WASTEWATER ENGINEERING: DESIGN OF WATER RESOURCE RECOVERY FACILITIES

MOP 8 SERIES

Edited by Christopher G. Schmit, Timothy G. Ellis,
Eric A. Evans, and Veera Gnaneswar Gude

2024

Water Environment Federation
601 Wythe Street
Alexandria, VA 22314–1994 USA
https://www.wef.org

ISBN 978-1-57278-440-6

About WEF

The Water Environment Federation (WEF) is a not-for-profit technical and educational organization of 30,000 individual members and 75 affiliated Member Associations representing water quality professionals around the world. Since 1928, WEF and its members have protected public health and the environment. As a global water sector leader, our mission is to connect water professionals; enrich the expertise of water professionals; increase the awareness of the impact and value of water; and provide a platform for water sector innovation. To learn more, visit www.wef.org.

This first edition of *Wastewater Engineering: Design of Water Resource Recovery Facilities (MOP 8 Series)* was prepared by editors Christopher G. Schmit, Timothy G. Ellis, Eric A. Evans, and Veera Gnaneswar Gude.

Review provided by:

Prof. Juan Antonio Baeza Labat
Jeanette A. Brown
Prof. Andrea G. Capodaglio, PhD, PE, Fellow IWA, BCEEM
Raj Chavan, PhD, PE, PMP
S. Rao Chitikela, PhD, PE, PEng, BCEE, F.ASCE
Natalie Cook, PE
Prof. Frederic Coulon
Christian Davies-Venn, PhD, PE, BCEE, BC.WRE, F.ASCE
Houssam B. El Jerdi
Val S. Frenkel, PhD, PE, D.WRE
John Friel
Ed Fritz
Vaughan Harshman
Mahmudul Hasan, PhD
Pat Herrick
Chip Kilduff
Stephanie Laughton, PhD
(Emy) Wenxin Liu, PhD, PE
Lu Liu, PhD
Indra Maharjan
Karthik Reddy Manchala, PE, PMP, ENV SP
Carol Martinson, PE
JB Neethling, PhD, PE
Kara Nelson
Tyler Robinson, PE
Steven I. Safferman, PhD, PE
Jianfeng Wen, PhD, PE
Brendan Wolohan, PE
Hong Yin, PhD, PE, PMP
Kai Zhang

Authors' and reviewers' efforts were supported by the following organizations:

Burns & McDonnell Engineering Co.
Cranfield University, UK
Greeley and Hansen, A TYLin Company
HDR Engineering, Inc.
Iowa Department of Natural Resources
Iowa State University
Johns Hopkins University, Whiting School of Engineering
RC-WEE Solutions LLC
Virginia Tech

Manuals of Practice of the Water Environment Federation

The WEF Technical Practice Committee (formerly the Committee on Sewage and Industrial Wastes Practice of the Federation of Sewage and Industrial Wastes Associations) was created by the Federation Board of Control on October 11, 1941. The primary function of the Committee is to originate and produce, through appropriate subcommittees, publications dealing with technical aspects of the broad interests of the Federation. These publications are intended to provide background information through a review of technical practices and detailed procedures that research and experience have shown to be functional and practical.

Water Environment Federation Technical Practice and Publications Advisory Panel

About the Editors

Dr. Christopher G. Schmit, professor of civil engineering and director of the Water and Environmental Engineering Research Center at South Dakota State University, received his bachelor's degree in civil engineering from the University of Wisconsin-Platteville and his master's and doctorate degrees in civil engineering from Iowa State University. Dr. Schmit's primary research areas include: high rate anaerobic digestion, biological nutrient removal in activated sludge, and stormwater treatment of nutrients and pathogens. Over the past 20 years, he has served as the primary investigator on a grant that provides undergraduate and graduate student operators to staff the Brookings Municipal Utilities Wastewater Resource Recovery Facility. He has also biennially taught an advanced wastewater treatment facility operations course for the past 20 years. Dr. Schmit has taught graduate courses in biological principles of environmental engineering, wastewater treatment facility design, industrial waste treatment, and hazardous waste management. He has been awarded the Patrick Hogan Teaching Excellence Award by South Dakota State University and has been recognized as the College of Engineering Researcher of the year and the College of Engineering Advisor of the year. Dr. Schmit is a licensed professional engineer in South Dakota and is active in the South Dakota Water and Wastewater Association, South Dakota Water Environment Association (SDWEA), and the Water Environment Federation (WEF). He has served on the board of trustees for WEF and has served as a chair for the Research and Innovation Community, as well as the Literature Review Community for WEF. He was awarded the Bedell award by the SDWEA for extraordinary personal service.

Dr. Timothy G. Ellis is an associate professor in the department of civil, construction, and environmental engineering at Iowa State University in Ames, where he has been teaching and engaging in environmental research since 1995. He received his bachelor's degree in civil engineering from Drexel University, his master's in environmental engineering from Georgia Institute of Technology, and his doctorate in environmental systems engineering from Clemson University. He has served as the executive editor of *Water Environment Research* and as a Fulbright Teaching Scholar at the University of Malta. Before academia, he worked internationally as a senior engineer in the United Arab Emirates on the retrofit of a wastewater treatment facility with full effluent reuse. His research area is primarily involved with biological wastewater treatment systems, and he is the inventor of the patented static granular bed reactor for high-rate anaerobic treatment and conversion of agricultural, industrial, and municipal wastes to renewable energy. He is a registered professional engineer in Virginia and Georgia.

Dr. Eric A. Evans is a senior process engineer at HDR specializing in wastewater planning and design with focus areas of process modeling and the application of innovative technologies. He received his bachelor's and master's in civil engineering and his doctorate in environmental engineering from Iowa State University. With more than 20 years of experience working in more than 50 facilities across the country, Dr. Evans has worked extensively to add nutrient removal capabilities to water resource recovery facilities. He has coauthored over a dozen technical papers and contributed to several books. He is a registered civil engineer in Iowa and Minnesota and a member of the Water Environment Federation.

Dr. Veera Gnaneswar Gude is NiSource-Meyer Charitable Foundation Professor of Energy and the Environment at Purdue University Northwest and director of Purdue University Northwest Water Institute. He is also professor (by courtesy) of environmental engineering at Purdue University. Previously, he was Kelly Gene Cook, Sr. Endowed Chair and professor of environmental engineering in Richard A. Rula School of Civil and Environmental Engineering at Mississippi State University. He received his bachelor's degree in chemical engineering from Osmania University and his master's and doctorate degrees in environmental engineering from National University of Singapore and New Mexico State University, respectively. His research interests include the energy–water–environment nexus, resource recovery from waste streams, advanced desalination and water purification technologies, and resilient and sustainable energy systems for rural economic development. His research efforts received funding from federal agencies such as NSF, USDA, USEPA, USGS, DOI, FHWA, and other state and industry sponsors. Dr. Gude has taught a wide range of civil engineering courses from freshmen to graduate levels, including water and wastewater engineering design courses. He received an early career award from ASEE's (American Society for Engineering Education) Environmental Engineering Division for the topic of Scientific Thinking and the Logic of Environmental Engineering Experiments. He also received an outstanding faculty research award from ASEE's Southeastern Section. He was awarded the James M. Robbins Excellence in Teaching Award at National and Southwest District levels by the Chi Epsilon Civil Engineering Honor Society. He was inducted into Bagley College of Engineering's Academy of Distinguished Teachers at Mississippi State University and was admitted as a distinguished member of the teaching academy at New Mexico State University. Dr. Gude is a licensed professional engineer and a board-certified environmental engineer and a diplomate water resources engineer. He serves as a member of the board of trustees, awards committee, and executive committee of AAEES. He also serves as an associate editor of *Water Environment Research*. He was named Engineer of the Year by ASCE's (American Society of Civil Engineers) Mississippi Section. He was elevated to the rank of fellow by both ASCE and the Environmental & Water Resources Institute.

Contents

Chapter 6 Suspended Growth Systems 171

Chapter 7 Attached Growth Processes 211

Chapter 11 Solids Thickening and Dewatering. 343

Chapter 12 Stabilization 389

List of Tables

List of Figures

Preface

Wastewater Engineering: Design of Water Resource Recovery Facilities serves several purposes. In its primary role as a textbook—intended for senior-level and graduate students in civil engineering—it provides the reader with the up-to-date knowledge and tools required to design water resource recovery facilities (WRRFs). The scope of coverage and the classroom-friendly features, such as learning objectives and example problems, make it an excellent tool for students transitioning from academia to practice, and it remains an essential resource for practitioners. Written by a team of professors and consultants and put through a rigorous industry review, *Wastewater Engineering* provides vital WRRF design information to an engineer at every stage of their education and career. The textbook also contains a sample three-credit course schedule to help instructors teach and organize an entire course on the design of WRRFs.

This book has its roots in WEF's *Design of Water Resource Recovery Facilities (MOP 8)*, a long-time fixture on the design engineer's shelf. As WEF approached how to update MOP 8, the organization decided to do so in a way that would address two significant needs in the water sector. The first was the need for current, nationally applicable, peer-reviewed design standards. The result was *Standards for Design of Water Resource Recovery Facilities (WEF 8)*. WEF 8 is a tool for seasoned design professionals, providing concise coverage of critical design criteria that has been formatted in quick-reference tables and bulleted lists. The second was the need for a wastewater design textbook and reference, as covered by this book. Within the MOP 8 series framework, this textbook serves as the entry point for new wastewater engineers and a primary reference work for practitioners, providing both theory and significant context behind the design parameters covered more briefly in WEF 8.

Sample Course Schedule

One of the most difficult aspects of using a textbook for an instructor is determining how to organize a course around the material in the textbook. With that in mind, a sample three-credit course schedule has been included in this text. This sample schedule is based on a 15-week semester with three 50-minute classes per week but can be adjusted as needed. The overall concept of this schedule is based on weekly topics, which are presented in bold for each week. The purpose of this concept is to give the user some freedom within each topic to modify the schedule to meet the needs of the course or to insert other material. Included at the end of the schedule are several notes that provide additional ideas for modifying the given schedule to accommodate alternative topics such as modeling, sustainability, or composting.

Sample 3-Credit Course Schedule

Week	Class Period	Topic	Sections
1		**Introduction to WRRFs**	**Chapter 1**
	A	Wastewater Characteristics	1.0–2.0
	B	Estimating Flows	5.0
	C	Regulations	6.0
2		**Design Fundamentals**	**Chapter 2**
	A	Design Process and Criteria	1.0–2.0
	B	Site Selection and Facility Layout	3.0
	C	Process Selection and Mass Balances	2.0, 4.0
3		**Preliminary Treatment**	**Chapter 3**
	A	Screening	2.0
	B	Grit Removal	4.0
	C	Grit Removal/Flow Measurement	4.0–5.0
4		**Sedimentation and Solids Separation**	**Chapter 4**
	A	Sedimentation Theory	2.0
	B	Primary Clarifiers	3.1–3.2, 3.8
	C	Secondary Clarifiers	4.0
5		**Biological Principles**	**Chapter 5**
	A	Biological Treatment/Microbiology	2.0–3.0
	B	Bacteria in Treatment Processes	4.0
	C	Principles of Biological Nutrient Removal	5.0–6.0
6		**Activated Sludge Processes**	**Chapter 6**
	A	Process Configurations	2.0
	B	Activated Sludge Design for BOD Removal	3.0
	C	*Optional (Oxygen Transfer Systems)*	4.0

(continued)

Sample 3-Credit Course Schedule (*Continued*)

Week	Class Period	Topic	Sections
7		**Biological Nutrient Removal in Activated Sludge**	**Chapter 8**
	A	Introduction/Biological Nitrogen Removal	1.0–2.0
	B	Phosphorus Removal Processes	3.0
	C	Finish Periods A and B	1.0–3.0
8		**Attached Growth Processes**	**Chapter 7**
	A	Biofilms	1.0–2.0
	B	Trickling Filter Design	3.0
	C	Integrated Biological Treatment	5.0
9		**Wastewater Solids**	**Chapter 10**
	A	Sources, Quantities, and Characteristics	2.0
	B	Mass–Volume Relationship	3.0
	C	Solids Storage, Transport, and Pumping	4.0–5.0
10		**Chemical Conditioning**	**Chapters 13**
	A	Conditioning Factors	2.0–3.0
	B	Conditioning Chemicals	4.0–5.0
	C	Thermal Conditioning	6.0
11		**Solids Thickening and Dewatering**	**Chapter 11**
	A	Solids Thickening	2.0
	B	Solids Thickening/Dewatering	2.0–3.0
	C	Solids Dewatering	3.0
12		**Stabilization**	**Chapter 12**
	A	Anaerobic Digestion	2.0
	B	Anaerobic Digestion	2.0
	C	Aerobic Digestion	3.0
13		**Tertiary Treatment**	**Chapter 15**
	A	Disinfection	1.0
	B	Filtration	2.0
	C	Membrane Processes	3.0
14		**Solids Disposal**	**Chapter 14**
	A	Incineration and Landfilling	1.0–2.0
	B	Land Application	3.0
	C	Start Week 15 Material	

(*continued*)

Sample 3-Credit Course Schedule (*Continued*)

Week	Class Period	Topic	Sections
15		**Hydraulics and Hydraulic Profiles**	**Chapter 9**
	A	Hydraulic Profile Principles	2.0–5.0
	B	Hydraulic Controls	6.0
	C	Unit Process Hydraulics	7.0

Notes.

- In week 1, the instructor could also add the sustainability section (7.0) in Chapter 1 if desired.
- In week 2, mass balances could be introduced if required (Chapter 2, Section 4.0).
- The Chapter 4 schedule combines primary and secondary sedimentation in one unit. Some instructors may prefer to move secondary sedimentation during the week in which Chapter 6 (activated sludge) is discussed.
- As noted in week 6, oxygen transfer systems is an optional topic; however, the topic might contain too much material for the week.
- Chapter 13 (chemical conditioning) (week 10) could be skipped in lieu of other topics, as those principles also are discussed briefly in Chapter 11.
- Week 15 includes hydraulic controls (Chapter 9, Section 6.0). This section can be omitted to save some time during that week.
- Composting (Chapter 12, Section 4.0) could be added as class C for week 14.
- Chapter 16 ("Modeling and Mass Balances") was omitted from the course schedule but can be added if another chapter is removed.

Notation Table

Symbol or Notation (used in equations)	Definition	Constant	Units
γ	Specific weight		kN/m^3
γ_w	Specific weight of water	1000	kg/m^3
ΔCOD	Removal rate of COD (mass basis)		kg COD/d
Δh	Hydraulic grade differential		—
ΔP	Pressure headloss		N/m^2
ΔP_d	Pressure drop		N/m^2·m
ΔT	Temperature difference		°C
ΔX	Biomass production rate (mass basis)		kg VSS/d
Δy	Change in water level		m
λ	Settling parameter		m^3/m^2·d
μ	Specific growth rate of the microorganism		1/d
μ	Dynamic viscosity of fluid		N·s/m^2
μ_{max}	Maximum possible growth rate of the microorganism		1/d
ν	Kinematic viscosity		m^2/s
ρ_f	Density of fluid		kg/m^3
ρ_p	Density of particles		kg/m^3
ϕ	Angle of bar screen		
ϕ_c	Confluence angle		
ϕ_{vw}	Angle of v-notch weir		—
A	Surface area		m^2
A/S	Air to solids ratio		
A_b	Area of bar screen opening		m^2
A_o	Cross-sectional area of gate or opening		m^2
A_s	Area of transfer surface		m^2
A_w	Area of wetted surface exposed to gas		m^2
b	Endogenous decay coefficient		d^{-1}
BA	Bicarbonate alkalinity		mg $CaCO_3$/L
BOD_{non}	Nonsettleable BOD (or COD) concentration		mg/L
BOD_{PE}	Primary effluent BOD concentration		mg/L
BOD_{PI}	Primary influent BOD (or COD) concentration		mg/L
b_{sf}	Bar screen shape factor		—
C	Hazen-Williams C-factor		

Symbol or Notation (used in equations)	Definition	Constant	Units
C_c	Cake solids concentration		mg/L
C_f	Feed solids concentration		mg/L
COD	Chemical oxygen demand		mg/L
C_p	Solids' specific heat or heat capacity		J/kg or MJ/kg
C_r	Filtrate/centrate/decant solids concentration		mg/L
C_s	Solids concentration in incoming sludges		mg/L
C_{sat}	Saturation concentration for dissolved oxygen		mg/L
C_w	Weir coefficient		—
d	Diameter of scoured particles		m
D	Detention time (solids stabilization per 40 CFR 503)		days
D	Pipe diameter or inside diameter		m
D_c	Critical depth		m
d_{DS}	Downstream water depth		m
d_{US}	Upstream water depth		m
E	Removal efficiency		Decimal
E	Solids capture		%
E_{BOD}	BOD removal efficiency		%
E_{BODmax}	Maximum BOD removal efficiency		%
E_{TSS}	TSS removal efficiency		%
E_{TSSmax}	Maximum TSS removal efficiency		%
f	Darcy-Weisbach friction factor		
f	Fraction of air dissolved at pressure *P*		
f_d	Fraction of active biomass that is biodegradable		—
f_p	Fraction of the digester influent $cBOD_5$ attributed to raw primary solids		—
g	Acceleration due to gravity	9.81	m/s^2
G	Root mean square velocity gradient or shear rate		s^{-1}
G_m	Volume of methane produced		m^3/d
G_{sgp}	Specific biogas production rate		m^3/kg
G_T	Volume of total biogas produced		m^3/d
h	Any height less than *H*		m
H	Height of ideal tank zone		m
H/L	Headloss per unit length		m/m
h_{cond}	Conductive heat transfer coefficient		$kJ/h \cdot m^2 \cdot °C$
h_{conv}	Convective heat transfer coefficient		$kJ/h \cdot m^2 \cdot °C$
H_f	Friction headloss		m
h_h	Height at which a particles with settling velocity v_h enters and is just removed		m
H_L	Headloss		m
HRT	Hydraulic retention time		s
H_v	Upstream velocity head		m

Symbol or Notation (used in equations)	Definition	Constant	Units
H_w	Head over weir		m
ISV	Intital settling velocity		m/h
k	Maximum specific substrate use rate		g COD/g VSS·d
k	Constant for type of scoured particles		
K_d	First-order reaction rate (solids destruction)		d^{-1}
K_{hl}	Headloss factor		—
K_{HW}	Hazen William's coefficient	0.849 for SI units 1.318 for U.S. customary units	—
K_{La}	Volumetric mass transfer coefficient		L/s
k_m	Manning's coefficient	1.0 for SI units 1.49 for U.S. customary units	—
k_s	Half saturation coefficient		mg/L
K_s	Half saturation constant of biodegradable substrate		g COD/L
K_v	Valve coefficient		
L	Length		m
m	Liquid mass flow rate (solids stream)		kg/h
M	Detention time multiplier (SOR divisor)		
M_c	Mass of solids within the cake		kg
M_s	Mass of dry solids		kg
M_{sgp}	Specific methane production rate per mass COD		m^3/kg COD
n	Manning's roughness factor		—
N	Pump rotational speed (Affinity Law)		rpm
NMHC	Non-methane hydrocarbons		
NOx	Nitrogen oxides (gas phase)		
P	Pressure		Pa or atm
P	Power		W or kW
PM	Particulate matter		
p_s	Percent solids (decimal percent)		
p_V	Weight fraction of volatile solids (decimal percent)		
P_v	Volatile fraction of TSS (digester)		—
p_w	Weight fraction of water (100-ps)/100)		
q	Heat load (digester heating)		J/h or MJ/h
Q	Flow rate		m^3/s or m^3/d
q_{cond}	Conductive heat transfer		kJ/h
q_{conv}	Convective heat transfer		kj/h
Q_g	Gas flow rate		m^3/s
Q_i	Influent flow rate		m^3/d
Q_R	Return activated sludge rate		m^3/d
Q_R	Pressurized recycle flow		m^3/d

Symbol or Notation (used in equations)	Definition	Constant	Units
Q_u	Underflow flowrate		m^3/hr
R	Particle radius		m
R_c	Rigidity coefficient		N·s/m^2
R_{min}	Minimum RAS rate		%
R_N	Reynold's number		—
r_{O2}	Oxygen transfer rate per unit volume		mg/(L·s)
s	Specific gravity of scoured particles		
S	Concentration of substrate		mg/L
S	Slope of energy grade line		—
s_a	Solubility of air at the required temperature		mL/L
S_{eff}	Effluent biodegradable substrate		g COD/L
SF	Safety factor		
S_F	Friction slope		—
SF_g	Solids flux resulting from gravity		kg/m^2·h
SF_t	Total solids flux		kg/m^2·h
SF_u	Solids flux resulting from underflow		kg/m^2·h
SLR	Solids loading rate		kg/m^2·d
S_M	Mass of primary sludge		kg/d
S_{NV}	Specific gravity of nonvolatile solids		
SOR	Surface overflow rate		m^3/m^2·d
SO_x	Sulfur oxides		
SRT	Solids retention time		d
SS	Specific gravity of dry solids		
SSV	30 minuted settled volume		mL/L
S_V	Specific gravity of volatile solids		
S_w	Specific gravity of wet solids		
Sy	Yield stress		N/m^2
t	Time (reaction time)		days
T	Temperature		°C
TA	Total alkalinity		mg $CaCO_3$/L
t_C	Pump cycle time		minutes
TDH	Total dynamic head		m
T_m	Temperature of heating medium		°C
TOC	Total organic carbon		mg/L
T_s	Temperature of solids at drying surface		°C
T_{sd}	Temperature at sludge gas interface		°C
TS_r	Total solids feed rate		kg/d
TSS	Total suspended solids concentration		mg/L
TSS_{non}	Nonsettleable influent TSS concentration		mg/L
TSS_{PE}	Primary effluent TSS concentration		mg/L
TSS_{PI}	Primary influent TSS concentration		mg/L
U	Overall heat transfer coefficient		J/(h·m^2·C)

Symbol or Notation (used in equations)	Definition	Constant	Units
U_b	Bulk downward velocity		m/h
v	Velocity		m/s
V	Volume		m^3
v_0	Sludge settling characteristic velocity		m/h
VA	Volatile acids		mg HAc/L
V_d	Rate of volatile solids reduction		kg/d
v_h	Settling velocity of any particle entering ideal tank zone at height *h*		m/s
V_H	Critical scour velocity		m/s
V_i	Hindered settling velocity due to gravity		m/h
v_L	Horizontal velocity through the tank		m/s
V_{lc}	Lower critical velocity		m/s
v_O	Settling velocity of critical particle		m/s
VS	Volatile solids concentration		%
VS_e	Digester effluent volatile solids fraction		
VS_i	Digester influent volatile solids fraction		
VSR	Volatile solids reduction		%
VS_r	Volatile solids feed rate		kg/d
VS_t	Volatile solids concentration at time, *t*		mg/L
v_u	Velocity upstream of bar screen		m/s
V_{uc}	Upper critical velocity		m/s
w	Width of ideal tank		m
w	Channel width		m
W	Power dissipated per unit volume (E/V)		W/m^3
w_b	Bar screen width		m
w_{dr}	Bar screen clear space		m
w_{US}	Width of the channel upstream of bar screen		m
X	MLSS concentration		g/m^3
$X_{e,v}$	Effluent VSS concentration		mg/L
X_i	Influent TSS concentration		mg/L
X_i	Concentration of settling flux solids		kg/m^3
$X_{i,v}$	Influent VSS concentration		mg/L
y	Depth of water above invert		m
Y	Yield		g VSS/g COD
y_{up}	Maximum depth in trough		m
Z	Elevation (invert elevation of pipe or channel from Bernoulli)		m

Chapter 1 | Introduction to Water Resource Recovery Facilities

LEARNING OBJECTIVES

At the end of this chapter, students should be able to:

- List the components of domestic wastewater and the significant sources of wastewater for a domestic WRRF.
- Describe the permitting requirements of the Clean Water Act (CWA) and the biosolids 503 regulations.
- Draw an example WRRF, clearly label the main unit processes, and give the function of each.
- Describe the mass balance concept and determine the boundaries.
- Describe the procedure for flow and loading data analysis.
- Describe sustainability principles involved in WRRF design.

1.0 INTRODUCTION

Water is an essential commodity for sustainable development. The extraction and use of both surface and groundwater sources in various domestic and industrial requirements results in the generation of wastewater that needs to be properly treated and disposed of before being discharged into the environment. The need for community wastewater collection and treatment systems globally has evolved over a period of more than 200 years, initially driven by the need to reduce human disease; then to eliminate gross water and environmental pollution effects, allowing native marine organisms to return to normal growth patterns and allowing full human recreational use; and, finally, to redefine wastewater as a resource with valuable products to be extracted through treatment. As a result, wastewater treatment facilities are now increasingly being recognized and designed as water resource recovery facilities (WRRFs).

About 238.2 million people are serviced by publicly owned treatment works (POTWs) in the United States (U.S. Environmental Protection Agency [U.S. EPA], 2016a). Of those serviced by POTWs, 127.7 million people are served by advanced wastewater treatment, 90.4 million people are served by secondary treatment, and 4.1 million are served by less-than-secondary treatment. Additionally, there are 2,281 non-discharging facilities serving 16.0 million people in the United States. Around 380 billion m^3 of municipal wastewater are produced annually in the world, taking only into account the shares of those countries with 87.5% of the global population. Of the wastewater that is accounted for, 55% (209 km^3) receives some kind of treatment. From those 209 km^3 annually undergoing treatment, about 10% of the wastewater is reused for some purpose (agricultural, industrial, etc.). For the most part, water reuse is based on water scarcity; 90% of the countries reusing wastewater have less than the average amount of renewable water resources per capita. The United Nations World Water Development Report (2021) estimates that there will be around US$ 1.5 trillion per year water infrastructure investment need across the world representing 20% of the total infrastructure investment. The majority of it (nearly 70%) will be in the global south which includes Asia, Latin America, and Africa.

Water conservation has become more common in water-limited areas. Water conservation needs result in beneficial recycling of treated wastewater for cooling, irrigation, agriculture, drinking water, and certain classes of industrial use. As water becomes scarce, intentional recycling of wastewater into drinking water supplies is becoming more prevalent. Increased efforts to control the discharge of toxins to the nation's waterways will continue in the future. Additionally, advanced wastewater treatment practices will become more prevalent. As climate change issues increase, municipalities will consider sustainability and carbon footprint as important criteria in evaluating alternative technologies. Changes in funding available to municipalities and wastewater management philosophy often drive the evolution and improvement of certain technologies. Along with water reuse, decentralized wastewater treatment and wet-weather flow management will play a role in future technology development and innovations (Burian et al., 2000).

2.0 WASTEWATER CHARACTERISTICS

2.1 WASTEWATER QUANTITY

Municipal wastewater primarily comprises domestic wastewater; industrial wastewater; infiltration/inflow from the collection system; and stormwater (in combined collection systems). Other sources may include septic tank waste (septage) generated in unsewered areas from septic tank cleaning contractors, and solids from sewer cleanings. Landfill leachate, water treatment residuals and, in some instances, contaminated groundwater—possibly with low concentrations of hazardous materials—can also be discharged to municipal WRRFs. All of these sources should be accounted for to the extent possible during the development of wastewater treatment facility design flows and loadings.

Domestic wastewater includes residential areas, commercial districts, institutions, and recreational facilities. Residential wastewater derives from water used in private residences for both indoor (e.g., drinking, cooking, bathing) and outdoor (e.g., landscape irrigation) purposes. Commercial, institutional, and recreational wastewaters come from a wide variety of facilities. Industrial contributions in any municipal wastewater may range from insignificant to many times the domestic contribution. Industrial operations and wastes may be continuous, or batch produced, and vary daily, weekly, and seasonally for any single industrial facility. Industrial discharges may also vary from one industrial facility to another for the same type of industry.

Some of the most significant components of municipal wastewater include infiltration, or unintentional water seepage or leaks through collection system pipes, house laterals, and manholes; and inflow, or surface and subsurface stormwater allowed to enter the collection system. These are typically considered together as I/I and vary from one community to the next. In combined systems, dry-weather flow includes domestic and industrial flows and I/I. Wet-weather flow occurs during storm or snowmelt events and is significantly higher than dry-weather flow. The quantity of wastewater depends on the geographical location, local climate, population demographics, industrial sector, water conservation practices, and water metering. Table 1.1 shows typical wastewater flowrates from urban residential sources in the United States. By increasing conservation efforts, the wastewater flowrates may be reduced by 30%.

Table 1.1 Typical Wastewater Flowrates From Urban Residential Sources in the United States (With data from Metcalf & Eddy, Inc./AECOM, 2014)

	Flowrate, gal/capita/d	Flowrate, L/capita/d
Household Size, No. of Persons	With Current Level of Conservation	With Current Level of Conservation
1	103	390
2	77	290
3	68	257
4	63	240
5	61	230
6	59	223
7	58	218
8	57	215

2.2 WASTEWATER QUALITY

Wastewater quality is described in terms of physical, chemical, and biological characteristics. The physical characteristics include suspended solids, color, odor, and temperature. Chemical characteristics may include organic and inorganic pollutants such as carbon, nitrogen and phosphorous elements, pH, dissolved solids, and other micro- and macro-pollutants. Flow and concentration of domestic wastewater reaching WRRFs vary hourly. Mass loading of pollutants to the WRRF depends on the hour of the day and other factors such as location, climate, socio-economic status of the community, and industrial discharges. Typical compositions of domestic wastewaters characterized in different strengths are shown in Table 1.2.

Table 1.2 Typical Composition of Untreated Domestic Wastewater (Adapted from Metcalf & Eddy, Inc./AECOM, 2014)

		Concentration[a]		
Contaminants	**Unit**	**Low Strength**	**Medium Strength**	**High Strength**
Biochemical oxygen demand, 5-day, 20°C (BOD_5, 20°C)	mg/L	100	200	400
Total organic carbon (TOC)	mg/L	100	160	320
Chemical oxygen demand (COD)	mg/L	300	500	1000
Nitrogen (total as N)	mg/L	20	30	70
Organic	mg/L	10	15	30
Free ammonia	mg/L	15	20	40
Nitrites	mg/L	0	0	0
Nitrates	mg/L	0	0	0
Phosphorus (total as P)	mg/L	3.5	5.5	11.0
Organic	mg/L	2	3	6
Inorganic	mg/L	1.5	2.5	4.5
Potassium	mg/L	10	15	30
Chlorides[b]	mg/L	35	60	120
Sulfate[b]	mg/L	24	36	72
Oil and grease	mg/L	50	75	150
Volatile organic compounds (VOCs)	mg/L	~100	100–400	~400
Total coliform	No./100 mL	10^6–10^8	10^7–10^9	10^7–10^{10}
Fecal coliform	No./100 mL	10^3–10^5	10^4–10^6	10^5–10^8
Cryptosporidum oocysts	No./100 mL	10^{-1}–10^1	10^{-1}–10^2	10^{-1}–10^3
Giardia lamblia cysts	No./100 mL	10^{-1}–10^1	10^{-1}–10^2	10^{-1}–10^3
Solids, total	mg/L	500	800	1600
Dissolved solids, total (TDS)	mg/L	350	550	1100
Fixed	mg/L	220	330	660
Volatile	mg/L	100	150	300
Suspended solids, total (TSS)	mg/L	125	200	400
Fixed	mg/L	30	40	80
Volatile	mg/L	100	150	300
Settleable solids	mL/L	8	12	24

Note. mg/L = g/m^3

[a]Low strength is based on an approximate wastewater flow rate of 570 L/capita/d (150 gal/capita/d). Medium strength is based on an approximate wastewater flow rate of 380 L/capita/d (100 gal/capita/d). High strength is based on an approximate wastewater flow rate of 190 L/capita/d (50 gal/capita/d).

[b]Values should be increased by amount of constituent present in domestic water supply.

3.0 WASTEWATER TREATMENT PROCESS

Water resource recovery facilities may contain preliminary treatment, primary treatment, secondary treatment, tertiary treatment, and/or disinfection, as shown in Figure 1.1. Most WRRFs also have solids processing facilities that further stabilize the materials that are removed from the incoming wastewater. In this example, an anaerobic digester is shown; however, smaller facilities typically use aerobic digesters.

A wastewater collection system comprises a network of pipes, conduits, tunnels, equipment, and appurtenances used to collect, transport, and pump wastewater. Typically, the wastewater flows through the network via conventional gravity sewers. Lift stations and pumping stations are used to move wastewater when gravity flow is not possible. There are three principal types of municipal wastewater conveyance systems: sanitary sewers, storm sewers, and combined sewers. Sanitary sewers convey wastewater from residential, commercial, institutional, or industrial sources, as well as small amounts of groundwater infiltration and stormwater inflow. Storm sewers convey stormwater runoff and other drainage. Combined sewers convey both sanitary wastes and stormwater.

Preliminary treatment takes place at the WRRF headworks. Headworks may also include flow measurement, flow equalization, pumping, and/or odor control. Larger materials such as wood, cardboard, rags, plastic, grit, grease, and scum that might damage downstream equipment or impair downstream operations are removed in this stage. These materials may be removed with chemical addition, bar racks, screens, comminutors, and/or grit chambers. Screens are typically followed by grit chambers, which are channels or small tanks where the wastewater velocity or speed is decreased to approximately 0.3 m/s (1 ft/sec). Here, heavier particles like sand, grit, eggshells, and heavier organic particles settle to the bottom of the tank where they can then be removed, while lighter organic particles remain suspended in the wastewater and pass on to the next process.

Primary treatment follows preliminary treatment. A primary clarifier is commonly used to slow the water further and remove heavier organic material and other particles by gravity. Grease and floatable material collect on the water surface and are skimmed off. Wastewater remains in a primary clarifier for approximately 2 hours (Water Environment Federation [WEF], 2017). Primary clarifiers may be

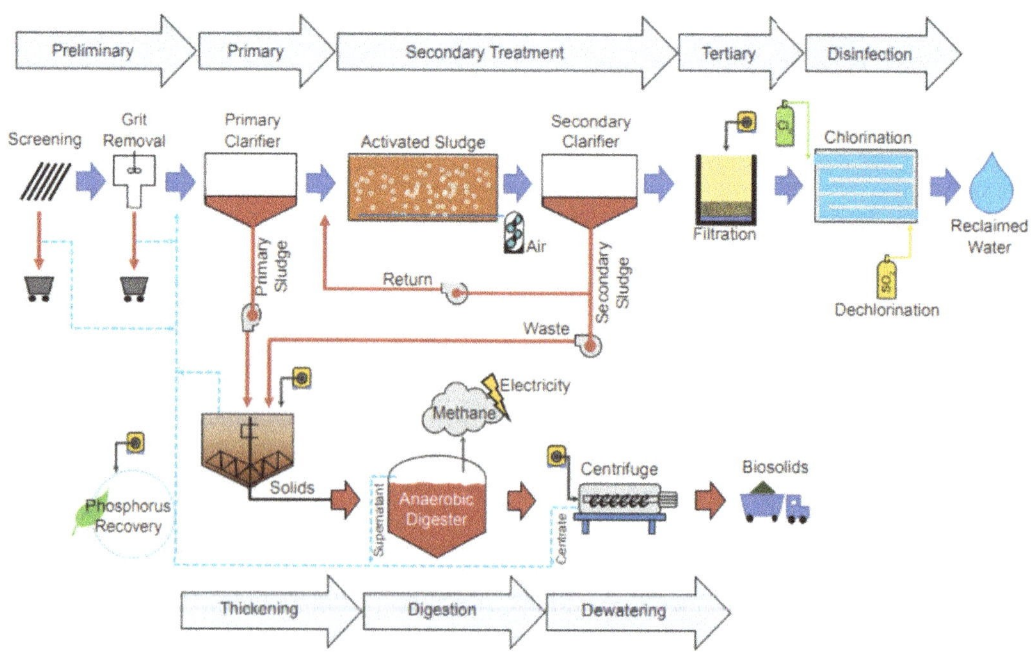

Figure 1.1 Treatment Process Involved in a Water Resource Recovery Facility (Reprinted with permission from Indigo Water Group)

rectangular or circular in shape. Primary clarifier can only remove settleable and floatable material, limited by the incoming wastewater characteristics. For domestic wastewater (residences and light commercial activities such as schools and stores), a primary clarifier can be expected to remove as much as 60% to 75% of suspended solids and between 20% and 35% of total 5-day biochemical oxygen demand (BOD_5). It does not, however, remove **colloidal solids**, dissolved solids, soluble BOD_5, soluble phosphorus, or ammonia. Chemically enhanced primary treatment is a method to increase the removal efficiencies of suspended solids and BOD_5.

Colloidal solids are small solid particles that are suspended in a fluid phase with a size range between 10 nm and several microns.

After the preliminary and primary treatment steps, the wastewater reaches the **secondary treatment** process, which is accomplished by biological treatment processes. Water resource recovery facilities with secondary treatment *remove at least 85% of influent total suspended solids (TSS) and BOD_5*, resulting in effluent concentrations between 10 and 30 mg/L. Secondary treatment processes may also remove ammonia, nitrate, and phosphorus even though, technically, any treatment that goes beyond the secondary treatment standards is considered tertiary treatment. Most secondary treatment processes involve biological treatment—typically, attached or suspended growth systems. Biological treatment processes rely on a mixed population of microorganisms, oxygen, and trace amounts of nutrients to treat wastewater. The microorganisms consume organic material in the wastewater for sustenance and reproduction. They are naturally present in the influent and do not need to be added to the treatment process. In attached growth systems, such as trickling filters, packed towers, and rotating biological contactors, the microorganisms form a **biofilm** that is attached to the supporting media. In suspended growth systems, such as ponds and activated sludge processes, the microorganisms are drifting throughout the wastewater. Hybrid treatment processes combine fixed film and conventional suspended growth biological treatment processes. Hybrid processes generally consist of a fixed media for the microorganisms to grow on within an aerated biological treatment tank providing greater treatment efficiencies per given reactor volume. Secondary treatment process effluent contains high levels of suspended biological solids that must be removed before the effluent is further treated or discharged to a receiving water body. Most WRRFs use secondary clarifiers to separate the solids from the liquid, as shown in Figure 1.2, although flotation, membranes, and other methods may be used.

Secondary treatment is any process designed to degrade the biological content of wastewater; typically follows primary treatment.

Biofilm is an accumulation of microbial growth on the surface of an object.

Physical and chemical treatment processes are used to remove oil, grease, heavy metals, solids, and nutrients from wastewater. For example, screening, sedimentation, and filtration are used to physically separate solids from wastewater. Chemical coagulation and precipitation are used to promote sedimentation. Coagulation is also used to improve capture and removal of colloidal solids. Activated carbon adsorption is used to remove organic pollutants. Breakpoint chlorination and lime are added to reduce nitrogen and phosphorus concentrations, respectively.

Advanced wastewater treatment processes are typically used to further reduce the concentrations of nutrients (nitrogen or phosphorus) and soluble organic chemicals in secondary treatment effluent. These processes may be physical, chemical, biological, or a combination. For example, membrane filtration—microfiltration, ultrafiltration, nanofiltration, and reverse osmosis—is used to remove organics, nutrients, and pathogens from wastewater. This method traditionally was used for industrial wastewater treatment but has been gaining popularity at WRRFs. Separate phosphorus reduction technologies are becoming more common because of more stringent discharge permit limitations. There are increasing requirements to both monitor and potentially treat microconstituents (sometimes termed **compounds of emerging concern**). Microconstituents may include pharmaceuticals and personal care products that may also be referred to as **endocrine-disrupting compounds**. This is of specific interest in areas where WRRF effluent is being considered for potable water applications.

Compounds of emerging concern, also called contaminants of emerging concern or microconstituents, are natural and man-made substances, including elements and inorganic and organic chemicals, detected within water and the environment, for which continued assessment of the potential effect on human health and the environment is a prudent course of action. Frequently mentioned examples include pharmaceutical and personal care products, pesticides, and industrial chemicals.

Endocrine-disrupting compounds are substances in the environment (air, soil, or water), food sources, personal care products, and manufactured products that interfere with the normal function of the body's endocrine system.

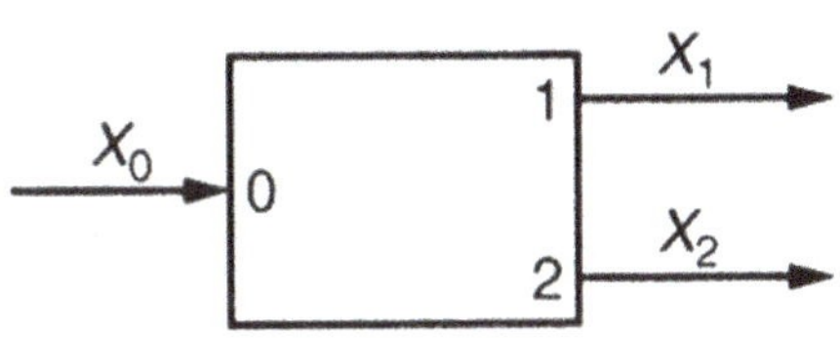

Figure 1.2 A Black Box for Calculating Mass Balances

Disinfection inactivates or destroys pathogenic bacteria, viruses, and protozoan cysts typically found in wastewater. These pathogens cause waterborne diseases such as bacillary dysentery, cholera, infectious hepatitis, paratyphoid, poliomyelitis, and typhoid. Disinfection reduces the number of bacteria and pathogens to safe levels to prevent the spread of waterborne illnesses and to protect the environment. Most WRRFs use chlorine gas, sodium hypochlorite (bleach), or ultraviolet radiation to disinfect their treated effluent; however, these are not the only disinfection alternatives available. Finally, the quality of effluent is influenced by the permit requirements set and the treatment processes needed to meet the discharge or reuse standards. The effluent from a WRRF can be discharged to a surface water body or wetlands, used to recharge groundwater aquifers via percolation through the ground or deep-well injection, or applied to land. It can be used (due to its nutrient content) to irrigate golf courses, parks, plant nurseries, and farms. It also could be the source water for a constructed wetland, adding nutrients that support the aquatic environment, thereby enhancing wildlife habitat and public recreation. In addition, effluent can be used by industries as cooling or makeup water for certain chemical processes. Sludge (or residuals) separated from primary and secondary clarifiers can be thickened to increase the solids content in the stream supplied to anaerobic digesters that convert the organic (volatile) solids into biogas containing 65% methane and 35% carbon dioxide and other gases. Sludge may also be physically treated (by drying and heating) to convert it into nutrient-rich and pathogen-free biosolids suitable for land applications.

Influent is water or wastewater flowing into a basin or water resource recovery facility. Biofilm is an accumulation of microbial growth on the surface of an object.

4.0 MASS BALANCE

Mass balance is a basic tool for understanding and designing a facility. A mass balance should be prepared to yield preliminary guidance concerning design quantities and significant differences between processing alternatives. The first mass balance that needs to be done is that of water; water coming into the system must leave the system (e.g., if the facility receives a total influent flow of 1000 m^3/d, then 1000 m^3/d must leave the facility through the various exiting streams).

Effluent is partially or completely treated water or wastewater flowing out of a basin or water resource recovery facility.

Consider a black box into which some material is flowing. All flows into the box are called **influents** and are represented by the letter M. If the flow is described as mass per unit time (e.g., kg/d or lb/d, calculated as concentration of a given component multiplied by the flowrate of the given current), X_0 is a mass per unit time of material X flowing into the box. Similarly, X_1 is the outflow or **effluent**. If no processes are occurring inside the box that will either make more of the material or destroy any, and if the flow is assumed not to vary with time (i.e., to be at **steady state**), then it is possible to write a **mass balance** around the box. This mass balance can be written in terms of mass/time or volume/time (for incompressible fluids).

Steady state is a condition in which all state variables are constant in spite of ongoing processes that strive to change them. For an entire system to be at steady state, i.e., for all state variables of a system to be constant, there must be a flow through the system. A steady state flow process requires conditions at all points in an apparatus remain constant as time changes. There must be no accumulation of mass or energy over the period of interest. The same mass flowrate will remain constant in the flow path through each element of the system.

$$[M_1 + M_2 + \ldots + M_m] = [M_1 + M_2 + M_3 \ldots M_n] \quad (1.1)$$

where M = mass flowrate. The subscripts in the above equation refer to the locations of the flows in or out of the black box.

If one stream enters the black box, a flow (mass or volume of M_0 per unit time), and the box splits the flow into two exit streams, 1 and 2, the flow from each is M_1 and M_2. A black box with one flow going in and two flows exiting is shown in Figure 1.2.

The two assumptions used in such mass balance equations are that the flows are in steady state (they do not change with time) and that no material is being destroyed (consumed) or created (produced), provided the density does not change. If these two assumptions are removed, and the amount of X can change with time (unsteady state) and this mass can either be consumed or created, then the mass balance is

Mass balance is a method for analyzing physical systems based on the law of conservation of mass. By accounting for all material entering and leaving a system, mass flows can be identified that otherwise might have been unknown or difficult to measure.

$$[M]_{\text{accumulated}} = [M]_{\text{in}} - [M]_{\text{out}} + [M]_{\text{produced}} - [M]_{\text{consumed}} \quad (1.2)$$

If the quantities do not change with time, the flows at one moment are exactly like the flows later; nothing can be accumulating in the black box, either positively (material builds up in the box) or negatively (material is flushed out); and the steady-state assumption holds and $[M]_{\text{accumulated}} = 0$. If nothing is being produced or consumed, $[M]_{\text{produced}} = [M]_{\text{consumed}} = 0$.

The above equations hold if only one material is considered in the mass balance. Mass balances become considerably more complicated (and useful) when several materials flow through the system. Mass and volume balances can be developed with multiple materials flowing in a single system. In some cases, the process is one of mixing, where several inflow streams are combined to produce a single outflow stream, while in other cases a single inflow stream is split into several outflow streams according to some material characteristics. Because the mass balance and volume balance equations are actually the same equation, it is not possible to develop more than one mass balance equation for a black box, unless there is more than one material involved in the flow.

This concept can be illustrated using a sludge thickening tank, shown in Figure 1.3.

The flow going into the tank is Q_0 at a solids concentration of C_0. The underflow concentration is C_u and the flowrate is Q_u, while the effluent flow and solids concentrations are Q_e and C_e, respectively. Two mass balances can be calculated: the volume balance and the mass balance. The volume balance is

$$Q_0 = Q_1 + Q_2 \tag{1.3}$$

That is, the flow (e.g., cubic meters per day, or gallons per day) going into the thickener must equal the flow out of the thickener.

The mass balance is calculated by recognizing that concentration multiplied by flowrate yields mass per unit time, or mass flowrate. For example, if the concentration is kg/m³ and the flowrate is m³/d, $C \times Q =$ kg/d. The mass balance is then

$$Q_0C_0 = Q_uC_u + Q_eC_e \tag{1.4}$$

Some unit operations in wastewater treatment separate one material from another. The effectiveness of this separation is known as *recovery*, calculated as

$$R = \frac{X_1}{X_0} \times 100 \tag{1.5}$$

where R = recovery of X (mass/time) from exit stream 1 with a feed stream of X_0 (mass/time), in percent.

Because concentration can be converted to mass flow by multiplying by the flowrate, this equation can also be written as

$$R = \frac{C_1Q_1}{C_0Q_0} \times 100 \tag{1.6}$$

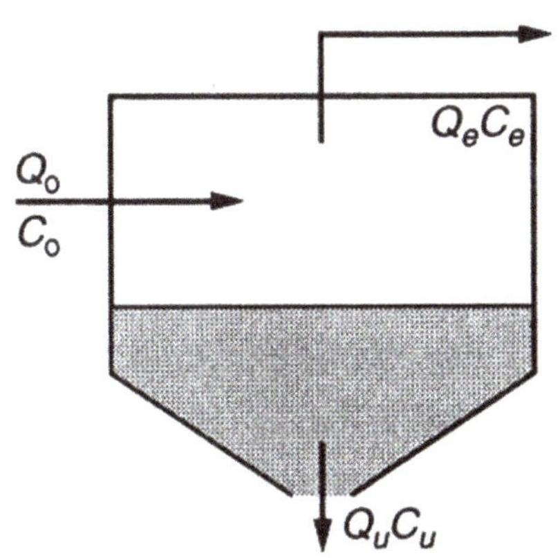

Figure 1.3 Sludge Thickening Tank

Where Q_0 is liquid flowrate of material entering the separator,
C_0 is concentration of material entering the separator,
Q_1 is liquid flowrate of material recovered or exiting the separator in the desired exit stream (exit stream 1), and
C_1 is concentration of material recovered.

The above equations apply to situations where the system is in steady state and there is no production or destruction of the material of interest. Consider next a system in which the material is being destroyed or produced in a reactor, but in which the steady-state assumption is maintained. That is, the system does not change with time, so that if the flows are sampled at any given moment, the results will always be the same.

An anaerobic digester destroys solids (converting them to gas and liquid), so that if a mass balance were to be used to describe the digester, the equation would be (in steady state):

$$Q_0C_0 = Q_sC_s + Q_dC_d + Z \tag{1.7}$$

Where Q_0 is inflow of raw sludge (m³/d),
C_0 is solids concentration of raw sludge (kg/m³),
Q_s is outflow of supernatant (m³/d),
C_s is solids concentration of supernatant (kg/m³),
Q_d is flow of digester sludge out of the digester (m³/d),
C_d is solids concentration of digested sludge (kg/m³), and
Z is destruction (digestion) of solids (mg/d).

Example 1.1

Two streams enter a mixing tank and one current emerges. One of the incoming streams has a flowrate of 340 m³/d and a solids concentration of 250 mg/L. The second incoming stream has a flowrate of 460 m³/d, with a solids concentration of 1500 mg/L. What is the flowrate in cubic feet per day of the stream leaving the mixing tank and what is its solids concentration?

Solution

Assuming steady state, by the principle of mass conservation, the amount of water flowing into the system must equal the water flowing out of the system (since no water is transformed within the mixing tank):

$$[\text{Water}_{\text{in}}] = [\text{Water}_{\text{out}}]$$

There are two streams flowing in and one stream flowing out, thereby:

$$[\text{Water}_{\text{in}}] = [340 + 460]\ \text{m}^3/\text{d}$$

Therefore, $[\text{Water}_{\text{out}}]$ = flowrate of the exit stream = 800 m³/d

A second balance can be calculated as the mass flow of the solids. Again assuming steady state and taking into account that no solids are destroyed or generated within the mixing tank:

$$[\text{Solids}_{\text{in}}] = [\text{Solids}_{\text{out}}]$$

The mass flow of solids can be calculated from the concentration and flowrate as:

$$[\text{Mass flowrate of solids}] = [\text{Concentration of solids}] \times [\text{Flowrate}]$$

Note that units will cancel out:

$$(C_AQ_A) + (C_MQ_M) = C_OQ_O$$

Where C is concentration of solids, Q = flowrate, A and M are influent streams, and O is the outflow.

Substituting

$$(250 \text{ mg/L} \times 340 \text{ m}^3\text{/d}) + (1500 \text{ mg/L} \times 460 \text{ m}^3\text{/d})] = [C_O \times 800 \text{ m}^3\text{/d}]$$

Note that the outflow is 800 m³/d calculated above. Thus,

$$C_O = 969 \text{ mg/L}$$

Example 1.2

A gravitational thickener, shown in Figure 1.4, receives a feed of 40 m³/h of precipitated metal plating waste with a suspended solids concentration of 5000 mg/L. The thickener is operated in a steady-state mode so that 30 m³/h of flow exits as the overflow, with a solids concentration of 25 mg/L. What is the underflow solids concentration and what is the recovery of the solids in the underflow?

Solution

Using subscripts i, u, and o for the influent, underflow, and overflow, respectively, the volume balance, assuming steady state and no production or consumption, is

$$40 \text{ m}^3\text{/h} = 30 \text{ m}^3\text{/h} + Q_u$$

$$Q_u = 10 \text{ m}^3\text{/h}$$

where Q_u = underflow. For the solids, the mass balance is

$$C_iQ_i = C_uQ_u + C_oQ_o$$

$$5000 \text{ mg/L} \times 40 \text{ m}^3\text{/h} = C_u \times 10 \text{ m}^3\text{/h} + 25 \text{ mg/L} \times 30 \text{ m}^3\text{/h}$$

$$C_u = 19\,900 \text{ mg/L}$$

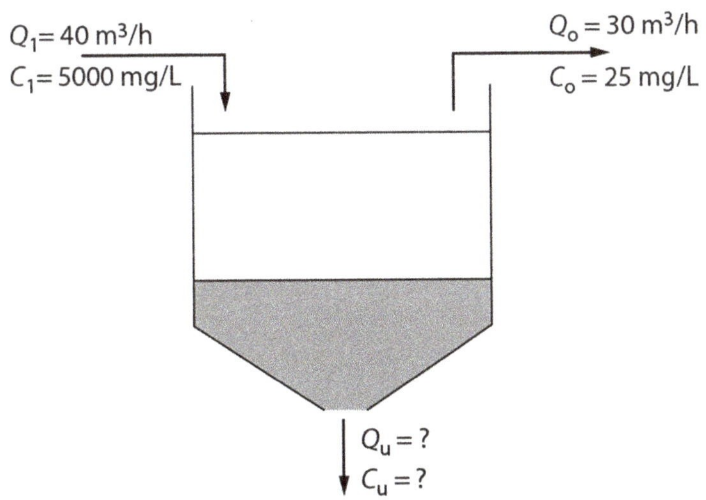

Figure 1.4 A Gravitational Thickener Receiving Precipitated Metal Plating Waste

where C_i is influent solids concentration,
Q_i is influent flowrate,
C_u is underflow concentration,
Q_u is underflow flowrate,
C_o is overflow solids concentration, and
Q_o is overflow flowrate.

The recovery of solids is

$$R = [(19\,900)(10) \times 100]/[(5000)(40)] = 99.5\%$$

Example 1.3

Consider a system shown in Figure 1.5a in which a sludge of solids concentration $C_O = 4\%$ is to be thickened to a solids concentration $C_E = 10\%$ using a centrifuge (note that 1% solids = 10 000 mg/L of

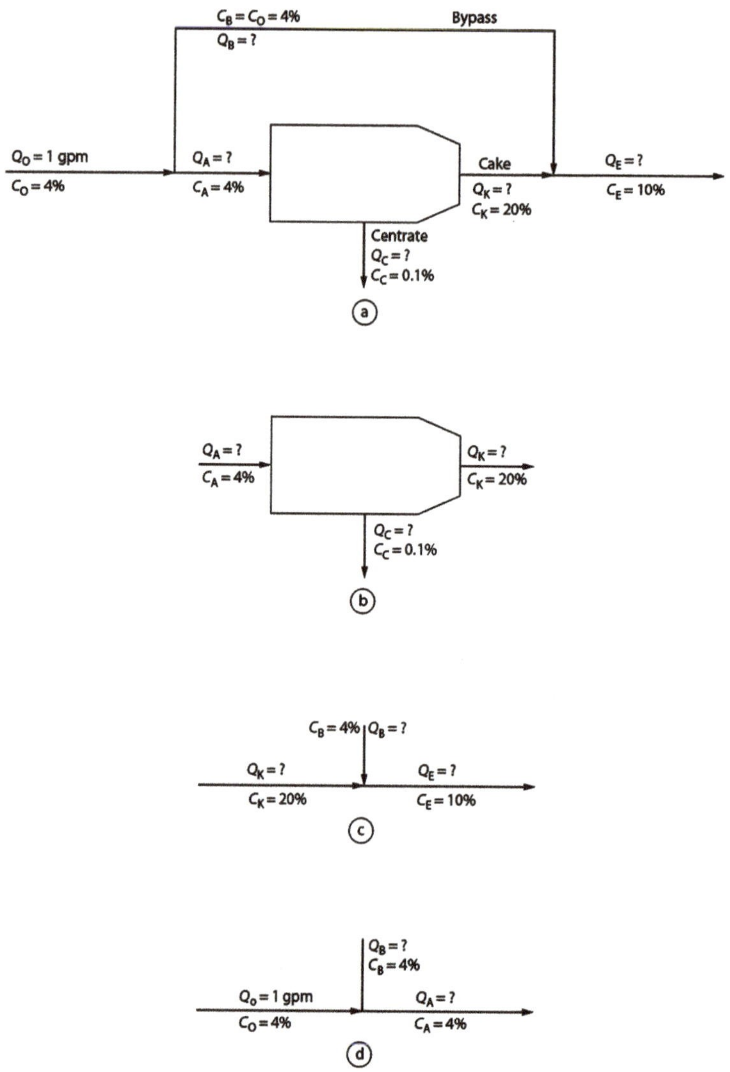

Figure 1.5 Mass Balance Around a Thickener Including a Bypass Stream (a); Mass Balance Around the Thickener (b); Mass Balance at the Point of Bypass (c); and Mass Balance at the Point of Mixing the Bypass Flow With the Thickener Outflow (d)

solids; therefore 4% solids = 40 000 mg/L of solids). However, the centrifuge produces a sludge at 20% solids from the 4% feed sludge. In other words, it works too well. Operators have decided to bypass some of the feed sludge flow and blend it later with the dewatered (20%) sludge so as to produce a sludge with exactly 10% solids concentration. The question is: How much sludge to bypass? The influent flowrate (Q_O) is 1 L/min at a solids concentration (C_O) of 4%. It is assumed that the specific gravity of the sludge solids is 1.0 g/cm³. That is, the solids have a density equal to that of water, which is usually a good assumption. The centrifuge produces a centrate (effluent stream of low solids concentration) with a solids concentration (C_C) of 0.1% and a cake (the high solids concentrated effluent stream) with a solids concentration (C_K) of 20%. Find the required flowrates.

Solution

Consider first the centrifuge as a black box separator (Figure 1.5b). A volume balance yields

$$Q_A = [Q_C + Q_K]$$

Note that the volume includes the volume of the sludge solids and the volume of the surrounding liquid. A solids balance on the centrifuge gives

$$[Q_A C_A] = [Q_K C_K + Q_C C_C]$$

$$Q_A\,(4) = Q_K\,(20) + Q_C\,(0.1)$$

There are only two equations and three unknowns. For future use, solve both of these in terms of Q_A, as $Q_C = 0.804\,Q_A$ and $Q_K = 0.196\,Q_A$. Next, consider the second junction in which two streams are blended (Figure 1.5c). A volume balance yields

$$[Q_B + Q_K] = [Q_E]$$

and a solids mass balance is

$$Q_B C_B + Q_K C_K = Q_E C_E$$

$$Q_B\,(4) + Q_K\,(20) = Q_E \qquad (1.8)$$

Substituting and solving for Q_K

$$Q_K = 0.6\,Q_B$$

From above, it was found that $Q_K = 0.196\,Q_A$. Hence

$$0.6\,Q_B = 0.196\,Q_A \text{ or}$$

$$Q_B = 0.327\,Q_A$$

Now consider the first separator box (Figure 1.5d). A volume balance yields

$$Q_O = Q_B + Q_A$$

From above,

$$Q_B = 0.327\,Q_A$$

Substituting

$$1 \text{ L/min} = 0.327\,Q_A + Q_A \text{ or}$$

$$Q_A = 0.753 \text{ L/min and}$$

$$Q_B = 0.327\,(0.753 \text{ L/min}) = 0.246 \text{ L/min}$$

which is the answer to the problem. Further, from the centrifuge balance

$$Q_C = 0.804\ (0.753\ \text{L/min}) = 0.605\ \text{L/min}$$

$$Q_K = 0.196\ (0.753\ \text{L/min}) = 0.147\ \text{L/min}$$

and from the blender box the balance is

$$Q_K + Q_B + Q_E$$

or

$$Q_E = Q_B + Q_K = 0.393\ \text{L/min}$$

There is also a check available. Using the entire system, a volume balance gives

$$Q_0 = Q_C = Q_E$$

1.0 L/min = 0.605 L/min + 0.393 L/min (calculation difference = 0.002) check.

5.0 ESTIMATING FLOWS AND LOADINGS

Wastewater flows can be estimated using water consumption data, if available, or with per capita or per source flow and loading rates. Consideration should be provided for future expansion or reduction of uses in current facilities, such as the reduction of flows from using low-flow toilets and other water conservation measures. Many municipalities have witnessed a drop in influent flow due to water conservation measures, despite the fact that loadings have remained constant or even increased.

5.1 DOMESTIC WASTEWATER

5.1.1 DOMESTIC WASTEWATER FLOWS

The residential flowrate portion of domestic wastewater is commonly based on population projections combined with per capita wastewater flow values. The population projections should take into account non-resident workers, non-permanent residents, and seasonal changes in populations (e.g., heavy tourist or commercial areas). Although the seasonal population may be present in a service area for only a portion of the year, it may have a significant effect on the wastewater flow treated by the WRRF. Similar consideration should be included for large schools or universities, for full-class relative to no-class months, and for large populations of transients who work and visit the area during the day but maintain their permanent residence elsewhere. Demographic projections developed for wastewater planning purposes should reflect other applicable planning estimates, such as those found in relevant zoning and master plans.

Flow per capita estimates for residential wastewater can be obtained from several available references if site-specific, per capita estimates are not available. Some U.S. state regulatory agencies use the *Recommended Standards for Wastewater Facilities* (Great Lakes–Upper Mississippi River Board of State and Provincial Public Health and Environment Managers [GLUMRB], 2014), which recommends 380 L/capita/d (100 gal/capita/d) for use as an average design flow. In general, about 60% to 90% of water consumption reaches the collection system, with the lower percentage applicable in semi-arid regions.

Flows from commercial sources are generally included within the allowance for domestic sources. This consideration becomes less appropriate for smaller service areas, however, where commercial operations such as laundromats, car washes, and sports venues may substantially affect the wastewater's character. If a commercial or institutional facility has many employees, for example, an estimate should be made for each contributing employee developed based on daily activities at the business location. Facilities that provide cafeterias, showers, and multiple handwashing requirements (e.g., food services) use more water per employee than those not including these water uses. Seasonal fluctuations in these flows must also be considered.

Where site-specific flow information is not available, representative commercial flows listed in Table 1.3 can be used. Some typical flowrates from institutional facilities, essentially domestic in characteristic, are shown in Table 1.4. Flowrates for these wastes also vary by region, climate, and type of facility. The actual flow records from the institutions are the best source of flow data for design purposes. Note that it is possible for a single institutional contributor to dominate WRRF design flows (or loadings).

5.1.2 DOMESTIC WASTEWATER LOADINGS

Typical major pollutant composition of untreated domestic wastewater was given in Table 1.2. Of the constituents listed, design loadings are typically developed for BOD, total suspended solids (TSS), nitrogen, and phosphorus. New WRRFs should be designed for a domestic load contribution of at least 0.077 kg/capita/d (0.17 lb/capita/d) BOD, 0.09 kg/capita/d (0.20 lb/capita/d) TSS, and 0.016 kg/capita/d (0.036 lb/capita/d) total Kjeldahl nitrogen (TKN) if nitrification is required, unless available information justifies other design criteria (GLUMRB, 2014). If garbage grinders are used in the service area, the design domestic loads should be increased to 0.091 kg/capita/d (0.20 lb/capita/d) BOD, 0.11 kg/capita/d (0.25 lb/capita/d) TSS, and 0.021 kg/capita/d (0.046 lb/capita/d) TKN. Water-saving efforts and water reuse will continue to increase the constituent concentrations in municipal WRRF influent. Depending on the conservation efforts planned, it might be preferable to analyze the influent for specific concentrations, rather than rely on historical data.

5.2 INDUSTRIAL WASTEWATER

The actual flow and constituent loading records from industries are the best source of data for design purposes. If data do not exist, and an industrial load is or may become significant, specific sampling

Table 1.3 Typical Wastewater Flowrates From Commercial Sources in the United States (Adapted from Metcalf & Eddy, Inc./AECOM, 2014)

		Flowrate, gal/unit/d		Flowrate, L/unit/d	
Source	**Unit Typical**	**Range**	**Typical**	**Range**	**Typical**
Airport	Passenger	2–4	3	8–15	11
Apartment	Person	30–45	38	120–170	145
Mobile home park	Unit	100–110	105	380–430	400
Office	Employee	6–12	10	23–45	38
Automobile service	Vehicle served	5–10	8	20–40	30
Station	Employee	7–11	10	26–42	38
Boarding house	Person	20–45	30	75–150	125
Conference center	Person	5–8	6	20–30	25
Department store	Toilet room	250–450	300	1000–1700	1100
	Employee	6–11	8	23–42	30
Shopping center	Employee	6–10	8	23–38	30
	Parking space	0.8–2.3	1.5	3–9	6
Hotel	Parking space	0.8–2.3	1.5	3–9	6
	Employee	6–11	8	23–42	30
Motel (with kitchen)	Guest	36–60	38	135–230	145
Motel (without kitchen)	Guest	32–53	34	120–200	130
Restaurant	Customer	6–8	6	23–30	24
Laundry (self-service)	Machine	320–413	338	1210–1560	1280
	Load	36–41	38	136–155	145
Public Restrooms	User	2.4–3.8	3	9–14	12
Theater (indoor)	Seat	1.6–3	2.3	6–11	9

Table 1.4 Typical Wastewater Flowrates From Institutional Sources in United States (Adapted from Metcalf & Eddy, Inc./AECOM, 2014)

Source	Unit	Flowrate, gal/unit/d		Flowrate, L/unit/d	
		Range	Typical	Range	Typical
Assembly hall or church	Guest	1.6–3	2.3	6–11	9
Hospital	Bed	125–250	150	500–900	600
	Employee	4–11	7.5	15–42	30
Institutions including prison other than hospitals	Bed	60–100	80	200–400	300
	Employee	4–11	7.5	15–42	28
School, day:		12–23	19	45–90	70
With cafeteria, gym, and showers	Student	8–15	11	30–60	42
With cafeteria only	Student				
School with residential halls	Student	30–60	40	120–200	150

programs and interviews can be used to establish the effect of present operations and anticipated changes. Daily, weekly, holiday, and seasonal variations of industrial releases should be expected, unless information to the contrary exists. Bench-scale or pilot facility evaluations may be necessary to develop or ensure the use of appropriate design criteria when industrial wastes are dominant. This is especially true if the industrial constituents are thought to have an adverse effect on treatment processes, including membranes. If membranes are being used, the industrial waste should be tested on specific membranes before design selection.

5.3 INFILTRATION/INFLOW

An allowable infiltration or exfiltration rate for new pipe construction is 9 L/d per meter diameter per meter length (L/d/m·m) (100 gpd/in. diameter/mile). Infiltration values can be 10 times higher in older sewers that have not undergone replacement or rehabilitation but can also be lower for newly constructed systems. In the absence of site-specific information, I/I can be estimated as a function of the area served by the collection system: 0.2 to 28 m^3/ha·d (Metcalf & Eddy, Inc./AECOM, 2014). Inflow can be very high in communities with older or combined sewer systems. Although combined wastewater service may represent only a small fraction of the influent service area, inflow derived from the combined wastewater service area often will dominate the design and operation of the WRRF. Precipitation-induced inflow may reflect low buffered, often acidic rainwater and the additional pollutants derived from rooftops, roadways, and land use of the service area. Inflow can be immediate or delayed; immediate inflow refers to rain entering the collection system directly during or immediately after the rainfall event. Delayed inflow refers to the runoff associated with the melting of an accumulated snow cover.

5.4 STORMWATER

In combined collection systems, oversized combined sewers and interceptors serve as traps for sediment and settleable solids. Older sanitary collection systems that receive high inflow when a rainstorm occurs after an extended dry period may have increased quantities of influent screenings, grit, and suspended solids that can be challenging for WRRF operations. Special consideration is needed for regulators, or overflow structures, which direct combined wastewater flow in excess of sewer or facility capacities to a receiving stream. These discharge locations often are referred to as combined sewer overflows. These systems can result in unintended reverse-flow conditions in cases where the receiving water elevation varies with the tidal pulse or high receiving stream elevations. Malfunctioning tide gates or backflow check gates can allow seawater to enter the collection system during both dry- and wet-weather conditions.

5.5 OTHER SOURCES

Septage is expected to have higher levels of organics, grease, scum, and grit relative to raw domestic wastewater. Septage characteristics vary widely from one load to the next; data for local septage should be used for design whenever possible. In the absence of available data, septage can be assumed to contain

15 000 mg/L TSS, 10 000 mg/L volatile suspended solids, 7000 mg/L BOD, 700 mg/L TKN, and 250 mg/L total phosphorus, with a pH of 6 (GLUMRB, 2014). Sewer cleanings are expected to exhibit highly variable characteristics of organically enriched grit. Sewer cleaning also can include high quantities of grease, rags, trash, and other debris. Management and treatment of the grease from sewer or wet-well cleaning need special consideration in the design of the treatment system and its components.

Landfill leachate characteristics can be observed in the form of varying soluble organic compounds and reflect the character and age of the material placed in the landfill and the amount of water that infiltrates the landfill from ground and surface sources. Leachate can contain various concentrations of heavy metals, volatile and semi-volatile organics, and color, nitrogen, phosphorus, and many other industrial chemicals including PFAS (per- and polyfluoroalkyl substances). Flows and concentrations vary, depending on rainfall events and integrity of the soil cover in the landfill.

5.6 ANALYSIS OF FLOW AND LOADING DATA

It is common to use historical data from the facility to formulate the required design flowrates and loadings. However, there are numerous opportunities for the introduction of sampling and analytical error into the historical data record. To assess the potential for errors associated with facility records, the designer should consider the location, method, and frequency of sample collection, especially relative to the point of introduction of in-facility recycle streams or septage (if applicable); methods of flow measurement or estimation; instrument installation and calibration frequency; level of accuracy of the laboratory's analytical methods and procedures; and the basis for any calculation of any process variables. In addition, specific calculations can be performed to check for data anomalies. Each of these considerations is described briefly below.

5.6.1 LOCATION, METHOD, AND FREQUENCY OF SAMPLE COLLECTION

Common locations for sample collection may include raw effluent, primary effluent, aeration system, return and waste activated sludge, secondary effluent, and stabilization processes. Samples can be taken as grab, interval, or composite samples, and the suitability of each sample type depends on the specific parameter being analyzed. A grab sample is a single sample taken over as short a time period as possible (usually instantaneously) and provides a "snapshot" representation of a parameter at a specific place and time. The sample is often analyzed immediately after collection. Grab samples are best for parameters that are time sensitive. Interval samples are a series of grab samples taken at various times throughout the day. The samples are taken at a preset time interval, transferred to an individual bottle, and analyzed. Interval samples are best for parameters that change over time.

Composite samples are a series of grab samples that are combined in a single container to provide concentrations over a flow- or time-paced quantity. Flow-paced composite samples are taken at a preset flow volume interval. Time-paced composite samples are taken at preset time intervals. Both types of composite samples involve collecting and transferring the sample into a common container where the constituents are combined to generate an averaged sample for the sampling period. Time-composite samples are best for parameters that are consistent over time and uniform (i.e., secondary clarifier/tertiary filter effluent). Flow-composite samples are best for characterizing parameters that are highly variable over the day (i.e., raw wastewater influent and periods of facility upset) and more adequately characterize loads. Consideration should be given to the location where in-facility recycle streams enter the main process stream. Measurements from facilities with flow equalization also should be verified to determine if corresponding flow and sample measurements are upstream and/or downstream of flow equalization.

5.6.2 METHODS OF FLOW MEASUREMENT OR ESTIMATION

Measurement error can be introduced in Venturi or magnetic flowmeters when there are changes in pipe size or abrupt changes in pipe direction near the device. The manufacturer should be consulted on the minimum upstream and downstream lengths of straight-run pipes for these units. Flow measurements can be considered accurate if the installed flowmeter is within 10% of a calibrated secondary flowmeter. Meters used for pacing chemical feed systems should have accuracy discrepancies that are less than 10%.

5.6.3 INSTRUMENT INSTALLATION AND CALIBRATION FREQUENCY

Instrument calibration and location affect the accuracy of data. Flow measuring devices and electronic sensors must be cleaned and calibrated on a regular basis. The physical location of flow measuring and sampling devices needs to be evaluated while reviewing facility operating data. Sampling locations should be in well-mixed areas to provide a representative sample. Sampling lines should be routinely maintained and cleaned.

5.6.4 LEVEL OF ACCURACY OF THE LABORATORY'S ANALYTICAL METHODS AND PROCEDURES

Samples must be preserved and/or homogenized to provide accurate results. *Standard Methods for the Examination of Water and Wastewater* (American Public Health Association et al., 2023) provides guidance on quality assurance and quality control procedures and contains sections on method development and evaluation, expression of results, and sample preservation. Data validity should include statistical screening of outliers, validation that the correct parameters are being tested, and verification that results are input with the proper date in operating records (date of sample collection, not the date of sample analysis).

5.6.5 METHODS TO CHECK VALIDITY OF DATA

The methods to validate facility data include use of constituent ratios for raw wastewater and primary effluent, determination of net yield from facility operating data, and mass balances around unit processes. Typically, constituent ratios for raw domestic wastewaters fall within the following ranges (EnviroSim Associates Ltd., 2016): BOD:TSS (0.82 to 1.43); chemical oxygen demand (COD):BOD (1.80 to 2.20); soluble BOD:BOD (0.20 to 0.40); BOD:TKN (4.2 to 7.1); and BOD total phosphorus (20 to 50).

Mass balances for conservative elements or pollutants allows the designer to gain a rapid understanding of the significant elements, in terms of recycle, solids capture, and validity of the sampling and measurement program. Inert solids (or total phosphorus) measurements and balances quickly can provide an overall assessment of the validity of a facility's monitoring program and determine whether the performance of a solid's destruction process is correctly defined. Percent accountable mass balance (out divided by in) closures of 100% plus or minus 10% are considered excellent; closures less than 80% and greater than 120% reveal suspect results from one or more processing points.

5.7 DEFINITION OF FACILITY CAPACITY

The hydraulic capacity of a WRRF typically is presented in terms of its annual average daily design flow. In some instances, the capacity is presented in terms of its peak hourly design flow, maximum month flow, or other flow condition, but this should be so noted. The average daily design flow is the average flow over the year when the facility is at design limits. This value is determined as the total flow for the year divided by 365 days, or more commonly as the average of the average monthly flows for 12 consecutive months. Influent solids, organic, and nutrients capacity is typically presented in terms of an annual average value, as well as maximum month and maximum day values. Facilities of equal design flow capacities can have different mass loading capacities. Changes in the effluent limitations or significant changes in the influent loading characteristics can result in significant reductions of the average daily flow capacity. For example, the inclusion of nitrogen or ammonia limits for a facility designed only for BOD removal requires an increase in the aerated solids retention time in the secondary biological process, by as much as three to four times that provided for BOD removal. This reduces the capacity of the existing facility, frequently requiring facility expansion of the biological process (bioreactors and clarifiers) to recover the initial design flow capacity. A reduction in flow to a facility does not necessarily result in a reduction in loadings because the flow reduction can result in higher influent constituent concentrations.

6.0 WASTEWATER REGULATIONS

Water quality and treatment objectives are criteria that must be established at the outset of WRRF design. The current and potential future regulatory requirements must be understood so that the design can meet current water quality objectives while also being flexible for future requirements. Figure 1.6 shows different sources of wastewater and U.S. EPA regulations. In the United States, the Clean Water Act (CWA, 1972) sets the regulations for water supplies and wastewater discharges including stormwater discharges. Regulations enforcing air quality standards and management of hazardous substances can also be important. The following subsections provide a general overview of key environmental regulations in the United States and internationally that often affect municipal WRRF design, including water quality objectives for discharge to receiving waters, the more stringent guidelines that apply to reclaimed water applications, air quality requirements, and regulation of hazardous wastes.

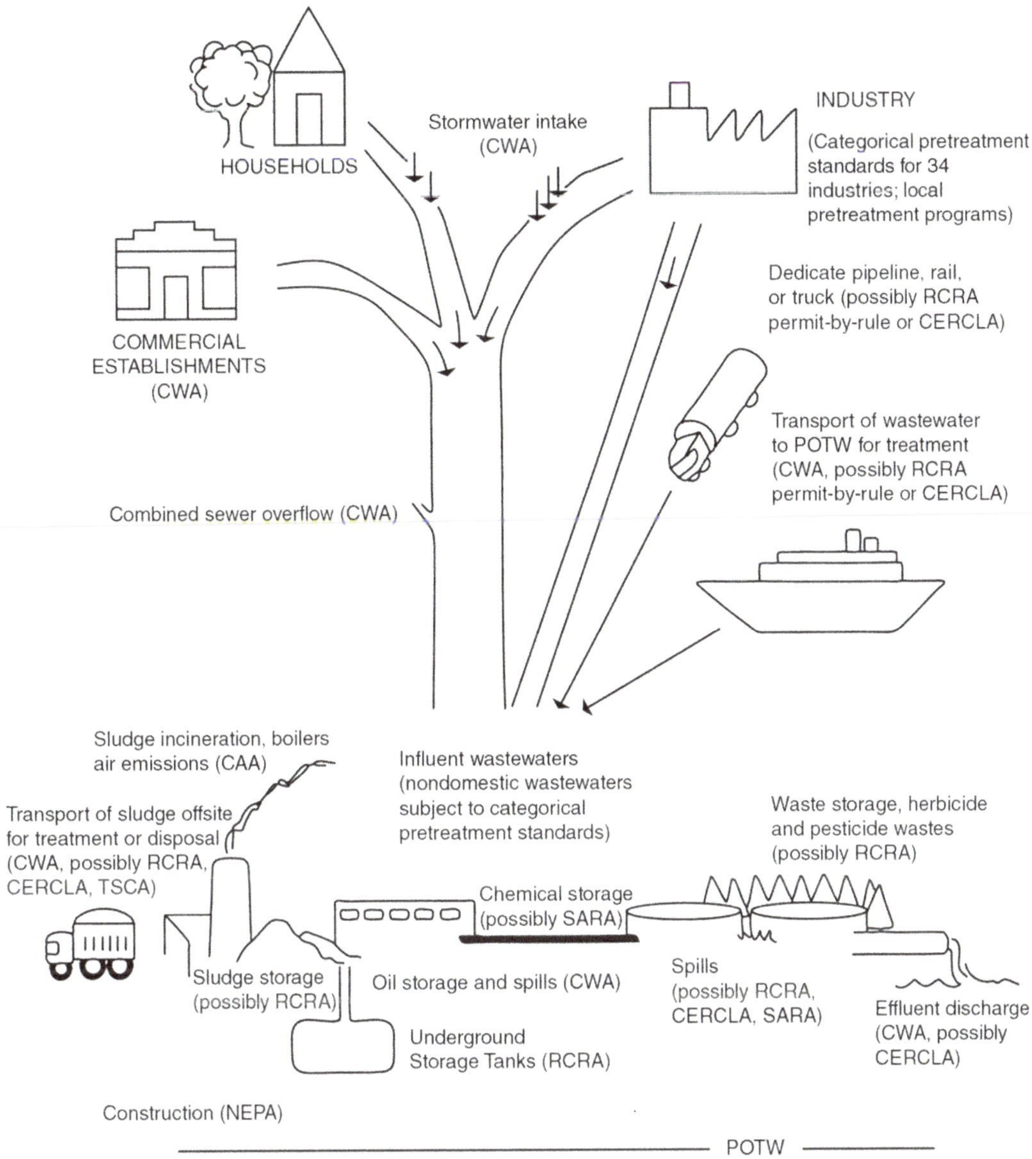

Figure 1.6 Activities and Sources of Pollutants Potentially Subject to U.S. EPA Regulations (U.S. EPA, 1989)

6.1 OVERVIEW OF UNITED STATES WASTEWATER DISCHARGE STANDARDS

The Federal Water Pollution Control Amendments of 1972 updated regulations that aim to improve water quality for human contact and recreation and eliminate pollution and introduction of toxic substances into the waterways of the United States. This legislation and updates are commonly referred to as the Clean Water Act (CWA, 1972), and amendments include the CWA of 1977, the CWA of 1987, and the CWA of 2002. The CWA of 1972 established technology-based effluent limits based on water quality standards developed by the various states and established a discharge permit system for point sources, exempting most nonpoint sources. The CWA of 1987 extended the discharge permit requirements to industrial dischargers and to municipal separate stormwater collection systems.

6.1.1 NATIONAL POLLUTANT DISCHARGE ELIMINATION SYSTEM

Discharge of treated wastewater to receiving waters of the United States and management of biosolids from municipal WRRFs are primarily regulated through National Pollutant Discharge Elimination System (NPDES) permits, which are generally issued at 5-year intervals. The U.S. EPA oversees the NPDES program, although the permitting process has been delegated to many states for management and control.

Regional offices of U.S. EPA administer and issue permits in states that have not accepted delegation. As of August 2016, 46 of the 50 states and one territory had received delegation to operate the wastewater component of their program under section 402.

Compliance standards are established on a national or regional perspective as the minimum quality of a facility discharge, both liquid stream and biosolids, that is acceptable under the permitting program. Water quality standards typically contain three components: designated use of a water body, water quality criteria for the type of water body, and an anti-degradation provision. If these compliance standards are not adequate to achieve the water quality standards of a receiving water body, more intensive limitations may be applied. Where technology-based effluent limits are not adequate to meet the uses established for a receiving water body, water quality–based effluent limits (WQBELs) can be applied. Implementation of the WQBELs can be in several steps. First, a total maximum daily load (TMDL) can be established, which identifies the maximum loading (often expressed in units of mass per day) of one or more pollutants that can be discharged to a receiving water body within a defined area. This is typically developed based on water quality models and field testing. The next step is to allocate the allowable discharge mass to the upstream point source and nonpoint source contributors to the water body. This process is often referred to as waste load allocation (WLA). In establishing the WQBELs, TMDL, and WLA, anti-degradation criteria are often applied to protect existing water quality from being degraded and to improve an existing water body when a higher beneficial use has been established than the current use of the water body. The result of these processes is a required effluent quality that is incorporated in a state or NPDES permit.

The NPDES effluent discharge standards are intended to protect and preserve beneficial uses of the receiving water body based on water quality criteria, technology-based limits, or both. National minimum standards, termed secondary treatment standards, for municipal wastewater dischargers are defined in Table 1.5. The secondary treatment regulations (40 CFR 403) include some exceptions and allow states to establish more stringent effluent quality requirements. Nitrogen and/or phosphorus limits are increasingly common in areas with receiving waters impaired due to eutrophication (i.e., low dissolved oxygen, excessive algal growth) (U.S. EPA, 2017). The NPDES effluent limitations can also be influenced by mixing zone rules that influence potential analysis of toxics. Discharges of contaminants of emerging concern, disinfection byproducts, and pharmaceuticals and personal care products are examples of wastewater constituents that are currently unregulated but are receiving increasing attention from the scientific community and regulatory agencies.

6.1.2 INDUSTRIAL PRETREATMENT PROGRAM

Section 403 of the CWA established an industrial pretreatment program for industrial dischargers to municipal WRRFs. General provisions were intended to prevent the pass-through of untreated industrial waste into the nation's waterways. Criteria for prohibited discharges to municipal WRRFs include those that are explosive, corrosive, obstructive, excessively variable, or excessively hot. Categorical discharges, defined in section 403.6, were identified in Title 40 of the *Code of Federal Regulations* (CFR) (U.S. EPA, 2016c) for 56 specific categories, applicable to various industries. The U.S. EPA estimates that as many as 45 000 facilities are regulated by these categorical standards. In addition to the categorical standards, industries that discharge to a municipal wastewater collection system must meet the municipal, state, or U.S. EPA pretreatment standards. More than 1600 municipal WRRFs are included in regulated pretreatment programs under section 307(b) of the CWA. Although the categorical standards are established, the program relies heavily on each municipality to identify its own discharge priorities and to propose solutions to account for site-specific factors. Designers are encouraged to identify the industries discharging to the WRRF during the design phase.

6.1.3 NUTRIENT TRADING

The concept of nutrient trading as a water quality tool began in the 1980s. Nutrient trading is a structured mechanism to provide treatment options and more cost-effective construction options to all pollutant contributors in a specific stream segment or receiving water body. The U.S. EPA has developed a water quality trading toolkit for permit writers (U.S. EPA, 2009). Design consultants may be able to reduce the structural components of a facility if nutrient trading effectively shifts loading to alternate locations.

Table 1.5 Minimum National Performance Standards for Publicly Owned Treatment Works (Secondary Treatment Equivalency) (40 CFR 133) (U.S. EPA, 2016d)

Parameter	30-Day Average	7-Day Average
Conventional secondary treatment processes		
5-day BOD[a] the most stringent of		
Effluent, mg/L	30	45
Percent removal[b]	85	—
5-day CBOD[a] the most stringent of		
Effluent, mg/L	25	40
Percent removal[b]	85	—
Suspended solids, the most stringent of		
Effluent, mg/L	30	45
Percent removal[b]	85	—
pH, units Whole-effluent toxicity	Within the range of 6.0 to 9.0 at all times, Site-specific	
Fecal coliform bacteria, MPN/100 mL[c]	200	400
Trickling filters and stabilization ponds (equivalent of secondary treatment)		
5-day BOD[a] the most stringent of		
Effluent, mg/L	45	65
Percent removal[b]	65	—
5-day CBOD[a] the most stringent of		
Effluent, mg/L	40	60
Percent removal[b]	65	—
Suspended solids[d] the most stringent of		
Effluent, mg/L	45	65
Percent removal[b]	65	—
pH, whole effluent toxicity and fecal coliform bacteria remain unchanged		

[a]Chemical oxygen demand (COD) or total organic carbon (TOC) may be substituted for the 5-day BOD when a long-term BOD:COD or BOD:TOC correlation has been demonstrated.

[b]Percent removal requirements may be waived on a case-by-case basis for combined sewer service areas and for separate sewer areas not subject to excessive inflow and infiltration where the base flow plus infiltration is 450 L/capita·d (~120 gpd/capita) and the base flow plus infiltration and inflow is 1041 L/capita·d (~275 gpd/capita).

[c]Not defined in federal secondary treatment equivalency regulations, but permits typically include cited levels, often only on a seasonal basis.

[d]The state may adjust the suspended solids limits for ponds subject to U.S. EPA approval.

6.2 OVERVIEW OF INTERNATIONAL WASTEWATER DISCHARGE STANDARDS

The World Health Organization (WHO) provides guidelines for drinking water, recreational water, wastewater for use in agriculture and aquaculture, and ship and aircraft sanitation. The WHO does not provide guidelines for wastewater effluent quality, but it promotes various public health programs in support of both centralized and decentralized treatment systems. Guidelines are provided for bacteriological quality of graywater in various agricultural applications. The high cost of centralized systems in the eastern Mediterranean region is regarded by the WHO as the primary constraint in expanding wastewater service, resulting in greater emphasis on decentralized systems (Bakir, 2000).

European Union (EU) member countries established Council Directives in 1980 and 1994, which were updated with Council Directives 98/83/EC in 1998 (Council of the European Communities, 1998) for water intended for human consumption. The EU developed legislation regarding chemical pollution of waters with Council Directive 76/464/EEC, which was amended and supported by several other directives between 1982 and 1990. Council Directive 91/271/EEC (Council of the European Communities, 1991) established standards of wastewater treatment with tertiary and secondary treatment standards based on population equivalents and location relative to the coast. The directive defines secondary treatment as treatment of urban wastewater by a process generally involving biological treatment with a secondary settlement. For sensitive areas, which are defined as "natural freshwater lakes, other freshwater bodies, estuaries and coastal waters which are found to be eutrophic or which in the near future may become eutrophic if protective action is not taken," removal of phosphorus and nitrogen must be implemented to achieve the required effluent quality (Council of the European Communities, 1991). Population equivalents are defined by the EU as "the organic biodegradable load having a five-day biochemical oxygen demand (BOD_5) of 60 g of oxygen per day" (Council of the European Communities, 1991).

This directive was amended in 1998 (Directive 1998/15/EC) to include refined administrative reporting and monitoring requirements. These directives were expanded by Directive 2000/60/EC, which is referred to as the Water Framework Directive (WFD), to include management of inland surface and groundwater and protective measures for the aquatic environment. The EU wastewater standards most recently were updated by Directive (397 final) presented by the commission on July 17, 2006 (Council of the European Communities, 2006). The proposed directive includes treatment requirements for 41 chemical substances, including heavy metals.

6.3 WATER REUSE

Increasing water demands in urban areas and more stringent effluent quality requirements have led to increased interest and reliance on reuse of reclaimed water (i.e., treated wastewater) worldwide. As the world's population continues to grow, and with the water usable for domestic purposes representing only approximately 2% of the earth's water, the available supply must be recycled faster than typically achieved in nature. Historically, most water reuse has been for agricultural and landscape irrigation. However, commercial, industrial, and potable reuses are becoming more prevalent as a result of changing climatic conditions and urban growth water demands.

6.3.1 RECLAIMED WATER QUALITY CRITERIA IN THE UNITED STATES

Reclaimed water quality criteria differ by state and by intended use of the water. Designers should confirm current regulations before the start of any design. Typically, the greater the perceived human exposure and health risk from the application of reclaimed water, the higher the water quality standard tends to be. For some uses with minimal human contact, lower quality water may be allowed. Where the application may affect sensitive species or areas, such as in rehydration of a wetland in a nutrient-limited area, discharge water quality that is higher than required by the entity's wastewater discharge permit may be appropriate.

Irrigation and agricultural applications typically allow lower quality reclaimed water where there is limited human exposure. Where golfers or athletes are likely to come in contact with turf irrigated with reclaimed water or where edible crops are irrigated and would be in direct contact with the reclaimed water, higher standards are typically applied. For example, irrigation of a root crop such as potatoes (agricultural reuse) (U.S. EPA, 2012) warrants higher water quality compared to irrigation of sod farms or commercial nurseries. Minimum water quality standards in the United States exceed those set by WHO (2006) and are typically 5 to 20 mg/L BOD and 5 to 20 mg/L TSS, with a limit of 2.2 to 23 coliform colonies/100 mL set in California.

Reclaimed water for environmental reuse can be used for stream augmentation, that is, creation of a flowing water feature where one does not currently exist; or, as is more common today, application to rehydrate a wetland area or create a new wetland area. Several wetland applications combine wetland rehydration, water quality improvement and detention through the wetland system, and groundwater recharge to augment an underlying aquifer that is used for public water supply.

Reclaimed water can also be used for most industrial nonpotable water applications. Some of the higher value applications include use as cooling water at power facilities; for industrial and commercial building cooling towers; and for central heating, ventilation, and air conditioning or chill water systems. High-quality reclaimed water can be used in the microchip manufacturing process or food and beverage industry.

Reclaimed water can be used for groundwater recharge, a process that takes advantage of soil aquifer treatment mechanisms, extended travel distances to sources used for potable purposes, and is often used where nonpotable use is ultimately made of the receiving aquifer. Augmentation of potable water supplies requires the highest levels of treatment and typically involves an extended public information period, pilot testing, and a study phase to implement a project.

6.3.2 INTERNATIONAL REUSE REGULATIONS AND GUIDELINES

World Health Organization recommendations for the use of reclaimed water rely on implementation of low-cost, low-technology systems, such as stabilization ponds with long detention times. The long detention periods facilitate the settling and removal of helminth eggs and protozoan cysts, which tend to pose the highest risk to the public from exposure to reclaimed water. The historical WHO standards originally issued in 1989 were based on agricultural irrigation and limited helminth eggs to 1 egg/L and fecal coliform bacteria to 1000 fecal coliforms/100 mL for unrestricted irrigation. For public lawn irrigation, a more restrictive 200 fecal coliforms/100 mL is applied. Compliance relies on effluent testing for fecal coliforms, and the standard for helminth removal is intended as a design guideline. In 2006, WHO updated its guidelines for the safe use of wastewater, excreta, and graywater (WHO, 2006). These guidelines expanded the bacteriological parameters in terms of type of irrigation, crop, and exposure, and included recommended logarithmic reductions to achieve recommended health-based disability-adjusted life year targets.

6.3.3 CONTAMINANTS OF EMERGING CONCERN IN RECLAIMED WATER

The term "microconstituents" was adopted by WEF in 2007 (WEF, 2007b) to describe a large number of elements and compounds that are being detected in water. Contaminants of emerging concern (CEC) is the term more commonly used to include a broad group of individual chemicals and classes of compounds present at trace concentrations that can include groups of compounds categorized by their end use; environmental and human health effect; or type of compound (U.S. EPA, 2012). Contaminants of emerging concern are ubiquitous in modern life, and all living organisms have experienced some exposure. Improvements in analytical technology now allow for detection of many of these CECs at sub-nanogram-per-liter concentrations. Current treatments known to reduce the concentration of most CECs include processes with a long sludge age, membrane treatment (WEF, 2007a), and advanced oxidation processes (Water Research Foundation, 2013). Designers should consider these processes if considering reduction of CECs.

6.4 AIR QUALITY REGULATIONS

In the United States, the Clean Air Act (CAA, 1963) establishes National Ambient Air Quality Standards for six pollutants known as criteria pollutants: carbon monoxide, particulate, lead, nitrogen dioxide, ozone, and sulfur oxides. The states establish U.S. EPA–approved State Implementation Plans, which address emission standards for stationary and mobile sources of criteria pollutants. Permits for major sources of air pollution are addressed through the Title V Operating Permit program, which is implemented by the states. The Title V program affects municipal WRRFs that operate incinerators and dryers, engines, and boilers. Specific emission reductions may be required through CAA programs, such as New Source Review/Prevention of Significant Deterioration, New Source Performance Standards, and National Emission Standards for Hazardous Air Pollutants. Some states and local air districts have also established more stringent requirements than the federal Title V program.

Toxic air pollutants, also known as hazardous air pollutants (HAPs), also are regulated under the CAA. The U.S. EPA is working with state, local, and tribal governments to reduce air toxics releases of 187 pollutants to the environment. Examples of toxic air pollutants include dioxin, benzene, toluene, and metals such as cadmium, mercury, chromium, and lead compounds. Examples of potential HAP sources at WRRFs include headworks, clarifiers, and the aerated zones of bioreactors. From a practical perspective, the three criteria for determining whether a Title V permit is needed are (1) 190 ML/d (50 mgd) approximate flow, (2) exceeding a 5 ppm concentration of volatile organic HAPs, and (3) exceeding an industrial contribution of 30% of the facility flow. A facility meeting two of these three criteria typically must commit to federally enforceable limits to maintain emissions below the major source level, modify the process, install control equipment for emissions, or achieve equivalent reduction through pretreatment.

6.5 RISK MANAGEMENT PLANS

Under the authority of section 112(r) of the CAA, the Chemical Accident Prevention Provisions require facilities that produce, handle, process, distribute, or store certain chemicals above a threshold limit to prepare a risk management plan (RMP), and submit the RMP to the U.S. EPA. A risk management plan includes prevention of release, process safety management, and emergency response. Chemicals requiring RMPs above a certain threshold quantity that are commonly used at municipal WRRFs include chlorine gas, aqueous ammonia, and methane. For methane, the 4540 kg (10 000 lb) RMP threshold applies to the total weight of the flammable mixture of digester gases, not just the weight of methane or flammables in the mixture. However, if a municipal WRRF uses methane (or a methane mixture) as fuel or sells it as fuel (as a retail facility), the amount of methane that is used or sold as fuel is not covered under 40 CFR Part 68 (U.S. EPA, 2016b). For aqueous ammonia, the threshold applies only to the weight of ammonia in the mixture. General guidance on risk management for chemical accidents should be considered for process modifications or features that limit the quantity and/or use of chemicals that would exceed the threshold amount. Sound design also accounts for chemical release prevention and control technologies.

6.6 REGULATION OF HAZARDOUS SUBSTANCES

Hazardous substances (termed priority pollutants) are subject to a variety of regulations, some based on considerations of water quality, and some based on the technologies available for pollution control in different industrial sectors. The Resource Conservation and Recovery Act (RCRA) (U.S. EPA, n.d.-a) regulates both hazardous and nonhazardous solid wastes. Disposal of nonhazardous wastes is allowed in underground storage tanks and municipal solid waste landfills. Hazardous substances are regulated from the point of generation, through transport, storage and treatment, and disposal. The RCRA exempts domestic wastewater and municipal WRRFs from its jurisdiction through the domestic wastewater exclusion, unless the municipal WRRF uses underground storage tanks or a municipal landfill for solids disposal. The domestic wastewater exclusion applies to industrial wastes, as long as they mix with domestic wastewater before entering the municipal WRRF headworks. Another act, the Comprehensive Environmental Response, Compensation, and Liability Act (CERCLA) (U.S. EPA, n.d.-b), primarily affects municipal WRRFs in the United States that are considering accepting wastewaters from Superfund cleanup sites. In these cases, municipal WRRFs will be required to ensure the Superfund discharger is following all requirements of the national pretreatment program. Liabilities of the CERCLA for a municipal WRRF can also be related to wastewater exfiltration and overflow in the collection system, storage leaks, effluent and solids disposal, and other aspects of municipal WRRF operation.

6.7 BIOSOLIDS REGULATIONS

The U.S. EPA's biosolids regulations (40 CFR 503) (1993) address the use and disposal of solids generated during the treatment of domestic wastewater and septage. They are organized into five subparts: general provisions, land application, surface disposal, pathogen and vector-attraction reduction, and incineration. Biosolids standards are typically incorporated into a WRRF's NPDES permit. Such permits are issued by U.S. EPA or by states with agency-approved biosolids programs. As of 2016, eight states are delegated authority to administer the NPDES biosolids program: Arizona, Michigan, Ohio, Oklahoma, South Dakota, Texas, Utah, and Wisconsin. Any facility that treats domestic wastewater (e.g., facilities that generate, treat, or provide disposal for solids, including non-discharging and "sludge-only" facilities) must have a permit. If the WRRF's solids are disposed of in municipal solid waste landfills or used as landfill cover material, they must comply with the requirements of municipal solid waste landfill regulations.

3.7.1 GENERAL REQUIREMENTS

Biosolids generators are responsible for complying with the regulation that establishes two sets of criteria for heavy metals—pollutant concentration and pollutant ceiling—and two sets of criteria for pathogen densities—Class A or Class B. It also allows for two approaches to reduce vector attraction: treating the solids, or using physical barriers. Biosolids that meet the higher quality criteria have fewer restrictions for land application. The minimum requirements for a biosolids to qualify for land application are a pollutant ceiling concentration, Class B pathogen requirement, and vector-attraction reduction requirements. Biosolids that meet pollutant concentration limits, Class A pathogen requirements, and vector-attraction reduction requirements can be land applied without the additional precautions (site restrictions) required for Class B biosolids. However, all land appliers must meet the minimum monitoring, recordkeeping, and reporting requirements.

3.7.2 POLLUTANT LIMITS

Before biosolids can be applied to land, the level of heavy metals therein cannot exceed either pollutant concentration limits or pollutant ceiling concentrations and cumulative pollutant loading rates (see Table 1.6). Bulk biosolids that will be applied to lawns and home gardens must meet pollutant concentration limits. Biosolids sold or given away in bags or other containers must meet either pollutant concentration limits or pollutant ceiling concentrations. Users should be directed to apply this material at rates based on annual pollutant loading rates.

3.7.3 PATHOGEN LIMITS

Part 503 labels biosolids as either Class A or Class B based on how they meet pathogen reduction requirements. Both classes have been treated to reduce pathogens and minimize their ability to attract vectors (e.g., rats). However, Class B biosolids still contain detectible levels of pathogens, while Class A biosolids are essentially pathogen free. When metals levels are also low (i.e., meet the high quality standard), then a Class A material is labeled as "exceptional quality" (EQ) biosolids.

Both Class A and Class B biosolids can be land applied per Part 503, but land application of a Class B material involves buffer requirements, public-access limits, and crop-harvesting restrictions. These rules are intended to protect public health and enable microorganisms in the soil to degrade remaining pathogens. Any biosolids being applied to lawns and home gardens, or sold or given away in bags or other containers, must meet Class A criteria.

3.7.3.1 Class A Requirements

To be considered Class A, biosolids must meet specific fecal coliform and salmonella limits at the time of use or disposal. The requirements of one of the following alternatives must also be met:

1. Thermal treatment;
2. High-pH, high-temperature treatment;
3. Prior testing for enteric virus/viable helminth ova to demonstrate process performance, then follow similar operating criteria;
4. Ongoing testing for additional pathogens at the time the biosolids are used/disposed of;
5. Treatment by a process to further reduce pathogens (PFRP); and
6. Treatment by a process equivalent to one of the PFRPs as determined by the permitting authority.

3.7.3.2 Class B Requirements

Biosolids must meet at least Class B pathogen requirements before being used or disposed of through one of three alternatives: (1) the monitoring of indicator organisms, (2) the use of one of the processes

Table 1.6 Pollutant Limits for Land-Applied Biosolids (All Limits Are on a Dry Weight Basis)

Pollutant	Ceiling Concentration Limits[a] (mg/kg)	Cumulative Pollutant Loading Rates (kg/h[a])	"High-quality" Pollutant Concentration Limits[b] (mg/kg)	Annual Pollutant Loading Rates (kg/ha·a)
Arsenic	75	41	41	2.0
Cadmium	85	39	39	1.9
Copper	4300	1500	1500	75
Lead	840	300	300	15
Mercury	57	17	17	0.85
Molybdenum	75	—	—	—
Nickel	420	420	420	21
Selenium	100	100	100	5.0
Zinc	7500	2800	2800	140

[a]Absolute values
[b]Monthly averages

to significantly reduce pathogens (PSRPs), or (3) the use of a process deemed equivalent to the PSRPs as determined by the permitting authority. Biosolids that do not meet Class B criteria cannot be land applied, but they may be placed in a surface disposal unit that is covered daily per Part 503. If Class B biosolids or domestic septage are land applied, site restrictions must also be met.

Municipal WRRFs typically use one of the following four processes to produce Class A or Class B biosolids: heat drying, digestion, composting, and alkaline stabilization. All the processes except digestion produce Class A biosolids. However, thermophilic digesters may produce Class A biosolids.

3.7.3.3 Vector-Attraction Reduction Requirements

Vectors (e.g., flies, rodents, and birds) are attracted to volatile solids. Materials with lower volatile solids concentrations are less likely to attract vectors, which spread infectious disease agents. There are many options for reducing vector attraction. All biosolids must meet at least one of them before they can be beneficially used. The options are

1. Volatile solids reduction;
2. Additional anaerobic digestion or aerobic digestion units at >40 °C for 14 days or longer;
3. A specific oxygen uptake rate for aerobically digested biosolids or use aerobic processes at >40 °C for 14 days or longer;
4. Alkali addition under specified conditions;
5. Dry biosolids with no unstabilized solids to at least 75% solids or dry biosolids with unstabilized solids to at least 90% solids; and
6. Inject biosolids beneath the soil surface or incorporate biosolids into the soil within 6 hours of application.

7.0 SUSTAINABILITY

The design and operation of WRRFs should look beyond the principal objective of meeting treatment goals to comply with regulatory requirements and address sustainability goals, such as minimizing costs, providing process reliability and flexibility, facilitating operations, maximizing safety and security aspects, and minimizing social effects during both construction and future operation. Sustainable design concepts are being increasingly adopted by WRRF designers and owners. Although there are many definitions of sustainability, the most commonly used definition is to "meet the needs of the present generation without compromising the ability of future generations to meet their needs in a sustainable manner" (United Nations, 1987). Sustainability goals will be different for each project and can be classified generally into four main groups of objectives: economic, environmental, social, and technical. Some of them can be easily quantified, while others require a qualitative approach to perform a comparison among alternatives.

7.1 ECONOMIC CRITERIA

When comparing design alternatives from the economic point of view, it is important to include criteria related to both construction and operation costs (including personnel, energy, chemicals, and maintenance). In particular, energy requirements (e.g., aeration, pumping, heating, and mixing) must be considered along with chemical requirements (e.g., metal salts for phosphorus precipitation, external carbon source to enhance denitrification efficiency, or chlorine for disinfection) and costs related to the collection and disposal of sludge. Any potential benefit from energy recovery (i.e., by means of methane production in the anaerobic digesters) should be included in the analysis. Any cost-related criterion chosen to compare alternatives, once quantified, must be normalized by expressing it as a percentage of the total budget.

7.2 ENVIRONMENTAL CRITERIA

Implicit in municipal WRRF projects is an objective related to the protection of environmental health, and the treated effluent quality generally ameliorates the quality of the receiving water body. However, there can be negative environmental effects associated with energy consumption, chemical reagents used, treatment residuals management, and atmospheric emissions of a project. Some of these negative effects can be offset by nutrient recovery, water reuse, and energy generation. Life cycle analysis is a useful tool for assessing the overall balance of environmental effects of a WRRF project. Life cycle analysis can be defined as a compilation and evaluation of inputs, outputs, and the environmental effect of a system through its life cycle, from the production of raw materials to the disposal of waste generated.

7.3 SOCIAL CRITERIA

Social criteria are becoming increasingly important when designing WRRFs. This category of criteria includes aspects related not only to facility personnel and external workers, but also to the relative effects of the facility on the neighbors and outside community. Noise, visual aesthetics (relative visual effect of the facilities, both near and distant views, from the perspective of blending in with the surroundings and having pleasing architecture and landscaping), community involvement, and odor related to air emissions are generally the distinguishing features for respective alternatives. As mentioned above, most of these criteria cannot be quantified easily, which requires a qualitative comparison between alternatives.

7.4 TECHNICAL CRITERIA

Technical aspects often are considered only when designing the units, equipment, and control strategies in detail. However, including technical criteria during conceptual design should ensure a reliable, flexible, robust, easy-to-operate, safe, and secure facility. Safety considerations could be considered a subset of technical criteria. Safety concerns include the potential for falls, confined space entry, exposed equipment or moving parts, and chemical transport, storage, and addition for facility operators. Safety also should consider the special precautions required to reduce the level of risk to the outside community, such as chemical deliveries and truck traffic. There are many numerical approaches based on simulation results concerning the evaluation of technical criteria. When this is not possible, a qualitative comparative ranking between alternatives could be enough, making the related uncertainty and subjectivity aspects explicit.

7.5 EVALUATION METHODOLOGY

There are several evaluation methodologies for decision analysis that involve multiple objectives. A multi-criteria decision analysis evaluation methodology can be considered to determine WRRF design (Flores et al., 2005). After a preliminary step to collect and analyze all available information, the next step is defining design objectives and evaluation criteria used to measure the degree of satisfaction of the objectives. Initial weight factors are assigned to determine the relative importance of the criteria. In the following step, there are several tasks related to the decision procedure: identification of the issue to be resolved, generation of the potential alternatives, selection of a subset of criteria defined for this specific issue, and evaluation of the proposed alternatives. This evaluation is approached as a multi-criterion method and comprises quantification; normalization of the evaluation criteria; and a weighted sum, where each alternative under evaluation obtains a score that is calculated by adding the product of each normalized criterion multiplied by its corresponding weight. The alternatives are ranked according to the score obtained. The alternative with the highest score is the one with the highest degree of satisfaction for the objectives considered, and the one recommended for implementation. The same methodology is applied iteratively to deal with each new issue that arises, until the conceptual design of the WRRF project is completed.

PRACTICE PROBLEMS

1. **A town has a population of 100,000. Estimate the following: (a) flow of wastewater to the treatment facility, (b) tons of BOD_5 per day, and (c) tons of suspended solids.**
2. **A pickle packing facility produces and discharges a waste brine solution with a salinity of 13 000 mg/L NaCl, at a rate of 100 L/min, discharged to a stream with a flowrate above the discharge of 1200 m^3/d and a salinity of 20 mg/L. Below the discharge point is a prime sport fishing spot; the fish are intolerant to salt concentrations above 200 mg/L. What must the level of salt in the effluent be to reduce the level in the stream to 200 mg/L?**
3. **Raw primary sludge at a solids concentration of 4% is mixed with waste activated sludge at a solids concentration of 0.5%. The flows are 20 and 24 L/min, respectively. What is the resulting solids concentration?**

This mixture is next thickened to a solids concentration of 8% solids. What are the quantities (in gallons per minute) of the thickened sludge and thickener overflow (water) produced? Assume perfect solids capture in the thickener.

4. A 10 000 m^3/d WRRF has influent and effluent concentrations of several metals as follows:

Concentration of metal		
Metal	Influent (mg/L)	Effluent (mg/L)
Cd	0.012	0.003
Cr	0.32	0.27
Hg	0.070	0.065
Pb	2.42	1.26

The facility produces a dewatered sludge cake at a solids concentration of 22% (by weight). Facility records show that 45 000 kg of sludge (wet) are disposed of on a pasture per day. The state has restricted sludge disposal on land to only sludges that have metal concentrations less than the following:

Metal	Maximum allowable metal concentration (mg metal/kg Dry Sludge Solids)
Cd	15
Cr	1000
Hg	10
Pb	1000

Is the sludge meeting the state standards?

5. A textile plant discharges a waste that contains 20% dye from vats. The color is too intense, and the state has told the plant manager to reduce the color in the discharge. The plant chemist tells the manager that color would not be a problem if they could have no more than 8% dye in the wastewater. The plant manager decides that the least expensive way of doing this is to dilute the 20% waste stream with drinking water so as to produce an 8% dye waste. The 20% dye wastewater flow is 900 L/min. How much drinking water is necessary for the dilution?

6. A capillary sludge drying system operates at a feed rate of 200 kg/h and accepts a sludge with a solids content of 45% (55% water). The dried sludge is 95% solids (5% water) and the liquid stream contains 8% solids. What is the quantity of dried sludge, and what is the liquid and solids flow, in the liquid stream?

7. A stream flowing at 11 355 m^3/d and 20 mg/L suspended solids receives wastewater effluent from three separate sources:

Source	Mile marker	Wastewater effluent quantity (m^3/d)	Wastewater solids (mg/L)
A	0	7570	200
B	1.2	22 712	50
C	1.9	11 355	200

What is the flow and suspended solids concentration at mile marker 2.0?

8. An industry produces a wastewater flow of 2000 m^3/d containing 120 000 mg/L salt. It is discharging to a river with a flow of 34 000 m^3/d and a salt concentration of 50 mg/L. The regulatory agency has told the company to maintain a salt level of no greater than 250 mg/L downstream of the discharge. What percent recovery of salt must be accomplished in the facility before the waste is discharged?

9. A desalinization facility produces a distilled water of zero salinity. To reduce the cost of water treatment and supply, the operator mixes this desalinated water with untreated groundwater that has a salinity of 500 mg/L. If the desired drinking water is to have a salinity not exceeding 200 mg/L, how much of each water—distilled water and untreated groundwater—is needed to produce a finished water flow of 100 000 m^3/d?

10. **A primary clarifier has influent suspended solids of 200 mg/L at a flowrate of 10 m^3/h. The effluent has a suspended solids concentration of 80 mg/L. What is the suspended solids removal in the primary clarifier?**

8.0 REFERENCES

American Public Health Association, American Water Works Association, & Water Environment Federation. (2023). *Standard methods for the examination of water and wastewater* (24th ed.). American Public Health Association.

Bakir, H. (2000). *Sanitation and wastewater management for small communities in EMR countries: Challenges and strategies for accelerated development within the water resources constraints* (Technical Note on Environmental Health). World Health Organization.

Burian, S. J., Durrans, S., & Rocky, S. (2000). Urban wastewater management in the United States: Past, present, and future. *Journal of Urban Technology*, 7(3), 33–62.

Clean Air Act. (1963). *U.S. Code*, Section 50–88, Title 40.

Clean Water Act. (1972). *U.S. Code*, Section 1251, Title 33.

Council of the European Communities. (1998). *Council Directive 98/83/EC of 3 November 1998 on the quality of water intended for human consumption* (Official Journal L 330, 0032–0054).

Council of the European Communities. (1991). Council Directive 91/271/EEC concerning urban waste water treatment. *Official Journal of the European Communities*, No. L 135, 40–52.

Council of the European Communities. (2006). *Proposal for a directive of the European Parliament and of the Council on Environmental Quality Standards in the field of water policy and amending directive* (2000/60/EC).

EnviroSim Associates Ltd. (2016). *BioWin Process Simulator.*

Flores, X., Bonmati, A., Poch, M., Rodríguez-Roda, I., & Bañares-Alcántara, R. (2005). Selection of the activated sludge configuration during the conceptual design of activated sludge plants using multicriteria analysis. *Industrial & Engineering Chemistry Research, 44*(10), 3556–3566.

Great Lakes–Upper Mississippi River Board of State and Provincial Public Health and Environment Managers. (2014). *Recommended standards for wastewater facilities.* Health Education Services. http://10statesstandards.com/wastewaterstandards.pdf

Metcalf & Eddy, Inc./AECOM. (2014). *Wastewater engineering: Treatment and resource recovery* (5th ed.). McGraw-Hill.

United Nations. (1987). *Sustainable development goals.* https://www.un.org/en/academic-impact/sustainability

United Nations. (2021, March 21). *UN world water development report 2021.* https://www.unwater.org/publications/un-world-water-development-report-2021

U.S. Environmental Protection Agency. (n.d.-a). *Resource conservation and recovery act (RCRA) laws and regulations.* https://www.epa.gov/rcra

U.S. Environmental Protection Agency. (n.d.-b). *Superfund: CERCLA overview.* https://www.epa.gov/superfund/superfund-cercla-overview

U.S. Environmental Protection Agency. (1989). *Overview of selected EPA regulations and guidance affecting POTW Management* (EPA-430/09-89-008).

U.S. Environmental Protection Agency. (1993). Standards for the use or disposal of sewage sludge. *Code of Federal Regulations*, Part 503, Title 40.

U.S. Environmental Protection Agency. (2009). *Water quality trading toolkit for permit writers.* https://www.epa.gov/npdes/water-quality-trading-toolkit-permit-writers

U.S. Environmental Protection Agency. (2012). *2012 guidelines for water reuse* (EPA/600/R-12/618). http://nepis.epa.gov/Adobe/PDF/P100FS7K.pdf

U.S. Environmental Protection Agency. (2016a). Chemical accident protection provisions. *Code of Federal Regulations*, Part 68, Title 40.

U.S. Environmental Protection Agency. (2016b). Designation, reportable quantities, and notification. *Code of Federal Regulations*, Part 302, Title 40.

U.S. Environmental Protection Agency. (2016c). General pretreatment regulations for existing and new sources of pollution. *Code of Federal Regulations*, Part 403, Title 40.

U.S. Environmental Protection Agency. (2016d). Secondary treatment regulation. *Code of Federal Regulations*, Part 133, Title 40.

U.S. Environmental Protection Agency. (2017). National water quality inventory report to Congress. https://www.epa.gov/sites/default/files/2017-12/documents/305brtc_finalowow_08302017.pdf

Water Environment Federation. (2007a). *Effects of wastewater treatment on microconstituents, Technical practice update.* Microconstituents Community of Practice, Technical Practice Committee, Water Environment Federation.

Water Environment Federation. (2007b). *Sources of microconstituents and endocrine-disrupting compounds, Technical practice update.* Microconstituents Community of Practice, Technical Practice Committee, Water Environment Federation.

Water Environment Federation. (2017). *Sustainability and energy management for water resource recovery facility design.*

Water Research Foundation. (2013). Optimization of advanced oxidation processes for water reuse. https://watereuse.org/wp-content/uploads/2004/12/WRF-06-012-20131107.pdf

World Health Organization. (2006). *WHO guidelines for the safe use of wastewater, excreta and greywater* (Volume 1).

9.0 SUGGESTED READINGS

European Union Common Implementation Strategy for the Water Framework Directive and the Floods Directive. (2016, July). *Guidelines on integrating water reuse into water planning and management in the context of the WFD.* European Union.

Water Environment Federation. (2013a). *Operation of Nutrient Removal Facilities* (Manual of Practice No. 37).

Water Environment Federation. (2013b). *Wastewater Treatment Process Modeling* (2nd ed.; Manual of Practice No. 31).

Water Environment Federation. (2018). *Financing and charges for wastewater systems* (4th ed.; Manual of Practice No. 27).

Water Environment Federation. (2020). *Control of odors and emissions from wastewater treatment plants* (2nd ed.; Manual of Practice No. 25).

Water Environment Federation & American Water Works Association. (2008). *Using reclaimed water to augment potable water resources* (2nd ed.). Water Environment Federation.

Chapter 2 | Design Fundamentals

LEARNING OBJECTIVES

At the end of this chapter, students should be able to:

- Describe the integrated design concept and identify similar concepts in existing water resource recovery facilities.
- Describe WRRF site selection criteria and facility layout requirements.
- Evaluate the effect of activities in the service area on flow patterns at the WRRF, and explain how the size of the collection system and population served affect diurnal flow patterns.
- Identify the economic parameters required for cost estimation.
- Evaluate different cost options for the construction of WRRFs.

1.0 THE DESIGN AND PLANNING PROCESS

Water resource recovery facilities (WRRFs) represent a significant investment by a community, and detailed planning is needed in advance of the WRRF design effort. Planning is generally completed through the development of a facilities plan to forecast influent flows, loadings, and community growth; to define other design criteria for new facilities; to incorporate current and potential future regulatory requirements and other performance goals; and to identify the recommended facilities necessary to achieve those goals. Collaboration with stakeholders early in the planning process helps to define the goals of the WRRF and to ensure the facility is designed, constructed, and operated as an integrated part of the community. This section describes general considerations for planning a WRRF design project.

1.1 PERFORMANCE GOALS

There are universal goals common to all WRRFs, as well as goals that are community- and site-specific. The overarching goal common to all WRRFs is to protect human health and the environment. Some beneficial uses that often apply to the water environment include domestic and industrial water supply; recreational uses such as swimming, boating, and fishing; wildlife and aquatic life; agriculture and aquaculture; navigation; and transportation. Environmental regulations and permits define the WRRF water quality and treatment objectives, depending on whether the facility will be discharging to receiving waters or will be conveying treated effluent for water reuse. The permits for WRRFs discharging to receiving waters are written to protect the intended beneficial uses of those receiving waters. In some areas, there are restrictions on new discharge flows to impaired water bodies or outright bans on discharge to promote water reuse and groundwater replenishment. Water resource recovery facilities that apply effluent to land (such as for irrigation or to groundwater infiltration basins) have quality requirements that protect human health. An overview of common regulatory requirements and water quality objectives for WRRFs were discussed in Chapter 1.

A variety of other community-specific performance goals may also influence WRRF design. Many areas have policies regarding work occurring within or near wetland, waterway, floodplain, other environmentally sensitive areas, and areas that may be contaminated with hazardous waste. Additional examples of locally adopted policies that influence WRRF design include but are not limited to: restrictions to service area coverage and associated constraints; intergovernmental agreements for shared treatment capacity; targeted economic development plans; adopted growth rates; industrial pretreatment programs; biosolids management programs; energy management policy and plans; sustainability plans or goals; water reuse programs; land use requirements, and noise and odor restrictions; and building, fire, and electrical codes.

Many communities have adopted standards for sustainability and environmental stewardship or goals for energy efficiency and on-site energy production (including Leadership in Energy and Environmental Design [LEED]–compliant designs). Using established rating frameworks during project planning can help define sustainability performance goals, which can minimize operating cost by maximizing energy production and recovering value through nutrient recovery and water reuse.

1.2 INTEGRATED DESIGN

The goal of integrated WRRF design is to produce a facility that achieves the defined performance goals and interfaces well with the local surroundings and other municipal systems of which the WRRF is a part. Using an integrated design approach encourages the facility owner, operation and maintenance (O&M) personnel, and designers to communicate and collaborate from the start of WRRF planning and design through construction and startup. Including community stakeholders in WRRF planning and design is also becoming more prevalent as more municipalities are focusing on recovery of resources (water, energy, and nutrients) from wastewater.

Collaboration and frequent communication between the owner, O&M personnel, and designers is at the heart of integrated WRRF design. Integrated design also demands attention to detail and coordination as the multidisciplinary design team progresses through the stages of the design from alternative evaluations through the production of contract documents.

1.2.1 PROJECT PARTICIPANTS

All projects begin with the identification of a need by the owner, regulator, public entity, or legal counsel. Projects then typically proceed through five phases—facilities planning, design, construction, startup, and operation. The roles of each participant vary with the project phases and delivery method. A municipal WRRF design project involves many participants including the owner, program manager, design professionals, O&M personnel, contractor, regulator, legal counsel, financier, process equipment vendors, community and other stakeholders, and state and local agencies. Table 2.1 summarizes the roles and activities of these participants.

The owner could be a public entity (such as a municipal government or government-authorized regional district), or private (such as a commercial land development firm). All project participants are responding to the needs of the owner except for the community and the regulators. Owners may choose to hire a program manager when multiple projects are occurring simultaneously and additional oversight and coordination among projects may be warranted. O&M personnel are usually employees of the owner, but alternatively could be hired through an operations contract. Integrated WRRF design encourages O&M personnel to participate in the goal-setting process during facilities planning and to review the design as it develops. This approach ensures the completed design considers operability and ease of maintenance (which are critical aspects that often lack a designer's full attention). The owner may also have other staff with designated roles in the organization, such as an energy manager, sustainability manager, or environmental manager. Including these staff in WRRF planning and design can help to achieve energy or sustainability performance goals.

1.2.2 COMMUNITY ENGAGEMENT

The level of engagement with community stakeholders during WRRF planning and design varies depending on the project. At a minimum, proactive engagement with the immediate neighbors is recommended because they are often affected by high traffic volumes, limited parking, as well as dirt, dust, and noise related to construction of the facility. These issues can be even more problematic if construction must be done outside of normal working hours or during normal periods of peak traffic. Community input is particularly valuable for topics such as facility siting and aesthetics, siting final effluent discharge locations, defining necessary levels of odor mitigation, and residuals management planning. Addressing

Table 2.1 Project Participant Roles in Integrated WRRF Design

Participant	Roles
Owner	Pays for the design, modifications, and construction. Responsible for regulatory compliance. For owners that are public entities, final project approvals of construction contract may be required from a board of directors or city council.
Program Manager	Hired by owner to oversee project activities, responsible for communication among all participants and for achieving local requirements and performance goals.
Design Professionals	Hired by owner to develop facilities plan and design facility. For design–bid–build projects, design professionals also may serve as owner's representative during construction and startup. For alternative project delivery methods, this role may differ (some design firms may also operate the facility).
Operation & Maintenance Personnel	Provide input at various project milestones, starting with alternative development and evaluations during facilities planning through various design submittals to provide input on use of specific equipment and layout.
Contractor	Typically selected through a public bidding process. Role varies depending on the procurement approach, but the contractor is always responsible for construction of the facilities in accordance with contract documents. May also have design and/or operational responsibilities.
Regulator	Approves planning and design documents and performs reviews during construction; reviews grant and revolving fund applications, payment requests, and project closeout documentation. Involved in outfall siting, effluent dispersion, and residual disposal planning. Involvement continues after project completion with receipt/review of discharge compliance reports, operator training, and periodic inspection of the operating facilities.
Legal Counsel	Required for projects under court order or consent agreement. Legal assistance continues until all terms of the orders are complied with and the owner is released from the terms of the order or consent agreement. Status reports are filed throughout planning, design, construction, and startup phases.
Financier	Required at the time of bonding. Many public owners use short-term construction borrowing, with final bonds acquired at the end of construction. Larger projects may benefit from financial advice during planning or early in design. Public/private partnerships are also becoming more prevalent.
Process/Equipment Vendors	Consulted for advice on functionality of their equipment to verify that equipment will work as designed by the design professional, and that they can bid on the specifications. For specialized applications, equipment may be selected or procured early during design phase so that the key facility components are customized specifically for the equipment.
Community	Involvement is beneficial during all phases, particularly for owners concerned with meeting good neighbor, sustainability, or resource recovery performance goals. Community input can be facilitated in many ways, including advisory committees, public meetings, or social media.
State and Local Agencies	Involvement by any of the following might be necessary depending on project: ■ Local building inspectors to determine building codes/interpretations ■ Fire department for rulings on chemical storage, routing of alarms, etc. ■ Planning or zoning officials for site-review, zoning rules, and setbacks ■ Conservation committees or others responsible for reviewing wetlands ■ Local floodplain administrator for delineation of floodplain and permits ■ Local utilities, to discuss electrical and water supplies, drainage, etc. ■ Agencies that protect use of sites with historical/archaeological significance

these topics during the planning process allows the design team to address concerns in a cost-effective manner.

Advocacy groups also can be stakeholders in a project and can have a similar role as neighborhood advisory committees. Public meetings should have two objectives: (1) to explain the specific steps that will be taken, and (2) to understand public concerns. To be effective, neighborhood meetings are not solitary events in advance of a public hearing, but regular events during the design and as construction progresses. These meetings provide an opportunity to identify and respond to issues early and effectively. Utilities have successfully used interactive websites and social media platforms to share information, design features, construction photographs, and contractor progress. These web-based tools have allowed utilities to maintain ongoing communication with the local neighborhood affected by a project.

1.3 INTEGRATED DESIGN FOR RESOURCE RECOVERY

Municipal WRRFs are critical elements of a community system that includes inputs and outputs of water, nutrients, and energy—all of which are finite resources. Figure 2.1 highlights many examples of how municipal WRRFs can integrate their operations with other community functions to recover resources from wastewater. Resource recovery not only improves the overall sustainability of a community but can also help reduce WRRF operating costs. When paired with energy conservation best practices, energy recovery can result in a WRRF becoming energy self-sufficient or even a net energy generator.

Many WRRFs provide reclaimed water and nutrients for various uses throughout the community, ranging in purposes from irrigation water for parks or golf courses to direct potable reuse as discussed in Chapter 1. Anaerobic co-digestion of source-separated organic materials (such as food wastes, industrial organic wastes, or airport deicing fluid) with biosolids at a WRRF allows a municipality to produce renewable energy from a waste stream that would otherwise be landfilled. Nutrients from biosolids or other high-nutrient sidestreams can be recovered to produce marketable fertilizer or soil amendment products for the agricultural uses.

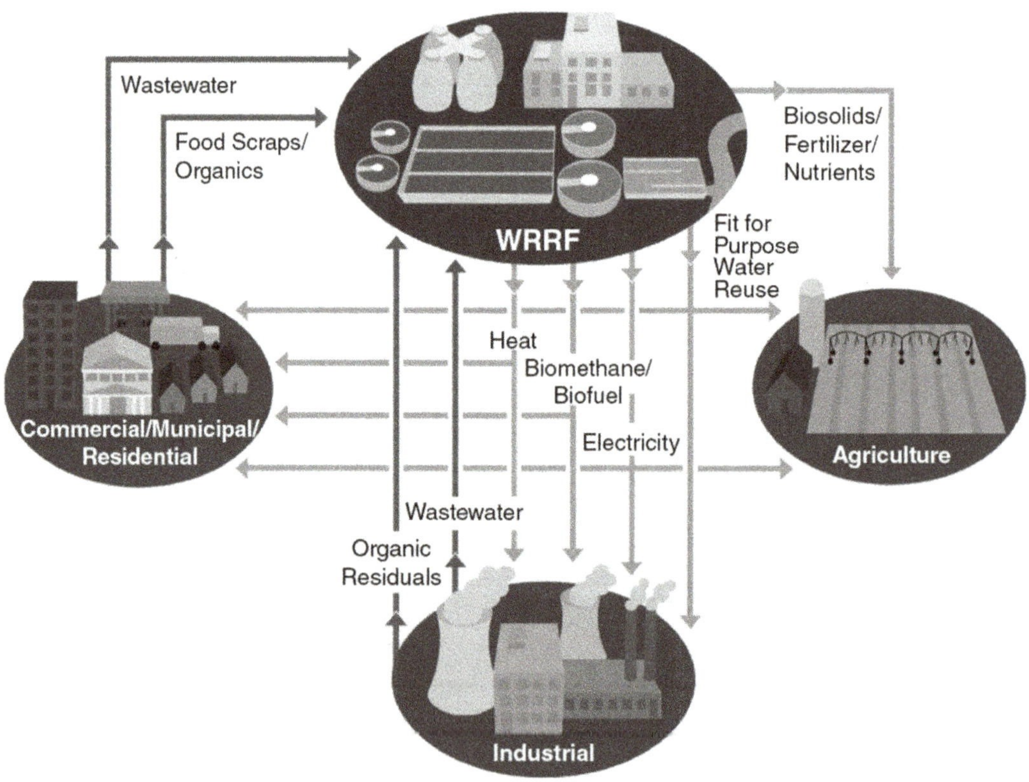

Figure 2.1 Opportunities for Integrated Resource Recovery at Municipal Water Resource Recovery Facilities (Reprinted with permission from CDM Smith, Inc.)

1.4 DESIGN STANDARDS

Many agencies in the United States and around the world have published guidelines or standards for the design of municipal WRRFs. In the United States, several federal agencies, states, municipalities, and cities have published WRRF design guidelines or standards. In addition, many states jointly have developed regional guidelines or standards for the design of WRRFs. It is the designer's responsibility during facilities planning and design to identify and use design standards appropriately to achieve compliance with regulatory-driven requirements applicable for a particular project. In the absence of such requirements, the designer can still use these resources as a source of guidance, while also considering site-specific conditions.

2.0 UNIT PROCESS DESIGN CRITERIA

Water resource recovery facilities involve unit operations (physical processes) and unit processes (biological and chemical processes) operating in series. Any given unit operation or process may have multiple units in parallel. Unit process design criteria—specifically, the design flows and loadings to each process—should be defined during facilities planning so that evaluation of alternatives and preliminary sizing of unit processes and equipment can be completed. Design flows and loadings should be used in conjunction with the design standards and design tools such as mass balances and process modeling to develop unit process criteria. General unit process sizing considerations are discussed in this section. Mass balances and process modeling are also described briefly, as are two other important unit process design considerations: redundancy and turndown.

2.1 MASS BALANCES

Mass balances prepared during planning and conceptual design yield guidance concerning design quantities and major differences between treatment alternatives. One of the most important applications of a mass balance is to define the design criteria for solids handling processes. Variability in solids production needs to be considered because it has a major effect on process performance. Most variability in solids production (net yield) from facility to facility is a result of differences in influent characteristics (inert suspended solids or the fraction of nonbiodegradable particulate organic matter), chemical use (phosphorus removal), and biological yield. The success of the entire facility and all unit processes depends on the ability to remove solids from the facility. The mass balance also provides the basis for the control logic of the process and instrumentation diagrams. The following steps can be used for mass balance preparation.

2.1.1 IDENTIFY PARAMETERS TO BE INCLUDED IN BALANCE

The mass balance should include each of the major pollutants that has a National Pollutant Discharge Elimination System (NPDES) limit or is significant for process control. At a minimum, flow, biochemical oxygen demand (BOD) or chemical oxygen demand (COD) and total suspended solids (TSS) should be included and, if applicable to the facility, nitrogen and phosphorus. It also may be desirable to balance the inert solids, especially if process modeling will be used for developing the mass balance. Balances should be performed for design average, maximum month, and maximum day conditions, although designers should remember that peak mass loads and peak flows rarely occur concurrently and that balances conducted with peak flows and masses will result in overly conservative design assumptions with limited physical significance.

2.1.2 ESTABLISH THE MASS BALANCE BOUNDARIES

It is critical to perform facility-wide mass balances that include all mainstream and sidestream processes. A whole-facility mass balance typically provides the best representation of the overall recycle characteristics, including COD, BOD, TSS, nitrogen, and phosphorus. Available facility data, such as solids capture and cake solids, can be used for mass balance development.

Endocrine-disrupting compounds are substances in the environment (air, soil, or water), food sources, personal care products, and manufactured products that interfere with the normal function of the body's endocrine system.

2.1.3 PREPARE INDIVIDUAL PROCESS BALANCES

To prepare a balance around an individual process, designers must use appropriate process performance data to calculate the incoming flow or mass, the outgoing flow or mass, any parameter removal or destruction (e.g., solids destruction in anaerobic digestion, ammonia removal in nitrification, solids capture in gravity thickening), and any parameter generation (e.g., inorganic solids from chemical addition, nitrate removal from denitrification) to provide a balance across the process for each parameter under consideration. Parameters and suggested values required for these balances for liquid treatment and solids handling processes should be used. Specific values for such parameters may be generated from historical facility data.

2.1.3 COUPLE MASS BALANCES FROM INDIVIDUAL PROCESSES

Using the process flow diagram, individual process mass balances must be coupled by using the balanced effluent from upstream processes as the influent to the next downstream process. Peaking factors for sludge flows and loads, and thus resultant sidestreams, should account for the actual expected hours of operation.

2.1.4 ITERATE

Because the outflows from one process will affect the mass balance of any downstream unit process, it is necessary to perform iterative calculations until all recycle and conversion conditions balance (achieve equilibrium) within reasonable limits. As mentioned above, because of the iterative nature of mass balances, they are best prepared using calibrated process models (Water Environment Federation[WEF], 2013).

2.2 FLOW PROJECTIONS

Population and flow projections, including nonpermanent residents and seasonal changes in populations, must be considered. The sum of the nonpermanent and the permanent population is considered the functional population and is the basis for WRRF flow. Most public wastewater treatment systems are designed with a phased approach, often using a minimum of 20 years of service life before an expansion is needed. However, in addition to the effects of lower startup flows, the designer should also consider potential additional future facility and site requirements in cases where the collection system service area is not expected to reach ultimate build-out levels within the design period adopted for the project.

Consideration of existing conditions and service area characteristics should be included when estimating flow per capita. When available, water consumption for an area should be used to estimate wastewater flow generation. At least 60% to 90% of the water consumption typically reaches the collection system (the lower percentage is applicable in semiarid regions) (WEF, 2023). If water consumption data are not available or an area is undeveloped, an estimation of flow per capita can be used to generate expected wastewater flow (as described in Chapter 1). Infiltration and inflow should also be considered, especially in areas with older collection systems. The most consistent component of municipal wastewater is the character of its domestic wastes. These wastes reflect the demographic makeup, character, and practices of the served population. Flowrates and concentrations will vary hourly in municipal systems. Generally, the smaller the system, the more variable the waste flowrates and concentrations, while larger systems may have smaller diurnal variations. Peaking factors are established and used in performance projections to evaluate the effects of higher flowrates, such as peak hourly and maximum day flows, especially during dry or wet weather conditions. These must be considered carefully both in terms of hydraulic loadings and pollutant loadings because treatment performance may be adversely affected with serious consequences such as exceeding effluent permit limitations.

2.3 UNIT PROCESS SIZING

Unit process design should be based on required performance at maximum process loading conditions for compliance, at a minimum, at the interval (e.g., maximum month or maximum week) included in a facility's NPDES permit. With the regulatory definition of the minimum compliance interval, design criteria may be developed to establish the controlling condition(s) for the facility's most stressed month and/or week (or more stringent intervals, if appropriate). Typically, the controlling condition reflects one or more of the following constraints: maximum flow, maximum constituent mass load, most stringent effluent limitation, or most restrictive processing condition (e.g., temperature). These constraints can occur simultaneously or at different times, depending on the circumstances of the facility, although maximum mass loads and maximum flows are not concurrent in separate systems.

Table 2.2 provides general guidance for the range of conditions that a unit process design should accommodate. Loading variability should be considered in terms of facility influent conditions, attenuation of these conditions through wet-weather management, and individual unit process conditions

Table 2.2 Design Guidelines for the Range of Conditions

Element	Guidelines
Hydraulics	Pipelines and channels must be carefully sized to avoid low-velocity conditions and associated solids deposition. Typically encountered problem areas are conduits common to both existing facility and planned expansion configurations, headworks, and primary clarifiers. Because troublesome deposits may be encountered even with careful design, means should be provided to facilitate cleaning. Pipelines and channels must also be sized large enough to maintain reasonable velocities and headloss at peak hour flows. Chapter 9 discusses guidelines for calculating hydraulic profiles.
Biological reactor sizing	Design for the controlling maximum month of the design life with the maximum solids retention time.
Biological reactor oxygen (air)	Design for the maximum daily demands in the project life and confirm equipment can turn down to meet minimum demands. Sensitivity to mixing supply and mixing needs is required and mixing often is limiting.
Biological nutrient removal systems	Check system performance for the minimum month (weakest) wastewater quality characteristics with the maximum and minimum recycle effects and provide appropriate safeguards (backup addition of supplemental carbon source and/or metal salt) to ensure compliance with effluent standards. The success of enhanced biological phosphorus removal is influenced strongly by the solids-processing train. Resolubilization of a fraction of the removed phosphorus is likely in solids storage and under solids stabilization.
Influent alkalinity	Design for minimum monthly or weekly value (typically high rainfall month) with maximum recycle.
Seasonal nitrification temperature	Design for average temperature of the minimum month or the monthly load/temperature conditions that result in the largest mass required under aeration. Remember that nitrification systems are limited by solids retention time, not nitrogen load.
Unit equipment sizing (particularly pumps)	Design equipment to meet both maximum and minimum processing needs where appropriate; for example, size pumps to handle peak hour or peak instantaneous flows depending on downstream processes and design pumping station wet wells and pumps for reasonable operation during the minimum month to week condition, and design minimum return solids capacity (and, if appropriate, waste solids capacity) from minimum monthly flows (and loads) in design life with reasonably anticipated minimum mixed liquor suspended solids concentrations and settled solids concentrations.
Waste solids	Design for minimum solids retention time conditions under at least the maximum monthly loading (may change with higher or lower storage) of the design life, with anticipated average or lower solids concentration under anticipated operating schedule. Provide appropriate storage with minimum solids retention time and anticipated average or lower solids concentration.
Solids handling unit processes	Design for the maximum day or maximum week conditions, taking into consideration the expected operational schedule for the solids handling or separation equipment (for example, 24/7 vs. 5 days per week.

with differing variability characteristics due to load modification by preceding unit process and recycle loads. The loading to any unit operation or process is dependent on the raw wastewater characteristics, performance of all preceding operations and processes, and characteristics of in-facility recycled sidestreams, which are a function of solids handling and disposal. Designers should consider the effects of varying process performance of upstream WRRFs and varying influent loads when sizing specific unit processes.

2.4 PROCESS MODELING

Process modeling is the preferred means for developing unit process sizing and unit operation mass balances. Process modeling can be used in concert with peaking factor analyses to assess whether the process design will perform as needed under the range of expected conditions. Simulation programs, such as BioWin (EnviroSim Associates, Ltd.), GPS-X (Hydromantis Inc.), or SUMO are available for process model development. These programs enable designers to consider, evaluate, and refine biological process configurations. Any process model should incorporate the whole WRRF to the extent possible, including sidestream treatment operations, so that the interconnection between liquid treatment processes (including both physical/chemical and biological units) and solids handling are captured in the simulations. Modeling the whole facility will capture the effect of processes that unavoidably solubilize pollutants and create special design and operating issues that vary with technologies and operating strategies used at the facility. The designer can use process modeling to evaluate different processes, especially when the facility's design objective includes the control of reintroduced pollutants. Chapter 16 includes details on process modeling.

2.5 REDUNDANCY REQUIREMENTS

Redundancy affects the overall reliability of the WRRF and cost. Water resource recovery facilities should have redundant machinery and equipment and process tankage to allow maintenance work or unplanned equipment failure. Designers should determine carefully the number of treatment units to be provided for each unit process. Multiple units are required for all critical components of the WRRF. Using multiple pieces of equipment of the same size and model and by the same manufacturer facilitates maintenance and reduces spare parts inventory. Typical practice for pumps and mechanical equipment, including solids processing and odor control equipment, is to provide one standby unit in addition to those required to handle peak design flow or load. Standby units are not required for process tankage, such as clarifiers, aeration tanks, fixed film reactors, or disinfection basins, but the need for redundancy depends on the size of the facility. However, hydraulic design should allow peak hourly flows, including associated sidestream flows, to be passed through the facility with the largest or longest flow path of each process unit removed from service. Redundant solids treatment trains should be considered when handling large quantities of solids—compared with the average daily production volume (i.e., with a digestion process)—to allow for anticipated failures and differing feed-solids characteristics in addition to maintenance. This would prevent solids handling from becoming a limiting condition or bottleneck to the entire system. Applicable state and other local criteria or guidelines should be consulted and applied.

3.0 SITE SELECTION AND FACILITY LAYOUT

There is no specific methodology for evaluating and selecting the most appropriate WRRF site. Although these general guidelines may apply to many site selection projects, it is important to develop a specific methodology to suit the community needs. The project manager and owner must have a basic understanding of the variables affecting site selection and work together to identify the methodology of evaluating these variables that works for their project. Variables that influence site selection may fall into one of the following categories: environmental, technical, institutional, and economic. The relative importance of some of these variables may not become evident until the site selection study is underway. For instance, the need for a buffer zone for odor and/or noise control may not be critical when comparing multiple rural sites but becomes critical when evaluating urban sites against rural sites. Consistency is critical to the process, particularly when it comes to public perception. Changing the quantitative and qualitative assessment once sites are identified may be perceived as altering the process to favor a political entity or influential owner(s). Thus, identifying key variables and developing qualitative and quantitative assessment frameworks early provides a consistent defensible evaluation process and avoids the possible perception of bias. Steps involved in site selection process include the following.

1. Develop a site selection work plan
2. Identify siting variables and develop evaluation criteria
3. Determine the WRRF technical components and minimum area requirements
4. Identify available candidate sites
5. Perform preliminary site assessments
6. Perform details (including costs) assessment
7. Site selection and public notification

Parallel steps to the above process involve public outreach and participation because the public must be involved in each step along the way. Legal, financial, and public relations work groups may also be involved in each step of the process. A summary list of the above steps is presented in Table 2.3.

Table 2.3 Site Selection Steps

Technical Steps	Nontechnical Steps
Develop site selection work plan	Develop public involvement plan (include public participation, education, relations, and involvement)
Determine variables and evaluation criteria	Inform and discuss process with stakeholders
Determine area requirements Determine future flow and loadings Determine area requirements from local WRRFs Determine WRRF area requirements by process and ancillary area requirements Determine preliminary cost estimates	Identify potential sites to stakeholders
Identify available sites within the minimum and maximum area requirements Assess environmental, technical, institutional, and cost conditions for each site	Identify elimination criteria and sites
Eliminate sites using Phase 1 qualitative criteria Eliminate fatally flawed or regulatory constrained sites Eliminate severe technically constrained sites	Identify elimination criteria and sites
Eliminate sites using Phase 2 quantitative evaluate criteria Quantify costs to correct/remediate environment, technical, and institutional issues Eliminate cost-prohibitive sites	Identify elimination criteria and sites
Recommend site(s) to owner	Hold town hall meeting for selected site and address technical merits of public opposition

3.1 DEVELOPING AND IMPLEMENTING A SITE SELECTION WORK PLAN

The first step in the site selection process is to develop a site selection work plan, which includes a consistent, technically sound methodology to site selection. The work plan should cover: (1) evaluation criteria including environmental, technical, institutional, and economic aspects; (2) area requirements; (3) site selection process; and (4) public participation. Wastewater treatment facilities are almost universally perceived by the public to be unacceptable neighbors. Public opposition can be strong; however, with early involvement of the public in the planning process and a sincere desire to listen and mitigate their concerns, much of the opposition can be minimized.

3.2 DEVELOPING EVALUATION CRITERIA

The four categories of site selection criteria—environmental, technical, institutional, and economic considerations—are detailed below.

- *Environmental* criteria are issues that would likely be examined in an environmental impact assessment for the project, including natural and man-made resources. They also may include regulatory constraints or aspects of the project that would require environmental permits. Adverse environmental impacts can render a site not acceptable for WRRF use.
- *Technical* criteria are constraints or objectives of project design and implementation. They include specific engineering requirements and site-specific conditions that may affect the technical feasibility of a project (and thus costs) on a particular site. Adverse technical impacts are generally overcome by more complex designs and higher costs.
- *Institutional* criteria are a "none-of-the-above" category for issues important to, or under the control of, other stakeholders. Site acquisition, conflict with public policy goals, and permits are some examples of this category of criteria.

- *Economic considerations* include the purchase price for the site, costs to mitigate unfavorable site conditions, capital costs for technologies or construction practices that are required for the sites, and 20-year (or more) operating costs of project implementation on a site.

It is important to note that when comparing multiple sites, only the differences between sites are important. If sites have the same issue, that issue does not affect the selection process.

3.2.1 ENVIRONMENTAL CONSIDERATIONS IN SITE SELECTION

3.2.1.1 General Considerations

The general rules when considering any kind of site development are (1) avoid environmental impacts by staying clear of the desirable or protected site feature(s); (2) minimize any impacts that cannot be avoided; and (3) mitigate unavoidable impacts. This three-step sequence is applied during site development planning but should be factored in earlier into the site selection process. The site selection process should result in eliminating sites that have major environmental constraints. For example, a small wetland may not necessarily preclude selection of a particular site but may significantly affect the layout of facilities on that site. The identification of site environmental features, such as wetlands, water bodies, floodplains, designated habitats, and historic/archaeological features, is critical in determining buffer zones, setbacks, and thus available area for each site.

3.2.1.2 Site Land Use

The current and previous uses of a particular site are important to its development potential. The project manager should obtain all available information on current and former site use, including former locations of buildings, structures, rail lines, underground tanks, and likely contaminants (often through the appropriate state regulatory agency that oversees removal of contaminated soils and groundwater). Possible sources of information include old aerial photos, soils maps, and land-use maps. If necessary, it is appropriate to perform a Phase 1 Environmental Site Assessment (ESA) to identify the presence of contaminated soil; buried, leaking fuel tanks; polychlorinated biphenyls (PCBs), buried herbicide and pesticide containers; and other dumping activities. When considering the demolition of existing structures, the presence of asbestos insulation and cement, flooring, and roofing materials should be assumed or identified. Phase 1 ESAs are typically inexpensive and can be performed quickly by experienced individuals before land purchase.

3.2.1.3 Surrounding Land Use

A determination of the effects of a WRRF on the surrounding land should consider area zoning regulations (including odor regulations), effects on adjacent property values, and compatibility with activities on neighboring properties. Constructing a facility in an industrial neighborhood rather than a residential area (particularly an affluent area) is generally more acceptable and less expensive. Water resource recovery facilities located near airports may require U.S. Federal Aviation Administration approval, because airplane glide slopes may control the height of some structures. Lagoons and ponds attract birds, which also affect air-traffic operations. If the selected site is surrounded by residences or zoned residential for future development, measures must be taken to minimize noise, odors, aerosols, air particulates, chemical hazards, insects, intrusive lighting, and traffic effects, and incorporate aesthetically pleasing architectural and landscaping features. However, all these factors affect costs, which should be planned.

3.2.1.4 Natural Resources

In general, sites within specially designated natural areas should be avoided. The development of areas designated as wild, scenic, or recreational, under the Wild and Scenic Rivers Act (1968), or habitats of rare, threatened, or endangered species may be prohibited or, at a minimum, very difficult. Similarly, shorelines often are reserved for public use; this is especially important in urban areas, where there is a shortage of open shoreline access.

3.2.1.5 Historical and Cultural Significance

Federal, state, and local entities that maintain catalogs of significant archaeological and historical areas and resources should be contacted during the site selection phase to determine whether the site has historical or cultural significance. If there is reason to believe that the site might have historical or cultural significance, a preliminary site investigation should be performed by qualified archaeologists or historians.

3.2.1.6 Setbacks and Buffer Zones
Setbacks (the distance between a building and the property line) are established by the building department having jurisdiction over the site. Buffer zones are recommended to reduce odor, noise, and light intrusion to the surrounding community. The amount of isolation and buffer area needed between facility processes and sensitive features and between facility processes and other property owners needs to be identified and included in the area requirements.

3.2.1.7 Air Quality
Water resource recovery facilities can be sources of odor, chemical emissions, particulates, and aerosols, all of which must be controlled and permitted. However, air quality (like seismic activity) is not often a differentiator; generally, all the sites will require the same air quality measures equally.

3.2.1.8 Noise Impacts
Noise management is an important consideration that can influence the site selection process. Communities may have noise ordinances that need to be incorporated into the comparative process. When comparing possible facility layouts, the difference in sound pressure can be quantified with the corresponding differences in distance. A survey to determine ambient levels of noise is recommended at any proposed WRRF site. A 3-dBA increase above ambient levels of noise by the WRRF has little or no effect on surroundings; noise level increases of 3 to 15 dBA have a moderate effect; and noise level increases of more than 15 dBA severely affect the surroundings (National Council of Examiners for Engineering and Surveying [NCEES], 2005).

3.2.1.9 Hazardous Chemicals
There is always a potential for chemical spills and leaks. For this reason, sites that are far removed from sensitive receptors, such as schools, hospitals, daycare centers, and convalescent homes, are preferable. If there are sensitive receptors near a proposed site, the use of chemicals and chemical safety must be addressed during the design process, leading to higher costs. These additional costs should be factored into the site evaluation process.

3.2.2 TECHNICAL CONSIDERATIONS IN SITE SELECTION

3.2.2.1 Proximity
An ideal site should be close to the raw wastewater collection system, treated effluent disposal point (or reuse application points), and biosolids disposal locations. It may be obvious that a site close to the existing collection system or the discharge point is preferred over a remote site. However, a remote site may be closer to the biosolids disposal location, thereby lowering biosolids hauling costs, so the proximity issue is not always clear, and a more detailed analysis may be necessary.

3.2.2.2 Discharge Permit Conditions
When multiple sites discharge to the same water body, the site selection process is unaffected by permit conditions. However, one potential site might be close to a receiving water body that requires tertiary treatment, while an alternative site might discharge to a different receiving water body that only requires secondary treatment. Under this scenario, the area requirements and costs for additional tertiary treatment, or conveyance back to the other receiving water body, must be factored into the selection process.

3.2.2.3 Elevation and Topography
Low-lying sites near the point of discharge facilitate the flow of wastewater from the service area by gravity and minimize the number of pumping stations in the collection system. However, such a site also may require flood protection. Constructing dikes, process tanks, buildings, entrances, or other structures, above the expected high-water levels provide flood protection, but these methods can be costly and may minimize the advantages of selecting a low-lying site. When contemplating the use of a site in a floodplain, a designer should contact the local floodplain management and local building authorities to identify the design flood event (100-year or 500-year floodplain) and any potential restrictions on property development.

A relatively flat site will better facilitate construction activities than a steeply sloped site. A site with significant topographic change may provide challenges for gravity flow, significant earth cuts and fill, vehicle access, and proximity between buildings and processes, all adding costs that need to be taken into consideration.

3.2.2.4 Geology, Hydrogeology, and Soils

Site geology, hydrogeology, and soil types significantly affect construction costs and are therefore important considerations in site selection. Unfortunately, adverse geology such as shallow rock formations and/or high groundwater conditions may not be readily apparent until soil borings are taken much later during the design process. Engineers can often get a qualitative understanding of soil conditions by working with a geotechnical consultant, or by reviewing well logs where available. If adverse geologic conditions are suspected, designers should plan for additional costs.

3.2.2.5 Seismic Activity

Seismic activity is generally not a differentiator and will not influence site selection when all potential sites are within the same seismic zone. However, if sites are within different seismic zones, additional design and construction costs to accommodate the seismic activity need to be factored into the evaluation. Generally, the building department for each community will confirm the seismic zone that must be accommodated.

3.2.2.6 Transportation and Site Access

A WRRF should be located close to all-weather roads for delivery of equipment and chemicals and off-site disposal of grit, screenings, and solids. Access roads that have reasonable gradients and curves of adequate radius allow the movement of large vehicles and heavy equipment. Service from a railroad spur offers the facility the option of accepting deliveries by rail, thus lowering the unit cost of bulk deliveries (i.e., chlorine and other chemicals). Costs for nonexistent or inadequate roads or bridges need to be included in the cost analysis. Access to public transportation supports worker satisfaction and reduces parking needs.

3.2.2.7 Utility Services

A WRRF should have potable water, redundant electrical power, and natural gas service, all of which can be expensive. Telephone and internet service are also important but easily supplied. If a site has strong potential but lacks nearby utility service, the design engineer should meet with representatives of utility companies to discuss utility sizing, extensions, and potential costs. Planning for utility extensions and capacity increases may take several years; and this process should be started early when planning and for site selection of a WRRF.

3.2.3 INSTITUTIONAL CONSIDERATIONS IN SITE SELECTION

Institutional criteria include issues that are important to, or under the control of, other stakeholders. Site acquisition, conflict with public policy goals, and permits are some examples of this category of criteria. State, regional, and local policies and interests also will play a role in shaping the criteria.

3.2.3.1 Sustainability

Consideration of sustainability goals in WRRF design has become important in recent years. This same focus on sustainability goals applies during the evaluation of alternative sites for WRRFs. Some states, and even regional and local agencies, have adopted sustainability policies and requirements that are relevant to site selection, building materials, and construction.

3.2.3.2 Land Acquisition and Ownership

The number and type of property owners of a proposed site (or adjacent sites) can significantly affect the degree of difficulty in acquiring the site. Ideally, one site has only one landowner who is a willing seller. As the numbers of sites, owners, and resistance to selling increase, so will the time and difficulty involved in acquiring the property.

Land acquisition also applies to off-site infrastructure (i.e., easements for collection mains). Ideally, pipelines can be installed within existing easements and rights-of-way; however, if extensive new easements are required, this consideration also should be factored into the site selection process.

3.2.3.3 Environmental Justice

Environmental justice, also referred to as environmental or social equity, represents the confluence of social and environmental movements and deals with the inequitable environmental burden borne by minority groups. The importance of environmental justice in the site selection of wastewater facilities has increased significantly in recent years. Today, the site selection of a WRRF is likely subject to federal, state, or regional environmental justice procedures, requiring documentation of the population characteristics of the host community in terms of minority and other less-represented groups, such as seniors, and evaluation of the effects of the proposed project on the well-being of those groups. Thus, evaluation and ultimate selection of a WRRF site should include evaluation of the relevant environmental justice factors.

3.2.3.4 Permit Requirements

The regulatory process(es) driving the site selection can influence site selection. If there are specific regulations that pertain to site selection of the project, the criteria should reflect these regulatory requirements. For example, if there are wetlands on a site, the site selection decision may be dependent, in part, on obtaining a U.S. Army Corps of Engineers permit to allow wetland filling. Local zoning ordinances are another example of an approval requirement that could heavily influence a site selection decision. It is imperative that all permits be identified during the planning process and a determination be made as to their relevance to a site selection decision.

The regulatory process(es) driving the site selection will strongly influence the choice of site selection criteria. If there are specific site selection regulations that pertain to the project, the criteria should reflect the regulatory requirements. For example, some states have, as part of their solid waste, hazardous waste, or groundwater protection regulations, specific criteria for evaluating sites for waste-handling facilities. Appropriate regulations should be reviewed when the list of criteria is compiled.

Local policies and interests also will play a role in shaping the criteria. One example of a local-interest criterion is "equitable distribution of regional resources." This criterion may or may not be important, depending on the service area of the proposed facility and the interests of the involved communities in making sure that public facilities are equally spread among the communities they serve. Input on these local criteria is best obtained from local officials and citizens.

3.2.3.5 Stakeholder and Public Involvement

In the context of WRRF siting, the term "stakeholder" generally refers to the group(s) or individual(s) that will play a significant role in the siting process. The project owner and/or representative plays a major role in determining the course of the project, ensuring that it remains on a predetermined course, schedule, and budget. Other stakeholders may include regulatory agencies, local officials, and members of the general public. Regulatory agencies must be convinced that the project meets regulatory requirement, whereas local officials and groups, as well as members of the public, will have interests in the project if they may be directly affected by the proposed facility.

All stakeholders have specific roles and appropriate opportunities for input regarding site selection; therefore, the stakeholders and roles should be determined early in the siting process. The public can play a crucial role in formulating, as well as implementing, the siting approach. This is especially true for controversial projects. The project manager should develop a public participation plan that describes the form and level of public involvement. Hiring a public relations expert to help guide the process is highly recommended, particularly for large and/or complex projects. When executing public participation programs, "more," "earlier," and "full-disclosure" are key to successful public involvement, particularly in the case of controversial siting studies for which project success may depend, in large part, on public approval.

3.2.4 COST CONSIDERATIONS IN SITE SELECTION

Costs of the actual WRRFs are likely the same regardless of the site, unless alternative discharge water bodies are available and thus alternative permit and WRRFs are required, or if topography requires an intermediate pumping station. Where costs are the same, they do not affect site selection. However, it is likely that additional treatment and features may be required at certain sites to be good neighbors. In some cases, compensatory/incentive measures may be needed for increasing the regulatory and public acceptability of projects. Compensatory or incentive measures are sometimes required to gain political and public acceptance of a project, and to make a whole project and its effects more acceptable to abutters and communities. Compensatory incentives generally are not legally required unless they are included in a contract developed between the project proponent and the host community specifically for the proposed project (i.e., a memorandum of agreement).

3.2.4.1 Capital Costs

Capital cost considerations for sites can include the cost difference for the following: land purchase; building and infrastructure demolition or relocation; brownfield/site/building remediation; wastewater conveyance (to the new site or discharge point); mitigation of wetlands, endangered species protections, archaelogical or historical concerns; additional effluent treatment (for differing NPDES permit limitations); geotechnical and/or topography hardship; electrical power, water, and gas utilities services; access road and/or bridge; flood protection; and good neighbor measures (aesthetically pleasing architectural features and additional landscaping, odor control, and/or noise abatement measures). The costs associated with most of these issues will initially be very difficult to predict. Where estimates can be reasonably predicted, they should be developed. However, where the planning budget is constrained or where not enough information is known, contingency factors can be applied to the overall cost estimate for each site.

3.2.4.2 Operations and Maintenance Cost Considerations

Operations and maintenance staff are likely to be identical for each site, and do not affect the site selection process. However, O&M costs for differing site conditions can include the following: conveyance of wastewater (pumping to the site or discharge point); land disposal cost differences (principally hauling costs); and O&M costs for additional treatment or mitigation. As with capital costs, O&M costs may be very difficult to estimate, and therefore, contingency factors can be applied to the overall cost estimate for each site.

3.3 DETERMINING WATER RESOURCE RECOVERY FACILITY SITE AREA REQUIREMENTS

It is difficult to derive an average area requirement for WRRFs because there are so many influencing variables, including odor control, effluent quality, biosolids handling, renewable energy systems, and many other factors. Project managers are encouraged to be conservative during this process, or to use practical and conservative assumptions to bound the site area requirements into minimum and maximum needs. To start this process, the following are identified as the major factors affecting area requirements: flow and constituent concentrations, effluent NPDES permit requirements, solids treatment processes, and ancillary facilities.

3.3.1 SOLIDS TREATMENT AREA REQUIREMENTS

Solids treatment area requirements are greatly affected by the ultimate treatment and disposal process selected. Thickening, dewatering, and landfilling will require lower area requirements, whereas digestion and/or land application (and liquid sludge storage) will require more land. Ancillary facilities include administration, maintenance, laboratory buildings, and parking areas for each. The area requirement will also include flow equalization roads and greenbelts for yard piping, splitter boxes, septage or vactor receiving station, substations and generators, and other utilities, space for buffer zones, setbacks, environmental controls, and spill containment areas. Additional area requirements may include off-site sewer maintenance and/or industrial pretreatment program equipment and staff, composting facility, and the like. Once the approximate processes and facilities are selected, area requirements for each can be approximated. A simple approach is to determine areas of local WRRFs similar in size and treatment and plot flow versus area. Google Earth or Google Earth Pro can be used to assist if the actual area is unknown. Figure 2.2 can be used to approximate the area requirements based on WRRF capacity. A more exacting approach is to size the various circles and boxes for each process area as described below.

3.3.2 LIQUID STREAM AREA REQUIREMENTS

The traditional processes to size include headwork (bar racks, influent pump station, fine screenings, and grit facilities), primary clarifiers, aeration tanks, secondary clarifiers, tertiary filters (if required), and disinfection. Hydraulic Institutes ANSI/HI 9.8 can be used to determine geometry and pump placement to determine overall pump station size. Dimensions required for different unit operations and processes involved in preliminary, primary, secondary, and advanced or tertiary treatment processes should be estimated. Considerable area needs to be added to the footprint of any building for process equipment, means of equipment removal, pipe galleries, electrical substations and motor control center equipment,

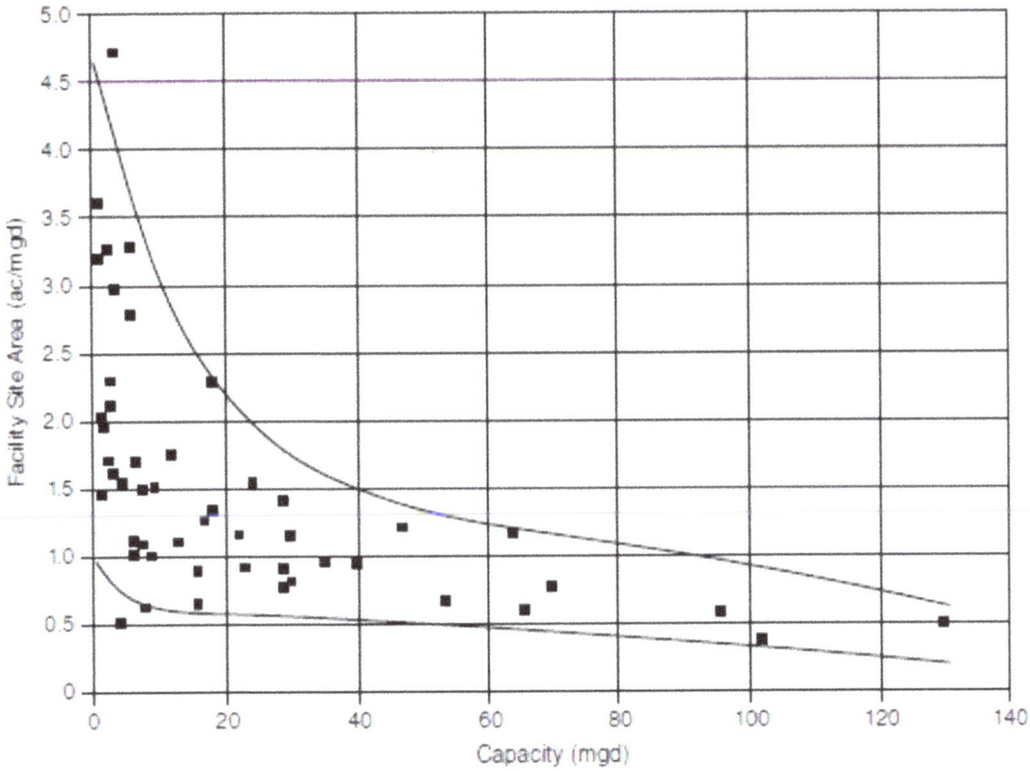

Figure 2.2 Facility Site Area Versus Facility Capacity (mgd × 3785 = m^3/d; ac × 0.4047 = ha)

control room, hallways, lavatories, and other structures. Yard piping does not need to be sized, but adequate room for pipe and splitter boxes need to be accounted for. An additional lift station may be necessary, depending upon site topography.

3.3.3 SOLIDS PROCESS AREA REQUIREMENTS

The type of solids treatment processes has a larger effect on area requirements than liquid treatment processes, and project managers are cautioned to use conservative estimates so that when a site is selected, there is room for alternative process types. Gravity thickeners require more area than mechanical thickeners, although the difference is not significant. Similarly, centrifuges take up less space than other equipment, but again the difference is negligible. More importantly, any preliminary equipment sizing should have ample room in the building envelope for maintenance access and removal. As previously noted, digestion should strongly be considered for energy recovery, and requires significant area. A loading facility with ample truck access and trailer storage should be added to the area requirements. Solar sludge drying beds, including sludge feed conveyors, require very large areas and could be assumed for a maximum area estimate.

3.4 CANDIDATE SITES AND SITE SELECTION

Once area requirements are determined, initial candidate sites meeting the minimum area requirements need to be identified. Initial candidate sites can be determined by identifying sites previously considered for similar uses or intensity of development; following active public solicitation for available land; and using criteria to map "exclusionary areas" where the facility should not be located and to identify the remaining areas as possible areas for consideration. Typically, geographic information system models and attributes are used to determine candidate areas and, sometimes, specific sites. Features such as wetlands, floodplains, steep slopes, groundwater recharge areas, poor soils, regulatory setbacks, and dedicated public land can be mapped as "exclusionary areas," which should be eliminated from further consideration in site selection investigations. Once exclusionary areas are mapped, the candidate sites would be selected from the remaining areas.

3.4.1 SITE EVALUATION PHASES

Generally, a qualitative and quantitative approach is used, typically accomplished in multiple phases, each resulting in the elimination of sites that do not meet the specified criteria defined for that phase. However, the number of phases will depend on the number of sites and the complexity of the project. The goal is to systematically narrow the field so that the least feasible sites are eliminated early in the process, leaving only the most feasible sites for detailed investigations. A two-phased evaluation process is described here.

3.4.1.1 Phase 1 Evaluation Criteria

Phase 1 evaluation criteria generally are of two types: critical constraints (including "fatal flaws" and mandatory regulatory constraints) and technical deficiencies. Fatal flaws are issues that have a high probability of prohibiting development of the project at a specific site. Mandatory regulatory constraints are nonnegotiable conditions that the facility is required to satisfy, or address protected areas that would be off-limits to the facility. Technical deficiencies are site obstacles that require significant engineering and design solutions conditions for the project to function properly. Some of the phase 1 critical constraints include land use compatibility, which should avoid developed residential areas and federal, state, and local parks; state or federally designated wetlands, floodplains, setbacks, buffer zones, and/or other officially protected areas; historic sites listed, pending, or recommended for inclusion in the National Register of Historic Places (state- and locally listed historic and archaeological sites also could be included here); state-catalogued endangered species habitat; state- or federally designated inactive hazardous waste disposal sites or corrective action sites or other sites known to be contaminated with hazardous wastes; and groundwater recharge areas.

3.4.1.2 Phase 2 Evaluation Criteria

At this stage, the facility could be constructed at any of the sites under consideration. All that remains is to determine which site is best from the perspective of environmental, technical, institutional, and cost criteria. It is critically important that this phase 2 selection process is quantitative (wherever possible), transparent, and the gauging criteria for nonquantitative criteria are established prior to the evaluation process. It will be easiest to base all decisions on costs or a cost-effective analysis, and therefore the differing environmental, technical, and institutional issues for each site should be translated to costs wherever possible.

For instance, under environmental criteria, adverse site land use conditions can be quantified by estimating the cost to remediate that site to a comparison site or to an assumed standard. Similarly, costs to mitigate noise, odors, or chemical hazards, and to create aesthetically pleasing architectural and landscaping features, must be estimated. Costs to mitigate technical issues can also be estimated. Conveyance (proximity), additional treatment (permit conditions), earthwork or intermediate pump station (elevation and topography), piling or groundwater pumping (geology, hydrogeology and soils), and electrical substations (utility services) can all be estimated, or factors applied to the overall costs for a given site.

3.4.2 EVALUATING THE SITES AND SITE SELECTION

At this point, costs for quantified criteria for each site are presumably established. Once the project team (including stakeholders) has selected the nonquantified criteria and defined them, the criteria must be applied objectively to each site. Inevitably, each site will have its own merits and limitations in each of the criteria categories, ideally resulting in a quantifiable cost-based assessment and a non-quantifiable ranking or level of impact. The final step in the process is to reach a site selection decision that is consistent with the established site selection criteria. The project manager can either recommend the final selection to the owner and stakeholders or present a few alternatives with the advantages and disadvantages of each. It is generally preferable for the project manager to present the "finalists" to the owner, thereby allowing the owner to make the final choice. One approach is to make a recommendation regarding the best site in each category of criteria (i.e., "best" technical site and lowest cost site) and then allow the owner to make the final selection. Once the site selection decision is made, agreements establishing site acquisition and development procedures (and often compensation) should be finalized with the host communities, as appropriate.

3.5 FACILITY ARRANGEMENT AND LAYOUT

Once the process treatment decisions have been made, WRRFs sized, and a site selected, the decisions on how to arrange the facilities are now made. Arrangement decisions are based on personnel and vehicular logistics as well as the physical connections between facilities. Facilities should be arranged to minimize the length of connecting and recycle piping, and to minimize staff personnel and vehicular travel around the site. In this manner, the arrangement of treatment processes on a facility site can affect the total life-cycle cost.

3.5.1 GENERAL CONSIDERATIONS

In general, liquid treatment processes can be arranged separately from solids treatment process. Likewise, the ancillary facilities (administration, laboratory, and maintenance buildings) can be arranged around the liquid and solids treatment facilities. The facility arrangement should start with isolating any areas of the property deemed less suitable for development, such as wetlandsor potential brownfield areas. Buffer zones, setbacks, 100-year floodplains, and other similar areas should also be identified as unavailable. Solids treatment facilities should be designated to areas less prone to odor complaints. From there, the facility arrangement can start with the location of the influent sewers and the ultimate discharge location.

The liquid stream facilities layout should progress sequentially from the influent sewer and lift station to the disinfection facility and outfall. This layout can follow a linear progression, or can progress out the back, particularly when the influent sewer and effluent discharge locations are nearby. Locating raw wastewater pumping stations near the point at which sanitary sewers enter the facility site minimizes the cost of building additional lengths of deep gravity sewers. Where possible, a designer should group together similar unit processes to facilitate operation, minimize piping, and allow for expansion. The designer should develop a preliminary hydraulic profile to establish the profiles of principal structures. Arranging treatment processes to follow site contours helps maximize the use of natural topography and reduces both pumping requirements and excavation costs for new structures. Flow splitting is another aspect that requires due consideration.

3.5.2 PERIMETER VERSUS INTERNAL FACILITIES

Many facilities should be located on the facility perimeter, within the WRRF or, for some facilities, cases can be made for both. Administration buildings and laboratories should be located on the site perimeter to facilitate visitors, sample deliveries, and the like. Facilities unlikely to cause odors, including tertiary treatment and/or ultraviolet disinfection facilities, can also be located on the perimeter. As previously stated, solids treatment facilities, particularly those likely to produce odors, should be located within the facility or in an area where odor complaints are less likely to occur. A significant exception is the biosolids disposal loading area; truck traffic should be kept to a minimum within the facility, so biosolids loading facilities should be located near the perimeter or near a major roadway. A separate access road for biosolids haulers should be considered. Chemical feed facilities have competing interests; they should be located near the application point, but a location that minimizes internal truck traffic should be considered. A maintenance facility should be centrally located within the WRRF because that will promote efficient logistics of personnel and parts. However, access for spare parts delivery must be considered. Chemical bulk storage facilities and chlorine storage areas should be located along the main service road to make deliveries more convenient. A separate set of ingress and egress gates and/or a loop road for chemicals and parts deliveries and for solids hauling should be considered.

In locating chemical storage and dosing facilities, rules, regulations, safety, compatibility of different chemicals, and O&M aspects should be considered. For example, there should be a minimum space between the areas where the public has access and where chlorine containers are stored. Incompatible chemicals should not be stored close to each other. A lime silo should be located as close as possible to the dosing point to minimize the O&M problems resulting from blockages.

Enclosures must be considered for blowers, compressors, large pumps, centrifuges, and other equipment that operate at high speeds in buildings of appropriate, sound-attenuating construction. If possible, facilities that generate noise should be located as far away as possible from potential receptors and the WRRF should plan for sound walls, berms, and heavy landscaping to be constructed in the surrounding area to minimize noise.

3.5.3 ADMINISTRATION, STAFF, AND SUPPORT FACILITIES

Adequate support facilities must be provided to ensure efficient O&M of a WRRF and accommodate the staff who operate and maintain the facility. The extent of provision of support facilities depends on various factors, including size and complexity of the treatment process, location, philosophy (how much is done by staff vs. outside contractors), and owner preference. Some WRRFs may require a high level of support provisions, including control rooms, administration offices, lunchrooms, training/conference rooms, locker rooms, maintenance workshops, storage rooms, laboratories, toilets for staff and visitors, reception areas, and visitor education facilities. Other facilities may want to contract out all but the routine maintenance and laboratory work. In the age of "doing more with less," combining WRRFs with Department of Public Works yards, potable WRRFs, or other municipal facilities should be considered and discussed with owners.

3.5.4 MAINTENANCE AND STORAGE FACILITIES

Provision of maintenance and storage room facilities depend on the WRRF and how much of the work is expected to be contracted out. The size and type (complexity) of the WRRF generally determines the size of the maintenance staff, the services to be performed in house, and the tools to be provided. As previously stated, maintenance facilities are best located near the center of a facility or in the most equipment-intensive area of the facility. Facilities with a high degree of instrumentation may require a separate instrumentation shop and maintenance staff. A designer should size the area that will house maintenance equipment manuals and shop drawings in addition to equipment records. In addition, facilities for personnel to review documentation and for computerization of maintenance records, inventory, and maintenance scheduling should be provided.

The storage area for the facility's spare parts and maintenance supply inventory is best located near the shop area and should be large enough to accommodate an array of shelves, bins, and drawers. Paints, lubricants, pesticides, herbicides, and similar toxic, flammable, and hazardous materials should be stored in an isolated, secure area with adequate ventilation. To facilitate deliveries, the storage area, which typically is secured, should be located adjacent to the roadway of the main facility. The building should be fully accessible to maintenance and delivery trucks and should include ramps for the passage of tractors, handcarts, and other mobile equipment.

3.5.5 LABORATORY FACILITIES

A laboratory is necessary to analyze samples for the purpose of operational control and regulatory monitoring and reporting. The size of the facility, type of treatment provided, need to analyze samples from off-site facilities (such as combined sewer overflow facilities or industrial pretreatment program samples), and extent of sample analyses to be performed on site determine the size and layout of the laboratory. For many small facilities, it may be cost-effective to contract all but the simplest analyses to outside laboratories. A small on-site laboratory can be cost-effective for conducting routine operational control tests, along with a large laboratory in a convenient location outside the facility for conducting complex tests for the facility and tests for outsiders. During the design stage, a designer should identify the type and anticipated frequency of tests and the analytical equipment required to conduct them.

3.5.6 OTHER LAYOUT CONSIDERATIONS

3.5.6.1 Roadways and Walkways

Roadways must provide access for all O&M needs (including access to double and overhead doors, access hatches, and roof- or yard-mounted heating and ventilation equipment), chemical and parts deliveries, vactor and septage receiving stations, and residuals loading areas. Pavement widths, adequate curve radii, and grades should be planned accordingly. Main roadways of 6 m (20 ft) in width and service roads of 4.9 m (16 ft) in width have been adequate at most facilities (J. M. Montgomery Consulting Engineers, 1985). However, 3.6 m (12 ft) is the minimum width recommended for one-way use. It is a good practice to provide looped roads to eliminate or minimize vehicle reversal. It is also advantageous to have two entrances—one for the staff and facility visitors, and another for deliveries, septage haulers, and the removal of residuals. Emergency access should always be provided for ambulance and fire trucks. The local fire department may have specific requirements for equipment vehicle turnarounds and width and location of fire lanes. Paved sidewalks at least 1.2 m (4 ft) wide should be provided in all areas of the facility, particularly for sampling stations, buildings, tanks, and other areas requiring frequent monitoring. Lighting should also be provided, particularly for outdoor equipment.

3.5.6.2 Security

Access to the site must be controlled. A perimeter fence and lockable gates, minimum 2.5 m high (8 ft), should be provided. In some areas, barbed wire will be required. Public entries may be controlled via gates that remain open during business hours, while access to the remainder of the facility may require on-demand access. Security gate systems may be remotely controlled using card access or phone systems. In larger facilities, closed-circuit television may be used to control facility access and maintain security. However, in designing security measures, access to the site by outside emergency response teams should be considered.

3.5.6.3 Site Drainage

Greenbelts, roadways, and parking areas must be designed for positive drainage. Where possible, grassed swales and infiltration should be encouraged to reduce hard piping that may conflict with other facility yard piping. Ponding of stormwater should be avoided because of the potential to attract undesirable species (i.e., geese and mosquitoes) and additional maintenance requirements. Stormwater from developed areas of WRRF sites generally cannot be discharged to receiving waters without a permit. The U.S. Environmental Protection Agency (U.S. EPA) (2021) states that a municipal WRRF with a design flow of 1 million gallons per day or more is considered an industrial regulated activity and requires an NPDES permit, 40 CFR 122.26(b)(14)(i)-(xi). Therefore, a designer should consult federal, state, and local requirements concerning stormwater handling.

3.5.6.4 Septage and Vactor Receiving Stations

The delivery of septage and vactor solids can create significant traffic, and a separate gate and access route should be provided. The unloading area should be located close to the headworks area, so wastes can drain by gravity from the holding tank to the influent raw wastewater pumping station or be pumped to the screening, metering, and degritting facilities. The unloading area should be designed to contain and control odors and provide wash-down facilities. A separate loading and weigh station typically is provided as well.

3.5.6.5 Visitors and Vehicle Storage and Parking

Parking should be provided for all personnel, disabled employees, and visitors. Visitor parking should be marked and placed close to the administration building. Large facilities or facilities anticipating visits from civic groups or schools should provide space for bus parking with adequate turnaround space or a looped drive. Where in-facility tours are provided, the agency should provide each visitor with a protective hard hat, protective glasses, and ear protection, as appropriate. Space should be provided to store this equipment in the administration building. Safety considerations concerning design issues can be found in the WEF Manual of Practice No. 1, *Safety, Health, and Security in Wastewater Systems* (WEF, 2021).

3.5.6.6 Architecture and Landscaping

Pleasing architecture and attractive landscaping greatly improve a WRRF's image and provide a pleasant atmosphere for the staff. This is particularly important if the facility is located in a scenic area or in a residential neighborhood. In such cases, the additional costs of special architectural treatment and landscaping are justified. In residential areas, the buildings and site should be designed to integrate into the surrounding neighborhood. In older industrial areas, a new, aesthetically pleasing WRRF can form the nucleus for the revitalization of the surrounding area.

3.6 LAYOUT EXAMPLES

Two layout examples are presented for discussion. Figure 2.3 presents a linear layout where the influent enters the facility at one end, the effluent at the other end, and the process flows linearly between the two. The site has a multiple-sites road providing ample room for truck access to each process area. The administration building, laboratory, and maintenance facilities are all located in one large building, which is efficient, but can complicate access when multiple vehicles are all accessing the same location. There are ample buffer zones around the perimeter and room for expansion. There is only one access road to the site, which is efficient and less expensive, but the single access point can get congested and requires a temporary access road when new pavement or pavement repairs are needed. Solids processing facilities are all located in one area, which is preferred.

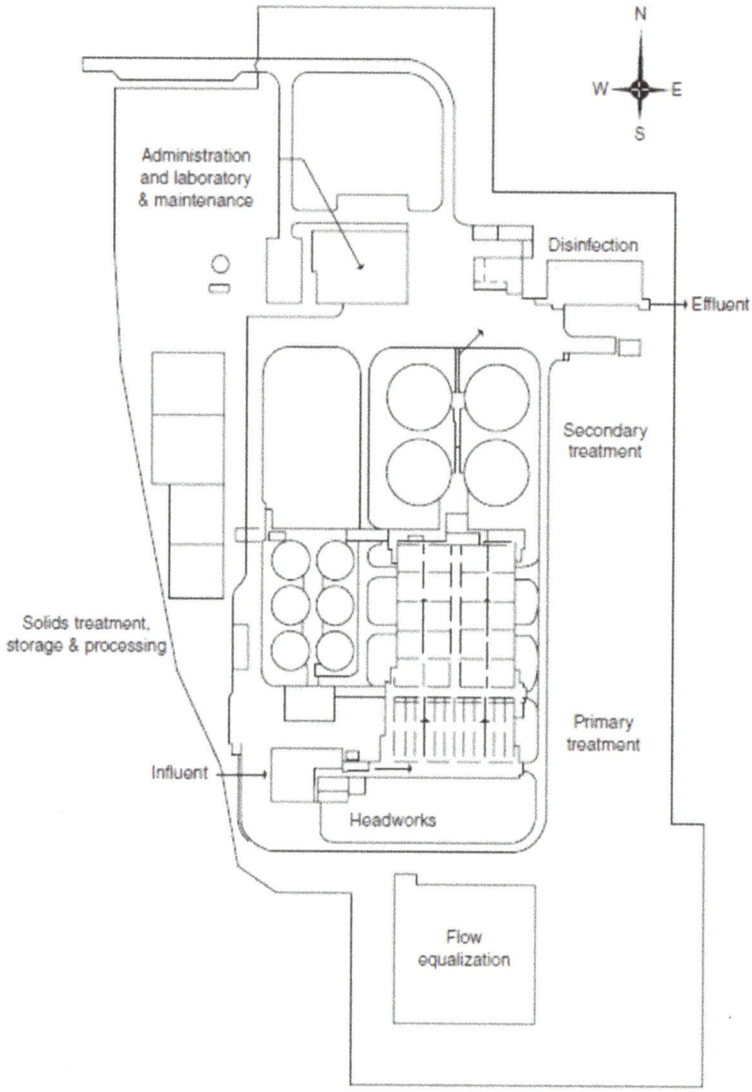

Figure 2.3 Linear WRRF Layout (Without Tertiary Treatment)

Figure 2.4, in contrast, presents a nonlinear WRRF on a triangular-shaped site. Like many older WRRFs, this site housed a primary WRRF initially, and grew with multiple expansions as population growth and permit limitations required it.

The flow follows a route that was dictated by site limitations. (Flow enters and leaves the site at the original locations, but progresses around the site as various process expansions were added.) The multiple-perimeter and access roads provide good site access and support isolated solids disposal truck traffic. The administration and laboratory buildings are combined and located on the perimeter with a parking lot, facilitating visitors. The maintenance building is separate and on the perimeter, which supports deliveries but requires longer travel for maintenance staff. There is little or no buffer zones or setbacks, and several of the buildings are on the edge of the right of way. A service building supporting O&M staff, locker room and shower, and parking lot are centrally located, which is ideal. Solids processing facilities are all located in one area, which is preferred.

4.0 COST ESTIMATION

A planning-level cost estimate typically is developed as part of a facilities planning effort and often is based on general facility footprints and conceptual process layouts. Although this estimate is updated

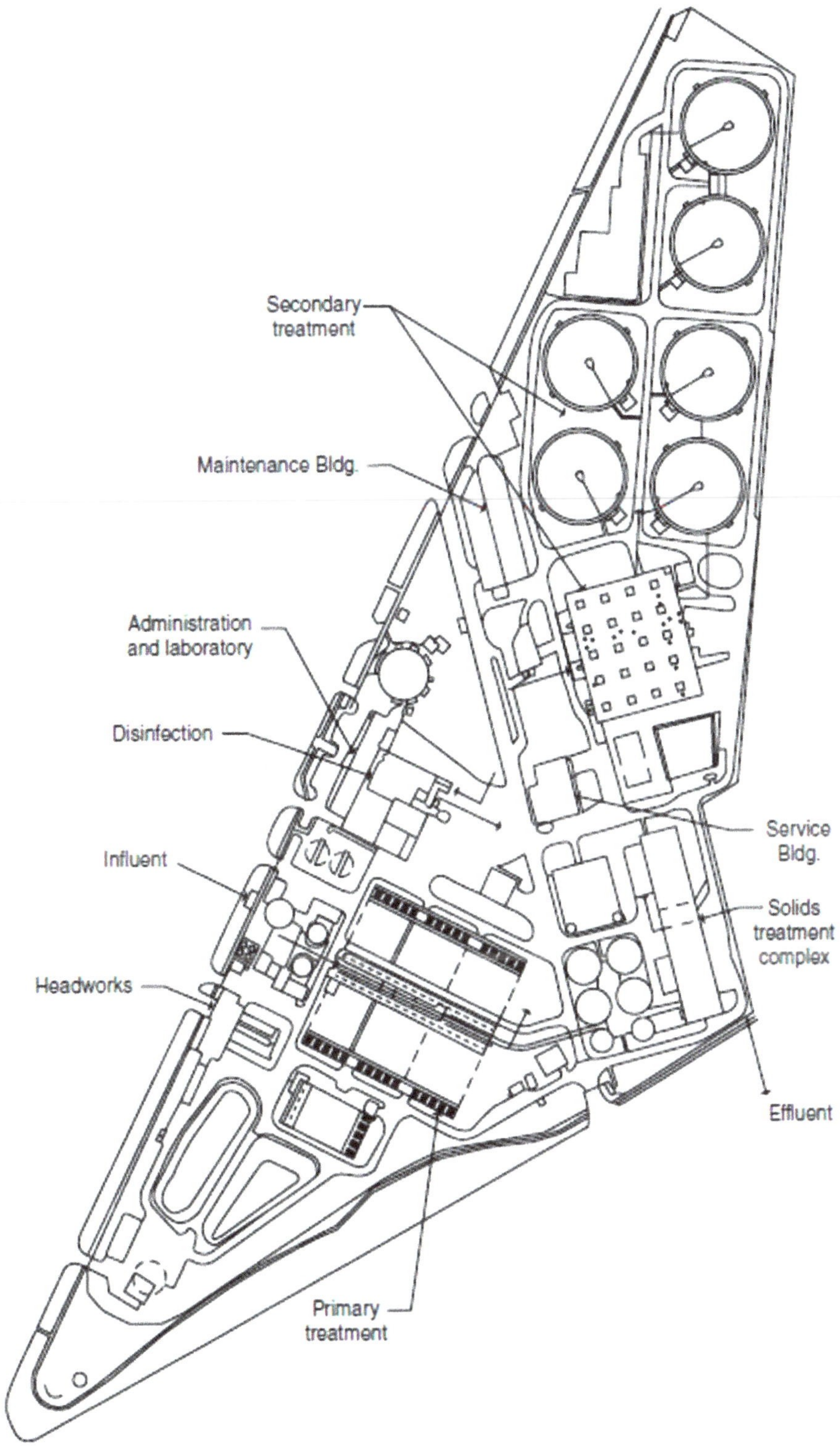

Figure 2.4 Nonlinear and Irregular Site

as the design progresses, it is critical to be aware of the initial cost estimate and how it was used in the development of a community's wastewater financial plan.

Water resource recovery facility projects represent large capital expenses for a community and typically are funded through a mix of funding sources. User charges or rate revenues typically fund the portion of the WRRF upgrade benefiting existing customers. Funds required for capital improvements compete with other rate-funded demands, including O&M of the wastewater utility and debt service. Connection or impact fees are another revenue source used to fund WRRF improvements. Impact fees are often developed based on adopted capital improvement plans and are available to fund only that portion of project costs attributable to growth. Rate revenue and impact fees are common revenue sources to retire municipal general obligation and revenue bonds that provide long-term financing of major capital improvements.

Miscellaneous revenues for funding a WRRF project include permit fees and fines, revenue from industrial surcharge fees, septage acceptance fees, food/organic waste acceptance fee, intergovernmental service agreements, and public/private partnerships. These can be significant sources of revenue for the construction of new facilities. State and federal block grants, state revolving loan funds, and rural community assistance programs also may be available to complement local funding sources. A more detailed discussion of rates and financing can be found in the WEF Manual of Practice No. 27, *Financing and Charges for Wastewater Systems* (WEF, 2018).

Estimating construction costs is an essential component during the planning and design of a WRRF. During the development of estimated project costs, it is important to understand the documents that will be used to develop the estimate as well as how the estimate will be used, for what purpose, and by whom. This section presents different types of cost estimates, the resources available to the designers developing cost estimates, and the application of markups. It also discusses the cost-estimating process as it relates to alternative delivery projects.

4.1 TYPES OF ESTIMATES

The Association for the Advancement of Cost Engineering (AACE) has developed definitions to help categorize cost-estimate types (AACE, 2020). The association's cost estimate classification system considers the most significant characteristics of a cost estimate, including degree of project definition, end usage of the estimate, estimating methodology, expected accuracy range, and the effort and time needed to prepare the estimate. Based on this characterization, the AACE has established five cost-estimate classes, which are defined in Table 2.4. A Class 5 estimate is based on the lowest level of project definition, and a Class 1 estimate is closest to full project definition.

End usage classification of a cost estimate can also be used to define estimate classification. However, the terminology used to define the different classifications (i.e., order of magnitude, budget, or definitive) tends to depend on the identity of the stakeholder (e.g., owner agency, designer, or contractor) and on the intended use for the estimate (e.g., capital planning or project funding, engineer's estimate, and bid development).

4.2 DIRECT AND INDIRECT COSTS

Construction cost estimates are composed entirely of direct capital costs. These costs include: land and site development costs; costs of utilities, transportation, and other services to the site; relocation costs; materials, equipment, and labor costs, including those needed for temporary facilities and systems required to maintain existing functionality of the facility; and all construction costs, including contractor's overhead, profit, mobilization, permits, bonds and insurance, and construction contingencies.

Indirect capital costs consist of engineering, permitting, and legal services during construction and any other associated costs. Additional project contingencies, beyond the construction contingency listed above as a direct capital cost, should be added depending on potential for significant changes in scope or schedule, bidding climate, risk tolerance, or other larger project factors. The sum of direct capital costs and indirect capital costs equals the total capital cost for a project. It is important to summarize direct capital costs and indirect capital costs separately and include contingencies for each category. Combining direct and indirect costs and contingencies could leave an unclear impression of the true individual estimates of construction and engineering costs.

4.3 LEVELS OF ACCURACY

The accuracy range of an estimate will depend on several factors related to the quality and depth of the input and the actual estimating process used. Other factors, such as the state of the technology being considered and the quality of reference cost-estimating data, also play an important role in defining

Table 2.4 AACE Cost Estimate Classification Matrix (AACE Recommended Practice 17R-97, 2020; used with permission from AACE International)

	Primary Characteristic	Secondary Characteristic			
Estimate Class	Level of Project Definition Expressed as % of Complete Definition	End Usage Typical Purpose of Estimate	Methodology Typical Estimating Method	Expected Accuracy Range Typical +/−-Range Relative to Best Index of 1[a]	Preparation Effort Typical Degree of Effort Relative to Least Cost Index of 1[b]
Class 5	0% to 2%	Screening or feasibility	Stochastic or judgment	4 to 20	1
Class 4	1% to 15%	Concept study or feasibility	Primarily stochastic	3 to 12	2 to 4
Class 3	10% to 40%	Budget, authorization, or control	Mixed, but primarily stochastic	2 to 6	3 to 10
Class 2	30% to 75%	Control or bid/tender	Primarily deterministic	1 to 3	5 to 20
Class 1	65% to 100%	Check estimate or bid/tender	Deterministic	1	10 to 100

Note. Used with permission of AACE International, 726 East Park Ave., #180, Fairmont, WV 26554. Email: info@aacei.org; Phone 304.296.8444; website: web.aacei.org.

[a]If the range index value of "1" represents +10/−5%, then an index value of 10 represents +100/−50% (at an 80% confidence interval).

[b]If the cost index value of "1" represents 0.005% of project costs, then an index value of 100 represents 0.5%.

accuracy range. Generally, estimate accuracy correlates with estimate classification and thus with the level of project definition, though this can be significantly affected by extended periods between the time of the development of the estimate and the expected schedule of construction. Table 2.4 includes expected accuracy ranges as a function of the estimate class as defined by the AACE (AACE, 2020).

4.4 QUANTIFICATION

The level of quantification needed for a specific estimate varies with the completeness of the design effort. For example, a Class 5 or Class 4 estimate, which rely on a conceptual and/or schematic design, will primarily use volumetric data such as facility capacity, pipeline flowrates, and tank volume requirements. A Class 1 estimate performed near the completion of contract documents will require detailed quantification of materials for all facilities, with vendor quotes and unit cost development that are specific to the area where the project is being constructed.

The construction industry typically refers to material quantities in ways that standardize the quantification process. As the estimate evolves from project definition through the contract document preparation phases of design, the quantities will change. The quantities of construction materials used in a construction cost estimate should be organized in an orderly manner. Standard forms make it easy to organize quantities on the forms by facility and then by specification section. This organization will allow quantities to be checked by project team members not directly engaged in the preparation of the cost estimate (i.e., design team).

4.5 COST RESOURCES

Pricing input to construction cost estimates is derived from various sources, including cost data books and standards or computer programs. It is important to know how the costs for a particular guide or program are structured and how they should be applied to the estimate being produced. Utilizing pricing input data specific to the geographic area where construction will take place will account for market-specific considerations such as labor and supply cost differences. Other cost resources include written and telephone quotations from manufacturers and vendors, quotes from contractors and subcontractors, estimates from similar completed projects and bid tabulations, cost curves, cost/capacity ratio formulas, construction indexes, and minimum wage rates.

4.6 APPLICATION OF MARKUPS AND CONTINGENCIES

The format of an engineer's construction cost estimate is similar to that of a general contractor's estimate for bidding. After the direct costs of material, labor, equipment, and subcontractors are subtotaled (including allowances to cover items that can be identified but not quantified at that level of design development), markups in the form of overhead, profit, mobilization, bonds and insurance, and construction contingencies are applied to arrive at the total bid price. Typical values of these markups are as follows: overhead (5% to 10%); profit (5% to 10%); mobilization (3% to 10%); bonds and insurance (1.5% to 2%); escalation for schedule (the time period between development of the estimate and construction), based on a local inflation rate compounded by the duration; and construction contingency (0 to 30%).

Markups are applied to the subtotal in a compounding manner. For example, the overhead markup is applied to the direct construction cost subtotal to obtain a subtotal that includes overhead. The profit markup is then applied to the subtotal that includes overhead to obtain the subtotal with overhead and profit. The compounding continues so that construction contingency is applied to the subtotal that includes overhead, profit, mobilization, and bond and insurance. Indirect costs such as engineering and implementation and project contingency are then added to the total bid price to calculate the total project cost.

Contingency is a true capital cost that accounts for future design definition and development as well as events that experience has shown will likely occur, including small changes in scope, variation in bidding climate, and other factors. The quantification of risks associated with these factors should be considered when developing the contingency. A contingency is different from an allowance: it varies with the level of an estimate. The greater the engineering detail provided, the lower the contingency. All estimates must have a contingency, including estimates prepared near or at the completion of contract document preparation.

4.7 ESTIMATE DOCUMENTATION

Complete support documentation provides defensible construction cost estimates to facility owners while helping to minimize the liability risk to the designer. The estimate should include the purpose, scope of work, assumptions, cost resources, markups, schedule, allowances, estimate accuracy, constructability, and statement of qualification and construction cost index, such as that established by *Engineering News-Record* (ENR) (http://www.enr.com).

4.8 ALTERNATIVE DELIVERY COST DEVELOPMENT

Alternative approaches for project delivery are becoming more common for the design and construction of WRRFs. The design–build model illustrates some of the special cost-estimating considerations associated with alternative delivery (Design–Build Institute of America, 2009). Elements of design development may occur differently with a single, integrated design and construction team. For example, procurement of long-lead items may occur much earlier in the project. This will allow definitive costs for this equipment to be known earlier, though connection of services to the equipment may still require design development. Depending on the scale of the equipment pricing, early knowledge of accurate pricing may result in reduction of construction contingency or elimination of escalation of costs due to schedule.

5.0 GUIDELINES FOR LIFE-CYCLE COST EVALUATIONS

This section presents guidelines for cost-effectiveness comparisons of WRRF project alternatives and general procedures for determining life-cycle costs (LCCs). These guidelines and procedures are based on the cost-effectiveness guidelines presented in Appendix A to 40 CFR Part 35, Subpart E (U.S. EPA, 2016).

5.1 PROCEDURES FOR PRESENT VALUE

Life-cycle cost comparisons for WRRF projects can be made using present-value or equivalent uniform annual value methodologies. The present-value method is a powerful tool because future expenditures are transformed into equivalent costs in today's dollars. Present-value costs provide a method to evaluate capital costs and the annual O&M costs of alternatives on an equivalent basis. All future capital and operating costs during the planning period are converted to an equivalent value during the base year. The present worth of an alternative is the amount of money in today's dollars that must be available during the base year, with a discount rate, to pay all anticipated capital and operating costs associated with the alternative through the end of the planning period. Alternatives with the lowest present worth are the most cost-effective over the life of the project. Common parameters used in all cost-effectiveness comparisons include design lives, discount rate, equipment and structure life expectancies, and the base year for analysis. Theories of present-worth analysis, equations used to calculate present-worth costs, and tables providing discount-rate factors for the equations can be found in any engineering economics text.

A present-worth analysis should be conducted using a base year that represents the time of the present-worth analysis or the initial year of operation of the facilities under construction. Capital costs should be referenced to an ENR construction cost index. Current costs typically are used for the analysis. Inflation rates are applied for detailed financial analyses undertaken to identify the financial effects of a selected alternative, but it is not necessary to include inflation rates when comparing alternatives during facilities planning.

The present worth is calculated as the amount of money, $Y, that must be invested for n years at interest rate i to have $X available every year, or

$$\$Y = \left[\frac{1}{(1 \div i)^n}\right]\$X \tag{2.1}$$

Where $Y is the amount that must be invested ($),
i is annual interest rate as a fraction,
n is number of years, and
$X is amount available every year ($).

The term

$$\left[\frac{1}{(1 \div i)^n}\right] \tag{2.2}$$

is called the *present worth factor*, or PWF.

Table 2.5 shows capital recovery, present worth, and sinking factors for a 6.125% interest rate. From Table 2.5, the PWF for 10 years is 7.316. If, again, $10 million is required, the funds needed every year would be

$$\frac{\$10\,000\,000}{7.316} = \$1\ 366\ 700$$

as before.

Example 2.1

A town wants to invest money in an account drawing 6.125% interest so that it can withdraw $20 500 every year for the next 10 years. How much must be invested?

Solution

From Table 2.5, the present worth factor for n = 10 is 7.316. Thus, the money required is 7.316 × $20 500 = $150 000.

5.2 DISCOUNT RATE

When a future amount of money is converted to its equivalent present value, the magnitude of the present amount is always less than the magnitude of the cash flow from which it was calculated. This is because for any interest rate greater than zero, all future value of the money is less than it is at present. For this reason, present-worth calculations are often referred to as discounted cash-flow methods. Regulations for the U.S. State Revolving Fund Program (40 CFR 35.2030[b][3], U.S. EPA, 2016) require that cost-effectiveness analyses performed in facilities planning are based on the

Table 2.5 6.125% Compound Interest Factors

Years	Capital Recovery Factor	Present Worth Factor	Sinking Fund Factor
1	1.061 25	0.942	1.000 00
2	0.546 39	1.830	0.485 14
3	0.374 97	2.666	0.313 72
4	0.289 41	3.455	0.228 16
5	0.238 20	4.198	0.176 95
6	0.204 16	4.898	0.142 19
7	0.179 93	5.557	0.118 68
8	0.161 83	6.179	0.100 58
9	0.147 82	6.764	0.086 57
10	0.136 67	7.316	0.075 42
11	0.127 60	7.836	0.066 35
12	0.120 09	8.326	0.058 84
13	0.113 78	8.788	0.052 53
14	0.108 41	9.223	0.047 16
15	0.103 80	9.633	0.042 55

discount rate established by the federal government. A discount rate for each federal fiscal year is established for water resources projects in the *Federal Register*. Owners also can choose to perform the cost-effectiveness analyses using a discount rate that reflects the owner's long-term, actual cost of money adjusted for inflation.

5.3 SALVAGE VALUE

Salvage value is the expected market value at the end of the useful life of the asset, which may be positive, zero, or negative. The salvage value may be negative if there are costs associated with decommissioning the asset, or the decommissioning costs are greater than the market value at the end of the useful life of the asset. Current depreciation methods approved for tax purposes typically assume a salvage value of zero, even though the actual salvage value may be positive. Note that this may force payment of extra income taxes when an asset is sold with a net realized value greater than the current book value (Blank & Tarquin, 2015).

5.4 LIFE EXPECTANCIES

Assets with expected useful lives shorter than the planning period have a cost associated with the replacement of the item during the planning period. Items with expected useful lives greater than the planning period have a salvage value at the end of the planning period. This can be accounted for by creating a cash flow of the costs and credits of specific items over a given time period and then performing a net present value analysis. Typically, life expectancies are taken to be 15 to 20 years for all equipment and 50 years for buildings, structures, and buried pipelines (U.S. EPA, 2002). Metal structures are given a shorter life expectancy than concrete structures. Certain jurisdictions may require specified expected useful lives in LCC analyses.

5.5 CAPITAL COSTS

Capital costs typically include land acquisition costs; estimated capital construction costs, including equipment, pipelines, buildings, and structures; engineering and project administration costs; and contingencies. Equipment costs should be separately identified from other costs because of the different life expectancies used in the present-worth analysis.

5.6 ANNUAL OPERATION AND MAINTENANCE COSTS

Annual O&M costs are those costs paid each year to keep the facilities in a good operating condition and to preserve the useful lives of structures and equipment. Annual O&M costs include wages, salaries, and benefits; maintenance repair and replacement; energy consumption; and chemicals.

Typically, O&M costs used in analyses are for the initial year of operation. These costs can be increased over the planning period if substantial increases are expected. O&M costs typically are based on costs at the time of the analysis, without future inflation. An average cost for wages, salaries, and benefits should be obtained from the owner and applied to the expected number of facility staff. Maintenance costs, which include lubrication oils, replacement parts, and other maintenance items, are often estimated to be 1% of the equipment capital costs. More detailed analysis of maintenance costs may be warranted if such information is available. Estimated maintenance costs should be increased beyond the roughly estimated 1% of the capital cost for any equipment expected to have abnormally high maintenance requirements.

Electrical costs should be based on the average power draw and the total cost for electricity (consumption plus demand). Chemical costs should be based on estimated average use and current unit costs. Solids disposal costs are often a significant portion of the O&M costs and should be carefully established.

5.6.1 CALCULATING ANNUAL COST

The capital costs of competing facilities can be estimated by calculating the cost that the municipality or agency would incur if it were to pay interest on a loan of that amount. Calculating the annual cost of a capital investment is exactly like calculating the annual cost of a mortgage on a house. The owner (municipality or agency) borrows the money and then pays it back in several equal installments. If the owner borrows $X and intends to pay back the loan in *n* installments at an interest rate of *i*, each installment can be calculated as

$$\$Y = \left[\frac{i(1+i)^n}{(1+i)^{(n-1)}}\right]\$X \tag{2.3}$$

Where $Y is the installment cost ($);
i is annual interest rate, as a fraction;
n is number of installments; and
$X is the amount borrowed.

The expression

$$\left[\frac{i(1+i)^n}{(1+i)^{(n-1)}}\right] \tag{2.4}$$

is known as the capital recovery factor, or CRF. The CRF does not have to be calculated because it can be found in interest tables or is programmed into handheld calculators. Table 2.5 shows CRFs if the interest rate is 6.125%.

From Table 2.5, if the number of annual installments *n* is 10 and if the interest rate is 6.125%, the CRF is 0.13667. If the $10 million was borrowed at the above terms, the annual payment would be $1 366 700 for each of the next 10 years.

5.6.2 CALCULATING CAPITAL PLUS OPERATION AND MAINTENANCE COSTS

Capital investments require not only the payment of the loan in regular installments, but also the upkeep and repair of the facility. The total cost to the community is the sum of the annual payback of the capital cost plus the O&M cost, or [total annual cost] = [O&M cost] + [annualized capital cost].

Example 2.2

A town wants to buy a new pumping station that has an expected life of 10 years. The town wants to borrow the $150 000 cost of the pumping station, paying it back in 10 annual payments. The interest rate is 6.125%. How much are the annual installments on this capital expense?

Solution

From Table 2.5, the capital recovery factor for $n = 10$ is 0.136 67. The annual cost to the town would then be 0.136 67 × $150 000 = $20 500. That is, the town would have to pay $20 500 each year for 10 years to pay back the loan on this pumping station (note that this pumping station does not cost 10 × \$20 500 = \$205 000 because the dollars for each year are different and cannot be added).

Example 2.3

A community wants to buy a pumping station that has an expected life of 10 years and costs \$150 000. They choose to pay back the loan in 10 annual installments at an interest rate of 6.125%. The cost of operating the pumping station (electric power, service, etc.) is \$20 000 per year. How much will this station cost the community every year?

Solution

From Table 2.5, the capital recovery factor for $n = 10$ is 0.136 67, so the annual cost of the capital investment is 0.136 67 × \$150 000 = \$20 500. The operating cost is \$20 000, so the total annual cost to the community is \$20 500 + \$20 000 = \$40 500.

5.7 INTEREST DURING CONSTRUCTION

U.S. EPA's cost-effectiveness guidelines (2010) present the following equation for estimating interest during construction:

$$I = 0.5PCi \tag{2.5}$$

Where I is interest during construction (%),
P is construction period (years),
C is total project cost, and
i is discount rate (%).

5.8 LAND COSTS

U.S. EPA's cost-effectiveness guidelines require that land values be escalated 3% annually to the end of the planning period. Land values also should be salvaged at the end of the planning period. The present worth for land acquisition is essentially the cost of interest through the planning period. The present worth of land costs is usually negligible in the final cost-effectiveness comparison and can be neglected. Costs should not be included if every alternative is using the same land (common factors excluded). By contrast, land cost should be included for alternatives associated with treatment by land application where the owner purchases the land. In cases where land is to be leased from another owner instead of purchased, the lease payments are included in the O&M costs or as a separate recurring annual cost of the alternative.

5.9 SUNK COSTS

Cost-effectiveness analyses can include only future expenditures. Any costs already incurred for an alternative must be considered as a sunk cost and excluded from the analysis. Alternatives benefiting from sunk costs will have reduced future costs—capital, O&M, or both. In some situations, the municipality or agency must save money by investing it so that it will have a reserve available at a later date. In environmental engineering, this most often occurs when the landfill owner must invest money during the active life of the landfill, so that when the landfill is full, there are sufficient funds available to place the final cover on the landfill. Such funds are known as sinking funds. Sinking funds may also be established for expanding WRRFs without having to issue new bonds. To determine sinking funds, it is necessary to calculate the funds, $\$Y$, necessary to be invested in an account at i interest rate, so that at the end of n years the fund has $\$X$ dollars accumulated. The calculation is

$$\$Y = \left[\frac{1}{(1+i)^n - 1}\right]\$X \tag{2.6}$$

Where $\$Y$ is the amount that has to be invested (\$),
i is annual interest rate, as a fraction,
n is number of years; and
$\$X$ is accumulated amount.

The term

$$\left[\frac{1}{(1+i)^n - 1}\right] \tag{2.7}$$

is known as the sinking fund factor.

Example 4.1

A community wants to have $500 000 available at the end of 10 years by investing annually into an interest-bearing fund that yields 6.125% interest. How much would they have to invest annually?

Solution

From Table 5, the sinking fund factor at 10 years is 0.074 52, and the required annual investment is therefore 0.074 52 × $500 000 = $37 260.

Note again that 10 × $37 260 = $372 600, which is considerably less than $500 000. This is because the investments during the early years are drawing interest and adding to the sum available.

PRACTICE PROBLEMS

1. An engineer has designed a pumping station for a wastewater collection system using the equipment from Flow Smooth pump manufacturers, which costs the following:

Hardware (from Flow Smooth)	$40 000
Land	$8000
Installation	$10 000

   The operating costs are as follows:

Electrical power	$500 per month
Labor	$500 per month
Parts/replacements (gaskets, valves, etc.)	$500 per month

   A city council member is approached by a salesperson for Suckemup Pumps who tells him that the city is being ripped off. She tells the council member that Suckemup can sell the *equivalent* pump to the city for only $30 000. While the pumps are not as efficient (She estimates power costs at $1000 per month and parts and replacements at $1000 per month), so this would still be a good deal for the city, and the council member should direct the engineer to use Suckemup instead of Flow Smooth.

   If you were called in to arbitrate, how would you analyze the situation? Would you side with the engineer? Why of why not? What other information must you know in order to make a qualified judgment?

2. A community of 5000 people wants to build a tricking filter WRRF and borrow money at 6.125% to pay for their share of the capital cost. The payback time for them would be 20 years. What is the annual cost to the community for paying back this loan?

3. The capital cost for a cast-iron, 20-cm sewer, 150 m long, is $300 000. The O&M costs (e.g., cleaning) will be approximately $10 000 per year. The pipe is expected to last 30 years. If a 25-cm pipe is used instead, the capital cost would be $350 000, and the O&M costs would be only $5000 per year (due to less chance of clogging).

   a. If both lines are hydraulically acceptable, and if the town is paying for the construction and O&M, which one should the town construct if the total cost is to be minimized? Assume an interest rate at 6.125%.

b. If the state is helping finance the project, and is contributing $250 000 for the construction (capital cost only—the town is responsible for the O&M cost), which sewer should the town construct if it wants to minimize the cost to the town?

c. If you are the design engineer on this project and are working for the town, which pipe would you recommend?

4. A town has established a surcharge of $10 per kg of BOD above 200 mg/L and $7 per kg of suspended solids over 150 mg/L. If an industry discharges 2000 m^3/d of wastewater at 300 mg/L of BOD and 1200 mg/L of suspended solids, (a) what will be the surcharge in dollars per day, and (b) what type of treatment would you advise the industry to install?
5. What is an integrated design for resource recovery? Provide two examples.
6. Describe the steps involved in WRRF site selection process.
7. Compare linear and nonlinear WRRF layouts.

6.0 REFERENCES

Association for the Advancement of Cost Engineering. (2020). *Cost estimate classification system* (rev. ed., AACE International Recommended Practice No. 17R-97).

Blank, T. B., & Tarquin, A. J. (2015). *Engineering economy* (7th ed.). McGraw-Hill.

Design–Build Institute of America. (2009). *Contract incentives and design-build acquisition* (Manual of Practice).

EnviroSim Associates, Ltd. (2016). *BioWin process simulator.*

Montgomery, J. M. Consulting Engineers Inc. (1985). *Water treatment principles & design* (pp. 237–261). John Wiley & Sons.

National Council of Examiners for Engineering and Surveying. (2005). *Fundamentals of engineering examination specifications.* http://www.ncees.org

U.S. Environmental Protection Agency. (2002). *The clean water and drinking-water infrastructure gap analysis.* Office of Water (4606M). EPA-816-R-02-020 September.

U.S. Environmental Protection Agency. (2010). *Guidelines for preparing economic analyzes.* National Center for Environmental Economics.

U.S. Environmental Protection Agency. (2016). State and local assistance. *Code of Federal Regulations,* Part 133, Title 40.

U.S. Environmental Protection Agency. (2021). *Stormwater discharges for industrial activities, Multi-sector general permit.* https://www.epa.gov/npdes/stormwater-discharges-industrial-activities

Water Environment Federation. (2013). *Wastewater treatment process modeling* (2nd ed.; Manual of Practice No. 31).

Water Environment Federation. (2018). *Financing and charges for wastewater systems* (4th ed.; Manual of Practice No. 27).

Water Environment Federation. (2021). *Safety, health, and security standards for water resource recovery* (7th ed.; WEF Manual of Practice No. 1).

Water Environment Federation. (2023). *Standards for design of water resource recovery facilities* (WEF 8).

Chapter 3 | Preliminary Treatment

LEARNING OBJECTIVES

At the end of this chapter, students should be able to:

- Characterize screens based on opening size.
- Estimate the quantity of screenings from municipal wastewater.
- Describe the operation and design of a coarse screen.
- Describe the operation and design of a fine screen.
- Describe the operation and design of an aerated grit chamber and a vortex grit chamber.
- Estimate the grit quantity from a grit chamber treating municipal wastewater.
- Describe the purpose and operation of a grit classifier and grit washer.
- Identify pumping technologies that are compatible with grit chambers.
- Explain the purpose and design considerations for flow equalization.
- Identify important flow measurement devices and their appropriate use in municipal wastewater treatment.
- Describe the process of neutralization as used in treatment of municipal wastewater.

1.0 INTRODUCTION

The purpose of preliminary treatment is to remove, reduce, or modify wastewater constituents in the raw influent that can cause operational problems with downstream processes or increased maintenance of downstream equipment. These constituents primarily consist of large solids and rags (screenings), abrasive inert material (grit), floating debris, and grease. This chapter presents descriptions of and design considerations for preliminary treatment processes including screening, coarse solids reduction, grit removal, flow equalization, flow measurement, and neutralization.

2.0 SCREENING

The main goal of screening wastewater flow is to protect downstream equipment and processes. Screening can be used to remove large objects that could damage influent pumps or block flow in raw wastewater channels and piping systems, or it can remove fine objects such as human hair, which protects sensitive downstream equipment including membrane systems, fabric filters, or suspended media used in integrated fixed film activated sludge (IFAS) and moving bed biofilm reactor systems. The passage of rags and debris into downstream processes is one of the most common causes of equipment maintenance and failure because of jammed pump impellers, clogged sludge and scum pipelines, and imbalanced operation of rotating equipment. Floating material in downstream processes or receiving streams is an aesthetic problem and a safety hazard to operators attempting removal. Several reports suggest that the effects of inadequate screening are significant and can impede proper functioning of oxygen delivery systems and mixers (Burbano et al., 2014) as well as result in excessive maintenance for other rotary equipment such as pumps in water resource recovery facilities (WRRFs) (Borys, 2014).

As wastewater processes continue to advance, damage associated with inert objects in wastewater becomes increasingly important. For this reason there is a trend toward installing screens with smaller openings. As screen openings become smaller, greater amounts of organic matter are removed, and, therefore, it becomes important to provide screening washer/compactors to WRRFs to return the organics to the wastewater flow stream. Washer/compactors are becoming standard design practice because these are able to return organic matter to the liquid stream, benefitting downstream nutrient removal processes that require soluble organic matter to function properly, and reducing odor from the storage of screenings.

2.1 TYPES OF SCREENING MEDIA

Typically, there are four types of screening media used in WRRFs: bars, wedge wire, perforated plate, and mesh.

2.1.1 BARS

Historically, bars have been the most used media because it is preferred for coarse screens and trash racks. Bars are available in a variety of shapes including rounded, rectangular, trapezoidal, and teardrop. Rounded bars have low capture efficiency and are only used on large opening bar racks. Trapezoidal bars have increasingly wide openings, allowing solids that pass through the narrowest opening at the front of the screen to pass through without getting trapped between the bars. Teardrop bars combine the benefits of the trapezoidal bars with better hydrodynamic flow characteristics, which minimize the headloss through the screen. Long vertical or horizontal gaps between bars can allow the passage of long, thin objects.

2.1.2 WEDGE WIRE

Wedge wire is a refinement of the trapezoidal bar screen used in much finer screening applications. The same narrow-to-wide opening profile is used to prevent trapping solids between the openings. The narrower openings of wedge wire screens result in much thinner media, which is why they are called "wires" instead of "bars." Wedge wire screens also have long, vertical gaps and are not allowed by some manufacturers of membrane bioreactor facilities because long thin objects such as hair must not be allowed to accumulate on the membranes. A close-up view of wedge wire bars is shown in Figure 3.1.

2.1.3 PERFORATED PLATE

Perforated plate media (Figure 3.2) is more effective at capturing solids than bars or wedge wire when fine screening (such as hair removal) is required. The technology for perforated plate media is constantly advancing, with the lower limit in size currently at 1 mm (0.04 in.). Perforated plate media exhibit higher headloss because the effective open area is decreased, orifice losses occur, and blinding is higher compared to bar or wedge wire media.

2.1.4 MESH

Mesh is used for fine screens 1 mm (0.04 in.) and smaller because of the manufacturing limitations of perforated plate media. Mesh media is more fragile and can result in "stapling" of solids within the media, which interferes with the release of captured solids by the removal mechanism. To avoid clogging the

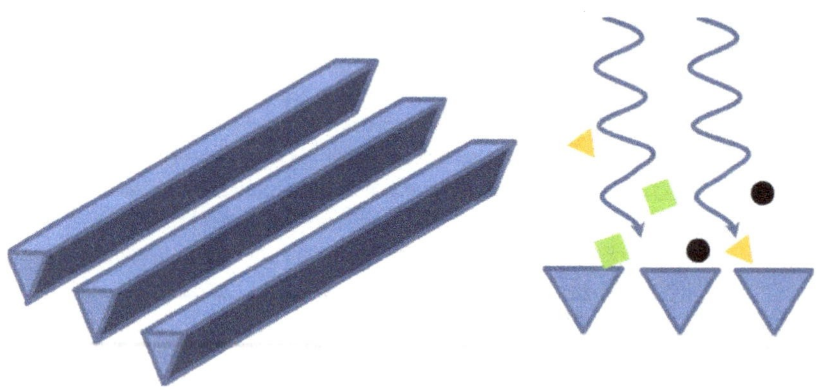

Figure 3.1 Wedge Wire Screen Material (Reprinted with permission from Indigo Water Group)

Figure 3.2 Perforated Plate Media (Reprinted with permission from Indigo Water Group)

mesh, high-pressure water jet cleaning is recommended. Openings in the mesh media can also be a source of confusion because the corner-to-corner distance of the square mesh opening is slightly longer than the side-to-side distance. Typically opening definition is provided for the side-to-side distance. (Refer to Figure 3.1 for an example.)

2.2 SCREENING CHARACTERIZATION

Screening of wastewater can be categorized according to screen opening size. Trash racks and bypass screens have openings greater than 36 mm (1.5 in.), whereas coarse screens have openings greater than 6 mm and less than or equal to 36 mm (0.25 to 1.5 in.). Fine screens have openings greater than 0.5 mm and less than or equal to 6 mm (0.02 to 0.25 in). Microscreens have openings less than or equal to 0.5 mm (0.02 in.).

2.2.1 QUANTITIES

The quantity of screenings removed can vary significantly depending on the wastewater flow, wastewater characteristics, type of collection system, screen type, screen opening, screen cleaning mechanism, and effectiveness of washer/compactor equipment. Actual operating data can be used when replacing screens with equal size openings, but it is of little value if an upgrade project involves putting in screens with smaller openings.

For screen openings between 25 and 50 mm (1 and 2 in.), each 13 mm (0.5 in.) reduction of clear opening size will approximately double the volume of screenings. For screen openings smaller than 25 mm (1 in.), the volume of screenings removed increases rapidly and becomes much more a function of wastewater characteristics and the efficiency of washer/compactor equipment. Removal by trash racks is more a function of wastewater characteristics than rack openings.

The quantity of screenings removed will depend on the sewer characteristics in the collection system, number and location of pumping stations, and whether or not the stations include screening. In short, gently sloping collection systems with low turbulence will contain more first flush screenings than lengthy, steep interceptor systems or systems with pumping stations. Slug screening loads corresponding to pumping stations startup periods have been reported in some facilities (Wodrich et al., 2005). Slug loads are common in combined collection systems after first flush conditions (especially after a dry period) and in areas with significant numbers of deciduous trees at the beginning of the fall.

Another critical factor in determining screenings quantity is the type of collection system feeding the WRRF. Previous experience indicates that combined systems produce several times the coarse screenings compared to separate systems. Peak wet weather removal from combined systems may vary by as much as 20:1 from average dry weather conditions on an hourly basis.

The Water Environment Federation (WEF) performed two surveys of utility members (WEF, 2016), obtaining data from 358 WRRFs across the United States. Figures 3.4 and 3.5 show the normalized data from the survey and differentiate between coarse and fine screens. Based on the survey data, extreme variations in screening quantities were reported, from less than 0.74 to 148 L/1000 m^3 (0.1 to 20 cu ft/mil. gal). Screenings quantity generation in the United States apparently is higher than in Europe. Recent vendor studies in Europe have found an average screenings production rate of 2.4 kg/person•d compared to 4.5 kg/person/d in the United States. This is based on U.S. screenings production of 40.7 L/1000 m^3 (5.5 cu ft/mil. gal); 378 L/person•d (100 gal/person/d) wastewater generation rate; and 800 kg/m^3 (50 lb/cu ft) screenings density. Some of this difference could be attributable to more widespread use of washer/compactors in Europe.

2.2.2 CHARACTERISTICS

The composition of the screenings removed will depend on the type of screen that is used. Coarse screenings consist of rags, sticks, leaves, food particles, bones, plastics, bottle caps, and rocks. These screenings are typically thought of as large debris that can immediately damage pumps and other equipment. Fine screens will remove coarse screenings along with a higher degree of rags, wipes, stringy material, and a large amount of organic matter. Fine screenings, if not removed, typically have a long-term effect on equipment and processes because it takes time for screenings to wrap around impellers or mixer blades. The smaller the opening size, the more organic matter will be removed with the screenings.

Unwashed, uncompacted screenings can contain 10% to 20% dry solids with a bulk density ranging from 600 to 1100 kg/m^3 (40 to 70 lb/cu ft). In the 2008 WEF member survey, screenings density averaged 825 kg/m^3 (55 lb/cu ft). There is currently no distinction made between coarse, fine, and microscreenings; however, it can be assumed that the finer the screening, the more water and organic material will be removed.

Typical performance specifications for washer/compactors are 90% reduction in organic content and an increase to 50% in dry solids. Recent European vendor studies have found that 50% dry solids are difficult

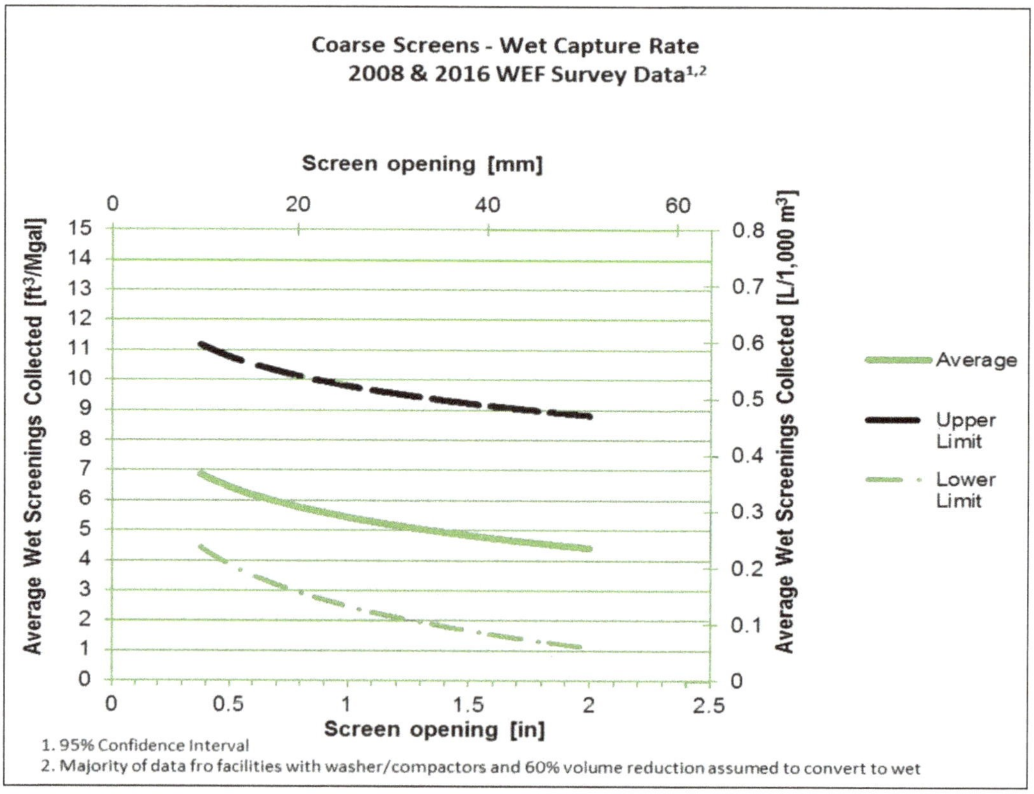

Figure 3.3 Screenings Quantities From Coarse Screens

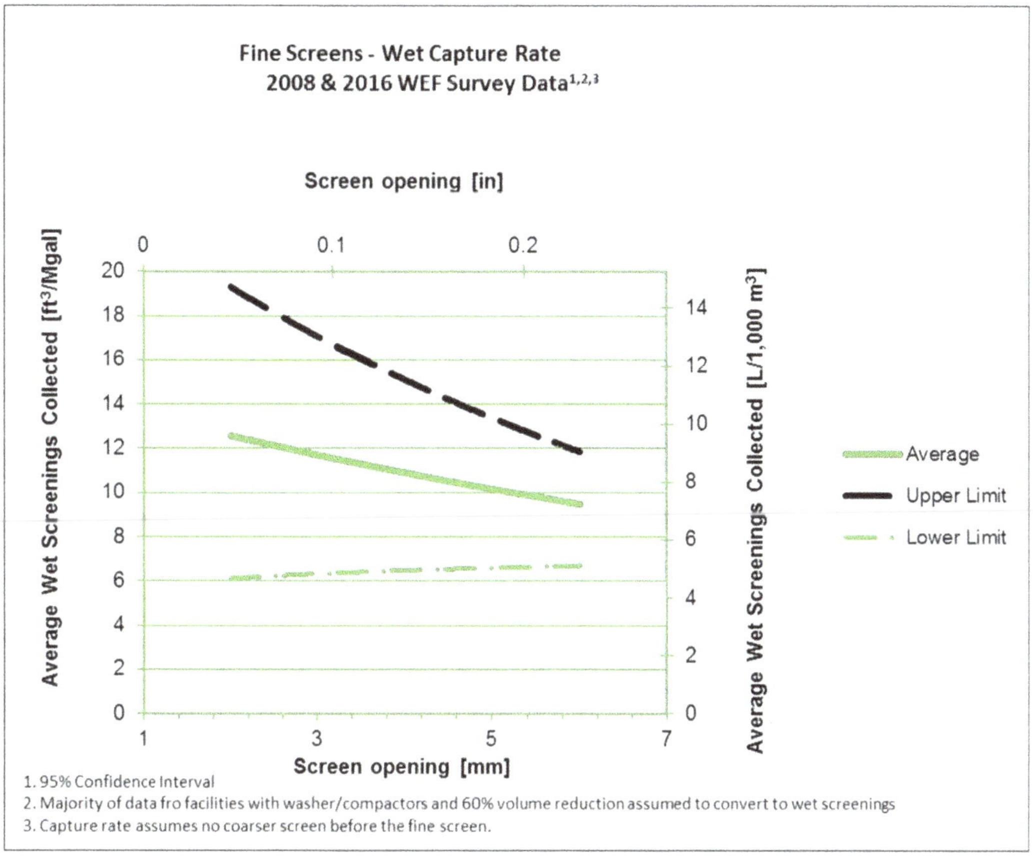

Figure 3.4 Screenings Quantities From Fine Screens

to achieve with standard washer/compactors. Based on the survey information, a range of 30% to 40% was estimated, with an average of 37%. From the inlet hopper to the discharge chute, it can be estimated that the screenings volume and weight is reduced by 50% to 80%.

Example 3.1

A WRRF with an average flow of 7560 m³/d (2 mgd) has installed fine screens with an opening of 4 mm (0.16 in.). The facility has also installed a compactor that is able to reduce the volume of screening by 60%. Estimate the volume of screenings that would be produced from this facility using Figure 3.4.

Solution

From the figure, the screenings produced is 8.5 L/1000 m³.

$$\text{Volume of uncompacted screenings} = 7560 \text{ m}^3/\text{d} \times 8.5 \text{ L}/1000 \text{ m}^3 = 64.3 \text{ L/d } (2.3 \text{ cu ft/d})$$

$$\text{Volume of compacted screenings} = 64.26 \text{ L/d } (1 - 0.60) = 25.7 \text{ L/d } (0.91 \text{ cu ft/d})$$

2.2.3 EFFECTS OF NONDISPERSIBLE PRODUCTS

Non-dispersible (flushable) consumer products were introduced into the market in the last decade. These newer personal hygiene products predominately consisted of nonwoven wipes that, because of their size and use, are frequently disposed of in toilets. Water utilities quickly learned that most of these products do not break up in the collection system, leading to problems in the collection system pumping stations and headworks of WRRFs.

Non-dispersible products have the potential to increase the screenable material carried to the facility due to non-dispersion in the wastewater, increase the instantaneous blinding of fine screens due to slug loads containing non-flushable material, and increase the wear and tear on screening equipment because more screenable material processing is required. These products also render certain types of screens inadequate to deal with the increased load of screenable material, which causes ragging issues in processes downstream of the screens when there is an inadequate capture of screenable material, and renders screening conditioning equipment (conveyance and conditioning) inadequate due to the higher peak load of screenable material, which reduces the facility's ability to deal with slug screening loads. Therefore, designers of headworks systems in WRRFs must take all of these factors into consideration for the proper selection and design of screening equipment that may be subjected to non-dispersible materials.

2.3 SCREEN TYPES

There are several screens designs and styles in the wastewater industry. The following sections summarize the various screen types based on their screening categories as listed below; however, some screens can be used in multiple categories. Table 3.1 summarizes advantages and disadvantages of various screen types.

Table 3.1 Summary of Coarse and Fine Screening Equipment

Type of Screen		Advantages	Disadvantages
Coarse Screens	Chain- or cable-driven screens	Simple channel construction High screenings loading rate Insensitive to fat, oil, and grease	Submerged components
	Reciprocating rake screens	No critical submerged components Widely used Allows for servicing the unit above the channel	Low screening loading rate High overhead clearance, particularly at deep channels
	Continuous self-cleaning screen	Medium to low headroom required Allows for servicing the unit above the channel	Several moving components Components subject to wear and tear
	Arc screens	Lower capital and operational cost No drive parts under water Uses 100% of channel width	Limited to small- to medium-flow facilities Not suited for deep channels
Fine Screens and Microscreens	Continuous element screens	Proven technology High screenings capture rate Allows a pivot design for servicing the unit above the channel	Numerous moving parts Some designs have components subject to wear and tear High headloss requirements High potential for grease blinding Potential carryover of screenings
	Multiple-rake screens	Widely used High screenings loading rate Low headroom requirements	Some designs have submerged components subject to wear and tear Medium headloss requirements High potential for grease blinding
	Stair screens	Low headloss requirements Allows a pivot design for servicing the unit above the channel	Stringy solids can pass through the screen Not recommended when rocks or excessive grit loads are expected Requires considerable footprint
	Band screens	Low screen carryover Well suited for expanding capacity without increasing the channel size	Requires pressure water for cleaning Difficult to access and to maintain Large solids removal may be an issue

(continued)

Table 3.1 Summary of Coarse and Fine Screening Equipment (*Continued*)

Type of Screen		Advantages	Disadvantages
Fine Screens and Microscreens (*continued*)	Drum screens	Low screen carryover Recommended for downstream processes sensitive to screenings (MBRs, IFAS, etc.)	Require considerable footprint Special hydraulic provisions required High headloss (pumping is usually required)
	Helical basket	Lowest screenings carryover potential Provides screenings washing/compaction in a single unit Recommended for downstream processes sensitive to screenings (MBRs, IFAS, etc.)	Requires significant footprint for installation Not suitable for deep channels
	Static screens	Minimal or no moving parts Well suited for smaller facilities	High headloss Operator intensive Susceptible to fast blinding
	Microscreens	Removal of very small solids	Requires special washing systems Should be preceded by larger opening screens High headloss

Note. MBR = membrane bioreactor; IFAS = integrated fixed film activated sludge

2.3.1 TRASH RACKS AND BYPASS SCREENS

Trash racks are used in older facilities and in facilities receiving wastewater from combined collection systems that can contain large objects. These are bar screens with large openings of 36 to 144 mm (1.5 to 6 in.) designed to prevent logs, timbers, stumps, bricks, and other large, heavy debris from entering treatment processes. Trash racks typically are followed by screens with smaller openings. Where space is limited, facilities sometimes have basket-type trash screens that are manually hoisted and cleaned.

Bypass screens are used for emergency screening purposes in the event the mechanically cleaned coarse or fine screen must be taken out of service. Openings range from 24 to 48 mm (1 to 2 in.). Manually cleaned trash racks and bypass screens typically are mounted on a 45° to 60° angle from the horizontal to facilitate cleaning using a rake and perforated plate drain pan. Depth should not be more than what can easily and safely be raked by operators. Mechanically cleaned trash racks are available and are mounted at an angle 75° to 80° from the horizontal.

2.3.2 COARSE SCREENS

Historically, coarse screens have been the most commonly used stand-alone screens in WRRFs because they provide sufficient screening without producing excessive volumes of organic materials, which eliminates the requirement for a washer/compactor. Coarse screens are characterized by openings that range from 6 to 36 mm (0.25 to 1.5 in.). Coarse screens with the smallest openings can still remove organic material and, therefore, the WRRF should provide a washer/compactor. The coarse screens most commonly used today are in the first stage of a two-stage screening process to provide protection and reduce blinding of the fine screens.

Coarse screens are cleaned mechanically, which allows the screening media to be mounted in a more vertical position, typically 70° from the horizontal. Mechanical cleaning reduces labor costs, improves flow conditions and screening capture, reduces nuisances and, in combined systems, better handles large quantities of stormwater debris and screenings. A mechanically cleaned screen is almost always specified for new facilities of all sizes. Many types of mechanically cleaned bar screens are manufactured, including, but not limited to, chain/cable-driven, multi-rake, single-rake, and continuous screens.

2.3.2.1 Chain-Driven Screens.

These types of screens are manufactured in several configurations: front clean/front return, front clean/rear return, and back (or through) clean/rear return. The front clean/front return type most efficiently retain captured screenings by minimizing carryover. Multi-rake screens (Figure 3.5) have become increasingly popular because the multiple rakes quickly clear accumulated material from the screen, allowing them to handle high screening volumes during peak flows.

2.3.2.2 Single-Rake Screens

The single reciprocating rake screen can be equipped with a back clean/back return mechanism or with a front clean/front return mechanism that minimizes solids carryover (Figure 3.6). Although the front clean design minimizes carryover, back clean is less vulnerable to jamming. The up-and-down reciprocating motion of the rake, similar to that of a person raking a manual bar screen, minimizes the possibility of jamming. The single biggest advantage of single-rake screens is that all moving parts are located out of the water. A disadvantage of reciprocating rake screens is the single rake, which limits the capacity to handle extreme loads, although this would typically only be a problem in deep applications in which cycle times are long.

2.3.2.3 Continuous Self-Cleaning Screens

Continuous self-cleaning screens consist of a continuous belt of plastic or stainless steel elements that are pulled through the wastewater to provide screening along the entire submerged length of the screen. Screen openings are designed with both horizontal and vertical limiting dimensions; the vertical spacing is slightly larger than the horizontal spacing. Continuous screens can be used in both coarse and fine screening applications, with openings as small as 1 mm (0.04 in.) ranging up to more than 72 mm (3 in.). The greater solids handling capacity of these screens allows smaller openings to be used, which results in greater capture of solids from the waste stream. Disadvantages of continuous screens include possible solids carryover resulting from the front clean/back return design and the difficulty of cleaning screen elements on the rear side.

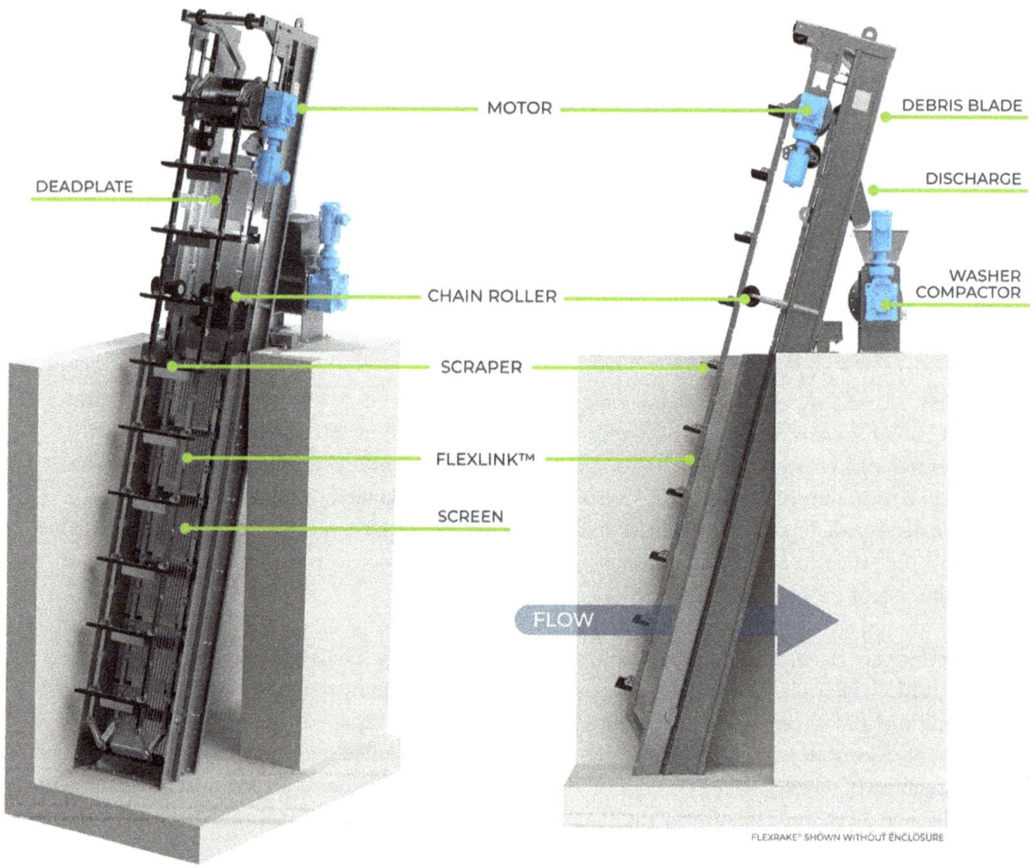

Figure 3.5 Duperon® FlexRake® Full Penetration Fine Screen (Reprinted with permission from Duperon Corporation)

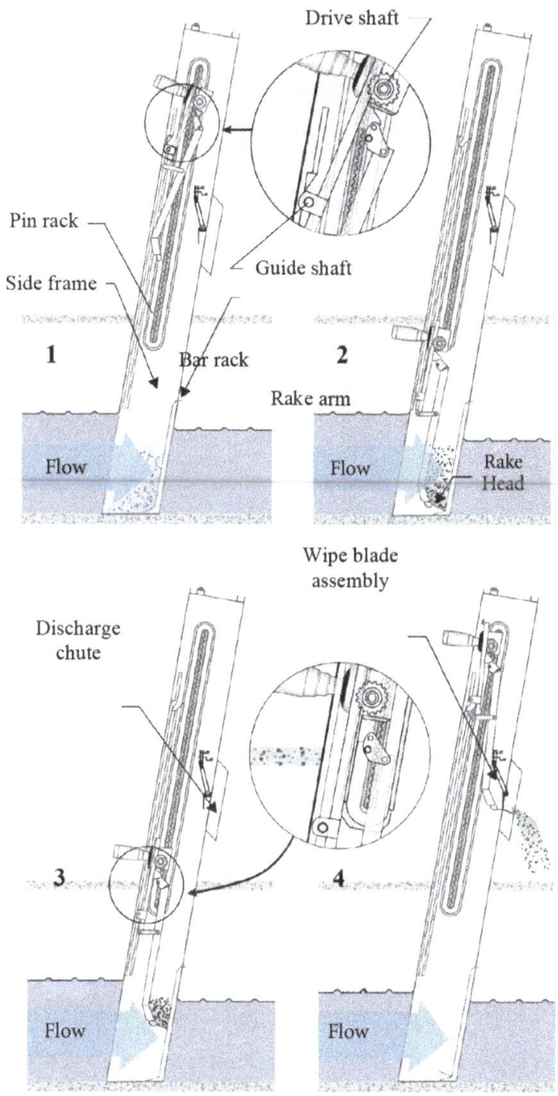

Bar Screen Operational Sequence

1. The bar screen cleaning cycle begins when the rake assembly travels downward after it is activated from the "park" position.
2. As the rake assembly rotates around the lower end of the pin rack, the rake arms force the rake teeth to engage the bar rack.
3. The rake assembly travels up the pin rack, cleaning debris from the bar rack and delivering the screenings to the point of discharge.
4. The rake engages the wiper blade to clean the rake head. The rake assembly then returns to the "park" position.

Figure 3.6 Reciprocating Rake Screen (Reprinted with permission from Vulcan Industries, Inc.)

2.3.3 FINE SCREENS

Fine screens are categorized by opening sizes ranging from 0.5 to 6 mm (0.13 to 1.5 in.). Mechanical cleaning of these screens is essential; the smaller the opening, the more critical cleaning performance is for proper operation. Water sprays or brushes are typically used for cleaning these screens. Hot water provides better results than cold water for cleaning fine screens because it helps to remove grease that has adhered to the surface. Brush-cleaned screens can have a lower capture performance because over time the bristles may bend or may become wrapped in screenings. Washer/compactors, either integral to the screen or stand-alone, must be used with fine screens because of the large amount of screenings and organic material removed. Fine screens typically involve mechanisms more complex than the ones found in coarse screens because of the need to remove smaller solids. Fine screens have more elaborate cleaning mechanisms compared to coarse screens, because attached organic material is more difficult to remove from smaller openings.

2.3.3.1 Continuous-Element Screens

Continuous-element screens consist of screens with an endless cleaning grid that is attached to a main drive via different configurations. Screenings are collected and conveyed to the top part of the screen and then discharged. Continuous-element screens can be provided in a variety of forms, but the most popular are the perforated plate and the belt type technologies. Perforated plate screens (Figure 3.7) are

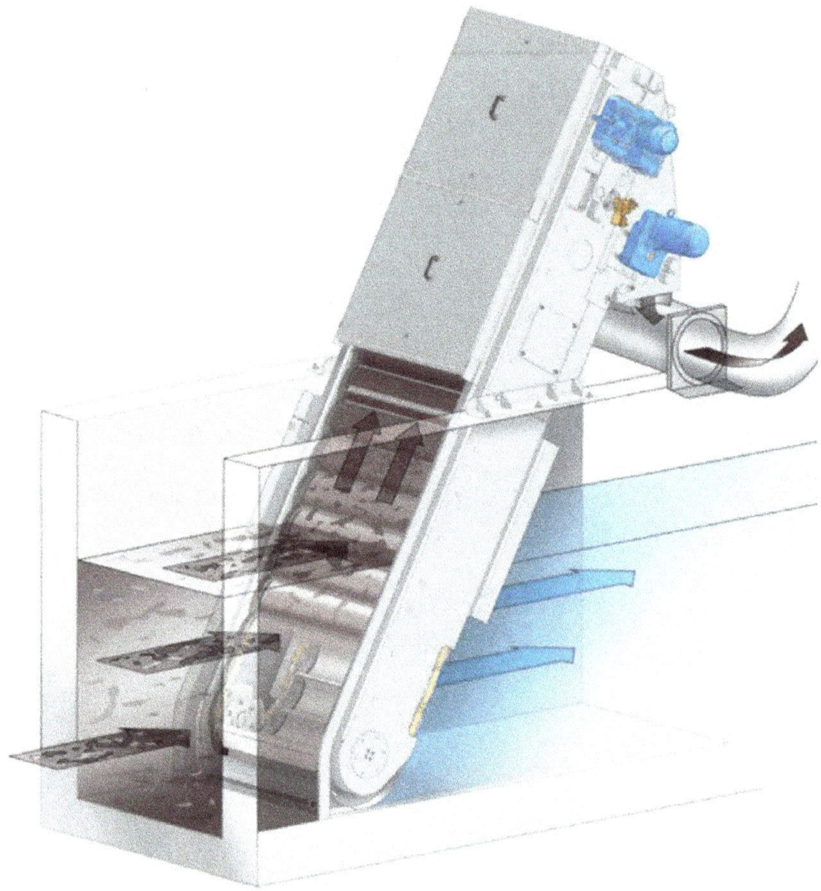

Figure 3.7 Perforate Plate Screen (Courtesy of Huber Technology, Inc.)

constructed of plastic or stainless steel panels with orifices throughout the panel, provided in a stacked panel configuration with rakes at regular intervals to prevent screenings rollback. Orifices are typically round for ease of construction. A proper cleaning mechanism (water spray and/or brushes) is required to remove buildup of organic material.

Perforated plate screens in a through flow configuration typically require an inclination angle between 35° and 75° from horizontal. Because the screening media travels downward on the back of the screen over the screened flow, screenings carryover can occur if the cleaning mechanism is not functioning properly. Continuous-element screens have one of the best screenings retentions in the industry and are often used where minimizing the screenings for downstream processes is important. Because the flow travels through the screen media twice, these screens have greater clean screen headloss than other screens and require a significant amount of flushing water.

2.3.3.2 Multiple-Rake Screens

The multiple-rake screens, discussed in the coarse screen section and shown in Figure 3.5, are being used with increasing frequency in fine screening applications. The basic construction is the same as for the multi-rake coarse screens; however, the raking system for finer screens is designed to operate at a higher cleaning frequency because of the higher screenings collection. The close tolerances of the multiple-rake mechanism allow this screen to be supplied for openings down to 4 mm (0.16 in.).

2.3.3.3 Stair Screens

Stair screens are also referred to as Step Screens® (Figure 3.8), and consist of thin, 2 or 3 mm (0.08 or 0.12 in.) long parallel lamellas of stainless steel with 3 to 6 mm (0.12 to 0.24 in.) clear openings. There are

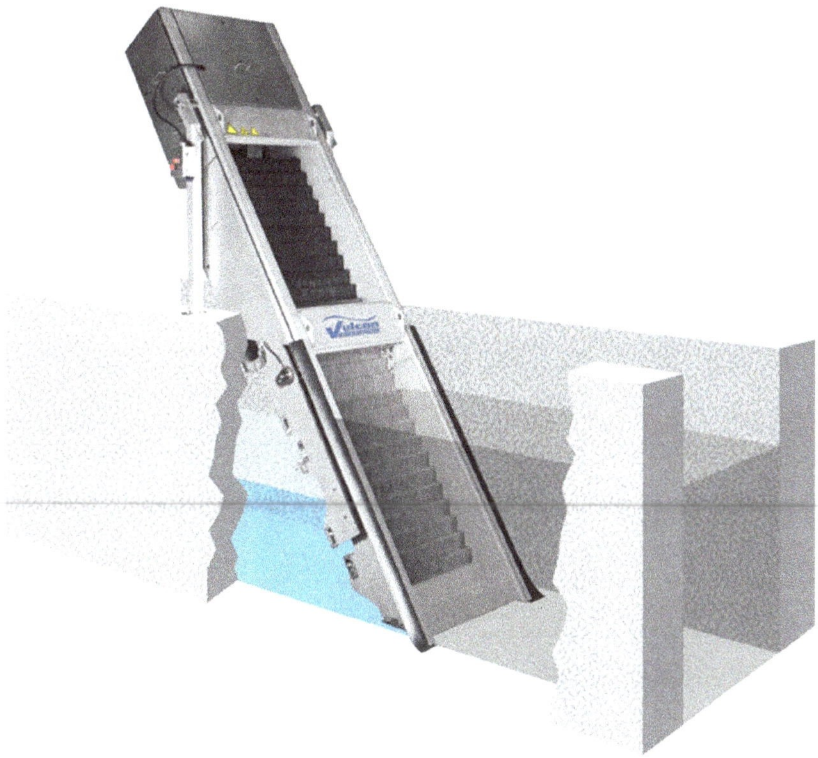

Figure 3.8 Stair Screen in a Channel (Courtesy of Vulcan Ind., Inc.)

two sets of lamellas in a stair screen; most designs have one fixed set and one moving set that rotates in and out of the screen to provide a step-motion pattern that lifts the collected screenings upward until they are discharged at the top of the screen. Sometimes wedging or larger objects have been a problem with this design. Stair screens provide a higher open area when compared with other fine screens, because they have open slots running from the bottom to the top of the screen grid. This slot configuration allows passage of stringy solids, which can be minimized through "matted operation" in which the screen is operated with a mat of accumulated screenings created by using differential head control. The larger open area of these screens helps to minimize headloss, which makes it suitable for retrofits where this is a significant concern. Stair screens typically require a significant footprint because it is recommended that they be inclined to 45° to prevent screenings rollback. To reduce footprint requirements, some designs use hooked steps, which allows the screens to be installed at a 75° inclination and allows deeper channels.

2.3.3.4 Band Screens

Band screens (Figure 3.9) are similar to perforated screens because both typically use perforated plate panels. Band screens are usually wider to accommodate the rotation of the screening plates, creating the "band" shape pattern. Lifting lips provided on each panel lift the screenings over the water and discharge them on a screenings collection trough. There are several configurations available for this technology depending on the way the wastewater flow approaches the screen, but only the centrally fed design is used in wastewater applications. Headloss through these screens is typically lower than with continuous-element screens as the water flows into the center of the screen and out two side panels. The benefit of this type of screen is that the screening debris remains on the upstream side of the screen and has no chance of passing over to the downstream side.

2.3.3.5 Drum Screens

There are three types of drum screens: aboveground, in-channel, and vertical. The aboveground drum screen is mounted on legs, requires pumped influent, and can be fed internally or externally. For the internally fed screen (Figure 3.10), wastewater is fed into a distribution pan inside the rotating drum, and screenings are conveyed to the front of the screen using large, internal auger flights. Screened wastewater is

Figure 3.9 Band Screen (Reprinted with permission from Headworks)

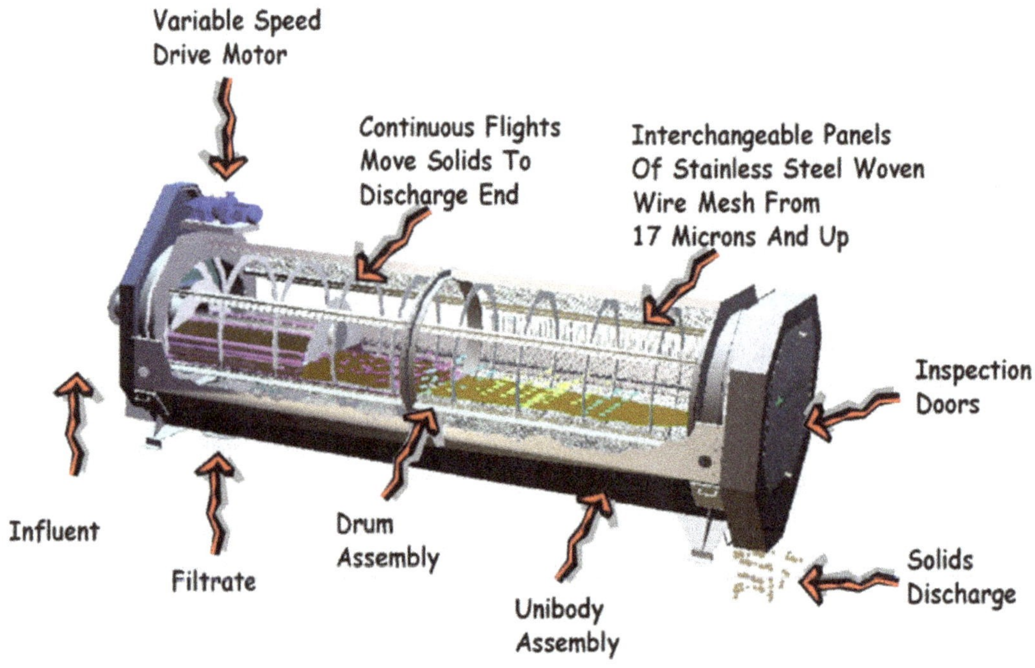

Figure 3.10 Internally Fed Drum Screen (Reprinted with permission from Chandco Systems Inc.)

then discharged out the back of the screen. The drum continuously rotates on wheels and screening media can be removed and replaced in sections. Wastewater is fed outside the rotating drum in the externally fed version (Figure 3.11) and solids are deposited to an auger below.

Gravity flow drum screens are available. These screens are similar to band screens except they are composed of perforated plate panels joined together in a circular pattern. The screenings are conveyed to the top of the screens where they are removed by gravity with the help of the washing system. Drum screens must have a diameter large enough to extend above the operating floor of the channel to allow screenings discharge. They should extend into the channel to a depth sufficient to submerge enough mesh to pass the required flow at low water level.

For large flows ranging from 8 to 35 m^3/s (180 to 800 mgd) in a channel, single-entry and double-entry drum screens have been successfully used (Figure 3.12). The drum screen rests with half of the screen in the channel as it slowly rotates around from wet side to dry side. Wastewater flows from the side into the center of the screen and then out through the screening panels. The screenings are rotated to a position above the concrete deck where they are removed by brush or spray nozzles. These massive circular screens are available in opening sizes ranging from 2 to 6 mm (0.08 to 0.24 in.) and designed to handle a large water differential across the screen.

2.3.3.6 Inclined Cylindrical Screens.

These screens (Figure 3.13) consist of a cylindrical screening basket, similar to the drum screens, with an internal screenings removal mechanism that is typically a helical screw. This screen is typically installed at a 30° to 45° incline. This type of screen can either be fixed with screenings removed by the internal helical screw or by an arm that rotates around the inside of the basket. The basket can consist of either perforated plates or bars. The screenings drop into an axially located hopper feeding a screw and are conveyed through an inclined washer/compactor pipe. Because of the central feed pattern, screenings carryover is low. One of the biggest advantages of this screen is that it is supplied with an integral washer/compactor. However, it requires a larger footprint and shallower influent channel compared to other screens.

2.3.4 *MICROSCREENS*

Microscreens are characterized by openings that are less than 0.5 mm (0.02 in.). Microscreens are devices designed to retain fine, screenable particles. They are typically used as a replacement for processes that

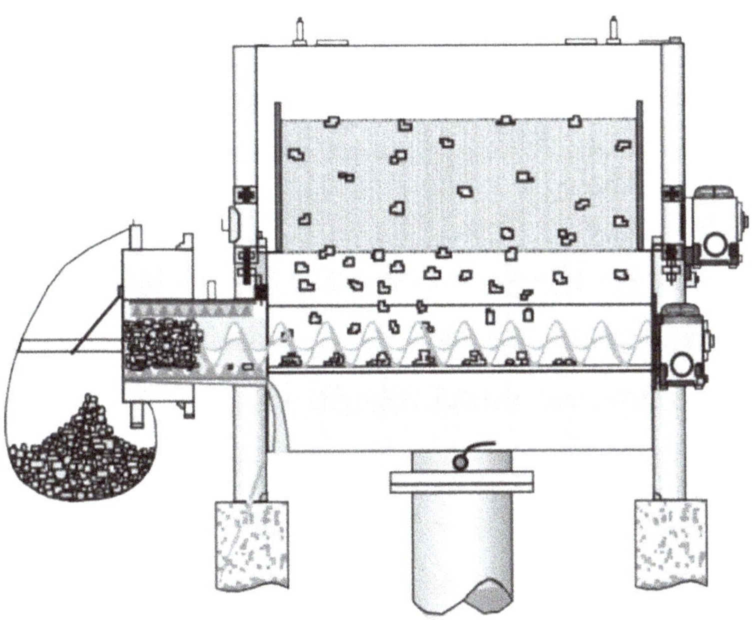

Figure 3.11 Externally Fed Drum Screen (Reprinted with permission from ANDRITZ Separation, Inc.)

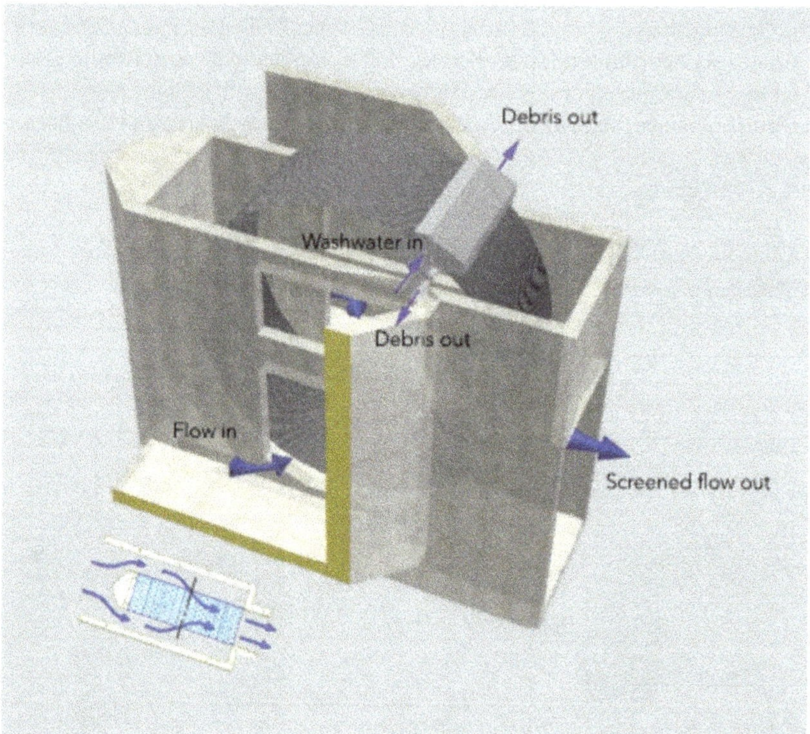

Figure 3.12 High Double-Entry Drum Screen (Reprinted with permission from Ovivo USA, LLC)

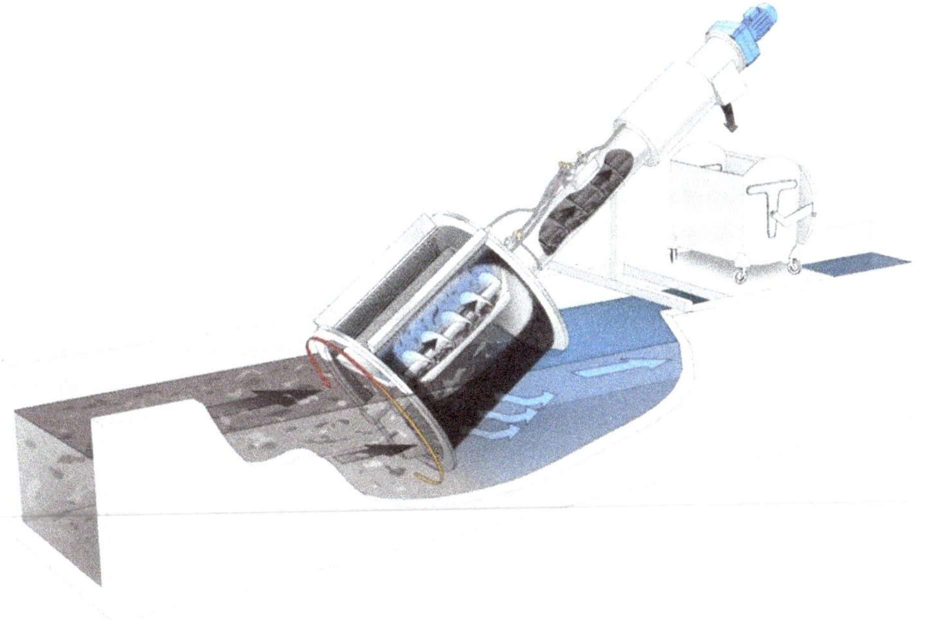

Figure 3.13 Inclined Cylindrical Screen (Courtesy of Huber Technology, Inc.)

require fine particle removal (such as certain membranes) or as a replacement for primary treatment processes. This section will briefly focus on using microscreens for the purposes of removing screenable material and grit from the wastewater.

Some microscreens have a configuration similar to drum screens but use a fine mesh fabric for their screening media and are capable of removing solids from 0.1 to 6 mm (0.004 to 0.24 in.) on average. If the wastewater contains high concentrations of grease or scum, a hot water unit should be considered as part of the cleaning cycle for the screen. It is recommended that a coarse screen be installed upstream to protect the microscreen.

Recently, the use of a specific type of microscreen, the filter screen, has resulted in the removal of grit particles because of the operating opening size that these units achieve. Filter screens are devices that use a rotating filter fabric as means to intercept solid particles, but can also be operated in a way to promote organic material buildup on top of the fabric to further reduce the operating opening size and increase solids capture. However, for grit removal applications, the primary mechanism for removal is fabric interception, so the selection of the fabric opening size becomes critical.

2.4 SCREENINGS PROCESSING

2.4.1 CONDITIONING

Washing and compaction of removed screenings is a critical function of the screening process, especially for fine screens. However, the amount of rock and grit removed with the screenings will need to be considered. If there is a high concentration of grit and rocks removed relative to the screenings, it may result in plugged compaction tubes and worn-out rotating equipment.

Washer/compactors have two equally important functions. Washing removes organic material from the screenings and returns it to the wastewater flow. Compaction reduces the volume of screenings, which reduces costs of storage and disposal as well as, in some cases, helping meet solid disposal requirements for screenings. Returning the organics to the wastewater stream reduces odor and handling hazards associated with the high concentrations of pathogens in the organic matter in screenings. For screens with a 6 mm (0.24 in.) opening or smaller, it is highly recommended that washer/compactors be provided to reduce the volume, weight, and odor of the large quantity of discharged screenings.

The most common form of washer/compactor is a screw auger, followed by a compression friction tube (Figure 3.14). Compactors can reduce organic content by 90% and increase solids to 50% (WEF, 2017). Washer/compactors can be equipped with grinders ahead of the auger to significantly reduce the quantity of screenings and to provide maximum washing.

2.4.2 SCREENINGS TRANSPORT, STORAGE, AND DISPOSAL

Conveyors are a widely used method of mechanically transporting screenings. Belt conveyors are the most common, but screw augers are also used. If the screen discharges directly to a washer/compactor, then the compaction friction tube can be extended to the screenings collection point. Friction tubes longer than 4.6 m (15 ft) should be avoided because the screenings can become too dry, requiring excessive force to push the screenings through the tube. Augers can be designed to serve both for collection/transport and as the first part of the washer/compactor.

Belt conveyors do not require significant maintenance but, depending on odor control requirements, they may need to be covered and ventilated. Belt conveyors are typically straight, but if direction changes are needed then curved conveyors can be used, eliminating the need for multiple units. Excessive slope on a belt conveyor should be avoided to prevent rollback and prevent screenings from falling off the conveyor.

Screw augers are used for screenings transport, particularly when an enclosed system is required for odor control. Both shafted and shaftless augers are used but, because objects are more likely to become wedged in shafted augers, shaftless augers are more common. The biggest benefit of augers is the ability to function as the first part of a washer/compactor system prior to discharge to a compaction tube.

Another method of screenings transportation is the sluice channel. Sluicing screenings is becoming more popular for a number of reasons but is only used in conjunction with a washer or compacter. As the need for cleaner screenings increases, requiring the use of washtubs on the washer/compactor, the sluice

Figure 3.14 Bagging System for Washed and Compacted Screenings (Reprinted with permission from Indigo Water Group)

essentially becomes an extension of the washtub, getting the washing process started as the screenings are discharged from the screen. Sluices have no moving parts and have a low capital cost compared to conveyors; however, the flushing water requirements will be higher when sludge gates are used. The use of sluice channels can also facilitate grinding and, in some cases, conveying to a site better suited for treatment and removal. The primary disadvantage to sluicing is that it requires a relatively large amount of water, especially problematic if a facility effluent water supply is not available.

To complete the handling of screenings, the material must be transported for disposal in accordance with local, state, and federal regulations. Landfilling is by far the most common method of disposal; however, increasingly strict European regulations for landfilling are designed to require incineration and other methods of disposal. On-site incineration, another method of screenings disposal, typically involves mixing the screenings with other treatment facility solids. Nevertheless, a few large facilities incinerate screenings separately because incomplete mixing of screenings with the primary sludge and waste biosolids creates unstable conditions in the incinerator. Screenings tend to clog feed mechanisms and should be ground or shredded and dewatered before being fed to the incinerator (U.S. Environmental Protection Agency [U.S. EPA], 1979).

2.5 DESIGN CONSIDERATIONS

Design considerations for screens include opening size; screen type; equipment height and footprint; channel depth, width, and approach velocity; headloss; provisions for removing the screen for maintenance; discharge height to accommodate screenings conveyed or washer/compactor equipment; angle of screen; screen cover for weather protection and odor control; construction materials and coatings for overall unit; drive unit service factor, motor size, environmental rating, and enclosure; spare parts; and provision of redundant screen or bypass manual screen.

2.5.1 EQUIPMENT SELECTION CRITERIA

There are several equipment selection criteria that must be considered when designing screens. Trash racks are seldom used ahead of coarse screens or grinders, except for combined systems with significant quantities of stormwater and at prisons with heavy screening quantities. A more common use of trash racks is for fine screens protection. Due to advances in technology, most new projects use screen openings from 3 to 12 mm (0.12 to 0.5 in.). Even finer screens with openings between 1 to 3 mm (0.04 to 0.12 in.) are used for advanced processes such as **membrane bioreactors** (MBRs). Along with finer screens, washer/compactors are now standard equipment. In the 2008 WEF utility members survey, the median screen opening is reported to be 9.5 mm (0.37 in.), and 53% of the U.S. facilities were operating washer/compactors.

Membrane bioreactor (MBR) is a wastewater treatment process that combines the activated sludge process with membrane filtration. The membranes are suspended in an activated sludge reactor to separate liquid from solids.

2.5.2 LOCATION OF SCREEN

Design practice for the location of screening devices at a WRRF varies. Screens can be located either upstream or downstream of influent pumps because most common raw wastewater pumps are capable of pumping screenings. However, with the recent introduction of non-dispersible products, preference has been given to screening prior to pumping. Screens are typically located upstream from the **grit** removal system. When fine screens are used, they could be located either upstream or downstream of grit removal, depending on the screen opening size provided and the anticipated effect on grit concentration equipment, which tends to plug with large solids.

Grit is sand, gravel, cinders, and other heavy solid matter with settling velocities substantially higher than those of organic (putrescible) solids in wastewater.

The question of whether screening equipment should be located in an enclosed structure depends on three conditions: equipment design, odor control needs, and climate. In climates with freezing temperatures, a heated enclosure is necessary. A structure not only protects the equipment but eases maintenance and improve aesthetics. Regardless of whether the screening equipment is housed, the drive mechanisms of a mechanically cleaned bar screen should be enclosed. In windy areas, the screen rake and discharge chute areas need covers to prevent screenings from blowing away.

The effects of backwater—the upstream water level—caused by headloss through the screen are important. Many installations include an overflow weir to a bypass channel to prevent upstream surcharging if the screen becomes blinded because of power failure or mechanical problems. Screens typically precede flow metering devices to prevent backup from affecting measurement accuracy.

2.5.3 HYDRAULIC CONSIDERATIONS

Hydraulic considerations described in this chapter apply to the design of a bar screen installation. For other screens and grinding devices, designers should follow manufacturer's recommendations for channel dimensions, capacity ranges, upstream and downstream submergence, power requirements, headloss information, and screening retention.

Velocity distribution in the approach channel has an important influence on screening equipment performance, regardless of screen type. A straight channel ahead of the screen ensures good velocity distribution across the screen and maximum device effectiveness. Use of anything other than a straight channel can divert too much flow to one side, resulting in accumulation of debris on that side of the screen. Hydraulic symmetry is important in the design of influent channels especially when multiple screens will be provided to prevent uneven distribution of screenings. In hydraulically sensitive areas, **computational fluid dynamics** (CFD) should be used to ensure the necessary hydraulic performance.

Computational fluid dynamics (CFD) are a series of algorithms and calculations used to predict or validate the behavior of liquid or gases flowing over or through constructed surfaces.

The design engineer must ensure that the wastewater approach velocity to the screen does not fall below a self-cleaning value or exceed velocities that may force material through the screen. Lower velocity will result in more solids and grit deposited in the channel. Ideally, the velocity in the channel ahead of the screens should exceed 0.4 m/s (1.3 ft/s) at minimum flows to avoid grit deposition. However, this is not always possible with the typical diurnal fluctuation in wastewater flows. As a reasonable compromise, the channel could be designed so that a velocity of at least 0.76 m/s (2.5 ft/s) for resuspending solids is attained during peak flow periods. Where significant stormwater must be handled, approach velocities of approximately 0.9 m/s (3 ft/s) should be used to prevent grit deposits from compromising screen operation during storms.

The flow velocity ahead of and through a bar screen affects its operation significantly. To avoid screenings pass-through, a peak flow velocity of no more than 1.2 m/s (4 ft/s) through the openings is recommended. However, there may be cases in which it is acceptable or even necessary to exceed the 1.2 m/s (4 ft/s) threshold at peak flowrates when high flow variability is observed. This should only be done when working closely with the equipment manufacturer. Because the screen open area is typically around 50% of the cross-sectional area of the flow channel, one of the most challenging aspects of screen design is balancing the need for grit suspension velocity upstream of the screen with maintaining low enough velocity through the screen openings to prevent screenings pass-through. If peak to average flow ratios are significant, peak flow velocities typically exceed recommended design values, causing higher than anticipated screening carryover. In these situations, hydraulic control structures, such as cutthroat flumes and partially submerged baffled walls (Figure 3.15), have been successfully used to control the water level inside the channel at peak conditions. This allows the velocities through the screen to be within the designed values. Provisions must be made for the added headloss in the hydraulic profile and for the scum that could collect in the channel.

Based on the velocities required for proper screen operation, the number of screens can be evaluated to help best meet the flow requirements. Typically, designs have resulted in at least one screen rated for 100% of the peak hour flow with either a fully redundant standby screen or a manual bar rack. Recently, more designs are including multiple screens to meet the design flow with at least two screens rated for 50% of peak-hour flow with a third channel with another 50% screen or manual bar rack.

Typical headloss allowed through a manually cleaned bar screen is 150 mm (6 in.), assuming frequent screen inspection. A nearly constant headloss for continuously cleaned mechanical screens can be maintained with a constant flow, but this headloss allowance should never be less than 150 mm (6 in.) because of intermittent cleaning cycles. Curves and tables for headloss through the screening device are available from equipment manufacturers.

Headloss through screens typically ranges from 50 to 635 mm (2 to 24 in.), depending on flow and screen opening size. Maximum headloss can reach 1 m (3 ft) or more in large units at peak flow with fine screens or when screen cleaning frequency is inadequate. The design engineer must take steps to ensure that headloss through the screening device at maximum flows will not surcharge influent sewers enough to impair upstream services.

Properly sizing screening equipment involves a careful balance between the channel width, available hydraulic grade line, the screen opening size, the maximum slot velocity through the screen at different

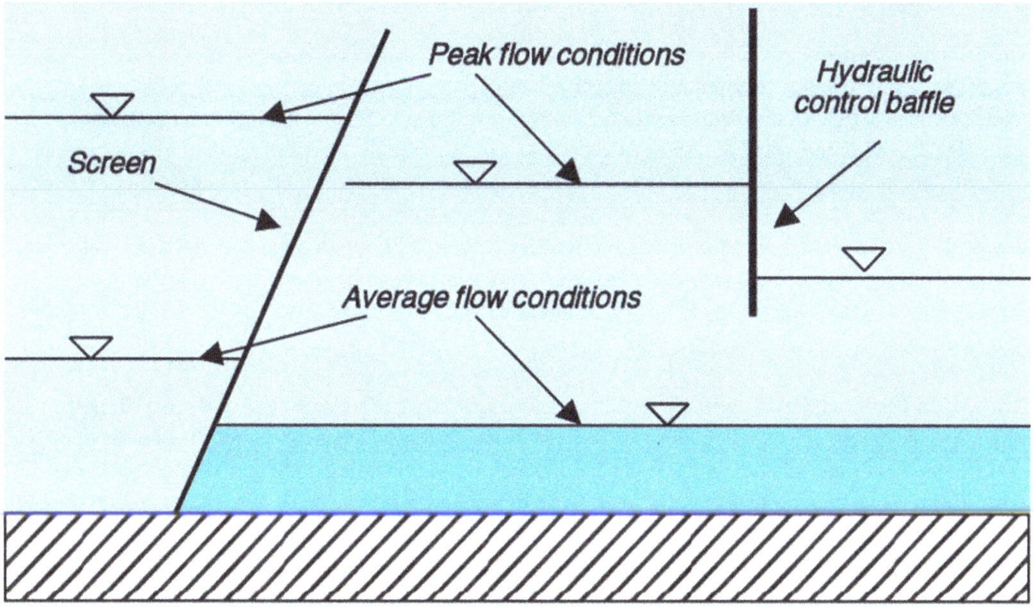

Figure 3.15 Hydraulic Control Structure Example

conditions, and the minimum velocity at low flows. Of paramount importance is the estimation of the headloss through the screen, as the screen size (screening area) is correlated to this number. The recommended practice for sizing screening elements should include estimation of the surface water elevation upstream and downstream of the screen.

2.5.4 SCREENING SUSTAINABILITY AND RESOURCE RECOVERY

With the increased focus on sustainability and resource recovery in WRRFs, design of the screening process should also consider the following principles. Sustainability is enhanced by a properly designed screening and screenings washing process by reducing the pathogen and organic content of the screenings; the direction of the highly soluble organic loading associated with the screenings to the liquid stream, enhancing the nutrient removal processes; and enhancement of the gas generation potential of the facility by keeping the downstream operating equipment working properly. When inadequate screening causes ragging issues, then the adverse effects on the different processes could range from increased energy demand for pumping, aeration, mixing, or other processes; to inadequate material for biosolids reuse. While reuse of screenings is not commonly practiced, some opportunity for resource recovery exists with facilities equipped with incinerators that practice heat recovery.

3.0 COARSE SOLIDS REDUCTION

Historically, in-channel grinders or **comminutors** have been used instead of screening to screen and shred material ranging in size from 6 to 19 mm (0.25 to 0.75 in.) without removing particles from the flow. Now comminutors are rarely used as stand-alone substitutes for screens. Solids from comminutors and screenings grinders have caused downstream problems, including deposits of plastics in digestion tanks and rag accumulations on air diffusers. Pulverized synthetic materials will not decompose in the digestion process and, if not removed, could bar public acceptance of biosolids for reuse as a soil amendment.

A comminutor is a circular screen with cutters that grind large wastewater solids into smaller particles.

Comminutors may create ropes or balls of material (particularly rags) that can clog treatment equipment (for example, mechanical aerators, mixers, pump impellers, pipelines, and heat exchangers). However, with the increased disposal of non-dispersible products many utilities have been adding comminutors ahead of pumping stations. There is also a grinding device used on raw wastewater in one screen design that uses an in-channel grinder ahead of a fine screen. Using grinders before screening greatly reduces the quantity and organic content of washed and compacted screenings, but can also result in passage of some material that would not have passed otherwise, which could affect the quality of biosolids produced.

4.0 GRIT REMOVAL

The primary purpose of grit removal is to reduce abrasion and wear of downstream mechanical equipment, prevent solids accumulation in channel and process tanks, and prevent loss of digestion capacity due to active volume decrease. If not removed, grit can accumulate in pipelines, channels, anaerobic digesters, and aeration basins. Settling of grit in primary clarifiers can result in excessively high concentrations of primary sludge, making it difficult to pump the sludge, especially in long suction lines.

Removed grit (grit slurry) is concentrated and washed to remove lighter organic material that is captured. The degree of washing and concentration of the grit slurry is dictated by several factors, including the degree of odor control and vector attraction required during storage, and costs and regulations associated with disposal.

A vector is an organism such as an insect or rodent that can carry or transfer pathogens, which spread disease.

Grit systems have traditionally been designed to remove 95% of particles larger than 200 μm (65 mesh) with a specific gravity of 2.65 (U.S. EPA, 1987). There is increasing evidence for higher percentages of grit in the influent wastewater smaller than 200 μm, and, therefore, an increasing demand to design grit systems capable of removing up to 95% of 100 μm (150 mesh) grit particles to avoid adverse effects on downstream processes.

A single grit removal unit with a bypass channel will suffice for small installations (average flow 15 000 m³/d or 4 mgd) or for facilities in which infrequent flows of wastewater containing grit can be tolerated in downstream processes. Facilities that are large, served by combined sewers, or with unit processes such as centrifuges that are more prone to abrasion from grit will require multiple grit removal units. This allows periodic removal of units from service for cleaning, maintenance, and repair.

Similar to screen design, wastewater flow extremes must be known so that chambers can be designed to efficiently remove grit from all flows. The quantity of grit entering the treatment facility is typically greatest during peak flows when scour velocities and transport rates are highest. Grit chambers are sized to remove grit effectively at peak flows but to avoid removing excessive organic material at lesser flows.

Typically, vortex grit removal systems are the most used grit processes in new facilities because of their compact size, mechanical simplicity, and lack of turbulence, which results in less release of odors from the process. Aerated grit systems, because of the use of larger tanks, are more capable of handling large, sudden quantities of grit.

4.1 GRIT QUANTITIES

Grit materials include particles of sand, gravel, other mineral matter, and minimally **putrescible** organics such as coffee grounds, eggshells, fruit rinds, and seeds. The quantity and characteristics of grit removed from wastewater will vary. Variables influencing grit quantity include the type of collection system (combined or separate), characteristics of the drainage area, use of household garbage grinders, condition of the collection system, sewer grades, types of industrial waste, and efficiency of grit removal. The perceived quantity of grit is affected by the efficiency of washing/dewatering equipment; therefore, upgrading washing/dewatering equipment can give the false impression that less grit is being removed because of elimination of organics and water. In most cases, grit accumulates in the collection system and is flushed into the WRRF in large quantities during high flow. Because these peak grit loading periods frequently overload poorly functioning grit systems, facility data can be of limited use when designing upgrades. In the 2008 WEF member survey, grit quantities reported were erratic. When correcting for outliers, however, quantities ranged from 3.7 to 148 L/1000 m³ (0.5 to 20 cu ft/mil. gal) and averaged 37 L/1000 m³ (5.0 cu ft/mil. gal).

Putrescible refers organic matter that is likely to become a rotten, foul-smelling product when it decays or decomposes.

4.2 GRIT PHYSICAL PROPERTIES

Grit solids characteristics will vary with solids content from 35% to 80% and volatile content from 1% to 55% (U.S. EPA, 1979). Well-washed grit should achieve a solids content of 90% with a minimum of putrescible solids. The moisture and volatile content will be influenced by the degree of washing the grit receives. The bulk density of dewatered grit will range from 1400 to 1800 kg/m³ (90 to 110 lb/cu ft). In the WEF member surveys (2016), reported grit density ranged from 1100 to 2200 kg/m³ (70 to 140 lb/cu ft) and averaged 1400 kg/m³ (90 lb/cu ft). While the effects of air voids in the removed grit must be accounted for, when comparing this data to clean sand (specific gravity of 2.65 or a density of 2600 kg/m^3 [165 lb/cu ft]), it can be seen that the characteristics of removed grit differ significantly from clean sand.

4.3 AERATED GRIT CHAMBER SYSTEMS

In aerated grit chamber systems, air introduced along one side near the bottom causes a spiral roll velocity pattern perpendicular to flow through the tank. The heavier particles with their correspondingly higher settling velocities drop to the bottom, and the roll suspends the lighter organic particles into the tank effluent. The rolling action induced by the air diffusers is independent of flow through the tank, allowing the aerated grit chamber to operate effectively with a wide range of flows. Heavier particles that settle are moved by the spiral flow of the water across the tank bottom and then into a grit trough or hopper.

When the wastewater flows into the aerated grit chamber, particles settle to the bottom at rates dependent on their size and specific gravity and on the velocity of roll in the tank. The rate of air diffusion and the tank shape govern the rate of roll and the size of the particle with a given specific gravity. Diffused air serves as a velocity control method with sufficient flexibility to accommodate varying conditions. Diffused air also requires only minimal headloss through the unit. Proper design depends on an understanding of the variables affecting the airlift pumping energy and its effect on the roll pattern in the tank. Figure 3.16 illustrates a typical aerated grit chamber.

Poor performance of some aerated grit chambers resulted in high percentages of organic matter in the removed grit, low removal efficiencies, and periodic abnormal grit loads on the collection equipment. There are several factors that can lead to poor performance in aerated grit chambers including improper tank geometry (most common); improper baffling; inconsistent air supply to diffusers, resulting in erratic roll patterns; and improper diffuser location.

Based on the work of Londong (1989), optimum aerated grit chamber design parameters are listed in Table 3.2. Air rates typically range from 5 to 12 L/s•m of tank length (3 to 8 cu ft/min/ft). Rates as low as 2 L/s•m (1 cu ft/min/ft) have been used for shallow, narrow tanks. Rates higher than 8 L/s•m (5 cu ft/min/

Figure 3.16 Typical Aerated Grit Basin (Reprinted with permission from Envirodyne Systems, Inc.)

Table 3.2 Aerated Grit Basin Recommended Design Criteria

Criteria	Design Values	Notes
Peak hour flow detention time	3–10 minutes	Use longer times for plants with heavy grit loads to improve fine grit removal
Width:depth ratio	0.8–1.0	
Depth	3.7–5 m (12–16 ft)	
Length:width ratio	3–8:1	Longer basins provide better fine grit removal
Bottom floor slope	30°	
Diffuser location	70% of total depth	

ft) are often used for deep, wide tanks. Providing valves and flowmeters is good practice for monitoring and controlling the air flow. Aeration is typically with coarse bubble non-clog diffusers. Separate, dedicated blowers are preferred. A minimum hydraulic detention time of 3 minutes at peak hourly flow of an aerated grit basin with proper geometry will provide a sufficient degree of coarse grit removal. Longer times should be considered if removal of fine (100 mesh) grit is desirable or if basins are covered and ventilated to odor control processes and to provide preaeration and volatile organic compound (VOC) removal. It is not unusual to size aerated grit chambers for 10 to 15 minutes' detention time. A chamber length-to-width ratio ranging from 3 : 1 to 8 : 1 is necessary, with longer tanks providing better grit removal. A width-to-depth ratio of 0.8 to 1.0 is necessary to prevent short-circuiting down the quiescent central core not affected by the roll pattern. A floor slope of 30° is recommended to ensure quick movement of grit into the longitudinal grit collection sump and to prevent re-entrainment of fine grit by the spiral roll. Inlet baffles are frequently used to turn the influent flow into the spiral roll. Flow through the tank should be perpendicular to the spiral roll pattern. Tank outlet is typically perpendicular to the flow through the tank.

Screw augers are the most frequently used method to transport grit along the bottom of the tank to the grit sump and pumps. Other methods include chain and flight collectors, chain and bucket collectors, clamshell buckets, and multiple sumps, each with dedicated grit pumps. Screw augers are used to transport grit to a grit pocket at the end of the tank and into flooded suction recessed impeller grit pumps. Both shafted and shaftless augers are used.

Chain and flight collectors can be used to drag grit up an inclined slope for removal. Chain and bucket conveyors are used; bucket systems typically require significant maintenance. In lieu of auger or chain-driven collectors, top-mounted self-priming or vertical cantilever pumps or flooded suction pumps mounted alongside the bottom of the tank can be used, but require formation of individual sumps along the length of the tank.

Example 3.2

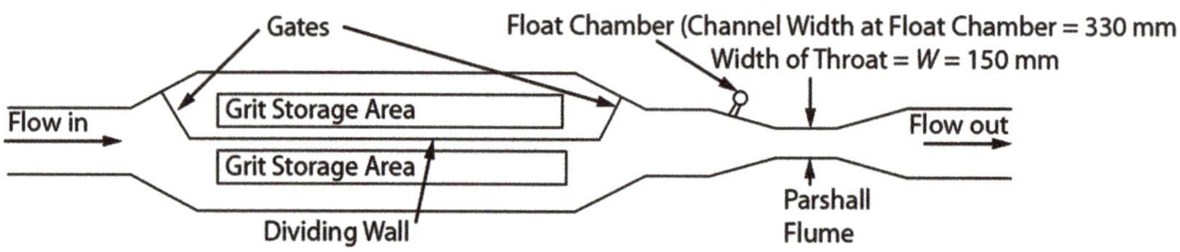

Diagram of grit removal chamber and Parshall flume (not to scale). One chamber used while other is cleaned

Design an aerated grit chamber to treat municipal wastewater for the following conditions.

Dual chambers will be required for cleaning one while the other is in operation. Each chamber will receive one-half of the flow.

Parameter	Value (SI Units)	Value (BG Units)
Average flowrate	0.6 m^3/s	13.69 mgd
Peaking factor	2.5	
Detention time (@ peak flow)	7.5 min	
Width-to-depth ratio	1:1	
Depth	4 m	13.1 ft
Number of chambers	2	
Air supply rate	8 L/s•m	5.2 cu ft/min/ft
Grit quantity	0.06 $m^3/1000\ m^3$	0.06 cu ft/1000 cu ft

Note. Values were determined from Table 3.2 and text.

Solution

Calculate the peak flowrate:

$$Q_{peak} = 0.6\ m^3/s \times 2.5\ \text{(peaking factor)} = 1.5\ m^3/s\ (3.42\ \text{mgd})$$

Determine the volume of each chamber:

$$V = \frac{Q_{peak}}{2\ \text{chambers}} \times t = \frac{1.5\ m^3/s}{2} \times 7.5\ \text{min} \times \frac{60\ s}{\text{min}} = 337.5\ m^3\ (11\ 920\ \text{cu ft})$$

Calculate the dimensions of the chamber:

$$\frac{\text{width (W)}}{\text{depth (D)}} = \frac{1}{1} = \frac{W}{4\ m} \qquad \text{Width} = 4\ m\ (13.1\ ft)$$

$$\text{Length} = \frac{\text{Volume}}{W \times D} = \frac{337.5\ m^3}{4\ m \times 4\ m} \qquad \text{Length} = 21\ m\ (68.9\ ft)$$

Calculate detention time (t) at average flow rate:

$$t = \frac{V}{Q_{avg}} = \frac{4\ m \times 4\ m \times 21\ m \times 2\ \text{chambers}}{0.6\ m^3/s \times 60\ s/min} \qquad \text{Detention time} = 9.3\ \text{min}$$

Calculate air supply:

$$\frac{\text{Air}}{\text{Chamber}} = 21\ m \times 8\ L/s \cdot m \times \left(\frac{1\ m^3}{1000L}\right) \times \left(\frac{60\ s}{\text{min}}\right) \qquad \text{Air supply} = 10\ m^3/\text{min}\ (353\ \text{cu ft/min})$$

Total air supply = 20 m^3/min (706 cu ft/min)

Estimate grit quantity (average per day):

$$0.06\ m^3/1000\ m^3 \times 0.6\ m^3/s \times \left(\frac{3600\ s}{hr}\right)\left(\frac{24\ hr}{d}\right) \qquad \text{Grit} = 3.11\ m^3/d\ (109.7\ \text{cu ft/d})$$

4.4 VORTEX GRIT REMOVAL SYSTEMS

4.4.1 *MECHANICALLY INDUCED VORTEX*

Mechanically induced vortex (MIV) systems combine the hydraulic forces associated with tangential incoming flow with mechanical paddles to create and enhance a vortex pattern. An MIV system comprises two circular chambers. The upper chamber is where the flow enters, allowing for the vortex creation that separates and settles the grit. The lower chamber is where the settled grit is stored until it is removed for further processing. Tanks are circular to encourage the vortex flow to force grit solids to the side walls where velocities are lower as a result of boundary layer effects. Once near the walls, the vortex and gravity continue to move the captured grit solids into the lower chamber. A straight flow pattern into the tank is critical to minimize turbulence at the inlet of the chamber. At the center of the chamber, rotating paddles maintain a constant circulation within the chamber and help separate organics from the grit path as it approaches the lower hopper. Grit solids are removed from the center hopper by top-mounted vacuum-primed pumps, flooded suction pumps, or (less frequently) air lifts.

Mechanically induced vortex grit systems include two basic designs: tanks with flat bottoms and a plate that separates the sump from the main tank and is surrounded by a narrow slot through which grit enters the sump; and tanks with a sloping bottom and a large opening to the grit hopper.

Ideally, flow into a vortex grit chamber should be straight and streamlined. Table 3.3 shows typical design criteria for vortex grit chambers. However, sizing of proprietary vortex grit chambers is based on recommended dimensions provided by equipment manufacturers, and the units are typically marketed in standard nominal sizes based on peak flow.

4.4.2 INDUCED VORTEX

Induced vortex grit removal systems (Figure 3.17) are fundamentally different from mechanical vortex systems because the vortex is created by the force of the incoming flow. Induced vortex basins are, therefore, smaller in diameter and require pumped flow. Induced vortex systems can be designed to capture a higher percentage of fine grit than mechanical vortex systems but result in significantly more headloss. Headlosses of 0.5 to 0.7 m (1.6 to 2.3 ft) are typical for 95% removal of 75 μm (200 mesh) grit. The use of an induced vortex system on raw wastewater has seen a decline in recent years and these systems are far more commonly used in the grit dewatering processes.

4.4.3 MULTIPLE TRAY VORTEX

The multiple tray vortex system (Figure 3.18) is a proprietary system that has rapidly increased in popularity. This system uses a flow distribution header for distributing influent over multiple conical trays. Tangential feed establishes a vortex flow pattern in which solids settle into a boundary layer on each tray and into a center underflow collection chamber for pumped removal. The system applies plate settler technology to grit removal by using the stacked trays to create a large, concentrated surface area and short settling distances. For new systems, the multiple tray system can be installed in a concrete basin with a smaller footprint than aerated systems. Multiple tray vortex systems can be designed to remove particles down to 75 μm in size and multiple installations have been proven to remove 95% of 100 μm particles.

4.5 OTHER SYSTEMS

Historically, other types of grit tanks have been used including the detritus tank and velocity-controlled tank. Detritus tanks, also know as square degritters, were one of the earliest grit removal tanks, characterized by having a constant level, short detention, and sized based on overflow rate. Velocity-controlled (horizontal flow) tanks were another early type of grit removal system that used proportional

Table 3.3 Vortex Grit Chamber Recommended Design Criteria

Criteria	Design Values	Notes
Peak hour flow detention time	20–30 seconds	
Influent channel velocity	0.5–1.1 m/s (1.6–3.5 ft/s)	Minimum acceptable velocity of 0.15 m/s (0.5 ft/s) for low flows
Target particle	50 mesh (> 300 μm)	Particle removal should occur between 40%–80% of peak flow
Influent channel	4–7 times inlet channel width	

Figure 3.17 Induced Vortex Grit System (Reprinted with permission from Hydro International)

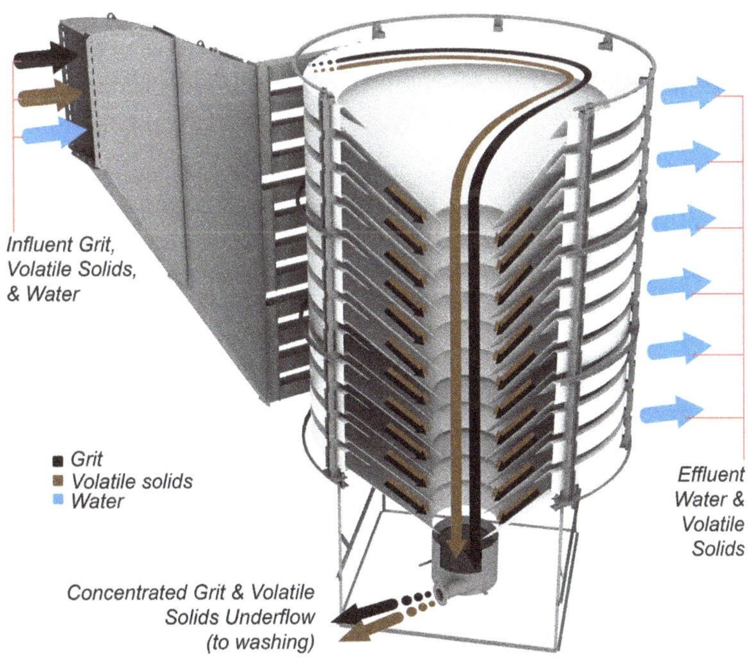

Figure 3.18 Multiple Tray Grit Separator (Reprinted with permission from Hydro International)

weirs or rectangular control sections (such as Parshall flumes) to vary the depth of flow and keep the velocity of the flow stream at a constant rate of 0.30 m/s (1.0 ft/s).

4.6 GRIT PROCESSING

One of the objectives of the primary grit removal process is to minimize the organic material removed. Varying degrees of organic material, however, will always be removed with the grit. Because of the effects of grease and detergent in wastewater, providing sufficient detention time for removal of fine grit results in increased capture of organic material. Design of each grit removal system must, therefore, weigh the need for fine grit removal versus the higher cost of more effective grit washing equipment. Another important consideration in grit processing is to avoid grit carryover, so the grit washing equipment must be designed to remove the same design particle as the primary grit system associated with it. Part of the process also includes pumping and transporting grit, which require extra design considerations compared to transporting water.

4.6.1 GRIT CYCLONE/CLASSIFIERS

Cyclone systems are used for separating grit from organics in grit slurries. Centrifugal forces that develop in the cyclone cause heavier grit and suspended solids particles to concentrate along the sides and on the bottom. Lighter solids, including scum, are removed from the center through the top of the cyclone. Cyclones operate best at constant flow and pressure. If flows depart from design flows, solids are lost to the centrate stream.

A cyclone separator concentrates the grit centrifugally, requiring a steady feed of grit slurry at an inlet pressure of 34 to 140 kPa (5 to 20 psi). The constant feed rate is within the range of 800 to 1900 L/m (200 to 500 gpm) depending on the size of the cyclone. Cycle times for intermittent operation can vary from 5 minutes to 8 hours; peak grit loadings may require continuous operation.

Grit classifiers, either the inclined screw or escalator type, wash grit by separating out putrescible organics. Classifiers should be sized based on settling velocity of the particles to be settled, feed flow capacity, and grit raking capacity. For a target particle size and flowrate, the design engineer selects a minimum pool area and overflow weir length. The design engineer checks the classifier slope to ensure removal of the desired marginal particle size. Flatter slopes will remove finer grit particles. An example of a typical grit cyclone and classifier is shown in Figure 3.19.

Traditional cyclone/classifiers are typically designed to capture down to 150 μm particles and are not designed to provide a high degree of washing with final organic content of greater than 25%. The more advanced proprietary induced vortex/escalator design is capable of capturing particles down to 75 μm and producing a final organic content of no more than 15%.

4.6.2 CONICAL GRIT WASHERS

Conical grit washer technology (Figure 3.20) represents an advancement in grit slurry processing resulting from strict European regulations for preliminary treatment residuals disposal. Stainless steel conical vessels are used to capture grit slurry and various systems are used to wash organics from the grit. Rotating arms within the vessel slowly mix the settled grit and a washing jet at the bottom of the unit, activated by a solenoid valve on a timer, vigorously washes it. Lighter organic material continuously overflows the unit and heavier organic material is blown off at regular intervals from a midlevel overflow. These washers create grit with low organic content. The surface flow velocity, including the wash water, should be less than 25 m/h (0.02 ft/sec) and weir overflow rate should be less than 15 $m^3/h{\cdot}m$ (0.045 cu ft/sec/ft). Design flow to these units is typically limited to 25 L/s (400 gpm), although higher flows can be used if the units are equipped with cyclone(s). Historically, target grit particle size has been 200 μm for these units. Washing provided by these units is superior to traditional cyclone classifiers with a final organic content of less than 5%.

4.6.3 GRIT TRANSPORT, STORAGE, AND DISPOSAL

Overhead clamshell buckets, chain and flight collectors, and augers can be used to remove grit from aerated grit basins, but most installations use mechanical pumping. This is because grit slurry can be pumped directly to washing and dewatering equipment.

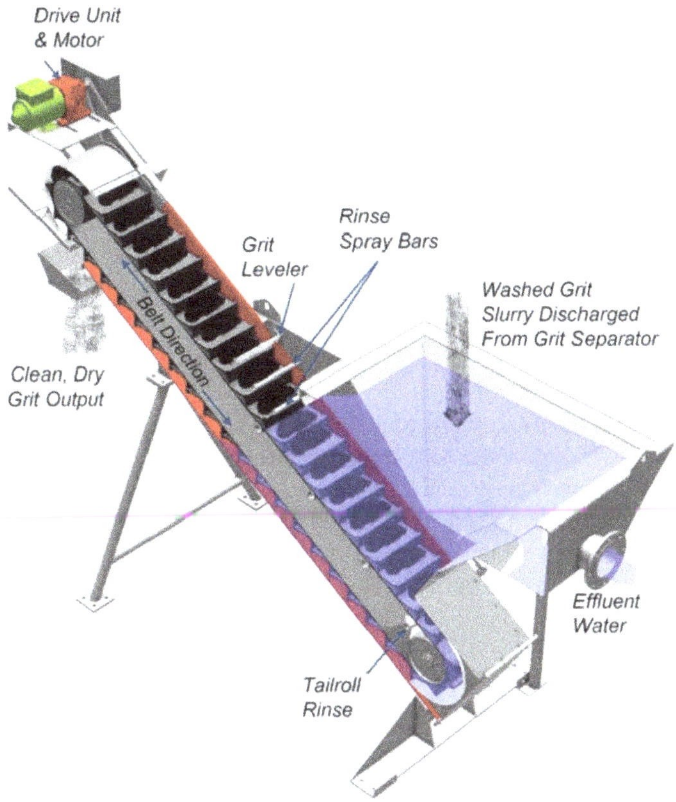

Figure 3.19 Grit Dewatering Escalator (Reprinted with permission from Hydro International)

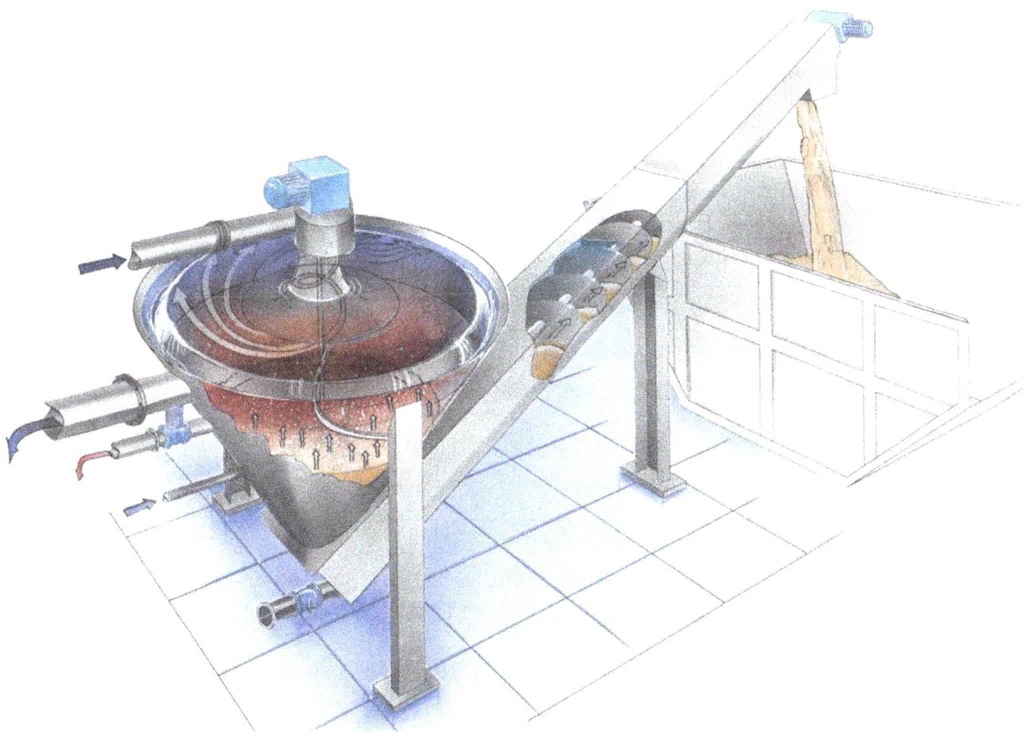

Figure 3.20 Conical Grit Washer (Courtesy of Huber Technology, Inc.)

Air lifts are a system in which compressed air is injected at the bottom of a pipe, displacing water and solids upwards through a discharge pipe.

Vortex or recessed impeller pumps are used to handle grit slurries. **Air lifts** have also been used, but have a history of operational problems. It is important that air release be provided before the grit enters the cyclone/classifier or washer. Grit can be conveyed directly to trucks, dumpsters, or storage hoppers. Containers should be covered to prevent odors during storage and hauling. Conveyors are frequently used to transport grit from handling facilities to containers. Overhead storage hoppers that discharge to truck containers avoid the need to keep a truck at the facility.

If grit is transported directly to trucks or dumpsters, then the container should have enough capacity to handle daily peak grit loads in the event that replacement trucks cannot arrive or leave (i.e., during inclement weather). Two bays are advisable to ensure continued loading if the loaded container must remain because of a truck breakdown.

4.6.4 GRIT PUMPING

One of the most important elements of successful grit removal systems is the design of the grit pumping system, which includes several key design considerations. Suction lines should be straight and provide flushing connections to minimize the potential for plugging. The suction and discharge lines should be sized to prevent grit deposition. The minimum pipe diameter to be used is 100 mm (4 in.) and a 1.2 to 2.1 m/s (4 to 7 ft/s) velocity should be maintained. If feasible, continuous grit pumping should be considered to minimize pipe clogging issues. Finally, pipe materials that are used in the pumping process must be abrasion resistant.

5.0 FLOW EQUALIZATION

Flow equalization is transient storage of wastewater for later release to a collection system or treatment process at a controlled rate to provide a reasonably uniform flow.

Flow equalization can accommodate wide variations in flowrates and organic mass loadings, which is one of the primary challenges in the design of WRRFs. Efficiency, reliability, and control of unit process operations within the facility can be adversely affected by the cyclic nature of waste generation, resulting in possible violations of effluent standards if treatment processes do not account for peak conditions.

Equalization can be used to minimize downstream process sizing in wastewater facilities. This section addresses the equalization processes for pretreatment facilities only. Equalization is designed to achieve more constant hydraulic or organic loading of downstream treatment processes. Equalization can result in some degree of treatment. This can be beneficial if, for example, a facility is receiving high concentrations of industrial waste. Treatment in equalization can be unintended, which must be taken into account when designing the downstream process. This can be particularly important in biological nutrient removal processes requiring maximum concentrations of soluble biochemical oxygen demand.

The benefits of hydraulic or organic loading equalization can include flow equalization, waste strength equalization, biological treatment enhancement, improved secondary clarifier performance, improved chemical feed and process reliability, preconditioning of wastewater for downstream processes, partial primary treatment, toxic substances dilution, and pH dampening.

Equalization usually occurs at the WRRF. However, in certain areas where the collection system has significant storage volume available, such as large gravity sewers (tunnels, collectors), or sufficient storage in the pumping station wet wells, equalization can also be practiced by carefully allowing storage to occur in the system (Heath et al., 2015).

Hydraulic equalization basins for WRRFs are constructed to dampen wet weather and diurnal peak flow variation. An additional benefit is a reduction in the variability of concentration and mass flow of wastewater constituents by blending in the equalization basin. This more uniformly loads downstream processes with organics, nutrients, and other suspended and dissolved constituents.

Waste strength equalization dampens the variability of the waste strength by blending wastewater in the equalization basin. For this purpose, the volume of wastewater in the equalization basin typically remains constant and flow remains variable.

Equalization reduces the size of unit processes at a new facility or relieves overloaded unit processes at an existing facility. By providing relatively constant loading, equalization can improve the efficiency, reliability, and operability of some types of facilities. For example, activated sludge and trickling filter facilities that are sized for peak organic loadings could be downsized. Equalization also protects from shock loading, which adds reliability to any biological process.

The use of equalization either to reduce the size of or relieve an overloaded condition on an individual treatment process is seldom economically justified. Nevertheless, if the cumulative effects of equalization throughout the facility are accounted for, equalization may be more cost effective than modifying existing unit processes.

Equalization operations have little resource recovery potential. However, they could provide significant sustainable features to a WRRF, as peak demand loadings and their associated larger energy requirements can be avoided when equalization is practiced.

5.1 DESIGN CONSIDERATIONS—METHODS

The design methodology for equalization involves determining the necessary volume, mixing, and aeration requirements, and the control of flows leaving the basin. The first step in estimating the volume involves determining the diurnal variations of wastewater. Whenever possible, this step should be based on actual operating data. Diurnal flow patterns will vary from day to day and seasonally, depending on the nature of the community (e.g., tourist areas, winter residences, agricultural food processing). It is, therefore, important to select a pattern that will ensure a large enough volume for equalization, taking into account conditions such as infiltration and storm-related inflow that influences variability of flow.

Several methods have been proposed for estimating the equalization volume required for typically varying wastewater flows and strengths (U.S. EPA, 1974). They include flow balance (mass diagram), concentration balance, combined flow and concentration balance, sine wave method, and rectangular wave method. These methods, sufficient for most WRRF applications, offer relatively simple procedures for estimating equalization volume requirements. If needed, other, more sophisticated methods of equalization analysis are available (DiToro, 1975; LaGrega & Keenan, 1974; McInnes et al., 1978; Novotny & Stein, 1976; Ongerth, 1979; Roe, 1951; Seidel & Bauman, 1961; Wallace, 1968). For all sizing methods, it has been common practice to add a 15% to 20% safety factor to the calculated equalization volume.

5.2 LOCATION

Equalization basins typically are located in the facility following the grit and screening units. To minimize solids accumulation, this is placed downstream of primary clarification. Equalization of solids handling sidestreams is also used frequently to regulate and reduce the effect of sidestream returns on the liquid treatment system. Figure 3.21 shows some typical equalization configurations.

5.3 OPERATION

Fill-and-draw mode is the most efficient method of operating an equalization basin for flow dampening. The basin is filled during the day when peak flows are occurring, and then it is drawn down at night when the facility is receiving low flows, so it is more capable of treating excessive flow. If an equalization basin is not operated in fill-and-draw mode, it acts as a mass loading equalization basin only, assuming the basin is completely mixed.

5.4 BASIN CONFIGURATION

Equalization basins are typically large concrete or earthen structures with sloping sides to facilitate cleaning. Earthen basins may have the side slopes lined with geomembranes or pavement to prevent erosion. Basin floors are sloped to one side to allow collection of settled solids along one wall. It is common practice to provide more than one equalization basin to allow for flexibility. Influent structures are typically designed to allow energy dissipation to prevent basin erosion, particularly in earthen basins. Effluent structures are built of concrete with a series of slide gates that serve as isolation or flow control, acting as weirs.

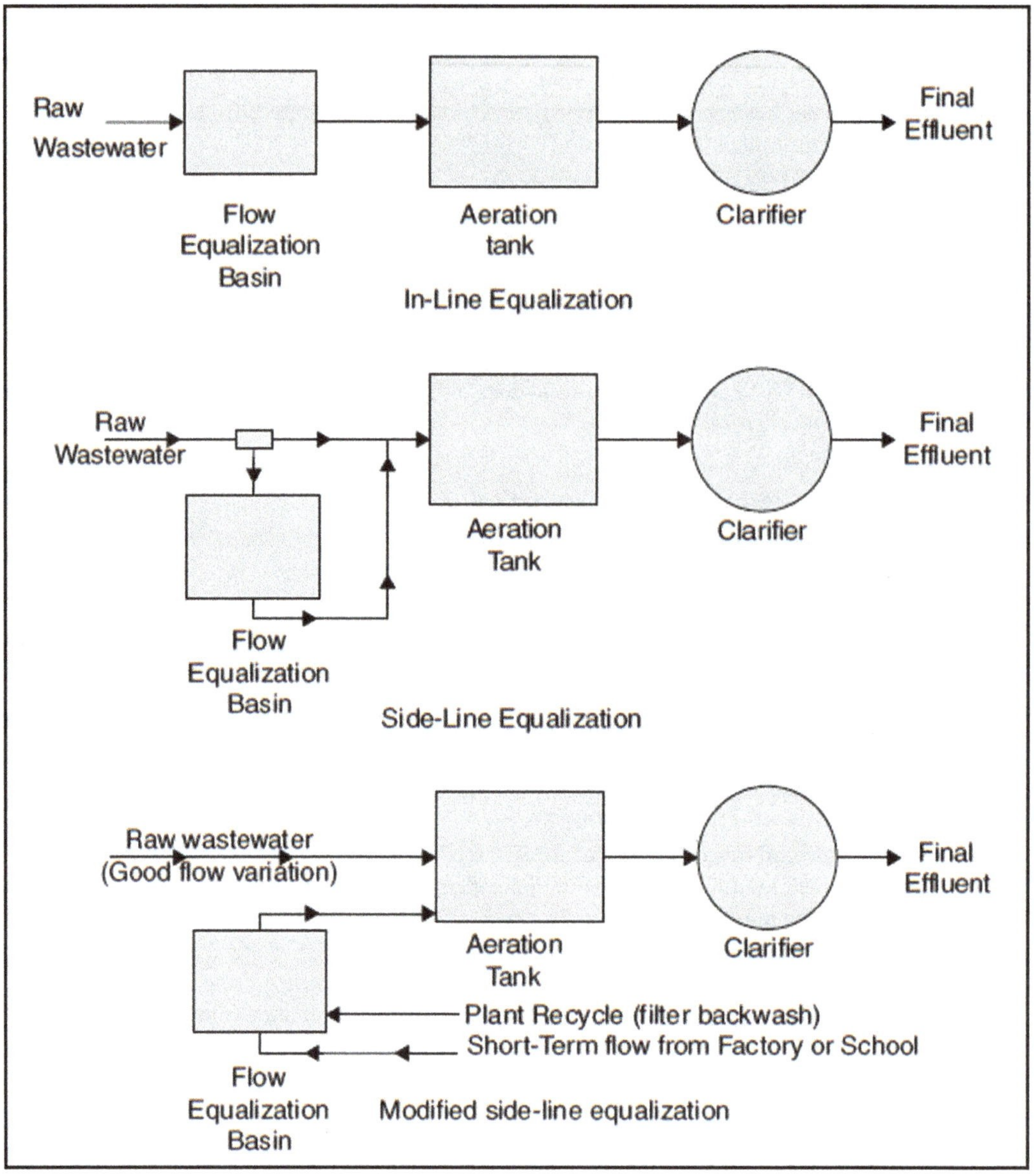

Figure 3.21 Equalization Configurations

Equalization basins are seldom covered because they are large. Facilities providing equalization for dry weather daily peaks and no wet weather inflows, however, have smaller basins that could be covered with domes similar to those used for primary clarifiers.

5.5 AERATION AND MIXING

The successful operation of equalization basins requires proper mixing and aeration. Design of mixing equipment provides for blending the contents of the tank and preventing deposition of solids in the basin. Ideally, equalization basins should be placed after screening and grit removal; however, the increased cost of facilities is generally not considered worth the relatively minor operational benefits to the equalization process. Aeration is usually provided to prevent septic conditions unless the equalization process is designed to treat only dilute wet-weather flows. Mixing requirements for blending municipal wastewater that has a suspended solids concentration of approximately 200 mg/L range from 0.004 to 0.008 kW/m^3 (0.02 to 0.04 hp/1000 gal) of storage (U.S. EPA, 1974). However, mixing requirements in equalization basins typically depend on basin geometry, mode of operation, and type of mixing system used. To maintain aerobic conditions, air should be supplied at a rate of 0.156 to 0.25 L/m^3•s (9.4 to 15.0 cfm/1000 cu ft) of storage.

Mechanical aerators, one method of providing both mixing and aeration, have oxygen transfer capabilities varying from 0.5 to 1.0 kg O_2/MJ (3 to 4 lb O_2/hp/hr) in clean water under standard conditions, but the oxygen transfer efficiency (OTE) in wastewater is lower. A reasonable OTE value for design would be approximately 0.16 to 0.39 kg O_2/MJ (1.0 to 1.5 lb O_2/hp/hr). Minimum operating levels for floating aerators typically exceed 1.5 m (5 ft) and vary with the power and design of the unit. If the equalization basin floor is subject to erosion (earthen basins), concrete pads on the basin floor are recommended. Baffling may be necessary to ensure proper mixing, particularly with a circular tank configuration.

Power requirements to prevent deposition of solids in the basin may greatly exceed what is needed for blending and oxygen transfer. In such cases, the most economical approach might be to provide mixing equipment to keep the solids in suspension and a diffused air system to supply the air requirements. Aspirating aerators have been used in equalization basins because these provide the horizontal mixing component.

Diffused aeration systems may use coarse or medium bubble diffusers. Membrane (not ceramic) fine bubble diffusers can also be used if increased OTE is required. With variable volume systems and, therefore, variable water depths, blowers must have pressure regulation controls. For varying level conditions, positive displacement blowers are typically preferred over centrifugal blowers especially because of surge conditions. Mixing requirements are 0.5 to 0.8 L/m^3•s (30 to 50 cfm/1000 cu ft) of basin volume.

Equalization basins located prior to primary clarification should always be designed based on the premise that some level of deposition will always occur. One method for containing solids deposition in small areas include the use of baffle walls within the equalization basins that allow settling of particles within the baffled areas and allow overflow of partially clarified wastewater to the remainder of the basin, which reduce the cleaning efforts after each use.

5.6 FLOW CONTROL METHODS

The use of flow equalization, in most cases, involves pumping either before or after the equalization basins. If flow is pumped to the equalization basin, then the basin effluent must be controlled by a regulating device. The design engineer must recognize the acceptable flow range of the device and the range of basin operating water levels that must be throttled. For pumped flow from the equalization basin, a variable speed pumping system will be required to carefully regulate the hydraulic loadings, providing a nearly constant flow to the downstream units. Influent pumping to the equalization basin requires a pump of sufficient capacity to handle diurnal peaks.

The use of waste strength equalization, typically requiring a constant volume, may require pumping to the equalization basin with a variable outflow equal to the input flow. Alternatively, if hydraulics allow, gravity discharge paced by a modulating device (valve or gate) tied to an effluent flow metering device may be used.

6.0 FLOW MEASUREMENT

Facility process control decisions depend on measurement of several parameters, including flowrate. Flow measurement can take place in several key areas including the influent, primary clarifier sludge, return and waste activated sludge, digested sludge and supernatant, return and recycle flows, disinfection process influent, and effluent. These areas present a wide variety of physical characteristics than can affect flowmeter selection and design. Flow measurement devices used to measure influent, sludges, and certain recycle flows must be designed to minimize problems associated with grease, grit, snails, rags, and solid materials. Final clarifier and facility effluent contain much smaller quantities of these problem materials.

All flow measurement devices used successfully in WRRFs must possess the following features. All flowmeters must control the waste stream in a manner that causes a known relationship between flowrate and some measurable parameter such as depth, velocity, or pressure drop. A sensor is used to measure the parameter and its value is then related to the flow. Flowmeters that control the waste stream (such as flumes) are called primary devices and those that measure the measurable parameter are called sensing devices. Flow measurement devices are further divided into two classes: those that operated under open channel flow conditions, and those that operate under pressure or full pipe flows.

Primary and sensing devices must be designed based on a number of considerations related to the type of flow being measured and the device that is being used. The flow condition—open channel or pressure—depends on the flow measurement device. Other design considerations include available headloss, physical characteristics of the stream, required accuracy and repeatability, spatial requirements, capital and operating costs, and maintenance requirements. A summary of primary devices is shown in Table 3.4. A sensing device summary is shown in Table 3.5. Actual calculations and applications of these devices are discussed in Chapter 9.

7.0 NEUTRALIZATION

One of the most common types of chemical processes used in WRRFs is pH adjustment. Adjustment of pH, frequently used for coagulation and phosphorus precipitation, simply raises or lowers pH to a more acceptable value. For example, wastewater that is excessively acidic or alkaline is objectionable in collection systems, WRRFs, and natural streams. Removal of excess acidity or alkalinity by chemical addition to provide a final pH of approximately 7 is called **neutralization**. Most effluents must be neutralized to a pH of 6 to 9 before discharge.

Neutralization is a chemical process that produces a solution that is neither acidic or alkaline.

There are three critical components of any pH control system: mixing intensity or turnover time in the reactor, response time of the control system, and the ability of the chemical metering system to match process requirements. If any one of these components is not properly designed, then significant problems in system performance may occur. Other factors that complicate the design of pH control systems include the amount of buffering capacity in wastewater, the change in mass flowrate of the hydrogen ions, and variations in wastewater flowrate or temperature.

Before designing a pH control system, a thorough wastewater characterization study should be performed. If characteristics are not clearly defined, additional sampling and analysis should be considered. However, in almost all cases, titration curves should be developed to provide definitive data with respect to chemical demands and consumption.

Methods used for pH adjustment are selected on the basis of overall cost because material costs and equipment needs vary widely. The volume, kind, and quantity of acid or alkali to be neutralized or partially removed are also variables that influence selection of a chemical agent. In those processes that demonstrate

Table 3.4 Primary Devices Summary

Primary Element	Advantages	Limitations
Open Channel		
Parshall flume	Self-cleaning	High cost
Palmer-Bowlus flume	Self-cleaning Suitable for installation in pipelines	High cost
Rectangular weir	Low cost Measurable wide flow range	Upstream solids accumulation May require ventilation
V-notch weir	Low cost High degree of accuracy	Upstream solids accumulation Narrow flow range
Compound weir	Accurate over wide flow range	Upstream solids accumulation
Pressure		
Venturi tube	High degree of accuracy	Subject to plugging High cost Pressure probes require periodic maintenance
Velocity flow tubes	Some types are more economical than are Venturi tubes	Lower accuracy than Venturi tubes

Table 3.5 Sensing Devices Summary

Device	Typical Accuracy	Advantages	Limitations
Gravity			
Mechanical float	± 1%	Simplicity of operation	Requires frequent cleaning
Bubbler	± 0.5%	Relatively maintenance free High degree of accuracy	Less accurate in low head applications
Capacitance probe	± 2%	Low cost	Requires frequent cleaning
Ultrasonic sensor	± 2%	Easy to calibrate Nonfouling	Affected by high winds, temperature, relative humidity, and foam
Pressure			
Ultrasonic velocity	± 1%	Simple to install Low cost	Requires periodic cleaning Suspended solids, grease, or gas bubbles may affect readings
Ultrasonic Doppler	± 1%	Simple to install Low cost Available as clamp on device for existing pipes	Suspended solids must be present in flow Pipe linings can affect results
Magnetic	± 1%	Available with wide range of lining materials High accuracy over a wide range	Requires proper grounding High cost Requires periodic cleaning

a significant variation in hydrogen ion mass flowrate, it is often necessary to use cascade neutralization processes of two to four stages. This is a problem most frequently encountered in industrial wastewater treatment, however, if tight pH control is required, cascade control offers significant improvements in process control and reliability.

PRACTICE PROBLEMS

1. Two aerated grit chambers are to be built for a facility with a maximum process design flow of 25 m^3/min (15 cu ft/s) combined sanitary and storm flow. Design the cross-sectional dimensions of the chamber, calculate the required airflow, and estimate the total quantity of grit per day for disposal.

2. Four aerated grit chambers are required for a facility with an average daily design flow of 63.1 m^3/min (24 mgd) and a peak hourly flowrate of 157.7 m^3/min (60 mgd).

   Find the volume and dimensions of the basins at peak hourly flow.

   Find the detention time of two basins operating at average daily flow.

   Estimate the amount of grit that would be produced by these basins at peak hour flow.

3. A grit chamber generates 8 m^3/d (2113 gpd) of grit. The grit is pumped through a 10 cm (4 in.) pipe with two pumps (one working, one on standby) with the pump cycling on once per hour. How many minutes each hour must the grit pump run to provide a minimum velocity of 1.2 m/s (4 ft/s) in the pipe?

4. A municipality has a flowrate of 0.44 m^3/s (10 mgd). Estimate the volume of screenings from a bar screen with an opening of 20 mm (0.75 in.).

   What would be the volume of the screenings if the facility switched to a fine screen with an opening of 5 mm (0.2 in.)?

8.0 REFERENCES

Borys, A. (2014, October 18–22). Influent screening upgrade to address process impacts at EBMUD. *Proceedings of the 87th Annual Water Environment Federation Technical Exposition and Conference* [CD-ROM], New Orleans, Louisiana, United States.

Burbano, M. S., Rife, J., Turriciano, A., Henne, R., Kirkman, R., & Harner, B. (2014, October 18–22). Modeling the effect of inadequate screening on nutrient removal. *Proceedings of the 87th Annual Water Environment Federation Technical Exposition and Conference* [CD-ROM], New Orleans, Louisiana, United States.

DiToro, D. M. (1975). Statistical design of equalization basins. *Journal of Environmental Engineering, 101*, 917.

Heath, G., Carr, J., Bailey, W., Karkhar, S., & Watson, K. (2015). Preliminary design of the DC water enhanced clarification facility. *New England Water Environment Association CSO/Wet Weather Issues Specialty Conference.*

LaGrega, M. D., & Keenan, J. D. (1974). Effects of equalizing wastewater flows. *Journal of the Water Pollution Control Federation, 46*, 123.

Londong, J. (1989). Dimensioning of aerated grit chambers. *Water Science Technology, 21*, 13.

McInnes, C. D., Middleton, A., & Adamowski, K. (1978). Stochastic design of flow equalization basins. *Journal of Environmental Engineering, 104*, 1277.

Novotny, V., & Stein, R. M. (1976). Equalization of time variable waste loads. *Journal of Environmental Engineering, 102*, 613.

Ongerth, J. E. (1979). *Evaluation of flow equalization in municipal wastewater treatment* (EPA-600/2-79-096). U.S. Environmental Protection Agency, Municipal Research Laboratory.

Roe, F. C. (1951). Pre-aeration and air flocculation. *Sewage Works Journal, 23*, 127.

Seidel, H. F., & Bauman, E. R. (1961). Effect of preaeration on the primary treatment of sewage. *Journal of the Water Pollution Control Federation, 33*, 339.

U.S. Environmental Protection Agency. (1974). *Flow equalization.* U.S. Environmental Protection Agency Technology Transfer.

U.S. Environmental Protection Agency. (1979). *Process design manual for sludge treatment and disposal* (EPA-625/1-79-011).

U.S. Environmental Protection Agency. (1987). *Preliminary treatment facilities design and operational considerations* (EPA-430/09-87-007).

Wallace, A. T. (1968). Analysis of equalization basins. *Journal of the Sanitary Engineering Division, Proceedings of the American Society of Civil Engineers, 94*, 1161.

Water Environment Federation. (2016). *Operation of water resource recovery facilities* (7th ed.; Manual of Practice No. 11).

Water Environment Federation. (2017). *Liquid stream fundamentals: Screening* [Fact sheet].

Wodrich, J., Winkler, T., Leaf, B., Clark, S., & Youker, B. (2005, October 29–November 2). Wet weather impacts on influent fine screening system design and operation. *Proceedings of the 78th Annual Water Environment Federation Technical Exposition and Conference* [CD-ROM], Washington, DC, United States.

Chapter 4 | Sedimentation and Solids Separation

LEARNING OBJECTIVES

At the end of this chapter, students should be able to:

- Describe sedimentation theory for settling types I through IV.
- Calculate the overflow rate, solids loading rate, detention time, and weir loading rate for primary and secondary sedimentation processes.
- Describe the operation and design a circular and rectangular primary sedimentation processes.
- Explain methods used to enhance sedimentation.
- Describe emerging technologies in sedimentation processes.
- Explain downstream considerations from the use of primary sedimentation.
- Describe the operation and implementation of membranes for the purpose of replacing sedimentation processes.

1.0 INTRODUCTION

Conventional sedimentation is an effective removal method for raw wastewater suspended solids, which range from a low concentration of nearly discrete particles to a high concentration of flocculent solids, as well as for solids in secondary treatment in processes like activated sludge and trickling filters. These particles tend to settle by gravity under quiescent conditions as grease and scum float to the surface. Clarifier design historically has been based on empirically derived design criteria. These criteria, when coupled with a theoretical understanding of the clarification process, may be used to design reliable, efficient clarifiers. Computational fluid dynamic (CFD) modeling can be used to refine designs.

This chapter discusses conventional sedimentation and theory as it applies to primary and secondary sedimentation as well as emerging technologies. Emerging technologies for solids separation use physical barriers and chemicals to aid in solids separation. This has led to new process improvements like enhanced sedimentation and membrane separation.

2.0 SEDIMENTATION THEORY

Gravity settling is one of the most important functions of a water resource recovery facility (WRRF) and is an effective method for the removal of suspended solids from wastewater suspensions. The behavior of solids ranges from discrete particle settling at low solids concentrations to flocculent settling at high solids concentrations. The effect of the concentration of solids on settling characteristics can be categorized by four distinct types of settling recognized in WRRFs.

These basic types of settling are categorized as types I through IV. Type I settling is also known as discrete settling in which particles are dispersed with no tendency to flocculate. The particles will settle independently (discretely). Although type I settling can occur in any sedimentation process, it is the primary mechanism for settling in grit chambers. Type II settling is known as **flocculent settling** and occurs with high solids concentrations in which particle interaction is occurring. In this type of settling,

Flocculent settling is a phenomenon referring to the sedimentation of particles in a dilute suspension as they coalesce (flocculate).

particles have a strong affinity to each other and combine (flocculate) into larger particles. These larger particles settle faster than the individual particles. Type II settling is the primary mechanism in primary sedimentation processes and sedimentation processes in drinking water treatment. Type III settling, also known as **zone settling**, occurs when particle concentrations are quite high (~>500 mg SS/L). At these higher concentrations the particles coalesce quickly, forming a matrix of particles. This matrix tends to settle together such that the particles retain their relative position to each other. Each individual particle may be hindered from settling by the matrix, so this is also known as hindered settling. Type III settling is the primary mechanism that occurs in secondary sedimentation processes, specifically those that follow activated sludge processes. Type IV settling is known as **compression settling** and occurs at the bottom of sedimentation tanks where particles are so close together that they actually touch each other, and the solids layer at the bottom is pressed down or compressed by the load of particles reaching the bottom of the clarifier or the sedimentation tank(s). The resulting compressive force from the weight of the particles compresses the matrix and squeezes water out of the sludge, which is thickened. Type IV settling occurs at the bottom of secondary sedimentation tanks and more substantially in gravity sludge thickening processes.

Zone settling is a sedimentation phenomenon in which the high concentration of particles cause the velocity fields of each particle to overlap, which in turn causes the solids to settle as one mass. Also known as hindered settling.

2.1 DISCRETE SETTLING

The primary mechanism of settling in grit chambers is type I or discrete settling. In primary sedimentation processes, type II settling is the main mechanism, with laminar flow conditions. To help in the design of these processes, the settling velocity of type I or type II settling particles can be predicted by Stokes' law (Equation 4.1):

Compression settling is a sedimentation phenomenon in which particles in a concentrated suspension will only settle if the existing structure of settled particles is compressed.

$$v = \frac{2}{9}\frac{\rho_p - \rho_f}{\mu} gR^2 \tag{4.1}$$

where v = settling velocity, m/s (ft/s);
ρ_p = density of particles, kg/m³ (lb/cu ft);
ρ_f = density of fluid, kg/m³ (lb/cu ft);
μ = viscosity of fluid, N·s/m² (lb·s/sq ft);
g = acceleration due to gravity, 9.81 m/s² (32.2 ft/sq ft); and
R = particle radius, m (ft).

From Stokes' law, it can be seen that the two main factors in settling velocity include the characteristics of the particle and the fluid in which the particle is being settled. Particles that are light (low density) or small (diameter) are going to settle slower than dense, large particles. Similarly, particles that are settling in light low-viscosity fluids will settle faster than in a fluid that is dense or has a high viscosity.

In design, a target particle for removal can be selected and the settling velocity of that particle can be calculated. The fluid properties of wastewater do not change significantly so the particle characteristics are the main determiner in settling velocity. The calculated settling velocity can then be used to design the solids separation process to assure removal of the desired particle.

Example 4.1

The target particles removed in grit chambers follow type I or discrete settling. The goal of most grit chambers is to remove 95% of 200 µm (65 mesh [0.0083 in.]) particles having a specific gravity of 2.65. What is the discrete settling velocity of that size particle.?

Solution

Use Equation 4.1 to find the settling velocity of the target particle. For the settling fluid, use water, which has a viscosity of 1.12×10^{-3} N·s/m² at 15.6 °C.

$$v = \frac{2}{9}\frac{\rho_p - \rho_f}{\mu} gR^2 = \frac{2}{9}\frac{[(2.65)(1000\ \text{kg/m}^3)]}{1.12\times10^{-3}\text{N·m}^2}(9.81\ \text{m/s}^2)(100\times10^{-6})^2$$

$$v = 0.032\ \text{m/s}\ (0.009\ \text{ft/s})$$

2.2 SETTLER DESIGN THEORY

The particle settling velocity represents the starting point for the design of a sedimentation process. The use of that velocity for design can be illustrated with the concept of the ideal settling tank. This concept is used to outline the mechanism that occurs in a sedimentation tank and will help identify the most important parameters for sedimentation design including the **detention time** and **overflow rate**.

Detention time is the time theoretically required to displace the contents of a tank or unit at a given rate of discharge.

Camp (1936, 1953) is credited with the idea of creating an "ideal tank" in which settling occurs. He suggested dividing the settling tank into four zones: the inlet, settling, outlet, and primary sludge zones (Figure 4.1). In the inlet zone, influent solids and flow are uniformly distributed over the cross-sectional area of the settling tank. In the settling zone, a uniform concentration of particles settles at the terminal settling velocity to the primary sludge zone at the bottom of the tank, where the settled solids are collected and removed. Effluent discharges over a weir in the outlet zone.

Consider a single particle, the *critical particle*, that enters the ideal settling zone at the far left and on the water surface and with a settling velocity, defined by the Stokes equation, of v_O. It is also traveling horizontally with the flow at a horizontal velocity of v_L. As shown in Figure 4.1, its trajectory will carry it just to the intersection of the sludge and outlet zones, so we can consider it just barely removed. All particles that start at the same place and have settling velocities, v_h, greater than v_O will enter the sludge zone and be removed. Similarly, all particles that enter the tank at the water surface and have settling velocities less than v_O will escape with the effluent.

Overflow rate is a measure of the upward water velocity in a sedimentation tank, expressed as flow per day per unit of basin surface area (L3/L2•/T, which is equal to L/T, a velocity). Also called surface loading rate.

If a particle enters that tank anywhere else below the water surface, say at some height h, its settling velocity will have to be less than v_O if it is to be removed. As seen in Figure 4.1, this velocity will have to be v_h if the particle is to be removed. By proportion,

$$\frac{v_h}{v_S} = \frac{h}{H} \tag{4.2}$$

Where v_O = settling velocity of the critical particle, m/s (ft/sec);
v_h = settling velocity of any particle entering the ideal tank zone at height h, m/s (ft/sec);
h = any height less than H, m (ft); and
H = height of the ideal tank zone, m (ft).

Consider now the triangle formed by the vectors v_O and v_L. If the ideal tank has proportions of height H and length L, then

$$\frac{v_L}{L} = \frac{v_o}{H} \tag{4.3}$$

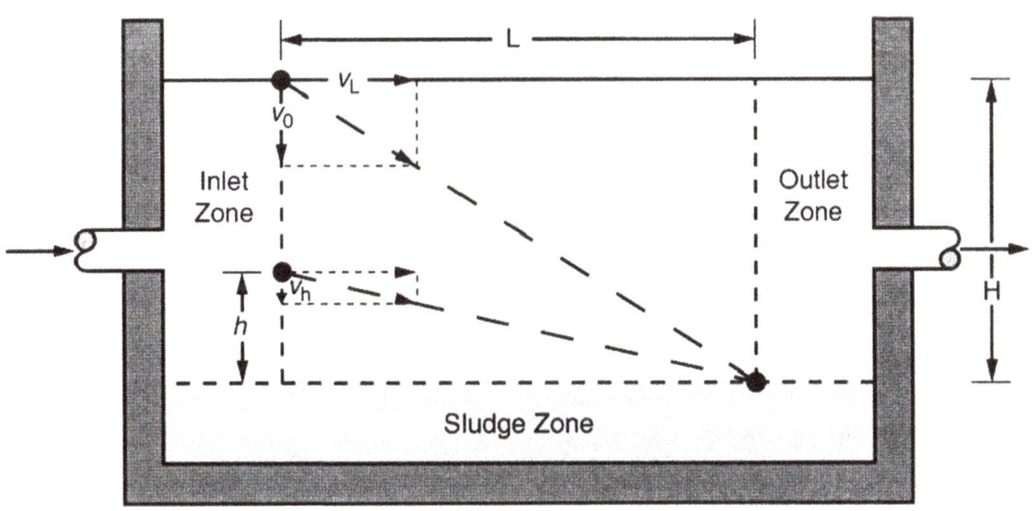

Figure 4.1 Ideal Settling Tank

Where v_L = horizontal velocity through the tank, m/s (ft/sec);
L = length of ideal tank, m (ft); and
H = height of ideal tank, m (ft).

The hydraulic retention time is:

$$HRT = \frac{V}{Q} = \frac{L}{V_L} = \frac{H}{V_o} = \frac{h_h}{V_h} \tag{4.4}$$

Where HRT = hydraulic retention time, s
V = tank volume, $L \times H \times W$, m³ (cu ft);
Q = flowrate, m³/s (cu ft/sec);
W = Width of ideal tank, m (ft);
h_h = height at which a particles with settling velocity v_h enters and is just removed, m (ft).

Starting with the above equation, through a series of substitutions:

$$V_O = \frac{Hv_h}{h} = \frac{H}{h} \times \frac{h}{HRT} = \frac{H}{HRT} = H\frac{Q}{V} = H\frac{Q}{L \times H \times W} = \frac{Q}{L \times W} = \frac{Q}{A} \tag{4.5}$$

Where A = surface area of the ideal tank, or $L \times W$, m² (sq ft).

This is a curious outcome. The critical particle velocity, also referred to as the surface loading rate, is defined as the flowrate divided by the surface area, and this also has dimensions of meters per second (feet per second). In the United States, engineers have called this relationship Q/A the *overflow rate*, although it has nothing to do with overflow. Instead, it is another way of defining the critical velocity.

If a tank is designed with a low critical velocity, then most of the particles, which have settling velocities greater than v_O, will be removed. If the tank is designed for a high critical velocity, this means that most of the particles will have settling velocities less than this critical settling velocity, and there is no time for them to be removed. The overflow rate is therefore another way of defining the critical settling velocity.

Example 4.2

Using the concept of an ideal settler, calculate the area and volume of a 3 m (9.8 ft) tank required to remove the particle from Example 4.1 when the flowrate is 0.02 m³/s (0.71 cu ft).

Solution

Use Equations 4.4 and 4.5 to determine dimensions of the tank.

$$V_O = \frac{Q}{A} = \frac{0.02\ \text{m}^3/\text{s}}{A} = 0.032\ \text{m/s} \qquad \textbf{Area} = \mathbf{1.6\ m^2}$$

The particle needs to settle 3 m to be collected at the bottom of the settler. Therefore, using a form of Equation 4.4:

$$HRT = \frac{H}{V_O} = \frac{3\ \text{m}}{0.032\ \text{m/s}} = 93.8\ \text{seconds}$$

Using the other form of Equation 4.4:

$$HRT = \frac{V}{Q} = \frac{V}{0.02\ \text{m}^3/\text{s}} = 93.8\ \text{seconds} \qquad \textbf{Volume} = \mathbf{1.9\ m^3}$$

If a width-to-length ratio is known, the dimensions of the tank can be computed.

2.3 SOLIDS FLUX

In secondary sedimentation, particularly sedimentation processes following activated sludge processes, type III settling dominates along with a significant regime exhibiting type IV settling. As described previously, these types of settling involve the interaction of particles that hinder the settling process. In the design process, the interaction of the particles must be accounted for in these secondary sedimentation processes. A concept known as solids flux was developed to account for the interaction. The solids flux is the mass of solids that are being settled per unit surface area (kg/m²·h). It is analogous to the overflow rate, which is a hydraulic parameter related to amount of liquid being settled per surface area (m³/m²·h).

Work by Coe and Clevenger (1916), Dick and Ewing (1967), Dick and Young (1972), and Yoshioka et al. (1957) advanced the solids flux approach to clarification. For a clarifier operating at a steady state, a constant flux of solids is moving downward (Figure 4.2). The total mass flux of solids is the sum of the mass flux resulting from hindered settling due to gravity (SF_g) and the mass flux resulting from bulk movement of the suspension (SF_u). The total solids flux, SF_t, in kg/m²·h (lb/sq ft·hr), is the sum of these two components:

$$SF_t = X_i V_i + X_i U_b \tag{4.6}$$

The total solids flux represents the maximum rate that solids can be continually applied to a clarifier for a given underflow rate, mixed liquor suspended solids (MLSS) concentration, and characteristic settling velocity at concentration X_i. The characteristic settling velocity is, in turn, a function of the settleability of solids.

2.4 PRIMARY TREATMENT

Primary treatment involves the separation and removal of suspended solids and floatables (fats, grease, oils, plastics, etc.) from wastewater by physical chemical methods. Much of the solids removed are organic and comprise a significant portion of the influent chemical oxygen demand (COD) and biochemical oxygen demand (BOD). Therefore, primary treatment results in reduced total suspended solids (TSS), COD, and BOD loadings to downstream biological treatment processes. Nutrients, such as nitrogen and phosphorus, and other constituents are also partly removed to the extent they are contained in the removed solids. The TSS, COD, and BOD reductions result in smaller reactor volume requirements and lower aeration demands for biological treatment. Removal of floatables minimizes operational problems

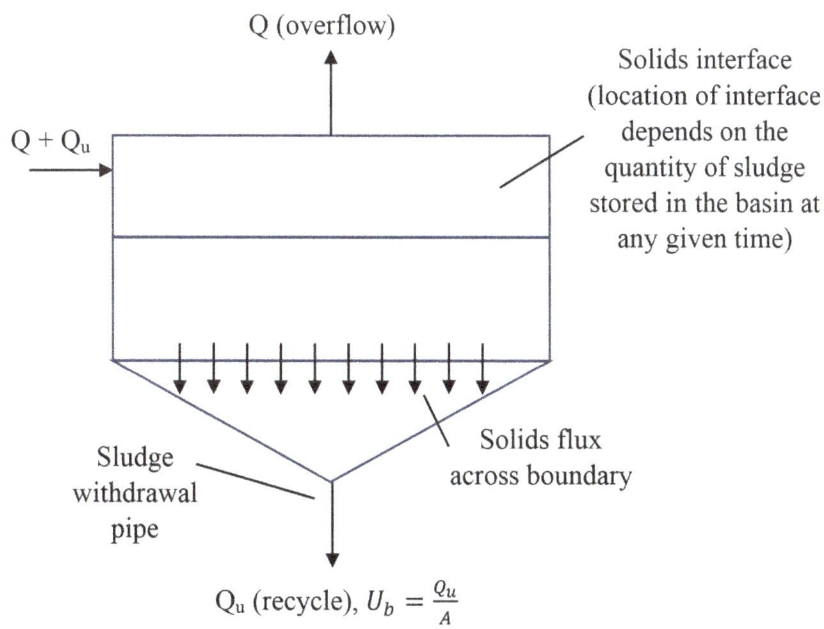

Figure 4.2 Settling Basin at Steady State (Ub = bulk downward velocity, m/h or ft/hr, and A = required area, m² or sq ft)

from the buildup of scum in downstream treatment processes and improves the facility's overall aesthetics by reducing visual blights and odors.

The selection and design of the primary treatment process should assess the economic effect of primary treatment on other processes throughout the WRRF. For example, primary treatment can reduce the costs (capital and operational) for biological treatment but increase the costs for solids handling facilities that receive the solids removed by primary treatment. Primary treatment is frequently coupled with anaerobic stabilization of waste solids, which produces methane gas that can be used to produce valuable energy. It is important to consider the net overall cost effect of primary treatment throughout the life of the WRRF.

There are several methods used for primary treatment. The most common form of primary treatment is quiescent (conventional) sedimentation, which includes skimming; collection; and removal of settled primary sludge, floating debris, and scum. Enhanced sedimentation methods, such as chemically enhanced primary treatment (CEPT) and high rate clarification (HRC) can be used to reduce the area required for primary treatment and/or to improve treatment performance. Microscreens and filters can also be used for primary treatment. Table 4.1 provides typical removal rates for TSS, COD, or BOD, phosphorus, and bacteria for some of the available primary treatment processes. The terms sedimentation and clarification are used interchangeably to describe the same process. Similarly, sedimentation tank, clarifier, and settler are synonymous.

2.5 SECONDARY TREATMENT

Secondary treatment involves the separation of biological solids from liquids after biological treatment. Biological solids that are separated are generally those that have been grown in a biological system and, thus, consist primarily of bacterial cells along with some nonbiodegradable end products and materials. The structural nature of the bacterial cells makes them difficult to stabilize biologically and to dewater.

There are several methods to achieve secondary treatment. Sedimentation is typically initially used to separate these solids, with further treatment required for significant increases in solids content. This treatment is typically performed in a separate set of sedimentation tanks (secondary clarifiers). Membrane filtration is a relatively new technique that is used to separate secondary solids and allows almost complete separation. These systems can be fit into existing biological treatment tanks and eliminate the requirement for secondary sedimentation tanks, saving space. However, membranes are more energy intensive than sedimentation tanks and can be more costly.

Table 4.1 Typical Primary Treatment Removal Rates

Primary Treatment Process	Total Suspended Solids Removal (%)	Chemical Oxygen Demand or 5-Day Biochemical Oxygen Demand Removal (%)	Phosphorus Removal (%)	References
Conventional primary clarifier	25–70	25–40	5–10	Metcalf & Eddy, Inc./AECOM, 2014; Steel 1979; U.S. EPA, 1987
Chemically enhanced primary treatment (CEPT)	60–90	40–70	70–90	Metcalf & Eddy, Inc./AECOM, 2014; Øedegaard, 2005
High-rate clarification	30–95	35–70	70–95	Metcalf & Eddy, Inc./AECOM, 2014; Stevenson et al., 2008
Primary effluent filtration[a]	60–80[b]	40–60[b]	5–10[b]	Caliskaner et al., 2015
Primary filtration	60–80[b]	40–60[b]	NA	Caliskaner et al., 2016
Rotating belt filtration/microscreens	30–60[b]	25–40[b]	NA	Sutton et al., 2008
Dissolved air floatation clarifier (with integrated biological treatment)	50–80	40–60	NA	Ding et al., 2015; Johnson et al., 2014

[a]Combined performance of primary clarifier and the primary effluent filtration systems.
[b]Without chemical addition.

3.0 PRIMARY CLARIFIERS

Conventional primary clarification is an effective removal method for raw wastewater suspended solids, which range from a low concentration of nearly discrete particles to a high concentration of flocculent solids. These particles tend to settle by gravity under quiescent conditions as grease and scum float to the surface. Primary clarifier design historically has been based on empirically derived design criteria. Computational fluid dynamic modeling can be used to refine designs.

Rectangular (Figure 4.3), circular (Figure 4.4), stacked, and plate and tube settlers (Figure 4.5) are four types of clarifiers. Although square tanks with circular sludge collection equipment ("squircles") are also seen occasionally, this configuration is prone to poor hydraulics and biological activity and should be avoided (Water Environment Federation [WEF], 2005). Rectangular and circular clarifiers are typically used for wastewater treatment.

3.1 RECTANGULAR CLARIFIERS

Rectangular clarifiers range from 15 to 90 m (50 to 300 ft) in length, 3 to 24 m (10 to 80 ft) in width, and 3 to 4.9 m (10 to 16 ft) in depth (Metcalf & Eddy, Inc./AECOM, 2014). Rectangular tanks with common wall construction are advantageous for sites with space constraints. The sludge removal equipment in rectangular clarifiers requires greater maintenance than similarly sized circular tanks because drive bearings in rectangular tanks are submerged; however, submerged bearings are not subject to corrosion.

3.2 CIRCULAR CLARIFIERS

Circular clarifiers range from 3 to 60 m (10 to 200 ft) in diameter and 3 to 4.9 m (10 to16 ft) in depth (Metcalf & Eddy, Inc./AECOM, 2014). Circular clarifiers can use relatively trouble-free, circular primary sludge removal equipment (drive bearings are not under water) (WEF, 2005). Walls of circular tanks are structurally designed as tension rings, which permit thinner walls than those for rectangular tanks. As a result of such advantages, circular tanks often have a lower capital cost per unit surface area than that for rectangular tanks. Circular tanks typically require more yard piping than rectangular tanks as well as separate structures for flow distribution and sludge pumping.

3.3 STACKED SEDIMENTATION TANKS

In areas where land for treatment facilities is not available or is extremely expensive, stacked sedimentation tanks can be double-decked or even triple-decked to increase the available clarifier area without increasing the clarifier footprint. Series flow and parallel flow are two types of stacked clarifiers. In the series flow unit, wastewater enters the lower tray, flows to the opposite end, reverses direction in the upper tray, and exits the effluent channel. In the parallel flow unit, pipes convey wastewater from the influent channel to both the upper and lower trays. The parallel tray unit is the most common stacked configuration used for primary sedimentation.

3.4 PLATE AND TUBE CLARIFIERS

In plate and tube clarifiers (commonly referred to as lamella or inclined plate clarifiers), bundles of parallel plates or tubes inclined at an angle of 45° to 60° to the water surface (Figure 4.5) are used to increase the

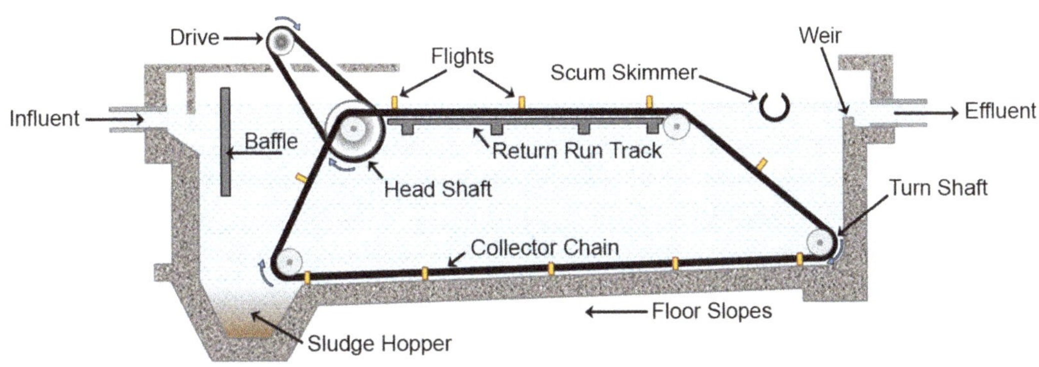

Figure 4.3 Typical Section View From a Rectangular Clarifier

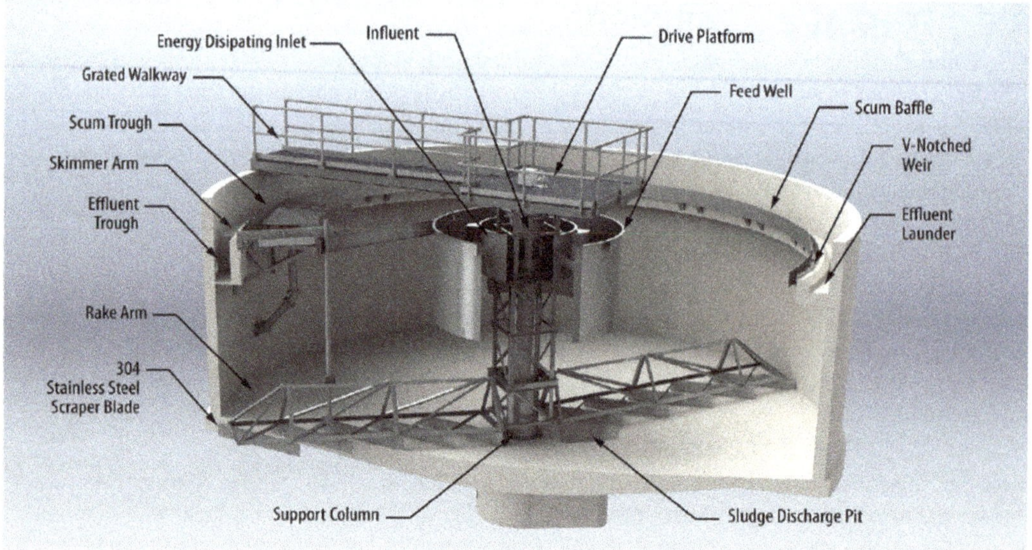

Figure 4.4 Circular Primary Clarifier Schematic (Reprinted with permission by Monroe Environmental [Monroe, MI] www.mon-env.com)

effective settling surface area of the tank. The spacing between plates is typically 40 to 120 mm (1.6 to 4.7 in.). The increase in settling surface area is equal to the total projected horizontal area for all plates or tubes installed. Depending on the height of the plates or tubes, the angle of inclination, and the spacing between adjacent plates or tubes, the effective settling surface area can be 6 to 12 times the plan area occupied by the plates or tubes (based on 100 mm [4 in.] spacing). The increase in effective surface area results in a smaller footprint and is beneficial when addressing site constraints. The performance of plate or tube settlers is similar to conventional primary clarifiers operating at the same overflow rate based on total projected horizontal area of the plates or tubes. Plate and tube clarifiers are used more frequently for high-rate and wet-weather clarifier applications because the intermittent nature of wet-weather influent and its characteristics tend to mitigate problems with settler cleaning.

3.5 DESIGN CRITERIA

Historically, primary sedimentation basin design has relied on criteria such as surface overflow rate, hydraulic detention time, depth, surface geometry, and weir loading rate. These criteria are helpful for design but are not accurate enough to permit prediction of actual sedimentation performance, which will depend on the wastewater characteristics (mainly settleability of suspended solids).

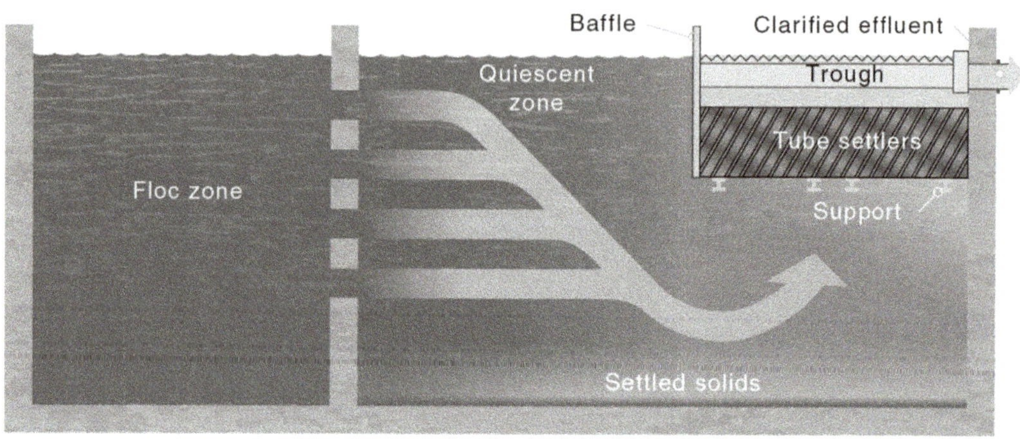

Figure 4.5 Typical Tube Settler (Courtesy of Brentwood Industries Inc., Reading, PA)

3.5.1 SURFACE OVERFLOW RATE

The surface overflow rate (SOR) of the clarifier can be expressed by the following equation:

$$SOR = \frac{Q}{A} \tag{4.7}$$

Where SOR = surface overflow rate, $m^3/m^2{\cdot}d$ (gpd/sq ft);
Q = flow to clarifier, m^3/d (gpd); and
A = surface area of clarifier, m^2 (sq ft).

Primary clarifiers historically have been designed on the basis of the SOR. Typical SOR values for conventional sedimentation range between 30 and 50 $m^3/m^2{\cdot}d$ (800 and 1200 gpd/sq ft) at average flow conditions and between 80 and 120 $m^3/m^2{\cdot}d$ (2000 and 3000 gpd/sq ft) with peak hourly flow (Metcalf & Eddy, Inc./AECOM, 2014).

As shown in Figure 4.6, there is considerable scatter in the dataset for SOR and corresponding primary clarifier removal efficiencies, but a lower SOR will typically increase TSS, COD, and BOD removal efficiencies. As a result, some regulatory agencies have established SOR requirements for primary clarifiers (Table 4.2).

3.5.2 HYDRAULIC DETENTION

The HRT is calculated from the required area and selected depth of the clarifier. Sufficient time for contact between solids particles is necessary for flocculation and effective clarification. As a result, some states have set limits (high and low) for detention times. Design considerations should include effects of low-flow periods to ensure that longer detention times will not cause septic conditions. Septic conditions increase potential odors, solubilization of **particulate matter**, and loading to downstream processes. Typical HRTs range between 1.5 to 2.5 hours for average flow conditions with a wastewater temperature of 20 °C (68 °F) (Metcalf & Eddy, Inc./AECOM, 2014). Detention times should be increased (SOR decreased) in cold climates to account for greater water viscosity at lower temperatures, which hinders particle settling.

Particulate matter is typically considered to be a solid particle larger than 1 μm; large enough to be removed from water or wastewater via filtration.

3.5.3 WEIR LOADING RATE

Proper clarifier operation also depends on outlet conditions. Effluent should be uniformly withdrawn to prevent localized, high-velocity gradients and short circuiting. If velocity gradients reach the scour velocity, then settled particles can be swept into the tank effluent. Density currents, rather than a high approach velocity, often cause primary sludge carryover of the effluent weirs. Therefore, effluent should be

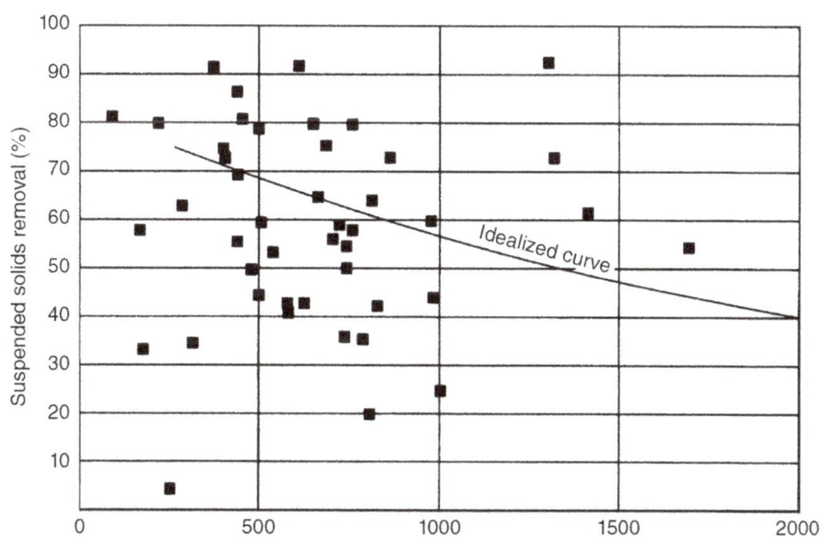

Figure 4.6 Primary Clarifier Suspended Solids Removals Versus Overflow Rates Showing Idealized Curve With Data (1 gpd/sq ft = 0.04075 $m^3/m^2{\cdot}d$)

Table 4.2 Primary Clarifier Surface Overflow Rate Requirements (Great Lakes-Upper Mississippi River Board of State and Provincial Public Health and Environmental Managers [2014] Recommended Standards for Wastewater Facilities. Health Research, Inc.)

	Surface Overflow Rates At:	
Type of Primary Settling Tank	Design Average Flow, m^3/m^2 d (gpd/sq ft)	Design Peak Hourly Flow, m^3/m^2 d (gpd/sq ft)
Tanks not receiving waste activated sludge	41 (1000)	61–81 (1500–2000)
Tanks receiving waste activated sludge (co-thickening)	29 (700)	49 (1200)

withdrawn from the tank in a manner that minimizes these currents. Typically, the effluent is withdrawn from a sedimentation tank by an overflow weir into a launder or effluent channel. The overflow weir must be leveled to control water surface elevation in the clarifier and promote uniform effluent withdrawal. Weirs may be either straight-edged or V-notched (Figure 4.7). V-notched weirs provide better lateral distribution of outlet flows than straight-edged weirs that are imperfectly leveled.

Weir loading rates have little effect on the performance of primary sedimentation tanks, especially with side water depths exceeding 3.7 m (12 ft) (Graber, 1974). Typical weir loading rates range between 125 and 500 $m^3/m\cdot d$ (10 000 to 40 000 gpd/ft) (Metcalf & Eddy, Inc./AECOM, 2014). In practice, weir loadings often do not exceed 125 $m^3/m\cdot d$ (10 000 gpd/ft) in facilities handling 3.8 ML/d (1.0 mgd) or less and 190 $m^3/m\cdot d$ (15 000 gpd/ft) in facilities handling more than 3.8 ML/d (1.0 mgd) (Merritt, 1983). Existing regulations sometimes govern weir loadings.

3.5.4 REDUNDANCY

Two or more clarifiers will allow the process to remain in operation while a tank is out of service for maintenance or repair. The U.S. Environmental Protection Agency (U.S. EPA) Design Criteria for Reliability states that there should be a sufficient number of units of a size, such that with the largest flow capacity unit out of service, the remaining units shall have a design flow capacity of at least 50% of the total average design flow to that unit process (U.S. EPA, 1999). Some states have increased this requirement to 75%.

Figure 4.7 V-Notch Weir Plate With Scum Baffle on a Secondary Clarifier (Reprinted with permission from South Platte Renew)

Example 4.3

Design circular primary settling tanks for the following conditions:

Design Parameters		
Average flowrate	30 000 m^3/d	7.93 mgd
Peaking factor	1.75	
Number of clarifiers	2	
Overflow rate @ Q_{avg}	41 $m^3/m^2 \cdot d$	1000 gpd/sq ft
Overflow rate @ Q_{pk}	102.5 $m^3/m^2 \cdot d$	2500 gpd/sq ft
HRT @ Q_{avg}	2 h	2 hr
Weir loading rate	190 $m^3/m \cdot d$	15 000 gpd/sq ft

Solution

Determine the required area of each clarifier using Equation 4.7 for average and peak flow conditions:

$$SOR = \frac{Q_{avg}}{A} = \frac{30\,000\ m^3/d}{(2\ tanks)A} = 41\ m^3/m^2 \cdot d \qquad Area = 366\ m^2 (3940\ sq\ ft)$$

$$SOR = \frac{Q_{pk}}{A} = \frac{(30\,000\ m^3/d)(1.7)}{(2\ tanks)A} = 102.5\ m^3/m^2 \cdot d \qquad Area = 256\ m^2 (2756\ sq\ ft)$$

Therefore, the average flow rate conditions govern; use area = 366 m^2 (3940 sq ft).

Determine the diameter of each clarifier:

$$Area = \frac{\pi d^2}{4} = 366\ m^2 \qquad Diameter = 21.6\ m\ (70.9\ ft)$$

Clarifiers are designed in 5-foot increments, so use a diameter of 22.9 m (75 ft) for each clarifier.

Determine the depth required to provide a 2 hour HRT at average flow:

$$HRT = \frac{V}{Q} = \frac{\left[\frac{(\pi)(22.9\ m)^2}{4}\right](Depth)}{(30000\ m^3/d)2\ tanks} = (2h)\left(\frac{1\ d}{24\ h}\right) \qquad Depth = 3.0\ m\ (\sim 10\ ft)$$

Depth is on the low range of 3 to 4.1 m (10 to 13 ft). The detention time could be increased to 2.5 hours to allow a deeper clarifier. A 2.5 hour detention time would allow a depth of 3.8 m (12.5 ft).

Check the weir loading rate (WLR) for the clarifiers, assuming the weir is on the periphery (circumference) of the clarifier:

$$WLR = \frac{Q_{peak}}{Weir\ length} = \frac{(30\,000\ m^3/d)2\ tanks}{(\pi)(21.6)} \qquad Weir\ loading\ rate = 221.2\ m^3/m \cdot d\ (17\,463\ gpd/ft)$$

This weir loading rate is higher than the design standards that were given. A designer could increase the weir length, which would increase the size of the clarifier, or provide a double weir configuration to increase weir length. Local standards would have to be applied to the situation.

3.6 DESIGN CONSIDERATIONS

3.6.1 DEPTH

The opportunity for contact between particles and flocculation increases with depth. Hence, removal efficiency should theoretically increase with depth. In actual practice, it is uncertain whether better removals can be obtained or higher overflow rates can be applied to deeper tanks. Clarifiers must be deep enough to accommodate mechanical primary sludge removal equipment, store settled solids, prevent scour and resuspension of settled solids, and avoid washout or carryover of solids with the effluent. Shallower depths may be acceptable with continuous primary sludge removal. Typical depths range from 3 to 4.9 m (10 to 18 ft) (Metcalf & Eddy, Inc./AECOM, 2014).

3.6.2 FLOW DISTRIBUTION AND INLET CONDITIONS

A splitter box is a chamber that divides incoming flow into two or more streams.

In facilities with multiple sedimentation tanks, practice has provided for equal distribution of flow and solids between the tanks. **Splitter boxes** and common channels sometimes are used for this purpose. Automatically controlled valves or gates, together with flow measuring devices, can also be used to split the flow equally to each tank. Manually controlled valves or gates are generally not effective for flow splitting because the fraction of flow directed to each tank can vary substantially with changing headlosses over the range of flows to be handled.

Inlet channel velocities should be high enough to prevent solids deposition. The inlet channel design typically allows a minimum velocity of 0.3 m/s (1 ft/sec) at 50% of design flow. Other alternatives to minimum velocities for prevention of solids deposition are inlet channel aeration or water jet nozzles (Yee & Babb, 1985). Provisions for scum removal from the inlet channel should also be included in the design.

Short-circuiting refers uneven flow through a vessel that causes the actual detention time to be shorter than the theoretical detention time. It occurs when density currents or inadequate mixing allows some of the flow to leave the vessel more quickly than the rest.

Sedimentation tank inlets should be designed to dissipate inlet port velocities, distribute flow and solids equally across the cross-sectional area of a rectangular tank or equally in all directions from the center feed area of a circular tank, prevent **short-circuiting**, and promote flocculation before quiescent settling.

Influent flow to rectangular clarifiers can be distributed by inlet weirs or by submerged ports or orifices. Inlet velocities to rectangular clarifiers typically are dissipated through some type of inlet baffle. Baffles typically are installed 0.6 to 0.9 m (2 to 3 ft) from the inlets. When full-width baffles are used, the top of the baffle should be 150 mm (6 in.) below the water surface to allow scum to pass, and the bottom of the baffle should be 0.3 m (1 ft) below the entrance opening (Metcalf & Eddy, Inc./AECOM, 2014).

There are several possible combinations to describe the flow pattern in a circular clarifier. The flow pattern can be center feed with center withdrawal, center feed with peripheral withdrawal, peripheral feed with center withdrawal, or peripheral feed with peripheral withdrawal. Center feed with peripheral withdrawal is the most common type of flow pattern.

Center feed circular clarifiers typically have a feed well with a diameter 15% to 25% of the tank diameter. Manufacturers' recommendations for submergence vary significantly. In practice, the feed well typically has been extended at least half of the tank depth. An energy dissipating inlet should be provided to dissipate inlet pipe and port velocities and promote flocculation as the influent flow enters the feed well area. Various configurations for energy dissipating inlets have been developed by clarifier manufacturers (Figure 4.8).

3.6.3 EFFLUENT LAUNDERS

Designers' opinions regarding launder spacing in rectangular clarifiers also differ. Some designers believe that launders should be spaced 5 to 6 m (16 to 20 ft) apart (American Water Works Association [AWWA] & American Society of Civil Engineers [ASCE], 2012). Launders typically are arranged transversely across the basin with chain and flight primary sludge collection equipment. With traveling bridge primary sludge collection equipment, launders must be arranged longitudinally as parallel finger weirs supported on piers.

Weirs and launders for circular clarifiers typically are mounted on the peripheral wall of the tank. Experience at some facilities has demonstrated that weirs and launders should be placed at least 15% of the basin radius inboard from the periphery of the tank (AWWA & ASCE, 2012). Such placement minimizes wall flow disturbance and draws effluent from a broader area.

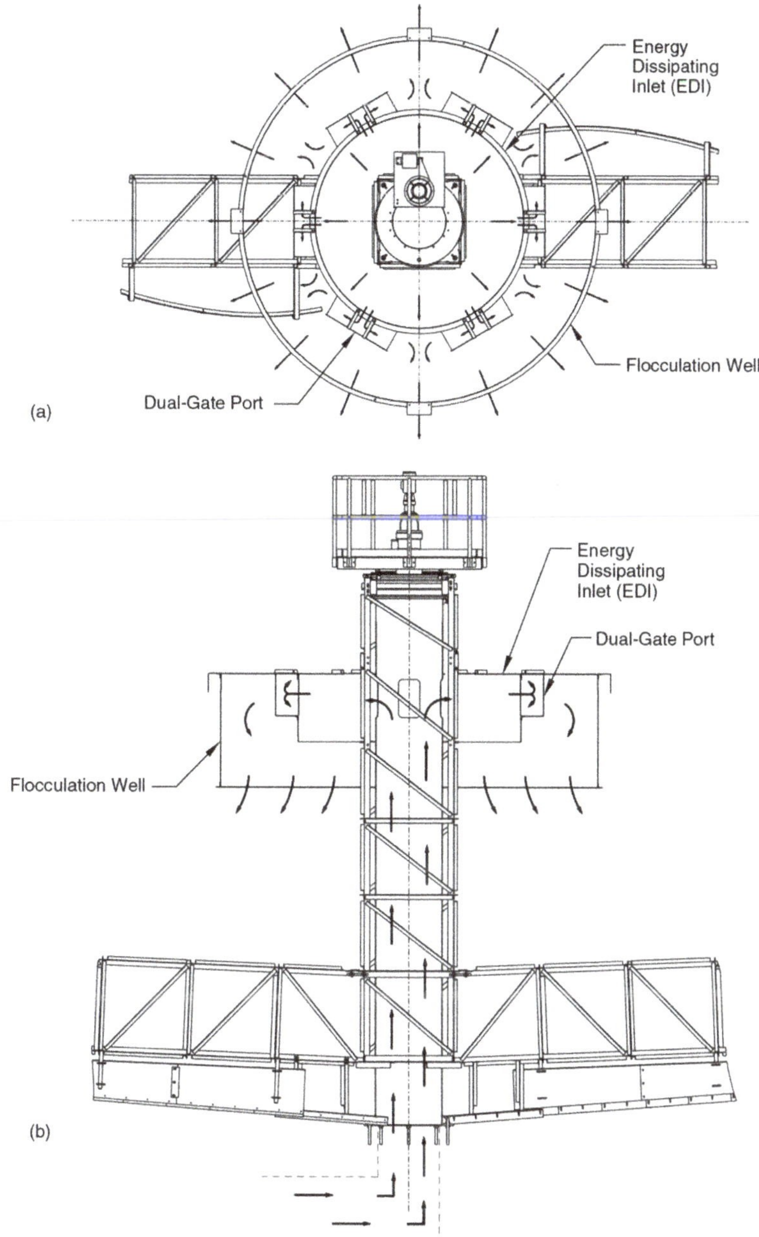

Figure 4.8 Energy Dissipating Inlet and Flocculation Well: (a) Plan and (b) Section (Reprinted with permission from WesTech Engineering, LLC)

3.6.4 LINEAR FLOW-THROUGH VELOCITY

Currents in the settling zone may hinder settling or result in scouring of settled solids. In practice, the linear flow-through velocity (scour velocity) has been limited to 1.2 to 1.5 m/min (4 to 5 ft/min) to avoid resuspension of settled solids (Theroux & Betz, 1959). The critical scour velocity may be calculated from the following equation (Camp, 1946):

$$V_H = \sqrt{\frac{8k(s-1)gd}{f}} \tag{4.8}$$

Where V_H = critical scour velocity, m/s (ft/s);
k = constant for type of scoured particles (unitless);
s = specific gravity of scoured particles;

g = gravitational acceleration, 9.81 m/s^2 (32.2 ft/sec^2);
d = diameter of scoured particles, m (ft); and
f = Darcy-Weisbach friction factor (unitless).

Typical k values are about 0.04 for rounded granular material and about 0.06 for nonuniform sticky and flocculent material. Typically, f values range between 0.02 and 0.03. Values for f are a function of the Reynolds number and characteristics of the settled solids surface. The particle diameter for scouring is typically 0.89 mm (150 mesh).

3.6.5 WEATHER CONDITIONS

Weather conditions can affect the performance of sedimentation tanks and must be considered in their design (Wells & Liberte, 1998). Wind may cause the water surface on the leeward side to be higher than on the windward side. Wind-caused turbulence may lead to unbalanced weir rates and short-circuiting (U.S. EPA, 1975). Design considerations for wind mitigation include orientation of tanks, installation of windbreaks or covers, increase of tank freeboard, and reduction of circular tank diameters to 37 m (120 ft) or less. Walls with a height of 1 to 1.2 m (3 to 4 ft) at the perimeter sometimes are used to provide wind protection for primary sedimentation tanks (Wall & Petersen, 1986).

Cold weather may also require freeze protection of surface sprays, insulation of piping, installation of underground piping at greater depths (below the freeze line), and provisions for auto-draining piping that conveys intermittent flow. Because scum collection equipment is prone to freezing, it needs adequate protection. Occasionally, these steps are insufficient in areas of severe cold and freezing; in this case, sedimentation tank covers may be required to avoid operational problems.

3.6.6 ODOR CONTROL

Odor control is a significant design consideration for primary treatment. Odors emanate from raw wastewater with high hydrogen sulfide concentrations and anaerobic conditions. The release of hydrogen sulfide and other similarly noxious odors occurs in areas of high turbulence such as sedimentation tank inlets, effluent launders, and weirs. Turbulence increases the water surface area for gas transfer with hydrogen sulfide release to the atmosphere. Scum and settled primary sludge handling systems can also be significant sources of odors. Available odor control strategies include source control, chemical treatment, preaeration, and containment (covers with ventilation to an odor control system).

3.7 ENHANCED SEDIMENTATION

Primary sedimentation can be enhanced by preaeration or chemical coagulation and flocculation, each of which is discussed in the following subsections.

3.7.1 PREAERATION

Preaeration of raw wastewater before primary clarification increases the settling of the wastewater by promoting flocculation of finely divided solids into more readily settleable flocs, thereby increasing suspended solids and BOD removal efficiencies. Other benefits of preaeration include scum flotation improvement, scrubbing of volatile organic chemical odor components, and prevention or mitigation of septicity.

Detention times of 20 to 30 minutes are necessary for floc formation and improved TSS, COD, and BOD removals. This range exceeds the range of 10 to 15 minutes suggested for odor control. The exact quantity of air required is a function of wastewater characteristics and tank geometry. The minimum air supply is typically 0.82 m^3/m^3 (0.11 cu ft/gal).

3.7.2 CHEMICALLY ENHANCED PRIMARY TREATMENT

Chemical coagulation and flocculation of raw wastewater promote clumping and agglomeration of finely divided solids into more readily settleable flocs, thereby increasing TSS, COD, BOD, and phosphorus removal efficiencies. Use of CEPT minimizes the required process footprint by reducing the size of primary clarifiers and aeration basins. It can efficiently remove 60% to 90% of TSS, 40% to 70% of COD or BOD, 70% to 90% of phosphorus, and 80% to 90% of bacteria loadings (Ødegaard, 2005; Metcalf & Eddy, Inc./AECOM, 2014; U.S. EPA, 1987). In comparison, conventional primary sedimentation may remove only 50% to 70% of TSS, 25% to 40% of COD or BOD, 5% to 10% of phosphorus loadings, and 50% to 60% of bacteria loading (McGhee & Steel, 2007; Metcalf & Eddy, Inc./AECOM, 2014; U.S. EPA, 1987). The CEPT process can also capture and remove heavy metals in the influent that could adversely affect the performance of downstream biological processes and effluent quality (Johnson et al., 2008).

Disadvantages of CEPT include an increased mass of primary sludge, production of solids that may be more difficult to thicken and dewater (particularly if lime or aluminum salts are used as coagulants), and an increase in operational cost and operator attention. Depending on downstream processes, there may be other site-specific disadvantages to CEPT. These include the removal of too much alkalinity, which may inhibit nitrification, or too much phosphorus, which may result in nutrient deficiencies in biological processes.

Historically, iron salts, aluminum salts, and lime are the chemical coagulants used for wastewater treatment. Iron salts typically are the most common coagulant used for primary treatment. Only a few facilities use lime as a coagulant because it is pH dependent and requires significant dosages to achieve the required pH. Lime also produces more primary sludge than metals salts and is more difficult to store, handle, and feed. To enhance sedimentation, some facilities use aluminum salts (alum). However, aluminum as a coagulant inhibits the specific methanogenic activity of bacteria needed for anaerobic digestion by approximately 50% to 70% (Noyola & Tinajero, 2005).

Anionic polymers sometimes are added during the flocculation step to promote floc formation. Anionic polymers should be added as a dilute solution to ensure thorough dispersion of polymers throughout the wastewater. Turbulence in the aerated headworks inlet channel is often used to provide mixing for flocculation, although this may not be optimal. Jar and full-scale testing are also recommended for selecting the flocculent and determining required dosages. Flocculent dosages should be assessed under high SORs (Peric et al., 2006).

3.8 HIGH RATE CLARIFICATION

High rate clarification combines techniques of chemically enhanced particle settling and solids contact/recirculation with lamella plate and tube settlers to achieve rapid settling. Very high settling velocities combined with rapid flocculation kinetics can reduce required process footprints to less than 10% of conventional primary treatment (Jolis & Ahmad, 2001). High rate clarificaton units are compact, have short startup times, and produce high quality effluent. High rate clarificaton facilities can be located at the WRRF or at remote satellite facilities that accept peak flow diversions from the main facility; however, the operator attention required at startup may be a disadvantage for satellite facilities that typically are unstaffed.

3.8.1 BALLASTED FLOCCULATION

Ballasted flocculation is a generic term for an HRC that adds a ballasting agent such as microsand along with coagulants to wastewater in a flash mixing zone. The ballasted flocculation process typically consists of influent screening, rapid mixing, flocculation, clarification with plates or tubes, sand separation, and recirculation (Figure 4.9). Fine screening is required to minimize plugging of the plates and hydrocyclone (WEF, 2005). Ballasted flocculation units are advantageous for wet-weather treatment because they can be offline, placed into operation, and meet performance requirements in 10 to 15 minutes (Leng et al., 2002).

The microsand is incorporated into the floc, which increases the particle density and settling velocity. Ballasted flocculation allows clarifier SORs between 700 and 1200 m^3/m^2 d (12–20 gpm/sq ft) at average flows and as high as 2300 to 3500 m^3/m^2 d (40–60 gpm/sq ft) at peak flows (Leng et al., 2002). Ballasted flocculation can achieve TSS removal rates of 75% to 90% and organic removal rates of 58% to 78% (Jolis & Ahmad, 2001; Leng et al., 2002).

3.9 PRIMARY EFFLUENT FILTRATION

A number of new wastewater filtration technologies, such as cloth depth filters (CDF) and compressible media filters (CMF), have emerged as popular solids removal processes within the wastewater treatment industry since about 2000. These technologies have been implemented successfully at hundreds of WRRFs as reliable tertiary filtration technologies to serve as a final treatment step before treated effluent is discharged to receiving water bodies or used as recycled water. These filtration technologies are now used to provide enhanced primary treatment.

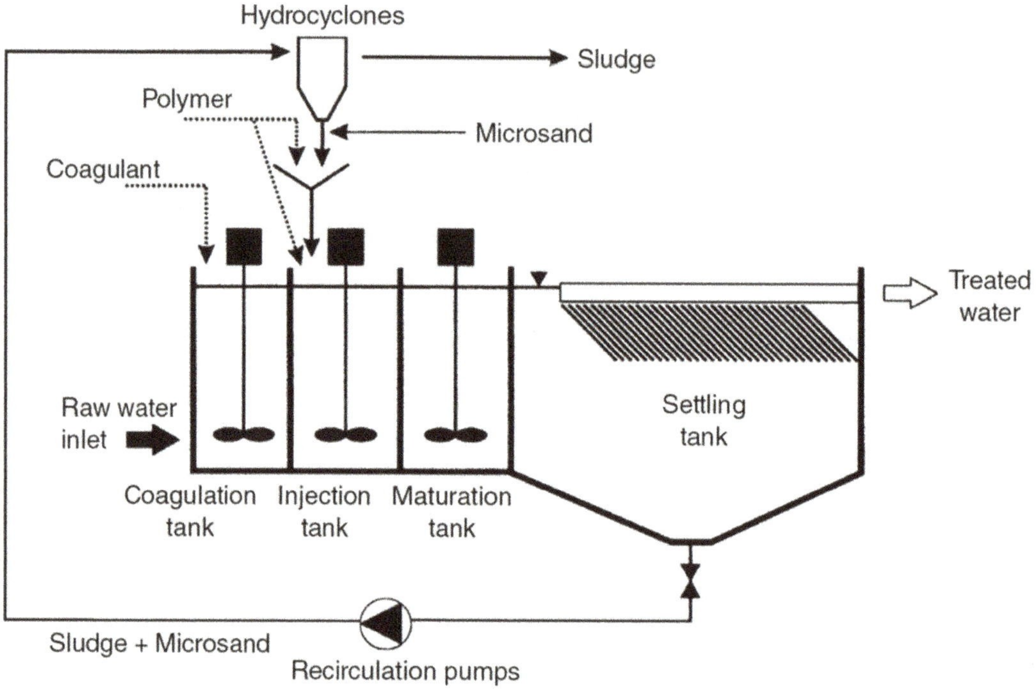

Figure 4.9 Ballasted Flocculation Process Schematic (WEF, 2005)

The effluent from the primary clarifier is filtered before secondary treatment. Reducing the primary clarifier effluent organic load will result in increased secondary treatment capacity (or decreased secondary treatment facilities requirements) and decreased aeration energy requirements. However, decreased organic loads may affect biological nitrogen removal. The filter backwash reject water (BRW) containing the solids captured from the primary effluent is diverted to the anaerobic digester after thickening. The high energy content of the captured volatile suspended solids in filter BRW will increase biogas production in the anaerobic digester.

3.10 EMERGING TECHNOLOGIES

Several technologies have recently emerged to provide a higher degree of primary treatment, reduced footprint, and decreased operational and maintenance requirements compared to conventional primary clarifiers. These emerging primary treatment technologies may also offer enhanced carbon diversion (i.e., increased removal of COD before biological treatment and routing of the COD to anaerobic digestion facilities) at WRRFs, resulting in additional energy and capital cost savings from decreased aeration energy requirements in secondary treatment, increased digester biogas production, and reduced sizing of secondary treatment facilities (or increased secondary treatment capacity of existing facilities).

3.10.1 PRIMARY FILTRATION

Filtering raw screened wastewater has recently emerged as a promising technology for primary treatment because of its significant energy and capital savings potential resulting from robust operation and high removal efficiencies (Caliskaner et al., 2016; Ma et al., 2015). Primary filtration reduces the footprint both for primary and secondary treatment facilities. Primary filtration in lieu of primary clarification can be used to increase the flow capacity of existing facilities and/or reduce the size and cost of new secondary treatment facilities by reducing the organic load to the secondary treatment system.

The filter BRW containing the solids captured from the raw wastewater is conveyed to the anaerobic digester after thickening. The high organic content of filter BRW increases biogas production in the anaerobic digester. The observed BOD and TSS removal efficiencies ranged from 50% to 55% and 80% to 85%, respectively, in a CDF.

3.10.2 ROTATING BELT FILTERS/MICROSCREENS

Rotating belt filters (RBF) and microscreens can provide treatment levels close to primary clarification in a reduced footprint. The use of RBFs and microscreens in lieu of primary clarifiers has increased since the early 2000s (particularly in Europe). Grit removal and/or coarse screening (e.g., with an opening size of 25–50 mm [1–2 in.]) are used upstream of the RBFs (especially in larger installations) to minimize operational and maintenance requirements. In the RBF, solids are removed by using a continuous loop fine mesh belt screen. The solids concentration of the reject stream is typically between 3% and 8%, allowing conveyance to solids digestion or dewatering facilities without further thickening. However, RBFs are also available with integrated screw press dewatering systems to increase the solids concentration to between 20% and 40%. Figure 4.10 illustrate the configuration of an RBF and microscreen system.

3.11 PRIMARY SLUDGE COLLECTION AND HANDLING

Settled primary sludge is typically scraped into a hopper where it is removed by gravity or pumping. The hopper for rectangular tanks is typically located at the inlet end of the tank to minimize travel time of particles to the hopper. For circular tanks, the hopper is typically located in the center of the tank. Common withdrawal pipes from two or more hoppers often result in unequal primary sludge removal from the hoppers. Therefore, multiple tanks and hoppers need separate pipes and pumps or valves on each outlet.

3.11.1 RECTANGULAR CLARIFIERS

Chain and flight or traveling bridge sludge collectors are typically used for rectangular primary sedimentation tanks. Chain and flight collectors (Figure 4.2) consist of two loops of chains with cross scrapers (flights) attached at approximately 3 m (10 ft) intervals. Revolving flights push the settled primary sludge to the hopper at the end of the tank. Most chain and flight collectors are within a 6.1 to 9.1 m (20 to 30 ft) width range (Green et al., 2007). Some installations, however, use side-by-side collectors without common walls in tanks as wide as 24 m (80 ft) or more (Metcalf & Eddy, Inc./AECOM, 2014). Flights travel along the long axis of the tank and, as the upper flights move away from the primary sludge hopper, they can skim the surface, pushing floating material toward the scum removal mechanism. At the end of the tank opposite from the sludge hopper, the flights drop to the floor and drag heavy, settled material to the primary sludge hopper for removal.

A traveling bridge clarifier consists of a scraper blade mechanism mounted on a bridge or carriage that travels approximately 1.8 m/min (6 ft/min) toward the hopper on tracks or rails mounted on top of the tank. As it travels away from the hopper at approximately 3.7 m/min (12 ft/min), the mechanism, largely out of the water, acts as a skimmer, pushing floating material toward the scum removal mechanism. As it reaches the end of the tank opposite from the hopper, the mechanism drops to the floor of the tank, reverses direction, and travels toward the hopper end of the tank, pushing settled primary sludge. In cold climates, the design of traveling bridge collectors should provide for control of snow and ice buildup on rails or tracks. Otherwise, this buildup could derail the bridge or reduce traction between the wheel and rail.

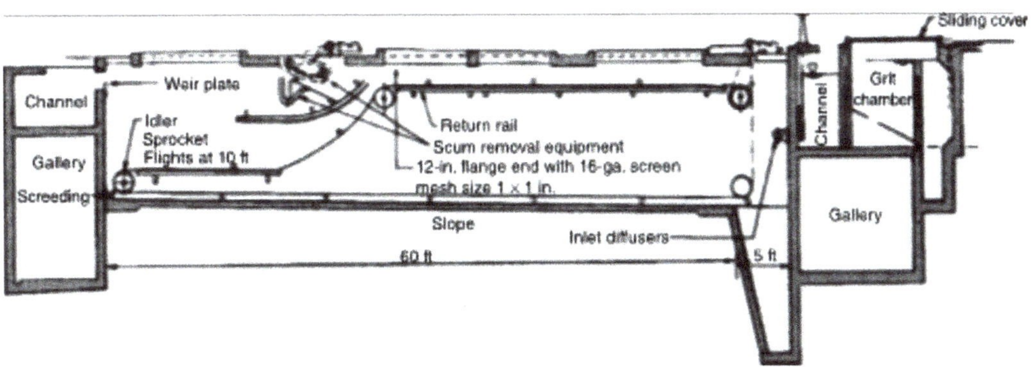

Figure 4.10 Typical Cross-Section of Primary Sedimentation Tanks at Valencia, California (ft = 0.30485 m; in. = 25.405 mm).

The sludge hopper, which can be up to 3 m (10 ft) deep, should have steep sides with a minimum slope of 1.7:1 (vertical:horizontal) and the bottom should have a maximum width of 0.6 m (2 ft) (Great Lakes–Upper Mississippi River Board of State and Provincial Public Health and Environmental Managers, 2014). Sludge hopper wall surfaces should be smooth with rounded corners to avoid any solids buildup. Rectangular sedimentation tanks with longer widths often need more than one hopper to reduce hopper depth.

3.11.2 CIRCULAR CLARIFIERS

Circular clarifiers typically have segmented rake- or plow-type primary sludge collection equipment. The conventional plow type consists of scrapers that drag the tank floor at a tip speed of approximately 1.8 to 3.7 m/min (6 to 12 ft/min) (AWWA & ASCE, 2012). Plows are located at an angle to the radial axis to force primary sludge towards the hopper, which is typically located at the center of the tank, as the device rotates. The center hopper is typically a vertical-sided sump where the primary sludge is removed by pumping. The rotating element of the device can be driven from either the center or the outside tank wall.

More recently, deeper, tapered, segmented, spiral-type sludge scrapers (Figure 4.11) are favored over the more conventional plow type. The continuously tapered spiral-shaped scraper blades allow faster operating tip speeds and move solids to the center hopper in as little as one revolution. This enables the facility operator to increase sludge transport capacity and improve sludge concentrations. Spiral-type scrapers have been successfully used in clarifiers up to 68 m (225 ft) in diameter to improve sludge collection/transport efficiency (Albertson & Okey, 1992; Kinnear, 2002).

3.11.3 QUANTITIES AND CHARACTERISTICS

Primary sludge production can be estimated from the following equation:

$$S_M = \frac{Q \times TSS \times E}{1000} \quad \text{(S.I.)} \tag{4.9}$$

Figure 4.11 Spiral Blade Clarifier (Reprinted with permission from WesTech Engineering, LLC)

where S_M = mass of primary sludge, kg/d (lb/d);
Q = primary influent flow, m³/d (mgd);
TSS = primary influent total suspended solids, mg/L;
E = removal efficiency, decimal;
1000 = units conversion factor ([1000 mg/L]/[kg/m³]); and
8.34 = units conversion factor ([8.34 lb/mil. gal]/[mg/L]).

The TSS removal efficiency in primary sedimentation tanks typically ranges from 50% to 70%. If actual removal data are unavailable, a removal efficiency of 60% can be assumed for estimating purposes. The CEPT process can increase primary sludge mass by 50% to 100% (Soap and Detergent Association, 1989). The addition of approximately 20 mg/L of ferric chloride and 0.2 mg/L of polymer to the headworks before primary sedimentation increased primary sludge production by approximately 45% (Chaudhary et al., 1989). Approximately 30% of this increase results from improved suspended solids removal and the remaining 15% stems from chemical precipitation and removal of colloidal material, via effective particle charge neutralization and bridging (Chaudhary et al., 1989).

Chemical sludge quantities can be estimated by the stoichiometric relationship between the raw wastewater constituents and coagulants. The stoichiometric quantity should be increased by approximately 35% for aluminum and iron salts to account for increased BOD, COD, and TSS removal (Mertsch, 1985; U.S. EPA, 1987).

The composition of primary sludge is variable and depends on the nature and degree of industrial development in the collection area. Table 4.3 lists typical characteristics. Some chemical sludges (particularly from lime and aluminum salts) are gelatinous, with a high water content, low suspended solids content, and high resistance to mechanical or gravitational dewatering. Feed solids composition merits careful consideration in the design of solids handling and processing units.

Example 4.4

The primary clarifiers from Example 4.3 are able to achieve a TSS removal of 53% with an influent TSS concentration of 185 mg/L. Estimate the amount of sludge produced on daily basis by the primary clarifiers.

Table 4.3 Primary Sludge Characteristics (U.S. EPA, 1979)

Characteristic	Range of Values	Typical Value
Volatile acids, mg/L as ascetic acid	200–2000	500
Total dry solids (%)	5–9	5
Volatile dry solids (%)	60–80	75
Specific gravity of individual solid particles	—	1.4
Bulk specific gravity (wet)	—	1.02
5-day biochemical oxygen demand:volatile suspended solids	0.5–1.1	—
Chemical oxygen demand:volatile suspended solids	1.2–1.6	—
Organic nitrogen:volatile suspended solids	0.5–1.1	—
Grease and fat, percent by weight of dry solids	7–35	—
Nitrogen, percent by weight of dry solids	1.5–4	2.5
Phosphorus, percent by weight of dry solids	0.8–2.8	1.6

Solution

Use Equation 4.9 to determine the primary sludge produced by the clarifiers:

$$S_M = \frac{(Q_{avg})(TSS)(E)}{1000\ L/m^3} = \frac{(30\ 000\ m^3/d)(185\ mg)(0.53)}{1000\ L/m^3} \qquad S_M = 2942\ kg/d\ (6471\ lb/d)$$

This is the amount of primary sludge that would have to be treated in the solids handling process.

3.12 DOWNSTREAM PROCESS CONSIDERATIONS

The selection and design of primary treatment systems should consider the objectives of the systems and the effects on downstream treatment processes (Bixio et al., 2000). Special downstream considerations are discussed below.

3.12.1 GENERATION OF VOLATILE FATTY ACIDS

Primary clarifiers that precede biological nutrient removal (BNR) systems may be operated to generate volatile fatty acids (VFAs) by fermenting primary sludge to enhance biological phosphorus removal (Barajas et al., 2002; Christensson, 1997; Christensson et al., 1998; Skalsky & Daigger, 1995). Online prefermenters are known as activated primary clarifiers (APCs). In addition, fermentation of primary sludge enhances its suitability as an electron donor and increases the quantity of soluble substrate (carbon) available to support denitrification in the downstream BNR processes (Baur et al., 2002; Lee et al., 1995; Millard, 2006). A primary clarifier can be operated as a prefermenter by increasing the sludge blanket depth and solids retention time (SRT) to create anaerobic conditions for **hydrolysis** and **fermentation**. Fermented sludge is recirculated from the bottom hopper to the top (inlet) of the clarifier where the influent wastewater flow elutriates generated VFAs. The APCs must have sufficient depth; otherwise, increasing the operating sludge blanket will be difficult and will increase the risk of sludge washout, particularly during wet-weather conditions. Sludge collection equipment must also be designed to withstand the additional torque resulting from a higher and thicker sludge blanket. Activated primary clarifiers are often plagued with erratic performance, are difficult to control, emit odors, and create less-than-optimal conditions for settling of suspended solids. Operation in a sequential cluster arrangement, where the feed is cycled between different primary clarifiers, provides better control but is limited by the number of available clarifiers. Typically, sidestream prefermentation in a separate covered reactor is preferable because it is more efficient, less prone to upsets and odors, produces more VFAs, and provides better SRT control (McCue et al., 2006).

Hydrolysis refers to enzyme-mediated reactions that convert complex organic compounds (i.e., particulate) into simple compounds or reduced-mass materials. Often refers to solids particle breakdown in the first step of anaerobic digestion.

3.12.2 PRESERVATION OF AVAILABLE CARBON

A primary treatment system upstream of biological nutrient removal processes designed for nitrification and denitrification might be specifically designed or operated for less than optimal removals of suspended solids and BOD to preserve available carbon needed for denitrification (Parker et al., 2001; Tang et al., 2004). The designer should carefully compare the benefits and disadvantages of such changes. For a more thorough discussion of BNR systems, see Chapter 8.

Fermentation is the extraction of energy from a carbohydrate in the absence of oxygen.

3.12.3 CO-THICKENING OF PRIMARY AND SECONDARY SOLIDS

Waste biological sludge is sometimes discharged to the influent end of primary sedimentation tanks for co-thickening (settlement and consolidation). In several fixed film facilities (i.e., trickling filters, biologically aerated filters, and solids contact), co-thickening of waste solids with raw wastewater has not adversely affected primary sludge settling or thickening (Kemp & MacBride, 1990). These facilities, which exist in areas with moderate climates, do not have excessive SORs, and are equipped with rapid sludge withdrawal systems to prevent increases in soluble BOD as a result of biological activity in the sludge blanket. Typical average SORs for co-thickening clarifiers range from 24 to 32 $m^3/m^2{\cdot}d$ (600 to 800 gpd/sq ft) with co-thickened sludge concentrations ranging from 3% to 5% (Metcalf & Eddy, Inc./AECOM, 2014). Several benefits of co-thickening include enhanced primary sedimentation because of increased flocculation with secondary solids and elimination, and simplification of separate secondary thickening facilities and solids handling processes (WEF, 2005). It should be noted that the nature of biological sludge from an activated sludge process (waste activated sludge [WAS]) is significantly different than that from fixed film processes. Separate WAS thickening has replaced the practice of WAS co-thickening for most activated sludge facilities.

3.12.4 DIGESTER HYDROGEN SULFIDE CONTROL

The primary treatment system might be designed to add ferric chloride, ferrous chloride, ferric or ferrous sulfate, or chlorine to control the content of hydrogen sulfide in digesters and digester biogas. Whereas chlorine oxidizes reduced sulfur compounds, iron salts react with sulfide to form an insoluble iron sulfide precipitate. This ties up the sulfide and prevents it from being released in either the primary or digestion processes. These chemicals sometimes are best added upstream where there is more turbulence for mixing.

3.12.5 SLUDGE DISPOSAL

Site-specific sludge disposal alternatives and the overall facility energy balance should be considered when weighing whether or not to provide primary treatment and in selecting the target TSS and COD or BOD removal efficiencies. For instance, incineration typically favors a high primary to biological solids ratio to maximize the caloric value of sludge to support **autogenic combustion**. Anaerobic sludge digestion also favors a higher primary to biological solids ratio, which results in higher biogas production rates.

Autogenic combustion refers to the burning that occurs when the heat of combustion of a wet organic material or solids is sufficient to vaporize the water and maintain combustion without auxiliary fuel.

4.0 SECONDARY CLARIFIERS

Gravity clarification traditionally has been used to separate MLSS from effluent in suspended growth systems. Table 4.4 lists many of the factors that affect clarifier performance. Design considerations to address these factors that are common to all shapes and sizes of clarifiers are discussed in the following paragraphs. These are followed by separate sections on rectangular and circular clarifiers. Further details and more in-depth analysis of the design of secondary clarifiers can be found in Water Environment Federation's *Clarifier Design* (Manual of Practice No. FD-8) (WEF, 2005) and *Recommended Standards for Wastewater Facilities* (Great Lakes–Upper Mississippi River Board, 2014).

Table 4.4 Factors That Affect Clarifier Performance (Ekama et al., 1997; reprinted with permission from IWA Publishing)

Category	Factors
Hydraulic and load factors	Wastewater (ADWF, PDWF, PWWF)*
	Surface overflow rate
	Solids loading rate
	Hydraulic retention time
	Underflow recycle ratio
External physical features	Tank configuration
	Surface area
	Depth
	Flow distribution
	Turbulence in conveyance structure
Internal physical features	Presence of flocculation zone
	Sludge-collection mechanism
	Inlet arrangement
	Weir type, length, and position
	Baffling
	Hydraulic flow patterns and turbulence
	Density and connection currents
Site conditions	Wind and wave action
	Water temperature variation
Sludge characteristics	Mixed liquor suspended solids concentration
	Sludge age
	Flocculation, settling, and thickening characteristics
	Type of biological process

*ADWF = average dry weather flow, PDWF = peak dry weather flow, and PWWF = peak wet weather flow.

4.1 SETTLEABILITY

It is not possible to predict the settling characteristics of a particular suspension of microorganisms from day to day. Microbial makeup is the primary factor affecting activated sludge settleability. A well-designed and operated activated sludge system provides an environment promoting the proliferation of desired microorganisms that readily flocculate and controlling the growth of organisms that can contribute to poor sludge settleability and foaming (e.g., filamentous bacteria).

Healthy mixed liquor includes a mixture of bacteria, protozoa, and metazoa. Filamentous bacteria are present in varying amounts and can hinder the ability of the sludge to settle and thicken. When viewed under a microscope, they are typically long and stringy in appearance. Ideal floc contains just the right mixture of filamentous microorganisms and floc formers, with the filaments forming the backbone of the floc (Figure 4.12a). If the floc lacks enough filaments, then it is likely to break up (Figure 4.12b) and effluent quality will deteriorate. If too many filaments exist, then bulking may develop that impedes zone and compression settling (Figure 4.12c) (Jenkins et al., 2003; Sezgin et al., 1978).

Two basic approaches are used in measuring sludge settleability: (1) volume of settled sludge after a given period of time, and (2) settling velocity of the sludge/liquid interface during zone settling. The sludge volume index (SVI) has long been a common measure of sludge settleability. It is the volume in milliliters occupied by 1 gram of the MLSS following 30 minutes of quiescent settling. The traditional

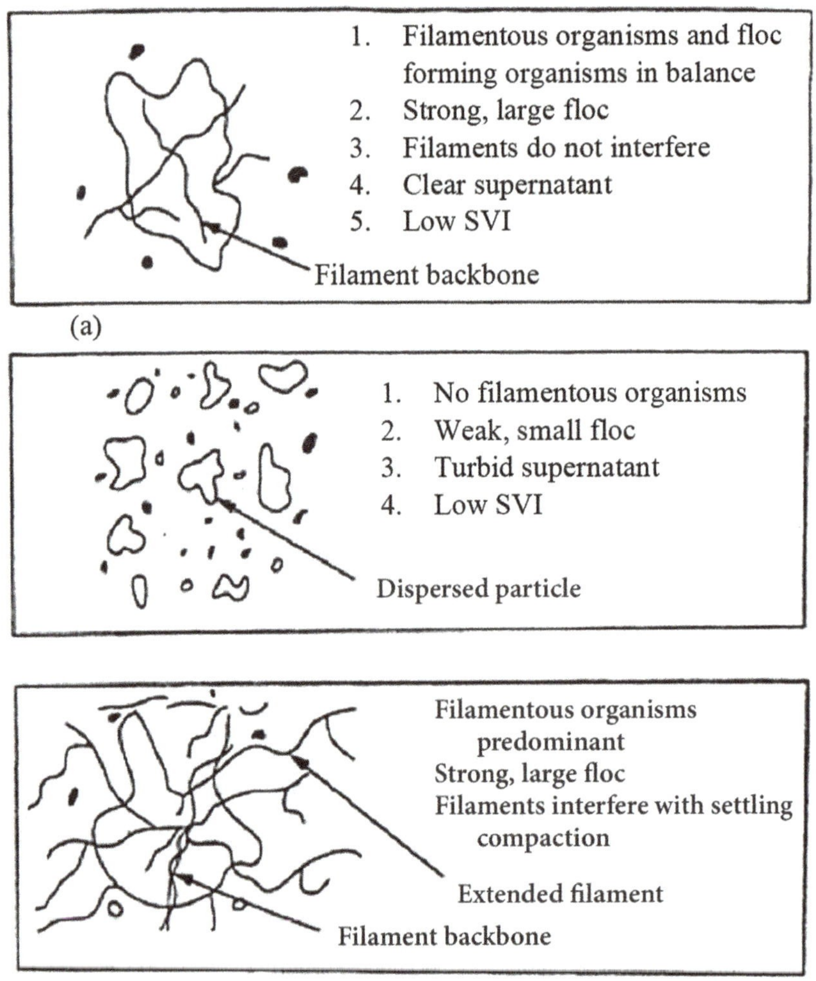

Figure 4.12 Effect of Filamentous Organisms on Activated Sludge Structure: (a) Ideal, Nonbulking Floc; (b) Pinpoint Floc; and (c) Filamentous, Bulking Floc (Ekama et al., 1997; reprinted with permission from the copyright holders, IWA)

method is carried out in a 1- or 2-liter settling column or graduated cylinder, but other methods produce more representative results. Standard methods specify gently stirring the sample during settling to eliminate or minimize wall effects (American Public Health Association et al., 2012). The dilute sludge volume index (DSVI) was developed to overcome concentration effects by diluting the suspension to result in a 30-minute settled sludge volume between 150 and 250 mL/L. Because of the relative insensitivity of $DSVI_{30}$ to solids concentration, it provides a common basis for comparing sludge settleabilities at different times and facilities.

4.2 DESIGN CRITERIA

Providing adequate surface area and depth is critical in clarifier design. Two criteria that define the area requirements are surface overflow rates (SORs) and solids loading rate (SLR). The SOR is the clarified effluent flow divided by effective surface area, which is defined as the dimensions to the inner tank wall and not the weir or orifice locations. The equation is presented in Equation 4.7. The SLR is the mass loading of solids (total flow to the clarifier, including return activated sludge (RAS), multiplied by the MLSS concentration) divided by the area and is presented subsequently. Some regulations specify numeric limits for surface area sizing; others accept justification calculations for loading. These in turn may include solids flux analyses and other such site-specific information.

The SOR used by design engineers, based on average dry weather flow (ADWF) and surface area, ranges from 0.5 to 2.0 m/h (300 to 1200 gpd/sq ft) for activated sludge secondary clarifiers. For peak hourly flow the SOR ranges from 1.7 to 2.7 m/h (1000 to 1600 gpd/sq ft). Some facilities are known to operate without difficulty at the upper end of this range and produce a high quality effluent. In many documented cases, diurnal or maximum pumping peak rates of 2.7 to 3.1 m/h (1600 to 1800 gpd/sq ft) do not exceed capacity. In other cases, poor clarification efficiency is encountered when the average and peak SORs are on the lower end of typical design ranges. A survey of consulting firms resulted in preferred SORs, shown in Table 4.5.

The Ten States Standards or similar guidelines may need to be applied (Great Lakes–Upper Mississippi River Board, 2014). In some cases, these capacity ratings were developed from clarifier designs in operation decades ago and do not reflect potential improvements in the design of inlet and outlet structures, depth, sludge collectors, and sludge removal that have been shown to increase allowable rates. It is projected that fully optimized clarifier designs will have 15% to 20% higher hydraulic capacity than pre-1970 clarifier designs having the same side water depth (SWD) (WEF, 2013).

4.2.1 SOLIDS LOADING RATE

The SLR can be expressed by the following equation

$$SLR = \frac{(Q+Q_R)X}{A(1000\ L \cdot kg/m^3 \cdot mg)} \quad \text{(SI)} \qquad (4.10)$$

$$SLR = \frac{(Q+Q_R)X(8.34)}{A} \quad \text{(U.S.)} \qquad (4.11)$$

Table 4.5 Preferred Overflow Rates ($m^3/m^2 \cdot h$ [gpd/sq ft])

	Circular Clarifiers		Rectangular Clarifiers	
Flow	Range	Average	Range	Average
Average	0.5–2.0	0.95	0.5–2.0	0.95
	(300–700)	500	(300–700)	500
Peak	1.70–2.72	2.09	1.70–2.72	2.09
	(1000–1600)	(1200)*	(1000–1600)	(1200)*

*Ten of 15 firms use 2.04 $m^3/m^2 \cdot h$ (1200 gpd/sq ft).

where SLR = solids loading rate, kg/d·m^2 (lb/d/sq ft);
Q = flowrate, m^3/d (mgd);
Q_R = return activated sludge (RAS) flowrate m^3/d (mgd);
X = MLSS concentration, g/m^3 (mg/L);
A = total plan area of clarifier, m^2 (sq ft); and
8.34 = conversion factor ([lb/mil. gal]/mg/L).

In establishing the maximum allowable SLR, most design engineers prefer to keep the average SLR (including full RAS capacity) in the range of 100 to 150 kg/d·m^2 (20 to 30 lb/d/sq ft) and peak SLR at 200 to 240 kg/d·m^2 (40 to 50 lb/d/sq ft). Rates greater than 240 kg/d·m^2 (50 lb/d/sq ft) or more have been observed in some well-operating facilities with low SVI, well-designed clarifiers, and effective solids removal. Approaches to determining the limiting SLR are presented below and include solids flux analyses and operating strategies.

Example 4.5

A flowrate of 40 000 m^3/d (10.6 mgd) is treated with conventional activated sludge followed by two clarifiers each with a diameter of 24.3 m (80 ft). The recycle ratio for the activated sludge system is 60% and the MLSS concentration is 2500 mg/L. Determine the solids loading rate on these clarifiers and comment on the values relative to standard values.

Determine the flow and recycle flow to each clarifier:

$$Q = \frac{(40\,000\ \text{m}^3/\text{d})}{2\ \text{tanks}} = 20\,000\ \text{m}^3/\text{d·tank}\ (5.3\ \text{mgd·tank})$$

$$Q_R = \frac{(40\,000\ \text{m}^3/\text{d})(0.60)}{2\ \text{tanks}} = 12\,000\ \text{m}^3/\ \text{d/tank}\ (3.2\ \text{mgd/tank})$$

Determine the area of each clarifier:

$$A = \frac{\pi d^2}{4} = \frac{\pi(24.3)^2}{4} = 463.5\ \text{m}^2\ \ (5024/\text{sq ft})$$

Use Equation 4.10 to determine the solids loading rate:

$$SLR = \frac{(Q+Q_R)X}{A(1000\ \text{L·kg/m}^3\text{·mg})} = \frac{(20\,000\ \text{m}^3/\text{d}+12\,000\ \text{m}^3/\text{d})(2500\ \text{mg/L})}{(463.5\ \text{m}^2)(1000\ \text{L·kg/m}^3\text{·mg})}$$

Solids loading rate = 172.5 kg/d·m^2 (34.5 lb/d/sq ft)

The solids loading rate is typically between 100 to 150 kg/d·m^2 (20 to 30 lb/d/sq ft); therefore, this solids loading rate is too high and the clarifier area would need to be adjusted for proper design.

Solids flux analyses are valuable in determining whether refinements to simple solids loading criteria are worthwhile and helping operators run the clarification process. The assumption of flux theory is that solids are continuously removed from the clarifier as they reach the design underflow concentration and that settling characteristics of the suspension are known. A detailed mathematical analysis of the flux theory is presented in *Clarifier Design* (WEF, 2005).

Solids concentration depth profiles consist of four zones: (1) clear water zone (h_1); (2) separation zone (h_2); (3) sludge storage zone (h_3); and (4) thickening and sludge removal zone (h_4) (Ekama et al., 1997). The fundamental premise of the flux theory is that during overloaded conditions (applied solids flux greater than the limiting flux), a critical zone settling layer (sludge storage zone, h_3) develops in the sludge blanket, which limits the conveyance of solids to the bottom of the tank. Consequently, not all solids that enter the storage zone from the separation zone are transferred to the thickening zone below, and the excess solids accumulate in the storage zone, causing it to expand. As it expands, the solids concentration remains constant in the storage layer. The depth of the separation and thickening zones (h_2 and h_4), however, do not increase substantially. The continued expansion of the storage layer will result in the sludge blanket reaching near the effluent structure level, causing a loss of solids with the effluent. At this point, the storage layer cannot expand further, and storage capacity of the clarifier is exhausted. The solids flux that could not be transferred through the storage layer is then lost with the effluent.

When the applied solids flux is less than the critical flux (underloaded condition), all of the applied solids can be transferred to the tank bottom, eliminating the need for solids storage. As a result, the sludge blanket is composed of the separation (h_2) and the thickening (h_4) zones only.

4.2.1.1 State Point Analysis

The state point analysis (SPA) is a graphical approach derived from solids flux theory (Keinath, 1985; Keinath et al., 1977). The SPA incorporates MLSS concentration and suspension settling characteristics, surface area available for thickening, and influent and RAS flowrates into the model. It can be used to assess different design and operating conditions. Figure 4.13 illustrates the components of the SPA. By definition, the state point is the point of intersection of the clarifier overflow rate (OFR) and underflow rate (UFR). The Clarifier Research Technical Committee Protocol of the Water Environment Research Foundation (WERF) provides guidance with respect to the development and application of the SPA (WERF, 2001).

4.2.1.2 Wilson Approach

Wilson (1996) presented a simplified method of evaluating secondary clarifier performance using the settled sludge volume (SSV or V_{30}) from a 30-minute settling test. He proposed that the SSV is correlated to the initial settling velocity (ISV), which also represents the required SOR provided that it is adjusted, where appropriate, for temperature, volatile solids content, and chemical addition. The relationships of this approach are:

$$R_{min} = SSV/(10^3 - SSV) \tag{4.12}$$

$$ISV = V_0 \times \exp(-4 \times SSV/10^3) \tag{4.13}$$

A volatile solid is an organic matter subject to decomposition; it is ignitable at 550 °C. Typically, this is used to represent the organic fraction of the sludge or other solids material.

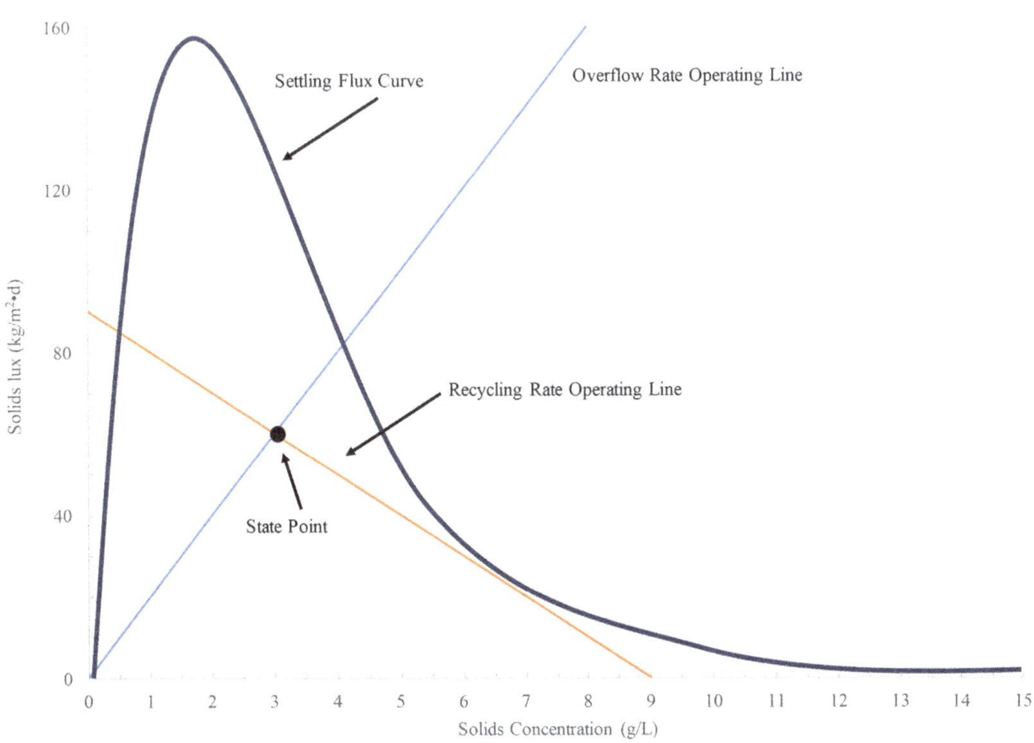

Figure 4.13 Elements of State Point Analysis of a Critically Loaded Clarifier (Adapted from Keinath, 1985)

where R_{min} = minimum RAS rate (%);
ISV = initial settling velocity, m/h;
SSV = 30 minute settled volume, mL/L; and
V_0 = sludge settling characteristic velocity, m/h.

Figure 4.14 presents a family of curves relating ISV (or clarifier SOR) to SSV for various values of V_0, assuming V_0 (m/h) is 0.3 to 0.5 times the temperature in degrees Celsius. Wilson concluded that the model compared well with the empirically validated German Abwassertechnische Vereinigung (ATV) approach as well as the model developed by Daigger (1995).

The Wilson approach entails determining ISV, which is also the maximum surface overflow rate (SOR_{max}), using Figure 4.14 or Equation 4.13; R_{min} can be derived from Equation 4.12. These values are then compared with SOR and RAS rates determined from facility operating data. Finally, the clarifier safety factor (CSF) and return safety factor (RSF) are calculated as follows:

$$RSF = \text{RAS rate}/R_{min} \tag{4.14}$$

$$CSF = SOR_{max}/SOR \tag{4.15}$$

A CSF value of less than 1.0 indicates clarifier overload. If CSF and RSF are both greater than 1.0, then the clarifier is underloaded. If CSF is more than 1.0 and RSF is less than 1.0, then the clarifier is most likely overloaded and the operating condition should be confirmed using other methods, such as the Daigger approach.

4.2.2 SIDE WATER DEPTH

Providing adequate tank depth is critical to good, consistent performance of any activated sludge clarifier. Selection of SWD is based on the size of the unit or the type of biological process preceding it. The trend in design practice is to make circular clarifiers deeper than in the past. Recommended values range from

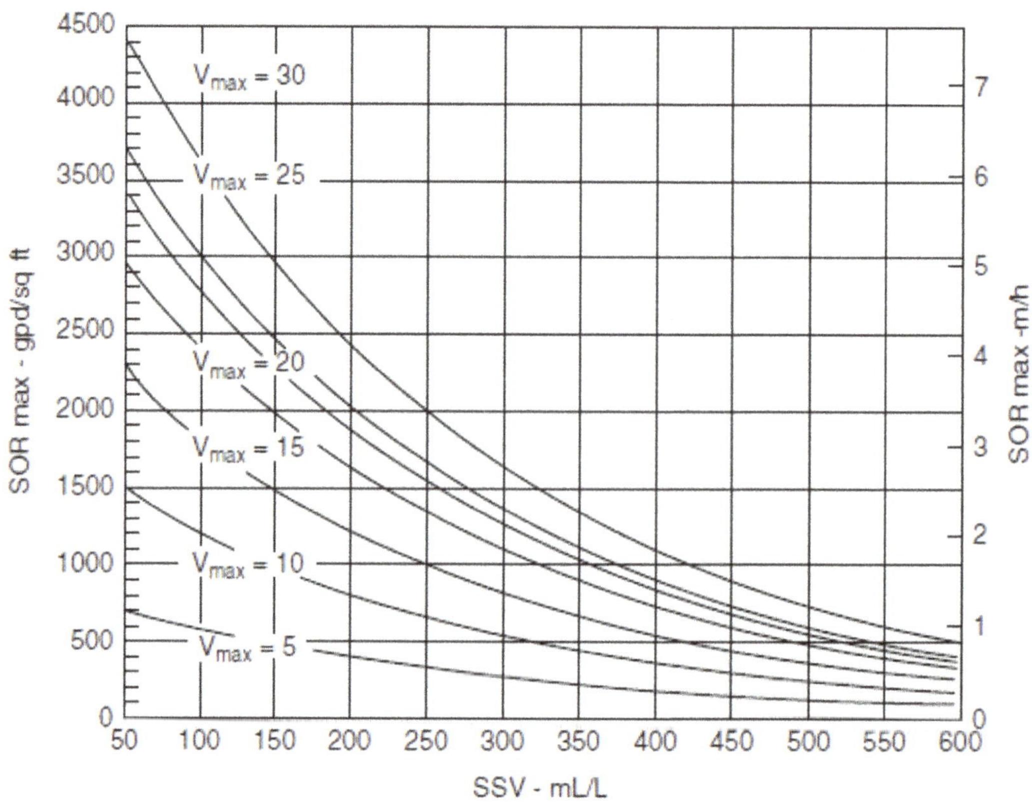

Figure 4.14 Wilson Model (Wilson, 1996)

3 to 3.5 m (10 to 18 ft) as a function of diameter. The distance of the sludge blanket from the effluent weir has a direct relationship to effluent quality (Miller & Miller, 1978). For circular tanks, based on historical operating data, Parker (1983) demonstrated the positive effect of SWD on effluent quality. At similar SORs, the average concentration of effluent TSS from a settler decreased as depth increased. Variability in effluent quality also decreased with increasing depth.

4.2.3 WEIR LOADING

Many regulatory statutes include weir loading in design criteria. Most design engineers believe that significantly higher weir loading rates would not impair performance and that placement and configuration have greater effects on a clarifier performance, particularly in the absence of excessive sludge blanket depths and high flow energies near the weirs. Misaligned weirs and those with excessive algae growth can cause flow imbalance within clarifiers. Many regulations limit maximum (peak hour) allowable weir loadings to 250 $m^3/m{\cdot}d$ (20 000 gpd/ft) for small WRRFs (less than 4000 m^3/d [1 mgd]) and 375 $m^3/m{\cdot}d$ (30 000 gpd/ft) for larger facilities.

4.2.4 SUMMARY OF SIZING STEPS

Many of the principles of design discussed above can be integrated into an approach that meets numerous objectives simultaneously. A step-by-step approach is given below.

1. Determine the operating MLSS concentration range of the facility to maintain an acceptable food to microorganism ratio (see Chapter 6), SRT, and effluent quality under various flow and mass loading conditions. For most municipal facilities, the range is 1000 to 4000 mg/L.
2. Determine the anticipated range of the MLSS SVI (or ISV). Select a statistically high value that would seldom be exceeded by the full-scale operating facility. This maximum design value should be based on an analysis of existing records, pilot facility data, or information from similar full-scale facilities. Where no data exist, SVI values of 150 to 200 mL/g for conventional and extended aeration systems and 120 to 150 mL/g for systems with selectors and BNR facilities with anaerobic or anoxic basins can be used.
3. Provide for 20% to 100% of ADWF RAS pumping rate capacity (up to 150% for extended aeration systems or others using high MLSS concentrations). The RAS rate increases higher than 80% of ADWF may be counterproductive because of the increased hydraulic load on the clarifier.
4. Determine the maximum theoretical SLR (function of SVI) using solids flux analysis to arrive at the solids loading limit and resulting surface area. This result should be checked against the governmental regulations that may place limits on solids loading.
5. Select an overflow rate to achieve the required effluent quality that is based on influent wastewater flowrate characteristics. This rate probably will not exceed the solids limiting value determined in Step 4. Design engineers should also make an allowance for taking a tank out of service for maintenance or repair.
6. Select a depth to provide adequate solids clarification, thickening, and storage. Allow 0.6 to 0.9 m (2 to 3 ft) for thickening; 1 m (3 ft) or more for buffering; and 2.4 m (8 ft) for clarification. More buffering depth is needed if diurnal or influent pumping flow variations or peak flow conditions are atypically large (e.g., greater than 2 : 1). Approximately 0.6 m (2 ft) of freeboard should be added to determine overall tank wall height.
7. Provide a reasonable weir length and place the weirs at strategic locations.
8. Select a mechanism for sludge removal. Plows, spiral curved blades, chain and flight systems, or hydraulic suction systems are available depending on shape.
9. Consider other details to improve secondary clarifier performance:
   - Flocculation inlet zones, preferably separated from the rest of the tank by baffles;
   - Midlength or midradius energy dissipation baffles (these may not be needed if a flocculating feedwell is provided);
   - Full radius skimmers (rotating trough or beach type for small tanks), full width for rectangular tanks, or partial radius beach type for large radial flow tanks (multiple blades, anti-rotation baffles, and spray nozzles may be added); and
   - Launder covers, algae removal mechanism, or chlorine addition equipment, if necessary, for algae control.

4.3 DESIGN CONSIDERATION

4.3.1 SHAPES

The current consensus of design engineers is that there are no significant shape advantages between circular and rectangular clarifiers if all of the design details are well done and in accordance with modern guidelines such as those of *Clarifier Design* (WEF, 2005). With that said, most design engineers and operators prefer circular clarifiers for activated sludge and specialty suspended growth systems treating municipal wastewater. The reliability of the mechanisms for circular tanks typically is cited as the primary reason.

4.3.2 RECTANGULAR DESIGN

Rectangular clarifiers have been used for activated sludge mixed liquor settling for nearly a century and frequently are found in large facility applications, although they are used for all sizes. Common-wall construction and space-saving footprints are attractive features. In addition, their galleries can accommodate piping and pumps.

4.3.2.1 Flow Patterns

Most activated sludge rectangular clarifiers have longitudinal flow patterns (Figure 4.15). A transverse flow option has been introduced recently, but few units of this type are built and operating. An important distinction is that longitudinal tanks can have concurrent, countercurrent, or cross-current sludge removal.

In concurrent flow in a longitudinal clarifier, the clarified liquid and sludge flow proceed down the length of the tank in the same direction. Inlet baffles or diffusers are designed to distribute the flow across the width of the clarifier and dissipate the inlet energy. The bulk of the mixed liquor solids settle to form a blanket interface. Settling action of the solids and removal of the sludge flow stream produce a density current along the bottom of the tank. The density current imparts a momentum that moves the sludge efficiently along the length of the clarifier to the hopper on the downstream end. Flights or scrapers assist the sludge movement.

The sludge hopper can be placed at the influent end so that the sludge is removed more quickly. In this case, the sludge flow reverses itself and is called a countercurrent sludge removal flow pattern. The sludge hopper can be placed at approximately mid-tank so that the sludge does not have to travel to the end of the tank for its removal. In most rectangular secondary clarifiers, the sludge hoppers are at the opposite effluent end or mid-tank.

4.3.2.2 Dimensions

Once the area and number of tanks needed for redundancy and normal size constraints have been established, specific tank geometric details can be defined. There are acceptable minimum ratios of length to width that effectively limit the maximum size of rectangular tanks. The length-to-width ratios of longitudinal rectangular clarifiers may range from 1.5 : 1 to 15 : 1. A minimum length-to-width ratio of 3 : 1 was recommended to prevent short circuiting, but it is typically greater than 5 : 1 (U.S. EPA, 1974). Some references recommend that the length of the rectangular clarifier should not exceed 10 times the depth (Metcalf & Eddy, Inc./AECOM, 2014). Typical depth ranges from 3.5 m (12 ft) to 6 m (20 ft). However, this length-to-depth ratio has been exceeded with success at larger facilities. The length, width, and depth

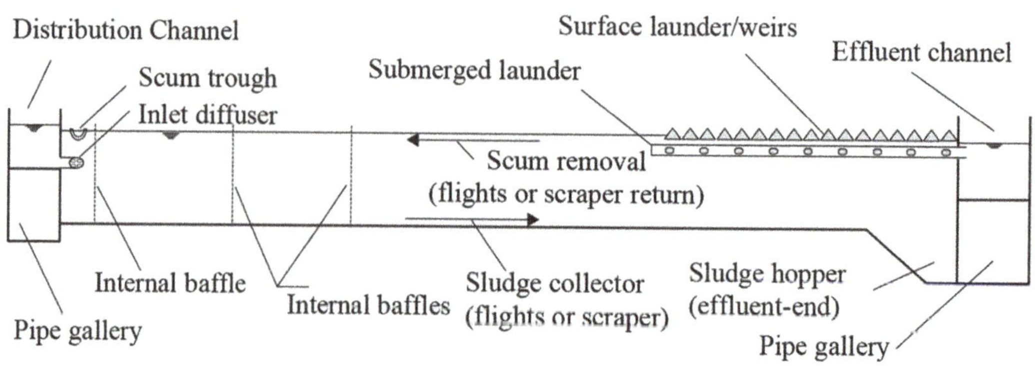

Figure 4.15 Rectangular Clarifier Design Features and Nomenclature (Hopper Locations May Vary) (Ekama et al., 1997; reprinted with permission from IWA Publishing)

dimensions should be proportioned so that horizontal flow velocities are not excessive. Rectangular clarifiers are seldom greater than about 100 m (300 ft) in length and are typically 30 to 60 m (100 to 200 ft) long.

4.3.2.3 Inlet Design

There are many different inlet designs for longitudinal flow rectangular tanks (Figures 4.17 through 4.19). Some include a flocculation zone baffled off from the quiescent portion of the tank and others do not. For concurrent flow tanks, density current problems can be minimized by positioning the inlet lower in the tank without placing it in the thickening zone, which is typically reserved for the bottom 1 m (3 ft) of the tank. Locating the inlet too low may scour the solids on the bottom and lead to resuspension. For countercurrent clarifiers with sludge hoppers at the inlet end, horizontal baffles to prevent density current flow into the hopper are recommended. Figures 4.19 and 4.20 show conceptual examples.

4.3.2.4 Scum Removal

The collection and removal of floatables in the clarifier concerns both inlet and outlet areas of the tank because foam passes through the middle of the tank unimpeded. If influent baffles are used, then they have slotted surfaces or include small downward opening gates to allow scum to pass through. For rectangular tanks with concurrent sludge removal, chain and flight collectors often are brought to the surface and move scum to the front of the tank for collection. For countercurrent designs, scum moves downstream and a collection barrier is needed upstream of the effluent launder area. A slotted roll pipe situated across the width of the tank can be used to remove scum. It is positioned at the point where scum is concentrated by the movement of the sludge flights on the surface of the clarifier. Scum flows into the horizontally mounted pipe when the slot is rotated below water level. The pipe can be rotated manually or automatically. Other designs use a separate set of chains and flights or spiral flight arrangement to move the scum up a beach and into a trough. Adjustable pan skimmers are also used on rectangular clarifiers. Water sprays can be used to move scum toward the pan and prevent the setup of foam.

4.3.2.5 Sludge Withdrawal

Most rectangular clarifiers have sloping floors and hoppers at one or more locations. Floor slope is typically 1%. Hoppers at either end of the tank are effective with shorter runs of piping through galleries to the recycle pumps. Mid-tank hoppers are sometimes used when internal baffles are provided, because the gap provided by the hoppers and between the sludge collectors are a convenient location for a baffle.

The typical hopper shape for rectangular clarifiers is an inverted pyramid with a rectangular opening on top. The sides are recommended to have the slope of at least 52° from horizontal to prevent solids from accumulating on the upper walls. A single rectangular tank may have two or more withdrawal hoppers, each equipped with a withdrawal pipe. Separate controls for each hopper are absolutely necessary.

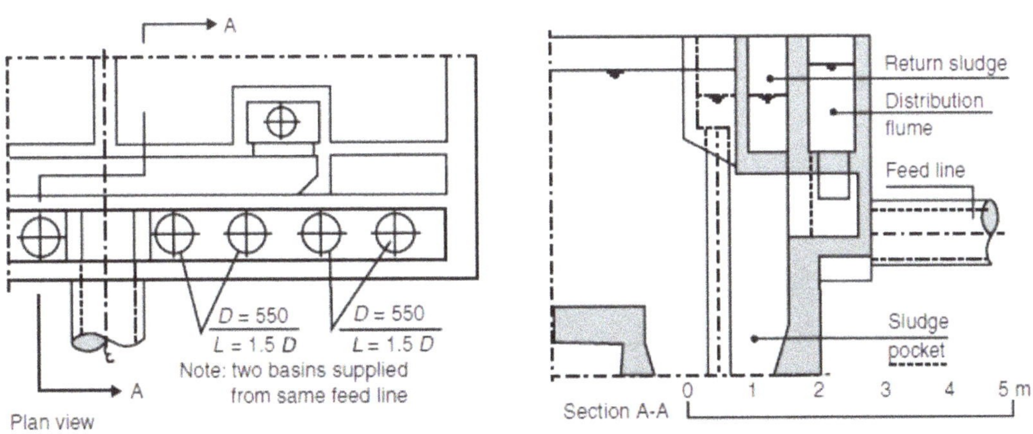

Figure 4.16 Inlet Design of Larsen (1977) to Avoid Floc Breakup (note that D is in millimeters)

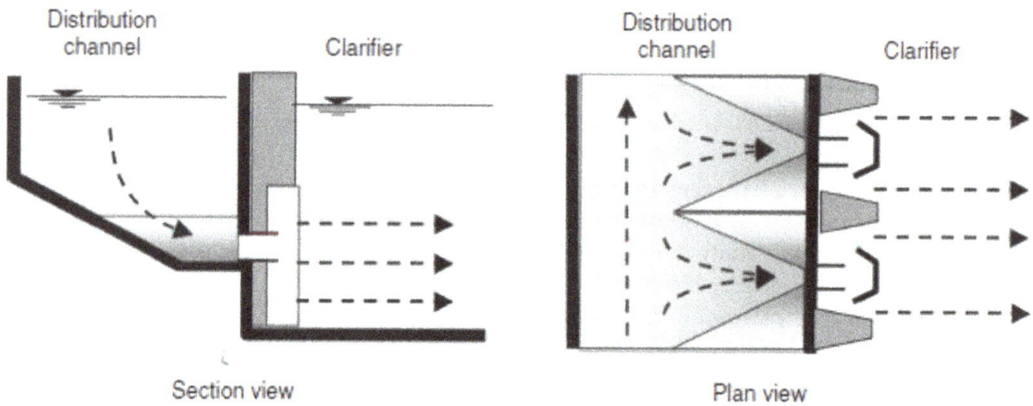

Figure 4.17 Distribution Channel With Funnel-Shaped Floor (Krauth, 1993) With a Stuttgart Inlet (Popel & Weidner, 1963)

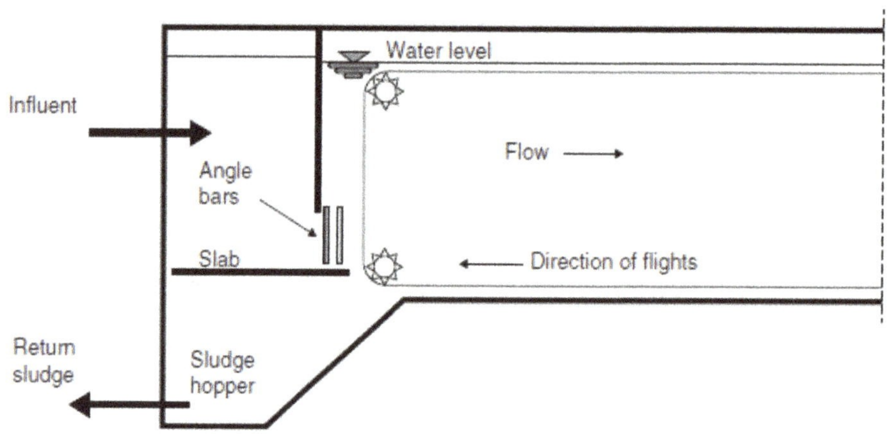

Figure 4.18 Aerated Distribution Channel With Horizontal Slab Deflecting Inlet Flow Energy From a Sludge Hopper at the Inlet End (Krebs et al., 1995)

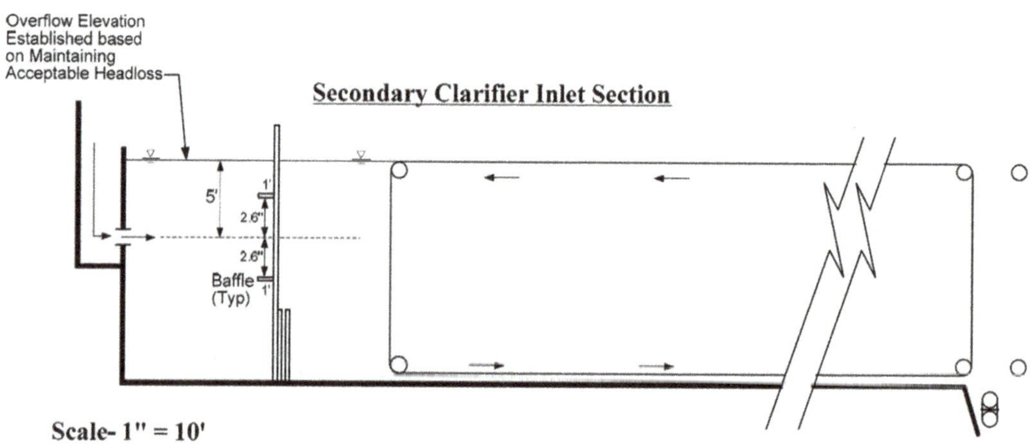

Figure 4.19 Inlet for New Secondary Clarifiers (Barnard et al., 2007; permission from J. L. Barnard)

Sludge removal systems for rectangular tanks typically feature chain and flight scrapers. Traveling bridge, hydraulic suction, and reciprocating flight collector systems have been used. In chain and flight scrapers, the flights are attached to two parallel chains driven by sprockets and move slowly along the clarifier floor, scraping the settled sludge to collection hoppers. On their return path, the flights can be designed to rise above water surface elevation and transport scum to collectors.

Historically, redwood or metal flights and steel chains have been used, but stainless steel or nonmetallic flights, chains, and sprockets have become more common. Typical flights measure 5 to 6 m long (16 to 20 ft), depending on the width of the clarifier. Newer flights can be 10 m (33 ft) long and entire systems can be up to 90 m (300 ft) long. Flights typically are spaced at 3 m (10 ft) intervals and travel at speeds of 5 to 15 mm/s (1 to 3 ft/min).

4.3.3 CIRCULAR CLARIFIERS

Clarifiers equipped with rotating mechanisms have gained in popularity through the years because they have a reputation for being mechanically the most trouble free and reliable. Most of the tanks used in suspended growth, secondary applications are circular in plan view and have radial flow patterns. Circular tanks have the disadvantage of taking up more footprint compared to equivalent capacity rectangular units built with common-wall construction. The former requires more inlet, outlet, and sludge piping.

The circular shape requires separate structures for flow splitting ahead of the tanks and for sludge pump stations. For flow splitting, the most common and effective method involves feeding the structure at a low elevation, causing flow to rise vertically, and then divide by flowing over two or more weirs. Another concept is to provide overflow weirs or orifices along an aerated channel that has low horizontal velocities. For some large facilities, modulating butterfly gates with computer controlled operators have been used successfully.

For sludge removal, it is important to have independently measured and controlled withdrawal for each clarifier. Many engineers will find that measuring hydraulic flows for this purpose is adequate. Some include solids concentration measurement that enables mass flux monitoring without sampling. Such instrumentation typically is located in a separate pump station serving multiple circular tanks.

4.3.3.1 Flow Patterns

Circular tanks are fed from a center inlet pipe or through ports or inlet baffles at the perimeter. Effluent is withdrawn over weirs or orifices near the surface at the tank wall or from launders supported away from the wall. Figure 4.20 presents a simplified illustration of these variations. The velocity vectors shown represent general movement when clarifiers have a thin sludge blanket at the bottom.

4.3.3.2 Dimensions

Clarifier mechanisms of more than 70 m (200 ft) in diameter are available, although the upper limit is considered to be 50 m (150 ft). For larger diameter clarifiers, wind can create surface currents that upset the radial flow pattern and concentrate scum downwind. Clarifier mechanisms are typically available in 5-foot increments. Before the early 1980s, circular secondary settling tanks often had depths of 2.4 to 3.0 m (8 to 10 ft) as measured at the wall. Performance data showed that a shift to greater depths resulted in lower effluent suspended solids and more resistance to upset from hydraulic peaking. Increased RAS concentrations also occurred with increasing depth, and guidelines were developed (Table 4.6).

4.3.3.3 Inlet Geometry

Most clarifiers are equipped with the mechanism drive located at the top of the center column for both center and peripheral feed styles. For center feed, the pipe also must bring influent into the tank and transmit rotational torque from the drive into the bottom foundation.

The influent pipe should be sized to keep material in suspension but to keep velocities low enough to avoid floc breakup and excessive headloss. Many manufacturers design the influent velocity at peak hour flow and maximum RAS flow, not to exceed approximately 1.4 m/s (4 ft/sec). Some other designers lower this to about half of this value to minimize floc breakup. For peripheral feed tanks, the velocity of inflow to the distribution feed trough or skirt should also be kept to less than this value.

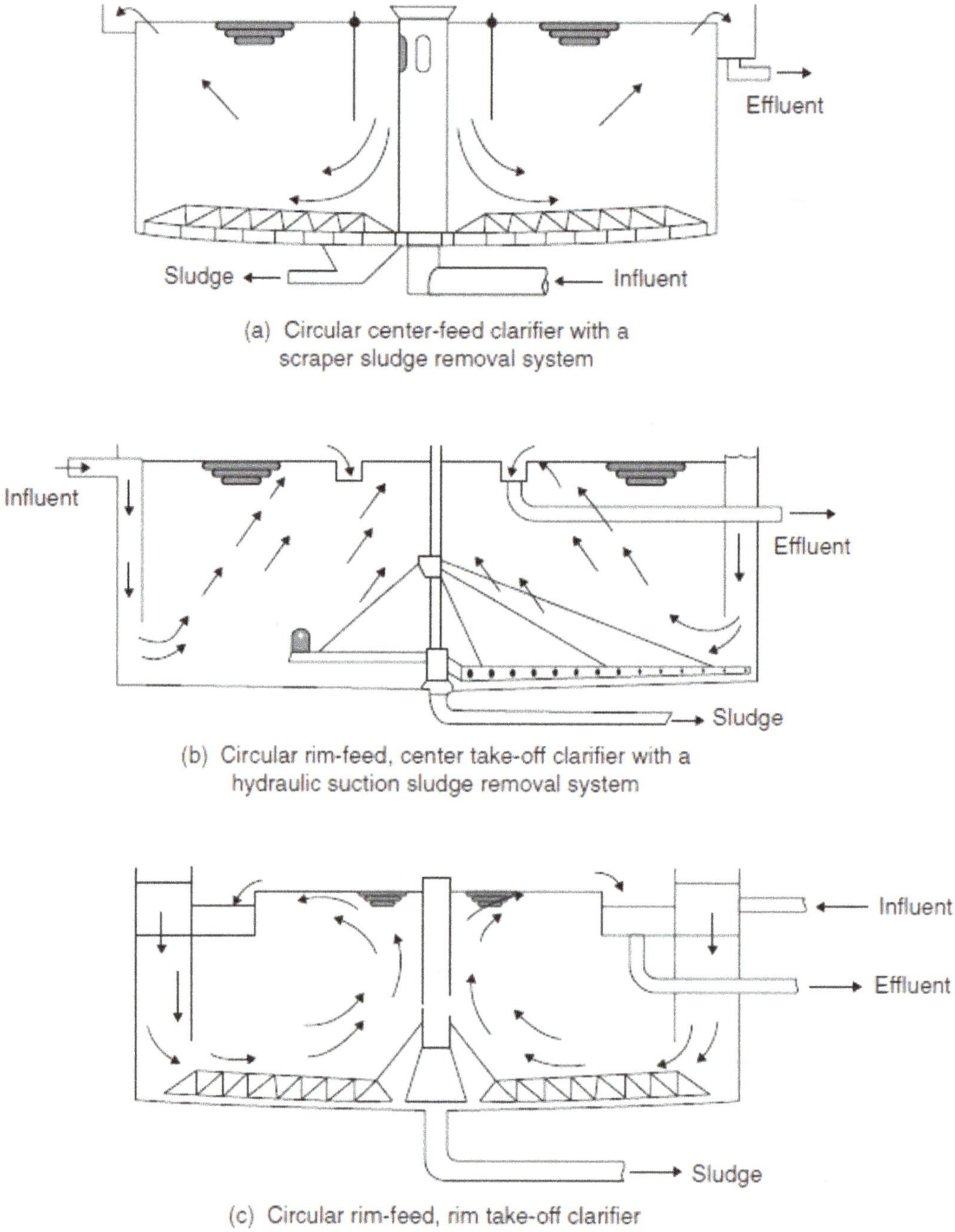

Figure 4.20 Typical Circular Clarifier Configurations and Flow Patterns

For center feed clarifiers with ports that transmit flow from the feed pipe into the feedwell, port velocities should not exceed feed pipe velocities discussed above. Most center feed columns have four rectangular opening ports. They are often submerged, although some designs may leave the top several centimeters of the ports exposed. Instead of ports, another popular feed pipe opening is to connect two segments of pipe with four vertical structural steel channels welded to each pipe exterior.

The standard center feed inlet design that is typically used for activated sludge is shown in Figure 4.21. For activated sludge facilities, the feedwell diameter is often 20% to 25% of the tank diameter. Criteria for downward velocity of flow determine the diameter of the simple feedwell. Some designers and manufacturers advise that feedwell diameters should not exceed 10 to 15 m (35 to 45 ft) regardless of tank

Table 4.6 Minimum and Suggested Side Water Depths for Activated Sludge Clarifiers

	Side Water Depth, m (ft)	
Tank Diameter, m (ft)	**Minimum, m (ft)**	**Suggested, m (ft)**
Up to 12 (40)	3 (10)	3.7 (12)
12–21 (40–70)	3.3 (11)	3.7 (12)
21–30 (70–100)	3.7 (12)	4 (13)
30–43 (100–140)	4 (13)	4.3 (14)
> 43 (140)	4.3 (14)	4.6 (15)

size. Likewise, downward flow velocities leaving the feedwell are often limited to about 0.7 m/min (2.0 or 2.5 ft/min).

A typical center feedwell extends down from as little as 30% to as much as 75% of the tank depth. Several manufacturers recommend that submergence be 25% to 50% of the SWD. It is also common that the center feedwell bottom edge be located about 0.3 m (1 ft) below the bottom of the center feed pipe ports. It must be low enough so that flow jetting out of the ports does not travel below the baffle and out into the settling zone.

4.3.3.4 Scum Removal

Activated sludge clarifiers often experience scum formation because of denitrifying sludges and foams (such as *Nocardia* filamentous bacteria). In the United States, it is common practice to remove floating materials from the surfaces of secondary clarifiers. For circular tanks, a variety of skimming mechanisms have been designed and are operated with varying degrees of capacity and success. The most common system used for center feed tanks is shown in Figure 4.22a. It features a rotating skimmer arm and wiper that travels around the outer edge of the tank next to a scum baffle. It moves the floatables onto a beach or egress ramp connected to a scum removal box. The skimmer blade is most effective if it is attached tangentially to the feed baffle, rather than perpendicular to it. The resulting pitch angle of the tangential design helps move floatables to the outer area of the tank.

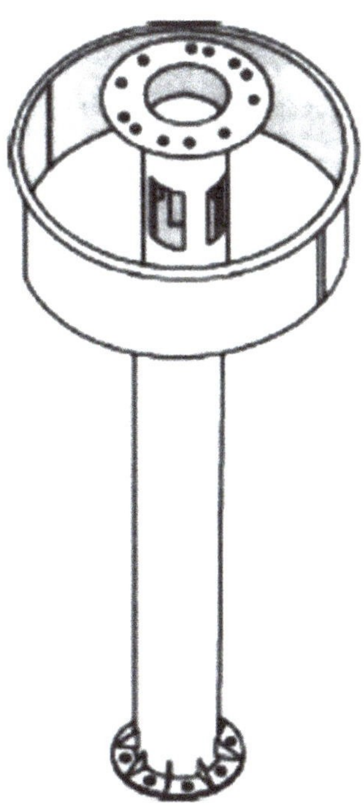

Figure 4.21 Standard Center Inlet Design

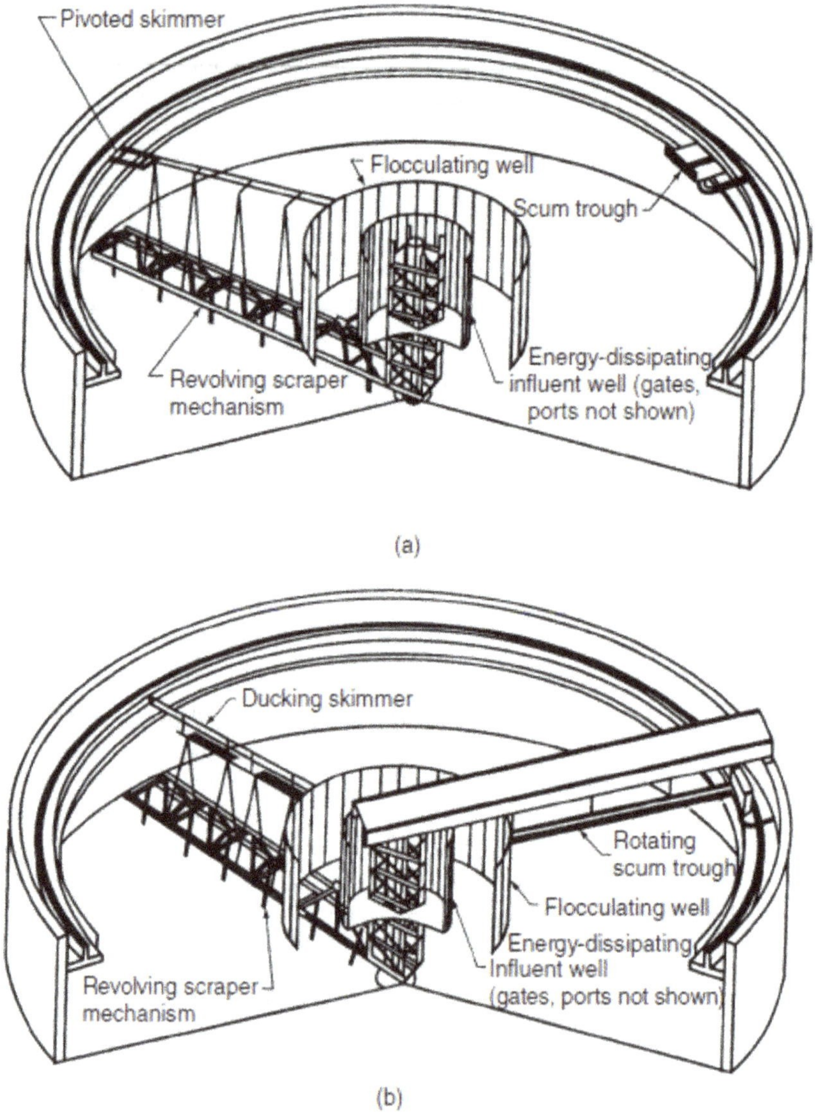

Figure 4.22 Alternative Skimming Designs for Circular Clarifiers: (a) Revolving Skimmers and Fixed Scum Trough and (b) Rotary Ducking Skimmers (Ekama et al., 1997; reprinted with permission from IWA Publishing)

Another skimming concept, known as the "ducking skimmer," is shown in Figure 4.22b. In this design, a skimmer board is connected to the sludge removal mechanism through a hinged, counterweighted assembly. It pushes the floatables toward a fixed, rotating trough that turns into position as the skimmer board approaches and trips a trigger switch. When the board reaches the rotating trough, it ducks under the trough, and its counterweights return it to the surface to continue rotation around the tank. This device has an advantage of offering full radius scum removal. Some installations have a reported a high amount of maintenance associated with the ducking skimmers. Issues have included controls, bearings, actuators, and binding of the rotating trough.

4.3.3.5 Outlets

Outlets for most circular center feed clarifiers consist of a single perimeter v-notch weir that overflows into an effluent trough. Alternatives to this include cantilevered or suspended double weir troughs and submerged orifice collector tubes. Another concept includes provisions of a square, octagonal, or circular double-sided launder suspended from the bridge or other structural support near the center of the tank.

4.3.3.6 Peripheral Weir

There are two common designs for peripheral weir outlets for circular tanks. In the first, a concrete trough is constructed on the inside of the tank wall. The weir plate is then bolted at the top of the inward face of the trough wall. In the second, the weir plate is bolted to the inside of the tank wall. A concrete effluent trough is then constructed outside the tank wall. This design can be less costly to build if soil elevation outside the tank can be used beneficially in formwork construction.

The most common type of weir plate involves placement of 90° v-notches at 150 or 300 mm (6 or 12 in.) intervals. This design allows a balance of relatively low increases in water surface elevation when flows increase with an allowance for imperfect leveling with reasonably good distribution. In contrast, a flat weir plate is susceptible to unbalanced withdrawal if the weir is not perfectly level or if wind effects on the surface are significant.

4.3.3.7 Cantilevered Double or Multiple Launders

In early years, and perhaps in some areas today, regulations that limit weir loading to sufficiently low values have enticed designers to go with multiple weirs, serpentine weirs, and other ways to increase the weir length for a given diameter of tank. Requirements in recent years have been relaxed in many design guidelines. Nevertheless, this cantilever double launder concept remains. It offers the opportunity for solids moving up along to the wall to resettle before inward flow takes them to the outer weir. Anderson (1945) and others recommend that the outer weir be at least 25% of the tank radius from the wall.

4.3.3.8 Sludge Withdrawal

There are two basic types of sludge removal mechanisms for activated sludge circular tanks: scrapers and hydraulic suction.

Several basic scraper designs are used; Figure 4.23 shows four different types. The multiblade plows, using straight scraper blades, have been used most extensively in the United States. The designs using

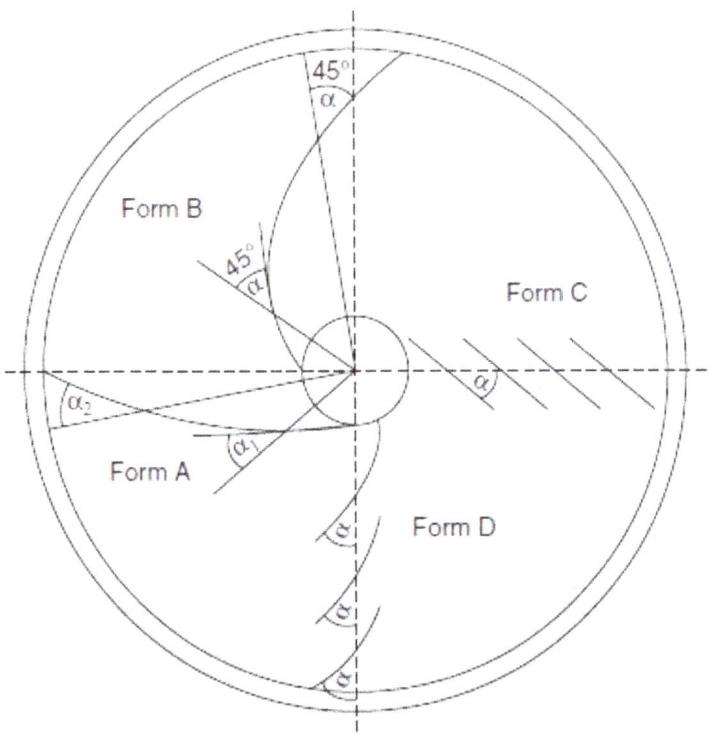

Figure 4.23 Scraper Configuration in Germany (Gunthert & Deininger, 1995). Form A is the "Nierskratzer" type, where $a1 > a2$. Form B is a logarithmic spiral with a constant at 45°. Forms C and D are "window shade" type scrapers.

curved blades are typically referred to as spiral plows and have been used for decades in Europe. Although blade angles of between 15° and 45° have been used, 30° has become popular in the United States. Suppliers report that the vast majority of new scraper mechanisms used for activated sludge are of the spiral design, as opposed to the multibladed "window shade" type. In early years, the tip speed of spiral scrapers was approximately 3 m/min (10 ft/min). Based on several facility improvement projects, Albertson (1995) and others have recommended values as high as 10 m/min (30 ft/min). These faster speeds, as well as deepening the spiral blades closer to the center of the tank, give this system a relatively high sludge transport and removal capacity. The higher speeds do induce some stirring of the tank contents, especially in smaller tanks, which may affect clarification. Some designs feature variable speed drives.

For activated sludge treatment with partial or complete nitrification, the occurrence of denitrification in clarifiers can cause solids to float and effluent quality to degrade. In the 1960s, the concept of hydraulic suction was engineered to assist in removing sludge more rapidly. Kinnear (2002) showed that this concept was more effective in maintaining lower sludge blankets.

To lift the sludge, a hydraulic head deferential is established by use of pumps or adjustable valves to move solids into the collector arms. There are two fundamentally different types of hydraulic suction removal mechanisms. The first, typically called an organ or riser pipe, has a separate collector pipe for each suction inlet orifice. The V-shaped plows direct the sludge to these pipes. The other type has a single or double arm extending across the full radius of the tank. The arm is tubular and has several orifices. It is typically referred to as a manifold design but is also known as header, tubular, or Tow-Bro in recognition of Townsend and Brower, who developed it.

4.3.3.9 Hoppers

Traditionally, most U.S. activated sludge clarifiers using scraper mechanisms were equipped with trapezoidal hoppers. Depending on tank size, these hoppers are typically 1 m (3 ft) deep and have walls with slopes of at least 50° above horizontal. Other types of hoppers have been developed to prevent "**rat holing**" and dilution of the RAS. One type consists of deep conical or annular sludge hoppers. The rotating mechanism has stirrups that reach into the annular hopper to prevent bridging.

Rat holing is the compaction on the side of a silo that occurs when the material being handled has sufficient cohesive strength to resist gravity flow through a silo. Typically, a channel will form where the material will fall through the silo; however, once the channel has emptied, all flow from the silo stops.

4.3.3.10 Floor Slopes

Most clarifiers with plow or spiral mechanisms in the United States have a constant floor slope of 1 on 12. Albertson and Okey (1992) have promoted the use of a dual slope floor that provides a steeper slope in the center for large tanks. This steeper slope provides for greater depth and sludge compaction. For hydraulic suction clarifiers, not much slope is needed. Because it is not necessary to move the sludge across the floor, relatively flat floors with slope of 1% or 2% for draining are common.

5.0 MEMBRANE BIOREACTORS

A membrane bioreactor (MBR) is a combination of suspended growth activated sludge biological treatment and membrane filtration equipment performing the critical solids liquid separation function that is traditionally accomplished using secondary clarifiers. **Low-pressure membranes** are typically used in MBRs (Metcalf & Eddy, Inc./AECOM, 2014).

Low-pressure membranes are membranes that operate at pressures ranging from 68.4 kN/m^2 – 205.2 kN/m^2 (10 – 30 psi). Microfiltration and ultrafiltration membranes are typically classified as low-pressure membranes.

There are two general types of membrane systems that can be used in MBRs: (1) pressure driven (in pipe cartridge systems that are located external to the bioreactor); and (2) vacuum driven (immersed systems that are designed for installation within the bioreactor). Immersed membrane technologies using hollow fiber or flat sheet membranes are the most popular because they operate at lower pressures (or vacuums); can more readily accommodate variations in solids; and typically provide a lower life-cycle cost, particularly for municipal facilities. Pressure-driven systems are more prevalent in industrial systems in which waste characteristics, such as high temperatures, require the use of ceramic membranes. In its simplest form, an immersed MBR can combine the functions of an activated sludge aeration system, secondary clarifier, and tertiary filtration in a single tank. In most cases, however, membranes are immersed in a tank separate from the bioreactor. A process flow diagram for a MBR is shown in Figure 4.24.

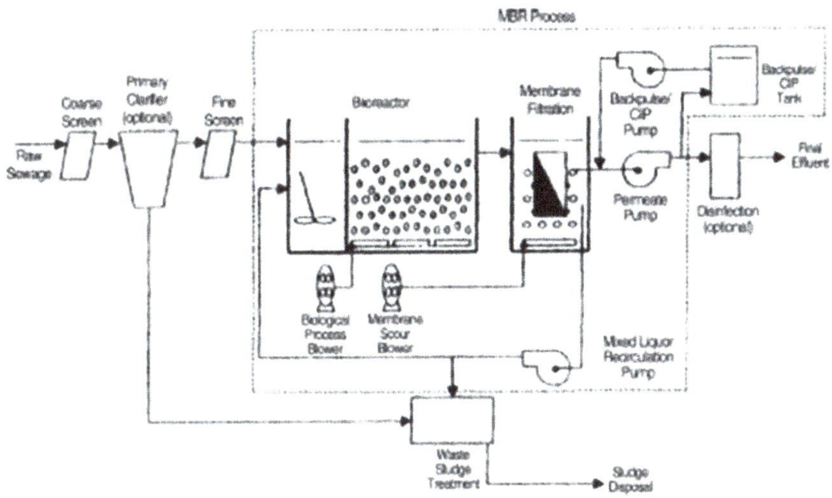

Figure 4.24 Process Flow Diagram of a Typical Membrane Bioreactor Facility (Reprinted with permission from Jacobs)

5.1 COMPONENTS AND CONFIGURATIONS

Unlike clarifier-based activated sludge processes, MBRs use membranes to separate biological solids from the mixed liquor. The membrane pore sizes are minute and often smaller than the pore sizes of filter papers used for laboratory analysis so the separation of solids from liquids is essentially complete, and all biological solids are retained in the process for use as return sludge or for wasting.

Although pore sizes are minute in the **microfiltration** or **ultrafiltration** range, they are not able to capture soluble organic compounds, metals, or trace contaminants such as pharmaceutical and personal care products, **priority pollutants**, or **endocrine-disrupting compounds**, or per- and polyfluoroalkyl substances. Although the biological process of an MBR may adsorb or reduce such contaminants, the filtration mechanism is not adequate to directly filter these materials from the wastewater (Maeng et al., 2013; Ternes and Joss, 2006; WERF, 2007).

5.1.1 APPLICATIONS

Membrane bioreactors are used to treat both municipal and industrial wastewaters and possess numerous potential benefits. In an MBR, the biomass is completely retained, resulting in consistently high quality final effluent with suspended solids concentrations of less than 1 mg/L and of suitable quality to be used as feed for reverse osmosis systems. Compared to conventional activated sludge (CAS) facilities with clarifiers, the effluent quality in an MBR is less dependent on the MLSS concentration and sludge properties. The incorporation of an MBR allows the elimination of secondary clarifiers and effluent filters, thereby reducing facility footprint. The aeration basin volume can be reduced because systems can operate at high MLSS concentrations. An MBR process can operate at longer SRTs than a CAS facility, reducing sludge production, and can also operate within a wide range of SRTs, resulting in increased flexibility and more options for optimization. The use of an MBR produces low turbidity effluent, which reduces downstream disinfection requirements by increasing the transmissivity of the effluent. This means less energy required for ultraviolet disinfection and a lower chlorine demand.

The use of MBRs often become viable for WRRFs requiring high quality effluent for reuse or for discharge to sensitive receiving waters and for facilities with significant land area restrictions (both new facilities and retrofits).

Although the advantages of MBR systems are numerous, MBRs are not suitable for every wastewater treatment application. The limitations of the membranes may cause problems for high flows that exceed the capacity of the membranes. Flows in excess of capacity must be diverted, which could compromise treatment. The life-cycle cost of MBRs may be high due to replacement costs of membrane components and the energy required to operate and maintain the membranes. Another potential disadvantage of

Microfiltration is low-pressure membrane filtration process that removes suspended solids and colloids larger than 0.1 μm from wastewater.

Ultrafiltration is a low pressure membrane filtration process that removes suspended solids and colloids large than 0.01 μm from wastewater.

Priority pollutants are a list of 126 chemicals that the U.S. EPA has designated for regulation due to environment or human health impacts.

Endocrine-disrupting compounds are substances in the environment (air, soil, or water), food sources, personal care products, and manufactured products that interfere with the normal function of the body's endocrine system.

MBRs is that the operating conditions within the system may promote the formation of foam due to retention of extracellular polymeric substances. This has been mitigated by implementing foam management options such as surface wasting to prevent foam accumulation.

5.1.2 EFFLUENT QUALITY

The effluent from an MBR process is essentially free of suspended solids and macrocolloidal material. An MBR facility also can be designed to remove nutrients from the wastewater, similar to conventional BNR processes, with minor modifications. Typically, effluent from MBR facilities contains less than 1 mg/L TSS, less than 5 mg/L carbonaceous BOD, less than 0.2 nephelometric turbidity units (NTUs), and low levels of bacteria. When membranes with pore sizes in the micro- and ultrafiltration ranges are used, MBRs alone will not remove dissolved solids, and the membranes themselves will have no effect on pH or alkalinity. Effluent from MBR facilities can be discharged to sensitive areas, reused on public access sites, or further treated by **nanofiltration** or **reverse osmosis**.

Nanofiltration is a high pressure membrane filtration process that removes particles greater than 1–10 nm from wastewater.

Reverse osmosis is a technological process (and high-performance membrane used to achieve this process) that uses forward osmosis phenomena to retain salts in the higher osmotic pressure solution when displacing water from this solution through the semipermeable high-performance membranes (reverse osmosis membranes).

5.2 PROCESS DESIGN

For successful operation, membrane equipment systems for MBRs include various combinations of air scour, backwash, relaxation, and chemical cleaning systems to maintain performance and permeability. Membranes can be operated across a range of flows, or flux, as long as an adequate differential pressure is provided across the membranes. Immersed MBR systems have limited differential pressure (less than 68.4 kN/m^2 [10 psi]) and, hence, flux rates are more restricted than with external, pressure-driven MBRs. Experience with membrane equipment suggests that average fluxes between 14 to 25 L/m^2·h (8 to 15 gpd/sq ft) are sustainable when the MLSS is 15 000 mg/L or less (WEF, 2023).

5.2.1 PRETREATMENT

It is normal practice that fine screening equipment with a maximum of 1 to 2 mm (0.04 to 0.08 in.) openings is provided to protect membranes from debris and fibrous materials. Typically, these screens are installed downstream of 6 mm screens, either at the headworks or following primary clarification. Additional measures include placing covers over the membrane tanks, or fine screening a portion of the mixed liquor as it is returned from the membrane tanks to the bioreactor.

Primary clarification is not specifically required for an MBR, although, just as for other activated sludge systems, this can reduce the total energy required for aeration and overall volume of the bioreactor. Primary clarification provides the additional benefit of settling out some of the undesirable trash and skimming off scum and floatables that would otherwise be removed by fine screening. Some membrane manufacturers will allow less stringent fine screening if primary clarification is included in the process.

PRACTICE PROBLEMS

1. **A municipality has two primary settling tanks receiving a suspended solids concentration of 250 mg/L, an average flowrate of 57 000 m^3/d (15.06 mgd), and a peak flow of 75 000 m^3/d (19.8 mgd). A sludge holding pond is to be constructed and the engineer needs to know the quantity of sludge that will be produced. The following design information is available. Calculate the daily mass of dry sludge produced and daily volume of wet sludge.**

   Design information:

Q_{avg} = average flowrate =	57 000 m^3/d	15.06 mgd
Suspended solids concentration	250 mg/L	
Sludge concentration =	4%	
E =	60%	
Specific gravity =	1.04	
Specific weight of water =	1000 kg/m^3	62.4 lb/cu ft

2. A primary settling tank was designed to treat 18 927 m^3/d (5 mgd). The hydraulic detention time is 2 hours and the depth is 3.0 m (10 ft). Is this design within accepted standards? A peaking factor of 2 has been observed. The width of the tank is 1.8 m (6 ft).

3. If settling tanks are designed only on the basis of surface area, why is depth important? Why not just make them very shallow tanks?

4. The ideal tank theory, as developed by Alan Hazen and Thomas Camp, that leads to the use of overflow rate as a design parameter, makes assumptions. What are these assumptions?

5. A WRRF operator is having trouble with the primary clarifier. Sludge is pumped out of the clarifier once a day. Sludge consistency is visually determined, and pumping is stopped when the sludge becomes "thin" (low in suspended solids). Gas bubbles out of the clarifier surface and floating solids are clogging the weirs. A consulting engineer recommends the installation of new sludge pumps to double the flowrate of the current sludge pumps. Is this recommendation correct? Will it solve the floating sludge problem? Why or why not?

6. A wastewater contains particles of only one size, shape, and density. These particles settle at a rate of 100 cm/h. The flowrate is 0.1 m^3/s.

   a. How long should a settling tank 2 m wide and 2 m deep be to remove 100% of these particles?

   b. What is the overflow rate?

   c. If a weir extends over the end of the tank (width), what is the weir loading?

7. If a primary clarifier in a water resource recovery facility has an influent of 4542 m^3/d (1.2 mgd), a depth of 3 m (10 ft), and a hydraulic retention time of 3.1 hours, what is the overflow rate? Is this within the design limits for your state's standards?

8. A primary clarifier receives a wastewater at 240 m^3/d (0.063 mgd) and a suspended solids concentration of 180 mg/L. It discharges an effluent with suspended solids concentration of 70 mg/L. What is the production of primary sludge in kilograms per day as dry solids? If the solids concentration of the settled solids is 4%, what is volume of sludge produced?

9. You are asked to design two secondary rectangular clarifiers following a conventional activated sludge system. The average flowrate to the activated sludge system is 15 142 m^3/d (4 mgd) with a 50% RAS return flow. The MLSS in the activated sludge system is 2 500 mg/L.

   a. Select a surface overflow rate and determine the required area for the clarifiers based on clarification.

   b. Select an appropriate solids loading rate and determine the area required for thickening.

   c. Using the results from the previous parts, determine the length and width for each clarifier using a length-to-width ratio of 8 : 1.

   d. Choose and justify a depth and calculate the HRT for the clarifier. The HRT should be between 2 and 4 hours.

   e. Design a weir system so that the weir loading rate is at or below standard values. Assume a peak hourly factor of 2.0.

10. Use the Wilson approach to determine the RSF and CSF for the following secondary clarifier parameters. Comment on the results and make any suggestions.

Design Parameters		
Average flowrate	11 500 m^3/d	3.0 mgd
Clarifier diameter	19.8 m	65 ft
Settled sludge volume	160 mL/L	
RAS rate	50%	
Temperature	10 °C	

6.0 REFERENCES

Albertson, O. E. (1995). Circular secondary sedimentation tanks. *Proceedings of the WERF/IAWQ Secondary Clarifier Assessment Workshop*, Dallas, Texas, United States. Water Environment Research Federation.

Albertson, O. E., & Okey, R. W. (1992). Evaluating scraper designs. *Water Environment Technologies, 4*(1), 52–58.

American Public Health Association, American Water Works Association, & Water Environment Federation. (2023). *Standard Methods for the Examination of Water and Wastewater* (24th ed.).

American Water Works Association & American Society of Civil Engineers. (2012). *Water treatment plant design* (5th ed.). McGraw-Hill.

Anderson, N. E. (1945). Design of final settling tanks for activated sludge. *Sewage Works Journal, 17,* 50.

Barajas, M. G., Escallas, A., & Mujeriego, R. (2002). Fermentation of a low VFA wastewater in an activated primary tank. *Water SA 28*(1), 89–98.

Barnard, J. L., Fitzpatrick, J. D., White, J., & Steichen, M. T. (2007, October 13–17). How important is the EDI in final clarifiers? *Proceedings of the 80th Annual Water Environment Federation Technical Exposition and Conference*, San Diego, California, United States.

Baur, R., Bhattarai, R. P., Benisch, M., & Neethling, J. B. (2002, October 2–6). Primary sludge fermentation—Results from two full-scale pilots at South Austin Regional (TX, USA) and Durham AWWTP (OR, USA). *Proceedings of the 77th Annual Water Environment Federation Technical Exposition and Conference* [CD-ROM], New Orleans, Louisiana, United States.

Bixio, D., van Hauwermeiren, P., & Thoeye, C. (2000, September 18.) Impact of primary treatment technologies on BNR: The case study of the STP of Ghent. *Watermatx Gent.*

Caliskaner, O., Tchobanoglous, G., Reid, T., Kunzman, B., Young, R., & Ramos, N. (2015, September 26–30) Evaluation and demonstration of five different filtration technologies as an advanced primary treatment method for carbon diversion. *Proceedings of the 88th Annual Water Environment Federation Technical Exposition and Conference* [CD-ROM], Chicago, Illinois, United States.

Caliskaner, O., Tchobanoglous, G., Young, R., Downey, M., & Kunzman, B., (2016, September 24–28). Advanced primary treatment via filtration to increase energy savings and plant capacity. *Proceedings of the 89th Annual Water Environment Federation Technical Exposition and Conference* [CD-ROM], New Orleans, Louisiana, United States.

Camp, T. R. (1936). A study of the rational design of settling tanks. *Sewage Works Journal, 8,* 742–758.

Camp, T. R. (1946). Sedimentation and the design of settling tanks. *TransActions of the American Society of Civil Engineers, 3*(2285), 895.

Camp, T. R. (1953). Studies of sedimentation basin design. *Sewage Industrial Wastes, 25,* 1–12.

Chaudhary, M. R., Shao, Y. J., Crosse, J., & Soroushian, F. (1989, October 16–19). Evaluation of chemical addition in the primary plant at Los Angeles' Hyperion Treatment Plant. *Proceedings of the 62nd Water Pollution Control Federation Annual Conference*, San Francisco, California, United States.

Christensson, M. (1997). *Enhanced biological phosphorus removal: Carbon sources, nitrates as electron acceptor, and characterisation of the sludge community* [Doctoral dissertation, Lund University, Lund, Scania, Sweden].

Christensson, M., Lie, E., Jonsson, K., Johansson, P., & Welander, T. (1998). Increasing substrate for polyphosphate-accumulating bacteria in municipal wastewater through hydrolysis and fermentation of sludge in primary clarifiers. *Water Environment Research, 70*(2), 138–145.

Coe, H. S., & Clevenger, G. H. (1916). Methods for determining the capacities of slime settling tanks. *Transactions of the American Institute of Mineral Engineering, 55,* 356.

Daigger, G. T. (1995). Development of refined clarifier operating diagrams using an updated settling characteristics database. *Water Environment Research, 67,* 95.

Dick, R. I., & Ewing, B. B. (1967). Evaluation of activated sludge thickening theories. *Journal of the Sanitary Engineering Division, American Society of Civil Engineers, 93*(SA4), 9.

Dick, R. I., & Young, K. W. (1972). Analysis of thickening performance of final settling tanks. *Proceedings of the 27th Industrial Waste Conference*, Purdue University, West Lafayette, Indiana, United States.

Ding, H. B., Doyle, M., Erdogan, A., Wikramanayake R., & Gallagher, P. (2015). Innovative use of dissolved air flotation with biosorption as primary treatment to approach energy neutrality in WWTPs. *Water Practice & Technology, 10*(1).

Ekama, G. A., Barnard, J. L., Gunthert, F. W., Krebs, P., McCorquodale, J. A., Parker, D. S., & Wahlberg, E. J. (1997). *Secondary settling tanks: Theory, modeling, design, and operation* (Scientific Technical Report No. 6; IWA Publishing). *Environmental Engineering ASCE, 121*(8), 558–564.

Graber, S. D. (1974). Outlet weir loading and settling tanks. *Journal of the Water Pollution Control Federation, 46*, 2355.

Great Lakes–Upper Mississippi River Board of State and Provincial Public Health and Environmental Managers. (2014). *Recommended standards for wastewater facilities*. Health Research, Inc., Health Education Division.

Green, W. D., Sangrey, K., Nowak, J., Foisy, M., & LaVergne, D. (2007, October 13–17). From theory to practice: Preliminary and primary treatment residuals handling improvements. *Proceedings of the 80th Annual Water Environmental Federation Technical Exposition and Conference* [CD-ROM], San Diego, California, United States.

Gunthert, F. W., & Deininger, A. (1995). Final settling tanks—New aspects and design approaches. *Proceedings of the 7th IAWQ Conference on Design and Operation of Large WWTPs*. Institute for Water Quality and Waste Management, Technical University of Vienna.

Jenkins, D., Richard, M. G., & Daigger, G. T. (2003). *Manual on the causes and control of activated sludge bulking and foaming and other solids separation problems* (3rd ed.). IWA Publishing.

Johnson, B.; Phillips, J.; Bauer, T.; Smith, G.; Smith, G.; Sherlock, J. (2014, September 27–October 1) Startup and performance of the world's first large scale primary dissolved air floatation clarifier. *Proceedings of the 87th Annual Water Environment Federation Technical Exposition and Conference* [CD-ROM], New Orleans, Louisiana, United States.

Johnson, P. D., Girinathannair, P., Ohlinger, K. N., Ritchie, S., Teuber, L., & Kirby, J. (2008). Enhanced removal of heavy metals in primary treatment using coagulation and flocculation. *Water Environment Research, 80*(5), 472–479.

Jolis, D., & Ahmad, M. (2001, October 13–17). Pilot testing of high rate clarification technology in San Francisco. *Proceedings of the 74th Annual Water Environment Federation Technical Exposition and Conference* [CD-ROM], Atlanta, Georgia, United States.

Keinath, T. M. (1985). Operational dynamics and control of secondary clarifiers. *Journal of the Water Pollution Control Federation, 57*, 770–776.

Keinath, T. M., Ryckman, M. D., Dana, C. H., & Hofer, D. A. (1977). Activated sludge unified system design and operation. *Journal of the Environmental Engineering Division, American Society of Civil Engineers, 103*, 829.

Kemp, F. D., & MacBride, B. D. (1990). Rational selection of design criteria for Winnipeg South End WPCC primary clarifiers. *Proceedings of the 63rd Water Pollution Control Federation Annual Conference*, Washington, DC, United States.

Kinnear, D. J. (2002, September 28–October 2). Evaluating secondary clarifier collector mechanisms. *Proceedings of the 75th Annual Water Environment Federation Technical Exposition and Conference* [CD-ROM], Chicago, Illinois, United States.

Krauth, K. H. (1993). *Abwassertechnik Vorlesungskript*. University of Stuttgart.

Krebs, P., Vischer, D., & Gujer, W. (1995). Inlet structure design for final clarifiers. *Journal of the Environmental Engineering Division, American Society of Civil Engineers, 121*(8), 558–564.

Larsen, P. (1977). On the hydraulics of rectangular settling basins; Report 1001. Department of Water Resources Engineering, University of Lund: Sweden.

Lee, S., Koopman, B., Park, S., & Cadee, K. (1995). Effect of fermented wastes on denitrification in activated sludge. *Water Environment Research, 67*(7), 1119–1122.

Leng, J., Strekler, A., Bucher, B., Gellner, J., Kennedy, K., & Neethling, J. B. (2002, September 28–October 2). High rate primary treatment—Emerging technologies. *Proceedings of the 75th Annual Water Environment Federation Technical Exposition and Conference* [CD-ROM], Chicago, Illinois, United States.

Ma, J., Reid, T., Caliskaner, O., Binder, D., & Castillo, C. (2015, September 26–30). Cloth depth filtration of primary domestic wastewater. *Proceedings of the 88th Annual Water Environment Federation Technical Exposition and Conference* [CD-ROM], Chicago, Illinois, United States.

Maeng, S. K., Choi, B. G., Lee, K. T., & Song, K. G. (2013). Influences of solid retention time, nitrification and microbial activity on the attenuation of pharmaceuticals and estrogens in membrane bioreactors. *Water Research, 47*(9), 3151–3162.

McCue, T. M., Randall, A. A., & Eremektar, F. G. (2006). Contrasting the benefits of primary clarification versus prefermentation in activated sludge biological nutrient removal systems. *Journal of Environmental Engineering, 132*(9), 1061–1067.

McGhee, T. J., & Steel, E. W. (2007). *Water supply and sewerage* (6th ed.). McGraw-Hill.

Merritt, F. S. (1983). *Standard handbook for civil engineers* (3rd ed.). McGraw-Hill.

Mertsch, V. (1985). *Characteristics of sludge from wastewater flocculation/precipitation. Schriftenreihe des Vereins fuerWasser, Boden und Lufthygiene* (Germany), *62*, 191.

Metcalf & Eddy, Inc./AECOM. (2014). *Wastewater engineering: Treatment, disposal, reuse* (5th ed.). McGraw-Hill.

Millard, S. (2006). *Fermentation of primary sludge for volatile fatty acids production* [Master's thesis, Cranfield University, Cranfield, Bedfordshire, United Kingdom].

Miller, M. A., & Miller, G. Q. (1978, October 1–6). Activated sludge settling in high purity oxygen systems—A full-scale operating data correlation. Paper presented at the 51st Annual Water Pollution Control Federation Conference, Anaheim, California, United States.

Noyola, A., & Tinajero, A. (2005, October 29–November 2). Anaerobic thermophilic digestion of sludge from enhanced primary treatment of municipal wastewater. *Proceedings of the 78th Annual Water Environment Federation Technical Exposition and Conference* [CD-ROM], Washington, DC, United States.

Ødegaard, H. (2005, September). Combining CEPT and biofilm systems. *Proceedings of the International Water Agency Specialised Conference on BNR*, Krakow, Poland. IWA Publishing.

Parker, D. S. (1983). Assessment of secondary clarification design concepts. *Journal of the Water Pollution Control Federation, 55*, 349.

Parker, D., Barnard, J., Daigger, G., Tekippe, R., & Wahlberg, E. (2001). The future of chemically enhanced primary treatment: Evolution not revolution. *Water, 21*(June), 49–56.

Peric, M., Riffat, R., Murthy, S. N., Cassel, A., Neupane, D. R., & Wilson, T. E. (2006, October 21–25). Laboratory clarifier test to predict and optimize full scale chemically enhanced primary treatment (CEPT) process. *Proceedings of the 79th Annual Water Environment Federation Technical Exposition and Conference* [CD-ROM], Dallas, Texas, United States.

Popel, F., & Weidner, J. (1963). Ueber einige Einflusse auf die Klarwirkung bon Absetzbecken. *GWF-Wasser/Abwasser, 104* (28), 796–803.

Sezgin, M., Jenkins, D., & Parker, D. S. (1978). A unified theory of filamentous activated sludge bulking. *Journal of the Water Pollution Control Federation, 50*, 362–381.

Skalsky, D. S., & Daigger, G. (1995). Wastewater solids fermentation for volatile acid production and enhanced biological phosphorus removal. *Water Environment Research, 67*(2), 230–237.

Soap and Detergent Association. (1989). *Principles and practices of nutrient removal from municipal wastewater.*

Steel, W. E. (1979). *Water supply and sewerage.* McGraw-Hill.

Stevenson, R., Nitz, D., Middlebrough, C., & Dyson, J. (2008, October 18–22). Operation of the largest high rate clarification system for CSO control in North America. *Proceedings of the 81st Annual Water Environment Federation Technical Exposition and Conference* [CD-ROM], Chicago, Illinois, United States.

Sutton, P., Rusten, B., Ghanam, A., Dawson, R., & Kelly, H. (2008, October 18–22). Rotating belt screen: An attractive alternative for primary treatment of municipal wastewater. *Proceedings of the 81st Annual Water Environment Federation Technical Exposition and Conference* [CD-ROM], Chicago, Illinois, United States.

Tang, C. C., Prestia, P., Kettle, R., Chu, D., Mansell, B., Kuo, J., Horvath, R. W., & Stahl, J. F. (2004, October 2–6). Start-up of a nitrification/denitrification activated sludge process with high ammonia side-stream: Challenges and solutions. *Proceedings of the 77th Annual Water Environment Federation Technical Exposition and Conference* [CD-ROM], New Orleans, Louisiana, United States.

Ternes, T. A., & Joss, A. (2006). *Human pharmaceuticals, hormones and fragrances: The challenge of micropollutants in urban water management.* IWA Publishing.

Theroux, R. J., & Betz, J. M. (1959). Sedimentation and preaeration experiments at Los Angeles. *Sewage and Industrial Wastes, 31*, 1259.

U.S. Environmental Protection Agency. (1974). *Process design manual for upgrading existing wastewater treatment plants* (Technology Transfer Series).

U.S. Environmental Protection Agency. (1975). *Process design manual for suspended solids Removal* (EPA-625/1-75-003a).

U.S. Environmental Protection Agency. (1979). *Process Design Manual for Sludge Treatment and Disposal,* EPA-625/1-79-011.

U.S. Environmental Protection Agency. (1987). *Design manual for phosphorus removal* (EPA-625/1-87-001).

U.S. Environmental Protection Agency. (1999). *Design criteria for mechanical, electric, and fluid system and component reliability* (EPA-430-99-74-001).

Wall, D. J., & Petersen, G. (1986). Model for winter loss in uncovered clarifiers. *Journal of Environmental Engineering, 12*, 1.

Water Environment Federation. (2005). *Clarifier design* (2nd ed.; Manual of Practice No. FD–8).

Water Environment Federation. (2013). *Operation of nutrient removal facilities* (Manual of Practice No. 37).

Water Environment Federation. (2023). *Standards for design of water resource recovery facilities* (WEF 8).

Water Environment Research Foundation. (2001). *Membrane bioreactors: Feasibility and use in water reclamation.*

Water Environment Research Foundation. (2007). *Fate of pharmaceuticals and personal care products through municipal wastewater treatment processes* (03CTS22UR).

Wells, S. A., & Liberte, D. M. (1998). Winter temperature gradients in circular clarifiers. *Water Environment Research, 70*(7), 1274–1279.

Wilson, T. E. (1996, October 5–9). A new approach to interpreting settling data. *Proceedings of the 69th Annual Water Environment Federation Exposition and Conference,* Dallas, Texas, United States.

Yee, L. Y., & Babb, A. F. (1985). Inlet design for rectangular settling tanks by physical modeling. *Journal of the Water Pollution Control Federation, 57*, 12.

Yoshioka, N., Hotta, Y, Tanaka, S., & Naito, S. (1957). Continuous thickening of homogeneous flocculated slurries. *Chemical Engineering* (Japan), *21*, 66.

Chapter 5 | Biological Principles

LEARNING OBJECTIVES

At the end of this chapter, students should be able to:

- Categorize the main groups of microorganisms and their functions in secondary treatment and nutrient removal processes.
- Compare and contrast carbon and energy sources for heterotrophic and autotrophic bacteria, and understand the terms yield, decay, maximum growth rate, half-saturation coefficient, and substrate.
- Define and explain the purpose of aerobic, anoxic, and anaerobic conditions in biological treatment processes.
- Predict the behavior of heterotrophic bacteria, nitrifying autotrophic bacteria, and phosphate accumulating bacteria under aerobic, anoxic, and anaerobic conditions.
- Explain the steps involved in nitrification, denitrification, and enhanced biological phosphorus removal (EBPR) and identify the microorganisms responsible for each transformation.

1.0 INTRODUCTION

Biological treatment depends on a healthy community of bacteria and other microorganisms to produce a final, treated effluent that is clean and clear with very low levels of biochemical oxygen demand (BOD) and total suspended solids (TSS) remaining. Some biological treatment processes also remove nitrogen and phosphorus by encouraging the growth of specialized bacteria. Several different types of biological secondary treatment processes such as lagoons, wetlands, trickling filters, rotating biological contactors, and activated sludge can be considered for wastewater treatment. This chapter describes some of the different types of microorganisms and environmental conditions present in those processes. The presence of microorganisms within biological treatment processes cannot be observed with the naked eye. Operators can only observe some of the physical effects of microbial activity such as color change, odor, bubbling of gases, floc size, sludge settling characteristics, and foaming. These direct observations of the treatment process are supplemented by measurements of physical and chemical characteristics, including BOD, TSS, pH, alkalinity, dissolved oxygen concentration, ammonia, nitrite, nitrate, phosphate, oxidation–reduction potential, and other parameters. An operator can run a water resource recovery facility (WRRF) effectively based on these observations and physical and chemical measurements alone. However, understanding why changes occur or do not occur in a biological treatment process requires understanding what different groups of microorganisms must grow as well as the conditions that can threaten their survival or decrease their activity. Understanding the microbiology behind the process makes it easier to optimize and troubleshoot secondary treatment processes.

2.0 BIOLOGICAL TREATMENT

Biological treatment was developed in the early part of the 20th century. Biological treatment relies on naturally occurring microorganisms to break down organic matter contained in wastewater into simple substances, mainly carbon dioxide and water. Some microorganisms use the organic matter present in wastewater as food to grow and reproduce. Others rely on ammonia or other materials to grow and reproduce. These microorganisms are retained at much higher concentrations in biological treatment processes than would occur in soil or natural bodies of water. The high concentrations of microorganisms maintained in treatment processes allow the waste to be broken down much faster than would otherwise occur in natural environments. Once treatment is complete, the **biomass** is separated from the treated wastewater, which can then be safely discharged as effluent to receiving waters such as a stream, river, lake, wetland, or sea.

Biomass is the mass of microorganisms in a biological treatment process.

2.1 PHYSICAL AND CHEMICAL REQUIREMENTS FOR BIOLOGICAL TREATMENT

Understanding the relationships between different wastewater components and whether or not they are biodegradable can help predict how wastewater characteristics will change as the wastewater moves through the treatment process. Chapters 3 and 4 discussed some of the mechanical processes (screens, grit basins, and primary clarifiers) that are used to physically remove larger debris and settleable solids. The particles that remain in the wastewater will not settle on their own within a reasonable amount of time. During secondary treatment, microorganisms consume biodegradable organic matter and nutrients to grow and reproduce, cleaning the wastewater in the process. Different groups of bacteria convert ammonia to nitrite and nitrate and then to nitrogen gas. Some other bacteria uptake phosphorus. The microorganisms form **biofilms** or **flocs**, depending on the type of secondary treatment process used. Biofilms and flocs are communities of microorganisms. Biofilms grow attached to surfaces, whereas floc particles grow suspended in the surrounding wastewater. These communities of microorganisms incorporate biodegradable matter and convert it into biomass. Some of the microorganisms in biofilms and flocs produce extracellular polymeric substances (EPS), a sticky substance that helps the bacteria stick together. Biofilms and flocs remove non-biodegradable particles from wastewater when they become trapped in the EPS around the microorganisms. Biofilms and flocs are denser than water and can be separated from the treated wastewater by gravity.

Biofilm is the accumulation of microbial growth on the surface of an object.

Flocs refer to (1) Small, gelatinous masses formed in water by adding a coagulant or in wastewater via biological activity; or (2) Collections of smaller particles agglomerated into larger, more easily settleable particles via chemical, physical, or biological treatment.

Figure 5.1 depicts some of the different wastewater components and typical percentages of various wastewater constituents found in the influent of WRRFs receiving domestic wastewater without significant industrial contributions.

2.1.1 *BIODEGRADABLE WASTE*

Bacteria are responsible for most of the biodegradation that takes place in WRRFs. Domestic wastewater contains organic matter consisting of carbohydrates, proteins, and fats that are biodegradable. Biodegradable material can be broken down into simpler components through the action of living organisms.

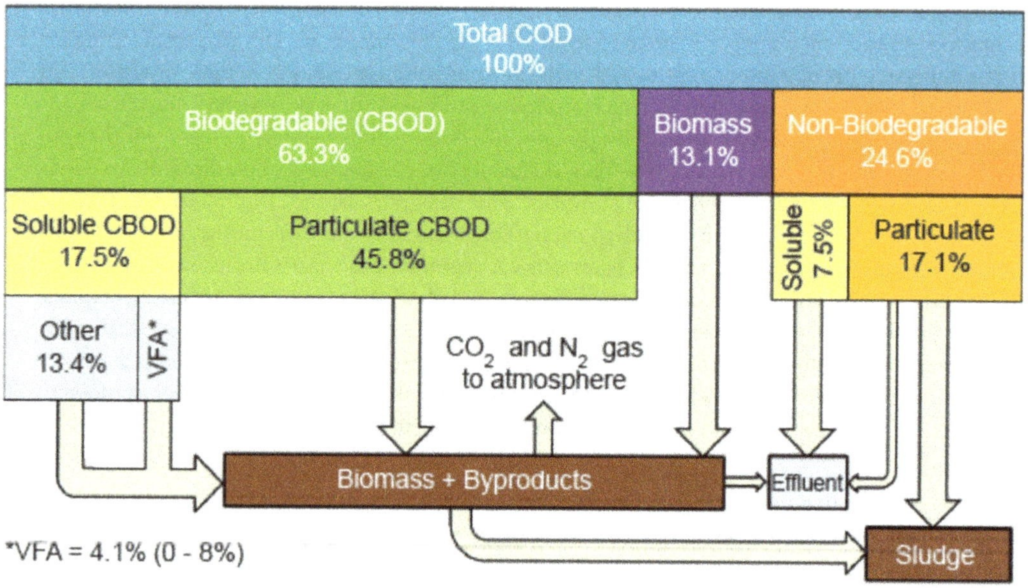

Figure 5.1 Organic Wastewater Fractions (COD: chemical oxygen demand, CBOD: carbonaceous biochemical oxygen demand, and VFA: volatile fatty acids) (Adapted from Pasztor et al., 2009)

2.1.2 NON-BIODEGRADABLE AND INERT WASTE

In addition to biodegradable organic matter, wastewater contains some organic matter and chemicals that degrade very slowly or are not biodegradable at all. Non-degradable organic chemicals are referred to as **recalcitrant**. Non-degradable inorganic matter, such as salt, silt, sand, and grit, are referred to as **inert**. Slowly biodegradable waste may not be completely treated by the microorganisms during normal detention times within biological treatment processes and may not be measured in the standard, 5-day biochemical oxygen demand (BOD_5) test. The slowly biodegradable waste and inert waste that is not physically removed in the treatment process is discharged with the treated wastewater effluent. Physical removal methods include settling and filtration.

Recalcitrant compounds cause inhibition effects that include phenols, short chain organic acids, and chlorinated aromatics.

Inert means having no inherent power of action, motion, or resistance. These materials are neither chemically nor biologically reactive.

2.1.3 BIOCHEMICAL AND CHEMICAL OXYGEN DEMAND

Biochemical oxygen demand measures how much oxygen is needed by the microorganisms in the process to break down or stabilize the waste materials present in the wastewater. In the United States, the BOD test is commonly conducted over 5 days and is referred to as the **BOD_5** test. Biochemical oxygen demand can be further divided into carbonaceous biochemical oxygen demand (CBOD) and nitrogenous oxygen demand (NOD), as shown in Equation 5.1. Carbonaceous biochemical oxygen demand is the amount of oxygen needed to stabilize organic compounds. Nitrogenous oxygen demand is the amount of oxygen needed to convert ammonia to nitrite and nitrate. Nitrogenous oxygen demand is not shown in Figure 5.1. The nitrifying bacteria exert this biochemical oxygen demand as they convert ammonia to nitrite and nitrate. For raw domestic wastewater (influent), the CBOD is typically about 85% of the total BOD (Water Environment Federation [WEF], 2022).

BOD_5 is five-day uninhibited biochemical oxygen demand.

$$BOD_{Total} = CBOD + NOD \quad (5.1)$$

Carbonaceous biochemical oxygen demand consists of both soluble and particulate fractions. Soluble CBOD includes any CBOD that can pass through a piece of filter paper. In practice, some very small particulates will also pass through the filter paper and be measured as soluble. An example of soluble CBOD is sugar dissolved in tea. Soluble BOD includes both soluble CBOD and NOD. Particulate CBOD remains on the filter paper after filtering. For domestic wastewater, including light commercial dischargers such as stores and schools, soluble CBOD is typically between 20% and 40% of the total CBOD (WEF, 2023).

Not all of the organic matter entering the treatment process is readily biodegradable. Some of the organic material is actually composed of live bacteria and other microorganisms (biomass). Moreover, some of the organic material cannot be broken down because the bacteria lack the necessary **enzymes**. Some of the organic material can be broken down; however, because it is so complex, the bacteria need more than the 5 days given in the BOD test to process the organic material. All of these organic components can be chemically **oxidized**. They may, given enough time, consume oxygen if they are discharged to the environment. Other substances in wastewater that can consume oxygen include ferrous iron and sulfides. Although these organic and inorganic substances are not measured in the BOD_5 tests, they can be measured in the chemical oxygen demand (COD) test.

Enzymes are a substance produced by a living organism which acts as a catalyst to bring about a specific biochemical reaction such as digestion and liver function.

Oxidized means being combined chemically with oxygen.

The COD test uses sulfuric acid, potassium dichromate, and high temperatures to break down organic matter. It does not matter if the organic matter is biodegradable or not. Chemical oxygen demand includes CBOD, non-biodegradable organic material, and biomass (Equation 5.2). The chemicals in the COD test do not react with ammonia and will not convert it to nitrite or nitrate, so the NOD is not included in the COD test result. There are a few types of complex organic compounds that cannot be broken down, even in the COD test.

$$COD = CBOD + Biomass + Non\text{-}biodegradable\ organics + Inorganic\ demand \quad (5.2)$$

The COD concentration will always be greater than or equal to the CBOD concentration because of non-biodegradable organic and inorganic matter that can be chemically oxidized. For domestic wastewater without a large industrial component, the ratio of COD to $CBOD_5$ is typically between 1.8 and 2.2, but can vary from one WRRF to another. In other words, the influent COD is typically about twice the influent

$CBOD_5$. The COD-to-CBOD ratio typically remains fairly consistent over time for a WRRF, which allows either parameter to be estimated based on measurements of the other. A facility may collect COD and $CBOD_5$ data for its influent and effluent over time to better refine the calculation for its wastewater. If a new industry begins discharging, the ratio of COD to $CBOD_5$ could change; therefore, it is a good idea to periodically run both COD and $CBOD_5$. The average percentages of biodegradable and non-biodegradable COD in wastewater are shown in Figure 5.1. Wastewater that has a high biodegradable content will have a relatively low ratio of COD to CBOD. A high ratio of COD to CBOD indicates that the wastewater contains relatively high amounts of non-biodegradable and slowly biodegradable wastes.

2.1.4 PARTICULATE ORGANIC MATTER

About 60% of the organic matter in wastewater is in particulate form. Slightly less than half of the particulate organic matter is large enough to settle out of suspension. If the WRRF has an efficiently functioning primary clarifier, the settleable solids and settleable organics will be removed upstream of the secondary treatment process. Particles between 1 and 100 μm remain in **colloidal** suspension. During treatment, colloidal particles become adsorbed onto biomass. Microorganisms can break down adsorbed biodegradable particles into soluble organic matter, which can then be used as a food source. Non-biodegradable particles, which are adsorbed onto biomass, are removed when the biomass is wasted or discharged from the system. In Figure 5.1, both the soluble and particulate biodegradable organic matter are turned into additional biomass. A small amount of biomass will escape into the final effluent; however, most of the biomass and adsorbed, non-biodegradable particulate matter will eventually end up in the sludge.

Colloidal refers to a suspended solid with a diameter less than 1 μ, which cannot be removed by sedimentation alone.

2.1.5 SOLUBLE ORGANIC MATTER

Some of the soluble organic matter in the influent wastewater is readily biodegradable and will be rapidly consumed by microorganisms. Soluble biodegradable organic matter can be measured in two ways: as soluble CBOD or as readily biodegradable COD (rbCOD). To determine rbCOD, the influent and effluent soluble COD are both measured. If soluble COD is present in the final effluent, it is assumed to be non-biodegradable. The difference between the influent soluble COD and effluent soluble COD is the soluble COD that the bacteria were able to consume (Equation 5.3).

$$\text{rbCOD} = \text{Soluble CBOD}_{\text{influent}} - \text{Soluble COD}_{\text{effluent}} \tag{5.3}$$

The influent rbCOD can vary significantly in temperate climates, with ranges from 25 to 125 mg/L. Microorganisms absorb and oxidize the influent rbCOD very rapidly. Readily biodegradable COD includes **volatile fatty acids** (VFAs), which are necessary for biological phosphorus removal. Volatile fatty acids are short chain organic acids such as **acetic acid**, propionic acid, butyric acid, isobutyric acid, valeric acid, and isovaleric acid. From a chemistry perspective, VFAs are organic acids that contain five or fewer carbon atoms. Municipal wastewater influent typically contains 30 to 70 mg/L of VFAs because of **anaerobic** biodegradation of organic matter in the wastewater collection system pipelines. Long, warm, flat collection systems will generate more VFAs than shorter, colder collection systems with steeper slopes. Essentially, the more time the wastewater spends in the collection system, the more VFAs will be generated. Some operators add sodium or calcium nitrate to their collection systems for odor control, which can reduce the concentration of VFAs and soluble CBOD in the WRRF influent.

Volatile fatty acids (VFAs), including acetic-, propionic-, and butyric- acid, are the essential intermediates of anaerobic fermentation.

Acetic acid is a chemical produced by fermentation that is used as a carbon source for biological processes. Chemical formula is CH_3COOH.

Anaerobic refers to the condition characterized by the absence of free oxygen and other electron receptors such as nitrate and sulfate. As a process, anaerobic implies the active presence of strictly anaerobic organisms. As an activated sludge process component, anaerobic implies the active presence of anaerobic and facultative organisms.

2.2 SOLIDS

Solid material in the influent can be classified by size and by whether it is organic or inorganic. Figure 5.2 shows the different types of solids and their relative proportions in the influent of WRRFs receiving domestic wastewater without significant industrial contributions. The different compounds that make up the volatile suspended solids (VSS) and volatile dissolved solids can also be measured as either COD or CBOD. Most of the dissolved, fixed solids are salts that cannot be removed by biological treatment processes. They pass through to the final effluent. Particulate and soluble volatile solids that are biodegradable will become biomass and end up in the sludge. Soluble, volatile solids that are non-biodegradable will also pass through to the final effluent. Non-volatile suspended solids will become part of the sludge.

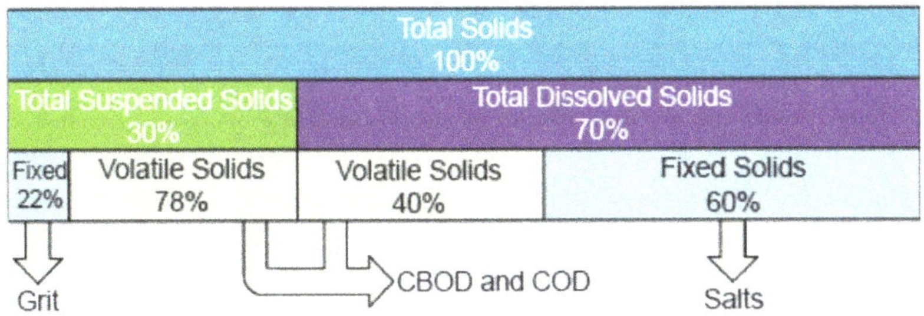

Figure 5.2 Wastewater Solids Fractions (Data from Metcalf and Eddy, Inc./AECOM, 2014)

3.0 MICROBIOLOGY

The biomass in biological treatment processes may contain as many as 300 different types of microorganisms. Microorganisms of greatest significance to wastewater professionals can be classified into four groups: bacteria, protozoa, metazoa, and viruses. Each of these groups plays a key role in the complex world of wastewater biology. Individual microorganisms that are important in particular secondary treatment processes will be discussed in the chapters covering those processes.

3.1 BACTERIA

Bacteria are *prokaryotes*, a diverse group of microorganisms. These small, single-celled organisms come in a variety of sizes and shapes and are the most prolific microorganisms on earth, essentially found everywhere. Each bacterium is a single cell varying in size from about 0.2 to 2 μm, which is too small to be seen with the naked eye. Most bacteria are spherical, but some are rod or spiral shaped. Filamentous bacteria are long chains made up of small bacterial cells, sometimes surrounded by a tubular sheath, and can be hundreds of microns long. Bacteria are responsible for nearly all the biological treatment that occurs in secondary treatment systems. Bacteria have a rigid outer coating that gives them structure and maintains their shape. This is the cell wall. Bacteria also have an inner, flexible membrane called the **periplasmic membrane or cell membrane**. The cell membrane is very important because it controls the intake of food and other nutrients and the discharge of waste products. The cell wall provides the structural support and maintains the shape of the cell. Much of the cellular contents are large protein molecules, known as enzymes, that are manufactured by the cell. Other cellular contents may include granules of polyphosphate, sulfur, or stored organic material.

Periplasmic membrane (or cell membrane) is a lipid bilayer that is semipermeable.

3.1.1 BACTERIA CELL ELEMENTAL COMPOSITION

The organic fraction of chemical elements in microbial cells can be represented as $C_5H_7O_2N$. A more complete chemical formula for the composition of a bacterial cell is $C_{60}H_{87}O_{23}N_{12}P$. As shown in Table 5.1, carbon, hydrogen, oxygen, nitrogen, phosphorus, and sulfur account for 96% of the dry mass of bacteria. Other elements account for 4%. The microbial cell average biomass contains 50% carbon, 12% nitrogen, and 2% phosphorus, by weight. If one or more of the components listed in Table 5.1 is not present in the influent wastewater in high enough concentrations, the bacteria will not be able to grow and reproduce at their maximum rate. Early researchers reported that a ratio of organic carbon (BOD_5) to nitrogen (N) to phosphorus (P) of 100 : 5 : 1 was needed for efficient treatment of wastewater. Later studies showed that these minimum ratios depended on the microorganism, process, type of wastewater being treated, and other environmental factors (WEF, 1994).

3.1.2 CHEMICAL TRANSPORT ACROSS THE CELL MEMBRANE

Bacteria absorb chemicals through their cell membrane. Waste products are eliminated from the cell by movement through the cell membrane to the external environment. Many chemicals cross the cell membrane by diffusion, moving from regions of high concentration outside the cell to regions of low concentration inside the cell. Diffusion is driven by the difference in concentration and does not require an expenditure of energy. Pores, or openings, in the cell membrane are approximately 0.008 mm in size. The pores in the cell membrane are so small that most molecules cannot pass through the membrane unless they are actively pumped across by the cell. Only water molecules; ions such as chloride (Cl^-), potassium (K^+), and sodium (Na^+); alcohols; and some small fat-soluble substances can pass through the pores passively (Gerardi, 2006; Madigan et al., 2003). The cell membrane also contains carrier molecules

Table 5.1 Elemental Composition of Bacterial Cells (Gerardi, 2006; WEF, 1994)

Elements	Percent	Elements	Percent
Carbon	45–55	Potassium	0.8–1.5
Oxygen	16–22	Sodium	0.5–2.0
Nitrogen	12–16	Calcium	0.4–0.7
Hydrogen	7–10	Chlorine	0.4–0.7
Phosphorus	2–5	Magnesium	0.4–0.7
Sulfur	0.8–1.5	Iron	0.1–0.4
		Trace elements	0.2–0.5
Total	96		4.0

that help transport substances across the cell membrane. Carrier molecules are proteins that can bind to specific substances and assist their diffusion through the cell membrane without an expenditure of cellular energy. Chemical transport by carrier molecules is known as **facilitated diffusion**.

Facilitated diffusion is the transport of substances across a biological membrane from an area of higher concentration to an area of lower concentration with the help of a transport molecule. As the transport is driven by concentration gradient, chemical energy is not directly required.

Soluble CBOD in the wastewater is rapidly transported across the cell membrane by the bacteria using active transport. To feed on particulate CBOD, the bacteria must first adsorb the particulate CBOD and then excrete enzymes. The enzymes convert particulate CBOD to soluble CBOD outside the cell. Only then can the organic molecules be transported across the cell membrane. Bacterial cells can absorb and metabolize soluble CBOD quickly, and they can adsorb particulate CBOD quickly. However, it takes time to convert particulate CBOD to soluble CBOD outside the cell. The degree of ionization of substrates, nutrients, and toxic wastes strongly affects chemical transport into bacterial cells. pH is a critical factor for microorganisms. Wastes produced by bacteria may significantly change the pH within a biological treatment process. A change in pH may result in different bacteria becoming dominant within a biological treatment process, which may affect treatment efficiency.

3.1.3 EXTERNAL SLIME LAYER

Many bacterial cells produce EPS, which are secreted through the cell wall onto the exterior surface of the cell. The EPS consist mainly of **polysaccharides**. Extracellular polymeric substances develop a protective coating on the outside of the cell wall (called a **glycocalyx**). The protective coating may be either a distinct gelatinous capsule or a diffuse and irregular slime or gel layer. The external slime layer or gelatinous capsule minimizes the loss of water and nutrients from the cell. It can also protect the cell from toxins. The slime or gel layer allows the bacteria to adhere to solid surfaces and to form colonies, biofilms, and floc particles. This ability to form larger particles, which can settle out of the water by gravity, is essential to many wastewater treatment processes, especially the activated sludge process. All bacteria produce multiple adhesive chemicals, which allow them to switch from attached growth to suspended growth forms under different environmental influences.

Polysaccharides are long-chain polymeric carbohydrates composed of monosaccharide units bound together by chemical linkages.

Glycocalyx is a polysaccharide or carbohydrate-rich lining that envelopes the outer layer of a cell.

3.1.4 CELLULAR STORAGE PRODUCTS

Bacterial cells store excess food and energy internally in the form of organic and inorganic molecules. The primary organic storage molecules include glycogen, starch, and poly-β-hydroxyalkanoate. Inorganic storage molecules include long chains of polyphosphate, known as **volutin granules**, and insoluble elemental sulfur that is stored as sulfur granules. Bacteria use these storage molecules to provide energy when external sources of food are limited or absent.

Volutin granules contain polymerized phosphate and represent a storage form for inorganic phosphate and energy.

3.1.5 MICROBIAL GROWTH AND REPRODUCTION

When a bacterial cell grows to full size, it reproduces by dividing into two daughter cells through **binary fission**. Rapid growth and cell division of the daughter cells result in exponential growth in the number of bacteria. Two daughter cells divide into four, four cells divide into eight, eight cells divide into 16, and so on. Under ideal conditions, some bacteria can double their numbers in as little as 20 minutes.

Binary fission is a form of asexual reproduction in some microbes in which the parent organism splits into two independent organisms.

Microorganisms generally exhibit a pattern of sequential growth phases, which have different growth rates (Figure 5.3). When a new food source becomes available, microorganisms often have a lag phase, during

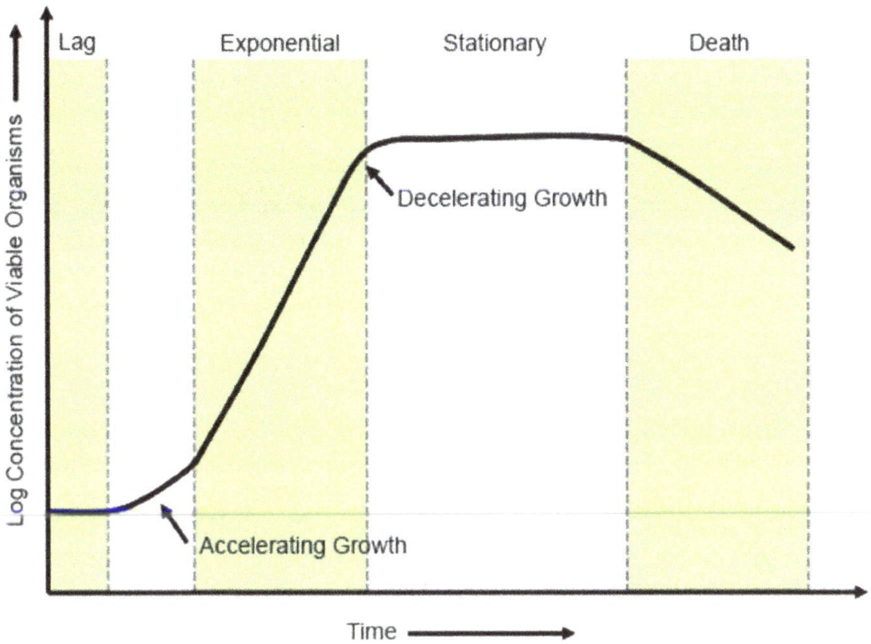

Figure 5.3 Bacterial Growth Curve

which no growth is observed initially. During the lag phase, microorganisms begin producing specific enzymes required to break down and oxidize the new food source. The lag phase is followed by a period of rapidly increasing growth called the acceleration phase. The initial lag and acceleration phases reflect the time required for the microorganisms to produce or activate the specific enzymes required to break down the new food source. Formation of a specific enzyme under the influence of its substrate is known as enzymatic adaptation. When the enzyme system for the specific food source is fully activated, microbial growth reaches the maximum growth rate, which remains constant until the food concentration decreases.

The lag phase is an adaptation period, during which the bacteria are adjusting to their new conditions.

Acceleration phase is a period of increasing growth before the log growth phase in a bacterial culture. The first division to occur during lag marks the end of the lag phase and the beginning of the acceleration phase, during which all cells capable of dividing will divide and all have acclimated to divide at the maximum rate afforded by the growth conditions.

Enzymatic adaptation is a change in the enzymatic constitution of an organism that is (a) induced by the presence of a specific substance, usually a substrate, in the environment and (b) unaccompanied by a change of genotype.

The acceleration phase is followed by an exponential growth phase, during which the microorganisms reproduce at their maximum rate. In this stage of development, resources are plentiful. The reproduction of microorganisms is by binary fission (i.e., each cell divides producing two new cells), and the increase in population follows in geometric progression: $1 \rightarrow 2 \rightarrow 4 \rightarrow 8 \rightarrow 16 \rightarrow 32$, and so forth. The population of bacteria (P) after the nth generation is given by the following expression:

$$P = P_0(2)^n$$

where P_0 is the initial population at the end of the accelerated growth phase. Taking the log of both sides of above equation:

$$\text{Log } P = \log P_0 + n \log 2$$

If the bacterial population is plotted on a logarithmic scale, this phase of growth plots as a straight line of slope n and intercept P_0 at t_0. This phase of growth is called the log growth or exponential growth phase.

Log growth (or exponential growth) phase refers to the state during which cell division proceeds at a constant rate. The exponential or log phase of growth is marked by predictable doublings of the population, in which 1 cell become 2 cells, becomes 4, becomes 8, etc.

As the number of bacteria increases, the amount of resources available to each bacterium gradually decreases and fewer bacteria will have enough resources to reproduce. Resources and energy are first used to maintain existing cell mass.

Bacteria can only reproduce if excess resources and energy are available. This period of gradually declining growth rate is called the deceleration phase. Eventually, the number of bacteria in the system will level out to match the available food and energy coming into the system. This is the stationary phase. The stationary phase is shown as a flat line in Figure 5.3. In the stationary phase, the number of new bacteria formed is

Deceleration phase refers to the state where the number of viable cells decreases in a predictable (or exponential) fashion. The steepness of the slope corresponds to how fast cells are losing viability.

roughly equivalent to the number of bacteria that die. The population stabilizes. The bacteria in secondary treatment processes will be in the stationary phase most of the time. If the amount of resources coming into the process decreases, there will be more live microorganisms than the available resources can support and some of them will begin to die. When cells die, they lyse (break apart), releasing contents that are then available as food for other organisms.

When a WRRF receives waste from a new source or receives a prohibited waste, the biomass in the treatment process may lack the type of bacteria or the specific enzymes required to break down the waste. Discharge permit violations may result as the new waste passes through the facility into the effluent. If the new waste is biodegradable, the biomass in the treatment process may develop the capability to break it down over a period of several days or weeks.

An indicator organism is a microbe whose presence indicates the absence or presence of a specific pollutant.

3.2 PROTOZOA

An operator can observe protozoa and metazoa under the microscope. These **indicator organisms** can also give insight to the health of the treatment system. Protozoa are unicellular eukaryotes that include amoebae, flagellates, and ciliates, which are shown in Figure 5.4. In WRRFs, protozoa account for over 90% of the total nonbacterial biomass. Protozoa, like bacteria, are typically found in natural systems like streams and lakes and are quite mobile. Protozoans common in secondary treatment processes include flagellates, amoebae, and ciliates. Ciliates can be crawling, free-swimming, or stalked protozoa. They are typically the dominant protozoa in wastewater systems in terms of number of species. Ciliates account for 10% of the biomass in suspended growth systems (104 ciliates per milliliter) and 20% of the total biomass in biofilms (Pauli et al., 2001). Protozoa are predators to bacteria and algae, which are their food sources. The large numbers of bacteria consumed by protozoa as food reduces the volume of waste biomass generated in WRRFs by approximately 20%. Absorption of colloidal particles by protozoa improves the settleability of suspended particles. In both activated sludge basins and clarifiers, the ciliate population has sufficient time to filter the entire wastewater volume several times, which helps clarify the water by removing bacteria and colloidal particles. Research has demonstrated that protozoa play an important role in suspended bacteria count in the effluent. Some protozoans are associated with waterborne disease. *Giardia* and *Cryptosporidium* are two protozoans responsible for significant outbreaks of illness in the United States.

3.3 BACTERIAL METABOLISM, NUTRITION, AND RESPIRATION

Nutrition is the biochemical and physiological process by which an organism uses food to support its life.

Microbial metabolism of interest in water and wastewater treatment is divided into two major categories, catabolism and anabolism. Catabolism is the breakdown or degradation of materials with the concomitant release of energy. Anabolism is the synthesis of new cellular materials using the energy released in catabolism and the breakdown products of catabolism. Metabolism also can be described by the type of energy source used (e.g., organic compounds, inorganic ions, and light). **Nutrition**, as defined here, is related to the chemical requirements associated with microbial metabolism. Most of the bacteria found in soils and natural waters and those used in wastewater treatment are not very fastidious with respect to nutrition. Inorganic nutrient sources, such as ammonia and nitrate for nitrogen or phosphate for phosphorus, are satisfactory, and a relatively wide variety of organic materials can be used as energy sources. A few microorganisms found in natural systems, such as the nitrifiers and the methanogenic bacteria, have tightly constrained nutrient requirements. Microbial respiration is defined by the type of terminal electron acceptor used in the energy-yielding reaction processes and can be classified generally as

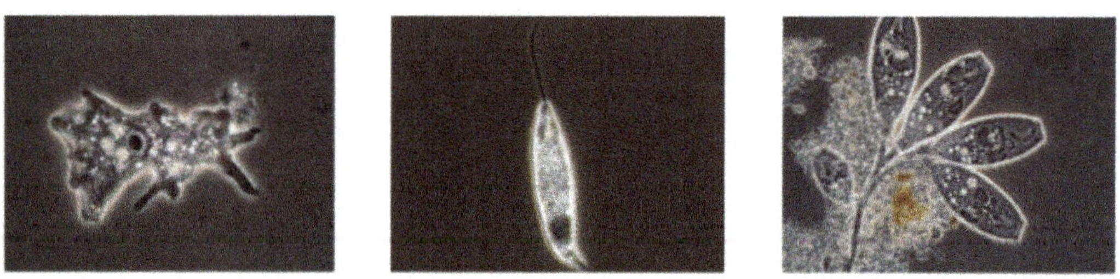

a) Amoeba b) Flagellate c) Ciliate

Figure 5.4 Protozoans in Wastewater (Reprin ted with permission by Steven C. Leach) aerobic, anoxic, or anaerobic and by the electron acceptor used (aerobic, anoxic, or anaerobic). A summary of bacterial metabolism and respiration is given in Table 5.2 (Lengeler et al., 1999; Madigan et al., 2003).

4.0 BACTERIA IN SECONDARY TREATMENT PROCESSES

The bacteria in secondary treatment processes can be broken into a few significant groups depending on their energy source, carbon source, oxygen requirements, and their preferred growth patterns. Bacteria oxidize organic compounds to carbon dioxide (CO_2) and water (H_2O). The bacteria require large quantities of carbon, oxygen, nitrogen, and phosphorus as well as a host of other materials like iron, manganese, and potassium in smaller quantities. Carbon, nitrogen, and phosphorus are called **macronutrients** because they are needed in large quantities. Iron, manganese, potassium, and other vital elements are called **micronutrients**. From Table 5.1, it is clear that bacterial cells contain about 50% carbon, 14% nitrogen, and 3% phosphorus by dry weight. The overall biochemical reaction that occurs for all of the bacteria in biological treatment processes may be expressed as follows:

Macronutrients are essential nutrients required in large quantities for energy generation. These include carbohydrates, fats, and protein.

$$\text{Carbon sources} + \text{Nutrients} + \text{Bacteria} + \text{Oxygen Source} \rightarrow \text{New Bacteria} + \text{Products} \quad (5.4)$$

Micronutrients are chemical elements required in trace amounts for the normal growth and development of living organisms.

Typical medium-strength domestic wastewater contains about 200 mg/L of $CBOD_5$, 40 mg/L of total nitrogen, and 7 mg/L of total phosphorus. This example wastewater has a C : N : P ratio of 100 : 20 : 3.5—an excess of nutrients (nitrogen and phosphorus). Wastewater produced by certain industries may be deficient in nitrogen and phosphorus. When the concentration of these nutrients is too low, nitrogen or phosphorus may need to be added to provide the correct ratio for microbial growth and achieve optimal biological treatment. Without nutrients, the bacteria are unable to reproduce.

4.1 ENERGY AND CARBON SOURCES

All living organisms can be grouped into broad categories according to how they obtain carbon and energy from the environment. Organisms that use carbon dioxide as their carbon source for growth are called **autotrophs**. Plants, algae, and many different kinds of bacteria are autotrophs. All other living organisms are heterotrophs, which derive the carbon they need for growth from organic compounds. Living organisms can further be divided into categories based on the way they obtain energy. All plants and algae and a few bacterial species obtain their energy from sunlight through photosynthesis

Autotrophs are organisms that derive their cell carbon from carbon dioxide.

Table 5.2 Classification of Living Organisms by Carbon and Energy Sources

Classification	Carbon Source	Energy Source
General		
Autotroph	CO_2	—
Heterotroph	Organic compounds	—
Chemotroph	—	Chemical compounds
Chemolithotroph	—	Inorganic compounds
Chemoorganotroph	—	Organic compounds
Phototroph	—	Light
Chemoautotroph	CO_2	Chemical compounds
Photoautotroph	CO_2	Light
Chemoheterotroph	Organic compounds	Chemical compounds
Photoheterotroph	Organic compounds	Light
Methylotroph	1-carbon compounds	1-carbon compounds
Other terms		
Eutroph	Use high concentrations for carbon and energy	
Oligotroph	Use low concentrations for carbon and energy	
Zymogenous	Grow rapidly when carbon and energy source added	
Saprophyte	Live off of dead organic matter	

and are called **phototrophs**. All other organisms obtain their energy from the oxidation of chemical substances and are called chemotrophs. If those chemical substances are organic, the organisms are **chemoorganotrophs**. If those chemicals are inorganic, the organisms are called **chemolithotrophs**.

Phototrophs are organisms that use light as their source of energy for certain metabolic functions such as ATP production.

Chemoorganotrophs are organisms that obtain energy from the oxidation of reduced organic compounds.

Chemolithotrophs are organisms that generate energy by the oxidation of inorganic molecules for biosynthesis or energy conservation via aerobic or anaerobic respiration.

Heterotrophs are organisms that derives their cell carbon from organic compounds.

4.1.1 HETEROTROPHIC BACTERIA

The bacteria feeding mainly on organic (carbonaceous) matter, are called **heterotrophs**. The word *heterotroph* means "other-nourishing" and derives from the Greek words *hetero* (other) and *trophe* (from trephine, meaning to nourish). *Heterotroph* literally translates to an organism that gets its nourishment from others. The bacteria do not literally consume one another whole, but they do get their energy and carbon from organic compounds that were produced by other microorganisms and not themselves. All animals and most bacteria are heterotrophs. This large group of different types of bacteria obtains its carbon and its energy from organic molecules such as carbohydrates, proteins, and fats in the wastewater. These compounds are measured collectively as CBOD. Heterotrophs require dissolved oxygen or an oxidized inorganic chemical such as nitrate (NO_3^-), nitrite (NO_2^-), or sulfate (SO_4^{-2}) to process CBOD. Heterotrophic bacteria are typically the dominant microorganisms in most environments, including in the biomass in wastewater treatment processes.

4.1.2 AUTOTROPHIC BACTERIA

Autotrophs are microorganisms that obtain their carbon from inorganic chemicals such as carbon dioxide (CO_2), carbonate (CO_3^{-2}), and bicarbonate (HCO_3^-). Most plants, algae, and many different kinds of bacteria are autotrophs. *Auto* means self and *trophe* means nourishing; an autotroph is literally self-nourishing. Autotrophs cannot use organic compounds (CBOD) as an energy or carbon source. Instead, they obtain their energy from inorganic sources. Autotrophs build or synthesize all of their own organic molecules. Although there are many different types of autotrophic bacteria, some of the most important ones in wastewater treatment are the nitrifying bacteria and the sulfate-reducing bacteria.

4.1.3 OXYGEN REQUIREMENTS

There are three different environments that can be present in wastewater treatment processes: aerobic (oxic), anoxic, and anaerobic. Aerobic environments contain dissolved oxygen and may or may not contain nitrite or nitrate. Anoxic environments contain nitrite and/or nitrate, but do not contain dissolved oxygen. Anaerobic environments do not contain dissolved oxyten, nitrite, or nitrate. The presence or absence of oxygen, nitrite, and nitrate influences which bacteria are active and how they process their energy source.

4.1.3.1 Obligate Aerobes

Obligate (or strict) aerobes are organisms that cannot live without oxygen in their environment for their growth and survival.

Respiration is the intake of oxygen and discharge of carbon dioxide as a result of biological oxidation.

Microorganisms that grow only in the presence of oxygen are called **obligate, or strict, aerobes**. Obligate aerobes cannot survive without oxygen. Some heterotrophs require an oxygen source for **respiration** and growth. For heterotrophs using dissolved oxygen, the overall biochemical reaction that occurs may be expressed as follows:

$$\text{Organic matter} + \text{Nutrients} + \text{Bacteria} + O_2 \rightarrow \text{New bacteria} + CO_2 + H_2O \tag{5.5}$$

Most of the nitrifying bacteria are obligate aerobes and must have dissolved oxygen. They function well in the aerobic or oxic environments, but cannot function in either anoxic or anaerobic environments. Most of the nitrifying autotrophs will become dormant for the time they are in either anoxic or anaerobic conditions.

4.1.3.2 Obligate Anaerobes

Microorganisms that grow only in the absence of oxygen are obligate, or strict, anaerobes. Examples of obligate anaerobes include *Actinomyces, Bacteroides*, and *Clostridium*. These microorganisms cannot use oxygen or nitrate for respiration. Instead, they partially break down organic matter to get some energy and then use the breakdown products for respiration. This type of anaerobic respiration is called fermentation. The byproducts of **fermentation** are organic acids and alcohols.

Fermentation is the conversion of organic matter into carbon dioxide, methane, and other low-molecular-weight compounds.

4.1.3.3 Facultative Anaerobes

Some heterotrophic bacteria are capable of surviving under different environmental conditions because they are adaptable. Facultative heterotrophic bacteria prefer to use dissolved but can switch to a chemically bound form of oxygen like nitrite (NO_2^-) or nitrate (NO_3^-) when oxygen is not available. When facultative heterotrophic bacteria use nitrite or nitrate as an oxygen source instead of dissolved oxygen, the byproducts are nitrogen gas (N_2), carbon dioxide (CO_2), and water (Equation 5.7). This process is called **denitrification**. Most species of denitrifying bacteria belong to three genera including *Alcaligenes*, *Bacillus*, and *Pseudomonas*. Using nitrate instead of oxygen is not as efficient for the bacteria. For facultative heterotrophs using nitrate as an oxygen source, the overall biochemical reaction that occurs may be expressed as follows:

Denitrification is a biological process in which nitrates are converted to nitrogen.

$$\text{Organic matter} + \text{Nutrients} + \text{Bacteria} + NO_X \rightarrow \text{New organisms} + N_2 + CO_2 + H_2O \qquad (5.6)$$

4.2 SPECIALIZED GROUPS OF BACTERIA

4.2.1 NITRIFYING BACTERIA

Nitrifying bacteria use ammonia (NH_3) or nitrite (NO_2^-) as their energy source and oxygen as an electron acceptor. The outputs from nitrifying autotrophic bacteria will be nitrite (NO_2^-), nitrate (NO_3^-), carbon dioxide (CO_2), and water (H_2O). Carbon is a necessary building block of all life on earth. The autotrophs use the inorganic carbon compounds, carbonate (CO_3^{-2}) and bicarbonate (HCO_3^-), as their carbon source instead of CBOD. Carbonate and bicarbonate are two components of alkalinity. For every 1 mg/L of ammonia that is converted to nitrate, 7.14 mg/L of alkalinity will be consumed. For the nitrifying bacteria, the overall biochemical reaction that occurs may be expressed as follows (not balanced):

$$NH_3 + O_2 + HCO_3^- + \text{Nitrifying bacteria} \rightarrow \text{New bacteria} + NO_3^- + CO_2 + H_2O \qquad (5.7)$$

The conversion of ammonia to nitrate is achieved by two different groups of nitrifying bacteria: the ammonia-oxidizing bacteria (AOB) and the nitrite-oxidizing bacteria (NOB). The ammonia oxidizers convert ammonia (NH_3) to nitrite (NO_2^-). The nitrite oxidizers then convert the nitrite to nitrate (NO_3^-). Both use inorganic carbonate and bicarbonate as a source of carbon to build the organic molecules needed for growth. Ammonia-oxidizing bacteria include the following: *Nitrosomonas, Nitrosococcus, Nitrosospira, Nitrosovibrio*, and *Nitrosolobus* (Koops & Pommerening-Röser, 2001). *Nitrosomonas* appears to be the most important genus in activated sludge and trickling filter systems receiving relatively high specific loadings of ammonia-N ($\geq$20 mg/L). *Nitrosospira* is the most important genus in biofilm systems receiving very low ammonia-N ($\leq$1 mg/L) concentrations (Grady & Filipe, 2000). Nitrite-oxidizing bacteria include the following: *Nitrobacter, Nitrospira, Nitrococcus*, and *Nitrospina*. Members of the genus *Nitrospira* appear to be the most important NOB in wastewater treatment systems, although *Nitrobacter* are also present. *Nitrococcus* and *Nitrospina* are marine species that do not live in wastewater treatment systems.

4.2.2 COMPARING HETEROTROPHS AND AUTOTROPHS

The heterotrophs use energy-rich organic compounds as their energy source, whereas the nitrifying autotrophs use relatively energy poor ammonia as their energy source. Heterotrophs grow much faster than the autotrophs. Under ideal conditions, a heterotrophic bacterium may only need 20 minutes to reproduce, whereas a nitrifying bacterium may need between 7 and 13 hours. For this reason, nitrifying bacteria will be present in the biomass in significant numbers only when the biomass is retained in a WRRF for several days. Fast-growing heterotrophs tend to outcompete autotrophs for nutrients and, therefore, are the dominant microorganisms in most environments. A comparison of bacterial metabolism, respiration, and the inputs and outputs of the processes are presented in Table 5.3 (Lengeler et al., 1999; Madigan et al., 2003).

4.2.3 PHOSPHATE-ACCUMULATING ORGANISMS

Phosphate-accumulating organisms (PAOs) are a specialized subgroup of facultative heterotrophic bacteria. When these bacteria are cycled between anaerobic and either anoxic or aerated conditions, they will pick up and store excess phosphorus. This process is called **luxury uptake** because the bacteria are incorporating more phosphorus than they need for growth. The PAOs are able to store up to 40% of their dry weight as phosphorus. Phosphorus is removed from the wastewater when excess biomass is removed from the system.

Luxury uptake is the uptake of phosphorus (as an example of a nutrient) by an organism from its phosphorus-enriched growth medium, where prior starvation by the species is not required.

Under anaerobic conditions, some of the CBOD is broken down through fermentation by anaerobic bacteria into soluble CBOD and simpler organic molecules, the VFAs. Volatile fatty acids are a preferred source of carbon and energy by heterotrophic bacteria, including the PAOs, because these compounds

Table 5.3 A Comparison of Bacterial Metabolism and Respiration

Type of Metabolism	Electron Donor	Carbon Source	Electron Acceptor	Principal Non-Cell Product	Microbial Group
Chemoheterotrophic	Organic compounds	Organic compounds			
			O_2	CO_2, H_2O	Aerobic heterotrophs
			$SO^{-2}{}_4$	CO_2, N_2, N_2O	Denitrifiers
			Fe^{3+}	CO_2, Fe^{+2}	Iron reducers
			$ClO^-{}_4$	CO_2, Cl^-	Perchlorate reducers
			CO_2	CO_2, CH_4	Methanogens
			Organic metabolites	Volatile acids, alcohols	Fermenters
Chemoautotrophic	Inorganic compounds	CO_2			
	NH_3		O_2	CO_2, No_2	Ammonia oxidizers
	NH_3		$NO^-{}_2$	N_2, H_2O, $NO^-{}_3$	Anaerobic ammonium
	$NO^-{}_2$		O_2	CO_2, $NO^-{}_3$	Nitrite oxidizers
	H_2		$NO^-{}_3$, $NO^-{}_2$	CO_2, N_2, N_2O	Denitrifiers
	H_2S		O_2	CO_2, $SO^{-2}{}_3$	Sulfide reducers
	H_2		CO_2	CH_4	Methanogens
	H_2		$SO^{-2}{}_4$	H_2S, H_2O	Sulfate reducers
Photoheterotrophic	Light (≈870 nm), H_2S	Organic compounds	CO_2		Phototrophs
Photoautotrophic				$SO^{-2}{}_3$, H_2O	
	Light (40 to 650 nm), H_2O	CO_2	O_2	O_2, H_2O	Phototrophs
	Light (≈870 nm), H_2S	CO_2	CO_2	$SO^{-2}{}_3$, H_2O	Phototrophs

Adenosine triphosphate (ATP) is an energy-carrying molecule found in the cells of all living things. It is a compound consisting of an adenosine molecule bonded to three phosphate groups. The breakage of phosphate linkage will release energy for physiological processes in living cells.

Polyphosphate is a phosphate compound used as sequestration agent to prevent the formation of iron, manganese, and calcium carbonate deposits.

are easily absorbed into the bacteria. The PAOs have a logistical problem: when they are under anaerobic conditions, they are exposed to VFAs, but without oxygen, nitrite, or nitrate present, they cannot access them. They also lack energy to capture VFAs. To address this problem, PAOs similar to other cells use a compound called **adenosine triphosphate** (ATP) to store energy. Adenosine triphosphate consists of three phosphate molecules chemically bound together. The chemical bond between phosphorus atoms stores energy until the cell needs it. The ATP is continuously formed and consumed within cells as energy is stored and recovered. The PAOs take ATP to the next level and form an energy-rich compound called **polyphosphate**, which strings together large numbers of phosphate molecules. Polyphosphate is a chemical battery. When the PAO needs energy, it breaks the chemical bonds between phosphorus molecules to recover the stored energy and temporarily releases phosphorus. This strategy allows the PAOs to pick up and store VFAs when they are under anaerobic conditions. Because they have a "battery pack," they are able to outcompete other types of bacteria for the available VFAs. When the PAOs are under anaerobic conditions, they use energy stored in polyphosphate to pick up VFAs. The VFAs are then combined into an internal storage product called poly-β-hydroxyalkanoate (PHA). The lack of oxygen, nitrite, or nitrate prevents PAOs from consuming the VFAs immediately. When the PAOs move into either anoxic or aerobic conditions, they immediately begin to consume the PHA and rebuild their polyphosphate battery. Cycling the PAOs between anaerobic and anoxic or aerated conditions is required for luxury uptake to occur. Without exposure to an anaerobic zone, there is no incentive for the PAOs to store phosphorus. Phosphate-accumulating organisms are discussed in more detail in the chapter on biological nutrient removal.

4.2.3.1 Sulfate-Reducing Bacteria

Sulfate-reducing bacteria (SRB) are a large group of microorganisms that live in many environments. There are 220 known species within 60 different genera. They can tolerate extreme pH values ranging from as low as 2 to as high as 10. Sulfate-reducing bacteria obtain energy by oxidizing organic compounds under anaerobic conditions using sulfate (SO_4^{-2}) as an oxygen source. Sulfate-reducing bacteria produce carbon dioxide (CO_2) and hydrogen sulfide (H_2S) as waste products. Hydrogen sulfide smells like rotten eggs and is extremely toxic. In addition to sulfate, SRBs can reduce other sulfur compounds including thiosulfate ($S_2O_3^{-2}$), sulfite (SO_2^{-3}), and elemental sulfur (S) to sulfide (S^{-2}). Most SRB are anaerobes, but some can switch to aerobic respiration when oxygen is available. *Desulfovibrio desulficans* is one SRB found in WRRFs that produces hydrogen sulfide.

4.2.3.2 Sulfur-Oxidizing Bacteria

The sulfur-oxidizing bacteria (SOB) are a diverse group of microorganisms that are capable of oxidizing sulfur as an energy source. Reduced sulfur compounds that are used by SOBs include hydrogen sulfide (H_2S), elemental sulfur (So), and thiosulfate ($S_2O_3^{-2}$). The final product of sulfur oxidation is sulfate ($S_2O_4^{-2}$) or sulfuric acid (H_2SO_4), which can corrode concrete and steel. Sulfur-oxidizing bacteria include *Thiobacillus* and *Beggiatoa*. Some sulfur-oxidizing bacteria can also oxidize ferrous iron (e.g., *Thiobacillus ferrooxidans* and *Sulfolobus* species). *Thiothrix*, common in wastewater treatment facilities, and *Sulfolobus*, confined to sulfur-rich hot springs, transform hydrogen sulfide to elemental sulfur. Sulfur-oxidizing and sulfate-reducing bacteria both live in wastewater collection systems and WRRFs. Both H_2S and H_2SO_4 can be formed within pipelines and treatment processes under alternating aerobic and anaerobic conditions.

4.3 MICROBIAL GROWTH RATES

Biological secondary treatment process design and operation are based on the growth kinetics of microorganisms. Kinetics is how fast or slow chemical and biological reactions take place. Kinetics determine how much time is required to treat wastewater.

4.3.1 MONOD GROWTH CURVE

The growth of bacteria is at its maximum rate only when the water surrounding the bacteria has an optimum pH and contains sufficient amounts of food, oxygen sources, nutrients, and trace elements. Microbial growth rates are often lower than the maximum growth rates because one or more of the substances they need to grow and reproduce is not available in sufficient quantities. When a substance is not present in sufficient quantities to support growth, it is said to be limiting. The most common limiting factors include (a) exhaustion of nutrients, (b) accumulation of toxic byproducts, and (c) changes in ion concentrations, especially pH (Monod, 1949).

Monod kinetics relates microbial growth rates to the concentration of a limiting substrate in an aqueous environment. Microbial growth rates are represented by an equation proposed by Jacques Monod (1942, 1949). The Monod equation states that when substrate concentrations are low, the growth rate of the microorganisms will be proportional to the concentration of the limited substrate. In other words, the growth and reproduction of the microorganism will be restricted by the availability of resources. The Monod equation also states that when resources are plentiful, the growth rate of the microorganisms will be maximized. However, the microorganism cannot grow faster than its maximum rate when resources are available in excess.

The Monod equation generates a characteristic growth curve, shown on the graph in Figure 5.5. The microbial growth rate, shown as μ, is determined by the concentration of the limiting substrate until the maximum growth rate (μmax) is reached. The growth rate has units of grams of VSS produced per gram of VSS added to the process per day (g VSS/g VSS·d). It is typically shown in mathematical equations simply as "per day" (or d^{-1}).

The Monod equation is written as follows:

$$\mu = \mu_{max} \times \left[\frac{S}{K_s + S} \right] \tag{5.8}$$

Where μ = specific growth rate of the microorganisms (pronounced mew), per day;
μ_{max} = maximum possible growth rate of the microorganisms, per day;
S = concentration of substrate, mg/L; and
K_s = half-saturation coefficient, mg/L.

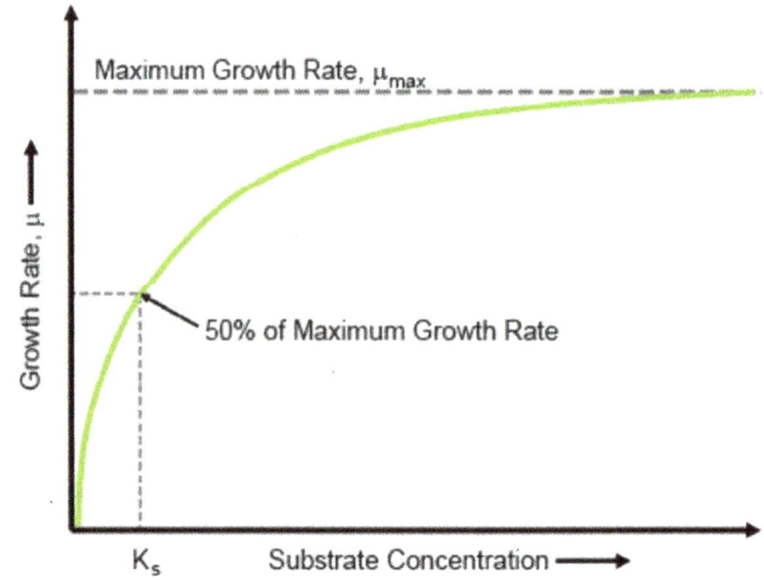

Figure 5.5 Graph of the Monod Equation

4.3.2 MAXIMUM SPECIFIC GROWTH RATE

The maximum specific growth rate is an important factor that, in part, determines biological treatment capacity. Microorganisms respond to higher concentrations of substrate (food) by increasing their growth rate up to the maximum growth rate (μ_{max}). When the substrate is scarce (low food concentration), microorganisms cannot grow as fast as they could if more food were available. In contrast, an abundance of substrate allows each microorganism to consume more food and increase its growth rate. Increasing the amount of substrate reduces competition for food and allows the microorganisms to increase their growth rate proportional to the available food source. In Figure 5.5, the growth rate curve increases rapidly from the origin in response to the increasing substrate concentration. It gradually flattens out and approaches the horizontal line, which indicates the maximum growth rate (μ_{max}). When the microorganisms are saturated by substrate and attain the maximum growth rate, the growth rate remains constant as substrate concentrations increase further. When surplus substrate is available, growth becomes limited by the number of microorganisms in the system instead of by the amount of substrate available.

The maximum growth rate (μ_{max}) and the half-saturation coefficient (K_s) used in the Monod equation are empirical coefficients, obtained by experimentation. Each species of bacteria has a different maximum growth rate (μ_{max}) and half-saturation coefficient (K_s) for every limiting nutrient and for the surrounding environmental conditions. The maximum growth rate for the biomass in a secondary treatment process is simply the average of the maximum growth rates for each of the different types of bacteria present in the biomass. The average maximum growth rate in a secondary treatment process can be determined for the biomass in the process using the same methods as those for a single type of bacteria and a single limiting substrate. In secondary treatment processes, μ_{max} is between 3 and 13.2 g VSS/g VSS·d for heterotrophic bacteria, while a μ_{max} between 0.20 and 0.90 g VSS/g VSS·d is considered for the nitrifying autotrophic bacteria (Metcalf & Eddy, Inc./AECOM, 2014). From this it is clear that the heterotrophic bacteria can reproduce much more rapidly than the nitrifying autotrophic bacteria.

4.3.3 SATURATION AND HALF-SATURATION COEFFICIENTS

The saturation coefficient is the concentration of a resource that a microorganism needs to grow at its maximum rate. Higher concentrations of the resource will not continue to increase growth rates. When resources are available at the saturation coefficient or higher, the microorganism will grow at its maximum rate. Below the saturation coefficient concentration, the assembly line will be limited and production of new microorganisms will slow down. The saturation coefficient is typically expressed as milligrams per liter (mg/L). The half-saturation coefficient, or half-velocity coefficient, shown as *Ks*, is the concentration of a resource needed to achieve exactly half of the maximum growth rate. The half-saturation coefficient is typically calculated from measurements of the maximum growth rate and is rarely measured directly. It is typically expressed in milligrams per liter (mg/L). The values of the maximum growth rate (μ_{max}) and half-saturation constant (K_s) are highly dependent on the microorganisms and substrates in the biological system. Readily biodegradable substrates have high μ_{max} values and low K_s values. In other words, soluble BOD, which is easy for the bacteria to take up, will help the bacteria to grow very quickly. They do not need to waste time and energy breaking down particulate BOD before they can bring it through the cell membrane. In contrast, slowly biodegradable substrates have low μ_{max} values and high K_s values. Therefore, the percentages of readily biodegradable and slowly biodegradable substrates in the wastewater will affect the μ_{max} and K_s values for each WRRF.

For facultative heterotrophic bacteria, the K_s ranges between 25 and 100 mg/L of BOD, with a typical value of 60 mg/L of BOD (Metcalf & Eddy, Inc./AECOM, 2014). This means that when the BOD concentration is in this range, the growth rate of the microorganisms is about half of maximum. Compare this to an effluent BOD limit of 30 mg/L for most mechanical treatment facilities. For the nitrifying bacteria, the K_s for ammonia-nitrogen is 0.50 mg/L (U.S. Environmental Protection Agency [U.S. EPA], 2010). When the ammonia-nitrogen concentration is higher than 1 mg/L in the activated sludge process, it becomes possible for the AOB to grow at their maximum rate. As the concentration of ammonia-nitrogen gets lower and lower, the AOB growth rate slows down. There is a similar effect with dissolved oxygen concentrations. The K_s for oxygen for the nitrifying bacteria ranges from 0.13 to 0.47 mg/L (Manser et al., 2005). Ultimately, the maximum growth rate of the microorganisms will be limited by the availability of any or all of the resources they need as well by water temperature.

4.3.4 EFFECT OF MULTIPLE LIMITING SUBSTRATES

Growth rates in wastewater treatment processes are often limited by two or more substrates at the same time. When growth rates are limited by more than one substrate or growth factor (e.g., organic matter and oxygen are both required by heterotrophic bacteria), multiple Monod terms can be multiplied together to estimate the combined effect of two or more substrates, as shown by the following equation:

$$\mu = \mu_{max} \times \left[\frac{S}{K_{s1} + S_1}\right] \times \left[\frac{S_2}{K_{s2} + S_2}\right] \tag{5.9}$$

Where K_{s1} represents the half-saturation coefficient for dissolved oxygen and S_1 is the dissolved oxygen concentration in the process. The K_{s2} term represents the half-saturation coefficient for ammonia and S_2 is the ammonia concentration in the process.

Example 5.1

The nitrifying bacteria need ammonia and dissolved oxygen to grow. The *Ks* for ammonia is 0.5 mg/L and the *Ks* for dissolved oxygen is 0.47 mg/L. What happens when one or both are limited? For the nitrifying bacteria, we will use a μ_{max} of 1 per day (U.S. EPA, 2010). In the first example, ammonia will be plentiful at 5 mg/L as N, but dissolved oxygen will be in short supply at 0.3 mg/L.

$$\mu = \mu_{max} \times \left[\frac{S_{NH_3-N}}{K_{sNH_3-N} + S_{NH_3-N}}\right] \times \left[\frac{S_{DO}}{K_{sDO} + S_{DO}}\right]$$

$$\mu = 1\ d^{-1} \times \left[\frac{5\ mg/L}{0.5\ mg/L + 5\ mg/L}\right] \times \left[\frac{0.3\ mg/L}{0.47\ mg/L + 0.3\ mg/L}\right]$$

$$\mu = 1\ d^{-1} \times \left[\frac{5\ mg/L}{5.5\ mg/L}\right] \times \left[\frac{0.3\ mg/L}{0.77\ mg/L}\right]$$

$$\mu = 1\ d^{-1} \times [0.91] \times [0.39]$$

$$\mu = 0.35\ d^{-1}$$

Because there was plenty of ammonia, the second term was calculated as 0.91 and is very close to 100% of the maximum growth rate. There was not enough dissolved oxygen, so the third term calculated out to be 0.39, which is pretty low. The overall growth rate is reduced from 1 per day to 0.35 per day. What if both ammonia and dissolved oxygen had been limiting? Imagine if the ammonia term had also calculated out as 0.39. Then, the overall growth rate would be reduced even further.

$$\mu = 1\ d^{-1} \times [0.39] \times [0.39]$$
$$\mu = 0.15\ d^{-1}$$

It seems odd that the growth rate would continue to drop, but it makes sense if considered in terms of the population of nitrifying bacteria instead of individuals. The further resources are reduced, the more likely it becomes that an individual bacterium will be lacking something essential. The percentage of bacteria in the whole population that have enough resources to reproduce goes down, and so does the overall growth rate. If a single bacterium is missing two essential resources, then the resource that is scarcest will determine the maximum growth rate.

4.3.5 BIOMASS YIELD

Yield is the amount of new microorganisms produced per amount of substrate added to the system (Equation 5.11). Substrates include organic compounds, ammonia, VFAs, oxygen, and nitrate. The availability of each resource affects how efficiently the microorganisms can grow and reproduce. Heterotrophic bacteria produce an average of 0.40 g of VSS, a surrogate measurement for new bacteria, for every 1 g of COD added to the process when oxygen is available (Metcalf & Eddy, Inc./AECOM, 2014). For secondary treatment processes receiving domestic wastewater, this is equivalent to about 0.6 g VSS/g BOD. The COD and BOD that do not become new biomass become carbon dioxide and water. Yields are very consistent for WRRFs accepting domestic wastewater. It is not possible to produce more new bacteria than the amount of food in the influent will support. Yields are somewhat lower when oxygen is not available and the facultative heterotrophs are forced to substitute nitrate or nitrite for oxygen (denitrification), requiring the bacteria to use more energy. Using nitrate or nitrite is less efficient for the bacteria than oxygen.

The AOB and NOB produce an average of only 0.15 and 0.05 g, respectively, of VSS for every 1 g of ammonia converted to nitrate (U.S. EPA, 2010). The yield is much lower for nitrifiers than it is for the heterotrophs, which means their growth is slower. The amount of new cells produced is essentially constant when the energy and oxygen sources are known. Yield can be calculated as follows:

$$\text{Yield} = \frac{\text{Mass of new cells formed}}{\text{Mass of substrate removed}} \tag{5.10}$$

4.3.6 BIOMASS DECAY RATE

Decay accounts for losses from the system caused by microorganisms consuming internal storage products, dying, and being eaten by other microorganisms. The amount of solids lost each day is a fraction of the total solids in the system (i.e., typically 0.10 g VSS lost per gram of VSS in the system per day for the heterotrophs and 0.17 g VSS lost per gram of VSS in the system per day for the nitrifying autotrophs) (Metcalf & Eddy, Inc./AECOM, 2014; U.S. EPA, 2010). In other words, about 10% of the heterotrophs in the process will die or be reduced in size each day. New microorganisms are produced each day (yield) and some get smaller and/or die each day (decay). The difference between the yield and the decay is the net yield or observed yield. This is an important concept for secondary treatment process control. The amount of COD or BOD entering the process each day dictates the amount of new cells that can potentially be produced. When the amount of COD or BOD decreases, fewer new microorganisms are produced. It is not possible to build up a population of live microorganisms without COD or BOD, or, in the case of the nitrifying bacteria, ammonia.

4.3.7 MONOD KINETICS

The yield, decay, maximum specific growth rate, maximum substrate utilization rate, and saturation coefficient are related to one another through the following equations (Metcalf & Eddy/AECOM, 2014):

$$\mu = Y \times \frac{kS}{K_s + S} - k_b \tag{5.11}$$

Alternatively,

$$Y = \left[\frac{K_s + S}{kS}\right][\mu + k_b] \tag{5.12}$$

Where μ = specific growth rate, g VSS/g VSS·d;
Y = Yield, g VSS/g BOD or COD;
k = maximum utilization rate, g COD/g VSS·d;
S = substrate concentration, mg/L BOD or COD;
K_s = substrate concentration at one-half the maximum specific substrate utilization rate, mg/L; and
K_b = decay, g VSS/g VSS·d.

This particular example relates the growth rate of heterotrophic bacteria to the availability of BOD or COD. Similar equations exist for the nitrifying autotrophic bacteria and are presented in the chapter on biological nutrient removal. Operators will not need to calculate growth rates for microorganisms, but it is useful to show the mathematical relationship so we can understand what happens to the growth rate as the available resources diminish.

If the amount of resources available to the microorganism is less than 100% of what is needed, then the potential growth will be reduced. It is possible for oxygen sources to be limited as well, or for there to be limits on the quantity and types of spare parts (carbon and nutrients). If the amount of resources available to each bacteria begins to decrease, they will not be able to grow or reproduce as quickly. If available resources continue to decrease, the bacteria will begin to use up stored resources. This is called endogenous respiration and is similar to what happens when you go on a diet. Eventually, the bacteria run out of stored resources and will lyse or break open. Their remains then become food for other bacteria in the process.

Endogenous respiration refers to the bacterial growth phase during which microbes metabolize their own protoplasm without creating more protoplasm because not enough food is available.

5.0 NITRIFICATION/DENITRIFICATION PROCESS PRINCIPLES

Biological nitrogen removal is a two-step process that requires nitrification in an aerobic environment followed by denitrification in an anoxic environment. An anoxic environment is defined as one with low oxygen levels but adequate electron acceptors present (e.g., nitrate). The oxidation reduction potential (ORP) in anoxic zones typically ranges from −50 to 50 mV. As with all biological activity, these reactions are affected by the specific environmental conditions in the reactor, including pH, wastewater temperature, dissolved-oxygen concentration, substrate type and concentration, and the presence or absence of toxic substances. Nitrification is the sequential oxidation of ammonium-nitrogen to nitrite-nitrogen and then to nitrate-nitrogen. Biological denitrification reduces nitrate-nitrogen or nitrite-N to nitrogen gas, which is sparingly soluble and can be released to the atmosphere. Denitrifying organisms are primarily facultative heterotrophs that reduce nitrate in the absence of molecular oxygen. A relatively broad range of heterotrophic bacteria can denitrify. However, the presence of organic matter under anoxic conditions is a must.

5.1 NITRIFICATION

Nitrogen contained in municipal raw wastewater occurs predominantly in organic and ammonia nitrogen forms. Typical concentrations of total nitrogen in domestic wastewaters range from 20 to 85 mg/L, with a medium strength of 40 mg/L (Metcalf & Eddy, Inc./AECOM, 2014). Approximately 40% of the total occurs as organic nitrogen and 60% as ammonia. Typically, less than 1% is present as nitrate or nitrite unless influenced by industrial waste contributions. Influent nitrogen concentrations, as with other constituents, have trended upward in areas with increasing water conservation efforts or reduced infiltration and inflow.

The growth of new cells will remove some of the influent nitrogen. This nitrogen removal will be approximately 12% of the net biomass generated. An additional fraction of the influent total nitrogen is non-biodegradable or removed as particulates (1% to 2% of influent total Kjeldahl nitrogen; can be higher with novel solids processing such as thermal hydrolysis). As a result, approximately 80% to 85% of the influent nitrogen may be available for oxidation. For given conditions of solids retention time (SRT) and CBOD removal, nitrogen removal by assimilation can be estimated as a percentage of the net biomass plus cell debris produced. Nitrogen assimilation depends on the ratio of BOD to nitrogen in the influent, and thus can be significant in systems treating wastewater with high concentrations of organic carbon.

Nitrification is the aerobic oxidation of ammonium to nitrate often through a nitrite intermediate. The reaction is often carried out by two distinct groups of autotrophic bacteria where nitroso-species (e.g., *Nitrosomonas*) oxidize ammonium to nitrite (AOB) and nitro-species (e.g., *Nitrobacter*) oxidize nitrite to nitrate (NOB). A member of the *Nitrospira* has been observed to perform complete oxidation of ammonium to nitrate in single cell (Daims et al., 2015), though this pathway is not described extensively in this text. Some members of the domain *Archaea* are known to oxidize ammonium aerobically (e.g., *Nitrosoarchaeum limnia*) but the stoichiometry is the same as the *Bacteria* nitrifiers. The net reaction is shown below.

$$NH_4^+ + 1.5O_2 \rightarrow NO_2^- + 2H^+ + H_2O \tag{5.13}$$

$$NO_2^- + 0.5O_2 \rightarrow NO_3^- \tag{5.14}$$

$$NH_4^+ + 2O_2 \rightarrow NO_3^- + 2H^+ + H_2O \tag{5.15}$$

Each gram of ammonia oxidized to nitrate (both expressed as N) will result in:

- 4.57 g of oxygen consumed,
- 7.14 g of alkalinity (as calcium carbonate) destroyed, and
- 0.15 g of new cells (nitrifiers) produced.

The synthesis of new nitrifying biomass results in oxygen and alkalinity consumption slightly less than that predicted by stoichiometry. The degree of biological nitrification will depend on the mass of nitrifying organisms allowed to remain in the system. Their presence depends on relative growth rates of the autotrophic populations involved, system SRT, and other conditions such as temperature, ammonia, organic substrate, and dissolved-oxygen concentrations. For a given maximum mixed liquor suspended solids (MLSS) concentration, the fraction of autotrophic biomass will be limited based on the heterotrophic biomass that is present. Denitrification, whether intended as part of the process or incidental—such as that which occurs in clarifiers or low dissolved-oxygen zones of aerobic reactors—can recover some of the consumed alkalinity and reduce consumption of oxygen due to anoxic heterotrophic growth.

5.2 DENITRIFICATION

Autotrophic denitrification has been reported using alternate electron donors (e.g., molecular hydrogen or elemental sulfur) (Sun & Nemati, 2012; Tang et al., 2012). This process frequently consumes alkalinity as opposed to heterotrophic denitrification, which produces alkalinity. Because heterotrophic mechanisms predominate in wastewater treatment, this discussion is restricted to pathways using organic compounds as an energy source. Nitrate can be used in lieu of ammonia as a nutrient for bacterial growth. Therefore, if ammonia levels are low in a reactor, and nitrate is present, then the system will not be nutrient limited.

Biological mechanisms and stoichiometry for biological denitrification are relatively well established. Basic reactions for the reduction of nitrate to nitrogen gas are as follows:

1. Reduction sequence:

$$NO_3^- \rightarrow NO_2^- \rightarrow NO \rightarrow N_2O \rightarrow N_2 \tag{5.16}$$

2. Overall reduction using a methanol carbon source:

$$NO_3^- + \frac{5}{6}CH_3OH \rightarrow \frac{1}{2}N_2 + \frac{5}{6}CO_2 + \frac{7}{6}H_2O + OH^- \tag{5.17}$$

3. Overall reaction, including cell synthesis ($C_5H_7O_2N$): CH_3OH carbon source, and nitrate nitrogen source:

$$\begin{aligned} &NO_3^- + 1.08\ CH_3OH + 0.24\ H_2CO_3 \rightarrow \\ &0.056\ C_5H_7O_2N + 0.47\ N_2 + 1.68\ H_2O + HCO_3^- \end{aligned} \tag{5.18}$$

4. Overall reaction, including synthesis: CH_3OH, carbon source, and ammonia nitrogen source:

$$\begin{aligned} &NO_3^- + 2.5\ CH_3OH + 0.5\ NH_4^+ + 0.5\ H_2CO_3 \rightarrow \\ &0.05\ C_5H_7O_2N + 0.5\ N_2 + 1.45\ H_2O + 0.5\ HCO_3^- \end{aligned} \tag{5.19}$$

5. Overall reaction, including synthesis: municipal wastewater carbon source and ammonia nitrogen source:

$$\begin{aligned} &NO_3^- + 0.345\ CH_{10}H_{19}O_3N + H^+ + 0.267\ HCO_3^- + 0.267\ NH_4^+ \rightarrow \\ &0.655\ CO_2 + 0.5\ N_2 + 0.612\ C_5H_7O_2N + 2.3\ H_2O \end{aligned} \tag{5.20}$$

There are many texts and references for more information (Barnes & Bliss, 1983; Ekama et al., 1984; Grady et al., 2011; McCarty et al., 1969; Metcalf & Eddy, Inc./AECOM, 2014; Parker et al., 1975; Pitter & Chudoba, 1990; Sharma & Ahlert, 1977; Stensel et al., 1973; U.S. EPA, 1993).

Overall denitrification results that interest the design engineer can be summarized as follows:

- Nitrate is converted to nitrogen gas in a stepwise manner; NO and N_2O are also gaseous and can be released from solution as alternate end products.
- Oxygen equivalency is 2.856 mg O_2/mg NO_3-N reduced to N_2. For the other steps, the oxygen equivalency is:
  - NO_3-N reduced to NO_2-N = 1.142 mg O_2/mg NO_3-N reduced to NO_2;
  - NO_2-N reduced to N_2 = 1.713 mg O_2/mg NO_2-N reduced to N_2;
  - NO_2-N reduced to NO-N = 0.571 mg O_2/mg NO_2-N reduced to NO-N;
  - NO-N reduced to N_2O-N = 0.571 mg O_2/mg NO-N reduced to N_2O-N; and
  - N_2O-N reduced to N_2 = 0.571 mg O_2/mg N_2O-N reduced to N_2.
- Alkalinity equivalency is 3.57 mg $CaCO_3$/mg NO_3-N.
- Heterotrophic biomass production is approximately 0.4 mg VSS/mg COD removed.

Because nitrification only oxidizes ammonium to nitrate and nitrite, denitrification must be used if total nitrogen removal is required. This step requires the presence of both a degradable carbon source and nitrate. This can be achieved in three ways:

1. Supplying an exogenous carbon source such as methanol or acetate to the denitrification zone or reactor;
2. Using cBOD in the wastewater as a degradable carbon source by either recycling a large amount of nitrified mixed liquor back to an anoxic reactor or diverting a portion of the raw influent or primary effluent flow to a zone containing nitrate; and
3. Using endogenous carbon present in cell mass as the degradable carbon source.

The amount of nitrification removed by secondary treatment systems is limited by the amount of refractory dissolved organic nitrogen (rDON) present in the facility influent and created in the biological treatment processes (Water Environment Research Foundation [WERF], 2008). The rDON is the organic nitrogen in soluble compounds that is not easily removed by biological treatment. Typical levels of rDON in WRRF effluent range between 0.5 and 1.2 mg N/L with most values in the 1 to 2 mg N/L (Pagilla, 2007). Several variables have been shown to significantly affect biological denitrification kinetics, including:

- Carbon substrate type and concentration,
- Dissolved-oxygen concentration,
- Alkalinity and pH, and
- Temperature.

5.3 NITRITATION AND DENITRITATION

Nitritation, or the conversion of ammonia to nitrite (NO_2^-) rather than to nitrate, reduces the amount of oxygen required to remove ammonia. For nitrogen removal, nitrite is reduced in anoxic zones to nitrogen gas but requires less carbon than the conversion of nitrate because of the lower oxidation state of nitrite.

$$NH_3 \rightarrow NO_2^- \rightarrow N_2 \tag{5.21}$$

This process is used primarily to treat nitrogen-rich and carbon-poor liquors from the dewatering of anaerobically digested sludge prior to recycle to the main treatment trains. The reaction to nitrite is controlled by operating at lower dissolved-oxygen levels, and denitritation is done using the endogenous decay of the biomass. Because dissolved-oxygen levels are low, this process is a version of the simultaneous nitrification/denitrification process discussed earlier.

5.3.1 ANAEROBIC AMMONIUM OXIDATION

Anaerobic ammonium oxidation (anammox) is the anoxic oxidation of ammonium to nitrogen gas carried out by members of the *Planctomycetes*, which utilizes nitrite as the terminal electron acceptor. The energy-yielding reaction is shown in Equation 5.22.

$$NH_4^+ + NO_2^- \rightarrow N_2 + 2H_2O \tag{5.22}$$

In this process, nitrite is concurrently oxidized to nitrate to yield reducing equivalents for carbon dioxide fixation and cell growth. The net reaction (including cell synthesis) increases the stoichiometry from 1 : 1 ammonium:nitrite. The stoichiometry of anammox metabolism has been found to vary from 1.11 to 2.00 nitrite-N reduced per ammonium-N oxidized (Dapena-Mora et al., 2004; Staunton & Aitken, 2015; Strous et al., 1999; van de Graaf et al., 1996) and the variation is thought to be due to the effect of SRT; with a longer SRT, the stoichiometry approaches 1 : 1 (Lee et al., 2013). The generally accepted reaction that includes cell synthesis is shown in Equation 5.23 (Strous et al., 1998).

$$\begin{gathered} NH_4^+ + 1.32NO_2^- + 0.066H\,CO_3^- + 0.13H^+ \rightarrow \\ 1.02N_2 + 0.26NO_3^- + 0.066CH_2O_{0.5}N_{0.15} + 2.03H_2O \end{gathered} \tag{5.23}$$

The low energy yield (357 kJ/mol) of Equation 5.23 coupled with the autotrophic nature of the bacteria involved leads to a very slow growth rate with an estimated doubling time of 11 to 14 days under warm sidestream conditions. Although not well established, the doubling time at the reduced temperature found in mainstream treatment has been estimated to be 35 days (Lotti et al., 2014). Achieving anammox in the mainstream treatment process requires a feed with the proper stoichiometry of ammonium to nitrite (alternatively, produce nitrite in situ), a low BOD-to-N ratio to prevent outcompetition of anammox bacteria by denitrifying bacteria for nitrite via shortcut denitrification, and efficient biomass retention in order to maintain an adequate SRT for the slow growing biomass.

5.3.2 NITRITATION/ANAMMOX

The dominant form of influent nitrogen to most influent wastewater treatment facilities is ammonium-N, and a fraction of the influent N must be selectively oxidized to nitrite prior to further treatment by anammox bacteria. In addition to the above requirements for anammox, NOB must be selectively inhibited. Similar control techniques for shortcut nitrogen removal can be used for nitritation/anammox systems. Nitritation and anammox can be designed to occur in the same reactor or in separate reactors.

Nitrogen removal by nitritation/anammox lends itself to a two-sludge AB-type system. In the A-stage, BOD is removed and is either oxidized or redirected to anaerobic digestion for biogas production; in the B-stage, nitrogen is removed by nitritation/anammox. It is important to note that anammox produces nitrate as a side product and a tertiary polish step may be necessary to reach very low nitrogen discharge permits. At the time of this writing, very few mainstream nitritation/anammox systems had been developed and all mainstream systems require an operational sidestream system.

6.0 ENHANCED BIOLOGICAL PHOSPHORUS REMOVAL PROCESSES

Nitrogen and phosphorus removal systems can achieve low levels of nutrients in facility effluents. Depending on the effluent quality required, treatment could be by biological means alone, by chemical means, or by a combination.

Phosphorus is essential for metabolism during biological growth. Therefore, all biological processes remove phosphorus from wastewater to varying degrees. The VSS in conventional activated sludge (CAS) processes contain 1.5% to 2.5% phosphorus. Assuming a phosphorus content of 2% (0.02 mg P/mg VSS), and 0.5 mg of VSS produced per milligram of BOD removed, then about 1.0 mg/L of phosphorus is converted to cell mass per 100 mg/L of BOD removed. Therefore, a CAS process reduces the influent phosphorus by 1 to 2 mg/L.

Phosphorus removal in excess of metabolic requirements can be achieved by using EBPR. This process relies on the selection and proliferation of a specialized heterotrophic microbial population capable of storing soluble phosphorus in excess of their minimum growth requirements. These PAOs can sequester up to 0.38 mg P/mg VSS and, as a result, mixed liquor from an EBPR system can contain 0.06 to 0.15 mg P/mg VSS (Wentzel et al., 2008). The higher the mixed liquor PAO fraction, the greater the phosphorus content of the waste sludge and the amount of phosphorus removed.

The EBPR processes consist of anaerobic and aerobic zones and, in combined nitrogen and phosphorus removal configurations, anoxic zones. The anaerobic zone provides a competitive advantage to PAOs and is defined as not intentionally containing usable dissolved oxygen or nitrate. The ordinary heterotrophic organisms cannot compete for the same substrate in the anaerobic zone because of the absence of an electron acceptor (oxygen or nitrate).

The PAOs do not grow in the anaerobic zone, but consume and convert readily available organic material (i.e., VFAs) to energy-rich carbon polymers, the PHAs. The energy required for this reaction is generated through breakdown of stored polyphosphate (poly-P) molecules, which results in phosphorus release and an increase in the bulk liquid-soluble phosphorus concentration in the anaerobic stage. Magnesium and potassium ions are concurrently released to the anaerobic medium together with phosphate, in the molar ratio $P:Mg^{2+}:K^{+} = 1.0:0.33:0.33$. Phosphate-accumulating organisms also break down glycogen, which is stored within the cells, to provide substantial reducing power for PHA production (Erdal et al., 2004; Filipe et al., 2001; Mino et al., 1987).

The PAOs metabolize the internally stored PHA in the anoxic and aerobic zones and use the energy to take up all of the soluble orthophosphates released in the anaerobic zone and additional phosphorus present in the influent to renew the stored polyphosphate pool. Phosphorus uptake in excess of metabolic requirement is possible because the energy released by PHA oxidation is significantly greater than the energy required for PHA storage. The PAOs also use PHAs as a carbon source for growth.

Effluent from the EBPR reactors is low in phosphorus because the removed soluble phosphorus is stored in the biomass that is removed from the system with the waste activated sludge (WAS). Return activated sludge containing stored polyphosphate is recirculated to the head of the anaerobic zone to seed the incoming flow. Some of the energy and carbon is used to restore the glycogen pool for reactions to continue when mixed liquor is recirculated to the head of the anaerobic zones. The events that take place in the anaerobic and aerobic stages are illustrated in Figure 5.6.

There are two basic guidelines designers should follow when designing phosphorus, or combined nitrogen and phosphorus, removal systems: (1) minimize oxygen and nitrate/nitrite to anaerobic zones, because the anaerobic environment is most beneficial to developing biological phosphorus removal populations when there are no or low levels of electron acceptors present in the anaerobic zone; and (2) minimize oxygen to anoxic zones, because high dissolved oxygen or entrained air in the nitrified recycle stream or reactor feed will reduce the carbon available for denitrification.

Internal facility sidestreams can contribute a significant nutrient load to secondary treatment system influent. The recycle load depends on both the nutrient and the solids-handling system at the WRRF. Typically, those facilities without solids digestion systems have low levels of nutrient return. Facilities with aerobic digestion may have low levels of ammonia return but higher nitrate returns that need to be considered in design. Higher levels of phosphorus in the dewatering liquor are also common, but typically not at high levels. Anaerobic digestion facilities typically have the highest levels of ammonia and phosphorus in the return stream, although nitrate is not present. All nutrient removal facilities with digestion must consider the effects of the recycle system on the design of the mainstream nutrient removal process. Because of the inherent complexity of phosphorus, and of combined nitrogen and phosphorus removal systems, the use of commercial process simulators that incorporate sophisticated stoichiometric and kinetic models has become standard practice for process design. Process models are discussed in Chapter 16.

Aerobic digestion is a solids stabilization process involving direct oxidation of biodegradable matter and oxidation of microbial cellular material.

7.0 AEROBIC DIGESTION

Aerobic digestion is a biological treatment process that is used to further process and stabilize the solids (sludge) that are removed from wastewater during primary, secondary, and tertiary treatment. The sludge is placed into a new basin, called a **digester**, where it typically remains for 40 to 60 days (WEF & Association of Boards of Certification [ABC], 2018). Oxygen is added to support the microorganisms and break down the sludge. The goals of aerobic digestion are to

A digester is a tank or other vessel used to store and anaerobically or aerobically decompose the organic matter in solids.

- Reduce the number of pathogens in the remaining solids to safe concentrations,
- Reduce vector attraction,

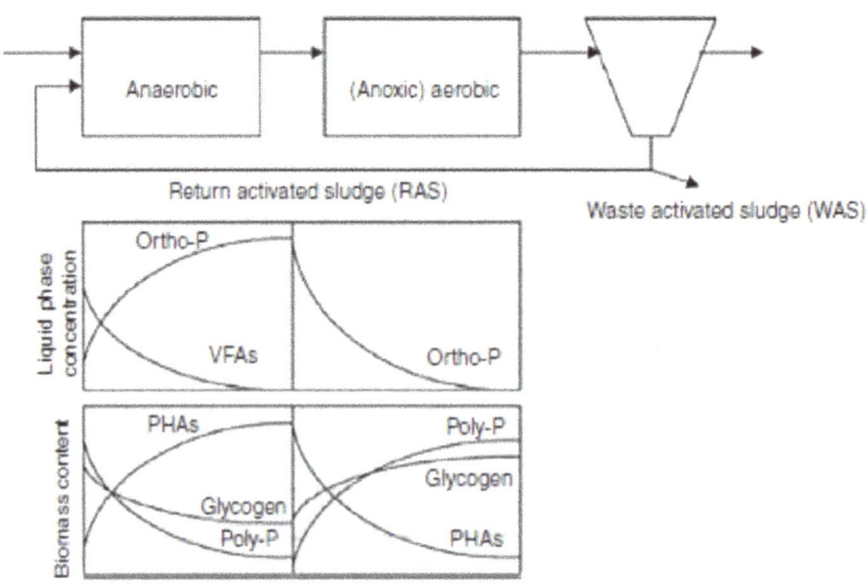

Figure 5.6 Typical Concentration Patterns Occurring in Enhanced Biological Phosphorus Removal Systems (WEF, 2013)

- Reduce the mass and percentage of volatile solids, and
- Reduce the total mass of solids.

Aerobic digestion is similar to the activated sludge process. Both processes contain the same types of microorganisms and use similar equipment. Aerobic digestion differs from activated sludge in two significant ways: (1) digesters receive sludge instead of raw influent or primary effluent, and (2) digesters have much longer SRTs than most activated sludge processes.

Aerobic digesters are typically fed with sludge produced during secondary treatment, although some receive a mixture of primary and secondary sludge. Unlike secondary treatment processes such as activated sludge, aerobic digesters do not receive a steady supply of BOD from the influent or primary effluent. They only receive sludge. Once in the digester, the microorganisms no longer have access to plentiful BOD. Even digesters that receive primary sludge have much less BOD available per microorganism than a secondary treatment process. Without access to BOD, they use up stored food, consume their own cell material, decrease in size, and eventually die. The dead microorganisms break open or lyse, releasing their contents to become food for the surviving microorganisms. This process of gradually consuming their own cell mass is called **endogenous respiration**.

Endogenous respiration is a bacterial growth phase during which microbes metabolize their own protoplasm without creating more protoplasm because not enough food is available.

Equation 5.24 shows endogenous respiration as a chemical reaction. The chemical formula for biomass (microorganisms) is shown as $C_{60}H_{87}O_{23}N_{12}P$ (Henze et al., 2008). Microorganisms also contain sulfur, iron, manganese, and other elements in lesser amounts. Two of the end products of endogenous respiration are carbon dioxide and water.

$$C_{60}H_{87}O_{23}N_{12}P + 62.5\ O_2 \rightarrow 48\ CO_2 + 12\ H_2O + 12\ NH_4HCO_3 + H_3PO_4 + \text{Energy} \tag{5.24}$$

From the above equation, aerobic digestion requires large amounts of oxygen: 1.5 kg of oxygen for every 1 kg of volatile solids destroyed. Volatile solids are converted to CO_2 and water. The total solids concentration in the digester will typically be lower than the concentration being fed to the digester. For every 1 kg of volatile solids destroyed, 0.12 kg of ammonia-nitrogen (NH_3-N) and 0.02 kg of orthophosphate as P (PO_4^{3-}-P) are released. Secondary sludge containing PAOs will release significantly more orthophosphate. The ammonium released by endogenous respiration reacts with dissolved carbon dioxide in the water to form ammonium bicarbonate (NH_4HCO_3). Because ammonium bicarbonate is a chemical buffer, it increases the alkalinity concentration in the digester.

Anaerobic is a condition characterized by the absence of free oxygen and other electron receptors such as nitrate and sulfate. As a process, anaerobic implies the active presence of strictly anaerobic organisms. As an activated sludge process component, anaerobic implies the active presence of anaerobic and facultative organisms.

Anaerobic digestion is a series of biological processes in which microorganisms break down biodegradable material in the absence of oxygen.

8.0 ANAEROBIC DIGESTION

Anaerobic digestion has been in use for nearly a century, but is only used in less than 5% of all WRRFs. As of 2020, there were 1269 WRRFs in the United States using anaerobic digestion to stabilize their biosolids (American Biogas Council, n.d.). There are an estimated 15 000 WRRFs throughout the United States (U.S. EPA, 2008). Anaerobic digestion is typically used at larger WRRFs that have permitted capacities greater than about 19 000 m³/d (5 mgd) and in WRRFs that produce primary sludge (WEF & ABC, 2018). Aerobic digesters are more common in smaller facilities and those that produce only secondary sludge.

Pathogens are highly infectious, disease-producing microbes typically found in sanitary wastewater.

Vectors are organisms such as insects and rodents that can carry or transfer pathogens, which can spread disease.

Anaerobic digestion is a biological treatment process that is used to further process and stabilize solids (sludge) that are removed from wastewater during primary, secondary, and tertiary treatment. The sludge is placed into a tank, called an **anaerobic digester**, where it typically remains for 15 to 20 days (Grady et al., 1999; U.S. EPA, 1979; WEF, 2023). The goals of anaerobic digestion are to

- Reduce the number of **pathogens** in the remaining solids to safe concentration;
- Reduce **vector** attraction;
- Reduce the mass and percentage of volatile solids;
- Reduce the total mass of solids;
- Produce biogas (methane) that can be used as fuel; and
- When certain requirements listed in the 503 regulations are met, convert sludge into **biosolids**.

Biosolids are solids that have been removed from wastewater and stabilized (e.g., digested or composted) to meet the criteria in the U.S. Environmental Protection Agency's 40 CFR 503 regulations and, therefore, can be beneficially used.

Anaerobic digesters may also produce electricity through co-generation. Phosphorus recovery is also possible from the dewatering centrate of the anaerobically digested sludge (due to phosphorus release under anaerobic conditions).

Hydrolysis refers to enzyme-mediated reactions that convert complex organic compounds (i.e., particulate) into simple compounds or reduced-mass materials. Often refers to solids particle breakdown in the first step of anaerobic digestion.

Acidogenesis is the formation of organic substrates to short-chain volatile fatty acids.

Acetogenesis is the metabolic process that converts volatile acids to acetate, the primary substrate of acetoclastic methanogenesis.

A rate-limiting reaction is the slowest reaction in a series of reactions. In other words, a rate-limiting step is a step that controls the rate of a series of biochemical reactions.

Anaerobic digesters are typically fed with a mixture of primary and secondary sludge, although some WRRFs feed only primary or only secondary sludge (Figure 5.7). Organic matter in the sludge is converted by three separate groups of anaerobic bacteria into methane (CH_4), carbon dioxide (CO_2), water (H_2O), hydrogen gas (H_2), ammonia (NH_3), and phosphate (PO_4^{-3}). A fourth group of bacteria, the sulfate reducing bacteria, convert sulfate to hydrogen sulfide. Anaerobic digestion consists of four distinct steps: **hydrolysis**, **acidogenesis**, **acetogenesis**, and methanogenesis. Acidogenesis and acetogenesis together are also referred to as fermentation (U.S. EPA, 2015). Inert material, shown as orange hexagons in Figures 5.7 and 5.8, passes through the digester unchanged.

Figure 5.8 summarizes the chemical reactions and organisms involved. In the first step, solid material physically breaks down and starts to become soluble. Next, extracellular *enzymes* (enzymes operating outside the bacteria) break down complex organic compounds like cellulose, proteins, fats, and other organic molecules into their smaller, soluble components. This step is crucial because bacteria can't "digest" solids or large organic molecules. It isn't possible for those large items to move through their cell membranes. Extracellular enzymes essentially predigest their food for them so it can be passed through the cell membrane.

Hydrolysis is thought to be the **rate-limiting** (slowest) step of anaerobic digestion (Pavlostathis & Gossett, 1988, 2004). Several factors can affect hydrolysis, including the organisms present, growth conditions,

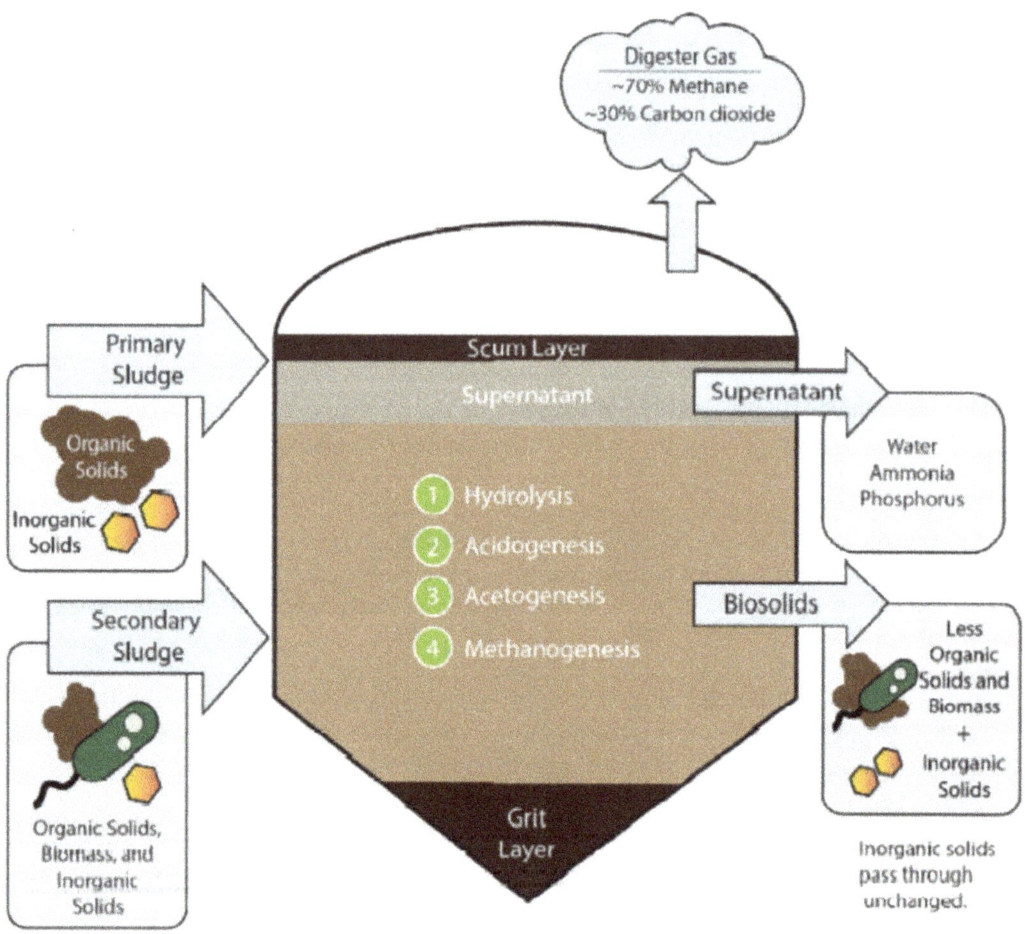

Figure 5.7 Overview of Anaerobic Digestion Process (Reprinted with permission from Indigo Water Group)

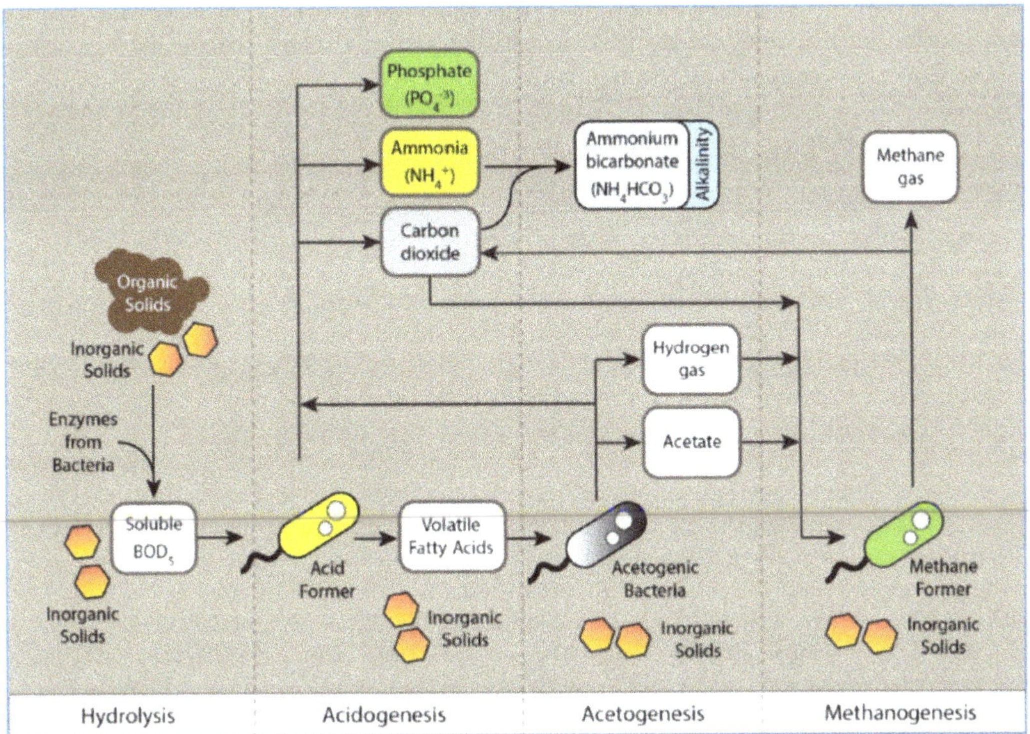

Figure 5.8 Four Stages of Anaerobic Digestion (Reprinted with permission from Indigo Water Group)

temperature, particle surface area (Sanders et al., 2000), and solids composition. This means the time required for hydrolysis can vary from one facility to the next. Not all organic matter can be hydrolyzed. The non-biodegradable fraction of municipal sludge varies. It may be as low as 35% to 40% or as high as 70% to 80% of the volatile organic matter (Parkin & Owen, 1986). The maximum amount of volatile solids reduction (VSR) that any digester can achieve will depend on the biodegradability of the feed sludge.

In the second step (acid formation), the acid-forming bacteria (acidogens) convert the soluble organic compounds created in the first stage into acetic acid (vinegar), propionic acid, hydrogen gas, carbon dioxide, and other low-molecular-weight organic acids (WEF, 2023). The concentrations and relative proportions of these organic acids (collectively termed VFAs) are an indicator of the overall condition of a digester.

Acidogenesis also produces ammonia, phosphate, and carbon dioxide (Figure 5.8). Note that most microorganisms contain between 12% and 16% nitrogen and between 2% and 5% phosphorus by weight. The PAOs can contain up to 40% phosphorus by weight. Organic matter in primary sludge also contains nitrogen and phosphorus. As the primary and secondary sludge break down in the digester, nitrogen and phosphorus are released. The ammonia reacts with dissolved carbon dioxide in the water to form ammonium bicarbonate (NH_4HCO_3). Because ammonium bicarbonate is a chemical buffer, it increases the alkalinity concentration in the digester. Phosphate can remain as dissolved orthophosphate, but may also form several different solid precipitates. Magnesium ammonium phosphate ($MgNH_4PO_4$), more commonly known as struvite, is a nuisance crystal that can form within the digesting sludge. Other phosphorus compounds include calcium phosphate, magnesium potassium phosphate, and iron phosphate (WEF, 2014). Iron phosphate is more commonly known as **vivianite**. Precipitated phosphorus leaves the digester in the solids.

Vivianite refers to particulate crystals or scale that is blue, green, or gray-black in color. It is soluble in hydrochloric acid or nitric acid (HNO_3) and turns opaque or dark when exposed to light. Also called hydrated iron phosphate.

In step 3, most of the proprionic acid and other organic compounds produced in step 2 are then converted into acetic acid by the acetogenic bacteria. More carbon dioxide, ammonia, and phosphorus are also produced along with a new byproduct, hydrogen gas.

In step 4, two different groups of methane-forming bacteria convert the acetic acid and hydrogen gas produced in steps 2 and 3 into methane and carbon dioxide (Figure 5.8). One group converts acetic acid into methane and carbon dioxide. About two-thirds of the methane is produced in this way (Bajpai, 2017). A second group of methanogens combines carbon dioxide with hydrogen gas to form methane. About one-third of the methane is produced this way.

8.1 ANAEROBIC DIGESTION PROCESS

In its simplest form, anaerobic digestion consists of a single, covered tank where feed sludge is combined with a mixed population of bacteria and other microorganisms. The bacteria break the sludge down in the absence of oxygen, nitrite, and nitrate. Many facilities operate two digesters in series. The facility shown in Figure 5.9 includes primary digesters with fixed covers (A), a secondary digester with membrane cover for gas storage (B), and a process for recovering phosphorus from the digester supernatant (C). Feed sludge is fed to the primary digester. Sludge is then transferred to the secondary digester when treatment is mostly complete. Larger facilities may have multiple primary and secondary digesters.

Covers prevent air from entering the digesters. This allows the desired mixture of anaerobic bacteria to grow and establish stable populations. Oxygen, nitrite, and nitrate are toxic to methane-forming bacteria. The relatively small amounts of dissolved oxygen, nitrite, and nitrate that enter the digester with the incoming sludge are used up almost instantaneously. The digester contents are truly anaerobic. Most anaerobic digesters are heated. Heating increases the activity of the bacteria and shortens the amount of time needed to digest the sludge. Most primary and some secondary anaerobic digesters are mixed. Mixing distributes the feed sludge throughout the digester, distributes waste products, minimizes the formation of grit and scum layers, aids in gas release, and helps to maintain a constant temperature throughout the digester. Where both a primary and secondary digester are in use, the primary digester is typically mixed and the secondary digester is not.

Biogas is the gas produced via the anaerobic decomposition of organic matter.

Anaerobic digestion produces several useful products: biosolids, **biogas**, and recovered phosphorus. Biosolids may be land applied as fertilizer. Biogas may be burned to heat the digester and other buildings, or can be converted into electricity with a generator. Phosphorus is an essential nutrient for agriculture. Anaerobic digestion generally requires more equipment, more process control sampling and monitoring, and more frequent process adjustments than aerobic digestion. This makes anaerobic digestion more expensive to construct and more complicated to operate and maintain. Despite these disadvantages, anaerobic digesters are gaining in popularity as the wastewater industry focuses more on reducing energy use and on resource recovery. Because anaerobic digesters do not require aeration, they use substantially less electrical energy than aerobic digesters.

8.2 ANAEROBIC TREATMENT OF WASTEWATER

Anaerobic treatment offers many advantages over conventional aerated activated sludge systems including lower energy consumption, the potential for energy recovery, low sludge production, operational simplicity,

Figure 5.9 Primary (Fixed Cover) Digesters (A) and Secondary (Membrane Gas Storage Cover) Digester (B) With Sidestream Treatment for Phosphorus Recovery (C) (Reprinted with permission from PASSARO ENGINEERING, LLC)

low land area requirements, improved sludge dewatering, ability to store sludge for long periods, simple designs, and less noise from mechanical equipment. While anaerobic treatment offers many benefits, significant limitations exist, including lower removal efficiencies for organics, suspended solids, and pathogens and essentially no nutrient removal with basic process configurations. As a result, anaerobic treatment requires aerobic posttreatment to meet standard effluent criteria for secondary and advanced treatment.

In anaerobic treatment, aeration is not required, thus eliminating the large energy demand needed to supply oxygen (through process air) in aerobic processes. Because organic matter is converted to methane, the process may produce energy for the facility. Anaerobic treatment processes typically are more energy efficient for higher strength wastewater (greater than about 300 mg BOD/L), because under these conditions, the fraction of methane dissolved in the effluent becomes insignificant relative to the total methane production (Cakir & Stenstrom, 2007).

The anaerobic decomposition of organic compounds yields less energy for the microorganisms, resulting in lower biomass yields. Typically, anaerobic treatment reduces the overall biomass yield by a factor of 6 to 8 when compared to aerobic treatment (Tchobanoglous et al., 2003). The reduced biomass results in lower sludge production, which decreases handling and hauling costs, yielding a savings of approximately 10% compared to an aerobic process (Speece, 1996). Because of low sludge production, nutrient requirements are less than for aerobic biological treatment. Anaerobic treatment can reduce the influent BOD and TSS by about 65% to 80%, yielding effluent concentrations of around 40 to 130 mg/L (Khalil et al., 2008; Noyola et al., 2006; Oliveira et al., 2006; von Sperling & Oliveira, 2008).

A survey published in 2008 identified about 3000 WRRFs (all industrial) using anaerobic treatment process (Totzke, 2008). Approximately 60% of the thousands of industrial WRRFs use the upflow anaerobic sludge blanket (UASB) process, making it the dominant anaerobic technology for industrial wastewater treatment (Latif et al., 2011). Due to the slow growth rate of anaerobic bacteria, anaerobic treatment of municipal wastewaters is now primarily used in warm climates where the wastewater temperature remains above 15 °C.

In search of ways to provide affordable wastewater treatment, communities in India, the Middle East, and Latin America have been advancing the use of anaerobic treatment for domestic wastewater treatment. Currently, numerous full-scale facilities with capacities of up to 164 000 m^3/d (1000 mgd) are in operation in these regions using the UASB process or variations of that process (Draaijer et al., 1992; Florencio et al., 2001; Giraldo et al., 2007; Heffernan et al., 2011; Monroy et al., 2000; Sato et al., 2006; Schellingkhout & Collazos, 1992; Seghezzo, 2004; Vieira et al., 1994).

Limited use of anaerobic treatment for wastewater in municipal facilities has been documented in the United States and Canada, but there is potential for its use in the U.S. South, where winter wastewater temperatures are greater than 15 °C to 20 °C. Advances in the anaerobic treatment of municipal wastewater, including use of two-stage reactors, coupled anaerobic reactor/digester combinations, UASB/hybrid reactors, and the expanded granular sludge bed reactor likely will expand use of this technology into more temperate climates (Switzenbaum, 2007).

Anaerobic treatment processes can be categorized as suspended growth, fixed growth, and hybrid processes. Most existing full-scale municipal WRRFs with anaerobic treatment of the liquid stream use the UASB process, which is considered a hybrid process (Malina & Pohland, 1992; Sutton, 1990; Van Lier, 2008). With high-strength wastewaters, dense granules that consist of a microbial consortium characterize the UASB process. When used for municipal wastewater treatment, however, granular sludge typically does not form, and the UASB process can be considered a suspended-growth process.

8.3 PROCESS CONFIGURATIONS

Anaerobic treatment eliminates three significant constraints on process loading that occur in aerobic processes: (1) oxygen transfer rates; (2) solids flux limitations, and (3) high-energy inputs for aeration that hinder floc formation (Speece, 1996). Although the lack of these limiting factors in anaerobic processes enables much higher mass loading rates, different constraints are imposed by low growth rates and high half-saturation coefficients of bacteria in the microbial consortia that is present. The slow growth rate of anaerobic bacteria and their sensitivity to temperature places a premium on biomass retention; the high half-saturation coefficients encourage use of staged reactors to increase performance. A wide range of reactor types exists for anaerobic treatment, and some of these are illustrated in Figure 5.10. For municipal applications, however, most installations are based on the UASB reactor. Also, approximately 60% of the thousands of full-scale anaerobic treatment facilities for the treatment of industrial wastewater are UASB-based reactors (Latif et al., 2011).

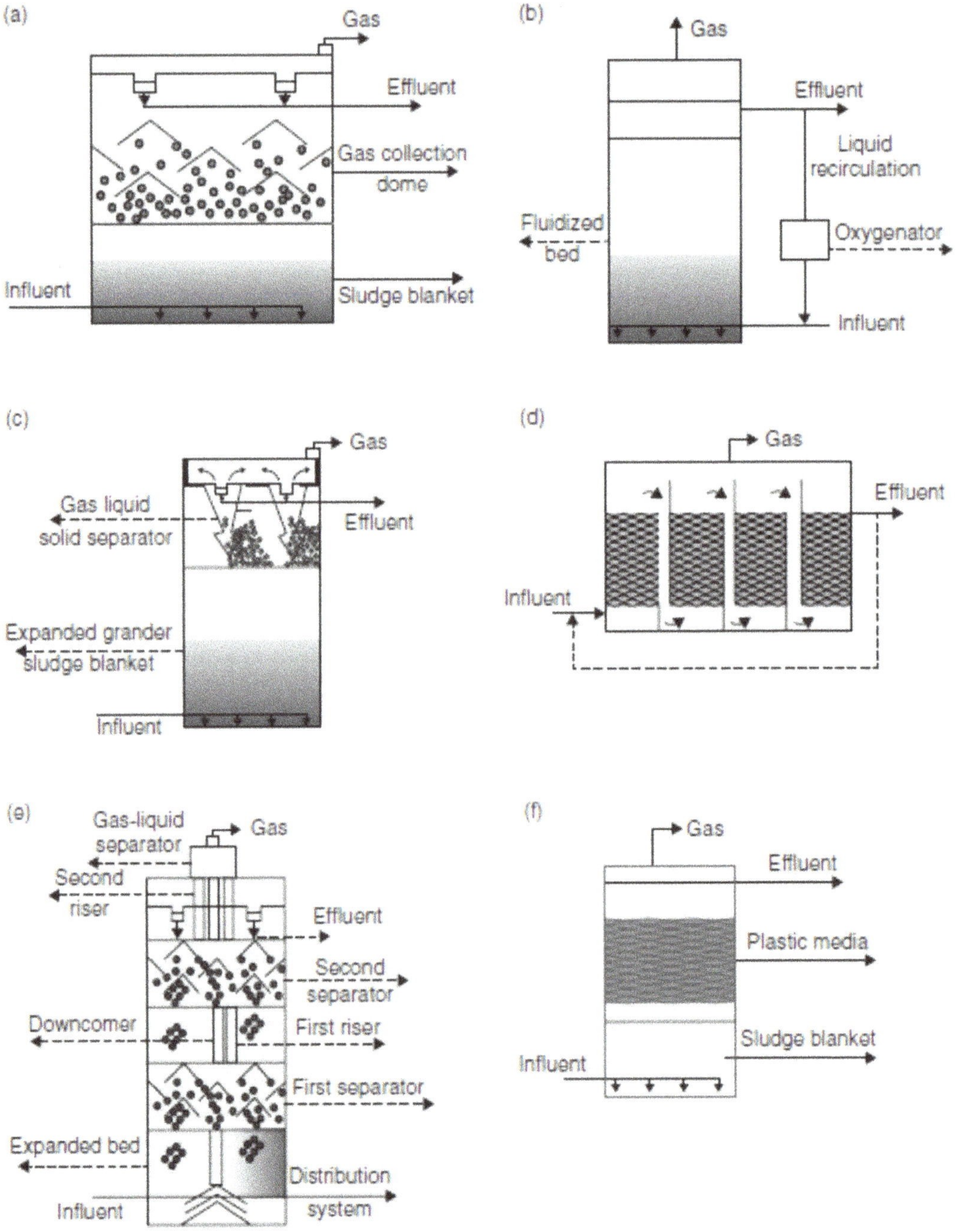

Figure 5.10 Schematic Illustrations of Several Types of Anaerobic Reactor Configurations: (a) Upflow Anaerobic Sludge Blanket (UASB); (b) Biofilm Fluidized Bed; (c) Expanded Granular Sludge Bed; (d) Anaerobic Baffled Reactor; (e) Internal Circulation; and (f) Anaerobic Hybrid Reactor (Nicolella et al., 2000)

PRACTICE PROBLEMS

1. Differentiate aerobic and anaerobic wastewater treatment processes.
2. Discuss the various microbiological processes involved in carbon, nitrogen, and phosphorous removal processes.
3. Explain the differences between oxidation/synthesis and endogenous respiration.
4. List the microorganisms based on their preference for carbon and energy sources and discuss their significance in WRRFs.
5. Certain bacteria cells divide and form two cells every 5 minutes. If you place 10^5 cells in a suitable environment, how many cells will you have in 30 minutes?
6. Determine the theoretical oxygen demand of 1 g (dry) of microbial cells using the two empirical cell formulas: (1) $C_5H_7NO_2$ and (2) $C_{42}H_{100}N_{11}O_{13}P$, without and with the essential nutrient phosphorus.

9.0 REFERENCES

American Biogas Council. (n.d.). *FAQs*. https://americanbiogascouncil.org/resources/faqs/

Bajpai, P. (2017). Basics of anaerobic digestion process. In *Anaerobic technology in pulp and paper industry* (Chapter 2). Springer. https://www.springer.com/gp/book/9789811041297

Barnes, D., & Bliss, P. J. (1983). *Biological control of nitrogen wastewater treatment*. E. and F. N. Spon.

Cakir, F. Y., & Stenstrom, M. K. (2007, October 10–13). Anaerobic treatment of low strength wastewater. *Proceedings of the 80th Annual Water Environment Federation Exposition and Conference* [CD-ROM], San Diego, California, United States.

Daims, H., Lebedeva, E. V., Pjevac, P., Han, P., Herbold, C., Albertsen, M., Jehmlich, N., Palatinszky, M., Vierheilig, J., Bulaev, A., Kirkegaard, R. H., von Bergen, M., Rattei, T., Bendinger, B., Nielsen, P. H., & Wagner, M. (2015). Complete nitrification by *Nitrospira* bacteria. *Nature*, *528*, 504–509.

Dapena-Mora, J., Campos, L., Mosquera-Corral, A., Jetten, M. S. M., & Méndez, R. (2004) Stability of the ANAMMOX Process in a Gas-Lift Reactor and a SBR. *Journal of Biotechnology*, *110*, 159–170.

Draaijer, H., Maas, J. A. W., Schaapman, J. E., & Khan, A. (1992). Performance of the 5 MLD UASB reactor for sewage treatment at Kanpur, India. *Water Science & Technology*, *25*(7), 123–133.

Ekama, G., Marais, G., & Siebritz, I. (1984). Biological excess phosphorus removal. In *Theory, Design and Operation of Nutrient Removal Activated Sludge Processes*. Water Research Commission: Pretoria, South Africa.

Erdal, U., Erdal, Z., & Randall, C. (2004, September). Biochemistry of enhanced biological phosphorus removal and anaerobic COD stabilization. *Proceedings of International Water Association, Biennial Conference*, Marrakech, Morocco. International Water Association.

Filipe, C., Daigger, G., & Grady, C. (2001). Effects of pH on the rates of aerobic metabolism of phosphate-accumulating organisms and glycogen-accumulating organisms. *Water Environment Research*, *73*, 213–222.

Florencio, L., Takayuki, M., & de Morais, J. C. (2001). Domestic sewage treatment in full-scale UASBB Plant at Mangueira, Recife, Pernambuco. *Water Science & Technology*, *44*(4), 71–77.

Gerardi, M. H. (2006). *Wastewater bacteria*. Wiley & Sons.

Giraldo, E., Pena, M., Chernicharo, C., Sandino, J., & Noyola, A. (2007, October 10–13). Anaerobic sewage treatment technology in Latin-America: A selection of 20 years of experiences. *Proceedings of the 80th Annual Water Environment Federation Exposition and Conference* [CD-ROM], San Diego, California, United States.

Grady, C. P. L., & Filipe, C. D. M. (2000). Ecological engineering of bioreactors for wastewater treatment. In S. Belkin (Ed.), *Environmental challenges.* Springer.

Grady, C. P. L., Daigger, G. T., & Lim, H. C. (1999). *Biological wastewater treatment.* CRC Press.

Grady, C. P. L., Jr., Daigger, G. T., Love, N. G., & Filipe, C. D. M. (2011). *Biological wastewater treatment* (3rd ed.). CRC Press.

Heffernan, B., Van Lier, J. B., & Van der Lubbe, J. (2011). Performance review of large scale up-flow anaerobic sludge blanket sewage treatment plants. *Water Science & Technology, 63*(1), 100–107.

Henze, M., van Loosdrecht, M. C., Ekama, G. A., & Brdjanovic, D. (Eds.). (2008). *Biological wastewater treatment.* IWA Publishing.

Khalil, N., Sinha, R., Raghav, A. K., & Mittal, A. K. (2008). UASB technology for sewage treatment in India: Experience, economic evaluation and its potential in other developing countries. *Proceedings of the Twelfth International Water Technology Conference*, IWTC12 2008. Egyptian Water Technology Association, Alexandria, Egypt.

Koops, H.-P., & Pommerening-Röser, A. (2001). Distribution and ecophysiology of the nitrifying bacteria emphasizing cultured species. *FEMS Microbiology Ecology, 37*(1), 1–9.

Latif, M. A., Ghufran, R., Wahid, Z. A., & Ahmad, A. (2011). Integrated application of upflow anaerobic sludge blanket reactor for the treatment of wastewaters. *Water Research, 45*(16), 4683–4699.

Lee, W., Kwak, J., Bae, P., & McCarty, L. (2013). The effect of SRT on nitrate formation during autotrophic nitrogen removal of anaerobically treated wastewater. *Water Science & Technology, 68*, 1751–1756.

Lengeler, J. W., Drews, G., & Schlegel, H. G. (Eds.). (1999). *Biology of the prokaryotes.* Thieme Verlag.

Lotti, T., Kleerebezem, R., van Erp Tallman Kip, C., Hendrickx, T. L. G., Kruit, J., Hoekstra, M., & van Loosdrecht M. C. M. (2014). Anammox growth on pretreated municipal wastewater. *Environmental Science & Technology, 48*(14), 7874–7880.

Madigan, M. T., Martinko, J. M., & Parker, J. (2003). *Brock biology of microorganisms* (9th ed.). Prentice-Hall.

Malina, J. F., & Pohland, F. G. (1992). *Design of anaerobic processes for the treatment of industrial and municipal wastes.* Technomic Publishing.

Manser, R., Gujer, W., & Siegrist, H. (2005). Consequences of mass transfer effects on the kinetics of nitrifiers. *Water Research, 39*(19), 4633–4642.

McCarty, P. L., Beck, L., & St. Amant, P. (1969). Biological denitrification of wastewaters by addition of organic materials. *Proceedings of the 24th Industrial Waste Conference.* Purdue University, Lafayette, Indiana, United States.

Metcalf & Eddy, Inc./AECOM. (2014). *Wastewater engineering: Treatment and resource recovery* (5th ed.). McGraw-Hill Education.

Mino, T., Arun, V., Tsuzuki, Y., & Matsuo, T. (1987, September 28–30). Effect of phosphorus accumulation on acetate metabolism in the biological phosphorus removal process. *Proceedings of the IAWPRC Specialized Conference*, Rome, Italy. IWA Publishing.

Monod, J. (1942). *Recherches sur la croissance des cultures bacteriennes.* Librairie Scientifique.

Monod, J. (1949). The growth of bacterial cultures. *Annual Review of Microbiology, 3*, 371.

Monroy, O., Famá, G., Meraz, M., Montoya, L., & Macarie, H. (2000). Anaerobic digestion for wastewater treatment in Mexico: State of the technology. *Water Research, 34*(6), 1803–1816.

Nicolella, C., van Loosdrecht, M. C. M., & Heijnen, J. J. (2000). Wastewater treatment with particulate biofilm reactors. *Journal of Biotechnology, 80*, 1–33.

Noyola, A., Morgan-Sagastgume, J. M., & Lopez-Hernandez, J. E. (2006). Treatment of biogas produced in anaerobic reactors for domestic wastewater: Odor control and energy/resource recovery. *Review of Environmental Science and Biotechnology, 5*, 93–114.

Oliveira, S. M. A. C., Parkinson, J. N., & von Sperling, M. (2006). Wastewater treatment in Brazil: Institutional framework, recent initiatives and actual plant performance. *International Journal of Technology Management & Sustainable Development, 5*(3), 241–256.

Pagilla, K. (2007, September 27–28). Organic nitrogen in wastewater treatment plant effluents. Presentation at the Water Environment Research Foundation and Chesapeake Bay Science and Technology Advisory Committee Workshop, Baltimore, Maryland, United States.

Parker, D. S., Stone, R., Stenquist, R., & Culp, G. (1975). *Process design manual for nitrogen control* (EPA 62511-75-007). U.S. Environmental Protection Agency.

Parkin, G. F., & Owen, W. F. (1986). Fundamentals of anaerobic digestion of wastewater sludges. *Journal of Environmental Engineering, 112*(5), 867–920.

Pasztor, I., Thury, P., & Pulai, J. (2009). Chemical oxygen demand fractions of municipal wastewater for modeling of wastewater treatment. *International Journal of Environmental Science and Technology, 6*(1), 51–56.

Pauli, W., Jax, K., & Berger, S. (2001). Protozoa in wastewater treatment: Function and importance. In *The Handbook of Environmental Chemistry* (Vol. 2, Part K: Biodegradation and Persistence), Chapter 3. Springer-Verlag.

Pavlostathis, S. G., & Gossett, J. M. (1988). Preliminary conversion mechanisms in anaerobic digestion of biological sludges. *Journal of Environmental Engineering, 114*, 575–592.

Pavlostathis, S. G., & Gossett, J. M. (2004). A kinetic model for anaerobic digestion of biological sludge. *Biotechnology and Bioengineering, 28*, 1519–1530.

Pitter, P., & Chudoba, J. (1990). *Biodegradability of organic substrates in the aquatic environment.* CRC Press.

Sanders, W. T. M., Geerink, M., Zeeman, G., & Lettinga, G. (2000). Anaerobic hydrolysis kinetics of particulate substrates. *Water Science & Technology, 41*, 17–24.

Sato, N., Okubo, T., Onadera, T., Ohashi, A., & Harada, H. (2006). Prospects for a self-sustainable sewage treatment system: A case study on full-scale UASB system in India's Yamuna River Basin. *Journal of Environmental Management, 80*(13), 198–207.

Schellingkhout, A., & Collazos, C. J. (1992). Full-scale application of the UASB technology for sewage treatment. *Water Science & Technology, 25*(7), 159–166.

Seghezzo, L. (2004). Anaerobic treatment of domestic wastewater in subtropical regions [PhD thesis, Wageningen University, Wageningen, Netherlands].

Sharma, B., & Ahlert, R. C. (1977). Nitrification and nitrogen removal. *Water Research (G.B.), 11*, 897.

Speece, R. E. (1996). *Anaerobic biotechnology.* Archae Press.

Staunton, E. T., & Aitken, M. D. (2015, September). Coupling nitrogen removal and anaerobic digestion for energy recovery from swine waste through nitrification/denitrification. *Environmental Engineering Science, 32*(9), 750–760.

Stensel, H. D., Loehr, R. C., & Lawrence, A. W. (1973) Biological kinetics of suspended growth denitrification. *Journal of the Water Pollution Control Federation, 45*, 249–261.

Strous, J., Heijnen J., Kuenen J. G., & Jetten M. S. M. (1998). The sequencing batch reactor as a powerful tool for the study of slowly growing anaerobic ammonium-oxidizing microorganisms. *Applied Microbiology and Biotechnology, 50*, 589–596.

Strous, M., Lan, C.-J., & Jetten, M. S. M. (1999). Key physiology of anaerobic ammonium oxidation. *Applied Environmental Microbiology, 65*, 3248–3250.

Sun, Y., & Nemati, M. (2012). Evaluation of sulfur-based autotrophic denitrification and denitritation for biological removal of nitrate and nitrite from contaminated waters. *Bioresource Technology.* doi:10.1016/j.biortech.2012.03.061.

Sutton, P. M. (1990, December 3–4). *Anaerobic treatment of high strength wastes: System configurations and selection.* Paper Presented at Anaerobic Treatment of High Strength Wastes, Milwaukee, Wisconsin; University of Wisconsin: Milwaukee.

Switzenbaum, M. S. (2007, October 10–13). Overview of challenges in providing anaerobic wastewater treatment. *Proceedings of the 80th Annual Water Environment Federation Exposition and Conference* [CD-ROM], San Diego, California, United States.

Tang, Y., Ziv-El, M., Meyer, K., Zhou, C., Shin, J. H., Ahn, C. H., McQuarrie, J., Candelaria, D., Swaim, P., Scott, R., & Rittmann, B. E. (2012). Comparing heterotrophic and hydrogen-based autotrophic denitrification reactors for effluent water quality and post-treatment. *Water Science Technology Water Supply, 12*, 227.

Tchobanoglous, G., Burton, F. L., & Stensel, H. D. (2003). *Wastewater engineering treatment and reuse* (4th ed.). McGraw-Hill.

Totzke, D. E. (2008). *Anaerobic treatment technology overview.* Applied Technologies.

U.S. Environmental Protection Agency. (1979). *Process design manual for sludge treatment and disposal* (EPA625/1-79-011).

U.S. Environmental Protection Agency. (1993). *Process design manual for nitrogen removal.*

U.S. Environmental Protection Agency. (2008, April 8). *EPA facility registry service (FRS): Wastewater treatment plants.* https://catalog.data.gov/dataset/epa-facility-registry-service-frs-wastewatertreatment-plants

U.S. Environmental Protection Agency. (2010). *Nutrient control design manual* (EPA-600/R-10-100). U.S. Environmental Protection Agency, Office of Research and Development.

U.S. Environmental Protection Agency. (2015). *Anaerobic digestion and its applications* (EPA 600/R-15/304).

van de Graaf, A. A., de Bruijn, P., Robertson, L. A., Jetten, M. S. M., & Kuenen, J. G. (1996). Autotrophic growth of anaerobic ammonium-oxidizing micro-organisms in a fluidized bed reactor. *Microbiology, 142*, 2187–2196.

van Lier, J. B. (2008). High-rate anaerobic treatment: Diversifying from end-of-the-pipe treatment to resource-oriented conversion techniques. *Water Science & Technology, 57*(8), 1137–1148.

Vieira, S. M. M., Carvalho, J. L., Barigan, F. P. O., & Rech, C. M. (1994). Application of the UASB technology for sewage treatment in a small community at Sumare, Sao Paulo State. *Water Science & Technology, 30*(12), 203–210.

von Sperling, M., & Oliveira, S. C. (2008, October 19–23). Comparative performance evaluation of full-scale anaerobic and aerobic wastewater treatment processes in Brazil. *Proceedings IX Latin American Workshop and Symposium on Anaerobic Digestion*, Easter Island, Chile. International Water Association.

Water Environment Federation. (1994). *Wastewater biology: The life processes.*

Water Environment Federation. (2013). *Operation of nutrient removal facilities* (Manual of Practice No. 37).

Water Environment Federation. (2014). *Moving toward resource recovery facilities.*

Water Environment Federation. (2022). *Basic laboratory procedures for the operator-analyst* (6th ed.).

Water Environment Federation. (2023). *Standards for design of water resource recovery facilities* (WEF 8).

Water Environment Federation & Association of Boards of Certification. (2018). *Wastewater treatment fundamentals I—Liquid treatment.* Water Environment Federation.

Water Environment Research Foundation. (2008). *Dissolved organic nitrogen (DON) in biological nutrient removal wastewater treatment processes.*

Wentzel, M., Comeau, Y., Ekama, G., van Loosdrecht, M., Brdjanovic, D., & Lopez-Vazquez, C. (2008). Enhanced biological phosphorus removal. In M. Henze, M. van Loosdrecht, G. Ekama, & D. Brdjanovic (Eds.), *Biological wastewater treatment principles, modelling and design*. IWA Publishing.

10.0 SUGGESTED READINGS

McCarty, P. (1964, October). Anaerobic waste treatment fundamentals, part two. *Public Works*, 123–126.

U.S. Environmental Protection Agency. (1974). *Sludge design manual for sludge treatment and disposal* (EPA625/1-74-006).

U.S. Environmental Protection Agency. (1976). *Anaerobic sludge digestion: Operations manual* (EPA 430/9-76-001). U.S. Environmental Protection Agency, Office of Water, Program Operations.

U.S. Environmental Protection Agency. (1987a). *Dewatering municipal wastewater sludges: Design manual* (EPA 625/1-87/014). U.S. Environmental Protection Agency, Center for Environmental Research.

U.S. Environmental Protection Agency. (1987b). EPA design information report: Anaerobic digester mixing systems (EPA/600/J-87/014). *Journal of the Water Pollution Control Federation, 59*(3), 162–170.

U.S. Environmental Protection Agency. (1987c). Sidestreams in wastewater treatment plants. *Journal of the Water Pollution Control Federation, 59*(1), 54–59.

Water Environment Federation. (2017). *Wastewater biology: The microlife* (3rd ed.).

Water Environment Federation. (2023). *Standards for operation of water resource recovery facilities* (Manual of Practice No. 11).

Water Environment Research Foundation. (2014). *Guidance document on anaerobic digester foaming prevention and control methods* (Final report INFR1SG10a). Water Environment Research Foundation; IWA Publishing.

Chapter 6 | Suspended Growth Systems

LEARNING OBJECTIVES

At the end of this chapter, students should be able to:

- Describe what constitutes a suspended growth process and diagram the individual basins and flows in a variety of activated sludge systems (e.g., completely mixed, plug flow, oxidation ditch, contact stabilization, step feed, and sequencing batch reactors).
- Explain the significance of filamentous bacteria in a suspended growth process and how selectors are used to limit the proliferation of filamentous growth.
- Given stoichiometric and kinetic parameters, be able to use the traditional model to design a basic activated sludge process for biodegradable carbon oxidation (i.e., cBOD removal) with respect to solids retention time, hydraulic retention time, aeration basin volume, sludge production, and oxygen requirements.
- Explain why safety factors are typically applied to SRT values calculated during design.
- Explain the effect of temperature, dissolved oxygen concentration, pH, toxic compounds, and nutrients on the performance of suspended growth processes.
- Discuss methods for aeration and how aeration performance is measured in suspended growth systems.

1.0 INTRODUCTION

Suspended-growth systems are biological treatment processes based on the growth and retention of a suspension of microorganisms. These microorganisms convert biodegradable, organic wastewater constituents and certain inorganic fractions into new cell mass and byproducts, both of which then can be removed by settling, gaseous stripping, and other physical means. Suspended-growth systems for wastewater treatment are predominantly aerobic processes, typically referred to as activated sludge, with a variety of reactor configurations and flow patterns. Strictly anaerobic suspended-growth processes for liquid-phase treatment are also in use.

This chapter presents fundamentals of aerobic activated sludge treatment and the process configurations that can be used. It covers the basic process design for carbon oxidation. The chapter concludes with a comprehensive process design example that illustrates several aerobic suspended-growth systems for different applications and performance requirements.

1.1 PROCESS DESCRIPTION

Figure 6.1 presents a typical flow-through suspended-growth activated sludge process. Wastewater is combined with biomass and other solids in the reactor, where mixing and aeration are provided. Typically, the process operates in a continuous-flow mode, but can also be operated as a batch process. Contents of the reactor, referred to as mixed liquor, consist of wastewater; microorganisms; microbial cell debris; and inert, biodegradable, and nonbiodegradable suspended and colloidal matter. The particulate fraction of the mixed liquor is termed **mixed liquor suspended solids** (MLSS). Suspended-growth anaerobic processes are similar in concept but without aeration or mixing of the solids resulting in a stratified **sludge blanket** of microorganisms.

Mixed liquor suspended solids is the suspended solids concentration in the mixture of wastewater and activated sludge being treated in the aeration basin.

Sludge blanket is an accumulation of solids hydrodynamically suspended in a tank.

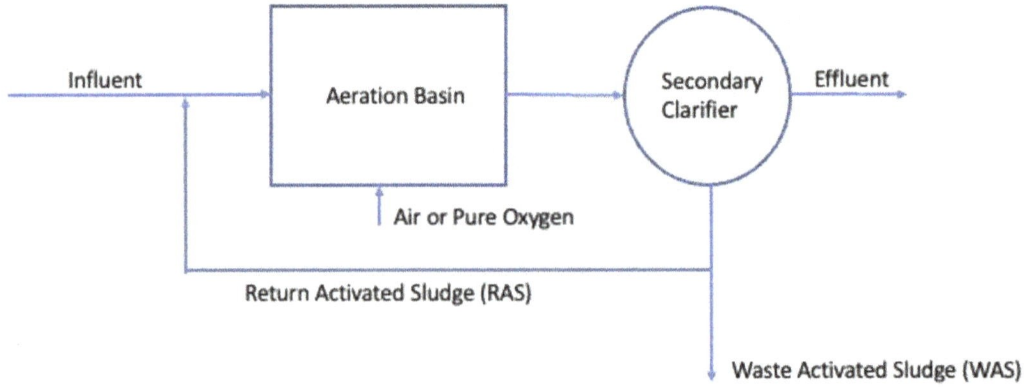

Figure 6.1 A Typical Activated Sludge Process Flow Diagram

After sufficient time for biological reactions, mixed liquor is conveyed to a separate settling basin (clarifier) or other solids-liquid separation step that separates MLSS from treated wastewater and produces clarified effluent. Settled MLSS are then recycled as return activated sludge (RAS) to the aeration basin to maintain a concentrated microbial population for efficient degradation of influent wastewater constituents in an economically sized reactor. Microorganisms and cell debris are generated continuously, which must be removed from the system through wasting to maintain a target **solids retention time** (SRT). Wasting can be continuous or periodic and typically is from the clarifier or return sludge line, although removal from the aeration basin is an alternative. Retention of MLSS and effluent clarification also can be accomplished with synthetic, microfiltration membranes. The MLSS recirculation and wasting requirements remain with this alternative.

Solids retention time is the average period of time that solids have remained in a process or system. Also called the sludge age or mean cell residence time.

1.2 HISTORIC OVERVIEW

The activated sludge process received its name from and was developed based on a series of experiments conducted in Manchester, United Kingdom, via a series of experiments with batch bottles conducted by Ardern and Lockett (1914) who discovered the value of recycling or reusing the biomass residue following biodegradation of wastewater. Widespread use of the process did not begin until the 1940s. Aeration basin hydraulic retention time (HRT = volume/flowrate) was one of the first design parameters used; short HRTs were chosen for what was considered to be weak wastewater and long HRTs for strong wastewater. Loading criteria eventually were developed, typically relating to the mass of biochemical oxygen demand (BOD) applied per day per mass of microbial solids present in the aeration basin.

Design equations were developed based on the concepts of microbial growth kinetics and mass balances back in the 1960s and 1970s. McKinney (1962), Eckenfelder (1966), Lawrence and McCarty (1970), and Ramanathan and Gaudy (1971) developed design approaches that yield similar results (Gaudy & Kincannon, 1977). These design approaches were based on the behavior and performance of microorganisms in the activated sludge process as characterized by measures such as BOD, total suspended solids (TSS), and kinetic and stoichiometric parameters representing a diverse biological population. More complex process simulation models that are based on detailed influent characterization and a variety of microorganisms were developed in the 1990s (Water Environment Federation [WEF], 2013). Process simulators are discussed in more detail in Chapter 16.

Although pilot- and full-scale studies can be used to determine reaction rates and parameters for a particular wastewater and process configuration, such studies typically are not conducted for municipal wastewater applications unless a new process or other special circumstance is involved. Alternatives to such studies are (1) to assume certain wastewater characteristics and embark on a semi-empirical design, (2) to use an entirely empirical approach relying on state or other guidelines, or (3) to develop a process model based on detailed wastewater characterization and calibrated or assumed constants and coefficients. This chapter explores the first alternative. This design approach is based on mass of organisms and SRT. Wastewater with a significant industrial fraction (more than 10% of the organic load) warrants special attention to establish values of coefficients.

Capabilities of the activated sludge process, through its ability to enrich biological populations to achieve specific objectives, have increased since the 1990s. The addition of unaerated zones to a nitrifying activated sludge process can result in effective removal of inorganic nitrogen and phosphorus through biological denitrification and enhanced biological phosphorus removal (EBPR), respectively. Over the last decade, the use of membranes with pore sizes of 0.1 mm allowed for higher reactor MLSS concentrations, resulting in smaller reactor volumes and effluent of exceptional clarity. In the wide variety of activated sludge process configurations and applications in use today, the fundamental biological processes at work are the same. However, advanced microbial sequencing techniques have allowed for a better understanding and, subsequently, improved isolation of environmental conditions to select for preferred microbial consortia.

1.3 ACTIVATED SLUDGE ENVIRONMENT

An activated sludge process uses a suspension of diverse microorganisms to treat wastewater. The dry weight of these microorganisms is 95% or more organic in composition. Suspension of microorganisms in an activated sludge process is typically 70% to 90% organic and 10% to 30% inorganic substances because of inert materials in the wastewater. The organic fraction is measured as volatile suspended solids (VSS), which is differentiated from the TSS that includes the inert matter. Composition of the organic fraction of biomass is approximated by the empirical formula $C_5H_7O_2NP_{0.2}$ (Grady et al., 2011). Successful facility performance depends on a microbial community that will oxidize the waste materials and form a flocculent biomass that is readily removed by gravity separation.

Heterotrophic organisms that require biodegradable organic matter for energy and new cell synthesis typically dominate the microbial population. Autotrophic bacteria that oxidize ammonia to nitrite and nitrate use inorganic materials for energy and cell synthesis. Such autotrophs typically are present in varying concentrations. A well-designed activated sludge system provides an environment that promotes growth of desired microorganisms and inhibits those that contribute to poor sludge settleability and foaming; it also can control nuisance organisms that may appear.

Most **bulking** microorganisms are filamentous bacteria. An abundance of filaments protruding from flocs are believed to prevent biomass compaction, which would lead to dense floc particles that settle quickly and produce a turbid effluent. Some researchers contend that an ideal floc contains just the right mixture of filamentous microorganisms and floc formers, with the filaments forming the backbone of the floc (Jenkins et al., 2003; Sezgin et al., 1978). Detailed discussions of the microbiology of biological treatment systems is presented elsewhere (Grady et al., 2011; Jenkins et al., 2003; Metcalf & Eddy, Inc./AECOM, 2014; U.S. Environmental Protection Agency [U.S. EPA], 1987).

Bulking is a phenomenon at activated sludge facilities in which solids do not readily settle or concentrate because of the predominance of filamentous organisms.

Biological nitrogen removal is achieved through ammonia oxidation followed by denitrification, which is the use of nitrate by biomass as an electron acceptor to oxidize carbon substrate. The end result of this process is the conversion of nitrate to nitrogen gas that is released to the atmosphere.

Excess biomass that must be wasted will remove the portion of the influent phosphorus that is incorporated into the biomass. Biological phosphorus removal can be enhanced by enriching the bacterial culture, which results in organisms that can retain greater amounts of phosphorus than a typical aerobic biomass. Given the expected phosphorus content of the biomass to be wasted, removal can be estimated from a mass balance across the system.

Aeration is an essential component of activated sludge process for two purposes. Oxygen serves as terminal electron accepter for the aerobic organic matter degradation by the heterotrophic bacteria, and ammonia oxidation to nitrate by nitrifiers. Aeration also provides mixing to keep the activated sludge solids in suspension.

1.4 SYSTEM COMPONENTS

A basic suspended-growth system consists of several interrelated components:

- Single or multiple reactors designed as completely mixed flow, plug flow, or intermediate patterns and sized to provide adequate SRT, organic loading, or other criteria resulting in an HRT between 2 and 24 hours or more.

- An oxygen source and equipment to disperse atmospheric, pressurized, or oxygen-enriched air to the aeration basin at a rate sufficient to keep the system aerobic.
- A means of mixing the aeration basin to keep solids in suspension.
- A clarifier, membranes, or other solids separation mechanism to separate the suspended solids from treated wastewater.
- A method of collecting and returning sludge from the clarifier or recycling concentrated solids from membrane zones back to the aeration basin for flow-through configurations (not needed for the sequencing batch reactor activated sludge process).
- A means of wasting excess biomass and accumulated nonbiodegradable influent solids from the system.

2.0 PROCESS CONFIGURATIONS AND TYPES

Suspended-growth reactors, used in activated sludge and biological nutrient removal (BNR) facilities, have been designed in many different configurations. They are categorized by basin shape, loading rates, feeding and aeration patterns, and other features. Combinations of features from several categories give design engineers an array of choices. Basic activated sludge facilities have often been called by such reactor descriptions as:

- Completely mixed activated sludge
- Plug flow or conventional activated sludge
- Sequencing batch reactor activated sludge
- Combination (capable of being operated in more than one configuration)

For smaller facilities, low-load processes (such as oxidation ditches and sequencing batch reactors [SBRs]) are common in part because of simplicity of operation and reliable performance. For larger facilities, conventional plug flow (some with configuration flexibility) is favored. Plug flow often is favored because completely mixed activated sludge (CMAS) reactors can promote growth of filamentous bacteria (due to low **food-to-microorganism** ratios) that hinder sludge compaction. This growth, however, is a site-specific issue related to dissolved oxygen concentration and other factors such as the rate at which substrate is applied and total available biomass. Plug-flow reactors generally offer more flexibility if they need to be converted for BNR. Specifically, anaerobic and anoxic zones can be created. Compartmentalization by using baffles or walls within a tank can provide flexibility for intermediate zones that can be used for more than one reactor. This compartmentalization may change the performance characteristics of the system, and system simulation as covered in Chapter 16 can be helpful to evaluate changes to the effluent quality.

Food to microorganism ratio (F:M) is typically calculated as the mass of substrate entering the aeration basin divided by the mass of biomass in the aeration basin. Low F:M ratios are desirable from the standpoint of lowering the effluent substrate concentration, but they can also lead to filamentous bulking and solids settling problems.

SBRs also are used widely, especially for smaller facilities. Several thousand SBR facilities are in operation around the world. Sequencing batch reactor designs can incorporate operational or cycle flexibility that supports removing nutrients to low levels (Young et al., 2008).

2.1 REACTOR TYPES

Categorization of reactor types leads to the definitions of completely mixed, plug flow, and SBR.

2.1.1 COMPLETELY MIXED REACTOR

By definition, a CMAS reactor has uniform characteristics throughout the contents of the entire reactor. In this configuration, shown in Figure 6.1, the CMAS influent wastewater is distributed rapidly throughout the basin, and operating characteristics of MLSS, respiration rate, and BOD are uniform throughout. Because the total body of basin liquid has the same quality as the basin effluent, only a low level of food is available at any time for the large mass of microorganisms present. This characteristic is cited as the primary reason why CMAS can handle surges in organic loading and toxic shocks (to a limited extent) without producing a change in effluent quality. As mentioned above, CMAS systems can promote the growth of filamentous bacteria that settle poorly. Nevertheless, many CMAS facilities produce excellent results if properly operated. Design of a selector tank before CMAS can help limit the growth of filamentous bacteria. Control of dissolved oxygen is simpler with CMAS geometry because of the approximately uniform oxygen demand throughout the reactor.

The CMAS basins are typically square, round, or rectangular. Tank dimensions may be controlled by the size and mixing pattern of the aeration equipment and local site considerations. Surface turbine units were popular for providing aeration for CMAS tanks, although diffused air has become the preferred technology today. Factors that influence mechanical aeration mixing effectiveness include length-to-width ratio, mixing power per unit of volume, and the locations of feed points and outlet structures. Achieving complete mixing in a real-world basin is difficult but can be attempted. Any square or circular basin with a reasonable detention time and level of mixing intensity can be considered a completely mixed reactor, regardless of the type of aeration system used. The length-to-width ratio of a basin typically should be maintained at less than 3 : 1 to remain primarily a completely mixed type using mechanical aeration and no baffles. Multiple mechanical aeration units in long, narrow basins—for example, length-to-width greater than 5 : 1—create a mixing pattern that starts to resemble plug flow. If diffused air is used, then full tank width influent feed and effluent removal weir structures typically are provided as good practice. Multiple feed points and withdrawal weirs along opposite sides of a rectangular aeration basin also could be used for this purpose. Oxidation ditches may be viewed as a completely mixed reactor even though they have some plug-flow characteristics. Some consider it to be completely mixed because the influent concentration of substrate is immediately diluted by the large mixed liquor flow to a value nearly equal to that of the aeration basin effluent.

2.1.2 PLUG-FLOW REACTOR

Plug flow and basins in series are discussed together because a plug-flow reactor can be viewed as several small, completely mixed basins in series. Plug-flow basins used for municipal activated sludge facilities are 5 to 9 m (15 to 30 ft) wide and up to 120 m (400 ft) long (length-to-width ratio of more than 10 : 1). Long basins may be constructed as single-pass tanks, side by side, or in a folded arrangement. The plug flow arrangement is commonly referred to as conventional activated sludge (CAS) and is shown in Figure 6.2.

The ideal plug-flow configuration has a relatively high organic loading at the influent end of the basin. The food-to-microorganism (F : M) ratio is reduced over the length of the basin as organic matter in wastewater is assimilated. At the downstream end of the basin, oxygen consumption shifts increasingly toward endogenous respiration. The high organic loading at the head end of this process (i.e., high F : M) discourages most types of filamentous bacteria growth and results in better sludge settling compared to a completely mixed reactor if sufficient dissolved oxygen concentrations are maintained. Keeping a low dissolved oxygen level too long, however, may encourage filamentous growth. As discussed elsewhere in this chapter, polysaccharide formation may result from high loading and low dissolved oxygen concentrations at the inlet end.

Whereas a completely mixed reactor can handle surges in loading, plug-flow configurations have a superior ability to avoid "bleed-through" of untreated wastewater during peak flows. Plug-flow reactors also have an advantage where high-effluent dissolved oxygen concentrations are desirable. In a completely mixed reactor configuration, the entire tank contents would have to be maintained at the elevated dissolved oxygen level to achieve that objective. Control of dissolved oxygen concentrations in a plug-flow system can be complicated if a wide range of oxygen demands at multiple locations are expected.

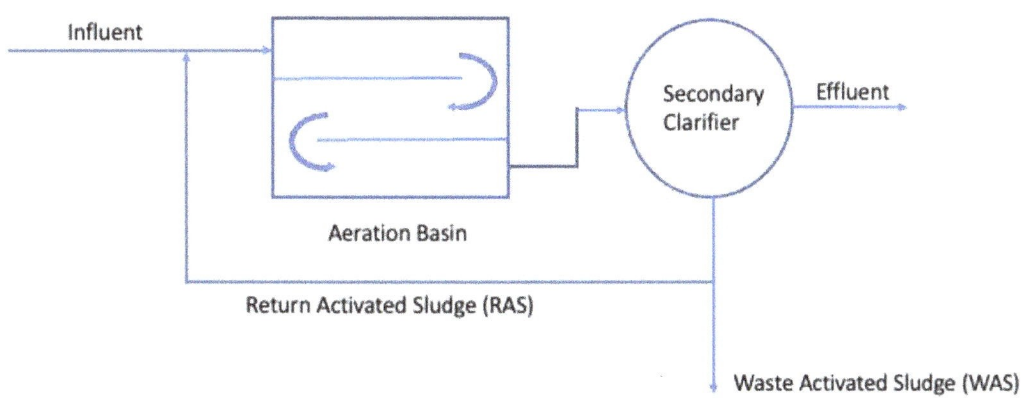

Figure 6.2 Conventional Activated Sludge

2.1.3 SEQUENCING BATCH REACTORS

The SBR process involves a fill-and-draw, completely mixed reactor in which both aeration and clarification occur in a single reactor. Settling is initiated when aeration is turned off. When settling time is up, a decanter device is used to withdraw supernatant. The sequential phases comprise a cycle with defined time intervals to achieve certain objectives. The bulk of MLSS remains in the reactor during the cycle, with periodic wasting. Specific treatment phases are illustrated as a percentage of reactor volume in Figure 6.3 as an example. The phases of each cycle include

- Fill (raw or settled wastewater fed to the reactor);
- React (aeration/mixing of the reactor contents);
- Settle (quiescent settling and separation of MLSS from the treated wastewater);
- Draw/decant (withdrawal of treated wastewater from the reactor); and
- Idle (delay period before beginning the next cycle; might include removal of waste sludge from the reactor bottom).

The idle phase may be omitted and sludge wasted at the end of the reactor draw phase. Cycles and phases may vary with each reactor. Because of the batch nature of the process, flow equalization or multiple reactors are required to accommodate the continuous and varying inflow of wastewater to the facility.

Advantages of SBR include elimination of a secondary clarifier and RAS pumping systems, high tolerance for short-duration peak flows and shock loadings, operational flexibility, and clarification that occurs under nearly ideal quiescent conditions. Disadvantages include the potential for sludge bulking

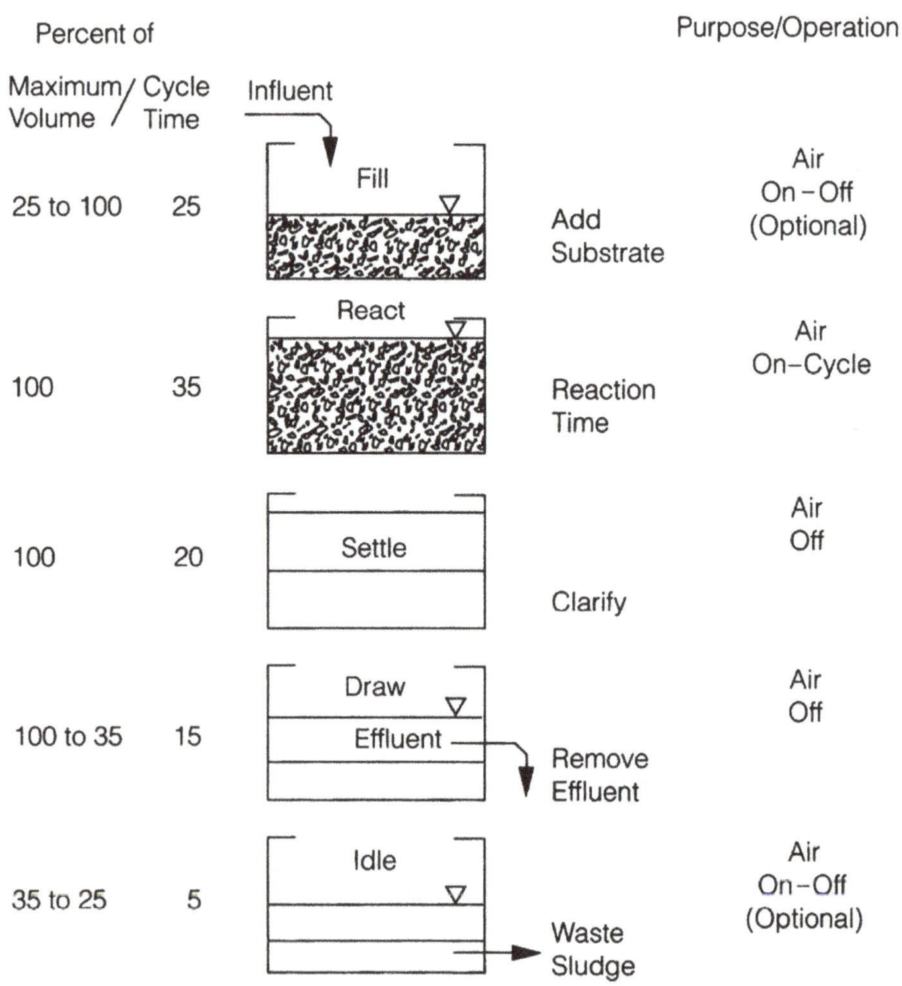

Figure 6.3 Sequencing Batch Reactor Typical Cycle

at low F : M ratios; the inability to effectively chlorinate RAS for filament control; and the need for multiple reactors for reliability, for adequate equalization, or to accommodate long-duration peak flows. Equalization of effluent decant also might be required for subsequent downstream treatment, conveyance, or discharge to small, hydraulically limited receiving waters. The intermittent cycle extended aeration system (ICEAS) was developed in Australia as a modification to the typical SBR (Goronszy, 1979). Influent feeds continuously to the reactor during all cycles as in a continuous-flow system; however, withdrawal is intermittent, similar to the SBR system. Continuous influent feed addresses some of the disadvantages of SBRs cited above. The pre-react zone of the ICEAS provides a higher F : M selector to limit the growth of filaments and also allows operation with a single reactor so that only two reactors are required for redundancy.

More than 1000 SBR-type facilities are operating in the United States. Approximately 80% of the facilities have flows of 4000 m^3/d (1 mgd) or lower; 70% have flows of 1900 m^3/d (0.5 mgd) or lower. Few, if any, are as large as 40 000 m^3/d (10 mgd). Larger facilities that have been constructed include the Kung Ming, in China (190 000 m^3/d or 50 mgd); Cardiff, in Wales (300 000 m^3/d or 80 mgd); Quakers Hill, in Australia (57 000 m^3/d or 15 mgd); and the Ringsend facility, in Dublin, Ireland (490 000 m^3/d or 130 mgd). One of the first large SBRs was constructed in Abu Dhabi, United Arab Emirates, in 1990 to provide 40% capacity expansion to the Al Mafraq water resource recovery facility (WRRF) by converting two emergency overflow basins to SBRs (Ellis et al., 1992). This two-basin SBR provides for 40 000 m^3/d (10 mgd) treatment and is still the largest two-basin SBR in the world (see Figure 6.4).

Sequencing batch reactors can be modified to provide carbonaceous oxidation, nitrification, and BNR. Nitrification takes place at the highest rates during the react phase and portions of the fill period when aeration is practiced. Because SBRs typically are designed and operated at long SRTs and low F : M, partial or complete nitrification is observed in nearly all facilities treating municipal wastewater. Denitrification can be achieved when aeration is reduced or stopped so that anoxic conditions form, but reaction rates are depressed if the reactor is not thoroughly mixed; therefore mechanical (or hydraulic) mixing might be warranted. Continuous-feed intermittent withdrawal SBR systems achieve higher denitrification rates compared to a batch feed process, including in the sludge blanket during the settle and decant phases, because carbon (BOD) is supplied continuously throughout the cycle. Conditions for EBPR can be created by incorporating a phase without aeration with readily available carbon substrate and low nitrate concentrations, such as at the beginning of a cycle. The flexibility of SBR allows for upgrades from regular carbon oxidation to BNR without costly construction.

Figure 6.4 Conversion of Two Overflow Basins to SBRs in Abu Dhabi, United Arab Emirates

2.2 LOADING RATES

Activated sludge processes also can be classified by loading or organic feed rate. Common terms are conventional, low rate, and high rate. Table 6.1 presents typical ranges for relevant design parameters.

2.2.1 CONVENTIONAL

Conventional loading rates apply to plug-flow or CMAS systems with an F:M loading of approximately 0.2 to 0.5 kg BOD/d·kg mixed liquor volatile suspended solids (MLVSS). These systems can obtain BOD removal efficiencies in the range of 85% to 95%. Conventional rate system MLSS design concentrations often range from 1500 to 3000 mg/L. Design MLSS concentrations increased considerably over the years because of improvements in the oxygen transfer capability of aeration devices, clarifier performance, and understanding of system concepts.

An important consideration in the design of conventional systems is that nitrification might occur, even when not desired. This often happens with low-loading conditions during summer months or high SRTs because of wasting practices. When nitrification occurs, denitrification may occur in the final clarifiers, resulting in "rising sludge" problems when nitrogen gas buoys and floats the biomass floc. Approaches used to limit nitrification and unwanted denitrification include reducing the SRT and the dissolved oxygen concentration to reduce nitrification, or increasing the dissolved oxygen before clarification. In warm climates, SRT reduction to prevent nitrification might adversely affect floc formation and secondary clarifier performance.

2.2.2 LOW RATE

Low-rate (also called extended aeration) facilities are characterized by the introduction of pretreated (e.g., screened and degritted) wastewater directly to an aeration basin with a long HRT, high MLSS concentration, high RAS rate, and low sludge wastage. This system, initially used in the United States for flows of approximately 4000 m³/d (1 mgd) or lower, often incorporated completely mixed reactors. During the past few decades, low-rate systems have been applied to larger sizes in the shapes of oxidation ditches and similar shapes.

Table 6.1 Design Parameters for Activated Sludge Processes (From Metcalf & Eddy, Inc./AECOM, 2014, with permission from the McGraw-Hill Companies)

Process Modification	SRT, d	F:M, kg BOD_5 Applied/ kgMLVSS·d[a]	Volumetric Loading (lb BOD_5/d/10^3 cu ft)	MLSS (mg/L)	HRT, h	Q_r/Q
Conventional	5–15	0.2–0.4	20–40	1500–3000	4–8	0.25–0.75
Complete mix	5–15	0.2–0.6	50–120	2500–4000	3–5	0.25–1.0
Step feed	5–15	0.2–0.4	40–60	2000–3500	3–5	0.25–0.75
Contact stabilization	5–15	0.2–0.6	60–75	(1000–3000)[b] (4000–10 000)[c]	(0.5–1.0)[b] (3–6)[e]	0.5–1.50
Extended aeration	20–30	0.05–0.15	10–25	3000–6000	18–36	0.5–1.50
High-rate aeration	5–10	0.4–1.5	100–1000	4000–10 000	2–4	1.0–5.0
High-purity oxygen	3–10	0.25–1.0	100–200	2000–5000	1–3	0.25–0.5
Oxidation ditch	10–30	0.05–0.30	5–30	3000–6000	8–36	0.75–1.50
Sequencing batch reactor	NA	0.05–0.30	5–15	1500–5000[e]	12–50	NA
Single-stage nitrification	8–20	0.10–0.25 (0.02–0.15)[d]	5–20	2000–3500	6–15	0.50–1.50
Separate stage nitrification	15–100	0.050–0.20 (0.04–0.15)[d]	3–9	2000–3500	3–6	0.50–2.00

[a]MLVSS = mixed liquor volatile suspended solids.
[b]Contact unit.
[c]Solids stabilization unit.
[d]Total Kjeldahl nitrogen/MLVSS.
[e]MLSS varies depending on the portion of the operating cycle.
Notes.
lb/10^3 cu ft × 0.0160 = kg/m³·d.
NA = not applicable.

A particular advantage of using long HRTs (typically 16 to 36 hours) is that they allow the facility to operate effectively over widely varying flow and waste loadings and lower overall solids production. Stable solids are often advantageous for subsequent solids-handling processes. Secondary clarifiers must be designed to handle variations in hydraulic loadings and high MLSS concentrations associated with this process.

One of the process goals is to maintain the biomass in a highly endogenous respiration phase. Because microorganisms are undergoing aerobic digestion in the aeration basin, more oxygen is required than for other single-stage systems. Many low-cost, low-rate facilities experience a dissolved oxygen deficiency when waste load is high because the design may not include automation to increase aeration in proportion to load. In some cases, the long SRT and excess dissolved oxygen at night will allow some nitrification, causing a daily, but noncoincidental, nitrification/denitrification cycle.

Some common problems with long extended aeration systems include continuous loss of pinpoint floc and the tendency to lose MLSS following short-term periods of low influent loading intensity, such as on weekends. Long HRTs combined with a long clarification time also can result in denitrification, leading to rising sludge in secondary clarifiers. This condition, combined with the lack of primary sedimentation to remove floatables, requires use of effective skimming devices in final clarifiers. Guo et al. (1981) suggest that the average MLSS concentration should not fall below 2000 mg/L. In cold climates, low temperatures likely will impair performance of the extended aeration process unless heat loss is controlled. Use of open surface aerators would be at a disadvantage compared to covered aerators or diffused air and mechanical velocity control mixers.

2.2.3 HIGH RATE

High rate is the term applied to an activated sludge system characterized by a short HRT and a high organic loading rate (OLR). The MLSS concentration may vary from 800 to as high as 10 000 mg/L; F:M ratios are higher than those used in conventional systems. Process integrity depends on maintaining the biomass in a relatively high rate of growth. Although high-rate systems can produce an effluent quality approaching that of a conventional system, they encourage a higher fraction of dispersed organisms than in higher SRT systems. This can result in turbid clarifier effluent. Therefore, high-rate systems must be operated with special care. For example, inadequate RAS flowrates, insufficient wasting, and high sludge flux rates make the clarifiers of these systems more sensitive to washout.

2.3 FEEDING AND AERATION PATTERNS

Changing the number and location of feed points of an activated sludge aeration basin can appreciably alter acceptable loading rates and quality of clarifier effluent.

2.3.1 CONVENTIONAL

Conventional activated sludge design typically would introduce influent to the head end of a rectangular basin. The RAS could be mixed with the influent before the tank or be added separately. Keeping the RAS separate facilitates subsequent conversion to other feed patterns (e.g., step feed or contact stabilization), if such flexibility is important. If RAS is blended with the influent ahead of multiple aeration basins operating in parallel, then care must be taken to ensure that the influent is well mixed before flow splitting occurs.

2.3.2 CONTACT STABILIZATION

Contact stabilization is a modification of the activated sludge process in which the feed point is moved downstream in the aeration tank (or into a separate tank). This provides a relatively short detention time for the MLSS to be in contact with the feed stream before mixed liquor leaves the reactor for solids separation. The RAS is added to the tank inlet separately and aerated before being blended with the mainstream influent. Because the upstream end of the aeration basin contains liquid at the RAS concentration instead of the MLSS concentration, a given volume of aeration tank would contain a larger mass of mixed liquor solids and, therefore, longer SRT. The longer SRT increases the time that the microorganisms are under aeration and allows for the metabolization of substrate that would otherwise not be removed because of the shortened HRT of the process. The reduced HRT results in fewer opportunities to oxidize ammonia and remove organic nitrogen. The process originally was developed to remove readily absorbed, soluble BOD, although particulate BOD is removed as well. Soluble organics absorbed and suspended organics adsorbed or enmeshed by the microorganisms in the short HRT zone subsequently are stabilized when returned to the reaeration zone. An existing aeration basin's capacity might, therefore, be readily increased by conversion to contact stabilization, if there is flexibility in feed point location, separate RAS return, and sufficient aeration capacity and distribution.

2.3.3 STEP FEED

The step-feed process, a modification of a plug-flow reactor, allows entry of influent wastewater at two or more points along the length of the aeration basin. With this arrangement, the oxygen uptake rate (OUR) becomes more uniform throughout the basin. Other operating parameters are similar to those of the conventional process. The RAS typically would be added to the aeration basin in a separate conduit at the inlet end of the basin. Step-feed configurations typically have diffused aeration. An existing plug-flow reactor can be modified for step feed by simply dividing the basin into compartments and redirecting the flow so that each compartment receives wastewater input. A step-feed tank configuration is shown in Figure 6.5.

One of the main advantages of the step feed process is decreased solids loading to secondary clarifiers. Because RAS enters separately at the head of the aeration basin, mixed liquor concentration decreases downstream as a function of the number and location of influent feed points, each of which further dilutes the mixed liquor. On the other hand, feeding influent near the head of the aeration basin increases the loading on the secondary clarifiers and, consequently, increases the RAS concentration. Shifting influent to downstream feed points lowers the solids loading rate and RAS concentration, thereby allowing higher hydraulic loading rates with less danger of clarifier solids overload.

2.4 SELECTORS

A variety of microorganisms affect settleability of MLSS, and environmental conditions can be altered to favor or select one type of bacteria over others (Jenkins et al., 2003). Mixed liquors that are low in nutrients, dissolved oxygen, or F : M tend to favor growth of filamentous bacteria that have high surface area-to-volume ratios. Many of these filamentous bacteria, which hinder settling, can be placed at a growth-rate disadvantage if mixed liquor is subjected to periods of high F : M. Organisms with the greatest ability to rapidly uptake soluble substrate and store it internally for use later during low concentration conditions tend to be those that are more flocculent and settle better.

Selector basins at the head end of main aeration basin can take advantage of these dynamics. A schematic drawing showing three selectors in series is shown in Figure 6.6. Selectors may or may not be compartmentalized like this. A single basin or even the head end of a long, narrow plug-flow aeration basin may be adequate to obtain improved results. To help reduce longitudinal mixing and overcome variations in waste flow and strength, it is recommended that three or more compartments be used to take full advantage. Albertson (2007) and the Water Environment Research Foundation (WERF) (2006) present case studies and general guidelines for selector implementation and design.

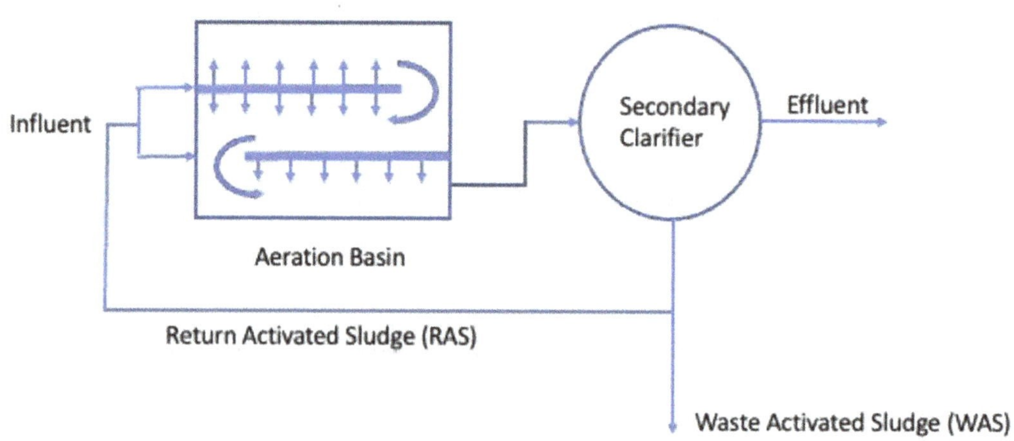

Figure 6.5 Step Feed Activated Sludge

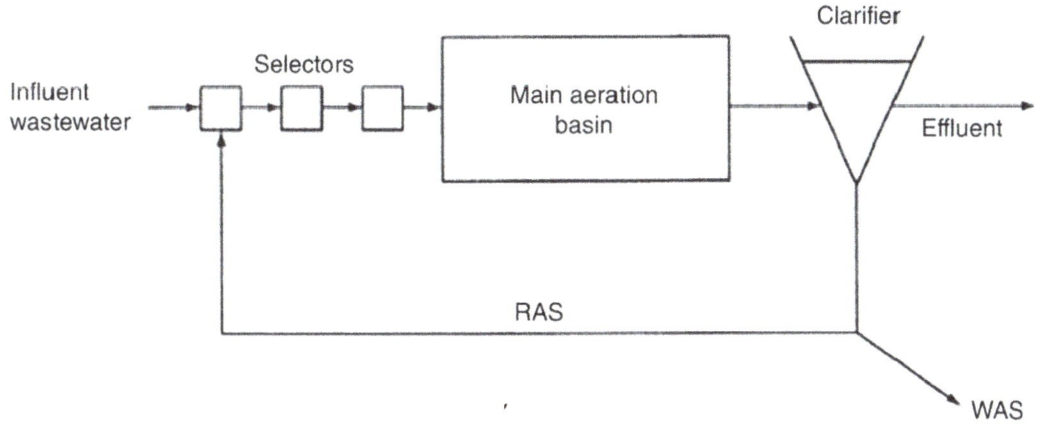

Figure 6.6 CMAS With Selectors

Selectors may be aerobic, anoxic, or anaerobic. Jenkins et al. (2003) suggest that selectors have at least three zones. For aerobic selectors, the first compartment F:M should be 10 to 12 kg COD/kg MLSS·d with an overall selector F:M of 3 to 6 kg COD/kg MLSS·d. For anoxic selectors, the first compartment F:M should be 6 kg COD/kg MLSS·d with an overall selector F:M of 1.5 kg COD/kg MLSS·d. For anaerobic selectors (often used in conjunction with enhanced biological phosphorus removal), design is typically based on detention time, and the HRT is typically in the range of 0.75 to 2.0 hours. In each case, the first two compartments should be equal in size and one-half the total selector volume when combined. The design of aerobic selectors should allow for dissolved oxygen concentrations of 1 to 2 mg/L.

Anoxic selectors can be mechanical or air mixed, with dissolved oxygen limited to low levels for the latter. If nitrate concentrations might interfere with anoxic selector performance, then denitrification might be required. Table 6.2 summarizes advantages and disadvantages of the three types of biological selectors.

2.5 OTHER VARIATIONS

2.5.1 PURE OXYGEN

Primary advantages claimed by manufacturers of the pure oxygen process include reduced power for dissolving oxygen into mixed liquor, improved biokinetics, ability to treat high-strength soluble wastewater, reduction in bulking problems from dissolved oxygen deficit stress and, with covered reactors, off-gas emissions and odors are contained.

Pure-oxygen systems in the past were characterized by high MLSS concentrations (3000 to 8000 mg/L) and relatively short HRTs (1 to 3 hours). Many of these systems for municipal wastewater are operated at MLSS concentrations of 1000 to 3000 mg/L while maintaining relatively short HRTs. This may be due, in part, to *Nocardia* foam accumulations in the reactors of some municipal facilities operating at high MLSS concentrations. Operation of older facilities at lower MLSS might also be a result of the solids loading limitations of secondary clarifiers that were sized based on overflow rate or without considering the effect of the sludge volume index (SVI) on solids-handling capacity.

For enclosed reactors, enriched oxygen gas is fed to the headspace concurrent with the wastewater flow. Mechanical mixers entrain the enriched atmosphere into the mixed liquor, and maintaining a constant gas pressure within the tanks maintains the oxygen feed. A dissolved oxygen concentration of 4 to 20 mg/L typically is maintained in the mixed liquor. Ideally, less than 10% of the inlet oxygen vents off from the last stage of the system, but 15% to 20% is not uncommon.

For open-reactor systems, oxygen is injected or entrained into a liquid stream entering the reactor (U.S. EPA, 1979). Many open-reactor systems are online worldwide, but most are small and used for municipal applications, although there are many large, open-tank systems for industrial wastewater treatment. An open-tank, pure-oxygen system can be used in combination with surface aeration to eliminate pH depletion due to carbonic acid formation. Typically, the open tank that receives the pure oxygen is the first aerobic cell that is 25% to 40% of the total aeration volume. In systems where all cells include surface or

Table 6.2 Comparison of Biological Selectors (Adapted from Sykes, 1993; Daigger et al., 2023)

Selector Type	Advantages	Disadvantages	Design Considerations
Aerobic	Simple process, no additional internal recycle streams other than return activated sludge	Does not reduce oxygen requirements	Contact time: 10–15 min. F:M: 12, 6, 3 kg COD/kg MLSS in ¼ vol first tank, ¼ vol second tank and ½ vol third tank, respectively
	Relies on basin geometry, not nitrification	Requires more complex aeration system design to meet maximum oxygen uptake rate in the initial high F:M zone	DO dissolved oxygen conc.: typically > 2 mg/L depending on F:M
Anoxic	Tends to buffer nitrification (recovers approximately 3.5 lb alkalinity as $CaCO_3$/lb of NO_3^--N denitrified)	Cannot be used with a process that does not nitrify Uses an additional recycle stream	Contact time: 45–60 min. F:M: 6, 3, 1.5 kg COD/kg MLSS in ¼ vol first tank, ¼ vol second tank and ½ vol third tank, respectively
	Lowers oxygen demand in a nitrification process (recovers approximately 2.86 lb O_2/lb of NO_3^- reduced)	Requires care in design and operation to minimize the introduction of oxygen in the anoxic zone; poor system design can induce low-dissolved oxygen bulking	Readily degradable COD to NO_3^- ratio: 7–9 mg rbCOD/mg NO_3-N
	The initial high F:M region occurs in the anoxic zone, with the high oxygen demand met by NO_3^- instead of oxygen		
Anaerobic	Simple design, no internal recycle other than return activated sludge	Does not reduce oxygen requirements May not be compatible with long solids retention times	Contact time: 45–120 min. F:M: 0.5–1 kg COD/kg MLSS in initial compartment where length: width ratio > 10:1
	The simplest of selector system to operate	Requires care in design and operation to minimize the introduction of NO_3^- and oxygen in the anaerobic zone	
	Can be used for biological phosphorus removal	Poor system design can induce low-dissolved oxlygen bulking	VFA: Orthophosphate ratio: 9–20 mg COD_{VFA}/mg P

diffused aeration to accommodate average and lower facility loading, pure oxygen can be injected to satisfy peak-loading aeration demands. Covered-tank pure-oxygen systems reduce pH values because of high partial pressure of carbon dioxide, which is not stripped out by the nitrogen gas in air leaving the liquid surface as in conventional aeration basins.

When the pH is depressed below 6.5 to 7.0, nitrification will attenuate, and the system can require a longer SRT, greater aeration tank volume and, perhaps, additional final clarifier capacity. These effects have led to consideration of separating the carbonaceous BOD (cBOD) removal and nitrification stages when using oxygen. Some design engineers suggest that the first stage of a two-stage system receive oxygen and the second stage receive air. Another option is to open up the last cell of the train to air rather than pure oxygen.

Covered tanks using oxygen have provisions for warning of potential explosions that could result from the presence of combustible, volatile hydrocarbons in influent wastewater. A detector system is used to purge the tanks automatically with air if the volatile hydrocarbon level becomes excessive. Covered tanks capture volatile organic chemical (VOC) emissions; therefore, off-gas volume from pure-oxygen systems is approximately 1% of that leaving a typical air system.

An atmosphere of high-purity oxygen and carbon dioxide in the reactor basin requires careful selection of construction materials. Compared with air, this atmosphere is more corrosive and reactive with organic compounds such as oils and greases. Some facilities have experienced corrosion of materials used in downstream conveyance channels and secondary clarifiers. Suppliers of high-purity oxygen systems have evaluated materials suitable for safe and reliable construction.

Mechanical surface turbines keep the reactor mixed. For deep tanks, submerged turbines or surface turbines with extended shafts are used to provide additional mixing blades closer to the bottom.

2.5.2 OXIDATION DITCH

In a classical oxidation ditch system, wastewater and mixed liquor are pumped around an oval pathway (racetrack) by brushes, rotors, or other mechanical aeration devices and/or pumping equipment located at one or more points along the flow circuit. Figure 6.7 shows oxidation ditches with alternative horizontal- or vertical-shaft aerators to maintain tank motion and aerate ditch contents. As mixed liquor passes the aerator, the dissolved oxygen concentration increases sharply and then declines as the flow traverses the circuit. Oxidation ditches typically operate in an extended aeration mode with long HRTs (12 to 24 hours) and SRTs (10 to 30 days). However, oxidation ditches can be designed for shorter HRTs and SRTs in warmer climates. Depending on the relative location(s) of the wastewater inlet, mixed liquor outlet, sludge return, and aeration equipment and control, oxidation ditches also can achieve nitrification and denitrification. For BOD removal or nitrification, the influent typically enters the reactor near the aerator and the effluent exits the tank upstream of the entrance. To enhance denitrification, influent should enter the ditch where the dissolved oxygen level is lower so that influent readily biodegradable organic matter can be used for denitrification.

Oxidation ditches have depths ranging from approximately 0.9 to 5.5 m (3 to 18 ft) and channel velocities from 0.24 to 0.37 m/s (0.8 to 1.2 ft/sec). Oxygen ditch depths can be as much as 7 m (24 ft) when surface turbines and draft tubes are utilized. Ditch geometry must be compatible with aeration and mixing equipment and should be coordinated with the manufacturers. Mechanical brushes, surface turbines, and jet devices are used to aerate and move the liquid flow. Combinations of diffused aeration and submersible mixers also have been employed (Christopher & Titus, 1983).

Since 1973, approximately 10 000 oxidation ditch facilities have been constructed in the United States. They are widely used in small- to medium-sized communities (5000 to 50 000 population) and with flows of 1900 to 19 000 m^3/d (0.5 to 5.0 mgd), although some are much larger. Advantages include simple operation, reliable performance, and cost-effectiveness.

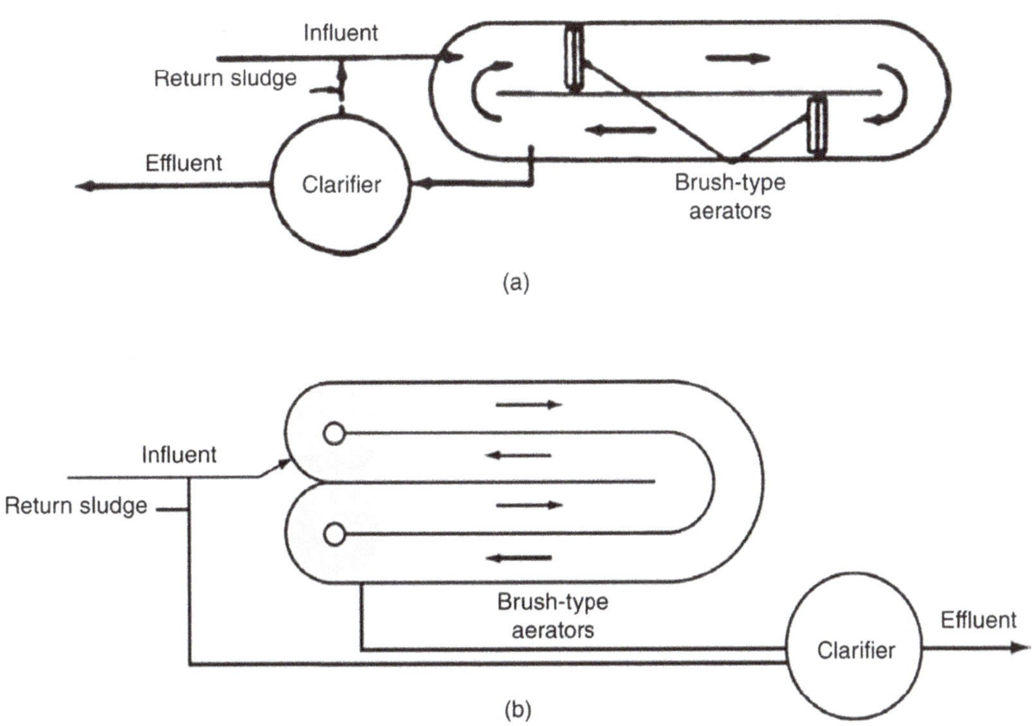

Figure 6.7 Oxidation Ditch Reactors: (a) Simple Loop and (b) Folded Loop

The vertical loop reactor (VLR) shown in Figure 6.8 is an aerobic activated sludge biological treatment process similar to an oxidation ditch. Wastewater in a VLR circulates in a vertical loop around a horizontal divider baffle. Proponents assert that overall oxygen transfer efficiency (OTE) for a VLR is higher than that of an equivalent conventional oxidation ditch due to the extended length of air bubble entrainment.

Another variation of the oxidation ditch concept is the concentric loop. There are more than 500 of these systems in the United States. In this process, mixed liquor is aerated and propelled around a series of concentric loops or a single loop by partially submerged disks that have honeycomb lattice surfaces. They carry air into the mixed liquor, which keeps moving because of drag. An opening in each of the interior walls allows flow to pass from one channel to the next in series. Different dissolved oxygen concentrations can be maintained in each loop, allowing for biological nitrogen removal.

2.5.3 AERATED LAGOON

Aerated lagoons are partially mixed, aerated reactors with long HRTs and without clarifiers or sludge return. Aerated, lagoon-style reactors consisting of lined earthen basins with HRTs of one or more days and with clarifiers and sludge return can be used to create activated sludge processes. Reactor shapes historically were square or rectangular, which is mixing limited, leading to development of oxidation ditches. Horizontal flow in a ditch keeps particles in suspension at lower energy input than mechanical or diffused aeration in lagoon reactors.

Another concept is to aerate and mix lagoon-style reactors with less energy by installing rows of diffused aerators near the bottom and then sequencing them on and off or suspending them from floating aeration piping. This extended aeration system was first introduced to the United States in 1986. Typically, these systems have lower capital costs and potentially lower operations and maintenance costs compared to conventional extended aeration basins with lower HRTs. The latter might be more cost-effective to construct where soil conditions are not favorable for a lined earthen basin. Design flows for facilities using aerated lagoon reactors are typically 400 to 190 000 m^3/d (0.1 to 50 mgd). The sequencing of aeration also can create conditions allowing for denitrification.

2.5.4 INTEGRATED SYSTEMS

Integrated fixed film activated sludge systems incorporate inert support media into the activated sludge reactors. This allows fixed film biomass to grow on the media and augment the microbial population of the mixed liquor.

2.5.5 GRANULAR SLUDGE

Granular sludge formation has been well studied under anaerobic conditions and used in anaerobic treatment processes since the 1980s; however, the first "full-scale" facility using granule formation under aerobic conditions was installed in 2005 and is patented as the Nereda® process. By 2020, there were over 80 worldwide installations of the Nereda® process for the treatment of municipal and industrial wastewater either in operation or in the planning stages. Aerobic granular sludge (AGS) processes reduce, on average, capital costs by 25%, operation and maintenance costs by 25%, and facility footprint by 75% in comparison to CAS treatment processes (Giesen et al., 2015). These savings are accomplished due to

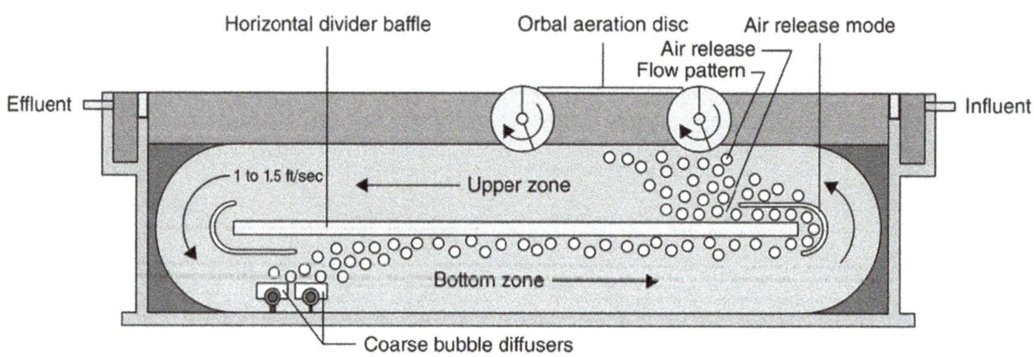

Figure 6.8 Vertical Loop Reactor

decreased volume requirements resulting from the granular biomass. The granules form distinct layers: the outer layer is the aerobic layer for heterotrophic growth and nitrification, an anoxic layer for denitrification lies below the aerobic layer, and an anaerobic layer is at the core of the granule. Operation of AGS is often at MLSS concentrations from 6 to 12 g and requires a short settling period. This enables simultaneous COD, nitrogen, and phosphorus removal in a single aerobic tank, and granule formation also improves sludge settling ability (Welling et al., 2015).

3.0 PROCESS DESIGN FOR CARBON OXIDATION

Eckenfelder (1966), Lawrence and McCarty (1970), McKinney (1962), and McKinney and Ooten (1969) led the way in developing a quantitative understanding of the activated sludge process. The work of Lawrence and McCarty (1970) was particularly significant in providing a more unified approach emphasizing the importance of SRT. This approach is the basis for carbon oxidation and nitrification process design and led to more sophisticated software models (WEF, 2013).

3.1 CARBON OXIDATION

Figure 6.10 illustrates a typical suspended-growth system flow diagram with the associated nomenclature of individual system parameters. Presentation of the fundamental relationships and development of design equations for carbon oxidation can be found in several references (Grady et al., 2011; Lawrence & McCarty, 1970; Metcalf & Eddy, Inc./AECOM, 2014; Rittmann & McCarty, 2001). Considering that we have two main events occurring in the activated sludge system, namely substrate removal (see Equation 6.1) and growth of biomass (see Equation 6.2), we can develop a set of mass balances around substrate and biomass using the general mass balance equation (see Equation 6.3).

$$\frac{dS}{dt} = -\frac{\mu X}{Y} \tag{6.1}$$

$$\frac{dX}{dt} = \mu X - bX \tag{6.2}$$

Where S = soluble biodegradable substrate concentration, mg $bCOD_5$/L or mg BOD_5/L;
μ = specific growth rate, h^{-1} or d^{-1};
X = biomass concentration, mg VSS/L;
Y = true growth yield, mg biomass formed/mg substrate utilized, mg VSS/mg bCOD or mg VSS/mg BOD_5/L; and
b = biomass decay coefficient, h^{-1} or d^{-1}.

$$in - out + generation = accumulation \tag{6.3}$$

Substituting the Monod equation for μ, the mass balance on biomass results in Equation 6.4, and the mass balance on substrate results in Equation 6.5:

$$X_o - (Q - Q_w)X_e - Q_w X_w + V\left(\left(\frac{\mu_{max} S}{K_s + S}\right)X - bX\right) = 0 \tag{6.4}$$

$$QS_o - (Q - Q_w)S_e - Q_w S_w - V\left(\left(\frac{\mu_{max} S}{K_s + S}\right)\frac{X}{Y}\right) = 0 \tag{6.5}$$

Where Q, Q_w = influent and waste flowrate, m^3/d;
X_o, X_e, X_w, X = influent, effluent, waste, and mixed liquor biomass concentration, mg VSS/L;
S_o, S_e, S_w, S = influent, effluent, waste, and mixed liquor substrate concentration, mg bCOD/L or mg BOD_5/L;

V = aeration basin volume, m^3;
μ_{max} = maximum specific growth rate, h^{-1} or d^{-1}; and
K_s = half saturation coefficient, mg bCOD/L or mg BOD_5/L.

At this point, a set of assumptions will simplify the equations. The first assumption is that the aeration basin is an ideal continuous stirred tank reactor (CSTR), meaning that the concentration of substrate is the same everywhere in the reactor as the effluent concentration (i.e., perfect and instantaneous mixing). The second assumption is that there is no biomass in the influent or effluent, and the third is that all the reactions occur in the aeration basin. When these assumptions are applied to our mass balance equations, they result in Equations 6.6 (from the mass balance on biomass) and 6.7 (from the mass balance on substrate).

$$\frac{\mu_{max} S}{K_s + S} = \frac{Q_w X_w}{VX} + b \qquad (6.6)$$

$$\frac{\mu_{max} S}{K_s + S} = \frac{QY}{VX}(S_o - S) \qquad (6.7)$$

There are some relationships that we may recognize from these two equations. One is the SRT and the other is the HRT, the definitions of which are given in Equations 6.8 and 6.9, respectively.

$$\frac{1}{SRT} = \frac{Q_w X_w}{VX} \qquad (6.8)$$

$$\frac{1}{HRT} = \frac{Q}{V} \qquad (6.9)$$

When we substitute SRT and HRT into Equations 6.8 and 6.9 and solve for the values of S and X, it results in Equations 6.10 and 6.11, respectively.

$$S = \frac{K_S(1 + b \cdot SRT)}{SRT(\mu_{max} - b) - 1} \qquad (6.10)$$

$$X = \frac{SRT \cdot Y(S_o - S)}{HRT(1 + b \cdot SRT)} \qquad (6.11)$$

These are our design equations for a single, completely mixed, activated sludge basin. When we have multiple basins, we apply the equations to each basin, resulting in completely mixed basins (or CSTRs) in series. Continuous stirred tank reactors in series is typically how existing, or proposed, treatment systems are modeled, and simulation programs as discussed in Chapter 16 are particularly helpful with this task.

One of the limits to biological treatment systems as modeled with this approach is how low a substrate concentration can be achieved. When we let the SRT approach infinity, Equation 6.10 reduces to Equation 6.12. At this point, the substrate concentration matches what is needed to support the maintenance requirements of the biomass. In other words, there is simply enough substrate to meet the biomass decay reaction, and biomass growth equals biomass decay. The other extreme situation is when we allow the specific growth rate, μ, to approach the maximum specific growth rate, μ_{max}. When we do this, we get a limiting SRT, or SRT_{min} as shown in Equation 6.13. At SRTs less than the SRT_{min}, the biomass will wash out of the bioreactor faster than they can grow and the system will fail. At most growth conditions, SRT_{min} will be less than 1 day, whereas typical SRTs are greater than 1 day, so biomass washout is not a concern. The one situation where SRT_{min} may be a concern is during startup. Without using seed biomass to start up an activated sludge system, it is important to not exceed the SRT_{min}. One way to eliminate this concern is to design the system with HRT greater than SRT_{min}, and the system will grow biomass faster than it is washed out of the system.

$$S_{min} = \frac{K_S \cdot b}{\mu_{max} - b} \tag{6.12}$$

$$SRT_{min} = \frac{K_S + S_o}{S_o(\mu_{max} - b) - K_S \cdot b} \tag{6.13}$$

Table 6.3 presents ranges and typical values for the coefficients. The units for substrate can be either COD or BOD as long as the coefficients selected are the appropriate units. The reader is cautioned that the units of mass for reported coefficients can be in terms of total or volatile suspended solids and be based upon active biomass, total biomass that includes cell debris, or total mass.

More recent computer simulations often divide influent COD into multiple fractions including:

rbCOD—readily biodegradable COD;
nbCOD—nonbiodegradable COD; and
sbCOD—slowly biodegrable COD.

Each fraction is further divided into soluble, colloidal, and particulate components with each fraction having different responses to the variable conditions in the treatment process (i.e., rbCOD is soluble and divided into acetate and non-acetate fractions, sbCOD is divided into colloidal and particulate, and nbCOD is divided into cellulosic and non-cellulosic). A detailed waste fractionation study can be performed to generate influent data for modern computer models, although these studies can be costly. The reader is referred to Chapter 16 for more detailed discussion of the benefits of calibrating models and the data required for the proper fractionation of the influent wastewater.

In design, it is important to account for all components of incoming wastewater that influence solids production and oxygen demand. Influent microorganisms are assumed to be negligible relative to those in the reactor, but can influence oxygen demand patterns in the aeration tank (Grady et al., 2011). Non-volatile suspended solids can be estimated as the difference between influent TSS and VSS. Nonbiodegradable VSS (organic) are approximately 40% of influent organic or VSS (Dague, 1983). Nonbiodegradable volatile solids cannot be directly measured but must be estimated.

Biodegradable VSS are presumed to be rapidly adsorbed into the biomass and subsequently hydrolyzed (solubilized) and are reflected in the BOD or COD of the influent and ignored as a specific component of suspended solids. Slowly hydrolyzable biodegradable volatile solids, however, can significantly affect system kinetics and mass balances.

Table 6.3 Activated Sludge Kinetic Coefficients for Heterotrophic Bacteria at 20 °C (From Metcalf & Eddy, Inc./AECOM, 2014, with permission from the McGraw-Hill Companies)

Coefficient	Unit	Range	Typical Values
μ_m	d^{-1}	3.0–13.2	6.0
K_s	mg bCOD/L	5.0–40.0	20.0
Y	mg VSS/mg bCOD	0.30–0.50	0.40
b	d^{-1}	0.06–0.20	0.12
Temperature coefficient, θ values			
μ_m	Unitless	1.03–10.08	1.07
k_d	Unitless	1.03–1.08	1.04
K_s	Unitless	1.00	1.00

The above equations allow for estimation of process volume, waste sludge production, RAS ratio, and oxygen required for an activated sludge system for cBOD removal. Design of a system requires determination of several items: number and volume of aeration basins, quantity of sludge to be wasted, oxygen demand (average and peak), return activated sludge flowrate, waste activated sludge flowrate, and sludge production.

By convention, the stated HRT of suspended-growth reactors typically excludes the RAS flow and other recirculation or recycle streams and is based only on the flow influent to the process. Alternative definitions of HRT or other parameters can be used, but their basis needs to be indicated.

The SRT of Equations 6.10 and 6.11 represents the theoretical SRT required to achieve a required effluent soluble substrate concentration. Typical designs are based on higher SRT values to provide for biological solids that have desired settling properties. For carbonaceous substrate removal by heterotrophic organisms, the theoretical SRT is used rarely. Rather, an SRT is selected based on experience. A safety factor can also be applied to establish a value for design (Rittmann & McCarty, 2001).

The SRT is the total mass of solids in a system divided by the rate at which solids are wasted. It is based on reactor volume only. Including secondary clarifier solids in the total would be an alternative approach, but would require an estimate of solids in the clarifier, which is an operational variable and beyond the designer's control. Additionally, sludge in clarifiers is often anoxic or anaerobic, and including it in an SRT definition could result in undersizing the system. For systems with low SRT and short HRT, such as pure oxygen activated sludge systems, the difference between the two methods of SRT calculation can be significant. Clear representation of the basis of the calculation is recommended in any case.

The true and net yield coefficients in the above equations characterize the production of biomass. An additional term, observed yield, Y_{obs}, is calculated from data gathered at operating activated sludge facilities and is equal to the sum of effluent and waste solids divided by the mass of substrate removed over the same period. Ignoring effluent solids, Y_{obs} can be determined from the relationship between the true growth yield, decay coefficient, and SRT as shown in Equation 6.14.

$$Y_{obs} = \frac{Y}{1 + bSRT} \tag{6.14}$$

From the observed yield, the biomass production, P_x, can be calculated using Equation 6.15.

$$P_x = Y_{obs} Q(S_o - S) \frac{kg}{1000\ g} \tag{6.15}$$

Oxygen requirements can be determined from Equation 6.16, which considers only the carbonaceous oxygen demand. In this equation, the second term is the oxygen equivalent of the biomass. An activated sludge system intended to remove cBOD that has a design or operating SRT equal to or greater than the theoretical minimum SRT for nitrification will nitrify to some degree, and therefore will experience additional oxygen demand, and Y_{obs} will include nitrifier biomass. In that case, the oxygen equivalent of the nitrifier biomass and cell debris produced should be subtracted from the additional oxygen demand resulting from nitrification. Nitrification oxygen demand and biomass production are discussed in Chapter 8.

$$O_{2req} = \left(\frac{Q(S_o - S)}{f} \cdot \frac{kg}{1000\ g} \right) - 1.42 P_x \tag{6.16}$$

Where f = conversion factor for BOD_5 to BOD_L, typically 0.45 – 0.68; f = 1.0 when S_o and S are in units of bCOD.

It is important to account for recycle streams from solids processing in the concentration and variability of influent. Although expected to be low in soluble BOD, return flows from dewatering of aerobically or anaerobically digested WAS could have significant particulate BOD and nonbiodegradable solids. Consideration also should be given to different potential modes of process failure, including an underloaded facility failing to perform when subjected to design conditions without adequate time for the system to acclimate.

The equations above are for a single, completely mixed reactor–fed soluble substrate (or assuming that solubilization of particulate substrate is inconsequential and not rate limiting). Plug-flow reactors are more efficient for first-order reactions that occur in the activated sludge process but are mathematically too complex to solve directly with a simplified model. Process modeling, as presented by the manual *Wastewater Treatment Process Modeling* (WEF, 2013) and discussed in Chapter 16, is required to assess plug-flow and other reactor configurations. Typically, a tanks-in-series approach of multiple completely mixed basins is taken to approximate the conditions in plug-flow or CAS systems.

3.2 DESIGN CONSIDERATIONS

Temperature, dissolved oxygen, nutrients, toxic and inhibitory wastes, pH, and the inherent variability of wastewater affect performance of activated sludge systems.

3.2.1 TEMPERATURE

Temperature will affect reaction rate, stoichiometric constants, and oxygen transfer rates (OTRs). Most temperature corrections used in biological treatment designs follow the modified van't Hoff–Arrhenius equation:

$$K_{T_2} = K_{T_1}\theta^{(T_2-T_1)} \tag{6.17}$$

Where K_{T_1} = a specific kinetic, stoichiometric, or mass transfer coefficient at temperature T_1;
K_{T_2} = a specific kinetic, stoichiometric, or mass transfer coefficient at temperature T_2; and
θ = temperature coefficient, dimensionless (see Table 6.3).

3.2.2 DISSOLVED OXYGEN

In systems designed for cBOD removal, a minimum average tank dissolved oxygen concentration of 0.5 mg/L is acceptable under peak loading conditions and 2.0 mg/L under average conditions. Using low values increases OTE but can lead to filament formation and poor settleability. In nitrifying systems, a minimum average tank dissolved oxygen concentration of 1.0 to 2.0 mg/L under all conditions is reasonable.

3.2.3 NUTRIENTS

An adequate nutrient balance is necessary to ensure an active biomass that settles well. Nutrients refer to nitrogen, phosphorus, and trace metals that are necessary for biological growth (Metcalf & Eddy, Inc./AECOM, 2014). Systems with higher SRT are expected to require fewer nutrients in the influent because nutrients released during endogenous respiration become available for growth of active biomass. Because nutrient requirements depend on SRT, they can be based on excess biomass and cell debris produced. The minimum nitrogen and phosphorus requirements are assumed to be 12% and 2% of the excess biomass and cell debris generated, respectively (based on the empirical formula for biomass of $C_5H_7O_2NP_{0.2}$). Normal domestic wastewater typically contains ample nutrients. Wastes with substantial industrial contributions may require nutrient addition.

3.2.4 TOXIC AND INHIBITORY WASTES

The presence of certain inorganic and organic constituents can inhibit or destroy suspended-growth system microorganisms. A listing of many of these (e.g., chlorinated phenols, quaternary ammonia) is presented elsewhere (Grady et al., 2011). Nitrification processes are particularly sensitive to toxic inhibition (U.S. EPA, 1993; Water Research Commission, 1984).

3.2.5 pH

The pH of mixed liquor should range from 6.5 to 7.5 for optimum cell growth in cBOD removal systems. Nitrifying systems are more sensitive to system pH because the rate of growth of these organisms is a function of pH over the range of 6.5 to 7.5 (U.S. EPA, 1993). Pure oxygen systems often depress pH more than air systems because the former traps carbon dioxide (formed in respiration) in the mixed liquor due to the enclosed headspace coupled with lower gas stripping rates. Unless stripped out by downstream channel aeration or a similar process, at least some carbon dioxide is recirculated through clarifiers back to the reactors. To avoid pH reduction, a residual alkalinity of at least 60 mg/L (as calcium carbonate) for either pure oxygen or conventional aeration systems should be provided. Operating at 50 mg/L level is minimal, and a value of 80 to 100 mg/L would better maintain a stable pH under varying conditions.

3.3 DESIGN APPROACH

3.3.1 INFLUENT CHARACTERISTICS

Municipal loads and wastewater characteristics typically vary with season, day of the week, and hour of the day. Influent characterization is discussed in Chapter 1. Unless these variations are addressed in the design of a facility, process performance can be affected significantly. The aeration basin/secondary clarifier combination is vulnerable to high levels of such change. Excessive hydraulic peaks shift aeration basin solids inventory to the clarifiers, which may not be able to contain the solids. Temperature changes may adversely affect the settling of solids in the clarifier, resulting in a loss of solids to the effluent and affecting the rates of reactions. Increases in organic load may lead to deterioration of mixed liquor settling and turbid effluent high in suspended solids. Periodically, toxic compounds in the influent may significantly reduce biological activity in the aeration basin and result in poor process performance. In some areas, long-term trends of increasing wastewater strength resulting from efforts to reduce infiltration and inflow into the collection system and from water conservation can significantly reduce nominal hydraulic capacity of a facility. Such trends of increasing influent concentrations might become apparent only through examination of data spanning many years.

Activated sludge facilities can include features to account for influent variations. Such features include flow equalization, larger clarifiers, alternative aeration basin feed patterns, and greater RAS capacities. Process models can be used to evaluate flow and load patterns to quantify the effects of peak flows, such as solids inventory shifting, and effluent quality variations. Furthermore, higher levels of in-facility sensing and automation facilitate flow diversion strategies that allow process units to be reduced in size from those with less sophistication.

3.3.2 VOLUME OF AEROBIC REACTOR BASINS

Sizing of aeration basins is based on several important factors. The first is sufficient time to remove soluble and particulate substrates (and oxidize ammonia nitrogen, if required) and to allow biomass activity to return to a declining growth or endogenous level. The second is maintenance of flocculent, well-settling MLSS that can be removed effectively by gravity settling. Another consideration is providing sufficient volume for oxygen transfer to occur. Lastly, the aeration basin volume will determine the mixed liquor concentration and, in effect, the solids loading on the secondary clarifiers. A larger volume (and consequently HRT) results in a lower solids loading.

For municipal systems, it is recommended that process design be based on S_e to meet effluent limits. Equation 6.10 is based on the Monod kinetic relationship for soluble biodegradable substrate. When it is used, the value of SRT predicted may be unrealistically low and is often scaled upward by a safety factor (Dague, 1983; Grady et al., 2011; Lawrence & McCarty, 1970; Metcalf & Eddy, Inc./AECOM, 2014; U.S. EPA, 1993; Water Research Commission, 1984). Alternatively, information from the literature, similar facilities, or pilot-facility studies may be useful for estimating SRT and other kinetic parameters.

From a practical point of view, selection of SRT for cBOD removal systems is not only based on kinetic considerations, but also on experience. Design is typically based on providing a high enough value of SRT for the system to yield a flocculent sludge that settles well and produces a clear effluent. Figure 6.9, which represents a non-filamentous sludge grown on a soluble waste (glucose plus yeast extract), shows that a minimum SRT value of approximately 3 days is required (Bisogni & Lawrence, 1971). In practice, an SRT of 1 to 5 days typically is used during warm weather and up to 15 days during cold weather. Nitrification may well occur in these ranges and should be taken into account during design. Values of SRT outside this range are selected in situations where environmental or performance conditions warrant lower or higher values.

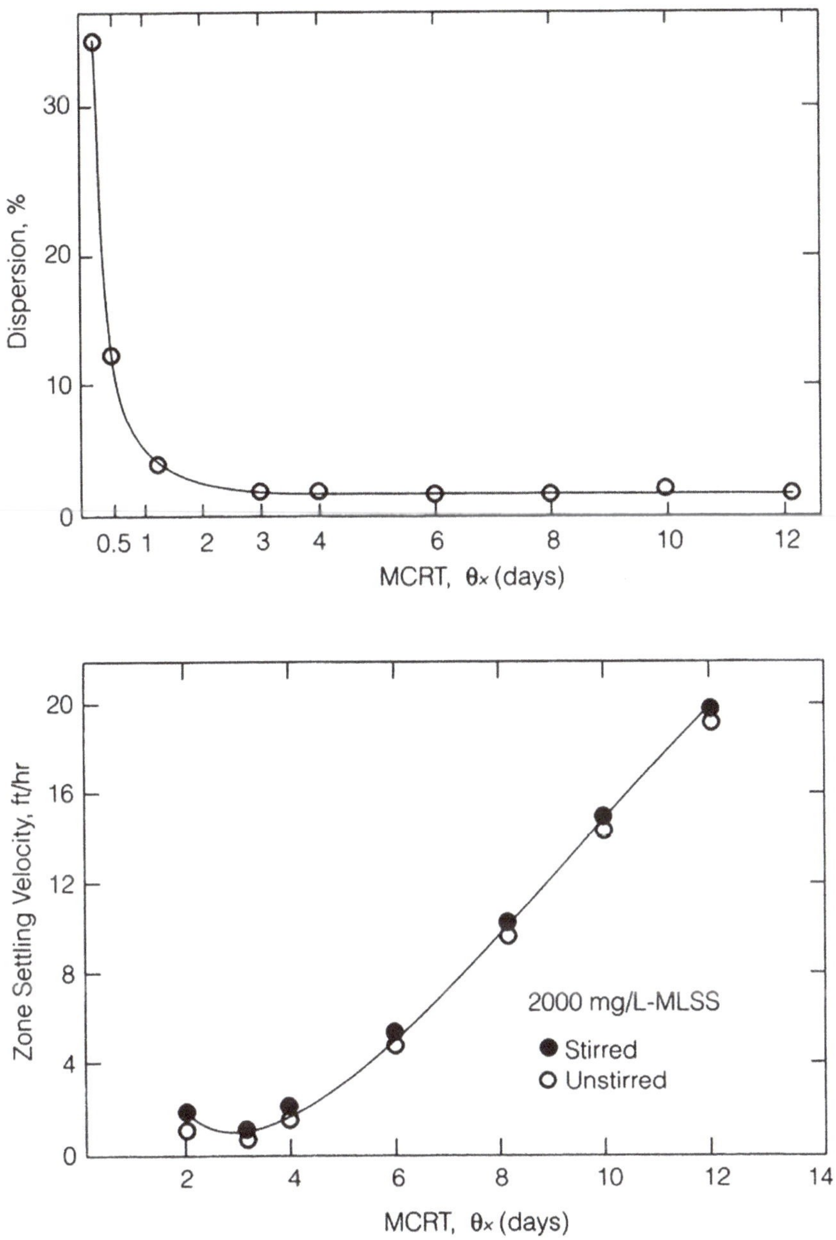

Figure 6.9 Effects of SRT (or Mean Cell Residence Time, MCRT, θ_x) on Dispersed Growth and Settling Velocity (Bisogni & Lawrence, 1971)

In warm climates where nitrification is not desirable, SRT values of 1 to 2 days are used. Also, long SRT values are often used in extended aeration systems in which secondary goals require minimization and stabilization of the excess sludge solids generated. Chapter 12 discusses the challenges of meeting federal regulations (Section 503) for stabilization by providing long SRTs. Figure 6.10 presents suggested ranges of SRT for BOD removal and nitrification at various temperatures (Bisogni & Lawrence, 1971).

Selection of MLSS may be determined by trial and error in the design process. Optimizing the aeration tank and clarifiers design should be based on the SRT required for wastewater treatment, oxygen transfer limitations, solids settling characteristics, and the allowable solids loading rate to the secondary clarifiers. Conventional air activated sludge system MLSS concentrations ranging from 1500 to 3000 mg/L often are used. It is possible, however, for these systems to accommodate higher concentrations. Extended aeration systems are frequently designed for up to 4000 mg/L MLSS. Membrane systems can be operated at MLSS concentrations of more than 10 000 mg/L, but foaming problems may be observed at values greater than 10 000 mg/L (2000 mg/L for systems with mechanical surface aerators).

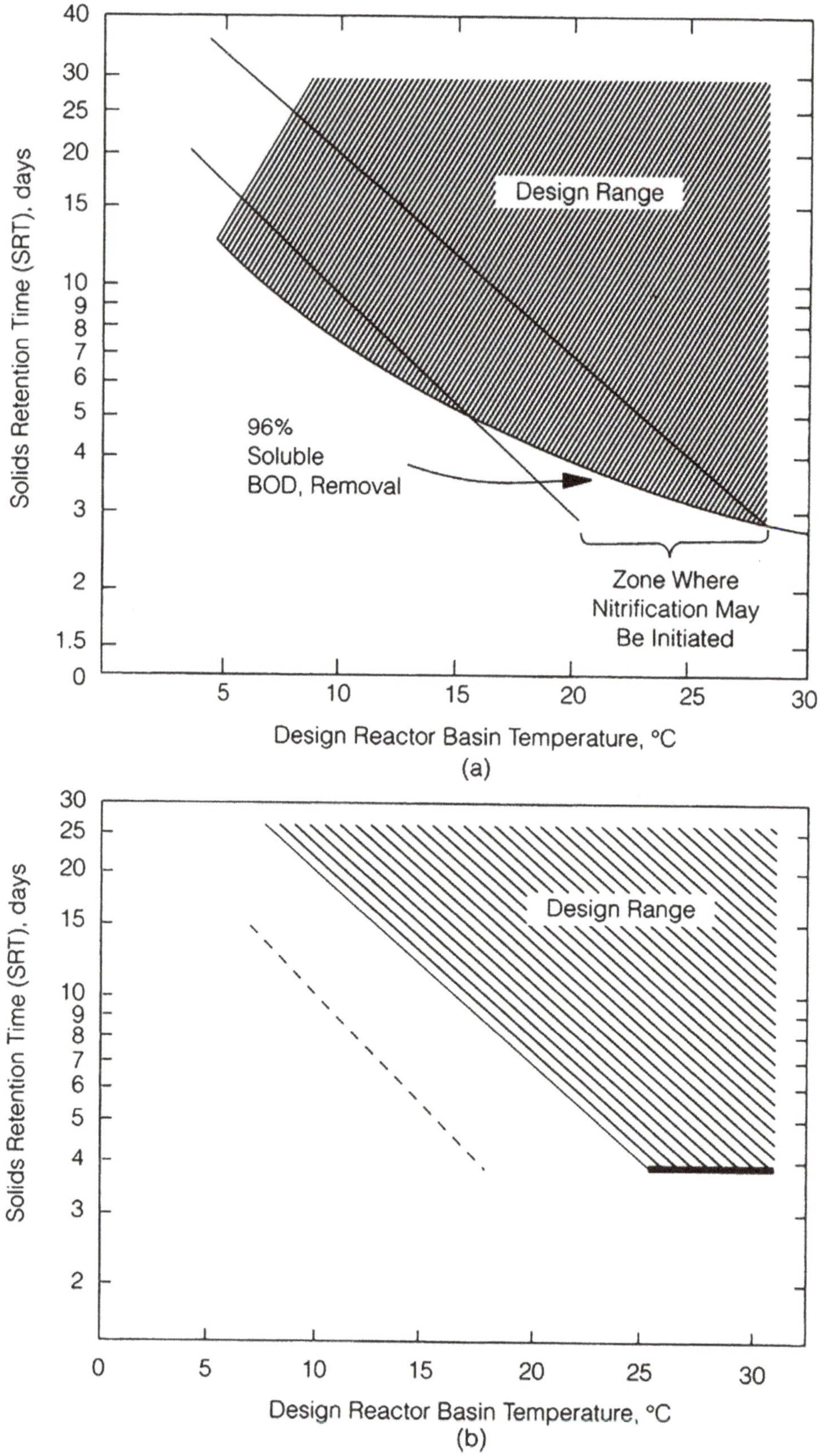

Figure 6.10 Design SRT for cBOD (a) Removal and (b) Nitrification (Bisogni & Lawrence, 1971)

Solids settling and thickening properties often dictate final selection of MLSS concentration for systems with secondary clarification. For air activated sludge systems, design for concentrations more than approximately 5000 mg/L is seldom economical (Eckenfelder, 1967). Figures 6.11 and 6.12 show suggested values as functions of SVI and temperature. Most facilities in the United States do not exceed 5000 mg/L MLSS. Operating MLSS levels for some of these facilities are 1500 mg/L or less in warm climates. Membrane bioreactors (MBRs) are designed for much higher MLSS concentrations, as discussed in Chapter 4.

Example Problem 6.1

Design an activated sludge system to meet 30 mg/L BOD_5 and 30 mg/L TSS effluent concentrations given the following:

$Q = 3785\ m^3/d$
$S_o = 220$ mg BOD_5/L (aeration basin influent concentration)
$\mu_{max} = 3\ d^{-1}$
$K_s = 40$ mg BOD_5/L
$Y = 0.5$ mg VSS/mg BOD_5
$b = 0.15\ d^{-1}$
BOD_5 of effluent TSS = 0.65
Design temperature = 20 °C

Solution

The first step in performing the process design is to calculate what the effluent soluble substrate concentration would be assuming that the clarifier is operated at its permitted effluent TSS concentration. Therefore, the effluent soluble substrate concentration for design can be calculated as the permitted BOD_5 minus the BOD_5 of the effluent TSS as shown in Equation 6.18.

$$S_{des} = \text{permit } BOD_5 - BOD_5 \text{ of effluent } TSS \tag{6.18}$$

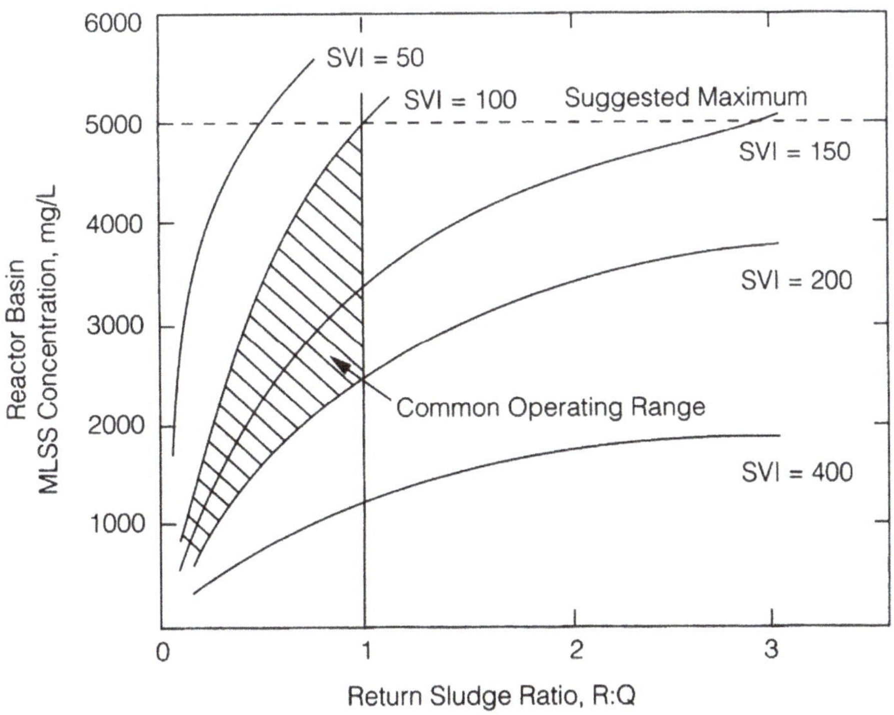

Figure 6.11 SVI as a Function of MLSS and Return Sludge Ratio at a Reactor Basin Temperature of 20 °C

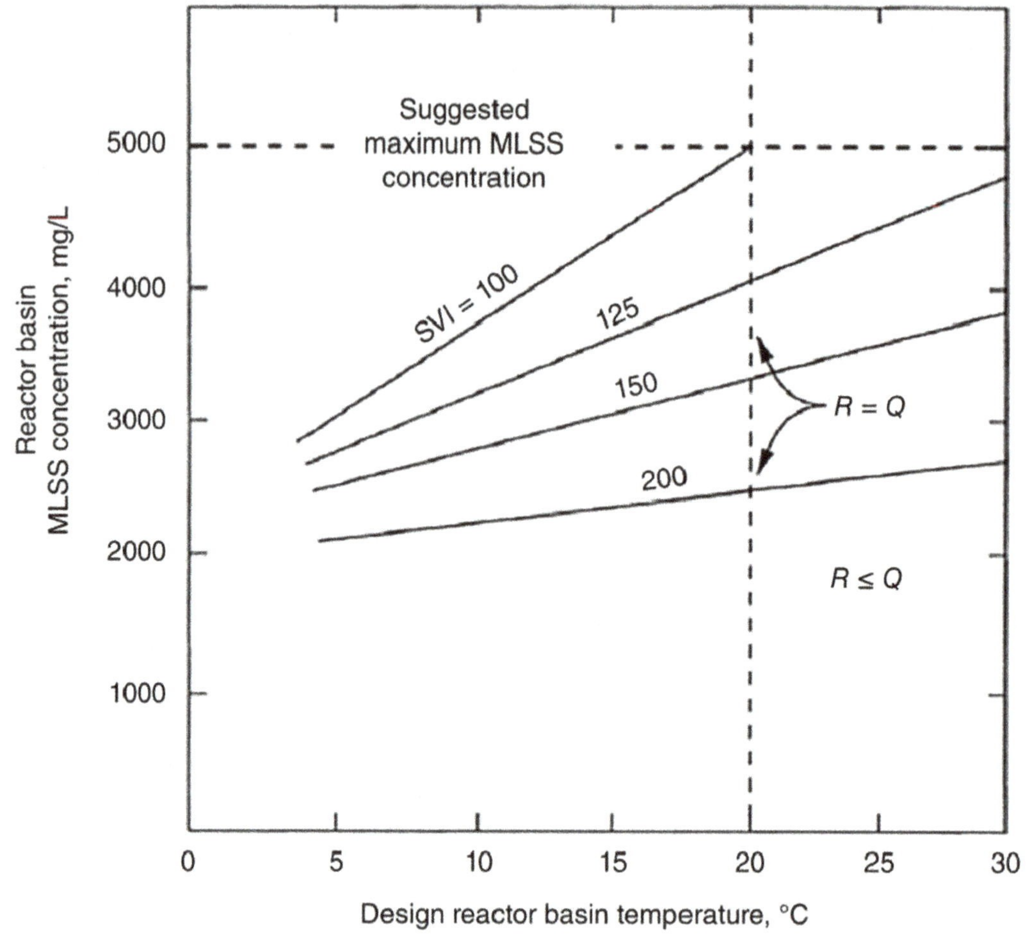

Figure 6.12 Suggested Maximum Mixed Liquor Suspended Solids Versus Temperature and Sludge Volume Index at Aerator Temperature

For our example, $S_{des} = 30 - (0.65)30 = 10.5$ mg/L as BOD_5. It is instructive to know what the minimum S_{des} achievable would be. Using Equation 6.12, we calculate:

$$S_{min} = \frac{40 \cdot 0.15}{3 - 0.15} = 2.1 \text{ mg/L as } BOD_5$$

If we get to this step in the design and realize that we cannot meet permit requirements even with a very long SRT, we still have options. The simplest is that we could stipulate that the clarifiers must be operated with effluent TSS concentration lower than the permitted level (thus allowing us an S_{des} that is achievable at reasonable SRT values). The next step is to calculate the required SRT to achieve this soluble effluent substrate concentration from Equation 6.10. Rearrangement of Equation 6.10 gives:

$$SRT = \frac{K_S + S}{S(\mu_{max} - b) - K_S \cdot b} \tag{6.19}$$

Substituting S_{des} in Equation 6.19 gives a value for SRT of 2.11 d:

$$SRT = \frac{40 + 10.5}{10.5(3 - 0.15) - 40 \cdot 0.15} = 2.11\ d$$

At this point in the design, the engineer has a degree of freedom with respect to the relationship between aeration basin volume (which is a function of HRT) and mixed liquor volatile solids concentration. If we select an HRT, we can calculate the resulting MLVSS concentration. Conversely, we can select an MLVSS concentration and calculate the resulting HRT. For our example, let's assume an HRT = 4 h (0.167 d). From Equation 6.11 we get the following MLVSS concentration:

$$X = \frac{2.11 \cdot 0.5(220 - 10.5)}{0.167(1 + 0.15 \cdot 2.11)} = 1005 \text{ mg VSS/L}$$

If we use the common assumption that $X' = 1.2 \cdot X$, where X' = MLSS, then X' = 1206 mg/L. Since this value is reasonably close to the common range of 1500 to 3000 mg/L, we can assume this a reasonable HRT and the resulting volume of the aeration basin can be calculated from Equation 6.9 as follows:

$$V = Q \cdot HRT = 3785 \frac{m^3}{d} \cdot 0.167 \ d = 632 \ m^3$$

Good engineering practice would dictate that we need multiple units of every treatment process, so we will design two aeration basins with a volume of 316 m^3 each. In is also good practice to see whether the HRT is greater than the SRT_{min} to ensure that the system can self-start during startup. Using Equation 6.13, we calculate

$$SRT_{min} = \frac{40 + 220}{220(3 - 0.15) - 40 \cdot 0.15} = 0.419 \text{ d}$$

Because this value is greater than our assumed HRT, we could either adjust HRT to 0.42 d, or stipulate during our design report that the system will either need to be started with seed sludge from another activated sludge system or we must be sure to start the system with a flow less than the design flow. For an HRT = 0.42 d, the flow should not exceed 1505 m^3/d during startup under these conditions.

3.3.3 WASTE SLUDGE GENERATION

The amount of biomass generated can be estimated using Equation 6.15 in terms of VSS. In reality, waste sludge will also include non-volatile, volatile biodegradable, and non-biodegradable VSS. A more detailed approach of estimated waste sludge generation uses an estimate of the fraction of biomass debris, f_D, resulting from decay. This value ranges from 0.08 to 0.2 and is commonly assumed to equal 0.15 (Metcalf & Eddy, Inc./AECOM, 2014). Inert, degradable, and biomass solids in the influent will also add to the waste sludge generation, and any precipitates that form from the addition of iron or aluminum salts to the activated sludge process for phosphorus removal or other purpose also should be included in this calculation. Figure 6.13 illustrates net secondary treatment system sludge production (to be removed as WAS and secondary effluent suspended solids) for the stated waste characteristics, both with and without primary sedimentation. Other texts (e.g., Grady et al., 2011) provide detailed extensions of the traditional model to account for inert and biomass solids and their effect on biomass yield, sludge production, effluent substrate concentration, and oxygen requirements. Chapter 16 in this text provides another approach to follow constituents in the activated sludge process for both steady-state and dynamic operation using simulation models based on the International Water Association's activated sludge model.

3.3.4 OXYGEN DEMAND

The oxygen demand of an activated sludge process can be estimated using Equation 6.16. Additional oxygen demand also can result from the presence of readily oxidizable compounds in the influent such as sulfide that has an approximate demand of 2 mg/L oxygen per mg/L sulfide (as S). Nitrification significantly increases the oxygen demand and is discussed in Chapter 8. Oxygen demand typically varies both spatially and temporally in a suspended-growth system. Temporal variations can be estimated from statistical analyses of data collected for influent loadings (cBOD and nitrogenous oxygen demand). Spatial variations depend on kinetic relationships between growth rates of the biomass and substrate removal rates and dissolved oxygen concentrations. They also depend on the flow regime and HRT of the process. Estimates of variation can be obtained from process computer models described by WEF (2013) or literature data.

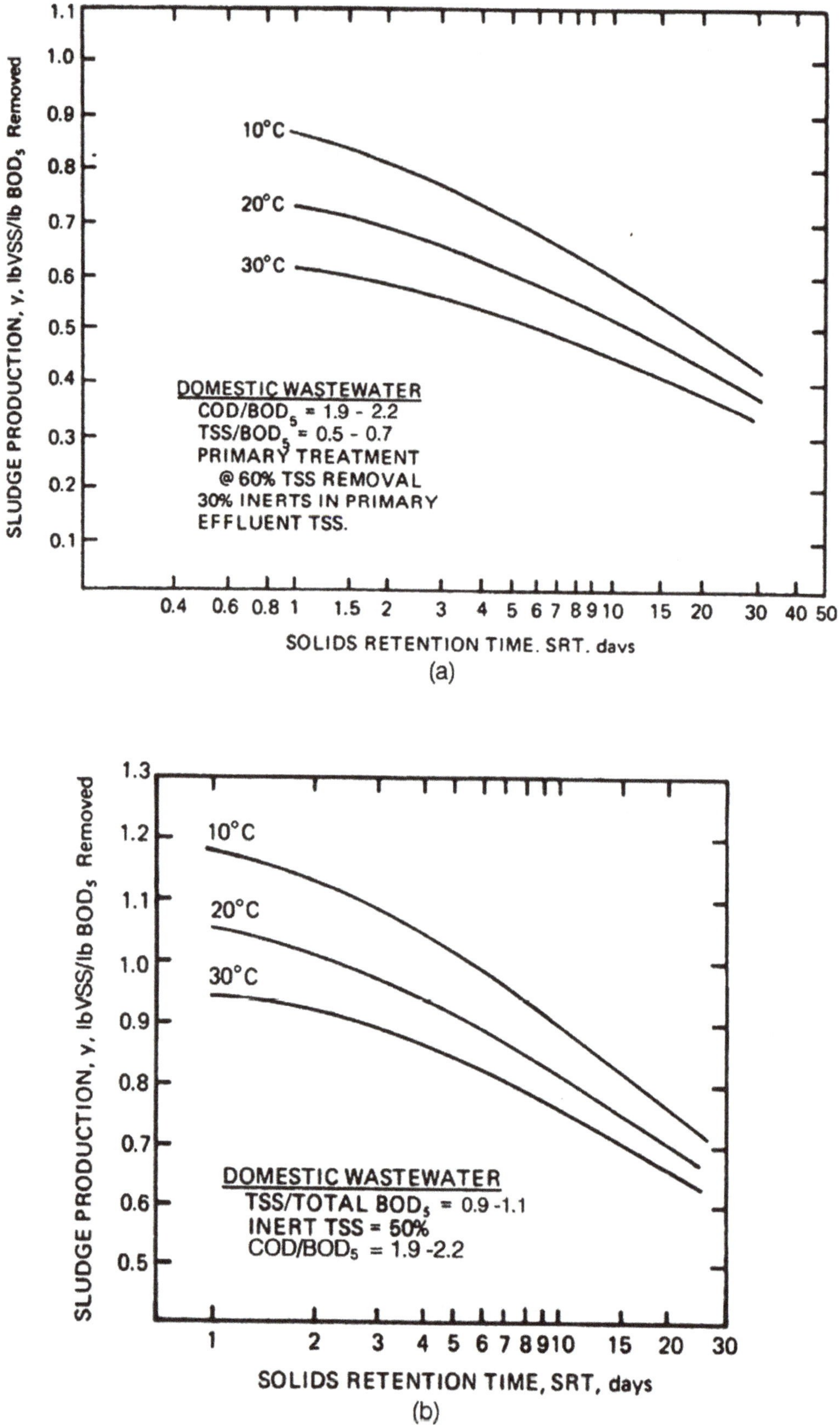

Figure 6.13 Net Sludge Production Versus SRT and Temperature (a) With Primary Treatment and (b) Without Primary Treatment

Total maximum oxygen demand for design should be based on peak loadings anticipated. At a minimum, requirements for a conventional system should be based on the 24-hour demand of the average day of the peak month. Some designers prefer to use the peak day or the peak 4-hour demand of the average day of the peak month. Basing the requirement on the peak day demand plus 50% of the peak 4-hour rate for the peak day has been suggested (Young et al., 2008). While the blowers must be sized to handle peak loads, multiple blowers should be provided to allow the airflow rate to be turned down to treat loads at or below average conditions.

3.3.5 RETURN ACTIVATED SLUDGE REQUIREMENTS

Requirements for RAS pumping capacity can be estimated for an assumed clarifier underflow concentration in the absence of site-specific solids settling characteristics. The ratio of RAS flow to influent flow (α_r) affects the size of secondary clarifiers without influencing the size of aeration tanks. As a guide, the design value of α_r should range from 20% to 100% of the average facility design flow for conventional systems and up to 150% for some systems depending on peak flow factors and expected clarifier performance.

Example Problem 6.2

In a continuation of the example design problem 6.1, the next step will be to calculate the RAS flowrate and concentration. Assuming a simple mass balance (ignoring the waste sludge), we obtain the following:

$$(Q+Q_r)X' = Q_r X'_r \quad (6.20)$$

Where Q_r = return activated sludge (RAS) flow rate; and
X'_r = return activated sludge concentration, mg/L as TSS.

The only unknown in Equation 6.20 at this point is the value of X'_r which can be derived with knowledge about the clarifier underflow concentration. In the absence of any historical or current field data, an estimate of the sludge volume index, SVI, can be used to derive the value of X'_r from Equation 6.21:

$$X'_r = \frac{10^6}{SVI} \quad (6.21)$$

An estimate of SVI for Equation 6.21 can be obtained from Figure 6.12. At an aerator temperature of 20 and MLSS of 1206 mg/L, the SVI can be approximated around 250, which divided into 10^6 gives an underflow concentration X'_r = 4000 mg/L. Using this value in Equation 6.20 results in a $Q_r = 0.43Q$, or $\alpha_r = 0.43$. This is an acceptable value (normal range is 0.2 to 1.0) and results in a Q_r = 1628 m^3/d.

3.3.6 SOLIDS LIQUID SEPARATION

Design of solids separation systems, whether membranes or secondary clarifiers, is an important function that is integral to the design of other components of a suspended-growth system. Details of membrane system design and clarifier sizing are covered in Chapter 4.

3.3.7 AERATION—MAINTAINING SOLIDS IN SUSPENSION

For coarse-bubble aeration with diffusers along one or two sides of the tank, inducing a spiral-roll pattern, a common criterion for aeration mixing, is 17 to 40 Nm^3/h·m (3 to 8 scfm/ft) header length (where Nm^3 is the volume of air under standard conditions of 0 °C and 1 atmosphere). An alternative criterion commonly presented in the literature is 0.9 to 1.5 Nm^3/h·m^3 (16 to 26 scfm/1000 cu ft) tank volume. An issue with this criterion is that for a given tank horizontal area, the air requirement for mixing a tank 6 m deep, for example, would be twice the air required for a tank 3 m deep.

One manufacturer provided an alternative basis for designing air mixing requirements by adjusting the unit volumetric airflow based on the tank depth. In this case, by recalculating the values provided, the criterion for mixing becomes approximately 4 $Nm^3/h \cdot m^2$ (0.23 scfm/sq ft) for spiral-roll mixing and for channel mixing in activated sludge systems. For channel mixing, other common criteria are in the range of 12 to 30 $Nm^3/h \cdot m$ (2 to 6 scfm/ft) header or channel length.

A commonly used design criterion for fine pore diffused aeration with full-floor coverage to maintain mixed liquor in suspension is 2.2 $Nm^3/h \cdot m^2$ (0.13 scfm/sq ft). This value has traditionally been applied irrespective of whether the flow entering the aeration basins is raw influent or primary effluent. However, some designers account for the removal of heavier material in primary clarifiers upstream of aeration tanks and adjust the minimum airflow required to maintain mixed liquor solids in suspension accordingly. With primary clarifiers, the minimum airflow required is within the range 0.9 to 1.5 $Nm^3/h \cdot m^2$ (0.05 and 0.09 scfm/sq ft) (Yunt, 1980). The shear rate may be a more fundamental approach to evaluating mixing intensity with a design range from 25 to 150 s^{-1}.

Other limitations for minimum airflow may override the above values, including the minimum airflow per diffuser to maintain air distribution throughout the grid and to prevent clogging of the diffuser, as well as the air requirements to meet process oxygen requirements.

In racetrack systems, some manufacturers use a mechanical mixer to maintain mixing and the aeration grid to provide aeration. To prevent the air bubbles interfering with the mixing pattern, a minimum distance between the mixer and the downstream aeration grid must be maintained. The mixer or diffuser vendor may provide this to the designer. Typical values are either the width of the channel or the water depth, depending on the size of the racetrack. To maintain solids in suspension, the velocity in the channel should be 0.3 m/s (1 ft/sec).

3.3.8 EFFECT OF MIXING ON FLOCCULATION

Mixing systems maintain solids in suspension by imparting velocity gradients in the liquid, which prevent settlement of individual floc particles. Velocity gradients can promote flocculation of mixed liquor floc particles or, if too large, can cause rupture of the floc particles. If floc breakup occurs to an excessive degree, then effluent quality could deteriorate due to dispersed solids that are incapable of settling during the retention times provided in secondary clarifiers.

3.3.9 ACTIVATED SLUDGE FOAM CONTROL

The incidence of foaming organisms can be a major activated sludge operating challenge. They manifest themselves as a problem particularly when the design of the activated sludge reactor is such that surface foam layers are trapped and foaming organisms can accumulate. In the absence of adequate design features, foam will eventually accumulate somewhere (or all over the surface) of the aeration tanks. Eventually the foam will bridge and harden and will not flow along the surface of its own accord. This becomes a burden for operations personnel as they try to remove the foam using high-pressure hoses. Often the best recourse at this stage is to remove the foam using a vacuum truck service, if access is possible. It is usually preferable to provide design features that avoid foam formation and accumulation.

Baffles that are commonly included in BNR facilities should promote the separation of distinct zones within the reactor but, at the same time, induce movement of foam through the reactor to a point where it can be wasted from the system in the secondary clarifiers, or within the reactor.

The foaming propensity of nocardioforms such as *Gordonia amarae* is mainly caused by free-floating dispersed filaments, more so than by floc-bound filaments (Narayanan et al., 2010). With subsurface withdrawal from reactors or inadequately designed baffles within the reactor that cause foam trapping, free-floating filaments tend to proliferate. Parker et al. (2014) posited that free-floating filaments are more likely to attach to the bubbles produced by fine pore diffused aeration and accumulate at the surface. Floc-bound filaments are heavier and would require more bubble attachment, which may require a dissolved-air flotation environment, not the turbulent diffused-air environment of aeration tanks.

Jenkins et al. (2003) discussed a number of measures to control foam in activated sludge reactors. For nocardioforms, these included aerobic selectors, anoxic selectors, anaerobic selectors, chlorination, cationic polymer addition, and surface wasting. For *Microthrix parvicella,* the measures included reducing SRT, providing plug-flow conditions, avoiding intermittent aeration or low dissolved oxygen, eliminating foam trapping, and adding a polyaluminum chloride coagulant. Several of the measures, such as lowering SRT or increasing dissolved oxygen, could be contrary to other fundamental objectives, such as nitrogen and phosphorus removal. It is also common to experience foaming issues despite the best design efforts to avoid them.

4.0 OXYGEN TRANSFER SYSTEMS

Transfer rates for diffused air systems are typically reported as oxygen transfer efficiency (OTE), expressed as a percentage, or oxygen transfer rate (OTR), expressed in units of mass/time. Table 6.4 summarizes the general characteristics of primary types of oxygen transfer equipment in use for 5 or more years. More detailed information is contained in the literature (U.S. EPA, 1989, 2010).

Mechanical devices are typically rated on the basis of specific oxygen transfer efficiency (SOTE) or standard aeration efficiency, expressed in units of mass/time/unit of power. As a secondary function, aeration devices furnish sufficient energy for mixing. Ideally, mixing energy should be sufficient to thoroughly disperse dissolved substrate and oxygen throughout a given segment of an aeration tank and keep MLSS suspended. This does not necessarily mean that both soluble and suspended materials should be uniformly mixed throughout the entire aeration tank. For example, plug-flow tanks and reactors with point-source oxygen addition do not rely on uniformity for proper operation.

The power required to satisfy oxygen demand depends on substrate and biomass concentrations, flowrate, and reactor volume. The power for mixing depends on aeration tank volume and, to a lesser degree, MLSS concentration. For certain combinations of biomass and substrate concentrations and other variables listed above, power requirements for mixing may exceed those for oxygen transfer. In systems with high biomass concentrations, oxygen demand will control power requirements; under low substrate loading, mixing may control power requirements; in plug-flow systems, mixing may control near the effluent end of aeration basin; for aerated lagoons, mixing often dictates power requirements. Additionally, some oxygen transfer equipment has a minimum airflow rate that must be maintained for proper air distribution or equipment function. Such equipment requirements must also be considered when determining the minimum airflow requirement of the system.

4.1 DIFFUSED AERATION

Diffused aeration, defined as the injection of a gas (air or oxygen) below a liquid surface, covers all equipment described in this section. Hybrid equipment that combines gas injection with mechanical pumping or mixing equipment is also classified herein as diffused aeration equipment. These hybrid devices include jet aerators and turbine-sparger aerators.

The wastewater treatment industry has witnessed the introduction of a wide variety of air diffusion equipment. In the past, the various devices were classified as either fine-bubble or coarse-bubble diffusers. In recent years, additional diffusion equipment including medium bubble and ultra-fine bubble have been introduced to the industry. Because the demarcation between air diffusion equipment based on bubble size has become increasingly difficult to define, air diffusion systems have been categorized herein by the physical characteristics of the equipment. In the following discussion, air diffusion devices are divided into three categories: porous diffusers, nonporous diffusers, and jet aerators.

4.1.1 POROUS DIFFUSER SYSTEMS

Porous diffusers are commonly used because of their relatively high OTE compared to other aeration technologies. An excellent reference on this subject was published by the U.S. EPA in cooperation with the American Society of Civil Engineers (ASCE) Committee on Oxygen Transfer (U.S. EPA, 1989). Much of the information presented in this section was derived from that source. Further information can be found on these specific topics in Groves et al. (1992) and U.S. EPA (1985, 1999).

Numerous materials have been used to manufacture porous diffusers. They typically can be divided into two categories: rigid materials of ceramic or plastic, and perforated membranes of thermoplastics or elastomers.

Table 6.4 Characteristics of Aeration Equipment (Arora et al., 1985; Boyle, 1996; Goronszy, 1979; Groves et al., 1992; Wilford & Conlon, 1957)

					Reported Clean Water Performance	
Equipment Type	Equipment Characteristics	Processes Where Used	Advantages	Disadvantages	SOTE (%)	SAE (kg/kW·h)
Porous diffusers	Ceramic, plastic, flexible membranes, dome, disk, panel, tube, plate configurations, total floor grids, single or dual roll, fine bubble	High-rate; conventional, extended, step, contact stabilization, activated sludge systems	High-efficiency; good operational flexibility; turndown approximately 5.1	Potential for air- or water-side clogging; typically require air filtration; high initial cost; low alpha	15–45	1.9–6.6
Nonporous diffusers	Fixed orifice; perforated pipe, sparger, slotted tube, valved orifice, static tube; coarse bubble; typically single or dual roll; some total floor grids	Same as for porous diffusers	Do not typically clog; easy maintenance; high alpha	Low oxygen transfer efficiency; high initial cost	9–13	1.3–1.9
Jets	Compressed air and pumped liquid mixed in nozzle and discharged fine bubble	Same as for porous diffusers	Good mixing properties; higher SOTE than nonporous diffusers	Limited geometry; clogging of nozzles; requires blowers and pumps; primary treatment required; low SAE	15–24	2.2–3.5
Mechanical surface radial flow, low speed (20–100 r/min)	Low output speed; large diameter turbine; floating, fixed-bridge, or platform mounted; used with gear reducer	Same as for porous diffuser	Tank design flexibility, high pumping capacity	Aerosols: some icing in cold climates; initial cost higher than axial flow aerators; gear reducer may cause maintenance problems unless adequate service factor provided	—	1.5–2.1
Axial flow, high speed (900–1800 r/min)	High output speed; small diameter propeller, direct, motor-driven units mounted on floating structure	Aerated lagoons and reaeration	Low initial cost; may adjust to varying water level; flexible operation	Some icing in cold climates; poor maintenance accessibility; mixing capacity may be inadequate	—	1.1–1.4
Horizontal rotor	Low output speed; used with gear reducer; steel or plastic bars, plastic discs	Oxidation ditch, applied either as an aerated lagoon or as an activated sludge	Moderate initial cost; good maintenance accessibility	Subject to operational variables, which may affect efficiency; tank geometry is limited	—	1.5–2.1
Submerged turbine	Units contain a low-speed turbine and provide compressed air to diffuser rings, open pipe, or air draft; fixed-bridge application; may employ draft tube	Same as for porous diffusers, oxidation ditches	Good mixing; high capacity input per unit volume; deep tank application; operational flexibility; no icing or splash	Require both gear reducer and blower; high total power requirements; high cost	—	1.1–2.1 (Typical) 2.0–3.0 (Drain tube turbine)
Aspirating	Same as axial flow; high speed	Aerated lagoons; temporary installations	Low cost, flexible operation	Same as axial flow, high speed	—	0.5–0.8

Note. SOTE = specific oxygen transfer efficiency; SAE = standard aeration efficiency

The oldest and most common rigid porous diffuser is produced from ceramic media, including alumina, aluminum silicates, and silica. Media consist of rounded or irregular-shaped mineral particles bonded together to produce a network of interconnected passageways through which compressed air flows. As air emerges from the diffuser surface, pore size, surface tension, and airflow rate interact to produce a characteristic bubble size. Currently, the most common rigid porous diffusers are manufactured from aluminum oxide. Porous plastics are made from thermosetting polymers. The two most common are high-density polyethylene (HDPE) and styrene-acrylonitrile. Porous plastic materials are lighter in weight; are inert in composition; and, depending on the actual material, may have greater resistance to breakage. Disadvantages include the brittleness of some plastics and fouling.

Membrane diffusers differ from rigid diffusers because they do not contain a network of interconnected passageways to diffuse gas. Instead, they use small, individual orifices (perforations) patterned across the membrane during fabrication to control gas passage. Perforation size, number, and pattern vary widely. Perforations are produced by slicing, punching, or drilling holes or slits in the membrane. Each hole acts as a variable aperture opening. The slit or hole size will affect bubble size (and, therefore, OTE) and back pressure. Typical slit or hole size is 1 mm, although perforated membrane panels of thin polyurethane are punched with significantly smaller slits (Boyle, 1996).

Perforated membranes are available in a variety of membrane materials, shapes, and perforation patterns. Two types of membrane materials commonly used are thermosplastics and elastomers. Typically, with elastomers, ethylenepropylene dimers (EPDMs) are mostly used for municipal applications. EPDMs contain carbon black, silica, clay, talc, oils, and various curing and processing agents. Oils make up a significant proportion of the mix and give the membrane its flexibility. Each manufacturer has its own formulation with distinctively different characteristics, including tensile strength, hardness, elongation at failure, modulus of elasticity, tear resistance, creep, compression set, and resistance to chemical attack. Thus, selection of the membrane for a particular application requires a thorough analysis of the material to ensure that it will perform effectively over the expected design life of the material. For membranes installed in industrial applications, the fouling potential, composition, and temperature of the wastewater must also be considered during membrane material selection.

Porous diffusers are available in plates, panels, domes, disks, and tubes. Dome diffusers are an older technology, consisting of a ceramic material mounted on a polyvinyl chloride (PVC) saddle attached by a center bolt. They are typically 180 mm (7 in.) in diameter and have downturned edges. Disk diffusers are currently the most commonly applied porous diffuser shape in aeration tanks. Disks vary in diameter from approximately 180 to 240 mm (7.0 to 9.5 in.) for ceramic and porous plastic materials and from 180 to 500 mm (7 to 20 in.) for perforated membranes. Like the dome, the disk is typically mounted on a PVC saddle but may be fastened by either a center bolt or a peripheral clamping ring.

Most tubular diffusers have the same general shape of 500 to 610 mm (20 to 24 in.) in length with an outside diameter of 63 to 80 mm (2.5 to 3 in.). Materials used include ceramics, porous plastics, and perforated membranes. Figure 6.14 shows some examples of porous diffusers.

Polyurethane membrane panels typically come in 1.2-m (4-ft) widths of variable lengths (1.2 to 1.8 m [4 to 6 ft] in 0.6-m [2-ft] increments). The base plate may be constructed of reinforced cement compound, fiber-reinforced plastic, or type 304 stainless steel. The panels are placed on the flat bottom surface of an aeration tank and fastened with anchor bolts.

With the exception of old plate designs, each porous diffuser is equipped with a flow control orifice to ensure uniform air distribution to each diffuser. Typical airflow rates for domes range from 0.014 to 0.071 m^3/min (0.5 to 2.5 cfm). For disks, the range is 0.014 to 0.08 m^3/min (0.5 to 3.0 cfm) for ceramics and 0.03 to 0.6 m^3/min (1 to 20 cfm) for perforated membranes, depending on their size. For tubes, the range is 0.03 to 0.14 m^3/min (1 to 5 cfm).

Plates, panels, domes, and disks are typically installed in a total floor coverage configuration, but plates also have been placed along the sides of aeration tanks to generate single- or dual-roll spiral mixing patterns. Disks and domes are arranged in a grid pattern with variable but uniform spacings to maximize OTE.

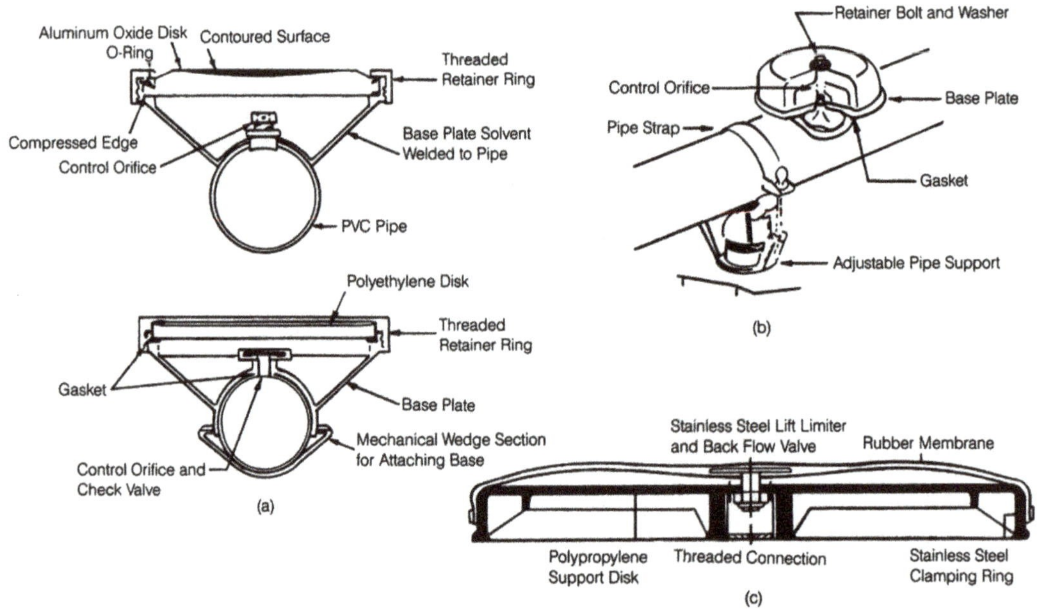

Figure 6.14 Selected Porous Diffusers: (a) Disks, (b) Dome, and (c) Perforated Membrane

Consideration should be given to a diffuser grid layout that allows for the addition of laterals and diffusers if conditions were to change. Tube diffusers are installed from removable drop pipes along one or both sides of an aeration tank. They also can be placed in a more efficient full-floor coverage configuration. Diffuser grids and configurations can be designed to allow removal of the diffusers for cleaning without process interruption.

4.1.2 NONPOROUS DIFFUSERS

Nonporous diffusers have larger orifices than porous devices and primarily produce coarse bubbles. They are available in a variety of styles and materials ranging from holes drilled in pipes to specifically engineered orifices in metal or plastic fabrications. Perforated piping, spargers, and slotted tubes are common examples of nonporous diffusers.

4.1.3 JET AERATION

Jet systems combine liquid pumping with air diffusion. The pumping system circulates mixed liquor in the aeration basin, ejecting it through a nozzle assembly. Air is typically supplied from a blower and introduced to the mixed liquor before it is discharged through the nozzles. Jets are often configured in either cluster or directional arrangements. Distribution piping and nozzles are typically made of fiberglass.

Because the recirculation pump is usually a constant-capacity device, turndown is accomplished by varying the air supply rate from the blower. A typical nozzle has a 30 mm (1 in.) throat through which air and mixed liquor pass. To overcome potential clogging problems, some systems are equipped with self-cleaning features.

4.2 MECHANICAL SURFACE AERATORS

Surface aerators can be grouped into four general categories: radial flow low speed, axial flow high speed, aspirating devices, and horizontal rotors. Each is used widely and has distinct advantages and disadvantages, depending on the application.

Surface aerators are typically float, bridge, or platform mounted. Platform and bridge designs should address torque and vibration. Bridges should be designed for at least four times the maximum moment (torque and impeller side load) anticipated. The aerator manufacturer can provide the magnitude of this moment. The efficiency and power draw of platform- and bridge-mounted aerators are sensitive to changes in the depth of impeller submergence. An increase in submergence results in increased fluid pumpage at an increase in power draw and can decrease gearbox life expectancy. High-speed (radial and axial) surface aerators are most often float mounted, providing portability and low cost.

4.3 AIR SUPPLY SYSTEM

Although selection and design of air diffusion systems typically receive the most attention, proper design of a supply system is critical to ensure that overall process goals are met. An air supply system consists of three basic components: air filters and other conditioning equipment (including diffuser cleaning systems), blowers, and air piping. Air filters remove particulates such as dust and dirt from the inlet air to protect blowers and diffusers from mechanical damage and clogging. Blowers are designed to develop sufficient pressure to overcome static head, diffuser, and line losses while delivering air at the required flowrate to the diffusion system. Piping conveys air to the diffusers.

4.3.1 BLOWERS AND COMPRESSORS

Many different types of blowers can be used to supply air for aeration processes. Historically, two main types of blowers were used at WRRFs: rotary positive displacement and centrifugal (multistage and single-stage) units. Within the last 10 years, two additional technologies have been introduced to the market: high-speed, direct-drive single-stage centrifugal blowers (turbo blowers) and low-pressure screw compressors. Information on each of these technologies, much of which is derived from Neville and Turriciano (2013) and Aerzen (2017), is presented in this section.

The positive displacement blowers are constant-volume machines capable of operating over a wide range of discharge pressures. Rotary lobe positive displacement blowers consist of bi- or tri-lobed impellers mounted on parallel, counter-rotating shafts. A finite volume of air is trapped as each impeller passes the inlet. This volume of air is displaced in a positive manner from the blower inlet to the blower discharge. The positive displacement blowers are a good selection for applications with significant variations in water level and temperature. They offer a low capital cost and high turndown, but can be less efficient and require more maintenance than other technologies. When operated at a constant speed, positive displacement blowers discharge at a constant flowrate. Discharge airflow can be varied using a variable frequency drive (VFD) to adjust motor speed in response to air demand. The positive displacement blowers are loud and require inlet and discharge silencers. Sound-attenuating enclosures can be added to further reduce noise pollution.

Centrifugal blowers are considered variable-flow, constant-pressure machines with one or more impellers mounted on a single rotating shaft. The rotating impeller(s) draws air into the blower along the axis of the shaft, accelerating the air radially outward and discharging it perpendicular to the shaft. Centrifugal blowers are not as loud as positive displacement blowers and can supply a wide range of airflows over a relatively narrow range of discharge pressures. Centrifugal blowers are categorized by the number of impellers and type of drive. Machines with one impeller mounted to the shaft are classified as single-stage centrifugal blowers and those with multiple impellers are classified as multistage centrifugal blowers. Multistage centrifugal blowers typically have good efficiencies at the design point, but efficiencies drop significantly as the machine turns down to the minimum airflow. Discharge airflow from multistage centrifugal blowers is controlled either by using an inlet throttling valve that restricts airflow on the suction side of the blower or by using a VFD to vary the motor speed. Variable frequency drive control is seldom feasible for systems with limited variation in water level and with discharge pressures that change only slightly in response to varying airflow.

4.3.2 TURNDOWN AND ENERGY EFFICIENCY

The selected blower system must be capable of supplying the volumes of air necessary to meet varying oxygen demands over the design life of the facility. To maximize operational flexibility and efficiency, blower systems should be designed to meet the maximum air requirement while still providing sufficient turndown to meet the minimum governing air requirement. In some instances, it may be necessary to install multiple blowers of smaller and/or varying capacity to efficiently or adequately meet the full range of air requirements. Therefore, in addition to site conditions, blower sizing and selection should take into account the following:

- Initial cold-weather minimum and hot-weather maximum air requirements based on existing influent loadings;
- Future minimum and maximum air requirements based on design influent loadings;
- Initial and future airflow requirements for mixing, depending on the number of diffusers and aeration tanks in use;

- Minimum airflow requirements for oxygen transfer equipment; and
- Energy efficiency of blowers to achieve greatest efficiency over the widest range and at the most common conditions expected.

The aeration process typically accounts for 25% to 60% of the total WRRF energy use (WEF, 2009). With rising power costs and the need for sustainable designs, improving aeration system efficiency has become increasingly important. A life cycle cost analysis (LCCA) including the equipment, operating, and maintenance costs projected over the life of the equipment can be used to compare multiple aeration system alternatives to identify the most cost-effective and/or efficient design. Guidance for developing a blower LCCA were developed by Neville and Turriciano (2013). Additional guidance for efficient aeration system design can be found in the U.S. EPA (2010) and the WEF (2009) publications.

As previously discussed, each type of blower has distinct operating characteristics. A blower must be compatible with the normal operating mode of the treatment system. Other factors such as noise, maintenance, and operator preference also are considered. If the aeration system is designed for operation with a fairly constant water depth, typical of most aeration basins, and with clean and maintained diffusers, then a centrifugal blower will be a good choice. Conversely, if the system will be operated over a wide range of depths, as in an SBR, then a positive displacement blower or compressor might be a better selection.

Both discharge pressure and the weight of air vary with inlet temperature. For this reason, blowers are sized to provide the required airflow rate at the maximum inlet air temperature anticipated. Blower motors are sized to deliver the required airflow rate at the highest inlet temperature expected.

4.3.3 AIR-PIPING MATERIALS

Primary considerations in piping material selection are strength and potential deterioration because of corrosion, thermal effects, and other environmental factors. Piping materials commonly used include carbon steel (painted), galvanized steel, stainless steel, and PVC. Piping materials less commonly used include ductile iron, fiberglass-reinforced plastic, and HDPE. Because blower discharge pressures are typically less than 100 kPa (15 psi), thin-walled pipe may be used with adequate provisions for protection from physical damage. Discharge air temperature in excess of 93 °C (200 °F) are common because discharge temperature increases with increasing discharge pressure. Therefore, the pipe and accessories (supports, valves, gaskets, and so on) must be designed accordingly. Because thermal stresses caused by pipe heating and cooling can be significant, provisions for pipe expansion and contraction including cooling loops or expansion joints are typically needed. Blower discharge piping is often insulated to protect workers from potential burns, help attenuate noise, and minimize heat loss to the blower room. Air intake piping may also be insulated to prevent condensation (in cold climates) and to help attenuate noise. Discharge piping should be sloped away from the blower with condensate drains installed at low points occurring upstream of the diffused aeration equipment.

Because of the potential for corrosion at the interface between the atmosphere and liquid in aeration tanks, stainless steel piping material is often used. For diffuser systems, stainless steel droplegs are connected to the diffuser manifolds and headers, typically stainless steel or PVC. The choice of material depends on the structural requirements of the diffusers, diffuser connection, temperature within the pipe, and whether a diffuser-cleaning system will be provided. For diffusers with cleaning systems, the designer must check the compatibility of the piping material with the cleaning gas or liquid. Polyvinyl chloride is typically selected because of its inert characteristics. However, PVC temperature limits need to be considered in warm climates, industrial applications with high water temperatures, and deep tank applications. Typically, PVC is only used below the water surface.

Stainless steel is typically selected for tube diffuser systems because of the cantilever load applied by the diffusers and its corrosion resistance. However, PVC also has been used successfully. The PVC is more typically used with disk or dome diffusers because they are mounted to the top of a header; thus, weight and buoyant forces transmitted through the connection to the header are minimal.

Because basins are periodically drained and left empty, PVC piping should be protected from sunlight deterioration. Titanium dioxide and carbon black are typically used in PVC piping for ultraviolet protection. Whenever possible, drained basins should be filled with clear water to protect PVC from ultraviolet deterioration. Other design considerations include provisions to protect against the effects of pipe freezing and thawing when tanks are empty. Standard practice at many facilities is to maintain a few feet of water depth in the basin with low airflow as long as no one is working in the basin.

4.3.4 AERATION SYSTEM CONTROL

Oxygen demand in aeration basins fluctuates in response to many factors, including diurnal influent loading; intermittent discharge of sidestreams from dewatering activities; dilution from storm events; and changes to air, water, and other wastewater characteristics. Fluctuations in oxygen demand result in changes in dissolved oxygen concentrations over time if the air supplied is not adjusted in response to fluctuating oxygen demands. Dissolved oxygen control strategies can be very simple or quite complex, depending on the system.

Manual control of dissolved oxygen requires operators to record dissolved oxygen measurements daily and to make changes to the airflow and flow splits between basins and zones daily, sometimes up to a few times a day. This method of dissolved oxygen control often does not match actual dissolved oxygen demand, providing too much or too little oxygen and leading to poor process performance and inefficient aeration system operation. Replacing manual dissolved oxygen control with automated dissolved oxygen control can provide considerable energy cost savings when designed and implemented correctly because it allows for airflow to aerated zones to adjust to the changing conditions within the basins rather than over-aerating for portions of the day. It has been estimated that by adding tight control of dissolved oxygen to the aeration process, facilities can save between 10% and 30% in total energy cost (U.S. EPA, 2010).

Automated dissolved oxygen control uses real-time dissolved oxygen measurements from dissolved oxygen probes as inputs to the process controller. The process controller provides control output to the aeration system. The aeration system responds to the control output by adjusting airflow from the blower and/or the valve position on the diffuser droplegs at the basin to maintain the target dissolved oxygen setpoint. Major components of an automated dissolved oxygen control system include:

- Dissolved oxygen probes: For control, dissolved oxygen probes should be installed near the center or close to the inlet of the aerated zones. For monitoring dissolved oxygen intrusion, dissolved oxygen probes can also be installed in anoxic or anaerobic zones.
- Blower airflow control: The total airflow supplied to the aeration system is controlled by modulating the airflow rate delivered by the blowers.
- Basin/zone airflow control: Total airflow is typically divided between multiple basins, aeration zones, and diffuser grids. The airflow split between basins, zones, and grids can be controlled by making manual adjustments to flow control valves or by using automated flow control valves. Implementing a mostly open valve control scheme with automated valves can improve efficiency by maintaining valves as open as possible and minimizing system pressure.
- Process control system: The process controllers receive feedback from the dissolved oxygen probes, compares the basin dissolved oxygen readings to dissolved oxygen setpoints, and sends signals to the air controller mechanism to make adjustments as needed.

Dissolved oxygen is typically used for aeration control because it is a good indicator of nitrification and is a reliable measurement. In recent years, alternative control parameters have been implemented in place of dissolved oxygen for aeration system control. For example, as real-time ammonia analyzers have become more reliable, there have been several WRRFs that have successfully implanted ammonia-based aeration system control. Ammonia-based aeration control uses ammonia concentrations as the input to the process controller as a more direct measurement of nitrification. Detailed information about specific dissolved oxygen control strategies, recent advancements in control strategies, and emerging technologies for using control parameters other than dissolved oxygen can be found in *Evaluation of Energy Conservation Measures for Wastewater Treatment Facilities* (U.S. EPA, 2010).

4.4 AERATOR DESIGN AND TESTING

Historically, many methods have been used to test and specify aeration equipment, which has led to confusion and misrepresentation of equipment performance. Furthermore, equipment suppliers, consultants, and users often use different nomenclature when they report capabilities.

The preparation of explicit equipment specifications is vital to installation of efficient, cost-effective devices. To enable a supplier to properly specify and quote aeration equipment, the prospective user or design engineer must provide accurate and detailed information about system requirements and constraints. The supplier can then provide equipment performance information based on reliable clean water test data and sound judgment based on experience from previous field applications.

Evaluation of reliable clean water oxygen transfer test data is only one step in understanding aeration system capabilities. Measuring oxygen transfer in the field with aeration equipment operating under actual process conditions can be used to provide additional information about the capabilities of installed oxygen transfer devices. Engineers must choose carefully among the various field test methods available and apply them with care.

The ASCE (2007) describes the measurement of OTR as the mass of oxygen dissolved per unit time in a unit volume of clean water by an oxygen transfer system operating at a given gas flowrate and power input condition. This method applies to both laboratory-scale devices with water volumes of a few liters and full-scale systems and is valid for many different mixing conditions and process configurations.

The method is based on dissolved oxygen removal from the test water volume by the addition of sodium sulfite with a cobalt catalyst, followed by transfer studies of reoxygenation to near-saturation level. Test water volume dissolved oxygen inventory is monitored during reoxygenation by measuring concentrations at several points that best represent tank contents. These dissolved oxygen concentrations can be measured in situ or on samples pumped from the tank. The method specifies a minimum number, distribution, and range of dissolved oxygen measurements at each determination point.

Data obtained at each sampling point are then analyzed by a simplified mass transfer model to estimate the apparent volumetric mass transfer coefficient (K_La) and the equilibrium spatial average dissolved oxygen saturation concentration (C^*) (ASCE, 2007; Jiang & Stenstrom, 2012). Nonlinear regression analysis is used to fit the dissolved oxygen profile measured at each sampling point during reoxygenation to the model's mathematical equation. In this way, estimates of K_La and C^* are obtained at each sampling point. After these estimates are adjusted to standard conditions, the system standard oxygen transfer rate (SOTR) is calculated from the aeration tank volume (V) and estimates of K_La and C^* at each of the n sampling points as follows:

$$\text{SOTR} = V\frac{\sum_{i=1}^{n} K_La_{20i}C^*_{\infty 20i}}{n} \tag{6.22}$$

Where K_La_{20i} = sampling point value of K_La corrected to 20 °C, time^{-1};
$C^*_{\infty 20i}$ = sampling point value of steady-state dissolved oxygen saturation concentration corrected to 20 °C, mass/volume;
C^* = equilibrium spatial average dissolved oxygen saturation concentration, mass/volume;
V = aeration tank volume, volume; and
n = sampling point.

Frequently, the SAE is calculated as the SOTR divided by power input. The SOTE in percent also can be estimated for diffused air systems by:

$$\text{SOTE} = \frac{\text{SOTR}}{W_{O_2}} \times 100 \tag{6.23}$$

where W_{O_2} = mass flow of oxygen in the gas feed stream, mass/time.

It is important to use consistent definitions during aeration testing, subsequent data analysis, and final result reporting. A consistent nomenclature has been established with more logical and understandable terminology, which will eliminate much of the difficulty in interpreting aeration literature (ASCE, 2007). Standard conditions for oxygen transfer tests are defined as water temperature of 20 °C, barometric pressure of 100 kPa (1 atm), and dissolved oxygen concentration of zero.

PRACTICE PROBLEMS

1. Calculate the theoretical carbonaceous oxygen demand (mg/mg or g/g) for biomass using the empirical formula: $C_5H_7O_2N$.
2. Calculate the theoretical nitrogenous oxygen demand (mg/mg or g/g) for NH_4^+.
3. What conditions will cause changes in the oxygen demand in an activated sludge system?
4. Discuss various options for aeration diffusers and what materials are typically used. What are some operational considerations for aeration diffusers?
5. Why is it important to know how the HRT of an activated sludge system compares to the SRT_{min} for that system?
6. What assumptions are used to develop the traditional activated sludge model for cBOD removal presented in this chapter?
7. Redo Example Problems 6.1 and 6.2 at a design temperature of 12 °C using the typical temperature coefficients given in Table 6.3.
8. Design a CMAS system for cBOD removal for a permit requirement of 20 mg/L BOD_5 and 20 mg/L TSS given the following:

   Q = 3785 m^3/d
   S_o = 150 mg BOD_5/L (aeration basin influent concentration)
   BOD_5 of effluent TSS = 0.65
   Design temperature = 15 °C
   Use typical values for kinetic coefficients and temperature correction factors from Table 6.3.
9. Design an aerobic selector to remove 80% of BOD_5 for the activated sludge system detailed in problem 6.

5.0 REFERENCES

Aerzen. (2017, January 23). How to select the most effective blower technology for wastewater applications. *Water Online*. https://www.wateronline.com/doc/how-to-select-the-most-effective-blower-technology-for-wastewater-applications-0001

Albertson, O. E. (2007, October 10–13). The case for staging biological reactors. *Proceedings of the 80th Annual Water Environment Federation Exposition and Conference*, San Diego, California, United States.

American Society of Civil Engineers. (2007). *Measurement of oxygen transfer in clean water* (ASCE/EWRI 2-06).

Ardern, E., & Lockett, W. T. (1914). Experiments on the oxidation of sewage without the aid of filters, *Journal of Chemical Technology and Biotechnology*, *33*, 523.

Arora, M. L., Barth, E. F., & Umphres, M. B. (1985). Technology evaluation of sequencing batch reactors. *Journal of the Water Pollution Control Federation*, *57*, 867–875.

Bisogni, J. J., & Lawrence, A. W. (1971). Relationships between biological solids retention time and settling characteristics. *Water Research (Great Britain)*, *5*, 753.

Boyle, W. C. (1996). Fine pore aeration: An update on its status. In *Enhancing design and operation of activated sludge plants* (Proceedings of the Central States Water Environment Association Seminar), Madison, Wisconsin, United States.

Christopher, S., & Titus, R. (1983). New wastewater process cuts plants cost 50 percent. *Civil Engineering*, *53*(5), 39.

Daigger, G. T., Kuo, J., Derlon, N., Houweling, D., Jiminez, J. A., Johnson, B. R., McQuarrie, J., P., Murthy, S., Regmi, P., Roche, C., Sturm, B., Wett, B., Winkler, M., & Boltz, J. P. (2023). Biological and physical selectors for mobile biofilms, aerobic granules, and densified-biological flocs in continuously flowing wastewater treatment processes: A state-of-the-art-review. *Water Research (Great Britain), 242*, 120245.

Dague, R. R. (1983). A modified approach to activated sludge design. Paper presented at 56th Annual Water Pollution Control Federation Conference, Atlanta, Georgia.

Eckenfelder, W. W., Jr. (1966). *Industrial Water Pollution Control.* McGraw-Hill.

Eckenfelder, W. W., Jr. (1967). Comparative biological waste treatment design. *Journal of Sanitary Engineering Division, American Society of Civil Engineers, 93*(SA6), 157.

Ellis, T. G., Soulsby, M. R., & Harper, P. C. (1992). The selection and design of sequencing batch reactors for treatment capacity expansion in Abu Dhabi. *Water Science and Technology, 26*, 1105–1114.

Gaudy, A. F., & Kincannon, D. F. (1977). Comparing design models for activated sludge. *Water Sewage Works, 123*, 66.

Giesen, A., van Loosdrecht, M., Robertson, S., & de Buin, B. (2015, September 24–28). Aerobic granular biomass technology: Further innovation, system development and design optimization. *Proceedings of the 89th Annual Water Environment Federation Technical Exposition and Conference*, Chicago, Illinois, United States.

Goronszy, M. (1979). Intermittent operation of the extended aeration process for small systems. *Journal of the Water Pollution Control Federation, 51*, 274–287.

Grady, C. P. L., Jr., Daigger, G. T., Love, N. G., & Filipe, C. D. M. (2011). *Biological wastewater treatment* (3rd ed.). CRC Press.

Groves, K. P., Daigger, G. T., Simpkin, T. J., Redmon, D. T., & Ewing, L. (1992). Evaluation of oxygen transfer efficiency and alpha factor on a variety of diffused aeration systems. *Water Environment Research, 64*, 691–698.

Guo, P. H. M., Thirumurthi, D., & Jank, B. E. (1981). Evaluation of extended aeration activated sludge package plants. *Journal of the Water Pollution Control Federation, 53*, 33–42.

Jenkins, D., Richard, M., & Daigger, G. (2003). *Manual on the causes and control of activated sludge bulking, foaming, and other solids separation problems* (3rd ed.). CRC Press–IWA Publishing.

Jiang P., & Stenstrom M. K. (2012). Oxygen transfer parameter estimation: Impact of methodology. *Journal of Environmental Engineering, 138*(2), 137–142.

Lawrence, A. W, & McCarty, P. L. (1970). Unified basis for biological treatment design and operation. *Journal of the Sanitary Engineering Division, American Society of Civil Engineers, 96*, 757.

McKinney, R. E. (1962). Mathematics of complete mixing activated sludge. *Journal of the Sanitary Engineering Division, American Society of Civil Engineers, 88*, SA3.

McKinney, R. E., & Ooten, R. J. (1969). Concepts of complete mixing activated sludge. *Transactions of the 19th Annual Conference on Sanitary Engineering* (Bulletin of Engineering and Architecture No. 60), University of Kansas.

Metcalf & Eddy, Inc./AECOM. (2014). *Wastewater engineering treatment and resource recovery* (5th ed.). McGraw-Hill.

Narayanan, B., de Leon, C., Radke, C., & Jenkins, D. (2010). The role of dispersed nocardioform filaments in activated sludge foaming. *Water Environment Research, 82*(6), 483–491.

Neville, M., & Turriciano, A. (2013). Don't blow it! Lessons learned from direct drive turbo blower designs for energy efficiency. *Journal of the New England Water Environment Association, 47*(1), 46–54.

Parker, D., Bratby, J., Espring, D., Hull, T., Kelly, R., Melcer, H., Merlo, R., Pope, R., Shafer, T., Wahlberg, E., & Witzgall, R. (2014). A critical review of nuisance foam formation and biological methods for foam management or elimination in nutrient removal facilities. *Water Environment Research, 86*(6), 483–503.

Ramanathan, M., & Gaudy, A. F., Jr. (1971). Steady state model for activated sludge with constant recycle sludge concentration. *Biotechnology and Bioengineering, 13*, 125.

Rittmann, B., & McCarty, P. (2001). *Environmental biotechnology: Principles and applications.* McGraw-Hill.

Sezgin, M., Jenkins, D., & Parker, D. S. (1978). A unified theory of filamentous activated sludge bulking. *Journal of the Water Pollution Control Federation, 50*, 362–381.

Sykes, J. C. (1993). Biological selector to minimize sludge bulking. *Waterwood Review*, September/October.

U.S. Environmental Protection Agency. (1979). *Full-scale demonstration of open tank oxygen activated sludge treatment* (EPA-600/2-79-012).

U.S. Environmental Protection Agency. (1985). *Summary report: Fine pore (fine bubble) aeration systems* (EPA-625/8-85-010). Water Engineering Research Laboratory, U.S. Environmental Protection Agency.

U.S. Environmental Protection Agency. (1987). *Summary report: Causes and control of activated sludge bulking and foaming* (EPA-625/8-87-012).

U.S. Environmental Protection Agency. (1989). *Design manual: Fine pore aeration systems* (EPA-625/1-89-023). Center for Environmental Research Information, Risk Reduction Engineering Laboratory.

U.S. Environmental Protection Agency. (1993). *Process design manual for nitrogen removal.*

U.S. Environmental Protection Agency. (1999). *Wastewater technology fact sheet: Fine bubble aeration* (EPA 832-F-99-065).

U.S. Environmental Protection Agency. (2010). *Evaluation of energy conservation measures for wastewater treatment facilities* (EPA 832-R-10-005).

Water Environment Federation. (2009). *Energy conservation in water and wastewater facilities* (Manual of Practice No. 32).

Water Environment Federation (2013). *Wastewater treatment process modeling* (2nd ed.; Manual of Practice No. 31).

Water Environment Research Foundation. (2006). *Develop and demonstrate fundamental basis for selectors to improve activated sludge settleability.*

Water Research Commission. (1984). *Theory: Design and operation of nutrient removal activated sludge processes.*

Welling, C., Kennedy, A., Wett, B., Johnson, C., Rutherford, B., Baumler, R., & Bott, C. (2015, September 24–28). Improving settleability and enhancing biological phosphorus removal through the implementation of a hydrocyclones. *Proceedings of the 89th Annual Water Environment Federation Technical Exposition and Conference*, Chicago, Illinois, United States.

Wilford, J., & Conlon, T. P. (1957). Contact aeration sewage treatment plants in New Jersey. *Sewage and Industrial Wastes, 29*, 845–855.

Young, T., Crosswell, S., & Wendell, J. (2008, October 18–22). Comparison of nitrogen removal performance in SBR systems. *Proceedings of the 81st Annual Water Environment Federation Exposition and Conference*, Chicago, Illinois, United States.

Yunt, F. (1980). *Results of mixing efficiency tests with Norton Dome Aeration System at LA Glendale Treatment Plant, Los Angeles County Sanitation Districts, Whittier, California* (cited in Mueller J., Boyle W., & Pöpel H. [2002]. *Aeration: Principles and practice.* CRC Press).

6.0 SUGGESTED READINGS

American Society of Civil Engineers. (1997). *Standard guidelines for in-process oxygen transfer testing* (ASCE-18-96).

Barnard, J. (1976). A review of biological phosphorus removal in the activated sludge process. *Water SA, 2*, 136.

Boyle, W. C., Stenstrom, M. K., Campbell, H. J., Jr., & Brenner, R. C. (1989). Oxygen transfer in clean and process water for draft tube turbine aerators in total barrier oxidation ditches. *Journal of the Water Pollution Control Federation, 61*, 1449–1463.

Doyle, M., & Boyle, W. C. (1985). Translation of clean to dirty water oxygen transfer rates. In *Proceedings of Seminar/Workshop Aeration Systems Design, Testing, and Operations Control* (EPA-600/9-85-005). U.S. Environmental Protection Agency.

Mino, T., Arun, V., Tsuzuki, Y., & Matsuo, T. (1987, September 28–30). Effect of phosphorus accumulation on acetate metabolism in the biological phosphorus removal process. *Proceedings of the IAWPRC Specialized Conference*, Rome, Italy. IWA Publishing.

Mueller, J. A., & Boyle, W. C. (1988). Oxygen transfer under process conditions. *Journal of the Water Pollution Control Federation, 60*, 332–341.

Mueller, J., Boyle, W., & Pöpel, H. (2002). *Aeration: Principles and practice.* CRC Press.

Reed, S. C., Crites, R. W., & Middlebrooks, E. J. (1995). *Natural systems for waste management and treatment* (2nd ed.). McGraw-Hill.

Rosso, D., & Stenstrom, M. K. (2006). Economic implications of fine-pore diffuser aging. *Water Environment Research, 78*(8), 810–815.

Stenstrom, M. R., & Gilbert, R. G. (1981). Effects of alpha, beta and theta factors on specification, design and operation of aeration systems. *Water Research (Great Britain), 15*, 643.

Sturm, B., Faraj, R., Amante, T., Wagner, B., Kiani, F., & Waybenais, D. (2015, September 24–28). Strategies for operating aerobic granular sludge reactors for low-strength municipal wastewater. *Proceedings of the 89th Annual Water Environment Federation Technical Exposition and Conference*, Chicago, Illinois, United States.

U.S. Environmental Protection Agency (1983). *Development of Standard Procedures for Evaluating Oxygen Transfer Devices* (EPA-600/2-83-102).

Chapter 7 | Attached Growth Processes

LEARNING OBJECTIVES

At the end of this chapter you should be able to:

- Describe what constitutes an attached growth process and diagram the individual basins and flows in a variety of attached growth processes (e.g., trickling filter, rotating biological contactor, trickling filter solids contact, external nitrification activated sludge).
- Explain why mass transfer limitations often control attached growth system performance.
- Describe trickling filter media selection and distribution system options.
- Given appropriate data, determine the size of a trickling filter to meet a desired effluent soluble BOD.
- Given appropriate data, determine the organic and hydraulic loading on a trickling filter.
- Explain what is meant by an integrated biological treatment process and give an example.

1.0 BIOFILMS

Biofilms are ubiquitous in nature and in engineered systems and can be used beneficially in municipal wastewater treatment. Biofilm and suspended growth reactors can meet similar treatment objectives for carbon oxidation, nitrification, denitrification, and desulfurization. In fact, the distinction between biofilm and suspended growth systems are becoming blurred as many installations now use a combination of the two (e.g., moving bed biofilm reactors, MBBRs). The root distinction is that biofilm systems rely on the growth of a microbial biofilm on an attachment media. The same microorganisms are responsible for biochemical reactions in both activated sludge and biofilm systems and respond in the same way to local environmental conditions (i.e., pH, temperature, electron donor, electron acceptor, and macronutrient availability, Morgenroth, 2008). Biofilm reactor designers should consider the effect of multiple substrates and biomass fractions and the way they are affected by mass transport limitations to evaluate system performance. Substrates typically considered during biofilm reactor design are soluble and particulate compounds, including electron donors and acceptors, nutrients, bacteria, and bacterial debris (e.g., extra polymeric substances, EPS).

1.1 BIOFILM REACTORS

Biofilm reactors have five primary compartments: (1) influent wastewater (distribution) system; (2) containment structure; (3) carrier with biofilm; (4) effluent water collection system; and (5) an aeration system (for aerobic processes and scour) or mixing system (for anoxic processes that require bulk-liquid agitation and biofilm carrier distribution). Because design is system specific, each is discussed relative to specific biofilm reactor types in subsequent sections. Five components determine the local environment of the biofilm: (1) carrier surface (i.e., substratum); (2) biofilm (including both particulate and liquid fractions); (3) mass transfer boundary layer (MTBL); (4) bulk liquid; and (5) gas phase (when significant). These components are illustrated in Figure 7.1.

1.2 BIOFILM PROCESSES AND STRUCTURE

Two processes—mass transfer and biochemical conversion—are characteristic of all biofilm reactors and influence biofilm structure and function. Mass transport inside the biofilm is controlled by molecular

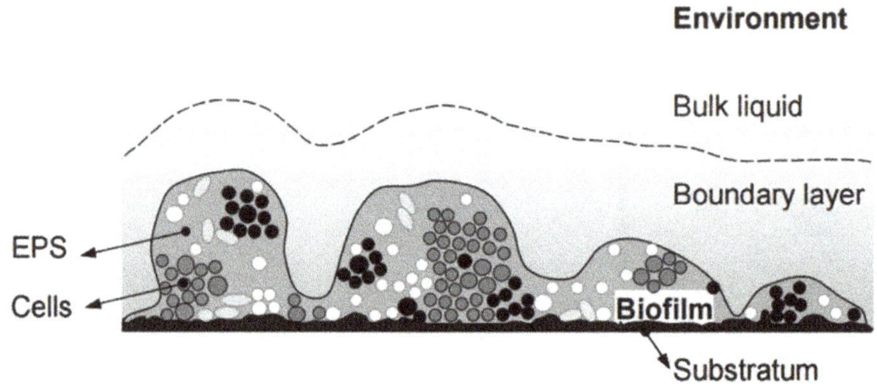

Figure 7.1 Schematic Representation of Biofilm Components: (1) Carrier Surface (i.e., Substratum) Medium; (2) Biofilm (Particulate and Liquid Components); (3) Mass Transfer Boundary Layer; (4) Bulk Liquid; and (5) Gas Phase (Wanner et al., 2006; reprinted with permission from IWA Publishing)

diffusion. If mass transfer to the biofilm is slow compared to biochemical conversion, the result is strong concentration gradients for substrates within the mass transfer boundary layer and inside the biofilm. These mass transport limitations result in inactive zones deep inside the biofilm and have implications for the design and operation of biofilm reactors and microbial ecology. Mass transport is the primary mechanistic difference between biofilm and suspended growth reactors. Typically, full-scale operating suspended growth systems are kinetically (i.e., biomass) limited, whereas biofilm reactors are typically diffusion (i.e., surface-area) limited. Therefore, it is necessary to understand the interactions between mass transport and substrate transformation processes to completely evaluate biofilm systems.

Microbial competition in a biofilm is based on local substrate concentrations and the location of different organism groups. Microorganisms closer to the biofilm surface have the advantage of higher local substrate concentrations but are also the most susceptible to shear- and abrasion-induced detachment and subsequent removal from the system with the effluent stream. Micororganisms living near the carrier, also called the substratum, are offered some protection from detachment but are subject to reduced substrate availability.

Biofilms in wastewater treatment applications typically are mass transfer limited, but for high bulk-liquid concentrations or low degradation rates, the limiting substrate can fully penetrate biofilms. Even systems that typically operate in a range of conditions that generate mass-transfer-limited biofilms may be periodically subjected to conditions that result in biofilms being fully penetrated. This is caused by a non-continuous detachment that is hydraulically triggered, induced by an organic shock load, or carrier/granule collision. Concentration gradients inside the biofilm allow for the development of different redox zones, and partial penetration can result in biofilms simultaneously having aerobic, anoxic, and anaerobic zones. The existence of aerobic and anoxic zones inside the biofilm can be beneficial, for example, in promoting processes such as simultaneous nitrification and denitrification. Once a biofilm is partially penetrated, increasing its thickness has no treatment benefit in removing the mass-transfer-limited, depleted substrate. Therefore, biofilm reactor performance is surface area dependent and not dependent on the total amount of biomass in the system. In some cases, biofilm penetration can be increased by increasing bulk-phase concentrations (e.g., by increasing bulk phase O_2 concentrations in aerobic biofilm reactors). Boltz et al. (2006) and Boltz and La Motta (2007) demonstrated that the removal of organic and inorganic particles is also a function of biofilm surface area.

A balance between microbial growth and detachment will result in biofilms with a range of thicknesses. Biofilm thickness that exceeds the rate-limiting substrate penetration depth typically hinders system performance. Excessive biofilm thickness can have two detrimental effects on full-scale reactors that rely on either passive or dedicated biofilm thickness control mechanisms. First, it can reduce biofilm surface area. Second, it can deprive biomass near the carrier of electron donor, electron acceptor, or macronutrients. As a result, the interior biofilm is likely to become anaerobic, may produce odors and result in uncontrolled

Table 7.1 Typical Biofilm Thicknesses for a Variety of Biofilm Reactors

	Typical Biofilm Thickness (mm)	
Type of Biofilm Reactor	Lower Estimate	Upper Estimate
Moving bed biofilm reactor	50	500
Biologically active filters	20	300
Fluidized bed biofilm reactors	20	400
Rotating biological contactors	200	2000
Trickling filters (carbon oxidizing units)	200	2000
Membrane biofilm reactors	20	500

detachment of biofilm segments that are equivalent in size to its thickness. The latter is known as sloughing. Passive biofilm thickness control mechanisms are inherent to normal biofilm reactor operating conditions, including, for example, mixing resulting in continuous carrier collisions in a moving bed biofilm reactor (MBBR). Dedicated biofilm thickness control mechanisms require function-specific operating cycles such as backwashing a submerged biologically active filter or flushing a trickling filter. A majority of the biofilm thickness control mechanisms are mechanically induced and hydrodynamically mediated. Table 7.1 lists, for relative comparison, the controlled biofilm thickness range typical of the reactor types.

Biofilm ecology in reactors affects their activity. Cold wastewater, for example, increases solubility (thereby increasing the rate of diffusion) but retards biochemical transformation processes. Soluble substrates may, therefore, penetrate deeper into the biofilm resulting in a reactor that can support more active biomass. Developing biofilms accumulate microbial cells, but all biofilms eventually lose particulate components. The loss of particulate components and their introduction to bulk liquid is called detachment. Bryers (1984) described four biofilm detachment processes: (1) abrasion, (2) erosion, (3) sloughing, and (4) predator grazing. Abrasion and erosion are illustrated in Figure 7.2. Abrasion, which is initiated by particle collision, and erosion, initiated by hydrodynamic shear near the biofilm surface, are the removal of small groups of cells. Predatory higher life forms such as macro- and micro-fauna graze biofilms (Boltz et al., 2008). Abrasion, erosion, and, to a certain extent, fauna grazing are associated with well-operating biofilm reactors. Sloughing and excessive predation is detrimental to

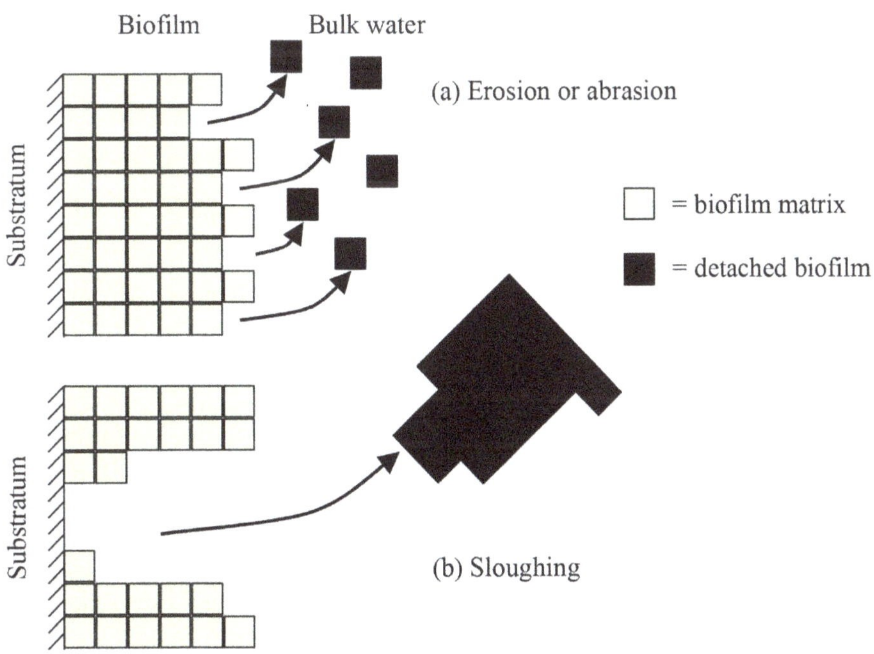

Figure 7.2 (a) Erosion and Abrasion and (b) Sloughing Biofilm Detachment Processes (Morgenroth, 2003; reprinted with permission from IWA Publishing)

biofilm reactor performance. Avoiding excess predatory fauna accumulation and promoting continuous detachment of small biofilm fragments results from proper thickness control, which occurs in a stable environment not subject to excessive mass transfer resistances.

2.0 BIOFILM REACTOR DESIGN APPROACHES AND CONSIDERATIONS

Several design approaches, biofilm reactor model, and biofilm model types typically are used in engineering practice. The primary objective of biofilm or biofilm reactor models is to predict soluble substrate flux (*J*) into the biofilm. This flux information can be used to obtain an estimate of the (1) overall biofilm reactor performance; (2) required active biofilm surface area; (3) electron acceptor (e.g., dissolved oxygen); (4) external electron donor (e.g., methanol or hydrogen); and (5) biosolids management requirements. This section discusses the relative benefits and limitations of general biofilm reactor design approaches. It is common practice to use more than one biofilm reactor design approach and base the final process design on a comparison of results.

2.1 BIOFILM MODELS USED IN ENGINEERING DESIGN

Typically, process design involves the use of mathematical biofilm models because (1) simulation models are efficient tools that allow rapid evaluation of a variety of scenarios; and (2) empirical procedures are inadequate for providing information that is now widely considered essential to biofilm reactor design (e.g., local flux as a function of bulk-liquid substrate concentrations, biofilm composition and competition for multiple substrates, and influence of individual processes on interaction between several microbial types).

Modern biofilm reactors facilitate the use of biofilm models in engineering design. For example, MBBRs do not present the hydrodynamic or operational complexities that historically have hindered use of mathematical biofilm model-based process design. This is because they are comprised of zones that are essentially completely mixed and are continuous flowing systems, respectively. Submerged completely mixed biofilm reactors allow for the application of modern biofilm knowledge and are conducive to simulation with existing biofilm models. As a result, most of the existing water resource recovery facility (WRRF) modeling programs have been expanded to include a submerged completely mixed biofilm reactor module that consists of a mathematical biofilm model. The process designer should understand the basis for the mathematical biofilm model, its supporting assumptions, and limitations before using these models in design.

Unfortunately, choosing a modeling approach that offers an appropriate level of complexity to meet the modeling objective can be difficult. An overview of different model approaches that are suitable for biofilm reactor design are as follows:

- One-dimensional homogeneous biofilm (single limiting substrate): This approach takes into account mass transfer limitations and the corresponding effects on concentration profiles and substrate flux into the biofilm. It is assumed that active microbes are homogeneously distributed across the biofilm thickness. The approach is valid only if the limiting substrate has been identified.
- One-dimensional homogeneous biofilm (multiple substrates and multiple biomass components): One key aspect of modeling biofilms is to evaluate the competition and coexistence of different bacterial groups (e.g., carbon-oxidizing heterotrophic bacteria, nitrifying autotrophic bacteria) and local process conditions (e.g., aerobic, anoxic, or anaerobic). Local process conditions can be determined by calculating biofilm penetration depths for different soluble substrates (e.g., COD, ammonia-nitrogen, oxygen, and nitrate-nitrogen). Growth of individual bacterial groups can be determined based on fluxes. To simplify calculations, it can be assumed that all bacterial groups are homogeneously distributed over the thickness of the biofilm (Boltz et al., 2009a, 2009b, 2009c; Rauch et al., 1999).
- One-dimensional heterogeneous biofilm: Different groups of bacteria are not distributed evenly and are competing for substrate and space. One-dimensional heterogeneous biofilm models must keep track of local growth and decay of different bacterial groups and detachment to biomass distributions over the biofilm thickness (Wanner & Reichert, 1996).

Different scales of heterogeneity are relevant for biofilm reactor design. Length scale of biofilm thickness, on the order of 100 to 1000 mm, is taken into account in one-dimensional and multi-dimensional biofilm models. Substrate fluxes from these simulations can then be integrated into models describing overall reactor performance in which the length-scale is on the order of 1 m. But, heterogeneities also can be observed in biofilm reactors in between these scales. For example, in some cases, patchy biofilms develop. This may occur in any underloaded biofilm reactor such as the lower reaches of some nitrifying trickling filters. These observations confirm that some sections of the biofilm carrier are bare, whereas others contain dense development. These heterogeneities between the micro and macro scale typically are not considered in biofilm models, and it is not clear to what extent they are relevant.

Wanner et al. (2006) provide a detailed description and comparison of different modeling approaches. Wanner et al. (2006) and Boltz et al. (2010) state that one-dimensional biofilm models are adequate for reactor design. Currently, most commercial software that is used for biofilm reactor design and evaluation takes into account multiple substrates and biomass fractions in either a one-dimensional heterogeneous or homogeneous biofilm. Some examples of software and references are summarized in Table 7.2.

Mathematical biofilm models are now standard design tools despite the fact that they are significantly more complex than graphical procedures and empirical and semi-empirical models.

2.2 LIMITATIONS OF BIOFILM MODELS FOR THE PRACTITIONER

Mathematical biofilm modeling has advanced biofilm reactor design. Furthermore, the models have been instrumental in explaining the mechanisms that result in mass transfer limitations, stratification of bacteria over the biofilm thickness (and throughout a biofilm reactor), competition between bacterial groups, and factors affecting the development of multi-dimensional heterogeneous biofilm morphologies. Adoption of biofilm models has been slow; additional research and development is needed to overcome limitations of existing models. There are several limitations of which process designers should be aware:

- Biofilm models are complex and they typically exclude important factors. Most models focus on small-scale heterogeneities inside the biofilm to better describe soluble substrate flux. Many approaches, however, fail to describe overall reactor operation (e.g., backwashing, flow distribution) as a function of system appurtenances (e.g., aeration system, mixing devices).
- Too many models are available. The choice of modeling approach depends on the specific question to be addressed. This is different from activated sludge modeling where mathematical models are directly

Table 7.2 Biofilm Models Used in Practice (Adapted from Boltz et al., 2009b)

Software	Company	Biofilm Model Type and Distribution	Reference
AQUASIM™	EAWAG, Swiss Federal Institute of Aquatic Science and Technology, Dübendorf, Switzerland (www.eawag.ch/index_EN)	1-D, DY, N; heterogeneous	Wanner & Reichert (1996) (modified)
AQUIFAS™	Aquaregen, Mountain View, California (www.aquifas.com)	1-D, DY, SE, and N, heterogeneous	Sen & Randall (2008a,b,c)
BioWin™	EnviroSim Associates, Ltd., Flamborough, Canada (www.envirosim.com)	1-D, DY, N, heterogeneous	Wanner & Reichert (1996) (modified)
GPS-X™	Hydromantis Inc., Hamilton, Canada (www.hydromantis.com)	1-D, DY, N, heterogeneous	Hydromantis (2002)
Simba™	ifak GmbH, Magdeburg, Germany (www.ifak.eu)	1-D, DY, N, heterogeneous	Wanner & Reichert (1996) (modified)
STOAT™	WRc, Wiltshire, England (www.wrcgroup.com)	1-D, DY, N, heterogeneous	Wanner & Reichert (1996) (modified)
WEST™	MOST for WATER, Kortrijk, Belgium (www.mikepoweredbydhi.com)	1-D, DY, N(A)[a], N[b], homogeneous[a], heterogeneous[b]	Rauch et al. (1999)[a], Wanner & Reichert (1996) (modified)[b]
SUMO	DYNAMITA (http://www.dynamita.com/)	1-D, DY, N; heterogenous	Wanner & Reichert (1996) (modified), Takács et al. (2007), Dynamita, 2016

Notes.
1-D = one-dimensional
DY = dynamic
N = numerical
N(A) = numerical solution using analytical flux expressions

applicable. The design engineer must be aware of the biofilm model type being used, its simplifying assumptions, and the resultant limitations. If this information is not clearly documented, then results from the model should be used with caution.

- There are significant deficiencies in approaches for calibrating biofilm models. Without reliable and transparent approaches for model calibration, the developed models will not be accepted as a robust design tool.
- Existing biofilm models adequately describe soluble substrate flux, but little basic research exists to allow for development of mathematical terms describing the fate of particulate matter.
- The mass transfer boundary layer that links the one-dimensional biofilm model with the bioreactor compartment is well understood.

3.0 TRICKLING FILTERS

During the first five decades of use, trickling filters included soil and rock biofilm carriers and their design was scattered and empirical in nature. During the 1950s and 1960s, Dow Chemical Co. (Midland, Michigan) began early experimentation with modular plastic packing media (Bryan, 1955; Bryan and Moeller, 1960). Other studies during that time that resulted in development of accepted design protocol (Atkinson et al., 1963; Eckenfelder, 1961, 1963; Galler & Gotaas, 1964; Germain, 1966; Howland, 1958; Schulze, 1960). In the early 1970s, when U.S. EPA issued its definition of secondary treatment standards, the trickling filter was regarded as being unable to produce water quality that consistently met published standards (Parker, 1999). This was partly because of poor secondary sedimentation tank (sometimes referred to as humus tank) design. In 1979 in Corvallis, Oregon, Norris and co-workers (1980, 1982) followed a rock-media trickling filter with a small aeration basin and a flocculator clarifier. The researchers demonstrated that WRRF effluent water quality could be significantly improved by bioflocculation in the solids contact basin and improved clarification. The researchers referred to the combined units as the trickling filter/solids contact process. The German definition of instantaneous hydraulic application rate, or Spülkraft (SK), was described by ATV-DVWK in 1983. Albertson (1995) advanced the use of Spülkraft with a timing mechanism that adjusted rotary-distributor speed with an electric drive.

3.1 GENERAL DESCRIPTION

The trickling filter is a three-phase system with fixed biofilm carriers. Wastewater enters the bioreactor through a distribution system and trickles down over the biofilm surface and air moves upward or downward in the third phase. Biofilm develops on biofilm carriers. Trickling filter components typically include a distribution system, containment structure, rock or plastic biofilm carrier, underdrain, and ventilation system. Figure 7.3 illustrates a modern trickling filter cross-section. Trickling filters that are treating wastewater produce total suspended solids (TSS), which means that liquid-solids separation is required. This is achieved with either circular or rectangular secondary sedimentation basins. The secondary segment of a trickling filter process typically includes an influent pumping station, trickling filter, trickling filter recirculation pumping station, and liquid-solids separation unit process.

3.1.1 DISTRIBUTION SYSTEM

Primary effluent (fine-screened and degritted wastewater) is either pumped or flows by gravity to a trickling filter distribution system. The distribution system uniformly distributes wastewater over the trickling filter biofilm carriers in intermittent doses. The distributors may be hydraulically or electrically driven. Intermittent application allows for resting, or aeration, periods. Efficient influent wastewater distribution results in proper media wetting. Poor media wetting may lead to dry media pockets, ineffective treatment zones, and odor. There are two types of systems: fixed-nozzle and rotary distributors. Because their efficiency is poor, distribution with fixed nozzles should not be used (Harrison & Timpany, 1988).

Hydraulic rotary distributors use retardant back spray orifices to slow rotational speed, while maintaining desired pump flowrate. Figure 7.4 illustrates both a modern, hydraulically driven rotary distributor that uses gates that either opens or closes distributor orifices to adjust rotational speed and an electrically driven rotary distributor. A typical dosing rate for conventional rotary distributor systems is 2 to 10 mm/passes at 0.2 to 1.5 min/rev. The rotary distributor typically is equipped with two to six arms. The distributed flow may be staggered for full coverage per arm. That is, each arm may provide 50% or 100%

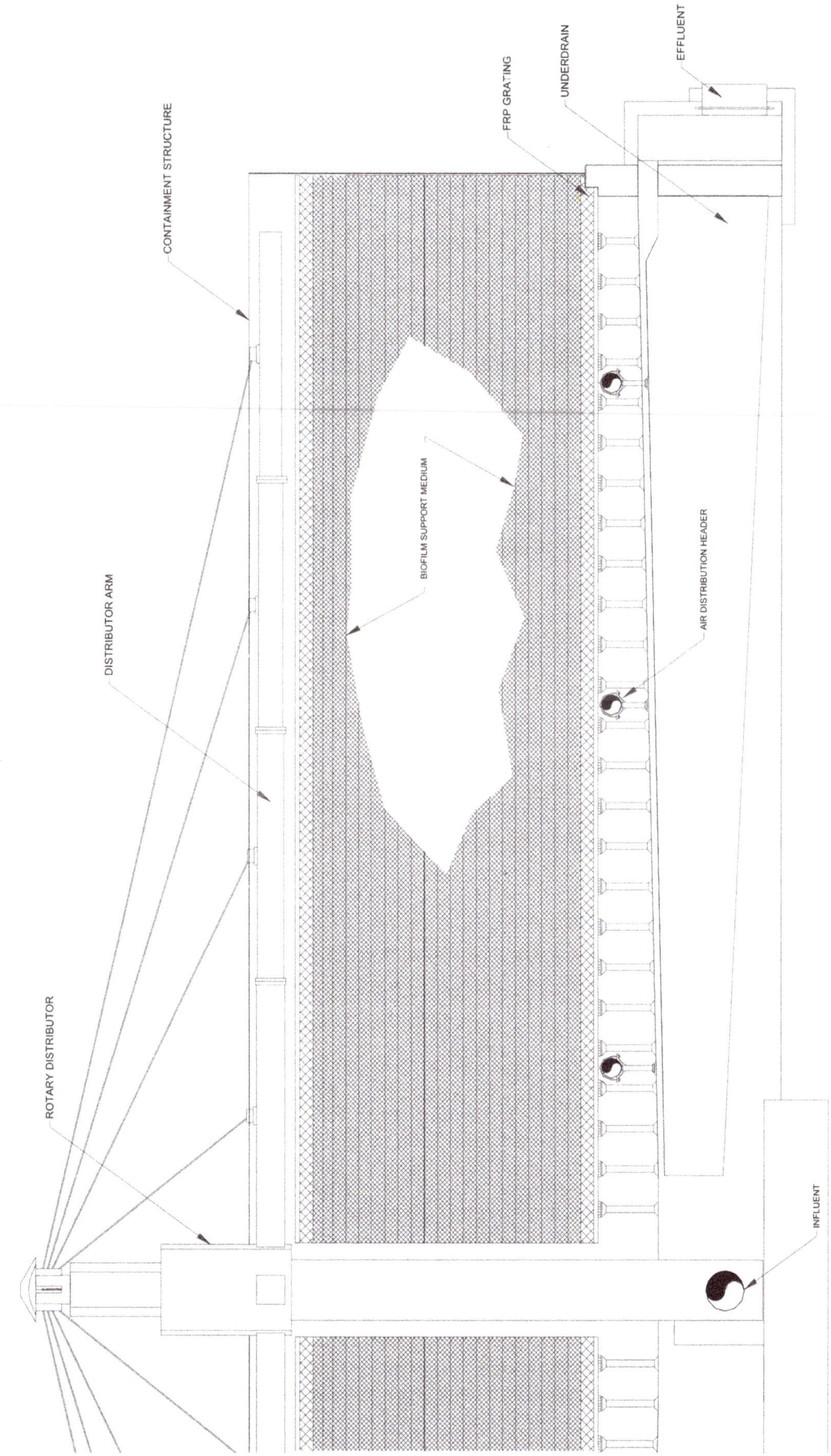

Figure 7.3 Typical Trickling Filter Components and Cross-Section (Daigger & Boltz, 2010)

Figure 7.4 Hydraulically Driven Rotary Distributors Use Variable Frequency Drive-Controlled Gates That Either Open or Close Distributor Orifices That Adjust With Varying Pumped Flows to Maintain a Constant Preset Rotational Speed (Left). On the Right Is an Electrically Driven Rotary Distributor. (Daigger & Boltz, 2010; reprinted with permission from WesTech Engineering, LLC)

coverage per revolution. Providing appropriate flushing intensity is difficult with rock media operating at typical application rates of 0.2 to 0.6 $m^3/m^2/h$ (0.08–0.25 gpm/sq ft). No minimum speed has been specified for a hydraulically propelled distributor. To increase flushing, distributor speeds have been reduced with reverse thrusting jets. Some hydraulically driven units have stalled or stopped rotating at speeds from 4 to 20 min/rev. The use of a variable frequency drive allows for more precise control of distributor-arm rotation. Electrically driven rotary distributors have motorized drive units that control distributor speed independent of the wastewater flow.

3.1.2 BIOFILM CARRIERS

Ideal trickling filter biofilm carriers, or media, provide a high specific surface area, low cost, high durability, and high enough porosity to avoid clogging and promote ventilation (Metcalf and Eddy, Inc./AECOM, 2014). Trickling filter biofilm carriers include rock, random (synthetic), vertical flow (synthetic), and 60° crossflow (synthetic) media. Both vertical-flow and crossflow media are constructed with corrugated plastic sheets. Some vertical-flow media is manufactured with corrugated sheets only while others have e-other sheet corrugated (the makeup are smooth plastic panels). Horizontal redwood or treated wooden slats also have been used as trickling filter media, but are no longer considered because of high cost and limited supply.

Modular plastic trickling filter media (i.e., self-supporting vertical flow or crossflow modules) is used almost exclusively for new trickling filter based WRRFs. Several trickling filters using rock-media exist provide good service when properly designed and operated. Figure 7.5 compares filter media; Table 7.3 presents physical properties of various media types. Figure 7.6 illustrates self-supporting crossflow modular plastic trickling filter media. Synthetic media allows for higher hydraulic loadings and enhanced oxygen transfer compared to rock-media because of the higher specific surface area and void space. Rock media has, ideally, a 50-mm diameter, but may range in size. Rounded rock trickling filter media helps mitigate issues associated with rigid, rock (slag) media. The slag-type rock contains numerous crevices that can retain water and accumulate biomass. Because of structural requirements associated with the large unit weight of the rock media, the trickling filters are shallow in comparison to synthetic media towers and are susceptible to excessive cooling. The water retained inside crevices in the slag-type rock media may then expand and sever rock fragments. This can result in the production and accumulation of fine material. The accumulation of both fine material and retained biomass is a primary contributor to rock-media trickling filter plugging (Grady et al., 1999). Typically, rock media has low specific surface area, void space, and high unit weight. Although recirculation is common, the low void ratio in rock-media trickling filters limits hydraulic application rates. Excessive hydraulic application can result in ponding, which results in limited

Table 7.3 Physical Properties of Commonly Used Trickling Filter Media (lb/cu ft × 16.02 = kg/m³ sq ft/cu ft × 3.281 = m²/m³)

Media Type	Material	Nominal Size, m (feet)	Bulk Density, kg/m³ (lb/cu ft)	Specific Surface Area m²/m³ (sq ft/cu ft)	Void space (%)
Rock (river)		0.024–0.076 (0.08–0.25)	1442 (90)	62.3 (19)	50
Rock (slag)		0.076–0.128 (0.25–0.42)	1601 (100)	45.9 (14)	60
Corrugated plastic modules[a]					
60° Crossflow	PVC	0.61 × 0.61 × 1.22 (2 × 2 × 4)	24.0–44.9 (1.5–2.8)	100 and 223.1 (30, 48, and 68)	95
Vertical flow	PVC	0.61 × 0.61 × 1.22 (2 × 2 × 4)	24.0–44.9 (1.5–2.8)	101.7 and 131.2 (31 and 40)	95
Random pack[b]	PP	0.185 ø × 0.051 H 7.3″ ø × 2″ H	27.2 (1.7)	98.4 (30)	95

[a]Manufacturers of corrugated plastic modules are (formerly) BF Goodrich, American Surf-Pac, NSW, Munters, (currently) Brentwood Industries, Jaeger, and Marley (SPX Cooling).
[b]Manufactures of random media are (formerly) NSW Corp. and (currently) Jaeger.
[c]Manufacturers of plastic strips are (formerly) NSW Corp. and (currently) Jaeger.

oxygen transfer and poor bioreactor performance. Existing rock-media trickling filters may sometimes be improved by providing forced ventilation, solids contact channels, or deepened secondary clarifiers that include energy dissipating inlets (EDIs) and flocculator-type feed wells. Replacement or deepening of the trickling filter using plastic media often is required if rock media quality is poor, space is limited, and WRRF expansion is expected. A well-designed and operated rock-media trickling filter can provide high-quality effluent. Grady et al. (1999) suggest that for organic loads of less than 1 kg BOD_5/d/m³, rock- and synthetic-media trickling filters are capable of equivalent performance. However, as organic load increases synthetic media will result in fewer nuisance problems and reduced plugging.

Drury et al. (1986) demonstrated that existing rock media trickling filters can be improved simply by using existing volume and replacing rock with synthetic media. This is because of increased specific surface area, ability to increase hydraulic loading, and improved ability to control biofilm growth. Changing the media can help address problems such as severe odor generation and deterioration of media or to expand capacity using existing footprint and assets. Crossflow media typically will perform better than vertical-flow media in low- to medium-organic loading scenarios. However, if the total organic load (TOL) becomes substantial, then biofilm accumulation will be so great that performance in the trickling filter will be hindered by the crossflow media. Parker (1999) suggested that this efficiency change illustrates the "switchover effect" in which efficiency switches over from crossflow to vertical-flow media at high TOLs that is not observed in the other studies.

Gullicks and Cleasby (1986) have demonstrated the importance of synthetic media wetting. Crine et al. (1990) have demonstrated that lava rock and random media wetting effectiveness decreased with increasing specific surface area. The researchers found that wetting effectiveness was only 0.2 to 0.6. In some cases, media can be combined such that media efficient in hydraulic distribution are in the upper layers and media less prone to excessive biofilm accumulation compose the remainder. Upper portions of the trickling filter will receive higher organic loading and lower layers will receive little organic loading. Many reports indicate that denser media, such as random and crossflow, that are effective for flow redistribution are more prone to solids retention and fouling (Boller & Gujer, 1986; Crine et al., 1990; Gullicks & Cleasby, 1986; 1990; Onda et al., 1968; Parker et al., 1989). Applications such as treating strong wastewaters, pretreatment with fine screens, and roughing tend to produce thicker biofilms. Vertical media types are preferred for these applications.

Synthetic trickling filter media has a high specific surface area and void space, and low unit weight. Because of the reduced unit weight, synthetic-media trickling filters can be constructed at depths more than three times that for a comparably sized rock-media trickling filter. Modular plastic media is manufactured with the following specific surface areas: 223 m²/m³ (68 sq ft/cu ft) as high-density; 138 m²/m³ (42 sq ft/cu ft) as

TRICKLING FILTER MEDIA COMPARISON

RELATIVE DIFFERENCE (A) IN TRICKLING FILTER MEDIA (B)

EFFLUENT QUALITY

	Low Load: Soluble BOD	Low Load: TSS	Low Load: Total BOD	Low Load: Nitrification	High Load: Soluble BOD	High Load: TSS	High Load: Total BOD
HORIZONTAL	●	■	●	●	●	■	●
VERTICAL	⊕	○	○	⊕	+	○	+
60° CROSSFLOW	+	+	+	+	⊕	+	⊕
45° CROSSFLOW	○	○	○	○	⊕	○	○
RANDOM	⊕	○	○	○	○	●	○
ROCK	○	○	○	■	■	○	■
COMMENTS	LITTLE SIGNIFICANT DIFFERENCE	LAMINAR FLOW INCREASES BIOFLOCCULATION	LITTLE SIGNIFICANT DIFFERENCE	CROSSFLOW GAVE BEST RESULTS	VERTICAL & CROSSFLOW NEAR EQUAL	LAMINAR FLOW INCREASES BIOFLOCCULATION	DIFFERENCES INCREASE AT HIGH LOADING

OPERATING CHARACTERISTICS

	Hydraulic Residence Time	Oxygen Transfer	Wetting	Filter Flies	Biogrowth	Plugging
HORIZONTAL	○	○	+	○	■	⊕
VERTICAL	■	○	○	+	+	+
60° CROSSFLOW	○	○	○	⊕	○	○
45° CROSSFLOW	○	○	●	○	○	●
RANDOM	⊕	○	■	○	●	○
ROCK	+	■	⊕	■	○	■
COMMENTS	GEOMETRICAL COMPLEXITY INCREASES TIME	MAY NOT CONTROL BIOKINETICS	POOR WETTING, RESULTS IN NUISANCES	PREVALENT W/POOR WETTING	POOR DISTRIBUTION IN SOME MEDIA	BLINDING OR SOLIDS BRIDGING

QUALITATIVE FACTORS

	Ease of Installation	Air Ventilation	Media Strength	Tower Construction	Media Height	Media Surface Area	Media Void Space	Weight of Media
HORIZONTAL	○	+	○	○	○	●	⊕	○
VERTICAL	⊕	●	○	○	⊕	⊕	⊕	⊕
60° CROSSFLOW	⊕	○	⊕	○	⊕	⊕	⊕	⊕
45° CROSSFLOW	+	○	⊕	○	⊕	⊕	⊕	⊕
RANDOM	○	○	●	●	○	⊕	⊕	⊕
ROCK	■	■	+	■	■	■	■	■
COMMENTS	SIGNIFICANT PLACEMENT DIFFERENCES	HORIZONTAL GIVES BEST CROSS VENTILATION	VERIFICATION BY TESTING ENCOURAGED	SELF-SUPPORTING MEDIA SIMPLEST	WEIGHT OR MEDIA COMPACTION PROBLEMS	BIOFILM THICKNESS ALSO IMPORTANT	HIGH RATE TYPE BEST	ROCK SIGNIFICANTLY HIGHER

GENERAL RECOMMENDATIONS

	General Recommendations
HORIZONTAL	SURFACE AREA LOW, NOT RECOMMENDED FOR TRICKLING FILTER USE. BEST USED WITH SLUDGE RECYCLE TO OPTIMIZE MEDIA.
VERTICAL	OPTIMUM USE IS AT HIGH LOADING TO THE FILTER ESPECIALLY WHERE PLUGGING MIGHT BE A PROBLEM WITH OTHER MEDIA.
60° CROSSFLOW	BEST AVAILABLE MEDIA FOR USE WITH FILTERS AT LOW TO MODERATE LOADING OR WHERE NITRIFICATION IS REQUIRED.
45° CROSSFLOW	NOT RECOMMENDED IN LIEU OF AVAILABLE 60° CROSSFLOW MEDIA
RANDOM	BEST USED WITH ROTARY DISTRIBUTORS AND AT LOW TO MODERATE LOADING.
ROCK	REPLACEMENT OF EXISTING ROCK MEDIA, WITH HIGH RATE TYPES USUALLY NOT JUSTIFIED. BEST USED AT LOW TO MODERATE LOADING.

NOTES:

(A) + BEST ⊕ GOOD ○ AVERAGE ● POOR ■ WORST

(B) ALL OPERATED IN THE TRICKLING FILTER PROCESS WITH 20 FOOT MEDIA DEPTH EXCEPT ROCK AT 8 FEET

IMPORTANT: COSTS MUST BE CONSIDERED ON A SITE SPECIFIC BASIS

CH2M HILL

Figure 7.5 Relative Comparison of Trickling Filter Media (CH2M HILL, 1984; reprinted with permission from Jacobs)

Figure 7.6 Self-Supporting Crossflow Modular Plastic Trickling Filter Media (Courtesy of Brentwood Industries, Inc., Reading, PA)

medium-density; 100 m^2/m^3 (30 sq ft/cu ft) as low-density. Both vertical and crossflow media are reported to effectively remove 5-day biochemical oxygen demand (BOD_5) and TSS (Aryan & Johnson, 1987; Harrison & Daigger, 1987). Research shows, however, that different synthetic media provide different treatment efficiencies despite being manufactured with virtually identical specific surface areas. The designer should carefully consider effects of media type and configuration on trickling filter effluent water quality.

Plastic modules with a specific surface area of 89 to 102 m^2/m^3 are well suited for carbon oxidation and combined carbon oxidation and nitrification. Parker et al. (1989) recommended a medium-density crossflow media and were against the use of high-density crossflow media in tertiary nitrification applications. Researchers claim that lower rates occur with high-density media because of the development of dry spots below the interruption points in the media. Higher density media has more interruptions and, therefore, is wet less effectively. Using medium-density media will reduce plugging. Vertically oriented modular plastic media is suited for high-strength wastewater (perhaps industrial) or high organic loadings such as with a roughing filter. Other advantages of vertical flow include more effective biomass flushing and less complicated geometry, which enhances air movement. In some cases, crossflow media has been placed in the top layer to enhance wastewater distribution.

3.1.3 CONTAINMENT STRUCTURE

Rock and random plastic media are not self-supporting when stacked and require structural support to contain the media within the bioreactor. Containment structures typically are precast or panel-type concrete tanks. When self-supporting media such as plastic modules is used, other materials such as wood, fiberglass, and coated steel are used as containment structures. The containment structure avoids splashing and provides media support, wind protection, and flood containment. Trickling filters are well known for the nuisance macrofauna such as filter flies and snails. A properly designed containment structure increases operator flexibility and allows control of nuisance macrofauna. It can include a variety of dosing alternatives and possibly provisions for flooding the filter.

3.1.4 UNDERDRAIN SYSTEM AND VENTILATION

The trickling filter underdrain system is designed to meet two objectives: collect treated wastewater for conveyance to downstream unit processes and create a plenum that allows for the transfer of air throughout the media (Grady et al., 1999). Clay or concrete underdrain blocks typically are used for rock media because of the required structural support. A variety of support systems, including concrete piers and reinforced fiberglass grating, are used for other media types. Figure 7.7 illustrates field-adjustable plastic stanchions and fiber-glass-reinforced plastic grating on the concrete floor of a trickling filter containment structure. The volume created between concrete and media bottom creates the underdrain.

The vertical flow of air through the media can be induced by either mechanical means (forced or fan ventilation) or natural air draft. Natural air ventilation results from a difference in ambient air temperature outside and inside the trickling filter. The temperature causes air to expand when warmed and contract when cooled. The net result is an air-density gradient throughout the trickling filter. Depending on the differential condition, an air front either rises or sinks, which results in continuous airflow through the bioreactor. Natural ventilation may become unreliable or inadequate in meeting process air requirements when neutral temperature gradients do not produce air movement. Such conditions may be daily or seasonal and can lead to odorous anaerobic biofilm conditions and poor performance. Therefore, mechanical forced air ventilation typically is included. Forced-air ventilation is accomplished by adding low-pressure fans to circulate air continuously. When using ventilation, the designer should ensure that the air is uniformly distributed to provide oxygen to all biofilm in the reactor.

3.1.5 TRICKLING FILTER PUMPING STATION

The pumping station lifts primary effluent and recirculates unsettled trickling filter effluent (also known as underflow) to the influent stream. Typically, trickling filter underflow is recirculated to the distribution system to achieve the hydraulic load required for proper media wetting and biofilm thickness control. The intent of recirculating bioreactor effluent is to decouple hydraulic and organic loading. Although effluent from the secondary clarifier can be recirculated, this not common practice because it may lead to the hydraulic overloading of secondary clarifiers. Influent pumping typically is selected to allow underflow

Figure 7.7 Field-Adjustable Plastic Stanchions and Fiber-Glass-Reinforced Plastic Grating on the Concrete Floor of a Trickling Filter Containment Structure. The Volume Created Between Concrete and Media Bottom Creates the Underdrain. (Courtesy of Brentwood Industries, Inc., Reading, PA)

to flow by gravity to the suspended growth reactor (or solids contact basin), secondary clarifier, or other downstream process.

3.1.6 HYDRAULIC AND POLLUTANT LOADING

Trickling filters are classified by the intended mode of pollutant degradation and loading, including carbon oxidation, combined carbon oxidation and nitrification, or nitrification. Organic loading is expressed as kg/d/m^3 (lb/d/cu ft) of filter media as BOD_5 or COD. General practice is to ignore the organic load imparted by recirculation streams, but the designer should account for the effects of recycle flow and pollutant loading (specifically ammonia-nitrogen) on treatment efficiency. The TOL may be calculated using:

$$\text{TOL} = \frac{\text{BOD}_5 \text{ apllied}}{\text{media volume}} = \frac{Q_{\text{in}} \cdot S_I}{V_M} \tag{7.1}$$

where V_M = trickling filter media volume, m^3;
Q_{in} = influent to the trickling filter (typically primary effluent), m^3/d; and
S_I = influent BOD concentration, g/m^3.

Nitrifying trickling filter data is expressed in terms of surface-based ammonia-nitrogen loading rate. Trickling filter hydraulic loading rate is calculated without and with recirculation. The wastewater hydraulic load (WHL) excludes recirculation and can be calculated using the following equation (m^3/d/m^2).

$$\text{WHL} = \frac{Q_{\text{in}}}{A} \tag{7.2}$$

where A = biofilm area in reactor (m^2), and
Q_I = influent flowing from upstream unit processes.

The total hydraulic load (THL) is used to gauge media wetting and biofilm thickness control, and considers trickling filter influent flowing from upstream unit processes, Q_I, and the recirculation stream, Q_R. The total hydraulic load can be expressed mathematically by Equation 7.3

$$\text{THL} = \frac{Q_I + Q_R}{A} \tag{7.3}$$

Where:

Q_I = influent stream
Q_R = recirculation stream.

Current practice is to reference hydraulic loading in units of cubic meter per day per square meter of plan trickling filter area (m^3/m^2/d) rather than referencing the required hydraulic application per unit biofilm growth area.

3.2 PROCESS FLOW SHEETS AND BIOREACTOR CONFIGURATION

3.2.1 PROCESS FLOW DIAGRAMS

The trickling filter process typically consists of preliminary treatment (including screening and grit removal), primary clarification, bioreactor, secondary clarification, and disinfection unit processes. Recirculation methods influence the process flow. There are two types of recirculation. The first allows for direct recirculation to the trickling filter, and the second passes flow through a primary clarifier before entering the trickling filter. Four trickling filter process flow diagrams, including both single- and two-stage trickling filters, are shown schematically in Figure 7.8. Recirculation dilutes influent wastewater, dampens organic variability in the influent because of diurnal fluctuations, and keeps the media properly wetted. Clarifying trickling effluent will enhance performance of a subsequent trickling filter in two-stage operation. The designer must ensure that the recirculation flow required for wetting and biofilm thickness control does not exceed the limiting hydraulic loading rate for the sedimentation tank. When

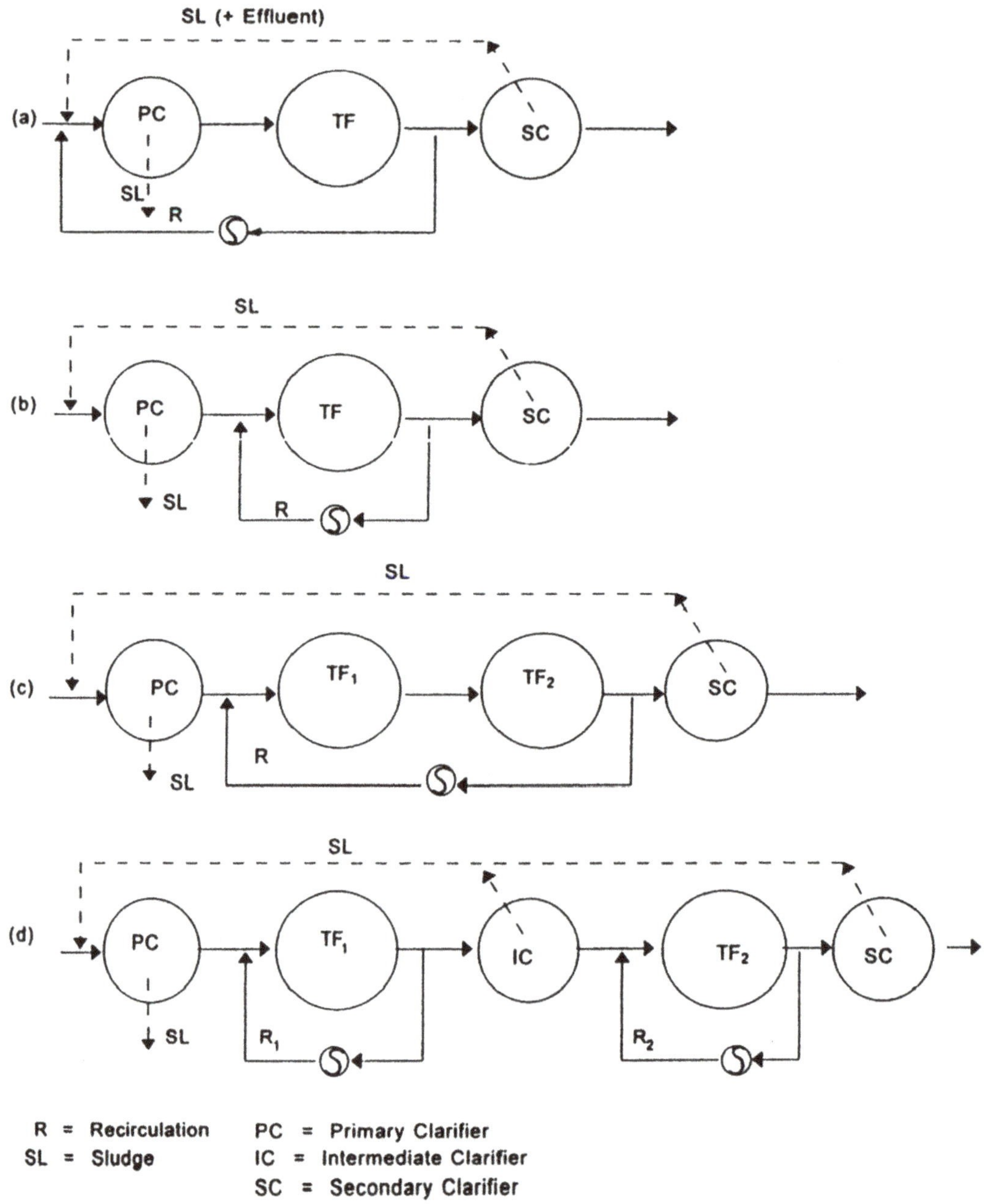

Figure 7.8 Flow Diagrams for Single- and Two-Stage Trickling Filter Facilities (R = recirculation; SL = sludge; PC = primary clarifier; IC = intermediate clarifier; SC = secondary clarifier)

two trickling filters are operated in series (rather than parallel) several studies have shown that the second stage trickling filter performance is not adversely affected by the absence of a clarifier between first and second stage units. However, there are indications that certain wastewaters containing high concentrations of soluble BOD_5 (likely from an industrial source) will result in excessive biofilm growth and subsequent excessive biomass production. These solids can adversely affect the second-stage trickling filter if not removed by intermediate settling. The design of settling tanks in two-stage trickling filter systems also is affected by the recirculation pattern. The designer should consider the most economical method of securing acceptable effluent quality.

The practice of alternating the lead trickling filter in a primary-secondary trickling system is referred to as an alternating double-filtration (ADF) system. This concept is beneficial when applied to modular plastic-media trickling filters. Gujer and Boller (1986) and Parker et al. (1989) observed patchy biofilm growth in the lower section of pilot-scale nitrifying trickling filters (NTFs). The researchers attributed the patchy growth to dry spots. Boller and Gujer (1986) advocated use of the ADF mode to enhance performance of the NTF. Aspegren et al. (1992) also observed improved nitrification using the ADF mode versus a single-stage NTF because of virtual elimination of biofilm patchiness. Use of the ADF approach with two trickling filters in series encourages full-depth biofilm development in both trickling filters. The lead trickling filter should be switched every 3 to 7 days to ensure that both contain a steady-state biofilm developed along the entire bioreactor length. Operating trickling filters in series (rather than parallel) may result in added nitrification without requiring ADF operation.

Sludge handling also affects the trickling filter process. Each of the process flow diagrams implies that waste biological solids are removed by co-settling the biological sludge with the primary sludge before withdrawal from the system. Many facilities exist that separately handle primary and secondary sludge. For example, primary sludge may be thickened by gravity thickeners and trickling filter humus by gravity belt thickeners. The benefits must be evaluated by the designer. In principle, however, sludge co-settling is sound practice. It does, however, require that operators consistently withdraw the solids from the process, and that designers provide equipment and means to maintain a near zero sludge blanket if necessary.

3.2.2 BIOREACTOR CLASSIFICATION

Within this section, trickling filters are discussed based on mode of operation or process application: (1) roughing, (2) carbon oxidation, (3) carbon oxidation and nitrification, and (4) nitrification (Daigger & Boltz, 2010). Table 7.4 summarizes typically accepted defining criteria for each operational mode. Roughing filters receive high-hydraulic and high-organic loadings and require the use of vertical-flow media to minimize plugging. Although they may provide a high-quantity organic load removal per unit volume, their settled effluent still contains substantial BOD_5. Roughing filters provide approximately 50% to 75% soluble BOD_5 conversion and may receive total loadings of 1.5 to 3.5 kg BOD_5/d/m^3. Carbon-oxidizing trickling filters provide settled effluent of 15 to 30 mg/L for both BOD_5 and TSS, and may receive BOD_5 loadings of 0.7 to 1.5 kg/d/m^3. Combined carbon oxidation and nitrification trickling filters provide effluent BOD_5 less than 10 mg/L and NH_3-N less than or equal to 0.5 to 3 mg/L (after solids separation). These trickling filters may receive BOD_5 loadings less than 0.2 kg/d/m^3, and TKN loadings of 0.2 to 1.0 g/d/m^2. Tertiary nitrifying trickling filters provide 0.5 to 3 mg/L effluent NH_3-N when receiving a clarified secondary effluent and NH_3-N loadings of 0.5 to 2.5 g N/d/m^2.

3.2.3 HYDRAULICS

Recirculation, distributor operation, biofilm thickness, and macro fauna accumulation affect wetting of trickling filter media. Albertson and Eckenfelder (1984) postulated that the active biofilm surface area in a trickling filter is dependent upon biofilm thickness and media configuration, and that increased biofilm

Table 7.4 Trickling Filter Classification (Daigger & Boltz, 2010)

Design Parameter	Roughing	Carbon Oxidizing (cBOD)	cBOD and Nitrification	Nitrifying
Media used	VF	RA, RO, XF, or VF	RA, RO, or XF	RA or XF
Wastewater source	Primary effluent	Primary effluent	Primary effluent	Secondary effluent
Hydraulic loading, (gpm/sq ft)	52.8–178.2 (0.9–2.9)	14.7–88.0 (0.25*–1.5)	14.7–88.0 (0.25*–1.5)	35.2–88.0 (0.6–1.5)
Contaminant loading, (lb BOD_5/d/1000 cu ft)	1.6–3.52 (100–220)	0.32–0.96 (20–60)	0.08–0.24 (5–15)	N/A
(lb NH_3-N/d/1000 sq ft)	N/A	N/A	0.2–1.0 (0.04–0.2)	0.5–2.4 (0.1–0.5)
Effluent quality, mg/L (unless noted)	50%–75% BOD_5 conversion	15–30 BOD_5 and TSS	<10 BOD_5 <3 NH_3-N	0.5–3 NH_3-N
Predation	No appreciable growth	Beneficial	Detrimental (nitrifying biofilm)	Detrimental
Filter flies	No appreciable growth	No appreciable growth	No appreciable growth	No appreciable growth
Depth, m (ft)	0.91–6.10 (3–20)	≤12.2 (40)	≤12.2 (40)	≤12.2 (40)

Notes. VF = vertical flow; RA = random pack; XF = crossflow; RO = rock; and TSS = total suspended solids.
*Applicable–shallow trickling filters; gpm/ft^2 = gallons per minute per square foot of trickling filter plan area gpm/sq ft × 58.674 = m^3/m^2·d (cubic meter per day per square meter of TF plan area); lb BOD_5/d/1000 cu ft × 0.016 0 = kg/m^3·d (kilograms per day per cubic meter of media); and lb NH_3-N/d/1000 sq ft × 4.88 = g/m^2·d (grams per day per square meter of media).

thickness reduces active surface area. The net result of excess biomass accumulation is reduced trickling filter performance. The researchers stated that for medium-density crossflow media with 98 m^2/m^3 specific surface area, a 4-mm biofilm thickness would cause a 12% reduction of active biofilm area, assuming that the entire media has been appropriately wetted.

Conventional practice is direct recirculation of the trickling filter underflow. Recirculation systems with recycle of settled effluent and direct recycle were compared at a facility in Webster City, Iowa (Culp, 1963). No significant differences in results were obtained when trickling filters operated simultaneously on the same settled wastewater under either winter or summer conditions. Researchers have concluded that it is the amount of recirculation and not the arrangement that is the most important factor for optimizing performance of rock-media trickling filters. The absence of intermediate settling between the first and second trickling filter operating in series does not adversely affect performance of the second-stage unit. The recirculation ratio (Q_r/Q) is typically in the range 0.5 to 4.0. Early studies demonstrated that vertical-flow corrugated media require an average application rate higher than 0.5 $L/m^2/s$ to provide maximum BOD_5 removal efficiency (Bryan, 1955; 1962; Bryan & Moeller, 1960). Shallow towers using crossflow media have used hydraulic rates of 0.39 to 1.08 $m^3/m^2/h$.

Slowed distributor operation benefits trickling filter facilities because of interrupted flow (periodicity of dosing), increased wetting efficiency (percent of media wetted), and controlled biofilm thickness. The designer should consider recirculation capabilities, the effect of reverse thrusting jets, or use of speed control on the distributor to enhance performance or improve operation. Another useful process control parameter is the dosing rate, Spülkraft. Methods for calculating dosing rate (mm/pass) are given by Equations 7.4 and 7.5.

$$DR = \frac{Q_I + Q_R}{A} \cdot \frac{1000\frac{mm}{m}}{N_a \cdot \omega_d \cdot 1440\frac{m\ in.}{day}} \tag{7.4}$$

or

$$DR = \cdot \frac{THL \cdot 1000\frac{mm}{m}}{N_a \cdot \omega_d \cdot 1440\frac{m\ in.}{day}} \tag{7.5}$$

Where N_a = number of arms on the distributor, and
ω_d = distributor rotational speed (rev/min).

Here, Spülkraft is in mm/pass. The typical hydraulically driven distributor in North America operates in the range of 2 to 10 mm/pass. Table 7.5 lists recommended operating and flushing dosing rates for rotary distributors. Higher dosing rates are recommended for higher organic loading rates to provide biofilm thickness control. Wastewater characteristics, temperature, or media type may influence dosing rate. It

Table 7.5 Operating and Flushing Dosing Rates for Distributors (Daigger & Boltz, 2010)

Total Organic Load ($kg/m^3 \cdot d$) ($lb\ BOD_5/d/1000\ cu\ ft$)	Operating Dosing Rate (mm/pass) (in./pass)	Flushing Dosing Rate (mm/pass) (in./pass)
<0.4 (<25)	25–75 (1–3)	100 (4)
0.8 (50)	50–150 (2–6)	150 (6)
1.2 (75)	75–225 (3–9)	225 (9)
1.66 (100)	100–300 (4–12)	300 (12)
2.4 (150)	150–450 (6–18)	450 (18)
3.2 (200)	200–600 (8–24)	600 (24)

Note. Actual values are site specific and vary with media type.

also may be beneficial to periodically use a higher flushing dosing rate for 5% to 10% of the 24-hour operating period. Work by Albertson (1989a, 1989b, 1995) and Parker (1999) and Parker et al. (1995, 1997) demonstrate that biofilm control measures enhance the trickling filter process when operating and flushing dosing rate values are used. These enhancements include improved performance, reduced odors, reduced power use for recycling, reduced nuisance organisms, and elimination of heavy sloughing cycles (Albertson 1989a, 1989b, 1995). Parker et al. (1995) described the use of both distributor speed control and variable frequency drive controlled recirculation pumps to maintain constant trickling filter hydraulic application. Pilot studies demonstrated mechanically driven distributor dosing did not improve performance of a nitrifying trickling filter. There is little research describing the effect of hydraulic transients on synthetic trickling filter media and their effect on media life.

3.3 OXYGEN REQUIREMENTS AND AIR SUPPLY ALTERNATIVES

Trickling filters require oxygen to sustain aerobic biochemical reactions. Several researchers have demonstrated that at least some portion of roughing, carbon oxidizing, combined carbon oxidizing and nitrification, and NTFs may operate under oxygen-limited conditions (Kuenen et al., 1986; Okey & Albertson, 1989; Schroeder & Tchobanoglous, 1976). Ventilation is essential to maintain aerobic conditions. Current design practice requires adequate sizing of underdrains and effluent channels to permit free airflow. Passive devices for ventilation include vent stacks on the periphery, extensions of underdrains through side walls, ventilating manholes, louvers on the sidewall of the tower near the underdrain, and discharge of effluent to the subsequent settling basin in an open channel or partially filled pipes. However, these methods may not be adequate if high performance is required or in the presence of low natural draft.

3.3.1 NATURAL DRAFT

One method to determine the amount of natural draft is to require 1 m^2 of ventilating area for each 3 to 4.6 m of trickling filter circumference. Another gauge is 1 to 2 m^2 of ventilation area in the underdrain area per 1000 m^3 of media. Inlet openings to rock media underdrains have a recommended nonsubmerged combined area equal to at least 15% of the trickling filter surface area. Drains, channels, and pipes should be sized to prevent submergence of greater than 50% of the cross-sectional area under design hydraulic loading. Forced ventilation should be provided for covered trickling filters. Benzie et al. (1963) studied 17 rock media based WRRFs in Michigan and demonstrated that airflow was stagnant when ambient and wastewater temperatures were equal. Researchers also concluded that, during winter months, recirculation has a cooling effect on natural draft.

Schroeder and Tchobanoglous (1976) proposed the following equation to determine the natural draft in synthetic media trickling filters.

$$\Delta P = 353 \cdot D \cdot \left(\frac{1}{T_C} - \frac{1}{T_H} \right) \tag{7.6}$$

Where ΔP = pressure head resulting from differential temperature, mm H_2O;
T_C = colder temperatures (ambient air or air inside trickling filter), °K;
T_H = hotter temperatures (ambient air or air inside trickling filter), °K;
$T_m = \dfrac{T_C - T_H}{\ln\left(\dfrac{T_C}{T_H}\right)}$ = average temperature inside the trickling filter, °K; and
D = media depth, m.

The log mean pore temperature, T_M can be replaced by the hotter temperature, T_H, to obtain a less conservative estimate of the average pore air temperature. Some trickling filters do not have adequate oxygen daily for at least part of the year. Air stagnation results in odors and performance variability. Furthermore, little data exists to provide guidance on defining the amount of natural airflow through rock or synthetic media trickling filters. The velocity of air is low except when there is a large difference between air and wastewater temperature or if the trickling filters are deep (6–12 m). In the case of natural draft, if the wastewater is colder than ambient air, then the air will flow down. Alternatively, if the ambient air is colder than the wastewater, then the airflow will be up. Because a constant temperature differential does not occur naturally, power ventilation by mechanical means is recommended.

3.3.2 FORCED VENTILATION

Most new and improved trickling filter based WRRFs use low pressure fans to force airflow. This practice, known as power ventilation, offers a wintertime benefit of limiting cold airflow and minimizing cooling of wastewater. Powered ventilation and enclosed trickling filters also can destroy odorous compounds in influent wastewater and prevent excessive ventilation during winter or during periods of high air–water temperature differentials. Trickling filters may be ventilated in either an upward or downward airflow pattern. Dow Chemical presented a calculation to determine differential pressure as a function of airflow rate ($m^3/m^2/h$) through VFM for natural draft and mechanical fans. Differential air pressure for natural draft typically is insufficient during some portion of the day. Whether the trickling filter is 1.5 m (5 ft) or 6 m (20 ft) deep, the driving force is low. With temperature differentials between ambient air and water of less than 62.8 °C (5 °F), airflow in the trickling filter can stagnate. Humidity differences also drive airflow, but ambient air values vary widely and are unpredictable. Downflow may reduce or eliminate trickling filter odors if incorporated with good flushing hydraulics.

Recommended minimum airflow for design of power ventilation has been developed. However, additional research is required to determine airflow rates necessary to maximize kinetics. Influent BOD_5 and effluent $rbBOD_5$ are used to determine the required airflow rate. The oxygen transfer rate used to determine airflow is 5% for carbon oxidizing trickling filters and 2.5% for both combined carbon oxidation and nitrification and NTFs. The higher air rate used for NTFs is to ensure that all areas of the trickling filter have airflow. Losses through vertical corrugated media can be described by using the following equation:

$$\Delta P = R \cdot \left(\frac{v^2}{2 \cdot g} \right) \tag{7.7}$$

Where v^2 = superficial air velocity, m/s;
g = acceleration because of gravity, 9.81 m/s^2;
R = tower resistance, velocity heads lost per unit tower depth, kPa/m; and
ΔP = total headlosses, kPa.

The term, *R*, is the total sum of individual headlosses through the trickling filter. Assuming adequate inlet and underdrain openings, the main loss through the trickling filter will be the packing loss (*Rp*) of the filter media. Dow Chemical proposed that *R* can be determined using Equation 7.8 for medium density (89 m^2/m^3) VFM (Metcalf & Eddy, Inc./AECOM, 2014):

$$R_p = 10.33 \cdot D \cdot e^{(1.36 \times 10^{-5})\left(\frac{L}{A}\right)} \tag{7.8}$$

Where R_p = packing headloss in terms of velocity heads;
L = liquid loading, kg/h;
A = trickling filter plan (slab) area, m^2; and
D = trickling filter media depth, m.

To estimate total headloss through the trickling filter, the value of *Rp* determined by Equation 7.8 should be multiplied by a factor of 1.3 to 1.5 to account for minor headlosses such as the inlet and underdrain. Because there are few data for other media types, the following multipliers should be used to determine R_p (Metcalf and Eddy, Inc./AECOM, 2014):

- Crossflow: 100 m^2/m^3 (30 sq ft/cu ft) = 1.3 3 R_p
- Crossflow: 138 m^2/m^3 (42 sq ft/cu ft) = 1.6 3 R_p
- Rock: 39 to 49 m^2/m^3 (12–15 sq ft/cu ft) = 2.0 3 R_p
- Random: 100 m^2/m^3 (30 sq ft/cu ft) = 1.6 3 R_p

3.4 TRICKLING FILTER DESIGN MODELS

Numerous investigators have attempted to delineate the fundamentals of the trickling filter process by developing relationships among variables that affect operation. Existing process models range from simplistic empirical formulations to complex numerical models. Analyses of operating data have established equations or curves and have led to development of various empirical formulae. Unfortunately, many models exist and there no industry standard. Trickling filter models may be classified as dissolved organic loading models, particulate organic loading models, hydraulic loading models, and mass transfer models. Although these models include many variables that affect operations, none can predict actual performance. Designers need to assess which equation best fits a particular situation, particularly when meeting discharge permit requirements.

3.4.1 VELZ EQUATION

Velz (1948) proposed the first formulation delineating a fundamental law, compared to previous empirical attempts that were based on data analyses. The Velz equation relates the BOD_5 remaining at depth D mathematically by:

$$\frac{S_D}{S_i} = 10^{-kt} \tag{7.9}$$

Where S_i = influent BOD_5 (mg/L);
S_D = BOD_5 (mg/L) removed at depth D, meters;
t = residence time (d); and
k = Velz first-order rate constant (d^{-1}).

The formula depicted in Equation 7.9 implies that k is constant for all hydraulic rates; however, Albertson and Davies (1984) presented evidence that k varies with the hydraulic rate. The Velz equation is presented because of its foundation in currently used design formulations, namely the Schulze and Eckenfelder equations. Several modifications of the Velz equation have been made.

3.4.2 SCHULZE FORMULA

Schulze (1960) postulated that the time of liquid contact with the biological mass is directly proportional to trickling filter depth and inversely proportional to the hydraulic loading rate. This is expressed by Equation 7.10.

$$t_c = \frac{c \cdot D}{THL^n} \tag{7.10}$$

Where t_c = liquid contact time (d);
c = constant (dimensionless);
D = trickling filter depth (m);
THL = hydraulic loading rate ($m^3/m^2/d$); and
n = exponent on hydraulic loading (dimensionless).

Combining the time of contact with the first-order equation for BOD_5 removal in an adaptation of the Velz theory, Schulze derived the following formula:

$$\frac{S_e}{S_i} = e^{-\left(\frac{k_S \cdot D}{THL^n}\right)} \tag{7.11}$$

Where S_e = soluble BOD_5 in trickling filter effluent stream (mg/L);
S_i = soluble BOD_5 in trickling filter influent stream (mg/L);
k_S = Schulze coeffecient (d^{-1} when n = 1);
D = trickling filter depth (m);
n = exponent on hydraulic loading (dimensionless); and
THL = hydraulic loading rate ($m^3/m^2/d$).

Equation 7.11 is similar to that proposed by Velz. However, Velz's constant, k, was not formulated to consider hydraulic load. For a given wastewater strength, the hydraulic rate is proportional to the loading rate. Thus, volumetric organic loading may still be the controlling process variable. The value of k published by Schulze (based on U.S. customary units) for a rock media trickling filter with a 1.8-m (6-ft) depth at 20 °C was 0.69 d^{-1}. The dimensionless constant characteristic of rock media trickling filter, n, was found to be 0.67. The common temperature correction value of $u = 1.035$ could be applied to determine k_t as follows by

$$k_t = k_{20} \times 1.035^{(T-20)} \quad \text{for} \quad k_s \tag{7.12}$$

and

$$k_t = A_s \times k_s$$

Where k_t = temperature-corrected coefficient value (d^{-1} when $n = 1$);
k_s = Schulze coefficient (d^{-1} when $n = 1$); and
A_s = clean surface area of the media, m^2.

3.4.3 ECKENFELDER FORMULA

Eckenfelder (1961) and Eckenfelder and Barnhart (1963) described a modified trickling filter formula to account for trickling filter media specific surface area (a, m^2/m^3). The formula proposed for soluble BOD_5 removal can be expressed mathematically by Equation 7.13.

$$\frac{S_e}{S_i} = e^{-\left(\frac{k_s' \cdot a^{(1+b)} \cdot D}{THL^n}\right)} \tag{7.13}$$

Where k_S' = overall treatability coefficient based on soluble BOD_5 $[(m^3/d)^{0.5}/m^2]$;
D = depth (m); and
THL = hydraulic loading rate ($m^3/m^2/d$).
b = surface area modifier for surface loss with increasing area.

With recirculation, Equation 7.13 can be extended and expressed mathematically by Equation 7.14.

$$\frac{S_e}{S_i} = \frac{e^{-\left(\frac{k_s' \cdot D}{WHL^n}\right)}}{\left(1 + \frac{Q_R}{Q}\right) - e^{-\left(\frac{k_s' \cdot D}{WHL^n}\right)}} \tag{7.14}$$

Using the Eckenfelder formula and $k_S a = a \times k_s$, Equation 7.14 can be rewritten as Equation 7.15, this is known as the Modified Velz Equation.

$$S_e = \frac{S_i}{\left(\frac{Q_R}{Q} + 1\right) \cdot e^{\left(\frac{k_s \cdot a \cdot D \cdot \theta^{(T-20)}}{\left[WHL \cdot \left(\frac{Q_R}{Q} + 1\right)\right]^n}\right)} - \frac{Q_R}{Q}} \tag{7.15}$$

Design engineers may use various empirical criteria and design formulations for sizing trickling filters. The Schulze equation (Equations 7.10 and 7.11) and Modified Velz Equation (Equation 7.15) are used for both rock and plastic media trickling filter design over a wide range of media specific surface areas and depths. The coefficients k and n vary, however. Bruce and Merkens (1970, 1973) conducted simultaneous testing of two trickling filters at the same flow and BOD_5 loading but at a 4 : 1 ratio in the application rate because one was 7.41 m (24.3 ft) deep and the other was 2.1 m (6.9 ft) deep. From this study it was concluded BOD_5 efficiency would be independent of depth. Further work showed that the BOD strength also affected BOD_5 removal efficiency. Thus, the value of k of the hydraulically driven equations could be modified as a function of media depth (D) and influent BOD strength (L) by Equation 7.16.

$$k_2 = k_1 \cdot \left(\frac{D_1}{D_2}\right)^{0.5} \left(\frac{L_1}{L_2}\right)^{0.5} \tag{7.16}$$

Where k = Velz's constant;
D = depth (m); and
L = length (m).

The variation of k with trickling filter depth is an important consideration. A k-value developed for specific depth should not be used for a different depth without proper modification. Using data from installations and simultaneous tests, Albertson and Davies (1984) showed that k could be used for any trickling filter configuration if corrected for depth. However, this research also indicated that inadequate wetting might produce lower k values than optimum.

The Eckenfelder, Schulze, and Velz equations are similar. Because the coefficients k (or K) and n must be, or have been, empirically derived, background data are influenced by a variables such as hydraulic rate, dosing mode, temperature, wastewater characterization, media configuration and depth, ventilation, and other unknown test-specific factors. The equations have proven effective in modeling specific WRRF data, but also have been proven to deviate significantly from observed results. When modifying the configuration, the value of the treatment constant, k (or K) changes, even when considering the same trickling filter media and wastewater.

Improved hydraulic application systems provide control of ponding, nuisance macrofauna, and odors. The designer may want to attempt to account for improved performance with increased dosing rates. A well-designed and operated rock media trickling filter can provide high quality effluent. Grady et al. (1999) suggest that for low organic loads (less than 1 kg $BOD_5/d/m^3$), rock and synthetic media trickling filters are capable of equivalent effectiveness at low to moderate organic loading.

3.4.4 FILTER MEDIA DEPTH

In North America, rock media trickling filters typically are 1- to 2-m deep, but may be as deep as 2.4 m. This depth limitation is associated with lack of adequate ventilation produced by natural draft and an increased tendency to pond. In Europe, deeper filters are common; units in Arnheim, Netherlands, were constructed at 4.9-m deep but equipped with power ventilation. Comparative data are lacking for deep, power-ventilated rock filters and shallow, natural draft rock media.

Synthetic media trickling filters typically are constructed between 5- and 8-m deep, although units up to 12.8-m deep exist. The limiting depth is because of height aesthetics, serviceability, pumping requirements, and structural design. The increased depth has no implication on biological treatment efficiency. Increasing the trickling filter depth is typically worthwhile to reduce the minimum flow required for high wetting efficiency. In taller filters that have high loadings, oxygen deficiency may occur in the uppermost layers. However, adequate ventilation and biofilm control measures can prevent problematic odors.

The effect of trickling filter media depth on bioreactor performance has been treated as a matter of controversy in previous design manuals. Several investigators suggest that volume, irrespective of depth, controls performance (Bruce & Merkens, 1970, 1973; Galler & Gotaas, 1964; Kincannon & Stover, 1982; National Research Council, 1946). Recent research indicates that performance is dependent on specific surface area (which may translate to bioreactor volume when considering identical media types), and not trickling filter depth. Essentially, performance is governed by substrate availability. In a combined carbon oxidation and nitrification trickling filter with a stratified depth, carbon oxidizing facultative heterotrophic dominate biofilms near top layer of media. Multispecies biofilms, including both facultative heterotrophs and autotrophic nitrifiers, live near the center, and autotrophic nitrifying biofilms may exist in greater numbers in the lower layers. The facultative heterotrophy dominated biofilm near the surface exists as a function of carbon-based substrate and oxygen; the autotrophic nitrifying biofilm exists as a function of ammonia-nitrogen and oxygen (in addition to the absence of carbon-based substrates). A given biofilm with a given bacterial density will have a capacity to oxidize a finite mass of carbon-based substrates. Therefore, the "layer" of trickling filter containing a carbon oxidizing biofilm will vary as a function of the influent load. This can theoretically be achieved with a tall or shallow trickling filter. Practical limitations are based on media wetting and site constraints for shallow trickling filters, and the aforementioned constraints for tall trickling filters. Most, if not all, of the improved performance with depth noted by some investigators is likely a result of improved hydraulic distribution. The average hydraulic rate should exceed 0.5 L/m²/s to ensure maximum performance.

3.6 DESIGN EXAMPLES

The following are examples of trickling filter sizing based on the equations and materials presented in earlier sections.

Example Problem 7.1

Determine the size of a trickling filter with plastic filter media for providing secondary effluent that will provide an average effluent soluble BOD (*Se*) of 15 mg/L.

Where q = 6630 m³/d or 76.7 l/s;
Li = 135 mg/L;
Specific surface area = 90 m²/m³ (27 sq ft/cu ft);
D = 5.49 m (18 ft);
R = 0.75;
Si = 75 mg/L or 75 g/m³;
K_{20} = 0.19 (l/s)$^{0.5}$/m²; and
Water temperature = 14 °C.

1. Determine the size of the trickling filter.
   (a) Determine the k_{20} using Equation 7.16:

$$k_2 = k_1 \cdot \left(\frac{D_1}{D_2}\right)^{0.5} \left(\frac{L_1}{L_2}\right)^{0.5}$$

$$k_2 = k_1 \left(\frac{6.1 \text{ m}}{5.49 \text{ m}}\right)^{0.5} \left(\frac{150 \text{ g/m}^3}{135 \text{ g/m}^3}\right)^{0.5}$$

$$k_2 = (0.19)(1.054)(1.054)$$

$$k_2 = 0.21 \ (1/\text{s})^{0.5}/\text{m}^3$$

   (b) Temperature correction for k_2 using Equation 7.12:

$$k_t = k_{20}/1.035^{(T-20)}$$

$$\theta_{14} = 1.035^{(14-20)}$$

$$= 0.814$$

   (c) Calculate the allowable total hydraulic application rate, THL, using Equation 7.15:

$$S_e = \frac{S_i}{\left(\frac{Q_R}{Q}+1\right) \cdot e^{\frac{k_t \cdot a \cdot D \cdot \theta^{(T-20)}}{\left[WHL \cdot \left(\frac{Q_R}{Q}+1\right)\right]^n}} - \frac{Q_R}{Q}}$$

2. Check to see whether the trickling filter BOD loading and size seem reasonable based on the experience-based criteria.
   (a) Calculate organic loading rate, TOL, to the trickling filter:

$$\text{TOL} = \frac{\text{BOD applied}}{\text{volume of media}}$$

$$\text{TOL} = \frac{(6{,}630\ \text{m}^3/\text{d})(135\ \text{g/m}^3)\ 1\ \text{kg}/10^3\ \text{g}}{(231\ \text{m}^2)(5.49\ \text{m})}$$

$$= \frac{895\ \text{kg/d}}{1268\ \text{m}^3}$$

$$= 0.706\ \text{kg/m}^3 \cdot \text{d}\ (44\ \text{lb BOD/d} \cdot 1000\ \text{cu ft})$$

Compare modeling results of TOL = 0.706 kg/m³/d with experience-based values in Table 7.4. Table 7.4 indicates that effluent of secondary quality standard may be possible with a TOL as high as 0.96 kg/m³/d. Consider local conditions and other trickling filter case histories in the area where treatment is to occur. It may be that assumptions for rational equations were not correct. Reassess assumptions, consider using a different equation, conduct a pilot test, or defer to actual trickling filter experience.

Example Problem 7.2

Determine the TOL, SOL, WHL and THL at an existing trickling filter WRRF.

Where,

Two, 25.9-m (85-ft) diameter plastic media trickling filters have a media depth of 4.27 m (14 ft).

q = 15 000 m³/d; r = 7500 m³/d;

$S_{i,total}$ = 140 mg/L; $S_{i,soluble}$ = 90 mg/L or 90 g/m³; and

Specific surface area = 98 m²/m³ (30 sq ft/cu ft); S_i − 90 mg/L.

1. Find the influent BOD load to the trickling filter.
   (a) Calculate influent total and soluble BOD loads in kg/m³/d.

$$\text{BOD applied} = q\,Li = (15\,000\ \text{m}^3/\text{d})(140\ \text{g/m}^3)\ 1\ \text{kg}/10^3\ \text{g}$$

$$= 2100\ \text{kg/d}$$

$$\text{SBOD applied} = q\,Si = (15\,000\ \text{m}^3/\text{d})(90\ \text{g/m}^3)\ 1\ \text{kg}/10^3\ \text{g}$$

$$= 1350\ \text{kg/d}$$

   (b) Determine trickling filter media area and volume.

$$\text{Area } A_1 = \pi\ (\text{dia})^2/4$$

$$A_1 = 3.14\ (25.9\ \text{m})^2/4$$

$$A_1 = 526.6\ \text{m}^2$$

$$\text{Total Area} = (\#\ \text{units})(A_1) = 2(526.6\ \text{m}^2)$$

$$A_T = 1053.2\ \text{m}^3$$

$$\text{Total Volume } V_T = A_T(D)$$

$$= (1053.2\ \text{m}^2)(4.27\ \text{m})$$

$$= 4497\ \text{m}^3$$

(c) Calculate total and soluble organic loading of the trickling filter, TOL and SOL, respectively.

$$TOL = \frac{\text{BOD applied}}{V_M} = \frac{2100 \text{ kg/d}}{4497 \text{ m}^3}$$
$$= 0.467 \text{ kg BOD/d} \cdot \text{m}^3$$
$$SOL = \frac{\text{SBOD applied}}{V_M} = \frac{1350 \text{ kg/d}}{4497 \text{ m}^3}$$
$$= 0.30 \text{ kg SBOD/d} \cdot \text{m}^3$$

2. Determine hydraulic loading rates to the trickling filter system.

(a) Calculate wastewater hydraulic loading to the trickling filter:

$$WHL = \frac{q(1000 \text{ m}^3/\text{ML})}{A}$$
$$WHL = \frac{(15\,000 \text{ m}^3/\text{d})}{1053 \text{ m}^2}$$
$$= 14.2 \text{ m}^3/\text{d} \cdot \text{m}^3$$

(b) Calculate total hydraulic loading to the trickling filter:

$$THL = \frac{\text{total pumped flow}}{A} = \frac{q + r}{A}$$
$$THL = \frac{(15\,000 \text{ m}^3/\text{d} + 7500 \text{ m}^3/\text{d})}{1053 \text{ m}^2}$$
$$= 21.4 \text{ m}^3/\text{d} \cdot \text{m}^2$$

(c) Calculate recirculation rate.

$$R = \frac{r}{q} = \frac{7500 \text{ m}^3/\text{d}}{15\,000 \text{ m}^3/\text{d}} = 0.5$$

(d) Determine surface loading rate to the filter:

$$SLR = TOL/SSA = 0.467 \text{ kg/d} \cdot \text{m}^3/98/\text{m}^3 = 4.76 \times 10^{-3} \text{ kg/d} \cdot \text{m}^3$$

4.0 ROTATING BIOLOGICAL CONTACTORS

Rotating biological contactor (RBC) design criteria has been presented in detail by the Water Environment Federation (2010). This discussion of RBCs is limited to a brief introduction with a few important design considerations in this section. As a secondary treatment process, RBCs have been applied where average effluent water quality standards are less than or equal to 30 mg/L BOD_5 and TSS. When RBCs are used in conjunction with effluent filtration, the process is capable of meeting more stringent effluent water quality limits of 10 mg/L BOD_5 and TSS. Nitrification RBCs can produce effluent having less than 1 mg/L ammonia-nitrogen remaining in the effluent stream. An RBC uses a cylindrical, synthetic media bundle that is mounted on a horizontal shaft. Figure 7.9 illustrates the shaft-mounted media. The media is partially submerged (typically 40%) and slowly (1.0–1.6 rpm) rotates to expose the biofilm to substrate in the bulk of the liquid (when submerged), and to air (when not submerged). Detached biofilm fragments suspended in the RBC effluent stream are removed by solids separation units.

Figure 7.9 Rotating Biological Contactor Cylindrical Synthetic Media Bundle Mounted on a Horizontal Shaft (Left) and Rotating Biological Contactor Covers (Right)

The RBC process typically is configured with several stages operating in series. Each reactor in series may have one or more shafts. Parallel trains are implemented to provide additional surface area for biofilm development.

Media-supporting shafts typically are rotated by mechanical drives. Diffused air-drive systems and an array of air-entraining cups that are fixed to the periphery of the media (to capture diffused air) have been used to rotate the shafts. The RBC process has the following advantages: operational simplicity, low energy costs, and rapid recovery from shock loadings. The literature has documented several examples of RBC failure resulting from shaft, media, or media support system structural failure; poor treatment performance; accumulation of nuisance macro fauna; poor biofilm thickness control; inadequate performance of air-drive systems for shaft rotation. State-of-the-art biofilm reactors such as the MBBR and biologically aerated filters (BAF) can provide equivalent or improved effluent water quality with reduced susceptibility to macrofauna infestation and reduced physical footprint.

5.0 INTRODUCTION TO INTEGRATED BIOLOGICAL TREATMENT

Integrated biological treatment (IBT) processes are two-stage, series, dual, or coupled processes. In this manual, the term "integrated biological treatment" is used to denote (1) conventional coupling in series of two or more different reactors, at least one of which is a fixed-biofilm reactor, and (2) the integrated fixed-film activated sludge (IFAS) process. It is recognized that the IFAS process differs from the conventional coupled-in-series processes, in that the dual suspended growth and biofilm processes occur within the same reactor.

5.1 PROCESS OBJECTIVES OF INTEGRATED SYSTEMS

Conventional IBT systems use a fixed-biofilm reactor (first stage) in series with a suspended growth biological reactor (second stage). The fixed-biofilm reactor typically consists of a biological tower, and the suspended growth reactor is typically an aeration basin or small contact channel. This combination results in a two-stage coupled unit process that has unique design parameters with treatment efficiency capabilities that often exceed those of the individual parent systems. The system may also consist of three stages, with the first and the last suspended growth systems, and the second a fixed-biofilm growth system. The IFAS process is yet another configuration, developed subsequent to conventional IBT systems, that combines both fixed film and suspended growth in the same reactor tank.

Designers have used integrated processes widely to use the advantages that these combined systems provide. Advantages of fixed-biofilm processes include their shock-load resistance, volumetric efficiency, low energy requirements, and low maintenance requirements. Advantages of suspended growth biological treatment processes include high-quality effluent and the flexibility to operate under various treatment modes to achieve different effluent quality objectives. By combining these two processes in IBT systems, designers capitalize on the advantages of each individual parent (single-stage) process.

Typically, in conventional IBT processes, the fixed biofilm primarily removes soluble five-day biochemical oxygen demand ($sBOD_5$). The suspended growth process, and the suspended growth/biofilm processes in IFAS systems, can provide a variety of functions, including flocculation to capture colloidal substrate and improve clarification, removal of residual $sBOD_5$, nitrification, denitrification, and enhanced biological phosphorus removal to meet advanced wastewater treatment requirements.

For design of IBT systems, the fixed-film process and the suspended growth process must be viewed as a combined system and developed accordingly. For example, in conventional IBT systems, removal of biodegradable organic matter drives design of the fixed-biofilm reactor. Design of the suspended growth process must then consider the amount of biodegradable organic matter removed in the fixed-biofilm process and establishment of conditions necessary for flocculation, as well as the treatment goals (e.g., nitrification or denitrification) of the combined process.

5.2 OVERVIEW OF INTEGRATED BIOLOGICAL TREATMENT SYSTEMS

Many combinations of treatment process trains are possible, depending on the type of parent process used, loading to the treatment units, treatment goals, and the point at which biological or recycle solids are reintroduced to the main flow stream (Harrison et al., 1984). Conventional integrated biological processes are divided into two groups: those that have low to moderate organic loadings applied to the fixed-film reactor and those with high (roughing) organic loadings.

The processes also are classified in terms of return of mixed liquor from the suspended growth stage. The process schematics in Figure 7.10 illustrate the various, common alternative methods for returning secondary sludge, or reaeration, and includes terms used by designers to distinguish process modes. The figure is not a comprehensive list of IBT systems; other process configurations are also discussed in this section.

5.2.1 TRICKLING FILTERS/SOLIDS CONTACT

The trickling-filter/solids contact (TF/SC) process uses a trickling filter that has low to moderate organic loads followed by a small, aerated solids contact tank or channel with secondary clarifiers that polish the trickling filter effluent through flocculation and additional removal of organics (Norris et al., 1982). Most of the organic substrate is removed in the trickling filter causing less filamentous microbial growth in the aerated solids contact tank or channel and, therefore, TF/SC flocs are more fragile than traditional activated sludge flocs. For that reason, flocculator clarifiers typically are used with the TF/SC process.

By combining the trickling filter with a solids-contact tank or channel, Parker and Matasci (1989) found the trickling filter reactor size is smaller than required when used alone. The magnitude of the size reduction will depend on the specific application and effluent requirements.

Conventional TF/SC does not include return solids reaeration before the solids contact tank or channel. However, it has been observed that reaerating the return sludge before the solids contact tank can enhance bioflocculation and may allow for a smaller tank or channel to be used. When both solids reaeration and solids contact are used, the acronym TF/SCR is used.

One significant benefit of the TF/SC process is the relatively low power requirements for the suspended growth system because of the ability of the trickling filter to remove the majority of the soluble substrate ($sBOD_5$) (Witzgall et al., 2013). Another benefit is the ability to upgrade existing rock trickling filters through polishing the fixed growth effluent by using return activated sludge (RAS) as a bioflocculating agent (Krumsick et al., 1984; Matasci et al., 1988). The TF/SC process has also been shown to achieve consistent nitrification even in cold climate conditions (Parker, 1998).

5.2.2 ROUGHING FILTER ACTIVATED SLUDGE

A common method of upgrading existing activated sludge facilities is to install a roughing filter ahead of the activated sludge process. As shown in Figure 7.10, both TF/SC and roughing filter activated sludge (RF/AS) processes have the same process flow schematic. However, with RF/AS, a much smaller trickling filter is used. This requires a larger suspended growth biological reactor to provide a significant amount of oxygen for $sBOD_5$ removal and biomass stabilization. This differs from the TF/SC process, in which

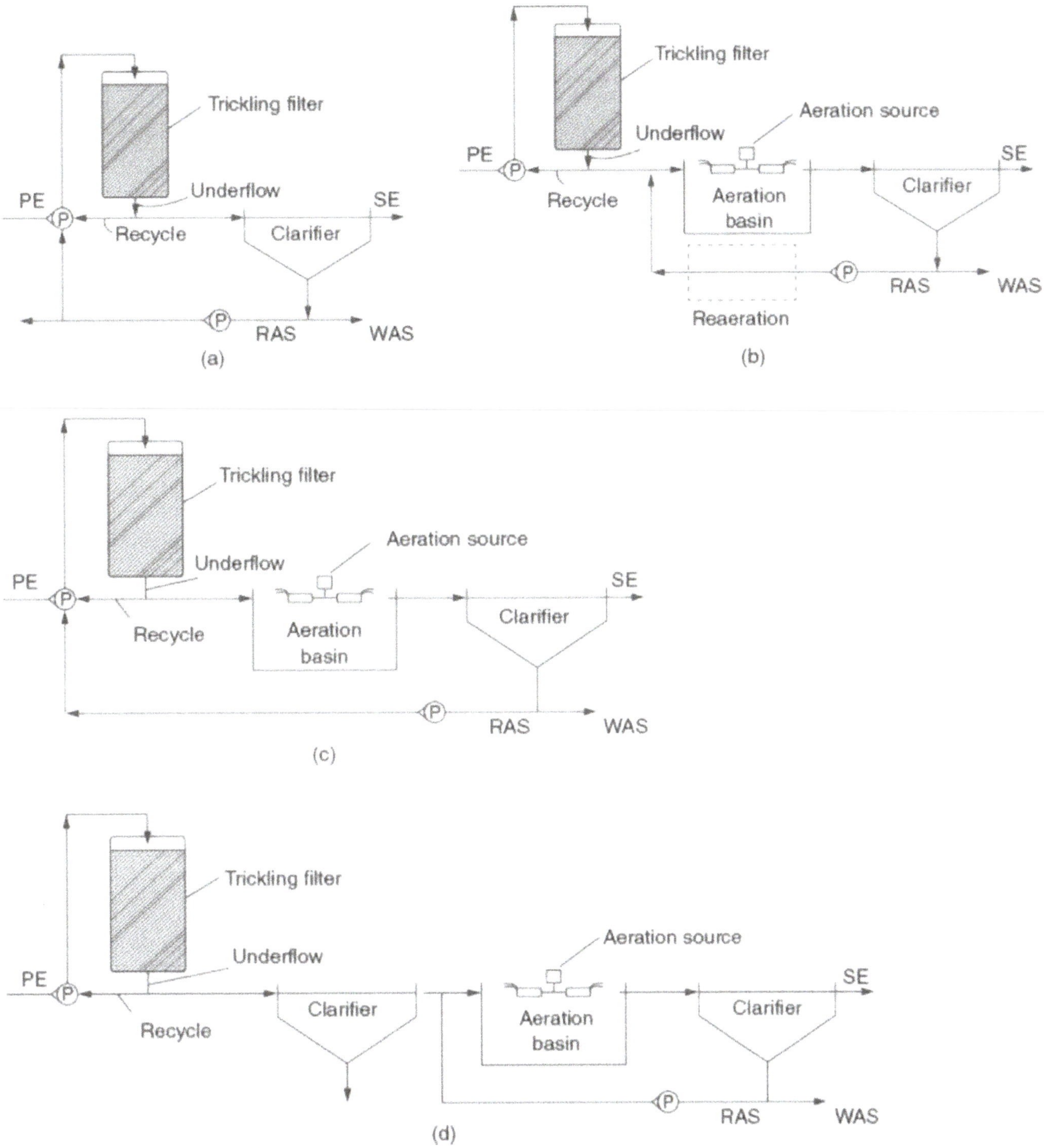

Figure 7.10 Combined Process Operations: (a) Activated Biofilter, (b) Trickling Filter Solids Contact and Roughing Filter Activated Sludge, (c) Biofilter Activated Sludge, and (d) Trickling Filter Activated Sludge (PE = primary effluent; SE = secondary effluent)

the trickling filter is larger and provides almost all of the $sBOD_5$ removal, thus allowing the solids contact channel or tank to be smaller and provide enhanced solids flocculation and effluent clarity. Because some of the $sBOD_5$ is left to be metabolized in the activated sludge tank in the RF/AS process, filament growth is usually sufficient, which results in a more stable floc.

Facility-specific circumstances that influence the balance between capital and operating expenses often will help to determine whether it is best to use the TF/SC or RF/AS process at a particular facility. During periods of reduced loadings, RF/AS facilities can be operated as TF/SC, thus using less suspended growth biological reactor volume and less aeration.

5.2.3 TRICKLING FILTER ACTIVATED SLUDGE

The trickling filter activated sludge (TF/AS) process is designed for high organic loads similar to those of RF/AS or biofilter activated sludge (which has an RAS flow back to the filter). A unique feature of TF/AS, however, is that an intermediate clarification step is provided between the biofilm and suspended growth biological reactors. The intermediate clarification removes sloughed solids from the biofilm reactor effluent before the effluent enters the suspended growth reactor. This solids removal step reduces the effects the sloughed solids may have on the suspended growth process dynamics. Because an intermediate clarification step is used, the TF/AS process is not a directly coupled system. The TF/AS schematic is illustrated in Figure 7.10.

The benefit of the TF/AS integrated process is that solids generated from carbonaceous biochemical oxygen demand ($cBOD_5$) removal in the trickling filter are removed before second-stage activated sludge treatment. This is often a preferred approach at facilities in which ammonia removal is required, as the suspended growth stage of the process can be designed to be dominated by nitrifying microorganisms. The intermediate clarification step can reduce required nitrifying aeration volume, which is a significant advantage for larger facilities or facilities processing higher strength wastewater (Daigger & Boltz, 2011).

The claim of significantly reduced oxygen demand or improved solids settleability from use of intermediate clarification is not universally accepted. This step is often eliminated because of the opposite claim: improved sludge settleability in the secondary clarifier when the trickling filter sloughs directly to the suspended growth reactor. The reduced oxygen demand afforded by intermediate clarification may not justify the capital and operating costs associated with intermediate clarification.

5.2.4 EXTERNAL NITRIFICATION BIOLOGICAL NUTRIENT REMOVAL ACTIVATED SLUDGE

The external nitrification biological nutrient removal activated sludge process is characterized by inserting a trickling filter midway through the suspended growth system, with the aim of feeding it an influent with reduced organic load so as to encourage nitrification (Hu et al., 2000). It is used at facilities in which both nitrogen and phosphorus removals are performed. In these facilities, the primary effluent (or potentially raw influent) is fed to a suspended growth system anaerobic zone to maximize phosphorus release and the uptake of readily biodegradable substrate. The effluent from the anaerobic zone, low in readily biodegradable substrate, is fed to an intermediate clarifier with the clarifier overflow fed to the nitrifying trickling filter for nitrification. Clarifier underflow is combined with nitrified trickling filter effluent and fed to an anoxic zone for denitrification, before polishing in an aerobic zone and secondary clarification (Muller et al., 2006).

5.2.5 INTEGRATED FIXED FILM ACTIVATED SLUDGE

The IFAS process is a different configuration of IBT systems, in which the attached and suspended growth environments are combined into one bioreactor system, so the attached and suspended biomass are in close proximity. The IFAS is the combination of an MBBR for attached growth and an RAS system (for suspended growth). The media allow the biomass in the bioreactor to be increased beyond that possible with suspended growth configurations alone, without increasing the solids load on the secondary clarifiers.

5.3 DESIGN OF CONVENTIONAL INTEGRATED BIOLOGICAL TREATMENT SYSTEMS

Design considerations for conventional IBT processes are similar to those for the parent processes. The designer should review appropriate parent processes and special design considerations for integration.

The design of integrated biological processes is a balance between two biological treatment reactors; neither the biofilm nor the suspended growth process can be sized separately. Good design includes consideration of the total integrated facility, including interactions among all units, especially the secondary clarifier (Harrison & Timpany, 1988).

Table 7.7 presents design criteria typically used by engineers for sizing integrated processes. In all instances, rational process design equations should be used to check the design because translating data from one situation to another may not always be successful.

The inability to control interdependent variables makes it impossible to precisely predict performance of integrated processes (Albertson & Eckenfelder, 1984). Design engineers often choose both conservative design loads and variable process modes for integrated processes that produce the desired effluent quality and solids characteristics under a wide range of operating loads.

5.3.1 SOLIDS RETENTION TIME

Use of the solids retention time (SRT) approach for sizing suspended growth reactors following fixed-biofilm systems has become the prevalent method for design (Harrison & Timpany, 1988). An SRT in the contact channel for TF/SC may range from 0.2 to 2 days. For a combined biofilter/activated sludge (BF/AS) facility, an SRT of 1 to 7 days may be more appropriate. Using high organic loading on the trickling filter will require increased SRTs in the suspended growth portion of the combined process.

Table 7.7 General Design Criteria for Fixed Film and Activated Sludge Systems With Segregated Return Activated Sludge

	Appropriate Design Criteria	
System	**Range**	**Commonly Used**
Activated biofilter		
Media type	High rate	High rate
Five-day biochemical oxygen demand loading, kg/m³ × d (lb/d/1000 cu ft)	0.2–1.2 (12–75)	0.5 (30)
Hydraulic loading, L/m² × s (gpm/sq ft)	0.5–3.4 (0.8–5.0)	1.4 (2.0)
Filter mixed liquor suspended solids, mg/L	1500–3000	2000
Trickling filter/solids contact		
Media type	Rock or high rate	High rate
Five-day biochemical oxygen demand loading, kg/m³ × d (lb/d/1000 cu ft)	0.3–1.6 (20–100)	0.8 (50)
Hydraulic loading, L/m³ × s (gpm/sq ft)	0.1–1.4 (0.1–2.0)	0.7 (1.0)
Channel mixed liquor suspended solids, mg/L	1500–3000	2000
Hydraulic retention time, hours	0.2–2.0	0.75
Solids retention time, days	0–2	0.1
Return activated sludge, mg/L	6000–12 000	8000
Diffused air, L/m³ × s (scfm/1000 gal)	250–500 (2000–4000)	370 (3000)
Mechanical, W/m³ (hp/mil × gal)	12–26 (60–130)	20 (100)
Food/mass, mg/kg × s (lb/d/lb)	2.3–23 (0.2–2.0)	11.6 (1)
Roughing filter, trickling filter, or biofilter with activated sludge		
Media type	High rate	High rate
Five-day biochemical oxygen demand loading, kg/m³ × d (lb/d/1000 cu ft)	1.2–4.8 (75–300)	2.4 (150)
Hydraulic loading, L/m² × s (gpm/sq ft)	0.5–3.4 (0.8–5.0)	0.7 (1.0)
Basin mixed liquor suspended solids, mg/L	1500–4000	2500
Hydraulic retention time, hours	0.5–4.0	2.0
Solids retention time, days	1.0–7.0	3.0
Food/mass, mg/kg × s (lb/d/lb)	5.8–13.9 (0.5–1.2)	10.4 (0.9)
Total available oxygen, g/kg (lb/lb)	600–1200 (0.6–1.2)	900 (0.9)
Oxygen typically used, g/kg (lb/lb)	300–900 (0.3–0.9)	600 (0.6)
Conventional activated sludge		
Solids retention time, days	5–15	10
F/M food to microorganism, mg/kg × d	0.3–0.5	0.4
Hydraulic retention time, hours	4.0–8.0	7.2 min (@12 °C)

Typically, the SRT calculation is similar to that presented earlier, i.e., the mass of solids under aeration divided by the rate at which solids are wasted from the system each day, either intentionally or unintentionally.

It is common to estimate solids production using a net process yield Y_{obs}. Even though soluble substrate may remain in the trickling filter effluent, it will typically be removed in the suspended growth bioreactor. The solids yield, Y_{obs}, for IBT systems is lower than for conventional activated sludge systems. However, because the retention of biomass in a trickling filter increases as TOL is decreased, the solids yield value typically is influenced more by the TOL on the fixed biofilm than by the SRT of the suspended growth biological reactor. Solids yield decreases as TOL is decreased. Solids yield values are approximately 0.7 to 0.9 kg TSS/kg BOD_5 removed. The SRT can then be calculated using the above solids wastage rate with the mass of mixed liquor suspended solids and the suspended growth reactor volume:

$$\mathrm{SRT}_{SG} = X_{SG} \times V_R / P_x \tag{7.17}$$

Where X_{SG} = net active solids in the suspended growth reactor (mg/L),
V_R = reactor volume (ML), and
P_x = mass of active solids wasted (kg/d).

The active biomass in the fixed-film reactor may be neglected in the calculations. The calculated SRT is an "effective" SRT in which residence time in the suspended growth reactor may be reduced to below 0.5 days for TF/SC processes; a value of 2 to 3 days may be sufficient with highly loaded filters such as RF/AS, TF/AS, or BF/AS. In the case of the activated biofilter (ABF), which does not have a suspended growth reactor, the SRT is for the fixed-biofilm portion only and since the liquid volume in the trickling filter cannot be accurately determined, the SRT cannot be accurately calculated.

5.3.2 SOLIDS PRODUCTION AND SETTLEABILITY

Solids yield or production from IBT systems often is dominated by biomass sloughing and storage occurrences within the fixed-biofilm reactor. The characteristics and amount of inert solids in the primary effluent will also be a factor in determining actual yields from the total integrated process.

Figure 7.11 shows various solids yields under several suspended growth SRTs and at low and high organic loads to the first-stage, fixed growth reactor (Newbry et al., 1988). The upper band in Figure 7.11 occurred when effluent suspended solids from the primary clarifier were approximately 50% higher than for the lower band. Figure 7.11 also demonstrates that nonbiodegradable solids in the primary effluent may affect net solids yield from an integrated process. It further illustrates that significant reductions in solids yield may not occur simply by increasing the SRT beyond 1 to 2 days. For example, one study indicated that, for integrated processes with an SRT greater than 1 day, little measurable benefit resulted from increasing the suspended growth reactor size to reduce sludge yield (Newbry et al., 1988). A comparison of 43 integrated WRRFs indicated that average solids yields were 0.7 kg TSS/kg primary effluent BOD, but values ranged from 0.15 to 4 kg TSS/kg BOD (Harrison et al., 1984). Case studies indicate that solids from integrated processes that use rock filter media in the first-stage trickling filter may yield 35% to 50% of those for systems with high-rate filter media (Harrison et al., 1984).

Many design engineers consider integrated processes to produce better settling sludge than conventional suspended growth systems. Although integrated processes are relatively resistant to shock loads, process variability and sludge-bulking problems may occur with integrated processes similar to suspended growth processes. This is particularly true for integrated processes that allow significant quantities of $sBOD_5$ to enter the suspended growth reactor. If, on the other hand, the process is designed where the $sBOD_5$ loading to the suspended growth reactor is consistently low, then a consistently good settling sludge will be produced and higher clarifier solids loading rates can be used (Parker et al., 1993). An air-induced minimum velocity gradient of 20/s has been shown to produce sludge of good quality (Motta et al., 2003)

The process designer should carefully evaluate sludge-settling characteristics when sizing secondary clarifiers. The concentration of RAS in an integrated process will be similar to that of a suspended growth

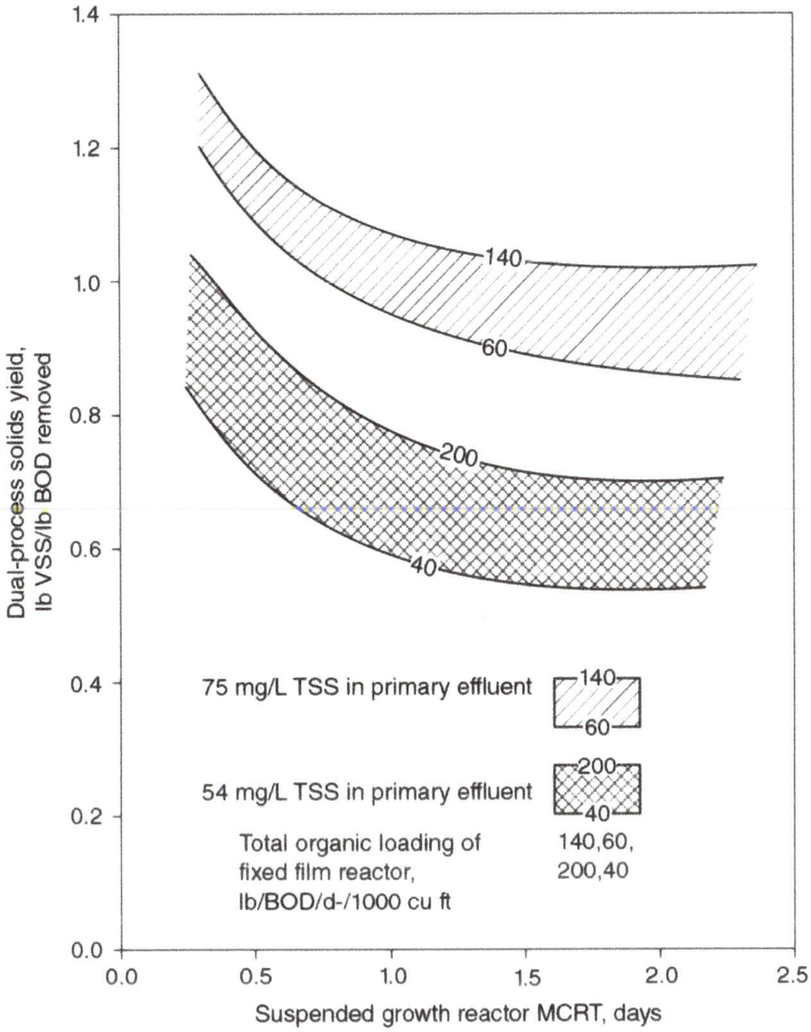

Figure 7.11 Combined Process Solids Yield as a Function of Suspended Growth Reactor Mean Cell Residence Time and Total Organic Loading of Fixed-Film Reactor With Crossflow Media. Solids Yield Varies With Wastewater Characteristics (lb BOD/d/100 cu ft, on the fixed-film reactor)

process. Return activated sludge concentrations ranging from 0.5% to 1.5% and averaging approximately 0.8% TSS typically will occur with integrated processes. After allowing for the expected sludge settleability, secondary clarifiers can be designed using the same approach as for suspended growth systems.

5.3.3 BIOCHEMICAL OXYGEN DEMAND AND TOTAL SUSPENDED SOLIDS REMOVAL

One study comparing various organic loads ranging from 0.74 to 3 kg BOD/m^3 × d (46 to 200 lb BOD/d/1000 cu ft) in the fixed-film reactor showed that sBOD$_5$ in treated effluent decreased and effluent quality improved with increased SRTs in second-stage suspended growth reactors (Newbry et al., 1988). Effluent sBOD$_5$ concentration, however, was determined to be independent of organic loading to the first-stage fixed-film reactor. Varying the organic load to the fixed-film reactor did not change the concentration of sBOD$_5$ in the effluent of the suspended growth step, as illustrated in Figure 7.12.

Researchers have shown that increasing organic loads to the fixed-biofilm reactor results in increased amounts of nonsettleable suspended solids as illustrated in Figure 7.13 (Newbry et al., 1988). With integrated processes, the second-stage suspended growth treatment reduces the amount of effluent nonsettleable suspended solids by increasing total system SRT. A sharp reduction in effluent suspended solids results from providing a minimum of a 1-day SRT.

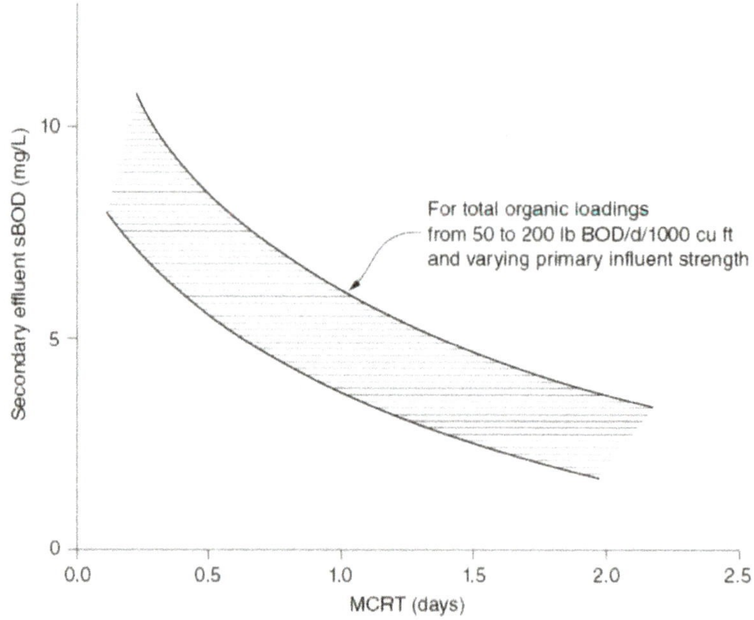

Figure 7.12 Effluent Soluble Biochemical Oxygen Demand of Combined Processes With Crossflow Filter Media for Various Mean Cell Residence Times and Total Organic Loadings. The sBOD Will Vary With Wastewater Characteristics.

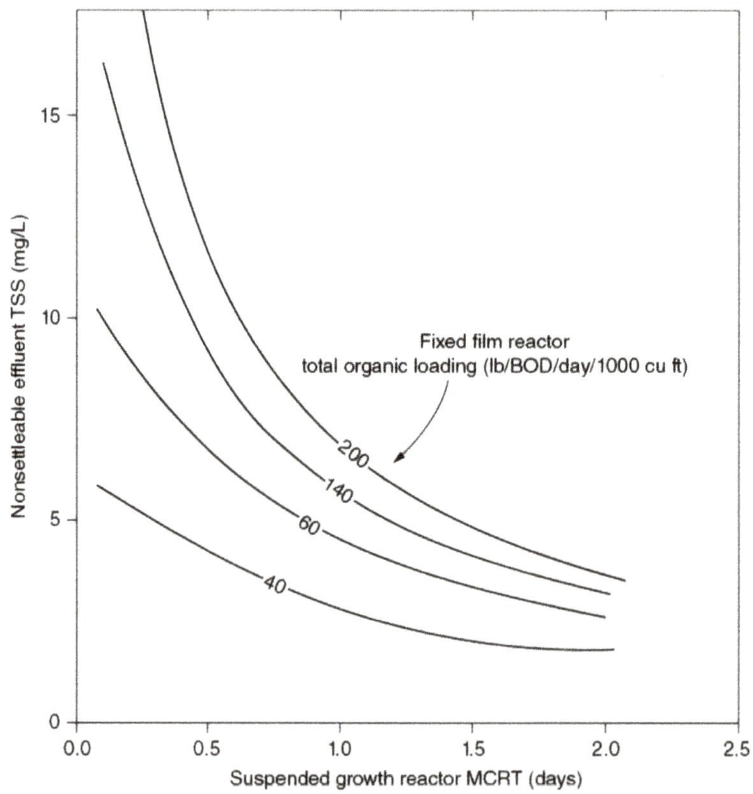

Figure 7.13 Non-Settleable Effluent Total Suspended Solids of Dual Process With Crossflow Filter Media for Various Mean Cell Residence Times and Total Organic Loadings. Non-Settleable TSS Are Measured in the Supernatant of the Settleable Solids Test. Secondary Effluent TSS Will Exceed the Non-Settleable TSS.

PRACTICE PROBLEMS

1. Discuss the factors that affect biofilm development.
2. Explain how distributor speed, ventilation, and recirculation in a trickling filter affects performance.
3. Explain the difference in solids production from an attached versus suspended growth process.
4. Redo Example Problem 7.1 assuming an effluent soluble BOD concentration, $S_e = 8.0$ mg/L.
5. Determine the size of a trickling filter with plastic filter media for providing secondary effluent that will provide an average effluent soluble BOD (S_e) of 10 mg/L given the following:

$Q = 38{,}000\ m^3/d$	$R = 1.0$
$Li = 135$ mg/L	$Si = 75$ mg/L
Specific surface area = 100 m^2/m^3	$K_{20} = 0.20\ (l/s)^{0.5}/m^2$
$D = 5.49$ m (18 ft)	Water temperature = 16 °C

6. Determine the TOL, SOL, WHL, and THL at an existing trickling filter WRRF given the following:

Two, 26-m diameter plastic media trickling filters have a media depth of 6.1 m.

$Q = 18{,}000\ m^3/d$;	$r = 7500\ m^3/d$;
$S_{i,total} = 135$ mg/L;	$S_{i,soluble} = 85$ mg/L
Specific surface area = 95 m^2/m^3	

6.0 REFERENCES

Albertson, O. E. (1989a). Slow down that trickling filter! *Water Environment Technology (Operator Forum)*, *6*(1), 15–20.

Albertson, O. E. (1989b). Slow motion trickling filters gain momentum! *Water Environment Technology (Operator Forum)*, *6*(8), 28–29.

Albertson, O. E. (1995). Excess biofilm control by distributor-speed modulation. *Journal of Environmental Engineering*, *121*(4), 330–336.

Albertson, O. E., & Davies, G. (1984, October). Analysis of process factors controlling performance plastic bio–media. *Proceedings of the 57th Water Pollution Control Federation Conference*, New Orleans, Louisiana, United States.

Albertson, O. E., & Eckenfelder, W. (1984). Analysis of process factors affecting plastic media trickling filter performance. *Proceedings of the Second International Conference on Fixed Film Biological Processes*, Washington, DC, United States.

Aryan, A. F., & Johnson, S. H. (1987). Discussion of a comparison of trickling filter media. *Journal of the Water Pollution Control Federation*, *59*, 915–918.

Aspegren, H. (1992). *Nitrifying trickling filters, A pilot study of Malmö, Sweden*. Malmö Water and Sewage Works: Malmö, Sweden.

Atkinson, B., Busch, A. W., & Dawkins, G. S. (1963). Recirculation, reaction kinetics and effluent quality in a trickling filter flow model. *Journal of the Water Pollution Control Federation*, *35*, 1307–1317.

Benzie, W. J., Larkin, H. O., & Moore, A. F. (1963). Effects of climactic and loading factors on trickling filter performance. *Journal of the Water Pollution Control Federation*, *35*(4), 445–455.

Boller, M., & Gujer, W. (1986). Nitrification in tertiary trickling filters followed by deep filters. *Water Research*, *20*, 1363.

Boltz, J. P., & La Motta, E. J. (2007). The kinetics of particulate organic matter removal as a response to bioflocculation in aerobic biofilm reactors. *Water Environment Research*, *79*, 725.

Boltz, J. P., La Motta, E. J., & Madrigal, J. A. (2006). The role of bioflocculation on suspended solids and particulate COD removal in the trickling filter process. *Journal of Environmental Engineering, 132*(5), 506–513.

Boltz, J. P., Goodwin, S. G., Rippon, D., & Daigger, G. T. (2008). A review of operational control strategies for snail and other macrofauna infestations in trickling filters. *Water Practice, 2*(4), 1–16.

Boltz, J. P., Johnson, B. R., Daigger, G. T., & Sandino, J. (2009a). Modeling integrated fixed film activated sludge (IFAS) and moving bed biofilm reactor (MBBR) systems I: Mathematical treatment and model development. *Water Environment Research, 81*, 576–586.

Boltz, J. P., Johnson, B. R., Daigger, G. T., Sandino, J., & Elenter, D. (2009b). Modeling integrated fixed film activated sludge (IFAS) and moving bed biofilm reactor (MBBR) systems II: Evaluation. *Water Environment Research, 81*, 555–575.

Boltz, J. P., Daigger, G. T., Johnson, B. R., Hiatt, W., & Grady, Jr., C. P. L. (2009c). Expanded process model describes biomass distribution, free ammonia/nitrous acid inhibition and competition between ammonia oxidizing bacteria (AOB) and nitrite oxidizing bacteria (NOB) in submerged biofilm and integrated fixed-film activated sludge bioreactors. *Proceedings of the Water Environment Federation Nutrient Removal Conference* [CD-ROM], Washington, DC, United States, June 28–July 1.

Boltz, J. P., & Daigger, G. T. (2010). Uncertainty in bulk-liquid hydrodynamics and biofilm dynamics creates uncertainties in biofilm reactor design. *Water Science & Technology, 61*(2), 307–316.

Boltz, J. P., Morgenroth, E., & Sen, D. (2010). Mathematical modeling of biofilms and biofilm reactors for engineering design. *Water Science & Technology, 62*(8), 1821–1836.

Bruce, A. M., & Merkens, J. C. (1970). Recent studies of high rate biological filtration. *Journal of Water Pollution and Control, 2*, 449.

Bruce, A. M., & Merkens, J. C. (1973). Further studies of partial treatment of sewage by high-rate biological filtration. *Journal of Water Pollution and Control, 5*, 499.

Bryan, E. H. (1955). Molded polystyrene media for trickling filters. *Proceedings of the 10th Purdue Industrial Waste Conference*, pp. 164–172. Purdue University.

Bryan, E. H. (1962). Two-stage biological treatment: industrial experience. *Proceedings of the 11th South Municipal Industrial Waste Conference*, p. 136. University of North Carolina, Chapel Hill, North Carolina.

Bryan, E. H., & Moeller, D. H. (1960). Aerobic biological oxidation using DowPAC. *Proceedings of the Conference on Biological Waste Treatment.* Manhattan College, New York.

Bryers, J. D. (1984). Biofilm formation and chemostat dynamics: Pure and mixed culture conditions. *Biotechnology and Bioengineering, 26*, 948–958.

CH2M HILL. (1984). A comparison of trickling filter media internal project report.

Crine, M., Schlitz, M., & Vandevenne, L. (1990). Evaluation of the performances of random plastic media in aerobic trickling filters. *Water Science & Technology, 22*(1/2), 227–238.

Culp, G. L. (1963). Direct recirculation of high-rate trickling filter effluent. *Journal of the Water Pollution Control Federation, 35*(6), 742–747.

Daigger, G. T., & Boltz, J. P. (2010). Trickling filter and trickling filter-suspended growth process design and operation: A state-of-the art review. *Water Environment Research, 83*, 388-404.

Drury, D. D., Carmona, J., & Delgadillo, A. (1986). Evaluation of high density cross flow media for rehabilitating and existing trickling filter. *Journal of the Water Pollution Control Federation, 58*(5) 364–366.

Eckenfelder, W. W. (1961). Trickling filter design and performance. *Journal of the Sanitary Engineering Division, American Society of Civil Engineers, 87*, 33–45.

Eckenfelder, W. W. (1963). Performance of a high rate trickling filter using selected materials. *Journal of the Water Pollution Control Federation, 35*, 1536.

Eckenfelder, W.W., & Barnhart, E. L. (1963). Performance of a high-rate trickling filter using selected materials. *Journal of the Water Pollution Control Federation, 35*(12), 1535–1551.

Galler, W. S., & Gotaas, H. G. (1964). Analysis of biological filter variables. *Journal of the Sanitary Engineering Division, American Society of Civil Engineering, 90*(6), 59.

Germain, J. E. (1966). Economical treatment of domestic waste by plastic medium trickling filters. *Journal of the Water Pollution Control Federation, 38*, 192.

Grady, L. E., Daigger, G. T., & Lim, H. (1999). *Biological wastewater treatment* (2nd ed.). Marcel Dekker.

Gujer, W., & Boller, M. (1983). Operating experience with plastic media tertiary trickling filters for nitrification. In V. de Emde & H. B. Tench (Eds.), *Design and operation of large treatment plants.* Pergamon: Oxford, United Kingdom.

Gujer, W., & Boller, M. (1986). Design of a nitrifying trickling filter based on theoretical concepts. *Water Research, 20*, 1353.

Gullicks, H. A., & Cleasby, J. L. (1986). Design of trickling filter nitrification tower. *Journal of the Water Pollution Control Federation, 58*(1), 60–67.

Harrison, J. R., Daigger, G. T., & Filbert, J. W. (1984). A survey of combined trickling filter and activated sludge processes. *Journal of the Water Pollution Control Federation, 56*, 1073

Harrison, J. R., & Daigger, G. T. (1987). A comparison of trickling filter media. *Journal of the Water Pollution Control Federation, 59*, 679.

Harrison, J. R., & Timpany, P. L. (1988). Design considerations with the trickling filter solids contact process. *Proceedings of the Joint Canadian Society of Civil Engineers, American Society of Civil Engineers National Conference on Environmental Engineering.* Canadian Society of Civil Engineers.

Howland, W. E. (1958). Flow over porous media as in a trickling filter. *Proceedings of the 12th Purdue Industrial Waste Conference.* Purdue University.

Hu, Z.-R., Wentzel, M. C., & Ekama, G. C. (2000). External nitrification in biological nutrient removal activated sludge systems. *Water SA, 26*, 2, 225.

Kincannon, D. F., & Stover, E. L. (1982). Design methodology for fixed-film reactors, RBCs and trickling filters. *Civil Engineering Practice and Design, 2*, 107.

Krumsick, T. A., et al. (1984). Trickling filter solids contact process demonstration, Salt Lake City, Utah. *Proceedings of the Annual Conference of the Utah Water Pollution Control Association.* Salt Lake City.

Kuenen, J. G., Jørgensen, B. B., & Revsbech, N. P. (1986). Oxygen microprofiles of trickling filter biofilms. *Water Research, 20*(12), 1589–1598.

Matasci, R. N., Clark, D. L., Heidman, J. A., Parker, D. S., Petrik, B., & Richards, D. (1988). Trickling filter/solids contact performance with rock filters at high organic loadings. *Journal of the Water Pollution Control Federation, 60*, 68.

Metcalf & Eddy, Inc./AECOM. (2014). *Wastewater engineering: Treatment and Resource recovery* (5th ed.). McGraw-Hill.

Morgenroth, E. (2003). Detachment: An often overlooked phenomenon in biofilm research. In S. Wuertz, P. Bishop, & P. Wilderer (Eds.), *Biofilm in wastewater treatment.* IWA Publishing.

Morgenroth, E. (2008). Modelling biofilm systems. In M. Henze, M. C. M. van Loosdrecht, G. Ekama, & D. Brdjanovic (Eds.), *Biological wastewater treatment—Principles, modelling, and design.* IWA Publishing.

Motta, E. J. L., Jimenez, J. A., Josse, J. C., & Manrique, A. (2003). The effect of air-induced velocity gradient and dissolved oxygen on bioflocculation in the trickling filter/solids contact process. *Advances in Environmental Research, 7*, 441.

Muller, A. W., Wentzel, M. C., & Ekama, G. A. (2006). Estimation of nitrification capacity of rock media trickling filters in external nitrification BNR. *Water SA, 32*(5), 611–618.

National Research Council. (1946). Sewage treatment at military installations. *Sewage Works Journal, 18,* 787.

Newbry, B. W., Daigger, G. T., & Taniguchi-Dennis, D. (1988). Unit process tradeoffs for combined trickling filter and activated sludge processes. *Journal of the Water Pollution Control Federation, 60,* 1813.

Norris, D. P., Parker, D. S., & Daniels, M. L. (1980). Efficiencies of advanced waste treatment obtained with upgrading trickling filters. *Journal of Environmental Engineering, 50*(9), 78–81.

Norris, D. P., Parker, D. S., Daniels, M. L., & Owens, E. L. (1982). High quality trickling filter treatment without tertiary treatment. *Journal of the Water Pollution Control Federation, 54*(7), 1087–1098.

Okey, R. W., & Albertson, O. E. (1989). Evidence of oxygen limiting conditions during tertiary fixed-film nitrification. *Journal of the Water Pollution Control Federation, 61,* 510.

Onda, K., Takeuchi, H., & Okumoto, Y. (1968). Mass transfer coefficients between gas and liquid phase in packed columns. *Journal of Chemical Engineering Japan, 1,* 56–62.

Parker, D. S. (1998). Establishing biofilm system evaluation protocols. WERF workshop: Formulating a research program for debottlenecking, optimizing, and rerating existing wastewater treatment plants. *Proceedings of the 71st Annual Water Environment Federation Technical Exposition and Conference* [CD-ROM], Orlando, Florida, United States.

Parker, D. S. (1999). Trickling filter mythology. *Journal of Environmental Engineering, 125*(7), 618–625.

Parker, D. S., Jacobs, T., Bower, E., Stowe, D. W., & Farmer, G. (1997). Maximizing trickling filter nitrification through biofilm control: Research review and full scale application. *Water Science & Technology, 36*(1) 255–262.

Parker, D. S., Lutz, M., Andersson, B., & Aspegren, H. (1995). Effect of operating variables on nitrification rates in trickling filters. *Water Environment Research, 67*(7), 1111–1118.

Parker, D. S., Lutz, M., Dahl, R., & Berkkopf, S. (1989). Enhancing reaction rates in nitrifying trickling filters through biofilm control. *Journal of the Water Pollution Control Federation, 61*(5), 618–631.

Parker, D. S., Brischke, K. V., & Matasci, R. N. (1993). Upgrading biological filter effluents using the TF/SC process. *Journal of Institution of Water and Environmental Management, 7,* 90.

Parker, D. S., & Matasci, R. N. (1989). The TF/SC process at ten years old: Past, present, and future. *Proceedings of the 62nd Annual Water Pollution Control Federation Conference*, San Francisco, California, United States.

Rauch, W., Vanhooren, H., & Vanrolleghem, P. A. (1999). A simplified mixed-culture biofilm model. *Water Research, 33*(9), 2148–2162.

Schroeder, E. D., & Tchobanoglous, G. (1976). Mass transfer limitations on trickling filter design. *Journal of the Water Pollution Control Federation, 48,* 772.

Schulze, K. L. (1960). Load and efficiency of trickling filters. *Journal of the Water Pollution Control Federation, 32,* 245.

Sen, D., & Randall, C. W. (2008a). Improved computational model (AQUIFAS) for activated sludge, integrated fixed–film activated sludge, and moving-bed biofilm reactor systems, part I: Semi-empirical model development. *Water Environment Research, 80,* 439–453.

Sen, D., & Randall, C. W. (2008b). Improved computational model (AQUIFAS) for activated sludge, integrated fixed-film activated sludge, and moving-bed biofilm reactor systems, part II: Multilayer biofilm diffusional model. *Water Environment Research, 80,* 624–632.

Sen, D., & Randall, C. W. (2008c). Improved computational model (AQUIFAS) for activated sludge, integrated fixed-film activated sludge, and moving-bed biofilm reactor systems, part III: Analysis and verification. *Water Environment Research, 80,* 633–646.

Velz, C. J. (1948). A basic law for the performance of biological filters. *Sewage Works Journal, 20,* 607.

Wanner, O., & Reichert, P. (1996). Mathematical-modeling of mixed-culture biofilms. *Biotechnology & Bioengineering*, *49*(2), 172–184.

Wanner, O., Eberl, H., Morgenroth, E., Noguera, D., Picioreanu, C., Rittmann, B., & Van Loosdrecht, M. C. M. (2006). *Mathematical modeling of biofilms, scientific and technical report no. 18.* IWA Publishing.

Water Environment Federation. (2010). *Biofilm reactors* (Manual of Practice No. 35). McGraw-Hill.

Witzgall, R., Parker, D., Waterman, N., Sen, S., & Hetherington, M. (2013, October 5–9). How a trickling filter/solids contact plant was designed and optimized for handling greater flow variability while requiring much lower energy than OCSD's activated sludge plants. *Proceedings of the 86th Annual Water Environment Federation Technical Exposition and Conference* [CD-ROM]; Chicago, Illinois, United States.

7.0 SUGGESTED READINGS

Abwassertechnische Vereinigung. (1983). *German ATV Regulations—A135*; Grundsätze für die Bemessung von einstufigen Tropfkörpern und Scheibentauchkörpern mit Anschluwerter über 500 Einwohnergleichwerten, D–5205; St. Augustine, Germany.

Boltz, J. P., Morgenroth, E., Brockmann, D., Bott, C., Gellner, W. J., & Vanrolleghem, P. A. (2011). Systematic evaluation of biofilm models for engineering practice: Components and critical assumptions. *Water Science & Technology*, *64*(4), 930–944.

Bosander, J., & Westlund, A. D. (2000). Operation of full-scale fluidized bed for denitrification. *Water Science & Technology*, *41*(9), 115–121.

Degremont. (2007). *Water treatment handbook* (7th ed.). Lavoisier SAS, France.

Gharagozian, A., Swanback, S., Narayanan, B., & Chan, R. (2011). *Solids yield surprise, research on trickling filter/activated sludge systems produces unexpected results.*

Le Tallec, X., Zeghal, S., Vidal, A., & Lesouef, A. (1997). Effect of Influent Quality Variability on Biofilter Operation. *Water Science & Technology*, *36*(1), 111–117.

Springer, A., & Green S. (2005). Colne Bridge BAFF process improvements. *Proceedings of Conference on The Design and Operation of Activated Sludge and Biofilm Systems*, Horan, Aqua Enviro Ltd.

Stenquist, R. J., Parker, D. S., & Dosh, T. J. (1974). Carbon oxidation-nitrification in synthetic media trickling filters. *Journal of the Water Pollution Control Federation*, *46*(10), 2327–2339.

Stensel, H. D., Brenner, R. C., & Lubin, G. (1984, July). Aeration energy requirements in sparged fixed film systems. *Proceedings of the International Biological Fixed Film Conference*, Washington, DC, United States.

Chapter 8 | Nutrient Removal Systems

LEARNING OBJECTIVES

Upon completing this chapter, you will be able to:

- Describe the processes used for biological nitrogen removal and discuss the microbiological principles behind these processes.
- Describe the processes used for biological phosphorus removal and discuss the microbiological principles behind these processes.
- Given wastewater characteristics, develop several alterative schemes for the simultaneous removal of nitrogen and phosphorus.
- Calculate stoichiometric doses of metal salts required for phosphorus removal and estimate actual dose based on treatment objectives; and select a metal salt and addition point based on treatment objectives.

1.0 INTRODUCTION

The two most prominent macronutrients in aquatic systems are nitrogen and phosphorus, which can act as limiting nutrients or result in phytoplankton production. Excessive phytoplankton production can result in **eutrophication**, a condition that causes decreased dissolved oxygen concentrations and a severe reduction in aquatic life diversity. The limiting nutrient is the nutrient that exists in the lowest concentration relative to what organisms need. Many estuarine and freshwater systems, such as rivers, streams, and lakes, tend toward phosphorus limitation, while marine water systems tend toward nitrogen limitation (Doering et al., 1995; Fisher et al., 1999). The limiting nutrient is the one that should be targeted for removal by wastewater treatment systems to control eutrophication. This chapter discusses various biological process configurations targeting nitrogen and phosphorous removal. Chemical precipitation methods for phosphorous removal and recovery are also discussed.

Eutrophication is a condition of excessive nutrients in a body of water that causes a dense growth of algae and plants and death of animal life (such as fish) from lack of oxygen.

2.0 BIOLOGICAL NITROGEN REMOVAL SYSTEMS

Biological nitrogen removal can be accomplished in both attached growth and suspended growth systems. Activated sludge or suspended growth system configurations may include single-sludge, dual-sludge, and triple-sludge systems.

Denitrification is a biological process in which nitrates are converted to nitrogen.

2.1 SINGLE-SLUDGE SUSPENDED GROWTH PROCESSES

2.1.1 WUHRMANN AND LUDZACK-ETTINGER PROCESSES

Wuhrmann (1954) proposed the single-sludge configuration for nitrogen removal shown in Figure 8.1. The Wuhrmann approach typically is referred to as post-denitrification. Without addition of an exogenous electron donor (i.e., a carbon source), the design relies on residual organic matter passing through the first stage or the endogenous respiration of biomass to provide the energy sources for **denitrification**. If complete **nitrification** (thus, complete carbon oxidation) is achieved, then endogenous respiration would provide the principal energy source. Nitrogen removals of 29% to 89% have been achieved in bench- and pilot-scale studies.

Nitrification is the process by which ammonia or ammonium is converted to nitrites and nitrates.

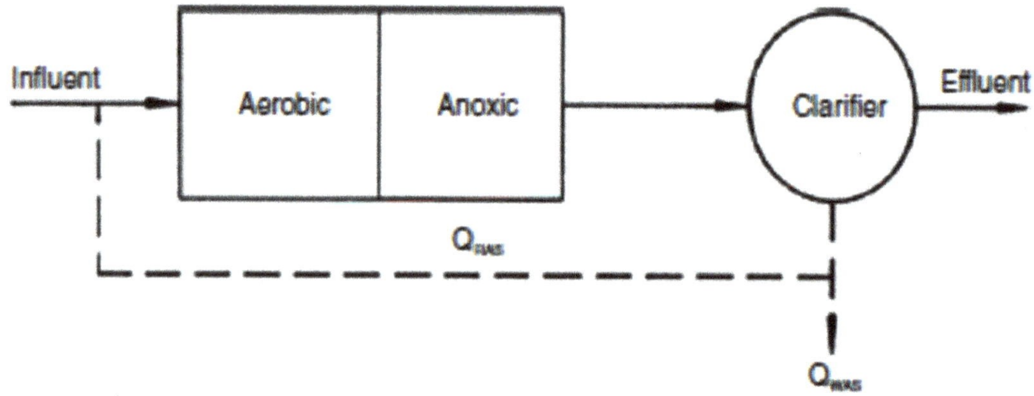

Figure 8.1 Wuhrmann Process for Nitrogen Removal (RAS = return activated sludge; WAS = waste activated sludge)

Variations of the Wuhrmann design have been developed to supply an exogenous electron donor to the anoxic stage. These consist of either bypassing the first stage with a portion (e.g., 15%) of the influent flow or supplying a suitable carbon supplement, such as methanol, directly to the anoxic zone.

The Ludzack-Ettinger configuration shown in Figure 8.2 reverses the sequence of anoxic and aerobic stages in the Wuhrmann design (Ludzack & Ettinger, 1962). The advantage of this design is the provision of influent biochemical oxygen demand (BOD) to the anoxic stage as an exogenous electron donor.

Barnard (1973a) proposed the modified Ludzack-Ettinger (MLE) configuration that incorporates an internal recycle (Q_{IR}) of nitrate rich mixed liquor from the aeration stage to the anoxic stage (Figure 8.3). This modification increases both the denitrification rate and overall nitrogen removal efficiency and provides control over the fraction of nitrate removed through variation of the internal recycle ratio. In addition, higher denitrification rates are attained because the anoxic reactor receives a source of readily biodegradable COD. This allows smaller anoxic volumes for a given nitrate removal requirement compared to the Wuhrmann and the Ludzack-Ettinger processes.

This process can be used when nitrification is occurring, and denitrification is required to recover alkalinity, lower overall oxygen demand, and provide a better settling sludge. The process effluent typically contains between 6 and 10 mg/L of nitrate nitrogen and is the most common method of achieving nitrogen removal.

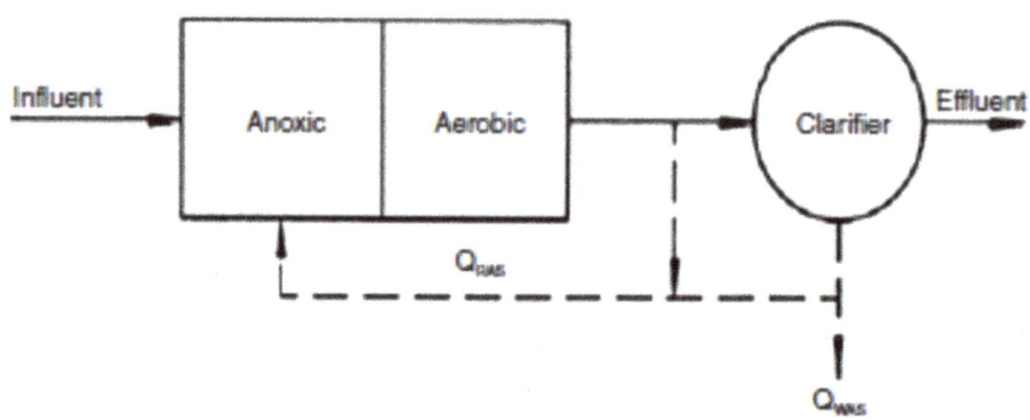

Figure 8.2 Ludzack-Ettinger Process for Nitrogen Removal (RAS = return activated sludge; WAS = waste activated sludge)

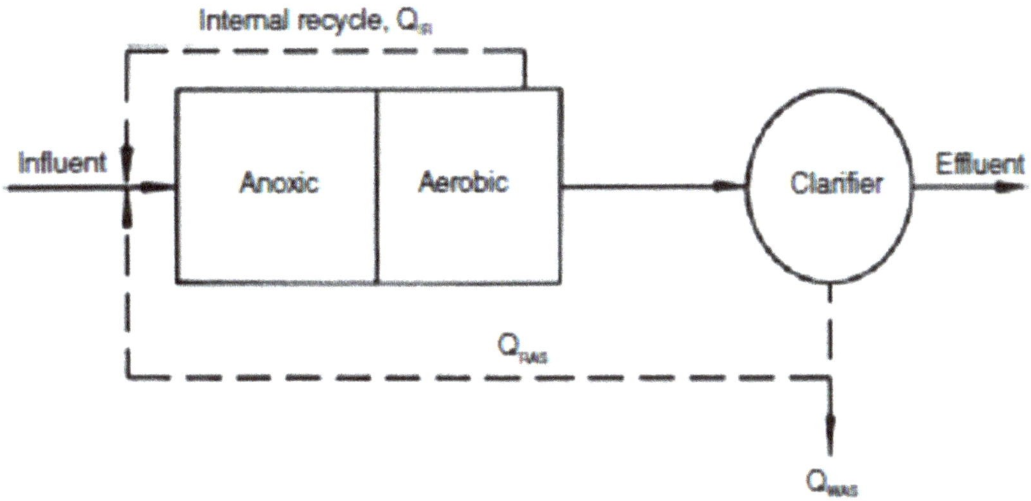

Figure 8.3 Modified Ludzack-Ettinger Process for Nitrogen Removal (IR = internal recycle; RAS = return activated sludge; WAS = waste activated sludge)

2.1.2 FOUR-STAGE BARDENPHO

The four-stage Bardenpho process consists of a series of four anoxic and aerobic zones with recycling of mixed liquor from the first aerobic zone to the first anoxic zone at a rate as high as four to six times the influent flowrate (Barnard, 1973b). This process (Figure 8.4) is intended to achieve more complete nitrogen removal than is possible with a two- or three-stage process. Complete denitrification cannot be attained with pre-aeration anoxic zones because part of the aerobic-stage effluent is not recycled through the anoxic zone. The second anoxic zone provides for additional denitrification using nitrate produced in the aerobic stage as the electron acceptor and endogenous or supplemental organic carbon as the electron donor.

The second anoxic zone is capable of almost completely removing the nitrate in the aeration tank effluent, provided the size is adequate and supplemental carbon is added. The final aeration stage strips residual gaseous nitrogen (N_2) from solution and minimizes phosphorus release in the final clarifier by increasing the oxygen concentration. The ability to successfully use the Bardenpho process to achieve an effluent concentration of total nitrogen as low as 2 to 4 mg/L depends on the ratio of oxidizable nitrogen to carbon in the influent to the activated sludge process and on the use of supplement carbon addition. Ekama et al. (1984) report that the total Kjeldahl nitrogen (TKN) to chemical oxygen demand ratio (TKN:COD) must be less than 0.08 to obtain complete denitrification.

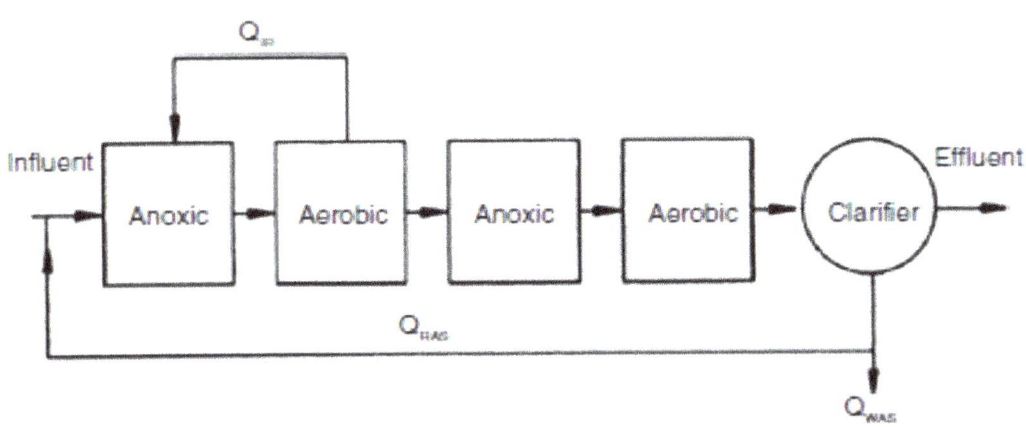

Figure 8.4 Four-Stage Bardenpho™ Process for Nitrogen Removal (IR = internal recycle; RAS = return activated sludge; WAS = waste activated sludge)

2.1.3 STEP FEED

This process is essentially the same as conventional step feed in which a portion of the influent to the process is fed to one or more points downstream of the head of the reactor. The difference in step-feed nitrogen removal is that each of the feed points has an anoxic zone for nitrogen removal. Step-feed nitrogen removal has been implemented at several full-scale facilities in a variety of configurations and number of feed points. Figure 8.5 shows a schematic of a three-pass step-feed system with an additional zone (similar to the Bardenpho process) (Johnson et al., 2003).

The primary benefits of the step-feed configuration include, as in conventional treatment, capacity improvement for a given volume or reduced reactor volumes for a specific capacity due to the higher mixed liquor suspended solids (MLSS) at upstream portions of the reactors. In addition, for nitrogen removal, the step-feed process reduces or eliminates the need to recycle nitrate back to the anoxic zones. Nitrate is supplied to the anoxic zones from the upstream aerobic reactor directly, except for the first pass. In the first pass, the only nitrate supply is from the return activated sludge (RAS) stream. For this reason, nitrified recycle from the end of the first pass to the head of the first pass is sometimes provided to make best use of the available carbon.

2.1.4 SIMULTANEOUS NITRIFICATION AND DENITRIFICATION

In this process, the dissolved oxygen level in an aeration tank is reduced to allow heterotrophs to denitrify and autotrophs to convert ammonia to nitrate/nitrite in a single basin. Typically, dissolved oxygen levels in these basins are less than 1 mg/L. This process has the advantage of eliminating the need for recirculation streams in the aeration basin and reduces aeration requirements due to increased oxygen transfer rate. Additionally, operation at a low dissolved oxygen concentration (< 0.5 mg/L) can result in nitrite shunting, whereby shortcut nitrogen removal is achieved with a lower net aeration and carbon requirement (Water Environment Federation [WEF], 2015). It does not, however, use influent carbon as efficiently as does the MLE or Bardenpho process. Also, there is some risk of reduced sludge settleability as a result of the low oxygen levels present simultaneously with elevated levels of soluble COD. Research has shown that these conditions can result in filamentous bulking (Jenkins et al., 2003).

2.1.5 OXIDATION DITCH

Extended aeration oxidation ditch systems are readily adaptable for carbon oxidation, nitrification, and denitrification as described above (Barnes & Bliss, 1983; Stensel, 1978; Van der Geest & Witvoet, 1977). Oxidation ditches are reactors that induce significant velocity and recirculation flows in basins with a racetrack or other type of configuration. Horizontal rotors, slow-speed mechanical aerators or rotating disks, or draft tube aerators provide the aeration and force to move the mixed liquor at one or more locations in the ditch. Also, submerged mixers and conventional diffused aeration can be combined to provide aeration and mixing power independent of each other. Dissolved oxygen concentration are highest at points of aeration and subsequently decrease because of oxygen uptake by the biomass as the mixed liquor moves around the looped reactor. After sufficient travel time, zones of simultaneous nitrification and denitrification form and fully anoxic conditions may occur upstream from aeration devices, as illustrated in Figure 8.6. Feed points typically are located in the anoxic zones to provide carbon for denitrification. The location and size of these anoxic zones will vary with time because oxygen uptake and transfer rates will vary with wastewater quality and flow. Therefore, reliance on this mechanism

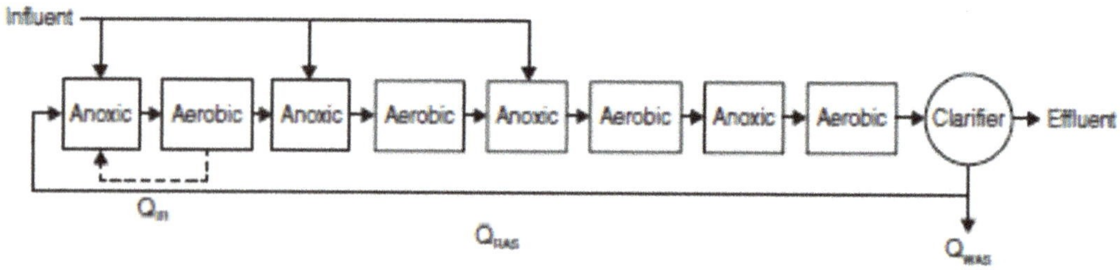

Figure 8.5 Three-Pass Step-Feed Nitrogen Removal System With Secondary Anoxic Zone (IR = internal recycle; RAS = return activated sludge; WAS = waste activated sludge)

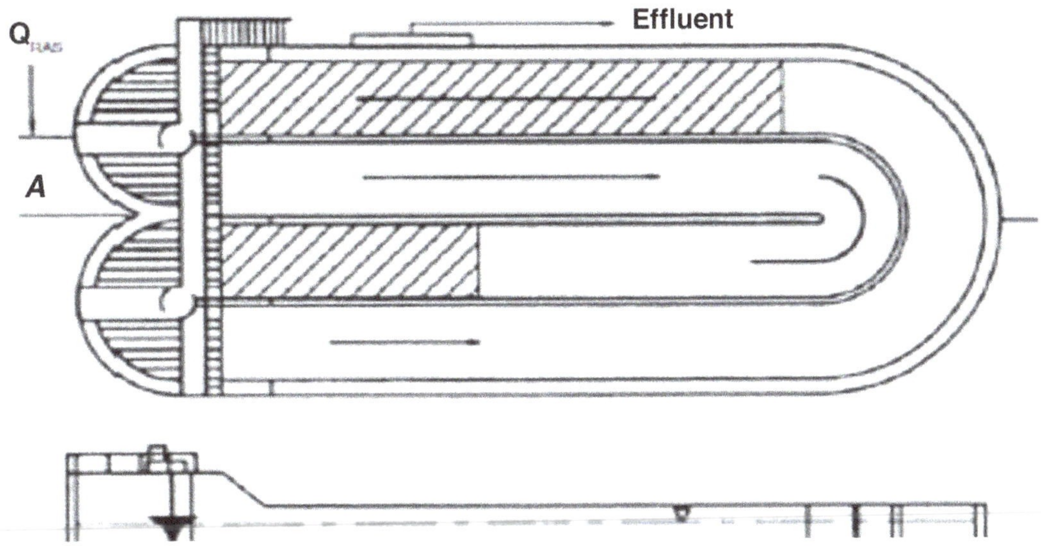

Figure 8.6 Oxidation Ditch Process for Denitrification

for denitrification requires a comprehensive control system to monitor and control dissolved oxygen throughout the basin.

The energy input for mixing and aeration must be controlled to maintain the mixed liquor in suspension. This system must afford sufficient operational flexibility with adjustable weirs, variable speed, or two-speed aerators for varying the oxygen input to match diurnal and seasonal changes in oxygen demand. Otherwise, during periods of low loading, necessary anoxic zones will not develop.

The variable levels of dissolved oxygen that are present in oxidation ditches can be used to promote simultaneous nitrification and denitrification or true anoxic zones within a single ditch. Additionally, oxidation ditches can be configured in series of concentric reactors with different oxygen levels in each, or simply as an aerobic reactor, as described previously in nitrogen removal processes.

In typical oxidation ditch reactors used for nitrogen removal, rates of both nitrification and denitrification may be low because of the relatively long solids retention times (SRTs) required for nitrification, low concentration of readily biodegradable COD, and marginal dissolved oxygen concentrations for either nitrification or denitrification. Oxidation ditch systems have the same limitations in removing nitrogen as other suspended-growth processes and can be designed to remove all but the lowest levels of nitrate/nitrite, subject to carbon availability. Large masses of mixed liquor in the system can compensate for low reaction rates. As with other nitrogen removal processes, highly variable influent flows pose a challenge to achieving a consistently low effluent nitrogen concentration. Nitrogen removals greater than 90% have been reported with oxidation ditch processes, but most operate at 5 to 10 mg/L effluent nitrate levels (Rittmann & Langeland, 1985).

An anoxic zone can precede the oxidation ditch, as in a typical MLE configuration using the ditch as the aerobic zone. The forward momentum of the moving mixed liquor is often sufficient that a small fraction can be diverted for the internal recycle without the use of an internal recycle pump as is required in a traditional MLE system.

2.1.6 TIME-CYCLIC PROCESSES

2.1.6.1 Sequencing Batch Reactor

Biological nitrogen removal (BNR) can be accomplished in sequencing batch reactors (SBRs) by creating, in one reactor, the proper cycle of aerobic and anoxic conditions in time sequence (Abufayed & Schroeder, 1986; Alleman & Irvine, 1980; Arora et al., 1985; Irvine et al., 1983; Palis & Irvine, 1985; Silverstein & Schroeder, 1983). Control strategies for BNR take into account reaction time, tank water level, and mixed liquor dissolved oxygen concentrations. The SBRs appear well suited for relatively small systems with highly variable wastewater flow and strength. Similar to conventional processes, successful operation depends on efficient clarification, which is accomplished in the same tank. For nitrogen removal, fill and react phases are subdivided into mixed fill, mixed react, and aerobic react. In this configuration, carbon oxidation and nitrification occur in the aerobic react phase, and denitrification takes place in anoxic conditions present during mixed fill and mixed react phases. A carbon source to support denitrification, needed in the anoxic react phase, is present in the beginning of each cycle from the feed. Nitrate is supplied from the previous aerobic cycle. Nitrification is attained in SBRs, as in any suspended growth biological treatment system, by designing for the appropriate aerobic SRT as discussed earlier in this chapter (also refer to Chapter 6). Denitrification results from selecting mixed fill and mixed react periods that are long enough to allow use of all dissolved oxygen, thus creating anoxic conditions.

A survey was done of 10 SBR systems in the Northeast United States using nitrogen removal (Young et al., 2008). It was found that facility effluent total nitrogen levels varied between 2.5 and 9.5 mg/L. None of these facilities added supplemental carbon, but all were operating at well below their design loads. This shows that time-cyclic processes can achieve low levels of nitrogen. Typically, for a given COD/total nitrogen ratio, conventional flow through activated sludge systems achieves lower total nitrogen levels because of their increased ability to change the amount of nitrogen brought back to the anoxic zones through recycle systems. Time-cyclic processes are generally limited in nitrogen removal capacity by the amount of nitrogen present at the beginning of their anoxic periods; however, advanced real-time control strategies may be applied to denitrify later in the cycle.

2.1.6.2 Continuous-Feed Intermittent Decant Systems

In these systems, feed continuously enters the reactor to help improve overall nitrogen removal (Peters et al., 2004). These systems typically are segregated into a first stage that continuously feeds downstream SBR tanks. Some variations do continuous feed within a single tank. Sludge from the SBR tanks is recycled back to the upstream tank continuously (from the SBR that is currently under aerate mode). In this way, these systems approach the carbon usage efficiency of conventional flow-through activated sludge systems.

2.1.6.3 Alternating Aeration

Nitrogen removal in a single-reactor activated sludge process using intermittent aeration is feasible (Barth & Stensel, 1981; Schwinn & Hotaling, 1988). Required equipment and operational requirements include (1) timers on blowers or aerators to provide aerobic/anoxic cycling; and (2) proper tankage and solids inventory for nitrification and denitrification.

Phased isolation ditch processes alternate aeration within multiple oxidation ditch reactors to create aerobic/anoxic cycling. The feed is also cycled between reactors so that the anoxic period coincides with the introduction of carbon for denitrification. Alternating aeration has been proposed as a method to selectively inhibit nitrite oxidizing bacteria (Kornaros et al., 2010) and promote shortcut nitrogen removal.

2.2 DUAL SLUDGE AND INTEGRATED SYSTEMS

Separate sludge systems, by definition, house various process stages in physically separate tanks, each with their own clarifier and return-sludge systems. Figure 8.7a–c shows dual-sludge configurations. In Figure 8.7a, the aerobic system first performs carbon oxidation and nitrification. Then, an external carbon source supplements the nitrate-laden stream before contacting the denitrifying biomass in the anoxic system. The system in Figure 8.7b uses the same configuration, except that a portion of influent wastewater fed to the second stage supplies the organic carbon to the anoxic system. Although this system eliminates the need for a carbon supplement, some TKN passes through because ammonia in the feed is not oxidized in the anoxic zone. A third configuration, shown in Figure 8.7c, also eliminates the need for supplemental carbon. In this configuration, the anoxic system precedes the aerobic system, thus providing sufficient BOD for denitrification. An additional recycle stream supplies nitrate. This flow scheme offers the possibility of reducing aeration needs because a substantial portion of the BOD can be oxidized in the anoxic system. Although some oxidized nitrogen is discharged, its magnitude is related to the recycle flow used.

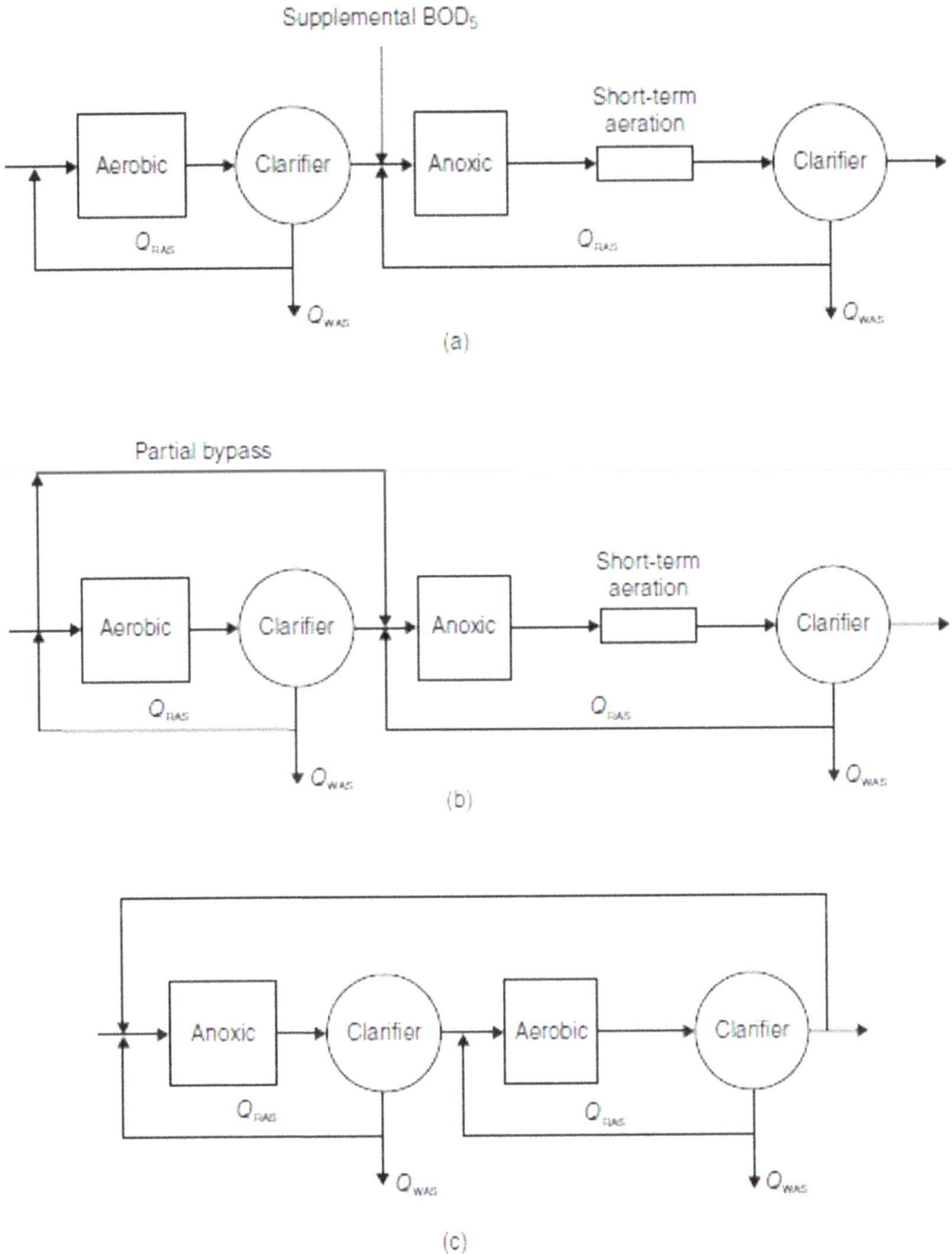

Figure 8.7 Dual-Sludge Processes for Nitrogen Removal (RAS = Return activated sludge; WAS = waste activated sludge) (Grady et al., 2011)

Biofilm processes such as integrated fixed-film activated sludge (IFAS), moving-bed biofilm reactors (MBBRs), denitrification filters, fluidized bed denitrification, and submerged rotating biological contactors can be considered in this application. The IFAS and MBBR processes use the same principles for nitrogen removal as suspended-growth systems. The IFAS media addition to anoxic zones will increase biomass available in a denitrification zone. The additional biomass will improve net denitrification rates within the zone. A suspended-growth SRT is maintained in the IFAS system through the recycle of RAS from the secondary clarifier system. In contrast to IFAS, MBBR systems do not have recycled activated sludge and do not necessarily include clarification. Moving-bed biofilm reactors systems sometimes are used as the primary secondary treatment process with nitrate recycle from a downstream aerobic zone.

2.3 NITRIC AND NITROUS OXIDE EMISSIONS

Nitrification and denitrification produce nitric oxide and nitrous oxide (N_2O) as byproducts and intermediates, respectively. Nitric oxide reacts in the atmosphere to form ground-level ozone (Chatfield et al., 2010), and N_2O is a potent greenhouse gas with a global warming potential approximately 300 times greater than CO_2 (Intergovernmental Panel on Climate Change, 2007). The mechanism of byproduct formation is unknown, but several factors have been associated with increased nitric oxide and N_2O production. Nitrifying systems have been shown to produce nitrous oxide when operated under low dissolved oxygen, with a high nitrite or ammonium concentration, low SRT, and low temperature, or with the presence of toxic compounds (Kampschreur et al., 2009). Release of nitrous oxide as an end product from denitrification has been linked to low concentration of dissolved oxygen, and limited availability of COD amendable to denitrification (Kampschreur et al., 2009). A recent survey of water resource recovery facilities (WRRFs) in the United States found 0.1% to 1.8% of influent TKN emitted as nitrous oxide (Ahn et al., 2010).

2.4 SIDESTREAM NITROGEN REMOVAL

2.4.1 NITRITATION AND DENITRITATION

Nitritation, or the conversion of ammonia to nitrite (NO_2^-) rather than to nitrate, reduces the amount of oxygen required to remove ammonia. For nitrogen removal, nitrite is reduced in anoxic zones to nitrogen gas but requires less carbon than the conversion of nitrate because of the lower oxidation state of nitrite.

$$NH_3 \rightarrow NO_2^- \rightarrow N_2 \qquad (8.1)$$

This process is used primarily to treat nitrogen-rich and carbon-poor liquors from the dewatering of anaerobically digested sludge prior to recycle to the main treatment trains. The reaction to nitrite is controlled by operating at lower dissolved oxygen levels, and denitritation takes place using the endogenous decay of the biomass. Because dissolved oxygen levels are low, this process is a version of the simultaneous nitrification/denitrification process discussed earlier.

2.4.2 ANAEROBIC AMMONIUM OXIDATION

Anammox is a nitrogen removal technique that uses anaerobic ammonium oxidation. Nitrite is used as an electron acceptor to convert ammonium to dinitrogen gas.

Anaerobic ammonium oxidation (**anammox**) is the anoxic oxidation of ammonium to nitrogen gas carried out by members of the *Planctomycetes*, which uses nitrite as the terminal electron acceptor. The energy-yielding reaction is shown in Equation 8.2.

$$NH_4^+ + NO_2^- \rightarrow N_2 + 2H_2O \qquad (8.2)$$

In this process, nitrite is concurrently oxidized to nitrate to yield reducing equivalents for carbon dioxide fixation and cell growth. The net reaction (including cell synthesis) increases the stoichiometry from 1 : 1 ammonium:nitrite. The stoichiometry of anammox metabolism has been found to vary from 1.11 to 2.00 nitrite-N reduced per ammonium-N oxidized (Dapena-Mora et al., 2004; Staunton and Aitken, 2015; Strous et al., 1999; van de Graaf et al., 1996) and the variation is thought to be a result of the effect of SRT; with a longer SRT, the stoichiometry approaches 1 : 1 (Lee et al., 2013). The generally accepted reaction that includes cell synthesis is shown in Equation 8.3 (Strous et al., 1998).

$$\begin{aligned} NH_4^+ + 1.32NO_2^- + 0.066HCO_3^- &\rightarrow N_2 + 0.13H^+ \\ &\rightarrow 1.02N_2 + 0.26NCO_3^- + 0.066CH_2O_{0.5}N_{0.15} + 2.03H_2O \end{aligned} \qquad (8.3)$$

The low energy yield (357 kJ/mol) of Equation 8.2 coupled with the autotrophic nature of the bacteria involved leads to a very slow growth rate with an estimated doubling time of 11 to 14 days under warm sidestream conditions. Although not well established, the doubling time at the reduced temperature found in mainstream treatment has been estimated to be 35 days (Lotti et al., 2014). Achieving anammox in the mainstream treatment process requires a feed with the proper stoichiometry of ammonium to nitrite (alternatively, produce nitrite in situ); a low BOD-to-N ratio to prevent out-competition of anammox bacteria by denitrifying bacteria for nitrite via shortcut denitrification; and efficient biomass retention in order to maintain an adequate SRT for the slow-growing biomass.

2.4.3 NITRITATION/ANAMMOX

The dominant form of influent nitrogen to most influent wastewater treatment facilities is ammonium-N, and a fraction of the influent nitrogen must be selectively oxidized to nitrite before further treatment by anammox bacteria. In addition to the above requirements for anammox, nitrite oxidizing bacteria must be selectively inhibited. Similar control techniques for shortcut nitrogen removal can be used for nitritation/anammox systems. Nitritation and anammox can be designed to occur in the same reactor or in separate reactors.

Nitrogen removal by nitritation/anammox lends itself to a two-sludge AB-type system. In the A-stage, BOD is removed and is either oxidized or redirected to anaerobic digestion for biogas production; in the B-stage, nitrogen is removed by nitritation/anammox. It is important to note that anammox produces nitrate as a side product, and a tertiary polish step may be necessary to reach very low nitrogen discharge permits. At the time of this writing, very few mainstream nitritation/anammox systems had been developed and all mainstream systems require an operational sidestream system.

2.5 DESIGN APPROACH FOR SUSPENDED GROWTH PROCESS FOR BIOLOGICAL NITROGEN REMOVAL

Biological oxidation of ammonia to nitrate can be achieved in combined carbonaceous biochemical oxygen demand (CBOD) removal and nitrification (single-stage) systems or in separate nitrification (two-stage) systems. The degree of nitrification in a combined, single-stage process depends on system SRT, provided that a population of nitrifying autotrophs can be maintained. The degree of nitrification, therefore, is governed to a large extent by design parameters (hydraulic retention time [HRT] and SRT for a nitrification system). Two-stage systems allow some separation of carbonaceous and nitrogenous oxidation processes. In the first stage (aeration basin with clarification and sludge recycle), most of the CBOD removal occurs and nitrification is limited. The second stage (separate aeration basin and clarifier with sludge recycle) maintains more favorable conditions for nitrification. Two-stage systems have been found to be costlier.

Presentation of the fundamental relationships and development of design equations for nitrification can be found in the References section (see Grady et al., 2011; Metcalf & Eddy, Inc./AECOM, 2014; Rittmann & McCarty, 2001; U.S. Environmental Protection Agency [U.S. EPA], 1993). A set of design equations for nitrification in complete-mixed, suspended-growth systems is presented below.

$$\mathrm{SRT}_{N,\min} = \frac{1}{\mu_N - b_N} \tag{8.4}$$

$$\mu_N = \mu_{N,\max}\left(\frac{N_e}{K_N - N_O}\right)\left(\frac{DO}{K_O + DO}\right) \tag{8.5}$$

$$\mathrm{SRT}_{\mathrm{design}} = \mathrm{SRT}_{N,\max}(\mathrm{SF}) \tag{8.6}$$

$$Y_{N,\mathrm{net}} = \frac{Y_N}{1 + b_N \mathrm{SRT}_{\mathrm{design}}} \tag{8.7}$$

$$M_{N,\mathrm{TSS}} = Q\left\{\frac{Y_{N\ \mathrm{net}(N_O - N_e)}}{f_v} + \left[f_d \times b_N \times \mathrm{SRT}_{\mathrm{design}} \times \frac{Y_{N\ \mathrm{net}(N_O - N_e)}}{f_v}\right]\right\} \tag{8.8}$$

$$R_n = 4.57Q(N_O - N_e) - 2.86(N_O - N_e - N_{3e}) - B \times f_v \times M_{N\,\mathrm{TSS}} \tag{8.9}$$

Where $\mathrm{SRT}_{N\min}$ = minimum SRT for nitrification, time;
μ_N = nitrifier specific growth rate, 1/d;
$\mu_{N,\max}$ = maximum nitrifier specific growth rate, 1/d;

N_o, N_e = influent and effluent oxidizable nitrogen concentrations, respectively, mg/L;
K_N, K_o = half-velocity constants for ammonia and oxygen, respectively, mg/L;
Y_N = nitrification yield coefficient;
$Y_{N\text{net}}$ = net nitrification yield coefficient;
DO = reactor dissolved oxygen concentration, mg/L;
$\text{SRT}_{\text{design}}$ = design SRT, d;
SF = safety or design factor, dimensionless ($\text{SRT}_{\text{design}} = \text{SF} \times \text{SRT}_{N\text{min}}$);
b_N = endogenous decay coefficient for autotrophs based on biomass in aerated zone, 1/time;
f_d = cell debris coefficient;
f_v = biomass volatile solids content, typically 0.85;
M_{NTSS} = mass of total autotrophic solids generated or removed as effluent total suspended solids or wasted, per day, kg/d;
R_N = mass of oxygen required per unit time to satisfy nitrification oxygen demand, kg/d; and
NO_3^- = effluent nitrate nitrogen, kg/L.

The basic approach to design of a suspended-growth nitrification process is the same as for carbon oxidation and begins with determining an appropriate design SRT. Table 8.1 presents ranges and typical values for the kinetic coefficients. The relatively slow growth rate of AOB causes nitrifying systems to be slow to recover following process upsets (e.g., low dissolved oxygen, depressed pH, and toxic inhibition) or to large changes in influent oxidizable nitrogen concentrations. A factor of safety is applied to the minimum necessary SRT for increased performance reliability based on consideration of variations in nitrogen loading, process performance requirements, and environmental factors. Equating the peak safety factor to the ratio of peak-to-average influent total nitrogen concentrations or to a combination of loading variations and an additional safety factor has been illustrated (Metcalf & Eddy, Inc./AECOM, 2014; U.S. EPA, 1993).

Equation 8.5 estimates the mass of total autotrophic solids generated or wasted. For combined carbon oxidation and nitrification, this quantity is added to carbonaceous waste mass for total mass generated.

It is important that in-facility recycle streams from solids processing be accounted for in the concentration and variability of influent nitrogen. Sidestreams are often generated from mechanical thickening and dewatering processes, which frequently occur during the day shift. This is the same diurnal period that corresponds to typically high loading to the WRRF and can cause a "double peak" of loading to the WRRF. By providing sidestream storage, the high load in recycle streams can be bled back to the system overnight when flows and loads are typically lower. This "load equalization" can significantly improve biological performance and stability and aid facilities needing to meet very low discharge permits (e.g., < 1.0 mg/L ammonium-N). Alternatively, the sidestream nitrogen load could be treated separately.

Table 8.1 Activated Sludge Nitrification Kinetic Coefficients 20 °C (Grady et al., 2011; Henze et al., 1987)

Coefficient	Unit	Typical Values
μ_{mn}	g VSS/g VSS·d	0.75
K_N	g NH_4·N/m^3	0.74
Y_N	g VSS/g NH_4·N	0.12
b	g VSS/g VSS·d	0.08
K_O	g/m^3	0.50
θ values		
μ_n	Unitless	1.07
K_N	Unitless	1.05
K_{dn}	Unitless	1.04

Consideration also should be given to potential modes of process failure that can include an underloaded facility failing to perform when subjected to design conditions without adequate time for the system to acclimate to higher nitrogen loads (Water Environment Research Foundation, 2006). For Equation 8.3, the dissolved oxygen concentration is considered to be a system average. It is recognized that some nitrification can occur in portions of an aeration basin where dissolved oxygen may be low (Albertson & Coughenour, 1995; Applegate et al., 1980; Smith, 1996). Conversely, the mechanism of dissolved oxygen penetration into biological floc will limit the dissolved oxygen available to entrained autotrophic organisms, which could result in the average reactor dissolved oxygen overpredicting nitrification performance, particularly when a relatively high amount of organic carbon leads to large biological flocs (Metcalf & Eddy, Inc./AECOM, 2014).

3.0 PHOSPHORUS REMOVAL PROCESSES

Phosphorus removal systems can achieve low levels of discharge in facility effluents. Depending on the effluent quality required, treatment could be by biological means alone, by chemical means, or by a combination. This section deals with biological removal of phosphorus followed by chemical treatment for phosphorus removal.

3.1 ENHANCED BIOLOGICAL PHOSPHORUS REMOVAL

The following sections describe selected enhanced biological phosphorus removal (EBPR) process configurations. Most variations of nitrogen and phosphorus removal systems are designed to minimize the return of oxygen and nitrate/nitrite to the anaerobic zone.

3.1.1 ANAEROBIC/OXIC (PHOREDOX AND A/O) SYSTEM

The anaerobic/oxic (A/O) process was first developed in the 1970s as the "Phoredox" system (Barnard, 1976) and later patented in the United States in the early 1980s as the A/O process. It entails a simple process configuration consisting of an anaerobic zone followed by an aerobic zone, shown schematically in Figure 8.8. Typically, the anaerobic zone HRT is between 30 and 45 minutes to select for phosphorus accumulating organisms (PAOs). Longer anaerobic HRTs often are required to promote fermentation in the zone to improve performance. The A/O configuration can be used with any type aerobic reactor and over the full range of aerobic SRTs.

In general, there has been mixed success with A/O EBPR facilities due to the difficulty of avoiding nitrification and the return of nitrates to the anaerobic zone with the RAS.

3.1.2 PHOSTRIP PROCESS

The PhoStrip process, illustrated in Figure 8.9, combines biological and chemical phosphorus removal. It diverts part of the phosphorus-rich RAS (typically 10% to 30% of influent flow) to an anaerobic stripper where sludge settles and phosphorus is released. The phosphorus-rich stripper supernatant is then precipitated with lime, while the biomass, stripped of phosphorus, returns to the aeration tank. The PhoStrip process combines biological and chemical phosphorus removal and is the first patented commercial system. It is not, however, typically used in modern EBPR facilities, although it is capable of achieving low effluent phosphorus concentrations.

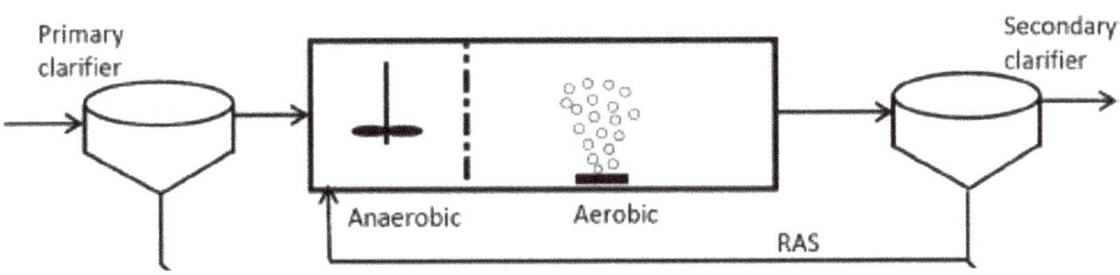

Figure 8.8 Schematic of Phoredox (A^2/O) Process (RAS = return activated sludge; WAS = waste activated sludge)

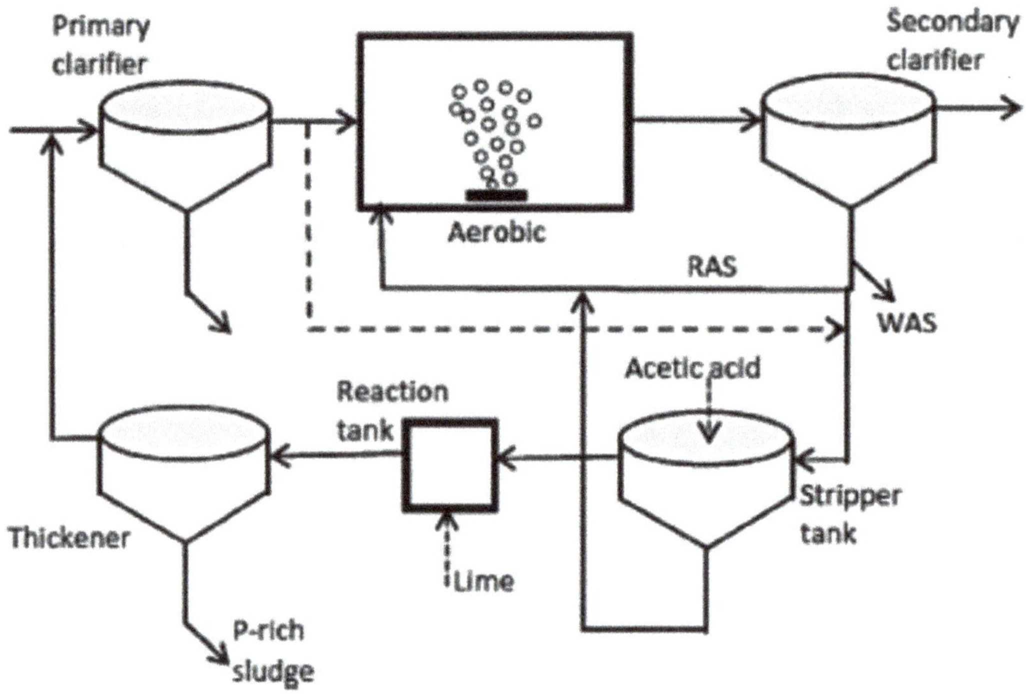

Figure 8.9 Schematic of the PhoStrip Process (RAS = return activated sludge; WAS = waste activated sludge)

There have been some modifications to the original concept, such as diverting a portion of primary effluent to the stripper tank to enhance phosphorus release; including a pre-stripper tank ahead of the anaerobic stripper, increasing the overall detention time for stripping; providing series anaerobic strippers; elutriating the stripped P by recycling settled sludge around the stripper tank; and incorporating anoxic zones for denitrification upstream of the aerobic reactors. When nitrification occurs, without a formal denitrification zone the pre-stripper tank would accept underflow from the secondary clarifier containing nitrates. Stripper underflow typically has high concentrations of soluble BOD and would provide the carbon source for denitrification. The pre-stripper tank hydraulic detention time is approximately 2 hours. Up to 70% denitrification has been observed (Kang, 1988; Matsch & Drnevich, 1987).

3.1.3 ANAEROBIC–ANOXIC–OXIC SYSTEM

Figure 8.10 shows the flow schematic of a typical three-stage anaerobic–anoxic–oxic (A^2/O) process. Mixed liquor is recycled from the end of the aerobic stage to the anoxic stage for denitrification at an internal recycle rate typically ranging from 100% to 400% of the influent flow. Clarifier underflow returns to the first stage of the anaerobic reactor with the reactor feed. This process was implemented in the Bushkoppies Wastewater Treatment Works in Johannesburg, South Africa, commissioned in 1985 (Pitman et al., 1982, 1988). Table 8.2 shows operating data with typical influent sewage, and with the discharge of a yeast factory effluent to the incoming sewer. This waste likely fermented in the sewer and in the anaerobic zone to form

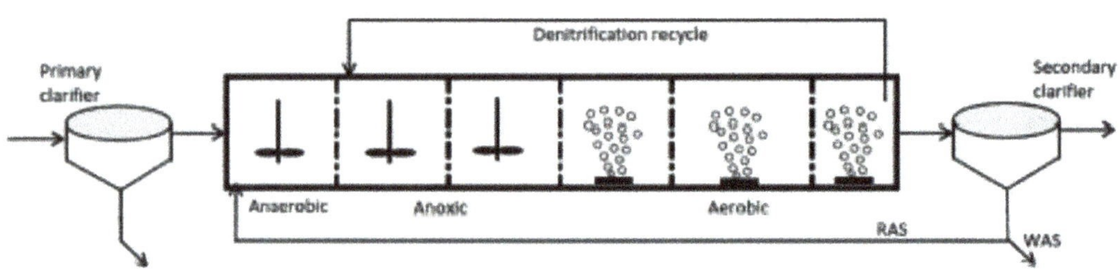

Figure 8.10 Schematic of (A_2/O) Process (RAS = return activated sludge; WAS = waste activated sludge)

Table 8.2 Performance of the Bushkoppie Facility Under Different Conditions (Pitman et al., 1988; reprinted with permission from IWA Publishing)

Operating Condition	Location	Temp (°C)	COD (mg/L)	Total P (mg/L)	PO_4-P (mg/L)	Nitrate (mg N/L)	TKN (mg N/L)	Ammonia (mg N/L)	TSS (mg/L)
No primaries	Secondary influent	22	500	9.0	5.4	0	42	26	230
	Secondary effluent	23	38	2.0	1.6	12	1.5	0.4	8
No primaries+	Secondary influent	22	640	8.0	4.2	0	36	23	330
Yeast factory effluent to sewer	Secondary effluent	23	53	0.7	0.3	8.6	1.4	0.5	7

Note. COD = chemical oxygen demand; TKN = total Kjeldahl nigrogen; TSS = total suspended solids

volatile fatty acids (VFAs). The results demonstrate the importance of influent wastewater characteristics. The effect of nitrates recycled back to the anaerobic zone is also apparent from the results because relative nitrate concentrations in the RAS are reflected by secondary effluent nitrate concentrations.

3.1.4 FIVE-STAGE MODIFIED BARDENPHO

The five-stage modified Bardenpho process (Figure 8.11) provides anaerobic–anoxic–aerobic–anoxic–aerobic stages for removal of phosphorus, nitrogen, and carbon. The early BNR facilities in South Africa (Goudkoppies and Northern BNR facilities, commissioned in the 1970s) were designed with this configuration (Barnard, 1976; Wentzel et al., 2008).

Table 8.3 summarizes the basic design and performance information for several five-stage facilities that have been in operation for several years. The Palmetto, Florida, WRRF, started in 1979, was the first in the United States to use this process (Burdick & Moss, 1980; Stensel, 1980). Most of these facilities have required supplemental chemical addition (metal salts and/or carbon) to meet effluent total phosphorus limits of less than 1.0 mg/L. Facilities using this process use a variety of aeration methods, tank configurations, pumping equipment, and methods of solids handling.

3.1.5 UNIVERSITY OF CAPE TOWN AND VIRGINIA INITIATIVE PLANT PROCESSES

Researchers at the University of Cape Town (UCT) developed the UCT and modified UCT (MUCT) processes shown in Figures 8.12 and 8.13. In the UCT process, both the RAS and aeration tank contents are recycled to the anoxic zone, and the contents of the anoxic zone are then recycled to the anaerobic zone. This recycle sequence decreases the amount of nitrates returned to the anaerobic zone. The internal recycle can be controlled to maintain near-zero nitrates in the effluent from the anoxic reactor, thereby ensuring that little nitrate will be returned to the anaerobic reactor. The Virginia Initiative Plant process is very similar to the UCT process (Daigger et al., 1988; Grady et al., 2011).

The anaerobic recycle (anoxic to anaerobic zone) is typically around 100% of the influent flow, and the actual anaerobic HRT is approximately twice the HRT of an A/O system. This is because in the UCT process, biomass is conveyed to the anaerobic zone via a mixed liquor recycle rather than RAS. Therefore, to maintain the required anaerobic SRT, larger anaerobic volumes are required because of the dilution effect of the anaerobic recycle.

In the MUCT process, the anoxic zone is divided into two reactors. The RAS enters the first reactor, and internal recycle from the aeration tank enters the second anoxic reactor. Internal recycle to the anaerobic zone comes from the first anoxic reactor. The MUCT process was intended to eliminate nitrate recycle to the anaerobic tank while limiting the actual HRT in the anoxic zone to one hour.

For wastewaters with low influent BOD-to-phosphorus ratios, the UCT processes can achieve both phosphorus removal and partial nitrogen removal up to 6 to 8 mg/L. It is reported that near-zero nitrate recycle can be maintained for TKN:COD ratios up to 0.14 (Ekama et al., 1983).

3.1.6 JOHANNESBURG AND WESTBANK PROCESSES

The Johannesburg (JHB) process, shown in Figure 8.14, also was developed in South Africa (Nicholls et al., 1987). The distinguishing feature of this process is the RAS denitrification zone before the anaerobic

Table 8.3 Basic Design Information for Existing Modified Bardenpho™ Systems

Parameter	Palmetto, Florida	Kelowna, Columbia, Canada	Orange County, Florida (phase III)	Fort Myers, Florida Central	City of Cocoa, Florida	Tarpon Springs, Florida	Johannesburg, South Africa (Goudkoppies)
Design flow, m³/d	5300	22 500	28 400	41 600	17 000	15 100	150 000
Final effluent standards Total nitrogen, mg N/L	3	6	3	3	7.3	6.3	3
Total phosphorus, mg P/L	1	2	1	0.5	0.7	3.1	1
Aeration mode	Submerged turbine	Submerged turbines	Carrousel™—mechanical surface aerators	Carrousel™—mechanical surface aerators	Fine-bubble	Carrousel™—mechanical surface aerators	Mechanical surface aerators
Primary settling	Yes	Yes	No	No	No	No	Yes
Sludge handling	Anaerobic digestion for primary sludge	DAF[b] thicken WAS[c]—land application gravity thicken primary	Belt press dewatering—landfill	Belt press dewatering—landfill	Belt press dewatering—landfill	Belt press dewatering—landfill	Anaerobic digestion of primary sludge
Anaerobic volume, m³	228	1760	2271	3680	1440	1500	6240
First anoxic volume, m³	614	3520	4013	3500	1500	2300	14 400
Aeration volume, m³	1060	7920	12 643	18 900	6000	5800	44 100
Second anoxic volume, m³	496	1760	2196	4100	1300	2600	14 400
Second aeration volume, m³	228	2640	379	1000	260	200	8100
Total volume, m³	2626	17 600	21 501	31 200	10 500	12 300	87 240
Clarification surface area, m²	230	1960	2100	3220	2100	930	12 350
Filtration	Yes	Yes	Yes	No	Yes	Yes	
Chemical addition	Yes	Yes	Yes	Yes	Yes		
Type	Alum	Alum	Alum	Alum	Alum		
Dose	90 gpd[a]		20–40 mg/L	50 mg/L			
Internal recycle	Vertical axial	Vertical axial	Horizontal axial	Vertical turbine	Horizontal axial	Vertical turbine	Archimedes
Type	flow	flow	flow		flow		screw
Ratio	4:1	4–6:1	4–6:1	4:1	4–6:1	4:1	4.5:1 to 18:1

Notes.
[a]gpd = 3.785 L/d = 0.0038 m³/d
[b]DAF = dissolved air flotation
[c]WAS = waste activated sludge

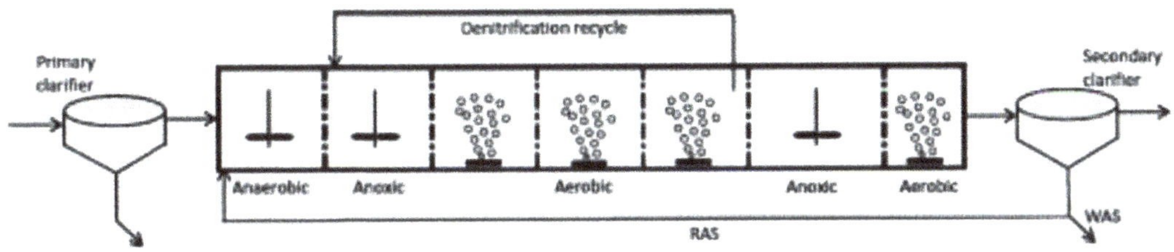

Figure 8.11 Schematic of Five-Stage Modified Bardenpho Process (RAS = return activated sludge; WAS = waste activated sludge)

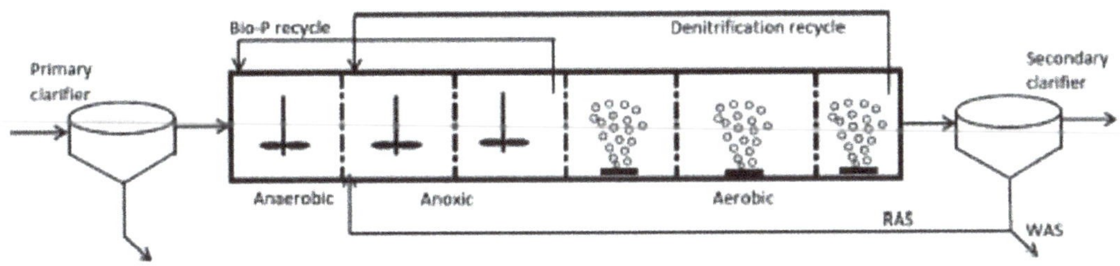

Figure 8.12 Schematic of University of Cape Town and Virginia Initiative Plant Process (RAS = return activated sludge; WAS = waste activated sludge)

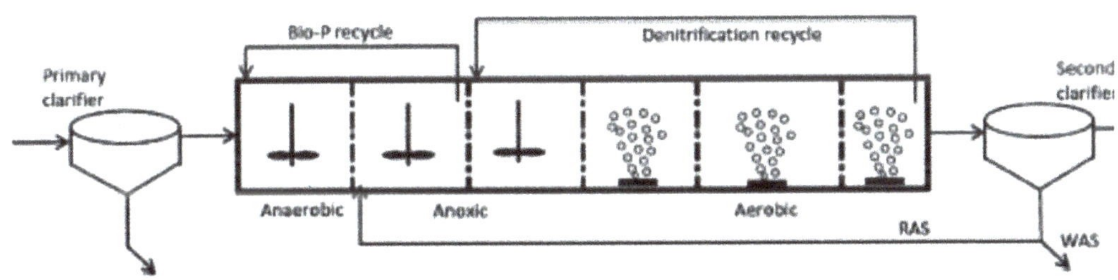

Figure 8.13 Schematic of Modified University of Cape Town Process (RAS = return activated sludge; WAS = waste activated sludge)

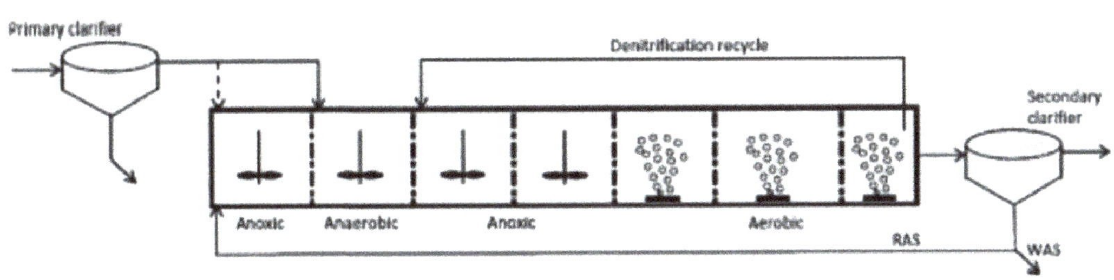

Figure 8.14 Schematic of Johannesburg Process; Dotted Line Signifies the Modification for the Westbank Process (RAS = return activated sludge; WAS = waste activated sludge)

zone. The concept behind this process is that endogenous respiration within the RAS provides the carbon needed to denitrify the RAS before it enters the anaerobic zone. This process has three primary benefits when compared to other combined removal systems: (1) anaerobic zone mixed liquor is at full concentration; (2) using endogenous respiration for nitrogen removal does not require carbon from the feed, thus resulting in efficient carbon usage; and (3) the denitrified mixed liquor recycle stream is eliminated.

A modification of the JHB process resulted in the Westbank Process (Oldham & Rabinowitz, 2001; Stevens et al., 1999). This process aims to improve the denitrification rate by adding either a small fraction of the process influent or supplemental carbon to the RAS denitrification zone.

3.1.7 TIME-CYCLIC PROCESSES

The SBRs can be operated to achieve combined carbon and nitrogen oxidation, nitrogen removal, and phosphorus removal by controlling the sequence and duration of cycles as shown in Figure 8.15 (Ketchum & Liao, 1979). Phosphorus removal is achieved by depleting the nitrate produced during the aerobic phase. Two ways in which this can be accomplished are by (1) adding an anoxic period following the aerobic period, and (2) cycling air on and off during the react phase to create several short aerobic–anoxic sequences. Either method will reduce the nitrates at the beginning of the fill cycle. This allows anaerobic conditions to develop during initial react phases when readily biodegradable substrate (VFAs) is available (Metcalf & Eddy, Inc./AECOM, 2014). These modifications typically reduce SBR cycles to three or four per day. If the SBR process is optimized for nutrient removal, then it can achieve less than 0.7 mg/L total phosphorus on a consistent basis depending on influent characteristics.

The time-cyclic, phased isolation ditch process configuration developed in Denmark uses a pair of oxidation ditches operated in an alternating mode to achieve EBPR. Phased isolation ditches and alternating aeration processes can be configured for combined nitrogen and phosphorus removal by adding an initial anaerobic zone for EBPR. Although it may be possible to maintain informal but localized A/O zones within an oxidation ditch or a similar looped reactor by carefully controlling dissolved oxygen levels, the most typically implemented EBPR configuration includes an external anaerobic basin for PAO selection followed by the oxidation ditch where phosphorus uptake occurs (WEF, 2013).

3.2 PHOSPHORUS REMOVAL THROUGH PRECIPITATION

Phosphorus may occur in the form of organic phosphorus found in organic matter and cell protoplasm; as complex inorganic phosphates (polyphosphates), such as those used in detergents; and as soluble inorganic orthophosphate (PO_4^{3}). As the final breakdown product in the phosphorus cycle, the orthophosphate

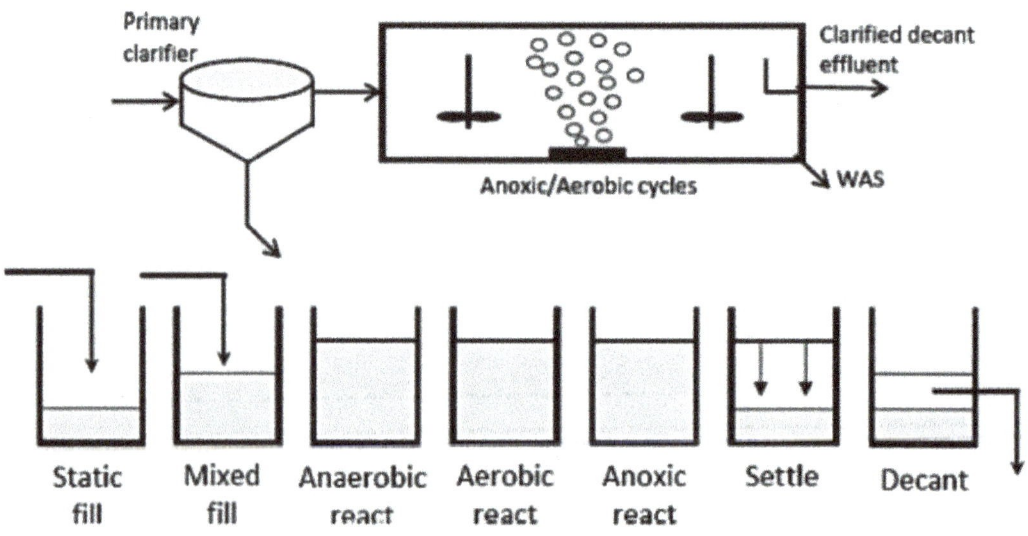

Figure 8.15 Schematic of Sequencing Batch Reactor Type Processes

form is most readily available for biological use or precipitation by a metal salt. When considering removal of phosphorus from wastewater, its form and solubility are important. Phosphorus enters a WRRF in all three forms. During the treatment process, most of the organic phosphate and complex phosphates are converted to inorganic orthophosphate. In wastewater that has been treated by biological processes, most of the phosphorus is in the soluble form, although a small amount of insoluble organic phosphorus also exists in the form of cell protoplasm. For example, the average amount of orthophosphate discharged in the effluent of municipal trickling filter facilities without chemical precipitation has been estimated at approximately 24 mg/L PO_4^{3} (8 mg/L phosphorus) (Griffith et al., 1973), which needs to be controlled in light of recent regulations. In activated sludge, the required minimum phosphorus-to-BOD_5 ratio has been estimated as 1:100, with similar ratios expected for other biological processes. Municipal wastewater has a BOD_5 in the range of 175 to 250 mg/L and a phosphorus content of 3 to 5 mg/L or higher, thus exceeding the nutrient requirements for aerobic biological treatment.

3.2.1 PHOSPHORUS REMOVAL METHODS

Phosphorus removal may be part of primary, secondary, or tertiary treatment processes. Physical and chemical treatment techniques include chemical precipitation and flocculation of phosphorus followed by sedimentation, flotation, or filtration. Incidental uptake by biological processes typically accounts for relatively little phosphorus removal unless the systems are designed for phosphorus removal, as discussed in the previous section. Phosphorus precipitation typically requires addition of chemicals and widely used coagulant aids (flocculant), and a coagulant (Daniels, 1975a, 1975b) can be used. Chemicals typically used for phosphorus precipitation are lime, alum, sodium aluminate, ferric chloride, ferrous sulfate, and polyaluminum chloride. Some examples on chemical reactions when using lime, alum, sodium aluminate, and ferric chloride are discussed in the following sections. These chemicals can be added at different locations in WRRFs. For example, as shown in Figure 8.16, chemical precipitation can be achieved in collection systems or facility headworks primary clarifier, prior to secondary treatment and secondary clarifier and a tertiary filter. It can be achieved in solids thickening, anaerobic digesters and supernatant, centrate, and filtration recycle streams. Chemical addition at points 3 and 4 might lead to increased MLSS and may affect the size of the activated sludge tank.

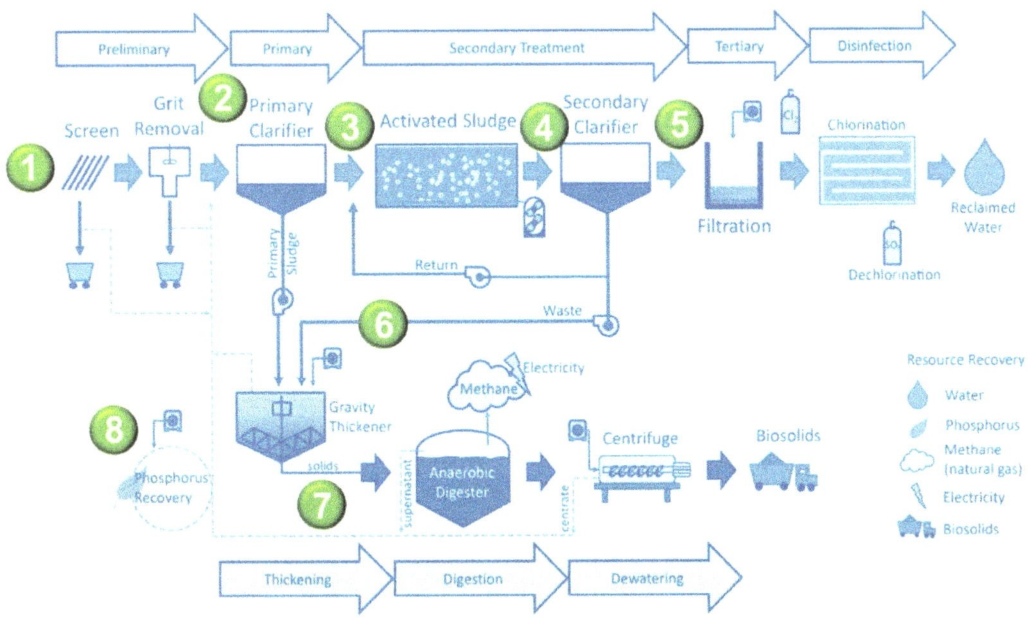

Figure 8.16 Chemical Addition Locations for Phosphorus Removal. 1. In the collection system or facility headworks; 2. Upstream of the primary clarifier; 3. Upstream of the secondary treatment process; 4. Upstream of the secondary clarifier; 5. Between the secondary clarifier and a tertiary filter; 6. Directly to the feed for solids thickening equipment; 7. Directly to anaerobic digesters; or 8. Directly to supernatant, centrate, and filtrate recycle streams (Reprinted with permission from Indigo Water Group)

3.2.2 LIME

In addition to its reactions with carbonate species, lime reacts with orthophosphate to precipitate hydroxyapatite, according to the following stoichiometric reaction:

$$5Ca^{2+} + 4OH^{-} + 3HPO_4^{-3} \rightarrow Ca_5OH(PO_4)_3 + 3H_2O \tag{8.10}$$

The theoretical molar calcium-to-phosphorus ratio is 5:3. The above equation is representative, but not exact, because the composition of the apatite precipitate varies. As a result, the calcium-to-phosphorus mole ratio may vary from 1.3 to 2.0. Hydroxyapatite solubility decreases rapidly with increasing pH and, typically, phosphate removal increases with increasing pH. Almost all orthophosphate converts to the insoluble form if pH is greater than 9.5. The actual pH required to precipitate a given amount of phosphate and the amount of lime required to raise the pH to the desired level vary with wastewater composition. These parameters should be determined by laboratory jar tests.

Wastewater alkalinity is the main variable that affects the lime dose for phosphorus removal. Unless a high pH is used, waters with low alkalinity (150 mg/L or less) form a poorly settleable floc because of the small fraction of dense calcium carbonate ($CaCO_3$) precipitate. Sufficient quantities of calcium carbonate act as a flocculant that enhances settling of the hydroxyapatite. For wastewater with high alkalinity, a pH of 9.5 to 10 can result in excellent phosphorus removal.

Magnesium hardness also affects the efficiency of phosphorus removal. At higher pH, magnesium hydroxide precipitates according to the following reaction:

$$Mg^{2+} + Ca(OH)_2 \rightarrow Mg(OH)_2 + Ca^{2+} \tag{8.11}$$

This reaction begins at a pH of approximately 9.5 and is complete at a pH of 11. The magnesium hydroxide precipitate is gelatinous and removes fine suspended solids as it settles. However, these same gelatinous properties can impair solids dewatering.

Because chemical requirements impose the principal operational cost associated with phosphorus removal, proper chemical dosages are critical to process economy. The lime dosage required to reach a given effluent phosphorus level is independent of influent phosphorus concentration. The degree of phosphorus removal is a function of pH. Lime requirements to reach the required pH correlate more closely with wastewater alkalinity than with any other variable. For other phosphorus precipitants (alum or ferric salts), chemical requirements are proportional to influent phosphorus concentration. In contrast, phosphorus removal is decidedly nonstoichiometric with other chemicals.

3.2.2.1 Process-Train Considerations

Various lime precipitation schemes exist for phosphorus removal (Figure 8.17). Lime may be added before the primary sedimentation tank in a biological WRRF. Because an excessively high pH interferes with the biological process, lime addition to the primary sedimentation tank ahead of an activated sludge system is limited to a pH of approximately 9.0, with a maximum phosphorus insolubilization of approximately 80%. At this limited pH, 2 to 3 mg/L of remaining soluble phosphorus is not unusual. Additional phosphorus may be removed, if necessary, by using aluminum or iron addition in aeration tanks or the final sedimentation tank. The use of lime in primary treatment has the added advantage of increasing organics and suspended solids removal efficiencies in the primary sedimentation tank, thereby decreasing the load on the aeration system.

3.2.2.2 Dosages and Removal

Lime precipitation of phosphorus may require filtration to ensure continuous compliance with effluent requirements. Even with a process pH as high as 11.4, the effluent may contain some residual phosphorus. Although high pH values ensure that virtually all phosphorus is insolubilized, effluent total phosphorus depends on suspended solids removal efficiency. In cases where the precipitate floc is difficult to settle, granular-medium filtration can ensure an extremely high degree of suspended solids and phosphorus removal. Typically, a residual phosphorus concentration of 0.1 to 0.2 mg/L may be achieved with granular-medium filtration.

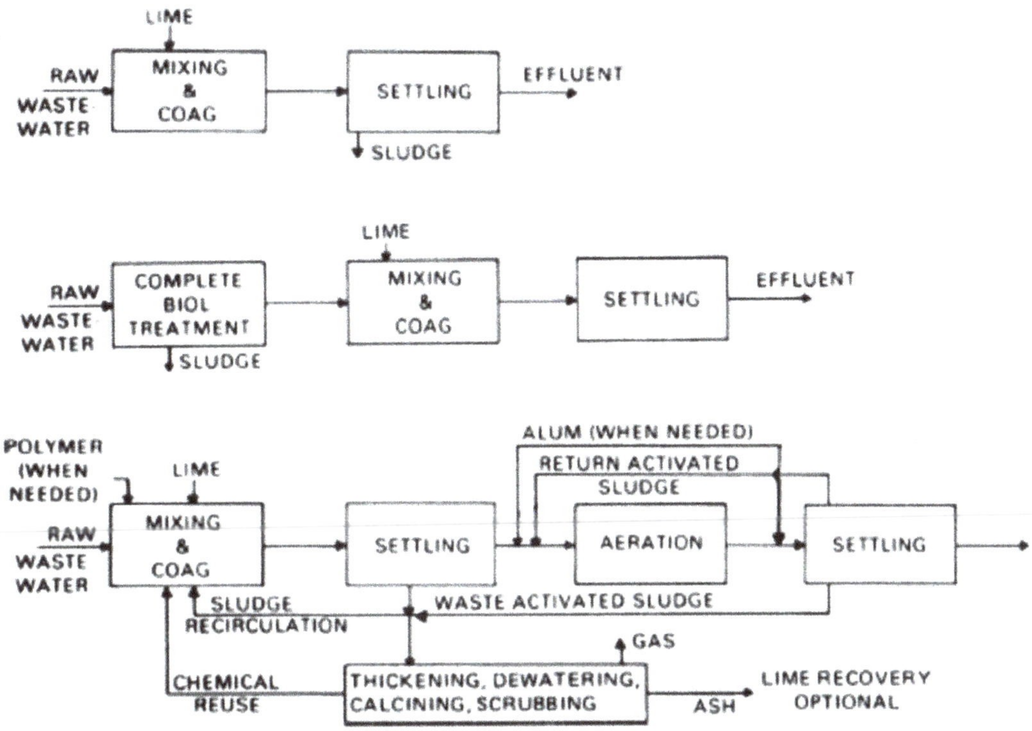

Figure 8.17 Various Schemes of Single-Stage Lime Precipitation for Phosphorus Removal

3.2.3 ALUM

Phosphorus may be precipitated from wastewater by chemical reaction with most mineral salt coagulants, including various salts of iron and aluminum. The principal source of aluminum in phosphorus precipitation is aluminum sulfate, typically known as alum. The theoretical stoichiometry of the chemical reaction is:

$$Al_2(SO_4)_3\cdot(14H_2O) + 2H_2PO_4^- + 4\,HCO_3^- \rightarrow 2AlPO_4 + 4CO_2 + 3SO_4^{2-} + 18H_2O \qquad (8.12)$$

The sulfate ion remains in solution, and pH is depressed. From the above equation, the calculated weight ratio of alum to phosphorus is 0.87:1.0. However, much more alum actually is required because of side reactions involving wastewater alkalinity and organic matter (). The actual weight ratio of alum to phosphorus is up to 3:1 and higher. The $Al_{0.8}(H_2PO_4)(OH)_{1.4}$ (molecular weight = 142.4 g/mol) is assumed to represent the precipitate formed after aluminum addition, and $Al(OH)_3$ (molecular weight = 78 g/mol) is the excess aluminum hydroxide formed.

The solubility of aluminum phosphate is a function of pH. The most efficient chemical use occurs at a process pH near the range of minimum solubility, approximately 5.5 to 6.5. Because alum use results in a small pH depression and most existing treatment systems operate at near-neutral wastewater pH, addition of alum almost automatically results in a pH in the range of minimum aluminum phosphate solubility. Sometimes excess alum is required to depress the pH sufficiently to reach the optimal pH range for phosphate removal. Excess alum then simply acts as an acid that might be replaced by another acid. Using alum following a lime treatment process is an example of such an application.

3.2.3.1 Process-Train Considerations

Alum addition offers flexibility in the location of chemical application points. Alum can be added before the primary settling tank, in the aeration tank, or between aeration and final sedimentation. Alum addition before the primary settling tank provides additional removal of suspended solids and organic matter in addition to phosphorus precipitation. These removals, however, result from reactions that compete with soluble phosphorus for available alum, thereby increasing the dosage required to obtain a given phosphorus residual level. In addition, some polyphosphates that are more difficult to treat may be present in raw wastewater. Alum addition in the activated sludge aeration tank allows use of the mixing already provided for that system. Another point of addition for alum in an activated sludge facility is the aeration basin effluent channel that carries mixed liquor to the final settling basin because of the turbulence in the channel. Addition of alum after the biological system takes advantage of wastewater stabilization with the attendant hydrolysis of the complex phosphates to the more readily reacted orthophosphate form. An added advantage is that dosage can be based on the phosphorus remaining after biological uptake.

3.2.3.2 Dosages and Removal

The stoichiometry of the reactions should be used to determine chemical dosages for alum and other mineral precipitants. In the case of lime, the degree of phosphorus removal depends directly on system pH. For aluminum and iron salts, phosphorus removal efficiency varies directly with chemical dosage up to the point at which mole requirements (molecular weight in grams of any particular compound) for phosphate precipitation and side reactions have been satisfied. Optimum dosages cannot be readily calculated because of ambiguity of the reactions involved. As a result, laboratory jar tests may be used to determine actual chemical requirements. Visible flocculation tends to occur when sufficient coagulant has been added to remove all phosphorus and lower the pH to less than 6.5.

3.2.4 *FERRIC CHLORIDE*

Both ferric and ferrous iron compounds may be used in the chemical precipitation of phosphorus. Although both types of compounds produce equivalent results, ferric chloride is used more often.

The dominant theoretical stoichiometric reaction between ferric chloride and phosphorus is believed to be similar to that of phosphate and alum:

$$FeCl_3 \cdot (6H_2O) + H_2PO_4 + 2HCO_3^- + FePO_4 + 3Cl^- + 2CO_2 + 8H_2O \tag{8.13}$$

This equation indicates that an iron-to-phosphorus mole ratio of 1 : 1 is required. This corresponds to an iron-to-phosphorus weight ratio of 1.8 : 1. However, similarly to alum, in practice much greater amounts of ferric chloride are required to satisfy side reactions with alkalinity that produce $Fe(OH)_3$. Alkalinity reactions occur because the well-flocculating ferric hydroxide precipitate aids settling of the colloidal ferric phosphate.

Experience has shown that efficient phosphorus removal requires the stoichiometric amount of iron to be supplemented by at least 10 mg/L of iron for hydroxide formation. Typically, iron requirements for municipal wastewater are 15 to 30 mg/L as Fe (45 to 90 mg/L as $FeCl_3$) to reduce phosphorus by 85% to 90%. Dosages vary with influent phosphorus concentration. The optimum pH range for iron precipitation of phosphorus is between 4.5 and 5.0. As with aluminum salts, iron salts may be added during primary, secondary, or tertiary treatment. The actual iron-to-phosphorus weight ratio is up to 10.0 : 1.0 to 15.0 : 1.0 and higher.

The $Fe_{1.6}(H_2PO_4)(OH)_{3.8}$ (molecular weight = 251 g/mol) is assumed to represent the precipitate formed after ferric iron addition, and $Fe(OH)_3$ (molecular weight = 106.8 g/mol) is the excess ferric hydroxide formed. Both aluminum and iron salts, when used as phosphorus precipitants, increase dissolved solids in facility effluent. In facilities with poor solids capture, ferric iron may impart a slight reddish color to the effluent.

3.2.5 *SOLIDS CONSIDERATIONS*

Addition of mineral salts for phosphorus precipitation can significantly increase the quantity of solids generated because of production of metal-phosphate precipitates and metal hydroxides and improved suspended solids removal (U.S. EPA, 1987).

As noted previously, phosphate precipitates of the form $Fe_{1.6}(H_2PO_4)(OH)_{3.8}$ and $Al_{0.8}(H_2PO_4)(OH)_{1.4}$ can be used to estimate the quantity of solids produced from precipitation of phosphorus. The remaining metal are precipitated as hydroxide according to these reactions:

$$Al^{3+} + 3H_2O \rightarrow Al(OH)_3 + 3H^+ \tag{8.14}$$

$$Fe^{3}+ + 3H_2O \rightarrow Fe(OH)_3 + 3H^+ \quad (8.15)$$

Addition of metals upstream of the primary clarifier results in a primary sludge mass increase of 50% to 100% because of phosphate and hydroxide precipitates and improved suspended-solids removal. The increase in solids production is almost equally attributable to improved suspended solids capture and additional chemical sludge generation (Knight et al., 1973). Overall facility solids mass increase is smaller because of decreased secondary sludge production from improved primary removals. For example, a 60% to 70% increase is typical across the entire facility.

For metal addition to secondary processes, waste mixed liquor solids mass may increase by 35% to 45%, and the overall facility solids mass may increase 5% to 25%. Metal addition to either primary or secondary treatment units not only increases solids mass, but also solids volume because settled solids concentration in clarifiers may decrease by as much as 20%.

As shown in the previous two reactions, each mole of cation beyond the quantity precipitated with phosphorus should react with three moles of water to produce one mole of metal hydroxide and three moles of hydrogen ions. Therefore, one milligram of alum, $Al_2(SO_4)_3.14H_2O$, reacts to produce 0.26 mg of insoluble aluminum hydroxide while consuming 0.5 mg/L of alkalinity as calcium carbonate. One milligram of ferric chloride, $FeCl_3.(6H_2O)$, will produce approximately 0.4 mg of ferric hydroxide and consume 0.56 mg of alkalinity as calcium carbonate. Alkalinity reductions are important design considerations for low-alkalinity waters or nitrified effluent. During nitrification, significant alkalinity reductions occur and therefore additional chemical treatment that further reduces alkalinity should be evaluated carefully. The metals dosing needs to be controlled to minimize potential negative effects in case of overdosing. As an example, the overdose of iron coagulants affects the performance of facilities, which are using ultraviolet disinfection because of the high ultraviolet light absorptivity by iron in case of overdose.

3.3 PHOSPHORUS PRECIPITATION AND RECOVERY

Phosphate precipitation and recovery is a sustainable alternative to manage sidestream loads with high concentrations of soluble phosphate. Phosphate recovery methods are typically based on precipitation and crystallization, or adsorption. This sidestream management strategy became popular with the development and implementation of magnesium ammonium phosphate (struvite or MAP) and calcium phosphate (apatite or CaP) precipitation and recovery systems. This section will focus on sidestream management aspects of phosphorus recovery. A more comprehensive review of phosphate recovery technologies can be found in the literature (Barnard et al., 2012; Morse et al., 1998; Sartorius et al., 2012).

3.3.1 ADSORPTION METHODS FOR PHOSPHATE SEPARATION AND RECOVERY

Processes based on adsorption, such as the REM NUT® process, are available; however, they are not commonly used for sidestream management applications. These processes include adsorbent packed columns, mixed reactors with media recovery, or ion exchange systems for phosphate removal. The phosphate recovery process follows downstream the adsorption process through media regeneration and phosphate recovery. These systems are typically more complex than precipitation and crystallization methods and have not been widely adopted by the industry. This section will focus on common precipitation and crystallization methods.

3.3.2 PRECIPITATION AND CRYSTALLIZATION METHODS FOR PHOSPHATE RECOVERY

Precipitation and crystallization methods aim to produce phosphate-based precipitates that can be crystalized and readily utilized as fertilizer products or can be post-processed to produce fertilizer. The most common precipitation products for recovery applications are magnesium ammonium phosphate—MAP (struvite) and calcium phosphate—CaP (e.g., hydroxyapatite). These systems provide multiple benefits including substantial reduction of sidestream phosphate load, scaling control, improved sludge dewaterability, and production of a sustainable fertilizer product. Increased dewaterability of the biosolids has also been achieved depending on the phosphorus recovery process configuration applied. Despite systems based on CaP recovery being in operation in Europe for several years, systems based on struvite recovery have dominated the North American market. Some of the proprietary systems available include: DHV-Crystalactor (CaP), P-RoC (CaP), Ostara Pearl

(MAP), Muiltiform Harvest (MAP), PhosNIX (MAP), PRISA (MAP), and AirPrex (MAP); however, non-proprietary systems also exist, particularly in configurations that induce struvite precipitation before dewatering.

Magnesium ammonium phosphate or struvite ($MgNH_4PO_4-6H_2O$) is a white crystal that contains equal molar amounts of magnesium, ammonium, and phosphorus. Struvite precipitation occurs when the concentration of magnesium, ammonium, and phosphate ions exceed the solubility product constant (K_{sp}). This reaction is affected by pH and ionic strength of the sidestream water. Therefore, a conditional solubility product is calculated from the K_{sp} value to take into account these parameters. In general, struvite solubility decreases as pH increases. However, as pH increases, the availability of the ionic species that are needed to form struvite changes. For example, for pH values higher than 9, ammonium ion concentrations will decrease as the equilibrium shifts toward ammonia. Optimum pH for effective struvite precipitation is reported to be between 8.0 and 10.7 (Doyle & Parsons, 2002).

Estimating struvite formation potential is complex because it requires an estimate of the concentrations of each ionic species available at a given pH. These estimates are often times performed using chemical equilibria software (Doyle & Parsons, 2002). Struvite precipitation occurs in two steps: (1) nucleation—the combination of ions to form embryonic crystals, and (2) growth—crystal development until equilibrium is attained.

Operating pH and temperature affect nucleation time. In general, an increase in pH and temperature decreases nucleation time. For pH values lower than 8.0, nucleation is slow and can take days to occur. For pH values higher than 8.0, nucleation occurs substantially faster, taking less than 1 hour for pH values higher than 8.5 (Doyle & Parsons, 2002). Mixing energy also affects nucleation time because more turbulent systems promote more release of CO_2 and therefore an increase in pH. Therefore, a higher mixing energy tends to reduce nucleation time (Doyle & Parsons, 2002). However, struvite recovery reactors typically operate at pH levels of 6.8 to 7.2 because struvite nuclei are either already present or are recycled into the reactor.

Calcium phosphate can crystalize as different compounds depending on the solution. Hydroxyapatite [$Ca_5H(PO_4)O$] is the most thermodynamically stable and is typically present in CaP recovery processes. The precipitation of CaP in wastewater requires a very high oversaturation of the solution and, for this reason, a seed is typically applied to the system to drive the crystallization process. The recovery of CaP may be attractive to specific applications depending on the target end-product market. Contrary to MAP-based products, CaP-based products can be used not only in the fertilizer industry but also in the lower value phosphate market because CaP is essentially phosphate rock (Nieminen, 2010).

3.3.3 STRUVITE PRECIPITATION AND RECOVERY PROCESSES

Struvite recovery systems are typically pre-engineered packaged systems, which may vary slightly from manufacturer to manufacturer. Two approaches are becoming predominant in water resource recovery applications: (1) systems that use sidestream liquors (e.g., solids dewatering recycled water) as a base source of nutrients (e.g., Ostara Pearl, Multiform Harvest, PHOSNIX); and (2) systems that use digested sludge liquors as a base source of nutrients (e.g., MagPrex®, NuReSys®, and Ostara®). In the former systems, struvite recovery is an integral part of the process. In the latter processes, some utilities chose not to recover the struvite separately, but rather allow it to be removed from the facility with the biosolids stream during dewatering.

3.3.3.1 Struvite Recovery From Sidestream Liquor

A typical struvite recovery processes from sidestream liquors is illustrated in Figure 8.18 and includes:

- Equalizations and pretreatment system: Dewatering reject streams have a high amount of impurities that can affect the performance of the recovery process or create operational problems. The removal of suspended organic and inorganic material benefits the downstream recovery process. Equalization and sedimentation may be an option for facilities that have intermittent dewatering schedules. Fine screens can be used as noted earlier for fibrous material removal.
- Fluidized bed reactors: In most processes, the nucleation and crystallization process takes place in a fluidized bed reactor. These conical shape reactors are generally tall with heights between 7.6 and 9 m

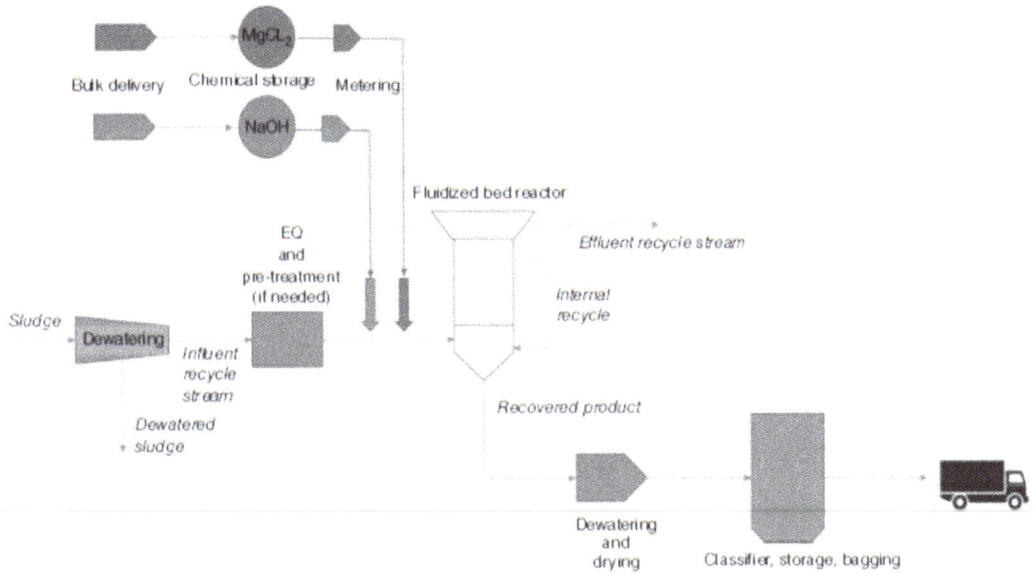

Figure 8.18 Typical Struvite Recovery Processes From Sidestream Liquors

(25 and 30 ft). Struvite is collected in the bottom of the reactor and conveyed for post-processing steps. Chemicals and the influent (e.g., centrate, filtrate) are added in the lower part of the reactor while effluent is collected at the top. Some designs will include a recirculation line to optimize hydraulic conditions of the reactor and promote formation and separation of larger particles.

- Chemical addition system: Two chemicals are commonly added to the reactor: (1) A magnesium source such as magnesium chloride or magnesium hydroxide, and (2) sodium hydroxide for pH control. Although magnesium hydroxide can be used as both a magnesium source and pH control agent, it may limit the opportunities for separate optimization of Mg addition and operating pH but may reduce overall chemical costs.
- Product dewatering and drying: Where recovery is practiced, the struvite product coming out of the reactor needs to be dewatered and dried for handling and post-processing. A dewatering or separation stage is added before post-processing. Product drying may be added to the process train.
- Classifier, storage, and bagging: Some systems may include on-site struvite handling and storage. This would typically include a conveyor system, a product classifier, storage silos, and bagging system. When selecting the process site, attention must be paid to the existing truck routes within the facility.

3.3.3.2 Struvite Recovery From Digested Sludge Liquor

A typical struvite recovery process from digested sludge is illustrated in Figure 8.19 and includes:

- Struvite precipitation reactor(s): The precipitation reactor is a tall cylindrical tank with a conical bottom typically of stainless steel construction. In this system, air via a coarse bubble diffuser is used to strip CO_2 and increase the pH of the sludge for optimal precipitation. The use of caustic for pH adjustment is typically not required, reducing the chemical handling requirements. Where only struvite precipitation is required, only one reactor will be used with a hydraulic retention time of 4 to 6 hours. Where recovery of the struvite is also targeted, a two-reactor system in series may be used, with an additional 4 to 6 hours of detention time to allow for growth of the struvite crystal, enhancing downstream separation efficiency. In some processes, air is introduced in the first reactor for both pH control and mixing. Others decouple pH control from the recovery reactor and introduce air in the first reactor and supplemental chemicals in the second reactor. Recirculation of struvite crystals can be provided to increase recovered crystal size.
- Chemical addition system: Where pH control can be accomplished with CO_2 stripping, only magnesium chloride is added to the recovery reactor as a magnesium source. Caustic or alternatively $Mg(OH)_2$ may be used where an alkalinity source is required. Similarly, chemical addition systems are comprised of bulk delivery system, storage tank, and metering system and pumps.
- Struvite product washing system: The recovered product is withdrawn from the bottom of the reactor and discharged into a classifier and washing system, similar to a grit washer and classifier commonly used in grit handling applications.
- Product handling system: The washed product is conveyed through a screw conveyor into a dumpster or storage container.

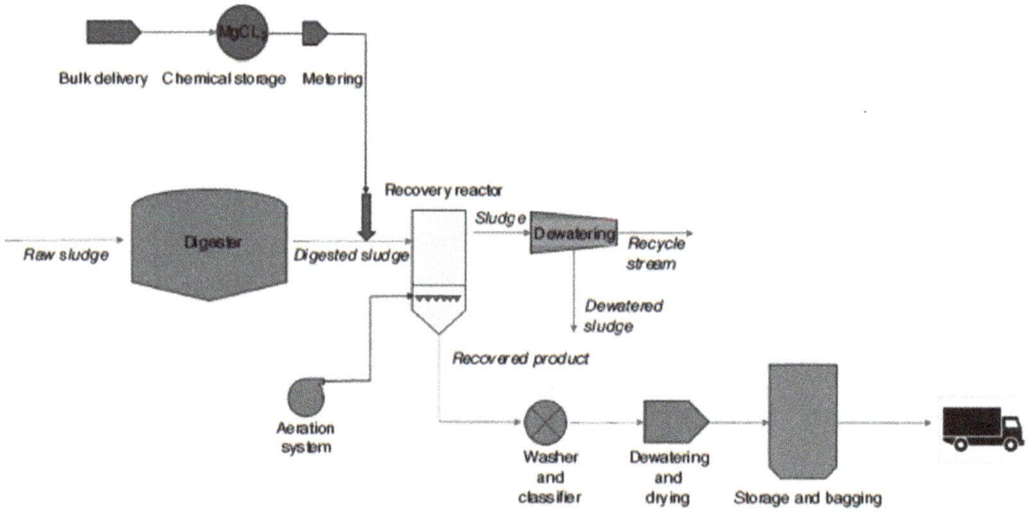

Figure 8.19 Typical Struvite Recovery Processes From Digested Sludge Liquors

3.3.4 CALCIUM PHOSPHATE RECOVERY PROCESSES

The general concept of a CaP recovery process is similar to MAP recovery processes. Some of the technologies available include the DHV-Crystalactor and the P-RoC processes. An important difference in these processes is the need for seed to drive the precipitation of CaP. A general CaP recovery process from sidestream liquors includes similar elements described for the MAP processes with the seed addition system.

3.3.4 BENEFITS OF SIDESTREAM MANAGEMENT THROUGH PHOSPHATE RECOVERY

The benefits of sidestream management through phosphate recovery go beyond the optimization of biological phosphorus removal in the mainstream by reducing the phosphate loading to the head of the facility. Some benefits are discussed below (Mayer et al., 2016).

3.3.4.1 Scaling Control and Mitigation

Struvite and vivianite can form and accumulate in pipelines, mechanical equipment, and process tanks. The scaling of mechanical equipment (e.g., pipes, heat exchangers, dewatering equipment, centrate handling equipment, and pipelines) is not only a process performance issue, but also a critical maintenance problem. Scaling removal is a labor-intensive process and is often costly as it involves equipment/pipeline disassembly, acid washings, and the use of special descaling solutions. The reduction of phosphate levels in the facility through controlled precipitation of MAP or CaP can considerably reduce deposits and scaling issues. Where struvite scaling control and dewaterability enhancement is the primary goal, the use of magnesium hydroxide upstream of dewatering in lieu of magnesium chloride is preferred because it has been shown to minimize unintended struvite precipitate formation on reactor internals and piping.

3.3.4.2 Reduction of Solids Disposal

Although the amount of solids recovered is small compared to the total solids disposed of, a reduction in solids disposal quantities is noted in processes that include phosphate recovery. The reduction in solids disposal will depend on the treatment process and can vary considerably from facility to facility.

3.3.4.3 Effects on Solids Dewatering Processes

The digestion of WAS from EBPR systems results in the release of PO_4^{-3}, Mg^{-1}, and K^+, which change the ionic balance of the sludge mixture. High ratios of monovalent to divalent cations are reported to reduce dewaterability and decrease cake solids (Higgins, 2014). High concentrations of phosphate result in a decrease of divalent ions (Mg^{-1} and Ca^{-1}) through precipitation and therefore increase the relative concentrations of monovalent cations (K^+) and consequent increase the monovalent to divalent ratio. The removal of phosphate in a controlled manner before the digestion process can alleviate some of the negative effects of EBPR sludge on dewaterability (Higgins, 2014).

3.3.4.4 Production of Sustainable Fertilizer Product

Phosphate, which is a key element for food production, is a limited resource and readily accessible global phosphate reserves are being depleted. The recovery of phosphate from human excreta is one of the measures proposed in the literature to overcome a potential future global phosphorus scarcity (Bird, 2015; Cordell & White, 2013). The recovery of phosphate as struvite produces a slow-release fertilizer product that requires minimum post-processing. Some manufacturers work with a business model whereby the fertilizer product is purchased from the WRRF for resale.

PRACTICE PROBLEMS

1. What are the functions of *ammonia oxidizing bacteria* in wastewater treatment? What are they supposed to do, and how does this work? Use chemical equations, words, and pictures.
2. Does phosphorus removal in a secondary treatment facility using biological processes such as Bardenpho increase or decrease with (a) total influent COD, (b) solid retention time, or (c) carbonaceous oxygen demand of the substrate in the anaerobic portion of the system?
3. An activated sludge facility has an aeration tank with a hydraulic retention time of 6 hours. The engineer wants to install baffles in the tank and convert the system to a modified Bardenpho process for removing both nitrogen and phosphorus. Would this be feasible? Why or why not?
4. A small WRRF has two trickling filters, which are run in parallel. The operator decides to run an experiment, operating these filters in series. That is, the effluent from the first filter is pumped into the second one. To the operator's amazement, the facility started to produce an effluent that is nitrified (the nitrogen is in the form of nitrate). Explain to the operator what happened.
5. Explain why denitrification requires a carbon source. What carbon sources are available in a typical WRRF?
6. What should be taken into consideration to convert an oxidation ditch, designed for BOD and suspended solids removal, into a biological nitrogen–phosphorus removal system?
7. Compare various phosphorus removal and recovery methods.

4.0 REFERENCES

Abufayed, A. A., & Schroeder, E. D. (1986). Kinetics and stoichiometry of SBR denitrification with a primary sludge carbon source. *Journal of the Water Pollution Control Federation, 58*, 398–405.

Ahn, J. H., Kim, S., Park, H., Rahm, B., Paguilla, K., & Chandran, K. (2010). N_2O emissions from activated sludge processes, 2008–2009: Results of a national monitoring survey in the United States. *Environmental Science and Technology, 44*, 4505–4511.

Albertson, O. E., & Coughenour, J. (1995). Aerated anoxic oxidation denitrification process. *Journal of Environmental Engineering, 121*, 720.

Alleman, J. E., & Irvine, R. L. (1980). Storage-induced denitrification using sequencing batch rector operation. *Water Research (Great Britain), 14*, 1483.

Applegate, C. S., Wilder, B., & Deshaw, J. R. (1980). Total nitrogen removal in a multi-channel oxidation system. *Journal of the Water Pollution Control Federation, 52*, 568–577.

Arora, M. L., Barth, E. F., & Umphres, M. B. (1985). Technology evaluation of sequencing batch reactors. *Journal of the Water Pollution Control Federation*, 57, 867–875.

Barnard, J. (1976). A review of biological phosphorus removal in the activated sludge process. *Water SA, 2*, 136.

Barnard, J. L. (1973a). Biological denitrification. *Water Pollution Control (Great Britain), 72*, 6.

Barnard, J. L. (1973b). Biological nutrient removal without the addition of chemicals. *Water Research (Great Britain), 9*(485), 15–104.

Barnard, J., Phillips H., & Steichen, M. (2012, September 29–October 3). State-of-the-art recovery of phosphorus from wastewater. *Proceedings of the 85th Annual Water Environment Federation Technical Exposition and Conference* [CD-ROM], New Orleans, Louisiana, United States.

Barnes, D., & Bliss, P. J. (1983). *Biological control of nitrogen wastewater treatment*. E. and F.N. Spon.

Barth, E. F., & Stensel, H. D. (1981). International nutrient control technology for municipal effluents. *Journal of the Water Pollution Control Federation*, *53*, 1691–1701.

Bird, A. R. (2015). Evaluation of the feasibility of struvite precipitation from domestic wastewater as an alternative phosphorus fertilizer resource. [Master's Projects, Paper 141].

Burdick, C., & Moss, O. (1980). Operating experience with the Bardenpho process at Palmetto, Florida. Paper presented at Annual Joint Technical Conference of the Florida Section American Water Works Association, Florida Pollution Control Association, & Florida Water Pollution Control Operators Association, Orlando, Florida, United States.

Chatfield, R. B., Ren, X., Brune, W., & Schwab, J. (2010). Controls on urban ozone production rate as indicated by formaldehyde oxidation rate and nitric oxide. *Atmosphere and Environment*, *44*, 5395–5406.

Cordell, D., & White, S. (2013). Sustainable phosphorus measures: Strategies and technologies for achieving phosphorus security. *Agronomy*, *3*(1), 86–116.

Daigger, G., Waltrip, E., Romm, E., & Morales, L. (1988). Enhanced secondary treatment incorporating biological nutrient removal. *Journal of the Water Pollution Control Federation*, *60*, 1833–1842.

Daniels, S. L. (1975a). *Coagulation/flocculation: State-of-the-art*. Dow Chemical Co.

Daniels, S. L. (1975b). Solid-liquid separation by coagulation and flocculation. In *Particle science and solid fluid separation*. Chemical Engineering Department, University of Houston.

Dapena-Mora, J., Campos, L., Mosquera-Corral, A., Jetten, M. S. M., & Méndez, R. (2004), Stability of the ANAMMOX process in a gas-lift reactor and a SBR. *Journal of Biotechnology*, *110*, 159–170.

Doering, P., Oviatt, C., Niwicki, E., Klos, E., & Reed, L. (1995). Phosphorus and nitrogen limitation of primary production in a simulated estuarine gradient. *Marine Ecology Progress Series*, *124*, 271–287.

Doyle, J., & Parsons, S. (2002). Struvite formation, control, and recovery. *Water Research*, *36*, 3925–3940.

Ekama, G., Marais, G., & Siebritz, I. (1984). Biological excess phosphorus removal. In *Theory, design and operation of nutrient removal activated sludge processes*. Water Research Commission.

Ekama, G., Siebritz, I., & Marais, G. (1983). Considerations in the process design of nutrient removal activated sludge processes. *Water Science and Technology*, *15*, 283–318.

Fisher T. R., Gustafson, A. B., Sellner, K., Lacouture, R., Haas, L. W., Wetzel, R. L., Magnien, R., Everitt, D., Michaels, B., & Karrh, R. (1999). Spatial and temporal variation of resource limitation in Chesapeake Bay. *Marine Biology*, *133*, 763–778.

Grady, C. P. L., Daigger, G. T., Love, N. G., & Filipe, C. D. M. (2011). *Biological wastewater treatment* (3rd ed.). CRC Press.

Griffith, E. J., Beeton, A., Spencer, J. M., & Mitchell, D. T. (Eds.). (1973). *Environmental phosphorus handbook*. Wiley and Sons.

Henze, M., Grady, C. P. L., Gujer, W., Marais, G. V. R., & Matsuo, T. (1987). Model for the single-sludge wastewater treatment. *Water Research (Great Britain)*, *21*, 505.

Higgins, M. J. (2014). Does Bio-P impact dewatering after anaerobic digestion? Yes, and not in a good way! *Water Environment Federation Biosolids and Residuals Conference*, Austin, Texas, United States.

Intergovernmental Panel on Climate Change. (2007, December). *Climate change 2007: The physical science basis*. Cambridge University Press.

Irvine, R. L., Ketchum, L. H., Breyfogle, R., & Barth, E. (1983). Municipal application of sequencing batch treatment. *Journal of the Water Pollution Control Federation*, *55*, 484–488.

Jenkins, D., Richard, M., & Daigger, G. (2003). *Manual on the causes and control of activated sludge bulking, foaming, and other solids separation problems* (3rd ed.). CRC Press–IWA Publishing.

Johnson, B. R., Daigger, G. T., Crawford, G. V., Wable, M. V., & Goodwin, S. (2003, October 11–15). Full-scale step-feed nutrient removal systems: A comparison between theory and reality. *Proceedings of the 76th Annual Water Environment Federation Exposition and Conference*, Los Angeles, California, United States.

Kampschreur, M. J., Temmink H., Kleerebezem R., Jetten M. S. M., & van Loosdrecht M. C. M. (2009). Nitrous oxide emission during wastewater treatment. *Water Research, 43*, 4093–4103.

Kang, S. (1988, October 3–7). Biological nitrification and denitrification in the PhoStrip™ process. *Proceedings of 61st Annual Water Pollution Control Federation Conference*, Dallas, Texas, United States.

Ketchum, L., & Liao, P. (1979). Tertiary chemical treatment for phosphorus reduction using sequencing batch reactors. *Journal of the Water Pollution Control Federation, 51*, 298–304.

Knight, C. H., Mondox, R. G., & Hambley, B. (1973, May 28–29). Thickening and dewatering sludges produced in phosphate removal. *Proceedings of the Phosphorus Removal Design Seminar*. Toronto, Ontario, Canada, Canada–Ontario Agreement on Great Lakes Water Quality; Training and Technology Transfer Division (Water), Environmental Protection Service, Environment Canada.

Kornaros, M., Dokianakis, S. N., & Lyberatos, G. (2010). Partial nitrification/denitrification can be attributed to the slow response of nitrite oxidizing bacteria to periodic anoxic disturbances. *Environmental Science and Technology, 44*, 7245–7253.

Lee, W., Kwak, J., Bae, P., & McCarty, L. (2013). The effect of SRT on nitrate formation during autotrophic nitrogen removal of anaerobically treated wastewater. *Water Science and Technology. 68*, 1751–1756.

Lotti, T., Kleerebezem, R., van Erp Tallman Kip, C., Hendrickx, T. L. G., Kruit, J., Hoekstra, M., & van Loosdrecht M. C. M. (2014). Anammox growth on pretreated municipal wastewater. *Environmental Science and Technology., 48*(14), 7874–7880.

Ludzack, F. J., & Ettinger, M. B. (1962). Controlling operation to minimize activated sludge effluent nitrogen. *Journal of the Water Pollution Control Federation, 34*, 920–931.

Matsch, L., & Drnevich, R. (1987). Biological nutrient removal. In *Advances in water and wastewater treatment*. Ann Arbor Science.

Mayer, B. K., Baker, L. A, Boyer, T. H., Drechsel, P., Gifford, M., Hanjra, M. A., Parameswaran, P., Stoltzfus, J., Westerhoff, P., & Rittmann, B. (2016). Total value of phosphorus recovery. *Environmental Science and Technology, 50*, 6606–6620.

Metcalf & Eddy, Inc./AECOM. (2014). *Wastewater engineering treatment and resource recovery* (5th ed.). McGraw-Hill Education.

Morse, G. K., Brett, S. W., Guy, J. A., & Lester, J. N. (1998). Review: Phosphorus removal and recovery technologies. *Science of the Total Environment, 212*, 69–81.

Nicholls, H., Osborn, D., & Pitman, A. (1987, September 28–30). Improvements to the stability of the biological phosphate removal process at the Johannesburg Northern Works. *Proceedings of the IAWPRC Specialized Conference*, Rome, Italy. IWA Publishing.

Nieminen, J. (2010). Phosphorus recovery and recycling from municipal wastewater sludge [Master's thesis, Aalto University, School of Science and Technology, Department of Civil and Environmental Engineering].

Oldham, W. K., & Rabinowitz, B. (2001). Development of biological nutrient removal technology in western Canada. *Canadian Journal of Civil Engineering, 28*(S1), 92–101.

Palis, J. C., & Irvine, R. L. (1985). Nitrogen removal in a low-loaded single tank sequencing batch reactor. *Journal of the Water Pollution Control Federation, 57*, 82–86.

Peters, M., Newland, M., Seviour, T., Broom, T, & Bridle, T. (2004). Demonstration of enhanced nutrient removal at two full-scale SBR plants. *Water Science and Technology, 50*, 115–120.

Pitman, A., Trim, B., & van Dalsen, L. (1988). Operating experience with biological nutrient removal at the Johannesburg Bushkoppie Works. *Water Science and Technology, 20*(4-5), 51–62.

Pitman, A., Venter, S., Nicholls, H., van Dalsen, L., & Soelbat, J. (1982). Operating experience with nutrient removal activated sludge plants in Johannesburg. *IMIESA, Journal of the Institute of Municipal Engineering of Southern Africa, 11*, 61–69.

Rittmann, B. E., & Langeland, W. E. (1985). Simultaneous denitrification with nitrification in single-channel oxidation ditches. *Journal of the Water Pollution Control Federation, 57*, 300–308.

Rittmann, B., & McCarty, P. (2001) *Environmental biotechnology: Principles and applications.* McGraw-Hill.

Sartorius C., von Horn J., & Tettenborn, F. (2012). Phosphorus recovery from wastewater–Expert survey on present use and future potential. *Water Environmental Research, 84*(4), 313–322.

Schwinn, D. E., & Hotaling, J. H. (1988). Modifying existing activated sludge plants to achieve nitrification and denitrification. Paper Presented at 62nd Florida Water Resources Conference.

Silverstein, J., & Schroeder, E. D. (1983). Performance of SBR activated sludge processes with nitrification/denitrification. *Journal of the Water Pollution Control Federation, 55*, 377.

Smith, G. (1996, October 5–9). Increasing oxygen delivery in anoxic tanks to improve denitrification. *Proceedings of the 69th Annual Water Environment Federation Exposition and Conference* [CD-ROM], Dallas, Texas, United States.

Staunton, E. T., & Aitken, M. D. (2015). Coupling nitrogen removal and anaerobic digestion for energy recovery from swine waste through nitrification/denitrification. *Environmental Engineering Science, 32*(9), 750–760.

Stensel, H. (1980). Performance of first U.S. full-scale Bardenpho™ facility. *Proceedings of U.S. EPA International Seminar on Control of Nutrients in Municipal Wastewater Effluents*, San Diego, California, United States.

Stensel, H. D. (1978). Carrousel activated sludge for biological nitrogen removal. In *Advances in water and wastewater treatment biological nutrient removal.* Ann Arbor Science.

Stevens, G., Barnard, J., & Rabinowitz, B. (1999). Optimizing nutrient removal in anoxic zones. *Water Science and Technology, 39*(6), 113–118.

Strous, J., Heijnen J., Kuenen J. G., & Jetten M. S. M. (1998). The sequencing batch reactor as a powerful tool for the study of slowly growing anaerobic ammonium-oxidizing microorganisms. *Applied Microbiology and Biotechnology, 50*, 589–596.

Strous, M., Lan, C.-J., & Jetten, M. S. M. (1999). Key physiology of anaerobic ammonium oxidation. *Applied Microbiology and Biotechnology, 65*, 3248–3250.

U.S. Environmental Protection Agency. (1987). *Design manual phosphorus removal* (EPA-625/1-87-001, Technology Transfer).

U.S. Environmental Protection Agency. (1993). *Process design manual for nitrogen removal.*

van de Graaf, A. A., de Bruijn, P., Robertson, L. A., Jetten, M. S. M., & Kuenen, J. G. (1996). Autotrophic growth of anaerobic ammonium-oxidizing micro-organisms in a fluidized bed reactor. *Microbiology, 142*, 2187–2196.

Van der Geest, A. T., & Witvoet, W. C. (1977). Nitrification and denitrification in carrousel systems. *Progress in Water Technologies, 8*, 653.

Water Environment Federation. (2013). *Automation of water resource recovery facilities* (Manual of Practice No. 21).

Water Environment Federation. (2015). *Shortcut nitrogen removal—Nitrite shunt and deammonification.*

Water Environment Research Foundation. (2006). *Effects of biomass properties on submerged bioreactor (SMBR) performance and solids processing.*

Wentzel, M., Comeau, Y., Ekama, G., van Loosdrecht, M., Brdjanovic, D., & Lopez-Vazquez, C. (2008). Enhanced biological phosphorus removal. In M. Henze, M. van Loosdrecht, G, Ekama, & D. Brdjanovic (Eds.), *Biological wastewater treatment principles, modelling and design*. IWA Publishing.

Wuhrmann, K. (1954). High-rate activation sludge treatment and its relation to stream sanitation. *Sewage and Industrial Wastes, 26*, 1–27.

Young, T., Crosswell, S., & Wendell, J. (2008, October 18–22). Comparison of nitrogen removal performance in SBR systems. *Proceedings of the 81st Annual Water Environment Federation Exposition and Conference*, Chicago, Illinois, United States.

5.0 SUGGESTED READINGS

Higgins, M. J., & Novak, J. T. (1997). The effect of cations on the settling and dewatering of activated sludges: Laboratory results. *Water Environmental Research, 69*(2), 215–224.

Oldham, W., & Stevens, G. (1984). Initial operating experiences of a nutrient removal process (modified Bardenpho) at Kelowna, British Columbia. *Canadian Journal of Civil Engineering, 11*, 474–479.

Chapter 9 Hydraulics and Hydraulic Profiles

LEARNING OBJECTIVES

At the end of this chapter, students should be able to:

- Read and understand a hydraulic profile of an existing water resource recovery facility (WRRF).
- Use fluid mechanics to properly size conduits, channels, flow splitters, pumps, and other hydraulic control features at a WRRF.
- Determine and properly set tanks walls, weirs, and equipment based on target water surface elevations.
- Develop a hydraulic profile that graphically shows water surface elevations in each unit process and throughout a WRRF.
- Develop hydraulic profiles for key design scenarios: low flow, average flow, peak flow.
- Document and conduct quality control on all hydraulic calculations.

1.0 INTRODUCTION

A key step to water resource recovery facility (WRRF) design involves evaluation of the volumetric flow and hydraulics throughout the treatment facility. Designers must evaluate and develop WRRF hydraulics with as much care as the treatment process itself to (1) provide the right wall heights (including effective freeboard) and elevations for tanks; (2) properly size control structures, pipes, and pump systems; and (3) support the treatment processes to achieve the treatment targets.

After a treatment concept is selected and a preliminary site layout determined, the designer can begin evaluating the hydraulic profile (water surface elevations) or **hydraulic grade line** (HGL) for the WRRF and the unit processes therein. The objective is to ensure there is adequate head to drive wastewater flow through the WRRF from one unit process to another in a controlled manner without overtopping. As part of the hydraulic design effort, the engineer establishes appropriate control weir elevations and water surface elevations (WSE) within each unit process, which, in turn, establishes tank wall elevations based after adding the desired **freeboard** above the WSE.

Hydraulic grade line is the sum of the elevation and pressure heads, excluding velocity head. The energy grade line is the HGL plus the velocity head. The HGL represents the water profile in an open channel or pipe flowing partly full.

Freeboard is the vertical distance between the normal maximum liquid level in a basin and the top of the basin. It is provided so waves and other liquid movements will not overflow the basin.

The hydraulic design of WRRFs includes making conduits, pipes, channels, and control structures large enough to support a wide range of flow conditions from a base low flow to the future peak wet-weather flow. The design must also consider the effects of future expansion beyond the capacity required during the design year. Sufficient hydraulic head, or driving force, should be provided to permit good distribution of the facility flow to all treatment processes over the range of expected flow conditions, without being excessive. During development of the hydraulic profile, the economics of building deeper facility structures may be considered as an alternative to pumping. Pumping stations result in increased operation and maintenance (O&M) costs and reduced WRRF reliability. Depending on the process(es) selected, however, intermediate pumping may be required. Pumping is common upstream of biotowers, trickling filters, and tertiary processes. The layout and elevations of the infrastructure at a WRRF relate closely to the hydraulics, and the hydraulic profile is the starting point for establishing how a WRRF is designed. Future expansion of the WRRF must also be considered, and appropriate gaps in the hydraulic profile may be desirable if a new unit process is planned for the near future.

This chapter reviews the procedures to develop the hydraulic profile. It is assumed that the designer has an understanding of basic fluid mechanics and hydraulics. There are numerous texts available to assist in this area. Several are included in the References section at the end of this chapter. Additional information on pumps and pumping hydraulics is presented in *Design of Wastewater and Stormwater Pumping Stations* (Water Environment Federation [WEF], 2022), and *Pumping Station Design* (Jones, 2008).

2.0 HYDRAULIC PROFILES

A hydraulic profile is a graphic representation of the hydraulic design that quickly and efficiently conveys the WSEs, the top of wall elevations, and weir or other hydraulic control elevations in unit processes throughout the WRRF. The hydraulic profile is developed based on sound hydraulic principles to support the water level required at each treatment process for wastewater to flow through the facility. The resulting water surface profile elevations typically are presented graphically on a drawing sheet in the WRRF construction and record drawings.

A hydraulic profile is typically prepared for the main liquid flow path extending from the facility inlet sewer to the receiving water, but a profile can also be prepared for ancillary flow trains, such as solids treatment and disposal facilities. Hydraulic profiles for ancillary flow trains are particularly useful when there is gravity flow between process units. The hydraulic profile is determined for the design peak flow for the facility (i.e., the peak wet-weather flow). However, it is useful to develop water surface profiles for maximum month, average, and minimum flowrates for current and design conditions. The hydraulic profile should present WSEs; hydraulic control devices, such as control valves and weirs; critical elevations of process structures, channels, and pipelines; and the top and bottom elevations of structures. The profile may include ground surface elevations, pipeline sizes, and other special features that enhance understanding of the system. Figures 9.1 and 9.2 show a WRRF layout and the corresponding hydraulic profile, respectively.

3.0 WATER RESOURCE RECOVERY FACILITY HYDRAULIC EVALUATION

The WRRF hydraulic evaluation follows a number of steps meant to work through different conditions or scenarios that may occur at the facility. A designer should work through the steps carefully and consider the most critical hydraulic conditions that may occur. This section discusses several of the key steps important to evaluation of WRRF hydraulics and development of the hydraulic profile.

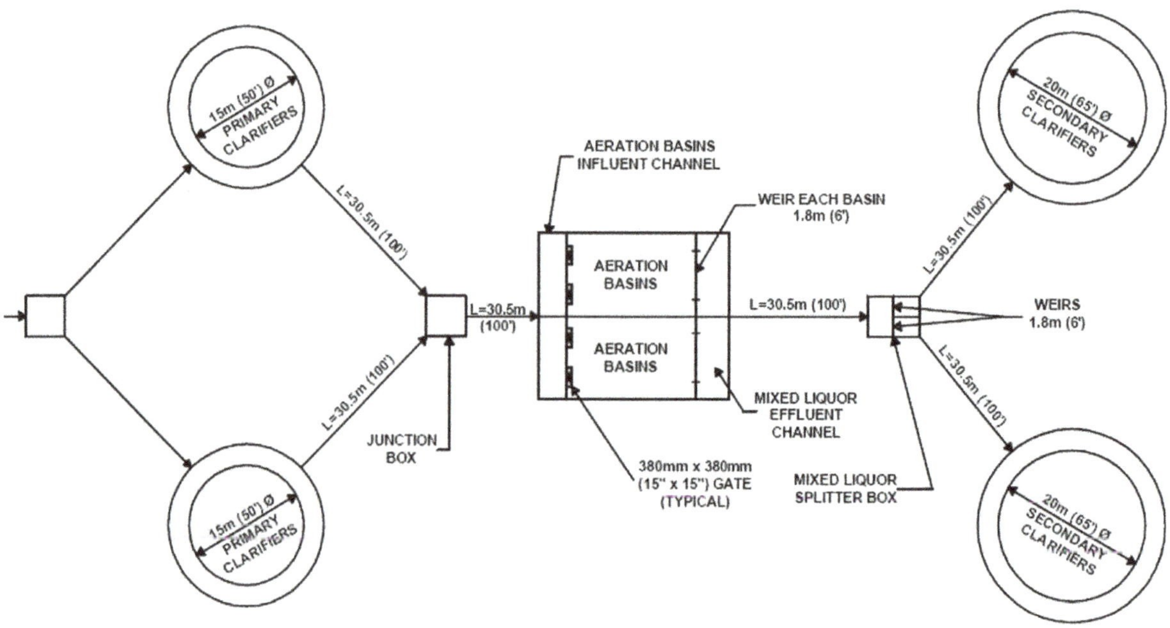

Figure 9.1 Example WRRF Site Layout

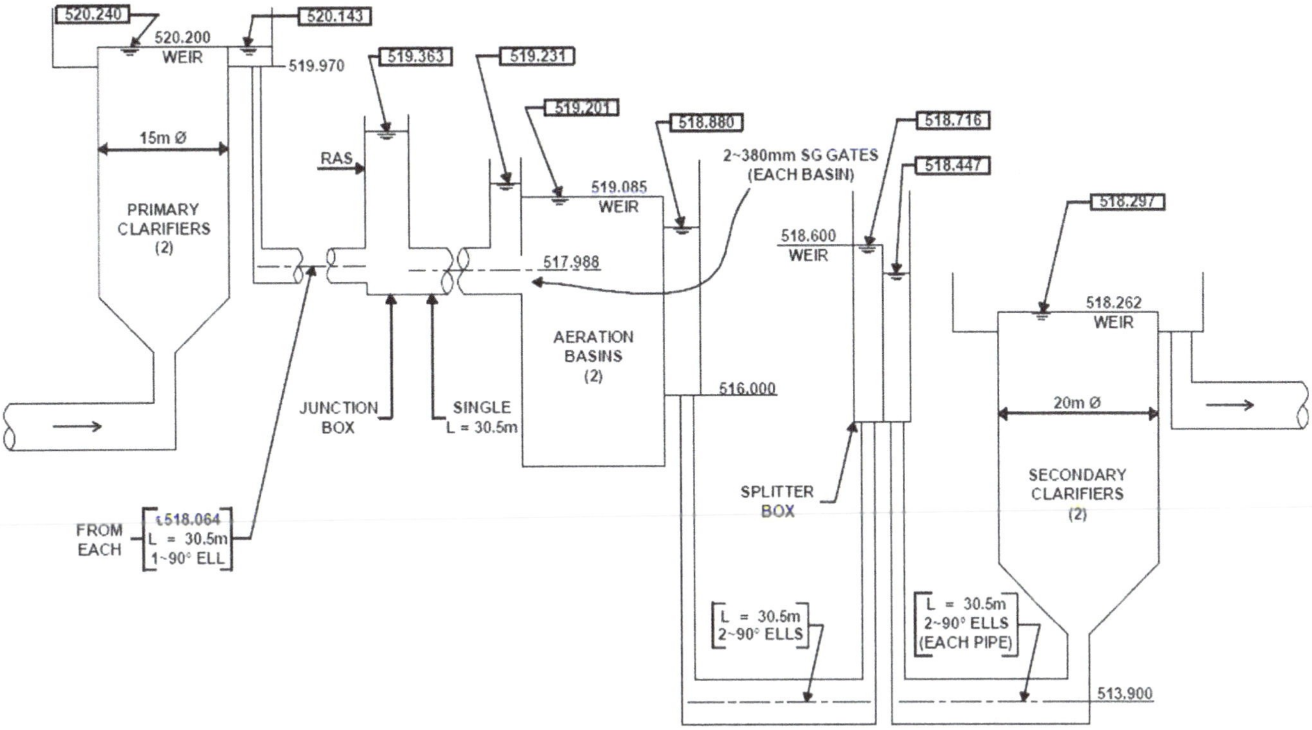

Figure 9.2 Example Hydraulic Profile Corresponding to WRRF Site Layout

In order to evaluate hydraulics, background information needs to be collected first, including basic WRRF site information, hydraulic control elevations, a big-picture overview of the WRRF hydraulics, design flowrates and water balance, initial conduit sizing, and establishment of calculation methods.

3.1 WATER RESOURCE RECOVERY FACILITY SITE INFORMATION

If designing modifications to an existing facility, the first step in developing the hydraulic profile is to obtain all the as-built drawings for the facility, particularly the civil, process, mechanical, electrical, and structural drawings. Key elevations, such as the tops of structures, weirs, and outlet pipe inverts, should be surveyed to verify actual versus plan elevations. Structure settlement over time and differences in survey data (as well as potential changes to weir adjustments over time) are common in older WRRFs, which require reconciliation before starting on the calculations.

For new facility construction, detailed drawings are not initially available, so preliminary sketches of process tank configurations, including rough dimensions and water level control concepts must be used. The design team must collect information on the proposed WRRF site including topography and groundwater levels. A preliminary site layout showing connecting conduits will need to be developed to provide a good starting point.

3.2 HYDRAULIC CONTROL ELEVATIONS

The hydraulic evaluation and development of the hydraulic profile starts with the **control elevation** or water surface elevation target. Typically, the effluent or discharge elevation requirements from the WRRF represent the primary target. In most cases, WRRFs discharge to a receiving water body. If the receiving water body is a river or stream, the controlling elevation is typically the required flood elevation (e.g., the 100-year flood level) as calculated by accepted hydrological methods or as obtained from a governmental entity such as the U.S. Army Corps of Engineers or the Federal Emergency Management Agency. The level of flood protection is established by the regulatory agency that governs the design and operation of wastewater facility.

Control elevation is a key level for water in a hydraulic profile maintained by control devices (such as weirs or flumes) that is often governed by the elevation at which water can safely be released at peak flow conditions.

Climate change may affect controlling elevations in the future, particularly in areas with encroaching water levels or where wet-weather event frequency is significantly different from historical standards. For a storage basin or pond, the controlling water level is the maximum water level in the pond. For a larger body of water, wind forces and seiches should be considered in addition to the highest water surface elevation. For discharge to an ocean or river subject to tidal action, the controlling water level is the high tide elevation based on the selected design storm. For ocean discharge, the designer also must consider the higher specific gravity of seawater (1.025) compared with wastewater (1.00). For every 12 m (40 ft) of depth below the seawater surface, an additional 0.3 m (1 ft) of head must be provided to overcome the heavier weight of seawater. For example, an ocean outfall discharging at a depth of 60 m (200 ft) below the ocean surface would require 1.5 m (5 ft) of head (over and above the headlosses typically encountered in the outfall pipe and diffuser orifices) to overcome the density differential.

Besides the receiving water body, additional control elevations may be important to consider as well. Elevations within an existing WRRF need to be reconciled with new flows or new structures. Topography of the site for a new or greenfield location may be a consideration for hydraulics as well. The hydraulic grade line of the influent stream(s) affects the need for pumping at the WRRF. Once the control elevation or elevations for WRRF hydraulics are established, basic hydraulic principles are applied to evaluate and determine water surface elevations throughout the WRRF.

3.3 OVERVIEW TO HYDRAULIC CALCULATION

Hydraulic calculations begin at the control point—which is typically the most downstream location where the water surface elevation is known. This is where the facility discharges to a receiving body of water, and specific regulatory criteria (such as the 100-year flood elevation) often determine the hydraulic elevation starting point. The calculation then proceeds upstream through all the unit processes to the headworks, accounting for the headlosses of the various components and hydraulic elements. Where there are multiple tanks, the hydraulic profile is calculated along the longest flow path. Although it is commonly assumed that flows to parallel tanks will split equally, this assumption may need to be validated by actual calculations or measurements. Adjustments to the hydraulic design may be needed.

If the hydraulic head at the facility inlet is not sufficient to meet the required head based on the hydraulic profile, revisions to the sizes and elevations of the hydraulic structures may be required. If available (or reasonable) revisions cannot produce the desired results, pumping may need to be considered. Pumping can be located at the facility inlet and/or elsewhere along the flow path through the WRRF. A cost-effectiveness analysis should be performed to determine the appropriate pumping location. From an O&M and reliability perspective, pumping primary or secondary effluent is preferred to pumping raw or screened wastewater.

3.4 FLOWRATES AND WATER BALANCES

Typically, the peak wet-weather flowrate is used for hydraulic design to establish maximum water levels and identify minimum freeboard levels in tanks and channels. The minimum flowrate is used to identify minimum velocities, maximum heights of freefall at weirs and channel outlets, and minimum submergences on equipment. This will identify locations where solids deposition may occur (low velocities) and where odors could be a problem (large freefall distances of raw or primary effluent).

The unit processes should accommodate the peak wet-weather flow (typically peak hourly, but in some cases may be considered maximum month), unless this flow would cause a hydraulic washout of the WRRF. In this situation, the designer should consider the use of equalization or storage basins to minimize any negative effect on the treatment process. In secondary WRRFs, the return activated sludge (RAS), trickling filter or biotower recycle flow, and other types of process recycle flows (which will vary in proportion to the influent flow depending on the process requirements of the facility) must be added to the peak flow.

A flow balance or water balance on unit processes may be useful to help verify flows through different flow elements. For example, activated sludge (aeration basins and clarifiers) will receive a blend of secondary influent flow mixed with RAS flow. The RAS flow may vary between 30% and 200% of the average influent flowrate. This means that the flowrate into and through the activated sludge process may be double or triple the flowrate through upstream processes. Recycle or recirculation flows are common at WRRFs, and evaluation of the balance through each process is important to establishing the design flowrates.

3.5 CONDUIT SIZING AND VELOCITY GUIDELINES

Generally, a minimum velocity of 0.6 to 0.76 m/s (2 to 2.5 ft/sec) at the design average flow for raw wastewater is required to prevent solids deposition in channels and pipelines. At low-flow conditions, however, velocities of 0.3 to 0.45 m/s (1 to 1.5 ft/sec) are needed to transport organic particulate matter. Achieving a particle resuspension velocity of 0.9 to 1.1 m/s (3 to 3.5 ft/sec) on a daily basis can be considered in situations in which it is not feasible to attain minimum flow velocity of 0.3 to 0.46 m/s (1 to 1.5 ft/sec) during low-flow periods. In some cases, the minimum velocity cannot be maintained because of specific process requirements, and consideration should be given to providing access for flushing and solids removal. The velocity in conduits carrying degritted wastewater typically is at least 0.45 m/s (1.5 ft/sec); the velocity in conduits conveying primary effluent typically is at least 0.3 m/s (1 ft/sec). If possible, the velocity should be approximately 0.6 to 0.9 m/s (2 to 3 ft/sec). Typically, conduits between process units are sized based on 1 to 2 m/s (3 to 6 ft/sec). Higher velocities are acceptable to minimize conduit size and cost, but must be balanced against headloss, which can become significant.

Maximum velocities due to gravity flow are not a typical concern at WRRFs, because conduit and channel slopes are flat, and headlosses are minimized. Velocities rarely are high enough to be a concern. For pumping systems, the maximum velocity in suction piping should not exceed 2.4 m/s (8 ft/sec) and keeping velocities below 1.5 to 2 m/s (5 to 6.5 ft/sec) is more typical. The maximum recommended velocity in force mains and pump discharge piping is 3.7 m/s (12 ft/sec) (Jones, 2008). However, to minimize the effects of water hammer and excessive headloss, velocities in discharge piping should be kept below 2 m/s (6.5 ft/sec).

Distribution channels are designed with low velocities (0.3 m/s or less [1 ft/sec]) to minimize headloss and ensure good flow distribution. Where low velocities result in solids deposition, aeration can be provided to keep the channel mixed. Nonaerated (pump) mixing could be considered if the channel is immediately upstream of an anaerobic or anoxic zone.

3.6 CALCULATION METHODS

The following sections describe the principal equations used to calculate the hydraulic profile. This is only intended to be a summary. The designer should consult the references cited previously for additional information and equations.

The **Bernoulli equation** is one of the fundamental equations of fluid flow and is founded on the conservation of energy. The classic Bernoulli equation has been rewritten, with the values of Z representing the invert of the conduit instead of the water surface, and $P_1/g = y_1$ and $P_2/g = y_2$ in:

The Bernoulli equation is the fundamental relationship governing dynamics for an incompressible fluid in a steady flow condition as derived from the conservation of energy.

$$Z_1 + y_1 + v_1^2/2g + H_L = Z_2 + y_2 + v_2^2/2g \quad (9.1)$$

Where Location "1" denoted by subscript 1 = downstream;
Location "2" denoted by subscript 2 = upstream;
Z_1, Z_2 = invert elevation of the channel bottom downstream and upstream, respectively, m (ft);
y_1, y_2 = depth of water in the channel or height of the piezometric surface above the invert of the conduit downstream and upstream, respectively, m (ft);
v_1, v_2 = velocity of flow in the channel downstream and upstream, respectively, m/s (ft/sec);
H_L = headloss between 1 and 2, m (ft); and
g = acceleration caused by gravity, 9.8 m/s² (32.2 ft/sec²).

The relationship between flowrate, velocity, and the cross-sectional area of flow can be expressed as follows:

$$Q = Av \quad (9.2)$$

Where v = velocity of flow in the channel, m/s (ft/sec);
Q = flow in the channel, m³/s (cu ft/sec); and
A = cross-sectional area of flow, m² (sq ft).

The equation for circular pipes flowing full can be expressed as follows:

$$Q = vD^2/C \tag{9.3}$$

Where Q = flow in the pipe, m^3/s (gal/min);
v = velocity of flow in the pipe, m/s (ft/sec);
D = pipe diameter, m (in.); and
C = constant = 1.2732 for SI units (0.4085 for U.S. customary units).

The hydraulic profile and associated headlosses can be calculated manually using the equations presented in this chapter and those found in hydraulics and fluid mechanics texts, such as Mays (1999), Benefield et al. (1984), Bergendahl (2008), Boulos and Nicklow (2005), Davis and Sorensen (1969), King and Brater (1963), and others. Practicing engineers often have access to proprietary software applications that facilitate the evaluation of WRRF hydraulics, provide documentation for the evaluation, and generate graphical outputs for communicating results. A thorough understanding of hydraulic principles and the procedure for the development of hydraulic profiles is important to maximizing the use of hydraulic software applications.

3.7 TOOLS AND ADVANCED HYDRAULIC EVALUATION

The hydraulic evaluation and design for WRRFs may be supported by tools to evaluate complicated flow schemes or to consider a number of flow scenarios efficiently. Tools are often built into digital applications that are available for commercial use. These tools require the user to build the flow structures (basins, biotowers, etc.), add the connectivity between the structures (pipes, channels, etc.), and layer in key control elements (weirs, flumes, etc.). The user must also develop the flow scenarios to be considered (low flow, average flow, peak flow). The tools output static and sometimes dynamic calculations including flow velocities, headlosses, HGL, and energy grade lines. Some examples of tools to consider include Visual Hydraulics (https://www.innovativehydraulics.net/visual-hydraulics-features), Water WorkSuite (https://www.bentley.com/software/hydraulics-and-hydrology/), and FLOW-3D (https://www.flow3d.com/products/flow-3d-hydro/water-treatment/).

The hydraulics within specific structures or processes may require advanced evaluation methods such as computational fluid dynamics in order to design properly. Computational fluid dynamic analysis may be useful or necessary when evaluating: two phase flow with suspended solids (aeration basins, clarifiers), pumping systems, areas of high flow turbulence, aeration systems, and mixing systems. For more discussion on computational fluid dynamics, the reader is referred to outside references (Nopens et al., 2012).

4.0 WATER SURFACE ELEVATIONS, FREEBOARD, AND FREEFALL VERSUS SUBMERGED

Each unit process should be designed hydraulically to prevent liquid from overtopping the walls of structures or tanks under all conditions. The top-of-structure elevation is set so that freeboard is maintained above the high-water elevation. Depending on the expected surface disturbance and relative frequency of the condition, freeboard can range from an extreme low of 150 mm (6 in.) to 1 m (3.3 ft) or more. The former may be acceptable in settling tanks, where the water surface is quiescent; the latter is appropriate in aeration tanks, where gases hold up in the liquid (with potential foaming) and around flow conduit bends and confluences (where surface waves and splashing can occur).

The typical minimum freeboard under maximum water level conditions is approximately 0.3 m (12 in.). Regulatory agencies may have established minimum freeboard requirements. Additional freeboard (or windscreens) may be necessary for locations where high winds would cause surface waves and inhibit scum removal, or where seismic conditions dictate. As a rule, there should be no submergence of control weirs under peak flow conditions. However, some submergence often is permitted when the peak flow condition is judged to be an infrequent occurrence (i.e., extreme peak wet weather), and extraordinary measures would be required to correct the condition. However, in any case, the effect on flow distribution and process performance should be considered when submergence occurs.

In contrast to freeboard, freefall represents a hydraulic drop from a control point, typically a weir. Figure 9.3 shows the difference between freefalling versus submerged flow for a weir. Many different design philosophies exist concerning the freefall allowance at weirs. A designer with a conservative approach may design to allow 80 to 150 mm (3 to 6 in.) of freefall between the weir elevation and the receiving weir trough water surface elevation. Another designer may establish a profile with no freeboard at the weirs, or

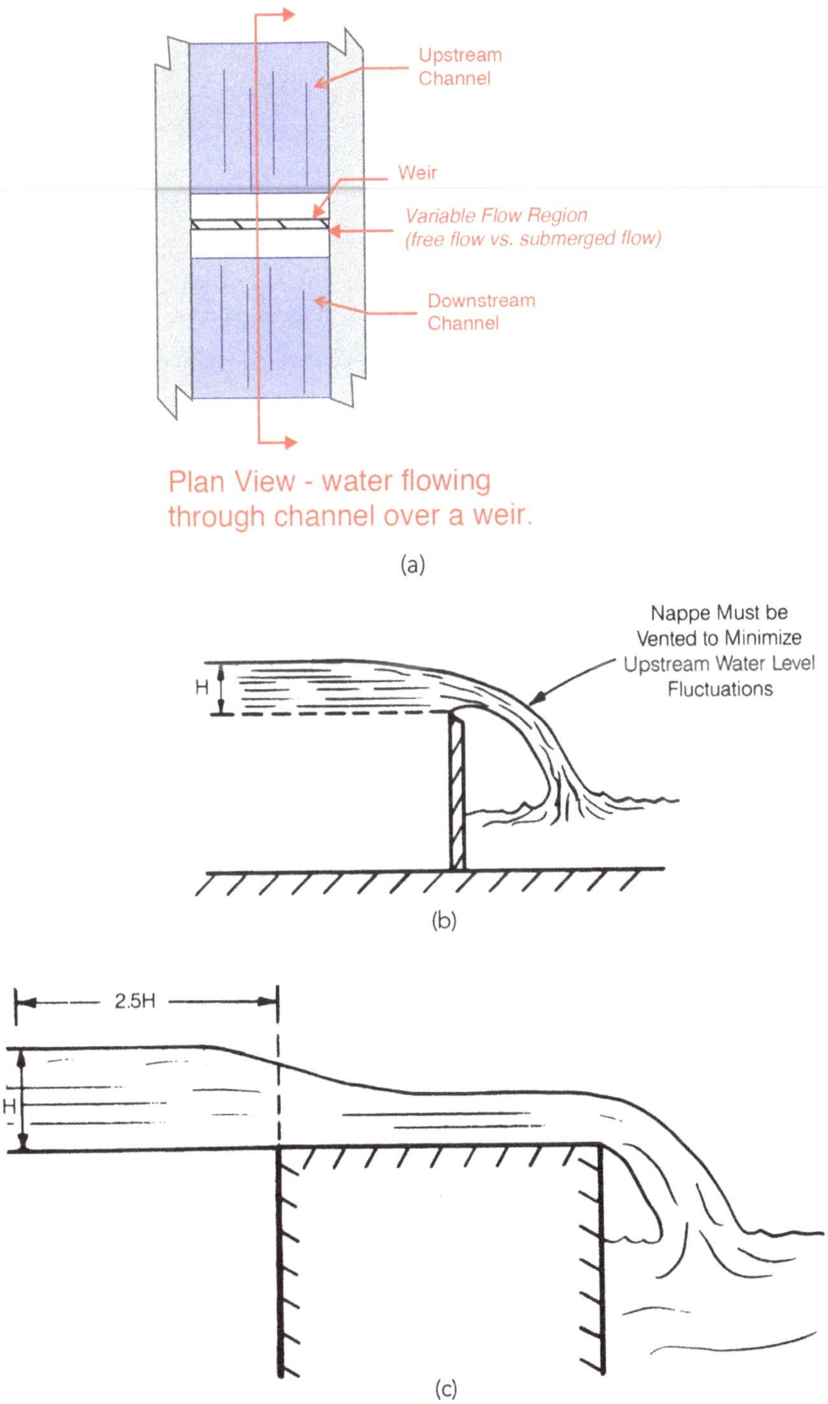

Figure 9.3 Weir Flow Conditions: (a) Plan View—Water Flowing Through Channel Over a Weir; (b) Section View—Free Flow; (c) Section View—Submerged Flow

even allow the weirs to flood at peak flow. These philosophies are established based on the available head in the facility and the cost to pump the liquid. If sufficient head exists without having to pump or construct excessively deep or high structures, a drop of at least 50 mm (2 in.) below the top of the weir (or the bottom of the "V" on a v-notch weir) is recommended. It should be noted, however, that excessive freefall in untreated or primary effluent will release hydrogen sulfide, which is odorous, potentially dangerous, and corrosive. Excessive freefall may entrain dissolved oxygen, generate foam, and/or cause floc breakup at mixed liquor splitting structures. On the other hand, additional freefall is sometimes necessary to accommodate future flows. Special control elements, such as ramps, may be used to minimize odor release, floc breakup, and other problems at freefall locations.

5.0 HYDRAULIC LOSSES

The hydraulic losses (or headlosses) in WRRFs consist of the following:

- Headloss through orifices (perforated baffle walls and gates),
- Head over weirs, and
- Friction and minor losses in channels and conduits.

These can be calculated using the equations found in hydraulic texts, several of which are mentioned at the start of this chapter. This section briefly reviews many of the important headloss calculations used in developing hydraulic profiles for WRRFs.

5.1 CONDUIT OR PIPE FLOW

Two types of flow conditions exist in conduits that connect unit processes: free surface flow, also referred to as open-channel flow, and pressure flow. Headlosses in conduits connecting unit processes consist of friction loss (in straight runs of conduit) and minor losses (in conduit bends and fittings, etc.). Friction losses in the conduit can be determined using the **Darcy-Weisbach equation**, Manning equation, or Hazen-Williams equation. The Darcy-Weisbach equation is applicable to any fluid (water, air, chemical, etc.) and is considered the most accurate. It requires iteration to solve, but most programs or apps support this with no difficulty. The Manning equation, commonly used in open channels, is applicable to full flowing conduits as well but is less commonly used for these, while the Hazen-Williams equation is used only for full-flow conduits.

The Darcy-Weisbach equation is an empirical equation that is used to determine headloss, or pressure drop, due to friction in a full flowing conduit or pipe resulting from the average velocity of fluid flow for an incompressible fluid.

5.2 PRESSURE FLOW

These equations and the friction characteristics for the various conduit materials can be found in hydraulics and fluid mechanics texts (Benefield et al., 1984; King & Brater, 1963; American Society of Civil Engineers [ASCE] & WEF, 2007).

Darcy-Weisbach equation:

$$H_L = f(L/D)(v^2/2g) \tag{9.4}$$

Where H_L = headloss, m (ft);
f = friction factor, which is based the conduit material and Reynold's number;
L = conduit length, m (ft);
D = conduit diameter, m (ft);
v = velocity of flow, m/s (ft/sec); and
g = acceleration caused by gravity, 9.81 m/s^2 (32.2 ft/sec^2).

The value for f is determined from the "Moody diagram," found in the texts referenced earlier and depends on the relative roughness of the conduit wall, e/D, and the Reynold's number, where e = roughness of the conduit wall, m (ft).

$$R_N = v{\cdot}D/n \tag{9.5}$$

R_N = Reynold's number (dimensionless) and
n = kinematic viscosity of the fluid, m^2/s (sq ft/sec).

Values for e can be found in the above-referenced texts. For noncircular conduits, the equivalent D can be determined as follows:

$$D \text{ equivalent} = 4R \tag{9.6}$$

Where R = hydraulic radius = A/P, m (ft);
A = cross-sectional area of flow m^2 (sq ft); and
P = wetted perimeter m (ft).

Manning equation:

$$Q = (k_m/n)\, AR^{0.67}S^{0.5} \tag{9.7}$$

Where Q = flowrate, m^3/s (cu ft/sec);
k_m = 1 (1.486 for U.S. customary units);
A = area of flow, m^2 (sq ft);
S = slope of the energy grade line, calculated as the vertical drop, m (ft) divided by the horizontal distance, m (ft), and sometimes called the friction slope S_F, m/m (ft/ft) or friction loss, m/m (ft/ft); and
n = Manning's roughness factor, which can be found in the referenced texts depending on conduit material; e.g., concrete = 0.013.

Hazen-Williams equation:

$$Q = K_{HW}CAR^{0.63}S^{0.54} \tag{9.8}$$

Where K_{HW} = 0.849 (1.318 for U.S. customary units); and
C = Hazen-Williams roughness coefficient.

Note that C can vary widely depending on the type of pipe material; and a value of 100 is often a default for concrete or ductile iron pipe. However, the value depends on the age and condition of the pipe. Also, a C value as high as 140 to 150 is common for new smooth wall polyvinyl chloride pipes. Other parameters are as defined above.

Minor losses reflect headlosses that occur because of valves, bends, other types of fittings, and entrances and exits from tanks to pipes. Minor losses can be determined in the hydraulic calculations using either the "K-factor" method or the "equivalent length" method. When evaluating minor losses with equivalent length method, the headloss is calculated as shown in Equation 9.9. When evaluating with the K-factor method, use the following:

$$H_L = K_{hl} \cdot v^2/2g \tag{9.9}$$

Where H_L = headloss resulting from the fitting, etc., m (ft);
K_{hl} = headloss factor, dimensionless;
v = velocity in the conduit, m/s (ft/sec); and
g = acceleration caused by gravity, 9.8 m/s^2 (32.2 ft/sec^2).

The equivalent length method converts the headloss to an equivalent length of pipe of the same diameter as the fitting or valve. Values for K and equivalent pipe length can be found in hydraulics texts cited above. The K-factor method is most common in practice.

5.3 OPEN CHANNEL FLOW

Open channel, or free surface, flow can be either steady or unsteady flow. In steady flow, the flowrate is constant with time and distance along the channel. This is the typical case. Unsteady flow (flow changes with distance or time) occurs in weir troughs and collection and distribution channels and will be discussed later in this chapter. In addition, steady flow can be uniform or nonuniform. In nonuniform flow, the depth of flow or the width of the conduit, or both, varies along the length of the channel. This is relatively common in WRRFs.

The calculation of the water surface profile in open channels uses the Manning equation (9.10) in combination with the Bernoulli equation, previously presented as Equation 9.1. In equation (9.10), headloss is as follows:

$$H_L = S_{ave} \cdot L \tag{9.10}$$

Where H_L = headloss between downstream and upstream locations, m (ft);
$S_{ave} = (S_{F,1} + S_{F,2})/2$ = average friction slope between downstream and upstream, m/m, (ft/ft);
$S_{F,1}$, $S_{F,2}$ = friction slope in the channel downstream and upstream, respectively, m/m (ft/ft); and
L = distance between downstream and upstream sections, m (ft).

The parameters $S_{F,1}$ and $S_{F,2}$ typically are determined using the Manning equation (9.7) based on the depth of flow downstream and upstream, respectively. For open-channel flow in WRRFs, typically the downstream values are known, and the upstream values are unknown and are to be determined. This is an iterative process in nonuniform flow, as the value for $S_{F,1}$ depends on the upstream depth. If there are any minor losses, they should be included. Minor losses can be determined using a K_{hl}-value approach similar to pipes.

When beginning to calculate the water depth in an open channel, it is very important to calculate the critical depth. Critical flow and critical depth are determined by the Froude relationship, and critical depth becomes important at flow constrictions. The critical depth, D_c, for rectangular channels is determined using Equation 9.11.

$$D_c = (Q^2/w^2 \cdot g)^{0.33} \tag{9.11}$$

Where D_c = critical depth, m (ft);
Q = flowrate, m³/s (cu ft/sec);
w = channel width, m (ft); and
g = acceleration caused by gravity, 9.8 m/s² (32.2 ft/sec²).

Critical depth for nonrectangular channels can be determined using methods described in King and Brater (1963). If the depth of water downstream is greater than the critical depth, the water depth in the channel matches the downstream depth. If the depth of water downstream is less than the critical depth, a constriction is occurring, and the downstream depth at the constriction will be the critical depth.

Aerated channels represent a special application whereby air is applied to maintain suspension of solids through the channel. For an unaerated channel, the Manning roughness coefficient used for formed concrete channels ranges from 0.013 to 0.015; however, the friction loss in aerated distribution channels is greater than that in unaerated channels due to gas holdup. The suggested value of Manning's n for aerated channels with a velocity of 0.4 m/s (1.3 ft/sec) is 0.035, and for a velocity of 0.3 m/s (0.9 ft/sec) is 0.0425 (Townsend, 1935).

5.4 ORIFICE

Headlosses through a submerged-control gate or an inlet gate to a treatment process, openings in baffle walls, submerged launders, perforated pipes, and filter underdrains typically are calculated using the orifice equation:

$$H_L = (Q/CA)^2/2g \quad (9.12)$$

Where H_L = headloss through the gate or orifice opening, m (ft);
Q = flow, m³/s (cu ft/sec);
C = gate or orifice coefficient, dimensionless;
A_o = area of gate or port opening, m² (sq ft); and
g = acceleration caused by gravity, 9.8 m/s² (32.2 ft/sec²).

The headloss is the difference in the water level between the upstream and downstream sides of the gate or orifice (Figure 9.4). Values of the gate, or orifice, coefficient for calculating headloss through ports under varying conditions can be found in Davis and Sorensen (1969), King and Brater (1963), and others. The value for C can vary, but typically is 0.6 for a sharp-edged opening and 0.9 for a rounded opening; a C value of 0.6 is used to be conservative.

The orifice equation, as defined in Equation 9.12, is applicable only when the gate or opening is completely submerged. When the gate or opening is submerged on the upstream side and the gate discharges free on the downstream side, the head is measured vertically from the centerline of the opening to the upstream water surface. If the opening is not fully submerged, the water surface upstream should be calculated as a submerged weir (King & Brater, 1963).

5.5 WEIRS

Although weirs sometimes are used for flow measurement in WRRFs, they commonly serve as control devices to maintain a required water level in a unit process. Weirs are classified based on the weir shape—rectangular, v-notch, trapezoidal, and proportional. Proportional weirs, sometimes called Sutro weirs, have the discharge head varying linearly with flow; these are used in grit channels to maintain constant velocity with depth. The upper edge of the weir plate (in the case of a rectangular weir) and the bottom of the notch in a v-notch or trapezoidal weir is the crest of the weir. The depth of the water over the crest is the head, which is measured some distance upstream of the weir (Figure 9.3). This distance can be as much as four to five times the head upstream (Ackers et al., 1978).

A weir is a baffle over which water flows.

Rectangular weirs are classified as either sharp-crested or broad-crested, as shown in Figure 9.5. Weirs also can be submerged or unsubmerged. Weirs typically are designed to be unsubmerged (i.e., not affected by downstream conditions).

Head over a sharp-crested rectangular weir under free-flow conditions can be calculated using the following equation:

$$H_w = [Q/(C_w L)]^{0.67} \quad (9.13)$$

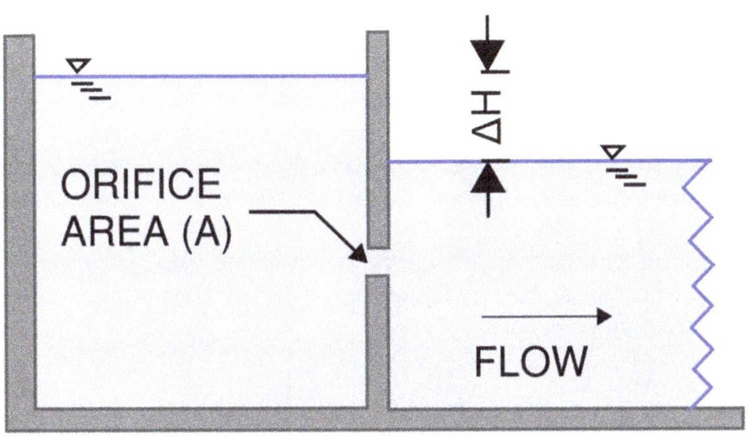

Figure 9.4 Orifice Loss

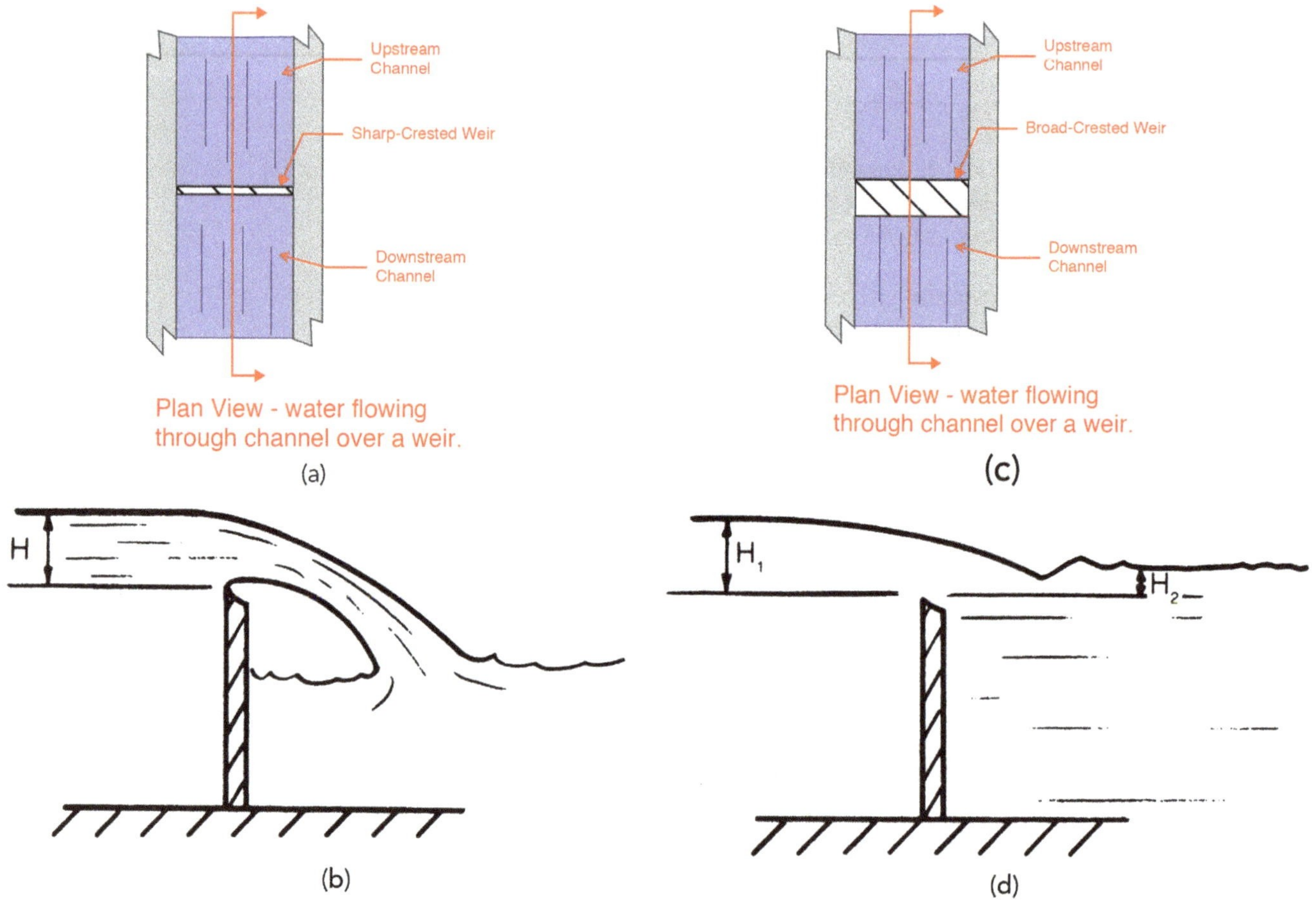

Figure 9.5 Weir Sections: (a) Plan View—Sharp Crested Weir; (b) Section View—Sharp Crested Weir; (c) Plan View—Broad Crested Weir; (d) Section View—Broad Crested Weir

Where Q = flow, m^3/s (cu ft/sec);
L = length of weir, m (ft);
H_w = head over the crest, m (ft); and
C_w = weir coefficient.

The weir coefficient accounts for the approach velocity, in terms of the ratio of the weir plate depth to the head over the crest. The values range widely depending on the approach velocity. The most commonly used value, 1.82 for SI units (3.3 for U.S. customary units), accurately represents deep-water conditions upstream, such as that found in basins, with an extremely low approach velocity toward a sharp-crested rectangular weir.

Hydraulic reference books often describe a rectangular weir with "end contractions." End contractions occur when the weir plate covers only a portion of the end of a channel. An adjustment is made to reduce the weir crest length, L, by $0.2H$, to account for the end contractions. This has the effect of increasing H slightly. In practice, this often is ignored, because the values of H typically are small compared with L.

V-notch weir angles range from 22.5° to 120°, with the 90° v-notch being the most common. Under free-flow (unsubmerged) conditions, as shown in Figure 9.6, the head over a sharp-crested v-notch weir can be calculated using the following equation:

$$H_w = [Q/(C_w \tan \phi_{vw}/2)]^{0.4} \qquad (9.14)$$

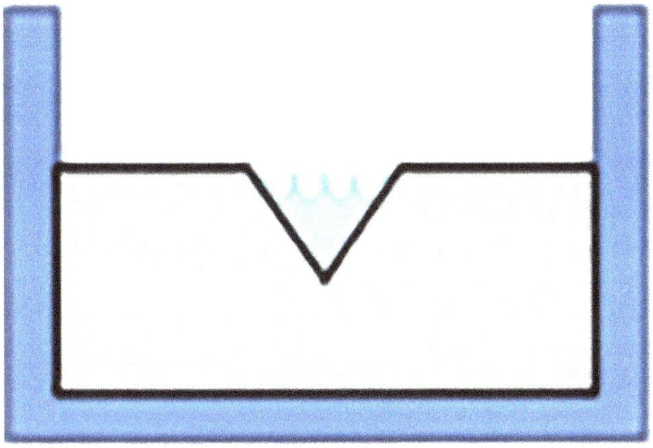

Figure 9.6 Sharp-Crested, V-Notch Weir (Grant, 1989) (Reprinted with permission from Teledyne ISCO. Teledyne ISCO. Isco Open Channel Flow Measurement Handbook, 3rd edition)

Where Q = flow, m^3/s (cu ft/sec);
ϕ_{vw} = angle of the notch, degrees;
H_w = head over the crest, m (ft); and
C_w = weir coefficient (1.38 for SI units [2.5 for U.S. customary units] for a 90° weir).

Equations 9.13 and 9.14 apply under free-flow conditions. When the downstream water surface elevation rises above the weir crest, the weir becomes submerged, as shown in Figure 9.3. Submerged weirs are not used for flow measurement because the weir equations do not directly apply to submerged conditions and thus produce sizable errors. The head over the crest can be calculated from Equations 9.13 and 9.14; however, the discharge flow, Q, must be corrected by using curves that have been developed experimentally (Chin, 2012; King & Brater, 1963; Street et al., 1996). The head upstream of a submerged weir is greater than for a free discharge condition.

5.6 LAUNDERS AND TROUGHS

Effluent **launders** typically are found in settling tanks and filters. Launders can convey water from freefalling weirs, where the flow passes over a weir and falls into the launder or trough. With free-flowing launders, water level in the tank is controlled by the weir. Alternatively, launders can be submerged, where there is no freefall across a weir and the tank water level is controlled by downstream constrictions.

A launder is a trough used to transport water.

When water flows through launders, or channels, it collects and accumulates along the length as the flow moves downstream toward the outlet end. Conveying flow along the launders requires head to overcome the friction loss along the channel and the exchange of momentum as the water falls perpendicular to the flow stream and accelerates along the direction of flow. The side overflow formula, described in detail in *Open-Channel Hydraulics* (Chow, 1959), is a differential equation for flow with increasing or accumulating discharge. Most designers prefer to use a computer program to solve this equation. Due to the low average value of the Reynolds number along the entire length of the launder, it is recommended to use a higher-than-normal friction factor (e.g., 50% higher). Because of the high degree of mixing caused by the falling flow and the corresponding distortion of the velocity profile, the velocity head value, $V^2/2g$, can be higher by approximately 1.3 times.

A free discharge occurs if the water level downstream of the end of the launder is less than the sum of the invert of the launder or trough plus critical depth. Critical depth is determined using Equation 9.11 for a rectangular trough. If this condition is not true, the trough is submerged. If the discharge from the trough or launder is "free," then the maximum water level in the trough at the upstream end is as follows:

$$y_{up} = 1.73 \cdot D_c \tag{9.15}$$

Where y_{up} = maximum water depth in trough, m (ft); and
D_c = critical depth determined from Equation 9.11 if rectangular, m (ft).

In the above equation, it is assumed that the trough or launder is level (typical), and friction loss is ignored (reasonable, as the troughs typically are short). If the trough or launder is not rectangular, or the discharge is not free, the designer should consult the reference texts.

Submerged launders may be used in primary sedimentation tanks in lieu of freefall launders to minimize odors. Primary effluent falling into a conventional launder strips hydrogen sulfide and other volatile, odorous compounds. Using a submerged launder eliminates the freefall and associated odors; however, they are a bit more complex. Headloss is determined with the orifice equation (9.12), based on the flow rate entering the submerged launder. Depending on the configuration, there could be an additional fitting loss, which would be determined using Equation 9.9. With submerged launders, the water level in the settling tank is ultimately controlled by a level-control valve; this is typically a butterfly valve or butterfly gate and a level control loop sensing the water level in the settling tank.

At facilities with large peak flows, a combination of submerged orifices and v-notches may be appropriate. When peak flow occurs, the tank level rises to overflow the weirs, thus avoiding the hydraulic restrictions through the orifices.

6.0 HYDRAULIC CONTROLS

Whereas the previous sections presented the fundamental hydraulic equations typically used in the calculation of the hydraulic profile, this section discuss specific hydraulic elements used to control flow at the WRRF.

6.1 FLOW DISTRIBUTION

To ensure proper treatment, it is essential to achieve equal flow distribution to each of the basins that make up a unit process. This requires splitting the total facility flow proportionally to the capacity of each basin or reactor. This can be accomplished through a wide variety of means.

Distribution boxes, channels, and header pipes are used for this purpose. Although modulating ports, weirs, gates, or valves can provide equal distribution, they require control systems that share the disadvantages inherent to any electrical and mechanical system (i.e., failure and high maintenance). Wherever possible, static devices such as fixed weirs, gates, and cut-throat flumes are preferred by facility staff for the distribution of flow rather than constantly modulating devices.

6.2 FLOW-SPLITTING STRUCTURES

The use of symmetry alone does not always ensure equal flow distribution, because small differences in headloss in the flow paths to the process units results in large differences in flow. A lack of symmetry may be caused by slight construction differences or structure settlement. Because of the long weirs associated with circular clarifiers, even slight differential settling of the clarifier structure impairs flow distribution and makes a large difference in flow entering the tank or clarifier.

Weirs provide good flow distribution without the need for mechanical equipment, such as rate-controlling valves, and are less subject to clogging. As a result, they generally provide the best technical solution for flow distribution or flow splitting. Where large flows are to be split, for example, those greater than 2 to 3 m^3/s (50 to 75 mgd), fixed weir splitting structures can become large and complex. Often a number of weir troughs, similar to those found in rectangular clarifiers, are used to achieve sufficient weir length.

A flow-splitting structure containing weirs is constructed at a location where the flows are to be split. Figure 9.7 illustrates this type of layout for two clarifiers. Figure 9.8 illustrates an example of flow control structures for facilities with asymmetrical layouts.

In the design of the splitting structure, the water surface upstream of the splitting weirs should be as quiescent as possible. One way to achieve this is to construct the influent pipe to enter vertically in the bottom. The "rise rate" or vertical velocity in the chamber upstream of the splitting weirs should be 0.3 m/s (1 ft/sec) or less, if possible. If heavy solids are being transported in the liquid (i.e., raw wastewater or raw or digested sludge), higher velocities may be required to prevent deposition, but this creates more turbulence, and flow splits may not be optimum. Design engineers should consider using baffles to improve flow distribution.

The top of the conduits leaving the splitting structure must be placed well below the water surface in the downstream chamber of the splitting box to ensure submergence at all times and avoid vortexing, which entrains air. This air accumulates in the pipe and causes backflushes, and is carried into the downstream process unit and causes surging.

Sizing the splitting weirs is based on Equation 9.13, assuming rectangular weirs are used. If there is a desire to have an equal split, all weir lengths must be the same. If all the downstream process units do not have identical capacities, then the weir lengths are proportionately adjusted. Ideally, there should be a freefall of approximately 75 to 150 mm (3 to 6 in.) between the weir crest and the downstream water surface elevation. For large or complex flow distribution structures, the use of three-dimensional computational fluid dynamics (CFD) modeling can be used to optimize the design.

6.3 DISTRIBUTION MANIFOLDS AND CHANNELS

An inlet pipe header or distribution channel also can be used to distribute flow to a unit process. However, in such cases, the inlet ports or gates to each basin should be designed with adequate headloss to ensure good distribution. As with flow distribution, equal distribution of solids to the treatment processes should be maintained. Unless provided for in the design, the equal distribution of solids to the treatment units may not occur coincidentally with the equal distribution of flow. This is especially common where flow is distributed by use of channels. Where channels are used, such as upstream of grit basins, the wastewater flow should be mixed well, such as with aeration, to ensure that the solids distribute evenly with the flow.

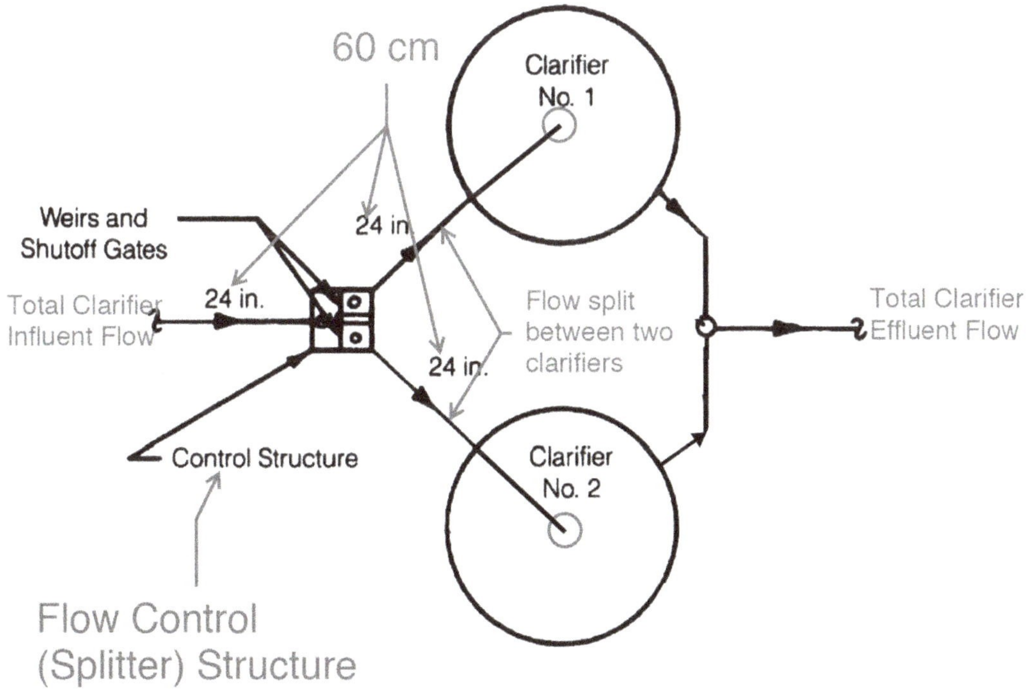

Figure 9.7 Flow Splitter for Symmetrical Layouts

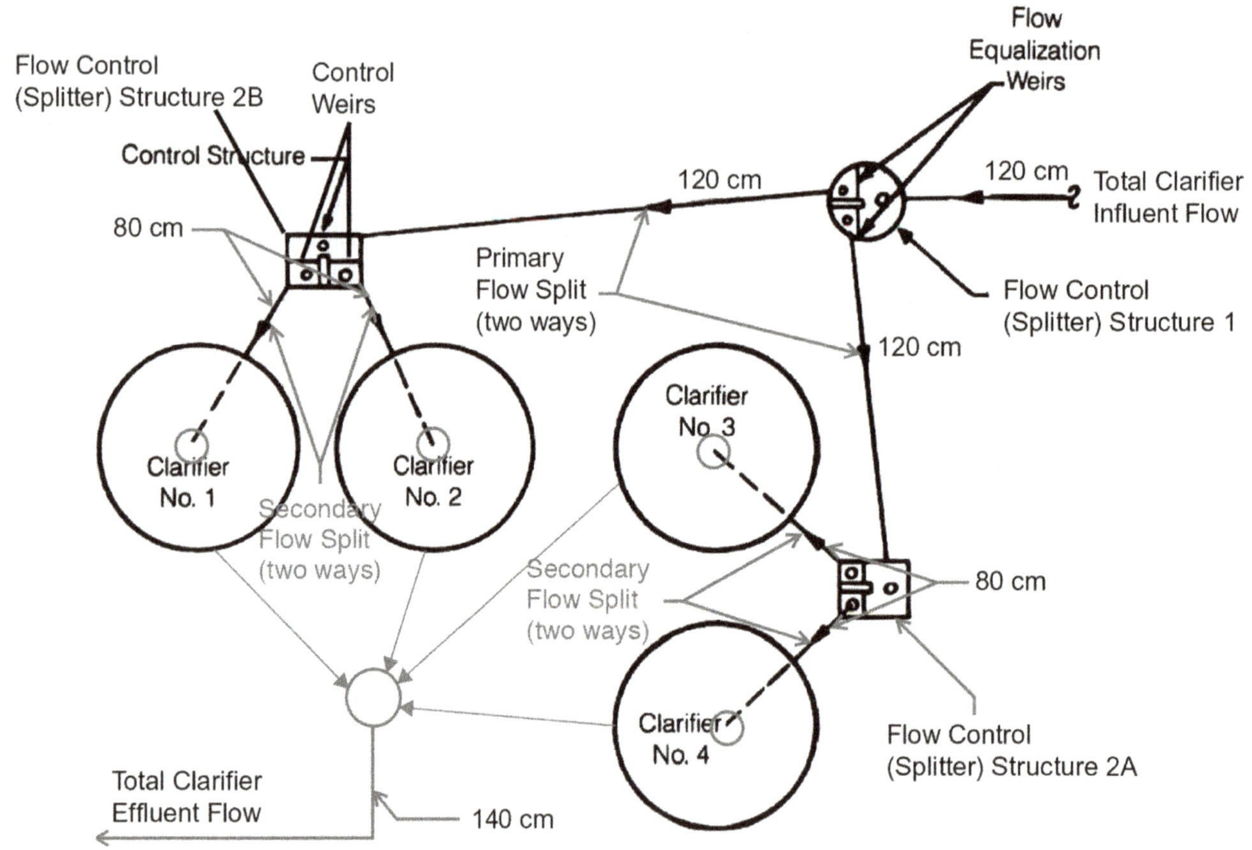

Figure 9.8 Flow Control for Asymmetrical Layouts

Ports or gates in the distribution channel or pipe header are sometimes used to equally distribute flow from or to unit processes. To split flow equally, the headloss through inlet openings must be much larger than the total headloss in the distribution channel or header pipe. Equation 9.16 (Camp & Graber, 1968; Fair et al., 1968) presents the hydraulic relationship between the friction loss in the header pipe or distributing channel and the outlet gates or ports. The equation assumes that the outflow occurs through numerous equally spaced ports or gates.

$$h_L = \Delta h/(1\text{-m}^2) \tag{9.16}$$

Where h_L = headloss through port, m (ft);

Δh = hydraulic grade differential along the distribution channel or pipe, m (ft); and

m = ratio of the lowest flow and the highest flow through the respective ports.

Equation 9.16 is used to determine the headloss that must be induced by the inlet gates or ports to ensure uniform distribution to all basins from a common inlet header or distribution channel. For example, to keep the flow to each basin within 5% of the other basins (m = 0.95), the headloss through the gate into the basin should be approximately 10 times the water level difference over the entire length of the distribution channel or pipe header.

When the outlet flow from a pressurized header approximates a uniform continuous outflow along the length of the header, the headloss along the header can be estimated by calculating the headloss as if the inlet flow to the header were to be conveyed along the entire length of the header and dividing the result by a factor of three. This estimation procedure is valid for m (ratio of the lowest flow and the highest flow through the respective ports) greater than 0.9 in Equation 9.16.

Ideally, the header or distributing channel must be made large enough so that the headloss from one end to the other is minimal. This can be accomplished by using a low velocity of less than 0.3 m/s (1 ft/sec); for very long channels, velocities less than 0.15 m/s (0.5 ft/sec) may be appropriate.

The distribution channel formula, described in Chow (1959), is a differential equation for flow with decreasing (or increasing) discharge. This formula can be used to analyze the hydraulic characteristics of a channel distribution header. However, most designers prefer to use a computer program to solve this equation.

Distribution channels may trap scum unless the discharge to the reactors or clarifiers is free and open. Installing a downward-opening scum gate at the end of the distribution channel facilitates scum removal. The use of weirs or cut-throat flumes minimizes scum accumulation. Flume fabricators make this as a standard product.

6.4 JUNCTIONS AND CONFLUENCES

Where the flow in channels and pipes join together, turbulence is created, and there is some loss of energy and consequently loss of head. A number of procedures can be used to determine the headloss resulting from converging flows. The simplest, though not necessarily the most accurate, is to estimate the loss using the K-factor Equation 9.9 with an estimate of K. Benefield et al. (1984) suggested a K of 1.8 for a 90° turn in a tee and 0.6 for the straight run of a tee. Montgomery (1985) provided charts to determine the headloss through junctions. A more accurate approach uses a pressure and momentum methodology, such as that developed by the City of Los Angeles, California (1968).

$$\Delta y = (A_1 + A_2)/2 = M_2 - M_1 = M_3 \cdot \cos\Phi \tag{9.17}$$

Where Δy = change in water level in junction, m (ft);
$\Delta y = z + D_1 - D_2$, m (ft);
z = change in invert elevation in confluence, m (ft);
D_1, D_2 = depth of flow or distance to HGL upstream and downstream respectively, m(ft);
A_1, A_2, A_3 = area of flow, m^2 (sq ft);
$M_1 = Q_1^2/A_1 \cdot g$, m^3 (cu ft);
$M_2 = Q_2^2/A_2 \cdot g$, m^3 (cu ft);
$M_3 = Q_3^2/A_3 \cdot g$, m^3 (cu ft);
Φc = confluence angle, degrees; and
g = acceleration caused by gravity, 9.81 m/s^2 (32.2 ft/sec^2).

6.5 FLOW AND PRESSURE CONTROL VALVES

Headloss through valves of any type can be determined by using K-factors and Equation 9.9 or by using the control valve equation presented below (Equation 9.18), which is more accurate. Care must be taken in the sizing of flow- and pressure-control valves. They must be sized so that the headloss through the valve is significant and the control valve operates within an acceptable opening range—not too close to its seat under minimum flow (maximum pressure) or too wide open under maximum flow (minimum pressure). Control valve sizing and headloss calculations can be determined using the flow coefficient K_v.

$$K_v = Q \cdot [1/\Delta p]^{0.5} \tag{9.18}$$

Where K_v = valve coefficient;
Q = flowrate, m^3/h (gal/min); and
Δp = pressure drop through the valve, bar (psi).

6.6 BAFFLES

Baffles are used to separate zones in unit processes or tanks. Baffles in unit process tanks can (1) be over–under, (2) be around the end, (3) contain openings or perforations across all or a portion of the baffle, or (4) be a combination.

For over–under baffles, the headloss for "over" baffle flow can be modeled or calculated using the submerged weir equations (King & Brater, 1963) and the headloss for "under" baffle flow can be calculated using the orifice equation, whereby the area is the total area of flow under the baffle.

For around-the-end baffles, the flow takes a 180° turn around a bend. This can be calculated as minor loss. The value for K varies from 2 to 3.5 (American Water Works Association, 2012). Values of 3.2 and 3.3 have been used successfully in practice in the design of baffled channel flocculators (Reichenberger, 1984). In perforated baffles, the perforations can be considered orifices.

Where over–under baffles are used in aeration basins, flow over the top is needed to avoid trapping scum. The over-the-top flow depth is typically 25 mm (1 in.). This occurs simultaneously with the flow under the baffle or through submerged openings in the baffle. Headlosses over and under (or through) the baffle are equal. Losses are calculated by simultaneously solving equations for a submerged weir and a submerged port or gate.

6.7 FLOW MEASUREMENT

Flow measurement can be applied to open channels and full-flowing pipes as previously introduced in Chapter 1. Although there are several types of flumes that can be used for open-channel flow measurement in WRRFs, the Parshall flume is the most common. Parshall flume head calculation equations can be found in a number of publications (e.g., Benefield et al., 1984; U.S. Department of the Interior, Bureau of Reclamation, 2001). The key to the design of Parshall flumes is to select the minimum throat width that provides the capacity range needed for the anticipated flow.

Submergence is the ratio of the downstream depth to the upstream depth, both relative to the flume's upstream invert elevation.

Most flumes have a limit on the amount of **submergence** that can be tolerated before metering accuracy deteriorates. Generally, the submergence should be less than 0.6 for throat widths less than 0.3 m (12 in.) and less than 0.7 for larger throats (Benefield et al., 1984). In setting the flume elevation, the designer should check the flume submergence under minimum, average, and peak flow conditions to ensure that the submergence is within the acceptable limits for all conditions. Although the flume may have satisfactory submergence under peak flow conditions, submergence may be problematic at minimum flow. To ensure metering accuracy, the approach to the flume should be as straight as possible, and the velocity distribution should be uniform.

Full-flow pipe meters include magnetic meters and sonic meters. Hydraulic considerations in full-flow pipes require the avoidance of upstream and downstream disturbances. These meters require a uniform velocity distribution across the entire cross-section in the upstream or approach piping. Valves and fittings immediately upstream of the meter are to be avoided. Typically, a straight-run pipe length of 10 diameters is required upstream and 5 diameters is required downstream. Flow conditioners may be used to reduce the required length of the straight run.

7.0 UNIT PROCESS HYDRAULICS

This section discusses hydraulics and flow considerations for unit processes found at WRRFs. The hydraulic designer must understand how the liquid flows through each unit process and the water depths that are required for each process. Within each unit process, devices are used to distribute flow, maintain a certain water depth, and control the flow. Typical devices include shutoff gates, weir gates, valves, ports, weirs, baffles, orifices, launders, and underdrains. Each of these devices imposes a headloss on the system that must be considered in the hydraulic calculations. The headloss through typical unit processes is shown in Table 9.1. Each unit process, the flow control devices for that process, and the interconnecting piping must be evaluated. Headloss calculations for unit processes should be evaluated for a number of conditions including low flow, flow peaking, and when units are offline. The following subsections outline fundamental hydraulic concepts and present some of the key points of consideration for many of the unit processes typically found in WRRFs.

7.1 SCREENS

Wastewater flowing through the screen creates headloss (Figure 9.9). The headloss depends on the type of screen (coarse bar rack, bar screen, or fine screen). For manufactured screens, such as comminutors,

Table 9.1 Typical WRRF Unit Process Headloss

Unit Process	Typical Headloss, mm (ft)
Bar screens or influent screening[a]	300–600 (1–2)
Vortex or aerated grit removal	300–750 (1–2.5)
Activated sludge reactor	450–900 (1.5–3)
Primary, secondary, or tertiary sedimentation	600–1200 (2–4)
Disinfection contact tank	450–900 (1.5–3)
Tertiary filtration[b]	1000–3000 (3–10)
Conduits between unit processes, including entrance and exit losses	150–1050 (0.5–3.5)
Flow distribution boxes	300–750 (1–2.5)

[a]Allows for partial clogging.
[b]Depends on whether low head type or conventional filter is used.

grinders, and rotary drum screens, the headloss can be obtained from the manufacturers. However, the designer is cautioned that the headloss provided may be the "clean screen" headloss. Headloss increases as the screen becomes partially clogged by debris or screenings. Mechanical screens use various types of motorized mechanisms to remove accumulated screenings (Figure 9.10). Allowances should be made for a partially clogged screen, because mechanical cleaning typically is intermittent. An allowance of 150 mm (6 in.) or more above the "clean screen headloss" typically is added.

Headloss through screens typically ranges from 50 to 635 mm (2 to 24 in.), depending on flow and screen opening size. Maximum headloss can reach 1 m (3 ft) or more in large units at peak flow with fine screens or when screen cleaning frequency is inadequate. The design engineer must take steps to ensure that headloss through the screening device at maximum flows does not surcharge influent sewers enough to impair upstream services. The headloss through coarse bar racks and manually and mechanically cleaned bar screens can be determined using the peak flow and the following equation (Metcalf & Eddy, Inc./AECOM, 2014):

$$h_L = k(v^2 - v_u^2)/2g \quad (9.19)$$

Where h_L = headloss through the screen, m (ft);
v = velocity through bar screen (i.e., between the bars), m/s (ft/sec);
v_u = velocity upstream of bar screen or approach velocity, m/s (ft/sec);
g = acceleration caused by gravity, 9.8 m/s^2 (32.2 ft/sec^2); and
k = headloss coefficient.

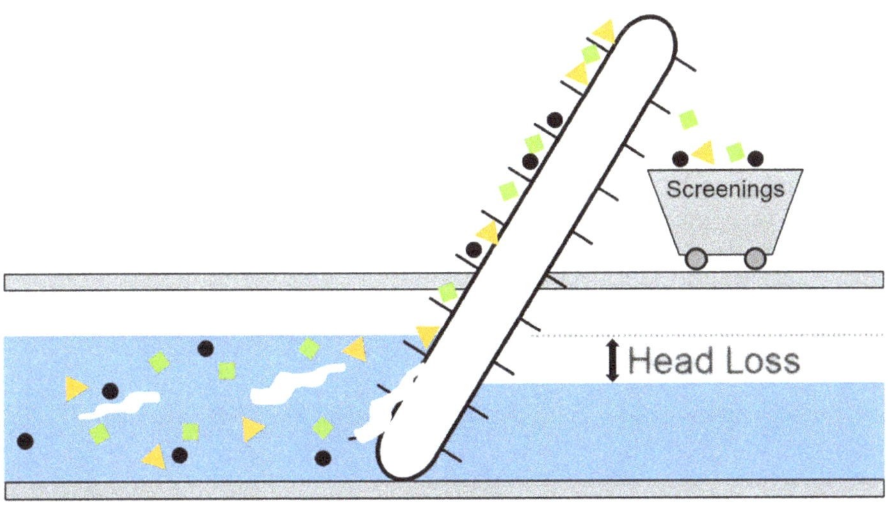

Figure 9.9 Headloss Across a Screen (Reprinted with permission from Indigo Water Group)

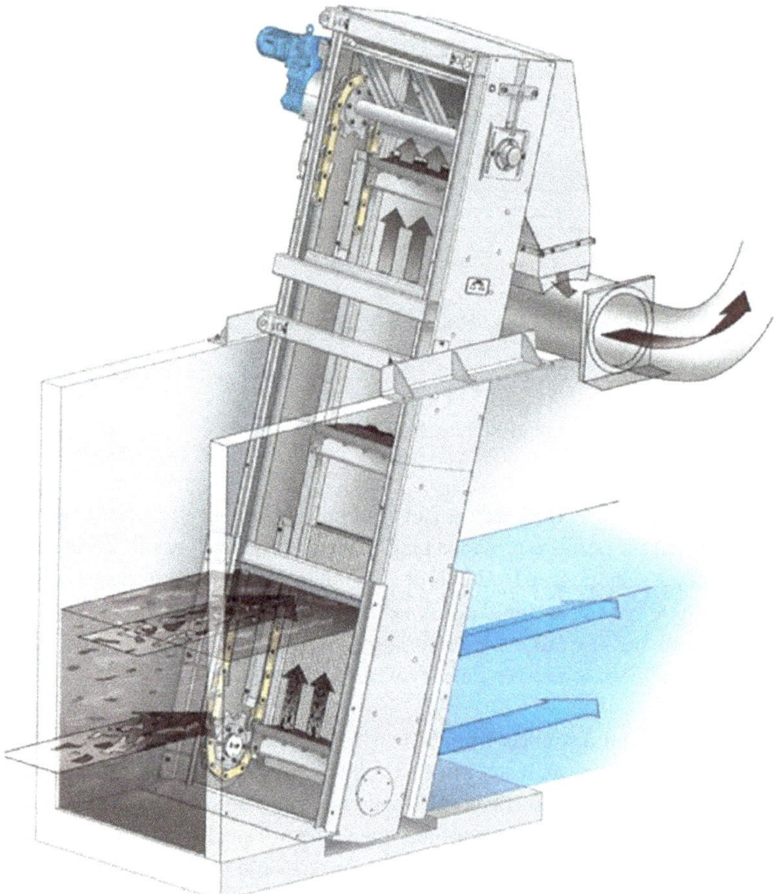

Figure 9.10 Front-Cleaning Screen (Courtesy of Huber Technology, Inc.)

The headloss coefficient k is a function of opening size and bar shape and can vary from 3.5 for a teardrop bar with 6 mm opening, to 1.4 for a trapezoidal bar with a 19 mm opening, clean. Historically, a value of 1.4 has been used for coarse bar screens.

This equation is recommended for use in preliminary stages of the project, when the screening components have not been fully selected. For more accurate results, however, Kirschmer's Equation 9.20 is recommended. This equation takes into consideration the angle of the screen, which significantly affects headloss through screens because the screening area is directly proportional to the incline of the screen. Headloss through a clean bar screen is indicated by Kirschmer's equation:

$$H_L = b_{sf}\,(w/b_{sf})^{1.33}\,H_u \sin\Phi \tag{9.20}$$

where H_L = headloss, m (ft);
b_{sf} = bar shape factor (see Table 9.2);
w = maximum cross-sectional width of bars facing upstream, m (ft);
W_{clr} = minimum clear spacing of bars, m (ft);
H_u = upstream velocity head = $v_u^2/2g$, m (ft); and
Φ = angle of bar screen with horizontal.

The velocity through the bar screen opening can be determined by the following equation:

$$V = Q/A_o \tag{9.21}$$

Where V = velocity through bar screen (i.e., between the bars), m/s (ft/sec);
Q = flowrate, m³/s (cu ft/sec); and
A = area of the bar screen opening m² (sq ft).

$$A_b = [W_{clr}/(W_b + W_{clr})] \times W \times (d_{DS}) \tag{9.22}$$

Where A_b = area of the bar screen opening, m² (sq ft);
W_b = width (thickness) of the bars, m (ft);
W_{clr} = clear spacing between the bars, m (ft);
W = total screen channel width, m (ft); and
d_{DS} = downstream water depth, m (ft).

$$v = Q/(W_{US} \cdot d_{US}) \tag{9.23}$$

Where v = velocity upstream of bar screen or approach velocity, m/s (ft/sec);
W_{US} = width of the channel upstream of the bar screen, m (ft); and
d_{US} = depth of flow upstream, m (ft).

In Equation 9.23, D_{US} typically is unknown and must be solved for iteratively using the Bernoulli equation (9.1) and knowing that y_1 is the same as d_{US}. To account for clogging of the screen, the value of A_o above can be reduced by the fraction or percent clogged. A 50% clogged screen is typically used. This increases the velocity through the screens, which increases the headloss.

The headloss for fine screens uses the orifice equation and is presented as follows (Metcalf & Eddy, Inc./AECOM, 2014):

$$H_L = \frac{1}{(2g)}\left(\frac{Q}{C \cdot A_e}\right)^2 \tag{9.24}$$

Where H_L = headloss through the screen, m (ft);
g = acceleration caused by gravity, 9.8 m/s² (32.2 ft/sec²);
Q = flowrate, m³/s (cu ft/sec);
A = area of the openings submerged in the wastewater, m² (sq ft); and
C = coefficient of discharge for the screen (typically 0.6).

In Equation 9.24, the headloss is the clear water, clean screen headloss. As with bar screens, an allowance should be made for partial clogging. The screen manufacturer should be consulted.

7.2 GRIT REMOVAL

There are three generic types of grit removal systems—aerated grit chamber, vortex grit chamber, and a constant-velocity grit channel. Regardless of the type of grit removal system used, equal distribution of both flow and grit to the individual units is important. Designers should consider flow distribution and flow separation points (i.e., those found at sharp corners) along the inlet distribution channels that can cause dead zones, where grit drops out. Once settled, the grit eventually tends to migrate to the closest basin or basins. The inlet distribution channel can be aerated to keep the flow well mixed and the grit in suspension. The use of air to keep the grit suspended and well mixed may require odor-control facilities to treat the offgases.

In an aerated grit chamber, the inflow direction and the outflow direction may be perpendicular to each other to help induce a spiral rotation through the grit chamber. Air is introduced on one side of the chamber to further induce spiral rotation (Figure 9.11). Typically, there is an overflow weir at the end of the chamber to control the water level, and there may be baffles installed in the chamber to improve performance, all of which must be accounted for in the hydraulics. Therefore, the designer needs to calculate headloss through the grit chamber based on the components inducing headloss (e.g., channels, baffles, weirs).

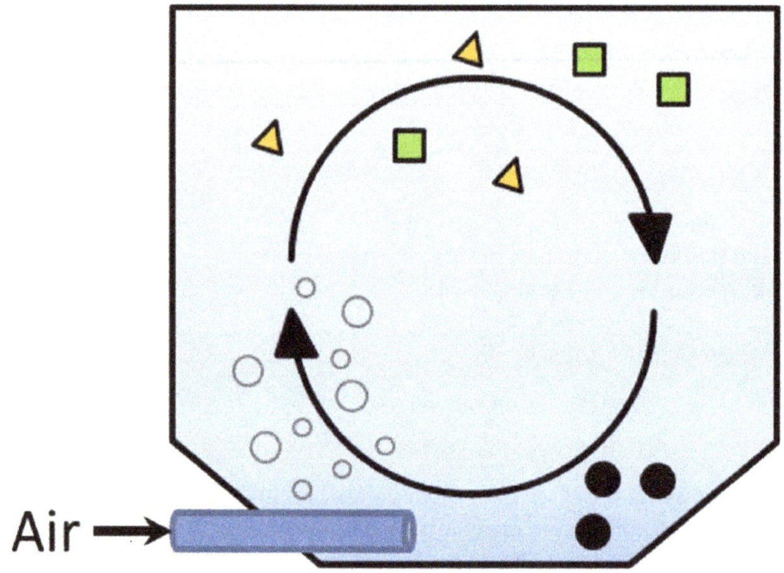

Figure 9.11 Aerated Grit Basin (Reprinted with permission from Indigo Water Group)

There are a number of proprietary vortex-type grit removal systems (Figures 9.12 and 9.13). Design data must be obtained from the supplier and must be carefully followed. Of particular importance is the approach configuration and velocity. The discharge must be a free discharge at all times.

7.3 SEDIMENTATION BASINS

Headloss through circular primary sedimentation basins, or clarifiers, consists of the headloss through the influent pipeline and inlet column, headloss through the orifices at the end of the inlet column pipe in the stilling well, headloss through any flow distribution orifices within the stilling well, and head over the effluent weir (Figure 9.14). If it is a rim-feed clarifier, the total headloss includes the headloss through the inlet trough orifices, headloss in the distribution trough, and head over the effluent weir.

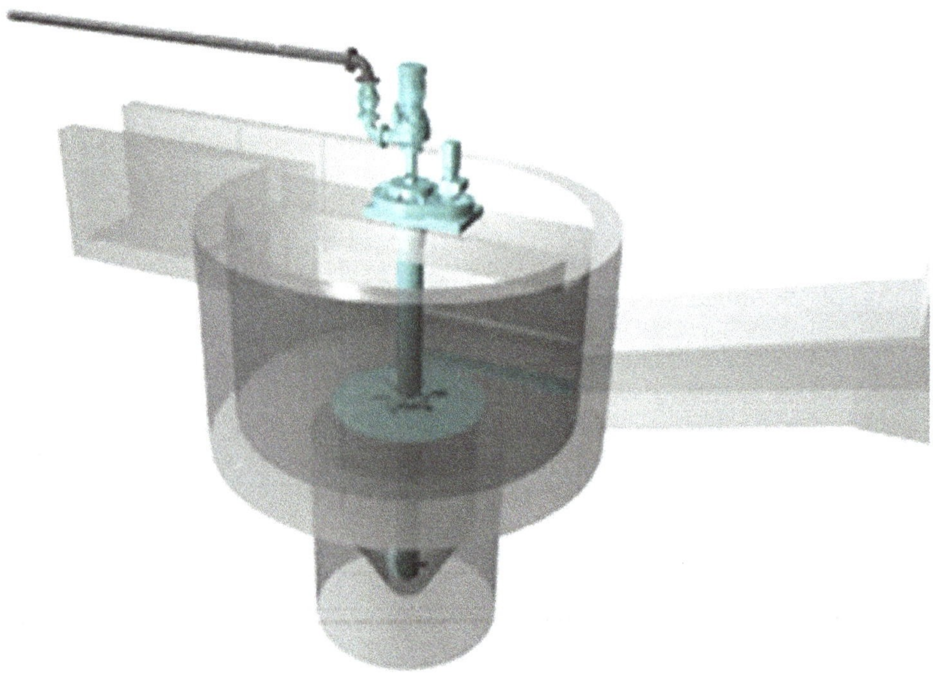

Figure 9.12 Flat-Floored, Hydraulic Forced Vortex Grit Chamber (Reprinted with permission from Smith & Loveless, Inc.)

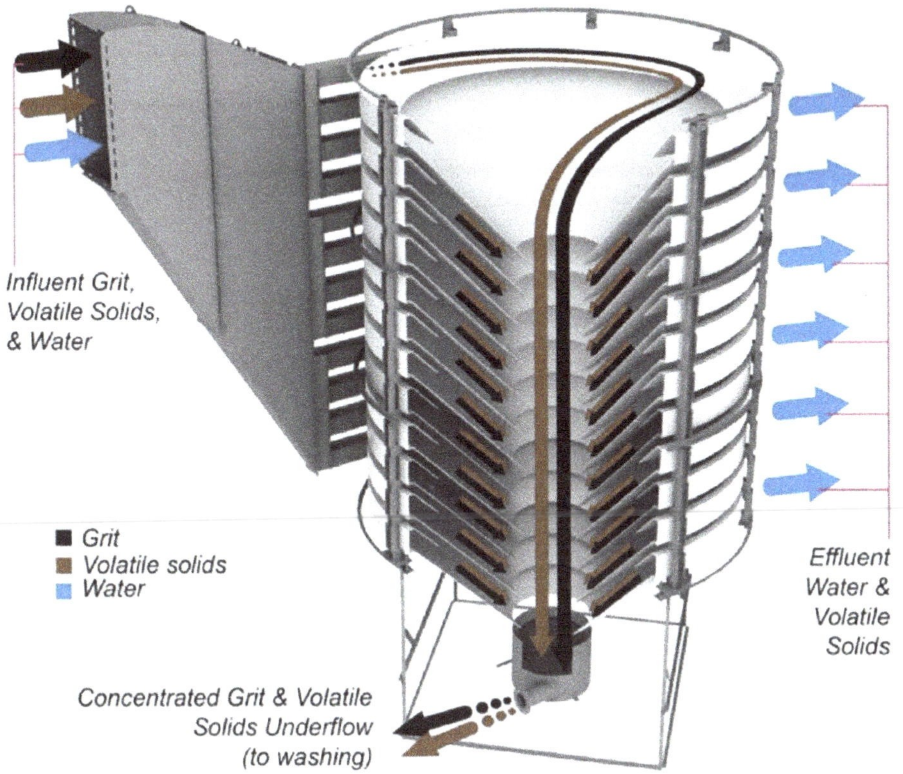

Figure 9.13 Multiple-Tray Grit Separator (Reprinted with permission from Hydro International)

For a rectangular basin, the headloss includes the inlet gate, energy-dissipating tee or other system, and head over the effluent weir. For the headloss through the inlet-energy-dissipating tee, a K-factor equal to 3.0 has been determined experimentally by the City of Los Angeles, California (Betz, 1981).

Headloss calculations for secondary sedimentation basins are similar to those for primary sedimentation basins. The flowrate into the secondary clarifiers must include the RAS for activated sludge facilities. For trickling filter and biotower facilities, the recycled flows typically are diverted away upstream of the

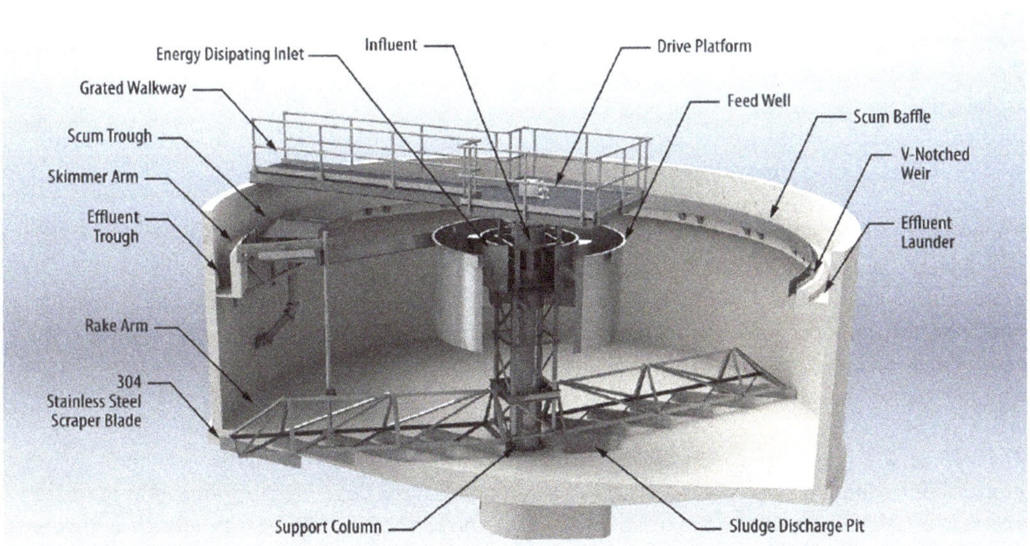

Figure 9.14 Circular Primary Clarifier Schematic (Reprinted with permission by Monroe Environmental [Monroe, MI] www.mon-env.com)

secondary sedimentation tanks; thus, they do not need to be included in the flow entering the secondary sedimentation tanks. However, this is not always the case.

For some activated sludge secondary clarifiers, the RAS is drawn off via a pipe or series of pipes situated within the inlet column. The headlosses in this system should be analyzed to ensure that the maximum anticipated RAS can be withdrawn and flow, by gravity, to the RAS pumps.

7.4 AERATION BASINS

As with any unit process, equal distribution of flow to the individual basins and within each unit process is important. As stated in previous sections of this chapter, headloss is used intentionally to help improve flow distribution. This can be generated by inlet gates (orifice loss), inlet weirs, or cut-throat flumes. To keep solids in suspension and well mixed, tanks and channels containing mixed liquor suspended solids should be aerated or mixed as appropriate. Aeration of the inlet channels is not appropriate in cases in which the treatment process would be affected (i.e., upstream of anoxic or anaerobic basins). Channels upstream of anoxic or anaerobic zones may be mixed using propeller mixers or jet nozzles.

The mode of operation for the aeration basin should be addressed during calculation of the hydraulic profile. Aeration basins can be operated as completely mixed flow, plug flow, stepped flow (feed), or a combination of flow patterns. Additionally, aeration basins may be baffled or partitioned to separate zones.

The designer calculating the water-surface profile should anticipate that the mode or modes of operation are planned during design, and that operational modes could change with future operational requirements. The hydraulic analysis should be performed along the path that provides the greatest total headloss. The headloss for flow through the aeration tank is small and typically is not calculated. If centrifugal aeration blowers are used, water level fluctuations between peak flow and initial minimum flowrates in the aeration basin should be kept to a minimum so that blower performance is not affected.

The flow used in the hydraulic analysis of inlet gates, aeration basin end weir(s), and internal baffles must include the RAS and internal mixed liquor recycle (often used in biological nutrient removal facilities), as appropriate.

7.5 TRICKLING FILTERS OR BIOTOWERS

For trickling filters to operate properly, a sufficient water surface elevation difference (hydraulic head) must exist to overcome the fixed-nozzle orifice loss and rotate the distributors. The head requirement at the inlet to the distributing arm typically is approximately 2 m (6.5 ft). The designer should consult with the manufacturer for actual headloss requirements. In the analysis, recycle flowrates must be included in the total flow.

7.6 DISINFECTION

Disinfection processes consist of long plug-flow channels, which may be baffled. Solids settling is not as much of a concern, and the basins can be designed to operate at low velocities with low headloss. Scum control, if desired, can be provided by a tilting pipe skimmer or similar system.

Minimizing short-circuiting in the design of chlorine contact chambers is paramount. Chorine contact basins should be designed as plug-flow reactors, with large length-to-depth or -width ratios (i.e., 40 : 1 or more). Internal perforated baffles may be installed just downstream from the inlet and at each "turn" in the tank. To ensure uniform flow distribution, a baffle should be installed upstream of the effluent weir (see Metcalf & Eddy, Inc./AECOM [2014] for additional design information). The principal headlosses in the chlorine contact tank occur as a result of the internal baffles (described above) and the end weir.

The water level in open channel ultraviolet systems must be maintained as constant as possible to ensure proper disinfection. This typically is done with a constant level gate, finger weirs, or similar device provided by the equipment supplier. It is important that this level control device have a free, unsubmerged discharge to operate properly. Also key to effective ultraviolet disinfection is uniform velocity distribution approaching the ultraviolet lamps. Some states, such as California, require velocity profiling in conformance with the National Water Research Institute (NWRI, 2012). This may require additional baffling at the inlet. Headloss data in ultraviolet systems can be obtained from the suppliers.

8.0 PUMPING SYSTEMS

The design of WRRFs confronts the engineer with a wide range of pumping applications, including raw wastewaters; treated wastewaters; mixtures of domestic and industrial wastes; raw sludges, thickened sludges, biosolids, and grit; scum containing a mixture of grease, floating solids, and trash; return and waste activated sludges (WASs); chemical solutions; flushing water; spray water and pump seal water; tank drainage; and sump pump water. Pumping of sludges and biosolids often results in higher pressure conditions, requiring modifications to the evaluation, and is not discussed in this section.

Pumping stations come in a variety of configurations depending on the application. Table 9.2 summarizes many of the most common wastewater pumping configurations. The table indicates applications for the configurations provided as well as key considerations.

Table 9.2 Summary of Wastewater Pumping Station Configurations at WRRFs

Configuration	Application	Key Considerations
Wet pit/dry pit	– All pumping systems	– Typical for large pumping capacities – Larger footprint, surface structure recommended for electrical equipment – Ventilation and lighting required for dry pit – Subject to flooding from broken pipes; needs sump pump – Seal water system generally required – Crane/hoist for maintenance
Submersible	– Influent, RAS/WAS – Intermediate and recirculation – Scum – Effluent	– Easily constructed, least costly – Less obtrusive, small footprint – Frequently used for smaller capacity stations – Maintenance more difficult – Requires crane access to remove pumps – Seal water not required – Moisture monitoring for motors
Submersible pumps in wet pit/dry pit	– All pumping systems	– Similar to wet pit/dry pit, except seal water not required – Pumps and motors safe from flooding
Self-priming	– Influent, RAS/WAS – Intermediate and recirculation – Effluent	– Pumps mounted at ground surface or shallow depth – Suction lift limited to approximately 6 m (20 ft) – Easy to maintain – Pumps may have reduced efficiency
Self-cleaning, trench wet well	– Influent	– Successful in larger lift stations – Easier to maintain than conventional design – Use in combination with wet pit/dry pit installations – Smaller footprint than conventional – Wet-well construction more complex – See Jones (2008) for details
Vertical pumps over wet well	– Influent, RAS/WAS – Intermediate and recirculation – Effluent	– Vertical turbine solids handling pumps for screened wastewater or primary effluent – Cost-effective – Small footprint – Requires crane access for maintenance of pumps
Archimedes screw	– Influent, RAS/WAS – Intermediate and recirculation – Effluent	– Low head application; limited to approximately 10 m (30 to 35 ft) lift – High efficiency – Essentially nonclog – Variable-speed pumping, even though at constant rotational speed – Turbulence in pumping and discharge can release odors – Open screws should be covered for odor control – Enclosed screw interior difficult to protect from corrosion – Bearing maintenance requires heavy equipment

Note. RAS = return activated sludge; WAS = waste activated sludge

The Hydraulic Institute is the global authority on pumps and pump systems that develops industry standards through a vast array of technical resources. (https://www.pumps.org/who-we-are/)

8.1 PUMP TYPES AND APPLICATIONS

There are two major pump classifications defined by the Hydraulic Institute—kinetic energy pumps and positive displacement pumps. Table 9.3 lists varying types in each of these classifications, including the major applications for these different types of pumps in WRRFs.

8.2 SYSTEM AND PUMP CURVES

When selecting pumps for a specific application, the design engineer must match the pump capabilities (flow and pressure or head) with the system properties (pipe size, fittings, needed flow), taking into account fluid properties (viscosity) and the range of expected operating flows.

The specific requirements of each pumping unit are determined by first calculating and plotting the system curve (Figure 9.15). The system curve is generated by plotting the sum of the static lift and friction

Table 9.3 Pump Classification and Applications in Wastewater Industry

HI - Major Classification	Pump Type	Pump Description	Major Pumping Applications
Kinetic	Centrifugal (volute)	– Separately coupled – Close coupled – Submersible – Axial split – Radial split	– Raw wastewater – RAS and WAS (nonclog) – Settled primary and thickened sludge – Secondary or tertiary effluent
	Peripheral (torque-flow)	– Separately coupled – Close coupled – Radial flow – Recessed impeller	– Raw wastewater – Scum – RAS and WAS – Dilute sludge – Dilute digested sludge
	Vertical (turbine)	– Lineshaft – Submersible – Horizontally mounted axial flow – Vertical turbine solids handling (VTSH)	– Screened wastewater (VTSH) – Primary effluent (VTSH) – Secondary or tertiary effluent
Positive displacement	Reciprocating	– Plunger	– Scum
		– Piston	– Primary, secondary, and settled sludges – Digested sludge – Thickened sludge – Chemical solutions
		– Diaphragm	– Scum – RAS and WAS – Digested sludge – Thickened sludge – Chemical solutions
	Rotary	– Lobe	– Raw wastewater
		– Screw	– Digested sludge (rotary-lobe) – Thickened sludge – Chemical solutions
		– Progressive cavity	– Scum – Sludge (when pumping primary sludge, grinder typically precedes pump) – Digested sludge – Thickened sludge
	Screw	– Spiral screw	– Raw wastewater – Settled primary and secondary sludges – Thickened sludge
	Pneumatic	– Airlift	– RAS and WAS
		– Ejector	– Raw wastewater at small installations – Scum

Note. HI = Hydraulic Institute; RAS = return activated sludge; WAS = waste activated sludge

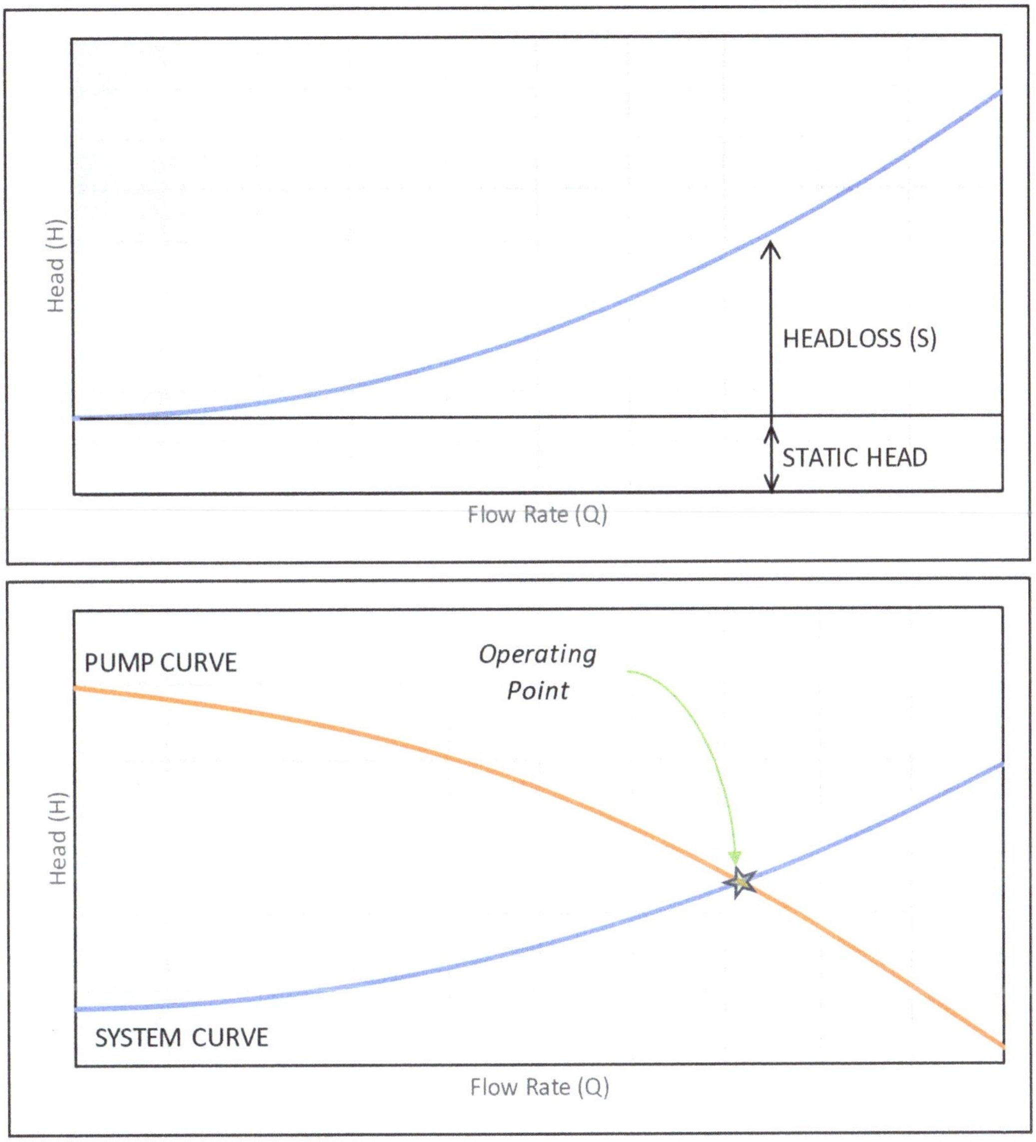

Figure 9.15 (Top) System Curve; (Bottom) Combining System and Pump Curves

headloss in the system at various (assumed) discharge flows. The static lift represents the elevation differential between the low-water level in the wet well and the high point of the discharge force main, or the high-water level at the discharge, whichever is higher. The friction headloss is the sum of friction headlosses through the suction pipe, suction fittings, discharge pipe, discharge fittings, and force main at various flowrates. The next step is to plot the pump performance curve, or pump curve as it is commonly called, on the same graph. This is unique to the pump and available from the pump manufacturers. It relates pump capacity and discharge head (or pressure); efficiency, net positive suction head (NPSH), and power are often plotted as well.

Jones (2008) describes many examples of typical system head curves, indicating portions of the **total dynamic head** that comprise static head and friction headloss. It is also important to note when plotting a system curve that there are the differences in the head capacity curves of one pump, two pumps, and three pumps in parallel operation. Pump curves for multiple pumps operating in parallel are obtained by adding the capacity of each pump at each of several given heads and plotting the results for each pump grouping.

Total dynamic head (TDH) is the total amount of pressure differential created by a pump for a given flow rate.

System and pump curves can be combined to solve a number of complex pumping problems. In the case of two remote pumping stations (with differing pump hydraulic characteristics) discharging to a common force main, the system head curve would be a compound curve with a segment for each of the following conditions:

- Pump A on, pump B off;
- Pump A off, pump B on; and
- Pump A on, pump B on.

When both pumps operate, the capacity contributed by each pump varies depending on static head, friction losses, and individual pump head capacity curves, and finds pressure equilibrium. Therefore, pump selection for such a system is an iterative process.

Head capacity curves for centrifugal pumps operated in a series are obtained by adding the operating head for each pump at a given capacity. An example of a series pumping arrangement would be where extreme static head conditions exist that exceeds the operating range of pumps that an owner might typically use. Series pumping increases the probability of a mechanical failure because all pumps operating in a series arrangement must be functioning properly in order to convey the desired pumping capacity. As such, series pumping arrangements should be avoided if the same objective can readily be achieved with a single reliable pump. Pump manufacturers should be consulted before designing series pump applications to ensure proper pump selection and thereby avoiding potential cavitation or motor overload conditions.

A family of curves is used to represent variable-speed pump head and capacity. Each individual curve corresponds to a discrete operating speed. The performance curves for different operating speeds can be obtained from the supplier or determined using the pump affinity laws, which relate speed, flow, head, and power. Superimposing the system head curve on the family of head capacity curves identifies specific operating points for the system.

Where variable- and constant-speed pumps discharge to a common header, a complete hydraulic analysis for all operating speeds is required to prevent cavitation. Pump cavitation creates unstable pumping conditions and can damage the pumps. Jones (2008) presents a discussion on pump cavitation and avoidance.

Variable-speed pumping applications have increased significantly over time, primarily because these are becoming more reliable and cost-effective while simultaneously providing greater operating range capability versus a constant-speed pump, thus reducing the overall number of pumps required for larger and more complex pump stations. Even so, careful consideration should be given to the use of variable-speed applications if a constant-speed pumping arrangement is sufficient for the purpose and objective of the pump station, as this does add cost and complexity to the station. Jones (2008) provides a description of the advantages and disadvantages of variable-speed pumping.

Affinity laws are governing relationships for centrifugal pumps used to predict pump performance in terms of flow rate and total dynamic head for changes in impeller diameter and impeller speed.

Affinity laws are used to assess pump performance under variable-speed conditions. These laws are used to predict centrifugal pump performance at different speeds and with different impeller diameters (WEF, 2022). For changes in speed, with the impeller diameter remaining constant, the affinity law becomes:

$$Q_2/Q_1 = N_2/N_1 \cdot (D_2/D_1) \quad (9.25)$$

$$H_2/H_1 = N_2^2/N_1^2 \cdot (D_2/D_1)^2 \quad (9.26)$$

$$P_2/P_1 = N_2^3/N_1^3 \cdot (D_2/D_1)^3 \quad (9.27)$$

Where Q_1 = pump flowrate, L/s (gpm) at speed 1;
Q_2 = pump flowrate, L/s (gpm) at speed 2;
N_1 = pump rotative speed, r/min (rpm);
N_2 = pump rotative speed, r/min (rpm);
D_1 = impeller diameter, m;

D_2 = impeller diameter, m;
H_1 = pump head, m (ft) at speed 1;
H_2 = pump head, m (ft) at speed 2;
P_1 = pump power required, kW (hp) at speed 1, and
P_2 = pump power required, kW (hp) at speed 2.

The power output necessary by a pump is determined by the energy needed by the contacting fluid, as follows:

$$P = \gamma QH/\eta \tag{9.28}$$

Where P = water power, kW;
γ = specific weight of the fluid, kN/m³;
Q = flowrate, m³/s;
η = efficiency; and
H = total dynamic head, m.

In U.S. customary units, the equation for power is the following:

$$P = \gamma QH/550 \tag{9.29}$$

Where P = water power, hp;
γ = specific weight of the fluid, lb/cu ft;
Q = flowrate, cu ft/sec;
H = total dynamic head, ft; and
550 = conversion factor from lb·ft/sec to hp.

Pumps work at varying efficiencies, and the pump efficiency must be factored in before selecting a motor. Brake power is defined as the power input required by a pump (the motor output power), which takes into account volumetric, mechanical, and hydraulic energy losses. The motor nameplate power rating, kW (hp), is its output power. Brake power is the water power (9.28) and (9.29) divided by the overall efficiency.

Because the brake power requirements vary with flow, head, and efficiency, changes in the pump discharge head will affect the brake power requirements. The pump brake power requirements should be checked at various points along the pump performance curve to ensure that the nameplate rating of the motor is equal to or greater than the maximum brake power required at all points.

To determine the total electrical power input to the motor for electrical power cost studies, brake power required should be divided by the motor efficiency.

8.3 WET WELLS

No single method for sizing wet wells applies to all design situations. Proper wet-well sizing considers three critical factors—detention time, pump cycle time, and turbulence at the pump intake. *Hydraulic Institute Standards* (Hydraulic Institute, 1983) provides guidance on the recommended configuration, sizing, and dimensions for wet well design.

As good practice, wet-well detention times generally should not exceed 30 minutes for average flow to minimize generation of unpleasant odors. In colder climates, longer detention times may be acceptable. Where such detention time limitations would be impractical, odor mitigation should be accounted for in wet-well design. Odor mitigation practice can include gas-tight covers, chemical feed systems (binding H_2S in solution), biofilters, or offgas scrubbing systems.

Pump cycle time refers to the elapsed time between successive motor starts (i.e., the time to fill the wet well plus the time to empty it). Excessive motor wear and shortened service life result from cycle times less than the manufacturer's recommendation. Minimum cycle times range from approximately 5 minutes for 4-kW (5-hp) motors to more than 30 minutes for 150-kW (200-hp) motors. The motor manufacturer should be consulted for minimum cycle time recommendations or special motor designs.

For constant speed pumps, the minimum wet-well volumes can be determined from the following equation:

$$V = T_C Q/4 \tag{9.30}$$

Where V = wet-well volume, m^3 (gal);
T_C = cycle time, minutes; and
Q = pump capacity, m^3/min (gal/min).

The minimum cycle time results if the influent flow equals 50% of the rated pump capacity. This limitation often determines wet-well volumes for both single- and multiple-pump installations. For multiple-pump installations, alternating the lead pump after each pumping cycle effectively doubles the cycle time and reduces wet-well volumes accordingly. Wet-well volumes also can be optimized with strategic pump "on" and "off" settings. In multiple constant-speed installations, the required wet-well capacity represents the sum of the wet-well capacities required for the individual pumps. Such an allowance prevents cycling when lag or standby pump units enter service.

Wet-well designs should allow adequate submergence and clearance between pump intakes to prevent turbulent currents and vortexes that could otherwise reduce pump efficiency or capacity. The pump manufacturer's recommendations and *Hydraulic Institute Standards for Centrifugal, Rotary, and Reciprocating Pumps* (Hydraulic Institute, 1983) should be consulted for sizing sumps and configuring wet wells. For large complex wet-well systems and very large pumps, where established design practice would be impractical, modeling may be necessary to ensure acceptable wet-well performance (Jones, 2008).

PRACTICE PROBLEMS

Concepts and Fundamentals:

1. What is the purpose of a hydraulic profile for a WRRF?
2. Name three hydraulic control elements or structures (i.e., elements use to control hydraulic elevation, flow splitting, or flow mixing).
3. What is the difference between the hydraulic grade line and the energy grade line?
4. How is headloss related to flow velocity, or how does a change in flow velocity affect headloss?
5. What are some of the key steps to evaluating hydraulics and developing a hydraulic profile for a WRRF?

Hydraulic Profiles:

1. What are two key reasons for developing a hydraulic profile?
2. Name three important elements of a hydraulic profile.
3. What are key steps to developing a hydraulic profile?
4. Draw a possible hydraulic profile for a flow condition of 150 000 m^3/d as described based on the following:
   a. Assumptions:
      - Top of structures are 1 m above water surface elevation
      - Structures have typical depths; e.g., flow control structures 3–4 m deep, clarifiers 5 m deep, aeration tanks 8 m deep

- It is appropriate to incorporate a hydraulic drop over a weir at flow control structures
- Grade is between 54 and 60 m
- Interconnecting pipes are 2 m below grade

b. River water elevation: 50 m
c. Headloss between outlet structure and river: 0.2 m
d. Final clarifiers: Six of six tanks in operation, water surface elevation: 51.2 m
e. Headloss between final clarifier splitter structure and final clarifier: 0.2 m
f. Aeration tanks: Four of four tanks in operation, water surface elevation (w/o aeration): 54.3 m
g. Headloss between aeration tank splitter and aeration tanks: 0.2 m
h. No primary clarifiers
i. Headworks headloss through screens and grit removal: 0.6 m

Flow and Headloss Calculations:

1. Calculate and show the flow balance for the following:

   a. Balance over a primary clarifiers
   b. WRRF influent flow is 5000 m^3/d
   c. WRRF plant recycle is returned upstream of primary clarifiers (facility recycle is effluent used for cleaning plus sidestream flow from thickening and dewatering)
   d. WRRF plant recycle is 10% of the influent flow
   e. Primary solids are withdrawn from the primary clarifiers with the following characteristics:

      - Primary solids mass rate: 7500 kg/d
      - Primary solids concentration: 15 000 mg/L

2. Calculate and show the flow balance for the following:

   a. Balance over secondary clarifiers (from activated sludge process)
   b. WRRF influent flowrate of 10 000 m^3/d
   c. WRRF plant recycle is 10% of the influent flow (facility recycle is effluent used for cleaning plus sidestream flow from thickening and dewatering)
   d. Return activated sludge flow is 50% of influent flow
   e. Waste activated sludge flow is 2% of influent flow

Energy Balance, Headloss Calculations:

1. What is the difference between the energy grade line and the hydraulic grade line?

2. What is a typical maximum velocity used for conduit sizing?

3. Determine the size (diameter) for a conduit conveying a peak flow of 100 m^3/d with a maximum velocity of 1.5 m/s.

4. What is the relationship between headloss and velocity?

5. Calculate the headloss in a 1 m conduit conveying conveying 2000 m^3/d of flow using Hazen Williams formula.

6. Calculate the headloss in an open rectangular channel conveying 4000 m^3/d of flow with a downstream depth of 2.5 m, width of 2.0 m, and length of 100 m.

7. Weir loss problem: If two, 1 m sharp-crested weirs are used to split 1000 m^3/d flow between two aeration tanks, what is the weir head when one and two tanks are in operation?

Unit Process Calculations:

1. Bar screen problem: What is the headloss in a bar screen with the following characteristics?

   a. Screening a total flow of 1000 m^3/d
   b. Two bar screens in operation
   c. Screen spacing is 6 mm
   d. Screen/channel width is 1.5 m

2. Clarifier problem: How many 90° v-notch weirs would be installed on a perimeter weir and launder for a clarifier treating 500 m^3/d of forward flow, and what is the water surface elevation in the clarifier?

Pumping Systems:

1. Pump power requirement: What is the power required to pump 5000 m^3/d of wastewater flow with a total dynamic head of 10 m?
2. Pump output at reduced speed: What is the speed requirement when reducing the pump flow from 1000 m^3/d at 900 rpm to 750 m^3/d?
3. Pump output at reduced impeller size: What impeller size is needed to pump 2500 m^3/d if the pump is rated for 4000 m^3/d with an impeller size of 0.4 m?
4. Wet well sizing: What is the minimum wet well volume required for constant speed pumps in a duplex pump station (one duty pump, one standby pump) with a cycle time of 10 minutes and a flow capacity of 1000 m^3/d?

9.0 REFERENCES

Ackers, W. R., White, W. R., Perkins, J. A., & Harrison, A. J. M. (1978). *Weirs and flumes for flow measurement.* John Wiley & Sons.

American Society of Civil Engineers & Water Environment Federation. (2007). *Gravity sanitary sewer design and construction* (2nd ed., ASCE Manuals and Reports on Engineering Practice No. 60, WEF Manual of Practice No. FD-5). American Society of Civil Engineers.

American Water Works Association. (2012). *Water treatment plant design* (5th ed.).

Benefield, L. D., Judkins, J. F., Jr., & Parr, D. A. (1984). *Treatment plant hydraulics for environmental engineers.* Prentice-Hall.

Bergendahl, J. (2008) *Treatment system hydraulics.* American Society of Civil Engineers.

Betz, J. M. (1981). Unpublished notes of a head loss test at the Los Angeles-Glendale Water Reclamation Plant (August 26).

Boulos, P. F., & Nicklow, J. W. (2005). *Comprehensive water and wastewater treatment plant hydraulics handbook for engineers and operators.* MWH Soft Press.

Camp, T. R., & Graber, S. D. (1968). Dispersion conduits. *Journal of the Sanitary Engineering Division, American Society of Civil Engineers, 94*(1), 31–39.

Chin, D. A. (2012). *Water resources engineering* (3rd ed.). Pearson-Prentice Hall.

Chow, V. T. (1959). *Open-channel hydraulics.* McGraw-Hill.

City of Los Angeles Bureau of Engineering. (1968). *Hydraulic analysis of junctions.* http://eng.lacity.org/techdocs/sewer-ma/haj.pdf

Davis, C., & Sorensen, K. (1969). *Handbook of applied hydraulics.* McGraw-Hill.

Fair, G. M., Geyer, J. C., & Okun, D. A. (1968). *Water and wastewater engineering, vol. 2: Water purification and wastewater treatment and disposal.* John Wiley & Sons.

Grant, D. (1989). *Isco open channel flow measurement handbook* (3rd ed.). ISCO.

Hydraulic Institute. (1983). *Hydraulic Institute standards for centrifugal, rotary, and reciprocating pumps* (14th ed.). Hydraulic Institute.

Jones, G. M. (Ed.). (2008). *Pumping station design* (3rd ed.). Butterworth-Heinemann.

King, H. W., & Brater, E. F. (1963). *Handbook of hydraulics.* McGraw-Hill.

Mays, L.W. (1999). *Hydraulic design handbook.* McGraw-Hill.

Metcalf & Eddy, Inc./AECOM (2014). *Wastewater engineering: Treatment and resource recovery* (5th ed.). McGraw-Hill.

Montgomery, J. M. (1985). *Water treatment principles and design*. Wiley-Interscience.

National Water Research Institute. (2012). *Ultraviolet disinfection guidelines for drinking water and water reuse* (3rd ed.).

Nopens, I., Batstone, D. J., Griborio, A., Samstag, R., Wicklein, E., & Wicks, J. (2012, September 29–October 2). Computational fluid dynamics (CFD): What is good CFD-modeling practice and what can be the added value of CFD models to WWTP modeling? *Proceedings of the 85th Annual Water Environment Federation Technical Exhibition and Conference*, New Orleans, Louisiana, United States.

Reichenberger, J. C. (1984). Unpublished notes for the design of Three Valleys Municipal Water District, Miramar Treatment Plant, Claremont, California.

Street, R. L., Waters, G. Z., & Vennard, J. K. (1996). *Elementary fluid mechanics* (7th ed.). John Wiley & Sons.

Townsend, D. W. (1935). Loss of head in activated sludge aeration channels. *Transactions of the American Society of Civil Engineers, 100*, 518.

U.S. Department of the Interior, Bureau of Reclamation. (2001). *Water measurement manual*, Chapter 8. Superintendent of Documents; U.S. Government Printing Office.

Water Environment Federation. (2022). *Design of wastewater and stormwater pumping stations* (Manual of Practice No. FD-4; 3rd ed.).

Chapter 10 Wastewater Solids: Characteristics, Processing, and Transport

LEARNING OBJECTIVES

At the end of this chapter, students should be able to:

- Identify sources of primary, secondary, combined and tertiary solids.
- Calculate the quantity of primary, secondary, combined, and tertiary solids generated from the treatment of municipal wastewater.
- List the characteristics of solids generated from the treatment of municipal wastewater.
- Determine mass, volume, and percent solids of a solids using the weight volume relationship.
- Estimate the specific gravity of solids.
- Perform a mass balance on solids processes.
- Describe the operation and design sludge storage process.
- Describe the operation and design sludge transport processes including sludge pumps and conveyors.

RESIDUALS TERMINOLOGY

The Water Environment Federation has adopted the following terminology for wastewater residuals.

Sludge is any residual produced during primary, secondary, or advanced wastewater treatment that has not undergone any process to reduce pathogens or vector attraction. Another common term for this is raw sludge. The term sludge should be used with a specific process descriptor (e.g., primary sludge, waste activated sludge, or secondary sludge).

Biosolids is any sludge that has been stabilized/treated to meet the criteria in the U.S. Environmental Protection Agency's (U.S. EPA's) 40 CFR 503 regulations for pathogen and vector attraction control and, therefore, can be beneficially used.

Solids and **residuals** are terms used when it is uncertain whether the material meets Part 503 criteria (e.g., during thickening, because stabilization may occur either before or after this process). In this textbook, the terms solids and residuals will be used as general references and for general descriptors (e.g., solids handling).

1.0 INTRODUCTION

The purpose of this chapter is to introduce the types of solids that are produced in a water resource recovery facility (WRRF) and how they are handled. This includes where solids are generated and how they are accounted for through mass balances. Solids also must be stored, conveyed, and transported, which are all topics covered in this chapter. Solids processing is a significant fraction of the cost of constructing and operating a WRRF. Methods and materials are based on the final disposal of the solids, which is limited to land application, landfilling, and incineration.

2.0 SOURCES, QUANTITIES, AND CHARACTERISTICS

Solids management is an important aspect of wastewater treatment design because of the interrelationships between the liquid and solids processes. It is impossible to completely separate liquid and solids handling processes, and engineers must consider their relationships when designing WRRFs. The liquid treatment processes chosen will affect both the amount of solids generated and their characteristics, which in turn will affect the choice of thickening, conditioning, and dewatering processes. Furthermore, the recycle streams from solids treatment processes can affect liquid processes. For example, dewatering anaerobically digested biosolids produces a sidestream with high ammonia and phosphorus concentrations, which will increase these loadings to the facility's liquid treatment processes.

The amount of solids generated during wastewater treatment is an important design parameter because it affects the sizing of solids treatment processes and all related equipment. Solids generation rates also affect the size of liquid treatment processes. For example, the size of a secondary treatment process to maintain a desired solids retention time (SRT) depends on how much solids will be produced.

In general, domestic wastewater typically produces about 0.24 kg/m^3 (2000 lb/mil. gal) of solids. The WRRFs using processes that destroy solids (e.g., digestion or heat treatment) will generate less, and those using chemical addition will produce more. That said, 0.25 kg/m^3 (2080 lb/mil. gal) is a convenient benchmark for cursory comparisons.

A good approach to estimating solids production is to provide a mass balance for the entire WRRF that relates solids production to design parameters for each treatment process (see Figure 10.1). The mass balance should show key constituents (e.g., flow, total suspended solids [TSS], and biochemical oxygen demand [BOD]) and the process assumptions used in the calculations. It should also include solids generated during nitrogen and phosphorus removal processes.

Recycle streams can be included in one of two ways. In the first approach, engineers assume that a fixed percentage of solids or BOD is recycled from downstream processes to the head of the facility. They then

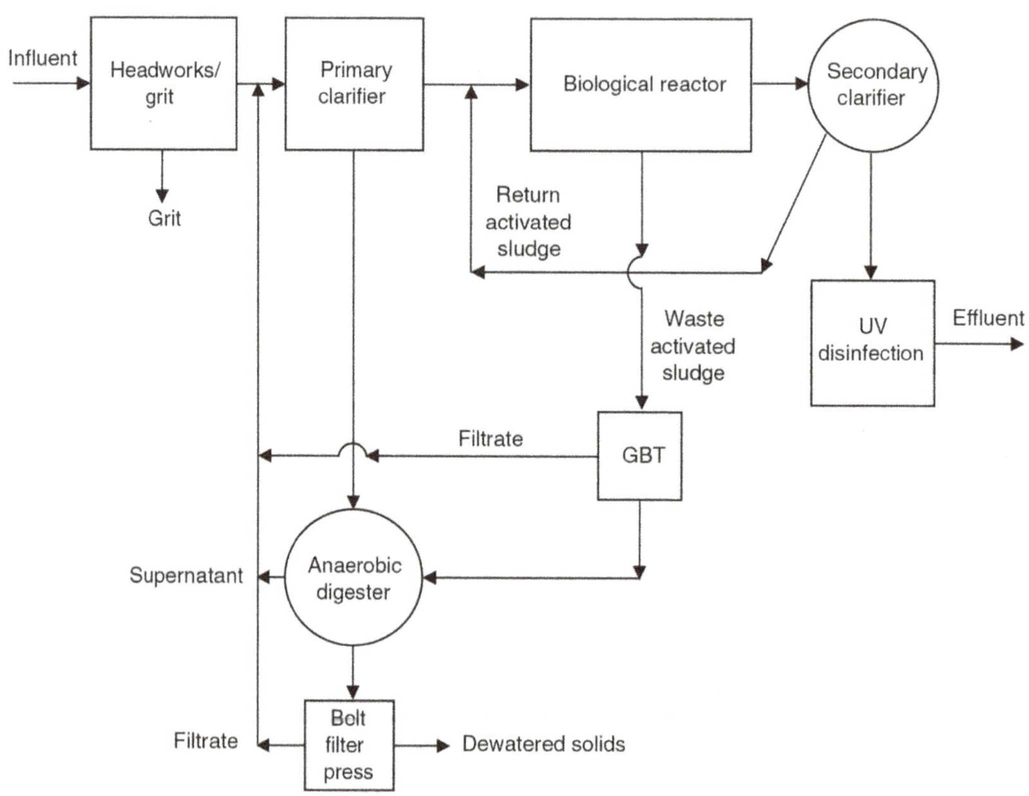

Figure 10.1 Process Flow Schematic for Mass Balance Example (GBT = gravity belt thickening)

iterate the solids balance until the recycled quantities assumed at the head of the facility equal the sum of recycled quantities computed for each process. The second approach is to estimate the WRRF's net solids production based on historical data, anticipated influent strength, or experience at similar facilities. Engineers then use this information to determine the amount of solids leaving the WRRF, and typically apply it to the output end of the dewatering process. They then back-calculate solids loading to a specific process via the mass balance.

In either approach, engineers must typically estimate the quantities of primary, secondary, and chemical solids separately. They must also take into account expected fluctuations in wastewater characteristics that result from changes in industrial contribution, stormwater flows, seasonal weather conditions, and an expanded collection area. Engineers need to understand peak solids production and diurnal variations to properly size solids handling processes.

When designing solids handling facilities (e.g., conveyance, conditioning, thickening, dewatering, or drying systems), engineers must know the characteristics and volumes of the solids involved. There are several types of solids (e.g., primary, secondary, mixed primary, and chemical), and their characteristics depend on many factors (e.g., liquid stream influent characteristics, the use of chemical precipitants and coagulants, process control, peak loads and weather conditions, and the treatment process chosen).

2.1 PRIMARY SOLIDS

Most WRRFs use primary sedimentation to remove settleable solids, which thicken via gravity. Called primary solids, this material consists of a combination of organic solids, grit, and inorganic solids. Primary solids typically are conveyed downstream for further processing (e.g., thickening, digestion, dewatering) before beneficial reuse and/or disposal. Most WRRFs use primary sedimentation tanks to remove settleable solids from wastewater. Primary sedimentation is a relatively efficient method for reducing BOD and TSS loading to secondary treatment processes. The amount of solids removed via primary sedimentation is typically related to either the surface overflow rate (SOR) or hydraulic retention time (HRT), which affect the removal rate of solids.

Figure 10.2 presents some typical curves relating primary solids production to the surface overflow rate (Great Lakes–Upper Mississippi River Board of State and Provincial Public Health and Environmental Managers [GLUMRB], 2014). This is the most common method for estimating primary solids production. However, other factors (e.g., hydraulic short-circuiting, poor flow distribution, density currents, and

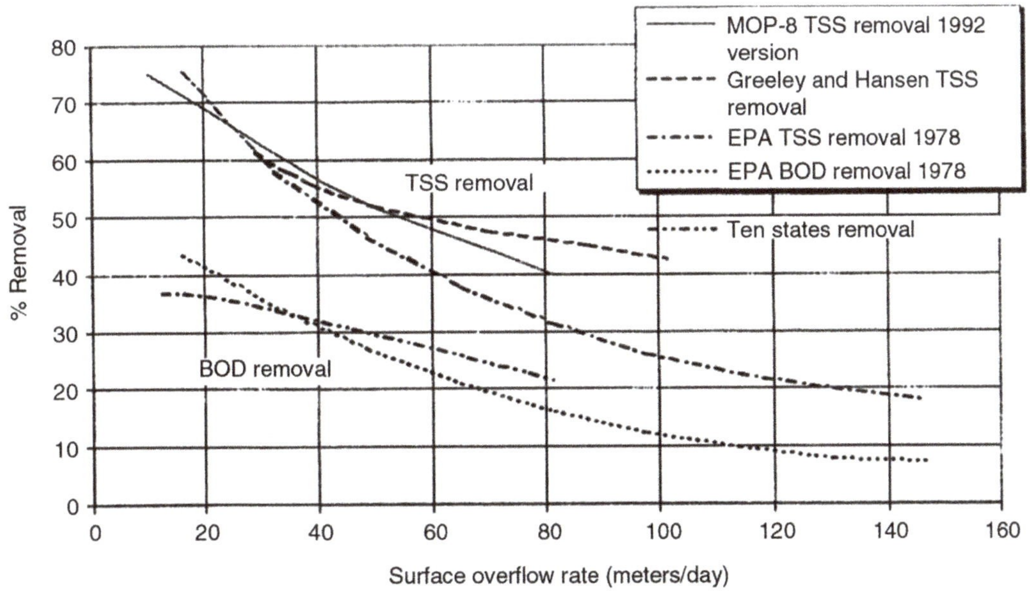

Figure 10.2 Primary Treatment Performance—Total Suspended Solids and Biochemical Oxygen Demand Removal

other mechanical factors) can affect performance significantly. Without proper clarifier design, actual facility data may indicate only a weak correlation between performance and either SOR or HRT. In fact, it is not unusual to briefly find occasional negative removal efficiencies for highly loaded primary sedimentation tanks.

The composition of primary solids varies widely from day to day, hour to hour, within a facility, and between facilities. Table 10.1 notes the typical composition. The heavy metal content of primary solids depends on the types of industries that discharge to the facility.

The total solids concentration depends on the characteristics of the solids, and the rate at which solids are removed from the primary sedimentation tank(s). If they are removed rapidly, a lower solids concentration can be expected. Some facilities remove solids more slowly or add chemicals to achieve any combination of type I, II, III, and IV settling (see Chapter 4). Furthermore, primary solids with more grit and heavier material will settle and thicken more quickly than other primary solids.

Facilities that receive both sanitary and stormwater flows or have a high contribution of infiltration and inflow will produce primary solids that vary greatly in both volume and volatile solids concentration. Those with inadequate grit removal may produce more primary solids, but they only contain 60% volatile suspended solids (VSS) because of all the inorganics and grit.

2.2 SECONDARY SOLIDS

Secondary solids are produced by biological treatment processes (e.g., activated sludge, biological nutrient removal, trickling filters, rotating biological contactors, and other attached growth systems) that convert soluble wastes or substrates (measured as BOD or chemical oxygen demand [COD]) into microorganisms or biomass. Secondary solids also include some of the particulate matter that remains after primary sedimentation and becomes incorporated into the biomass. The quantity of secondary solids produced is a function of many factors (e.g., the efficiency of the primary treatment process, the ratio of TSS to 5-day BOD [BOD_5], the amount of soluble BOD or COD in the wastewater, and the design parameters of the secondary treatment process). Typically, biological sludges are more difficult to thicken or dewater than primary solids and most chemical solids.

Table 10.1 indicates the typical composition of secondary solids. The WRRFs with high food: microorganism ratios tend to produce secondary solids with higher nitrogen levels than conventional activated sludge facilities do because less endogenous respiration occurs. Facilities with biological phosphorus removal processes produce solids with higher phosphorus levels. Facilities that treat significant amounts of industrial wastewater can produce solids with higher heavy metal concentrations.

2.3 TERTIARY SOLIDS

Tertiary treatment systems vary dramatically based on the intended treatment. In some cases, systems can be implemented to treat for metals or pharmaceuticals, but for purposes of this chapter, tertiary treatment refers to nutrient treatment (nitrogen and/or phosphorus). These systems are typically biological, chemical,

Table 10.1 Typical Characteristics of Residual Solids (Adapted from American Society of Civil Engineers, 2000; U.S. EPA, 1979)

Parameter	Primary Solids	Secondary Solids	Tertiary Solids	Combined Solids	Mesophilic Anaerobic Solids	Aerobic Digested Solids
Total solids, %	1.0–8.0	0.4–2.5	0.1–1.0	3–10	2.0–5.0	2.0–5.0
Total volatile solids, % of total solids	60–85	60–85	10–85	50–80	30–58	38–67
Grease, %	5.0–8.0	5–12	0–5	5–10	0–5	0–5
Phosphorus, % of total solids	0.8–2.8	1.5–3.0	—	—	—	—
Protein, % of total solids	20–30	32–41	—	—	—	—
Cellulose, % of total solids	8–15	—	—	—	—	—
Nitrogen, % of total solids	1.5–4.0	2.4–7.0	—	—	—	—
Volatile solids reduced, %	—	—	—	—	55–65	38–50
pH	5.0–8.0	6.5–8.0	6.5–8.0	6.0–8.0	6.8–7.2	6.0–7.2

or a combination of the two. As such, the sludge derived from these processes varies greatly. Table 10.1 indicates typical composition of sludge found in some of the widely used tertiary systems for nitrogen and phosphorus removal.

When treating for phosphorus, the primary mechanism for removing phosphorus from the liquid stream is to precipitate it in the sludge. These tertiary systems typically produce much higher quantities of sludge than do nitrogen treatment systems. When treating for nitrogen, the primary mechanism for removing nitrogen is by converting it to N_2 gas, so these systems typically have lower quantities of sludge.

2.4 COMBINED SOLIDS

The WRRFs without primary treatment processes generate combined solids. The characteristics of combined solids are dependent on the proportions of each type and their compositions. Table 10.1 indicates typical characteristics of combined, unthickened sludges. Design engineers can estimate how much combined sludge will be produced by adjusting the yield coefficients used to estimate secondary solids production to account for the additional solids.

2.5 CHEMICAL SOLIDS

Chemical solids are the result of adding metal salts, polymer, or lime to wastewater to improve suspended solids removal or precipitate phosphorus. Typically, these chemicals improve thickening and dewatering performance. Modern polymers have become more effective and thus more popular than metal salts for improving thickening and dewater performance. The characteristics of chemical solids are affected by wastewater chemistry, pH, mixing, reaction time, and opportunities for flocculation. That said, these typically contain more heavy metals than other solids because of the heavy metals in the coagulant and those that coprecipitate with iron and aluminum.

Design engineers can estimate the quantity of chemical solids produced based on anticipated chemical reactions. Solids production typically increases in direct proportion to the amount of chemical added; however, competing reactions must be considered. For example, adding ferric chloride will generate more solids than estimated by considering the reaction of ferric chloride to form ferric hydroxide, which preferentially reacts with phosphate, yielding more precipitates than the hydroxide reaction alone. As a general rule, engineers can assume 1 g more solids per 1 g of ferric chloride added. Similarly, adding lime can significantly increase solids production. Both chemicals are also used in the chemically enhanced primary treatment process. Adding inorganic chemicals for chemically enhanced primary treatment or phosphorus removal increases the mass and characteristics of primary solids.

2.6 THICKENED SOLIDS

Solids concentrations and constituents vary greatly depending on the treatment processes used upstream, strength of the wastewater, and other characteristics of the wastewater influent. Typically, storing, conveying, treating, and/or disposing of solids at low concentration is not economical, because the media is primarily water. Therefore, many facilities thicken their solids from the concentrations listed in Table 10.1, to 4% to 10% total solids, depending on the thickening technology and polymers used. Thickening systems are described in more detail in Chapter 11. These systems generally take advantage of gravity to concentrate solids and use types III and IV settling to remove the free water. Some of these systems are passive, like a gravity thickener, and some are mechanical, like a gravity belt thickener or rotary drum thickener. In many cases polymer is used to enhance settling and capture rate with effective particle charge neutralization and bridging.

2.7 DIGESTED SOLIDS

Some common forms of stabilization are anaerobic digester, aerobic digestion, and lime stabilization. New methods of stabilization are growing in popularity, and are discussed further in Chapter 12. Digestion is a biological process that stabilizes organic matter, reducing the total solids inside the digesters by volatilizing a portion of those solids into the gaseous state. Digested sludge is also often less odorous and has few vector attractants than raw sludge. Table 10.1 indicates typical characteristics of traditional high-rate mesophilic digested sludge; this does not include advanced digestion and predigestion processes such as thermal hydrolysis and/or recuperative thickening. Table 10.1 also shows typical characteristics of traditional aerobic digestion.

2.8 DEWATERED SOLIDS

Dewatering solids is generally intended to remove as much water as possible to prepare solids for the most economical means of disposal. In some instances, downstream processes (incineration, drying, soil amending, etc.) require some degree of water to remain in the solids, but in most cases the more water removed results in the least water transported and thus more economical disposal. Dewatering is typically done by mechanically enhancing type III and IV settling. For this reason, dewatering processes typically require some level of thickening upstream. Polymer is typically added to aid dewatering and capture. Similar to other solids handling processes, dewatering is highly dependent on the upstream treatment systems and characteristics of the solids. High ratios of primary to secondary sludge can typically achieve higher percent solids, while solids from phosphorus removal processes are generally more difficult to dewater. One could expect to see 18% to 28% total solids from traditional processes, but in some instances facilities have been able to achieve 30% to 40% total solids.

2.9 THERMALLY DRIED SOLIDS

Thermal drying is the process in which heat is applied to solids to promote the evaporation of water, increase the total solids content, and further reduce pathogens in the solids. Methods of thermal drying include convection (direct drying), conduction (indirect drying), radiation, or a combination therein.

3.0 MASS–VOLUME RELATIONSHIP

The relationship between mass and volume in solids is important for both design and operation calculations. Wastewater solids contain both dry solids and water, but mostly water. A 10% solids mixture contains 90% water and 10% dry solids. Furthermore, the solids within the mixture can contain either volatile or non-volatile solids with each presenting a different specific gravity that affects the final specific weight and gravity of the solids–water mixture. As solids are processed through the WRRF, the weight and volume of solids can change depending on the fractionation of the water and solids within the solids handling devices. For example, solids capture in a dewatering process or volatile solids destruction in a digester can affect the weight and volume of solids leaving a process.

3.1 SPECIFIC GRAVITY OF SOLIDS

To determine the specific gravity of wet solids (S_w), the specific gravity of water and dry solids (S_s) must be known. The specific gravity of water is given as 1.0, whereas the specific gravity of the dry solids depends on the fraction of volatile and non-volatile solids. The specific gravity of the dry solids is given by

$$\frac{100}{S_S} = \frac{p_V}{S_V} + \frac{100 - p_V}{S_{NV}} \qquad (10.1)$$

Where S_S = specific gravity of dry solids;
p_V = weight fraction of volatile solids, (%);
S_V = specific gravity of volatile solids; and
S_{NV} = specific gravity of nonvolatile solids.

Rearranging Equation 10.1 for S_S gives

$$S_S = \frac{100 S_V S_{NV}}{p_V(S_{NV} - S_V) + 100 S_V} \qquad (10.2)$$

The nonvolatile solids typically have a specific gravity of 2.5 while the volatile solids have a specific gravity of 1.0. Substitution into Equation 10.2 then gives

$$S_S = \frac{250}{1.5\, p_V + 100} \qquad (10.3)$$

Using a similar method, the specific gravity of the wet solids can then be determined as follows:

$$S_W = \frac{p_w + (100 - p_w)}{p_w + (100 - p_w)/S_s} = \frac{100S_s}{p_w S_s + (100 - p_w)} \tag{10.4}$$

Where p_w = weight fraction of water $(100 - p_s)/100)$; and
p_s = percent solids.

Example 10.1

Estimate the specific gravity of belt filter press solids that have 22% solids, of which 72% are volatile.

Solution

Determine the specific gravity of the dry solids using Equation 10.3:

$$S_S = \frac{250}{1.5\,p_v + 100} = \frac{250}{(1.5)(72) + 100} = 1.20$$

Determine the weight fraction of water:

$$p_w = (100 - p_s)/100 = (100 - 22)/100 = 78$$

Determine the specific gravity of the wet sludge using Equation 10.4:

$$S_w = \frac{100S_s}{p_w S_s + (100 - p_w)} = \frac{(100)(1.20)}{(78)(1.20) + (100 - 78)} = 1.04$$

Note: Even at 22% solids, the specific gravity is only 4% different than the specific gravity of water. Thus, if solids are 10% or under, we generally assume a specific gravity of 1.0.

3.2 MASS–VOLUME RELATIONSHIP

The relationship between mass and volume of solids can be used to complete mass balances around processes to determine flow characteristics entering and leaving a process. The relationship depends on the percent dry solids, specific gravity of the wet sludge, and knowledge of either the volume of solids or the mass of dry solids. The relationship is given by the following equation

$$M_s = V S_w \gamma_w p_s \tag{10.5}$$

Where M_s = mass of dry solids, kg (lb);
V = volume of solids, m^3 (gal); and
γ_w = specific weight of water, 10^3 kg/m^3 (8.34 lb/gal); and
p_s = *percent* solids (decimal percent).

Example 10.2

If the belt filter press in Example 10.1 has a daily solids production of 1200 kg/d (2640 lb/d), determine the volume of those solids.

Solution

Use Equation 10.5 to determine the volume of solids produced along with the specific gravity calculated in Example 10.1:

$$M_s = VS_w\gamma_w P_s = 1200 \text{ kg/d} = (V)(1.04)(1000 \text{ kg/m}^3)(0.22)$$

$$\text{Volume} = 5.2 \text{ m}^3/\text{d (1374 gpd)}$$

Note: If the specific gravity is ignored (S_w = 1.0), the volume increases to 5.4 m³/d. This shows the effect of the specific gravity on calculations.

3.2.1 MASS BALANCE SOLIDS PROCESSES

Wastewater sludge dewatering processes generate two products: a solids cake that concentrates the solids in a smaller volume, and a liquid stream that consists of water and dissolved substances removed from the feed sludge stream and minimal residual solids (see Figure 10.3). The liquid stream removed from the sludge, which can be called many things based on the type of dewatering process used (e.g., supernatant, decant, underdrainage, filtrate, centrate, or evaporated moisture), is typically recycled back to some point within the main wastewater treatment process and is treated alongside the incoming facility flow or routed to a separate sidestream process before reintroduction into the main wastewater process stream. The performance of the dewatering process is measured using indicators common to all dewatering processes: cake solids content, solids capture, and cake bulk density.

3.2.1.1 Cake Solids Content

Cake solids content is a measure of solids and moisture content of the dewatered cake; it is the weight of total dry solids in the cake divided by the total weight of the cake (solids plus remaining water in the cake matrix), expressed as a percentage. Because the amount of dissolved solids in the water trapped in the cake matrix is minimal when compared to the amount of suspended solids in the dewatered cake, a cake's total solids and TSS are often considered equal. Depending on the characteristics of the particular sludge being dewatered, the type of dewatering used, and the type and amount of pretreatment, the solids content of dewatered sludge cake typically ranges from around 12% to as high as 50%.

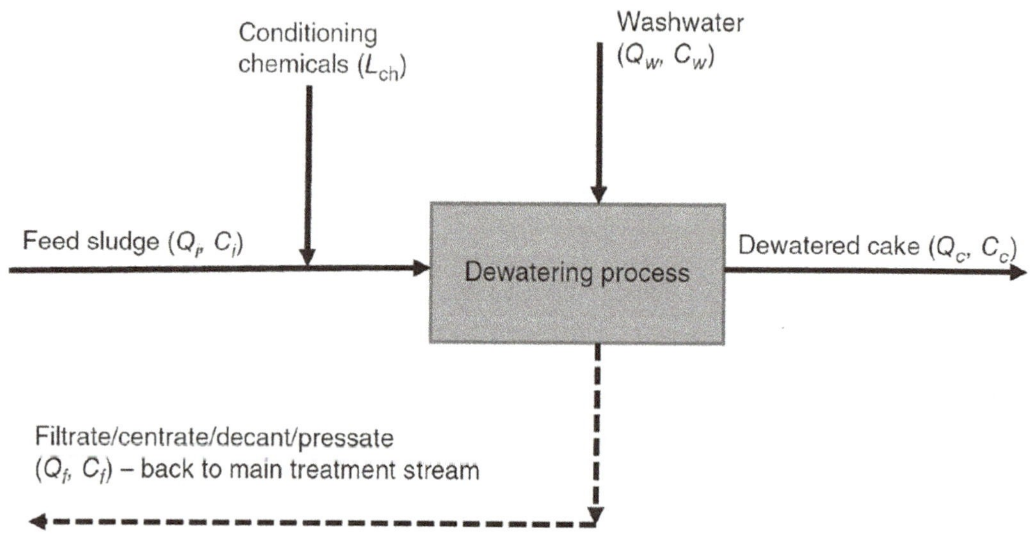

Figure 10.3 Dewatering Solids and Water Balance

A short-cut conversion of percent solids to concentration estimates that each 1% solids is 10 000 mg/L. This assumes that the solids have a specific gravity of 1.0. The percent concentration can be modified to account for specific gravity, as shown in Example 10.3.

3.2.1.2 Solids Capture

Solids capture rate represents the percentage of suspended solids in the sludge feed to the dewatering process that remain in the dewatered cake. If a chemical is added to condition the solids before dewatering, then the weight of the added chemical should be included in the weight of the feed solids when calculating the solids capture rate. The solids capture rate is calculated performing a solids and water balance around the dewatering process (see Figure 10.3). The solids capture rate for a dewatering process can be expressed by the following equations, derived from the mass balance around the dewatering process:

$$E = M_c / M_f \times 100 = C_c \times (C_f - C_r) / [C_c \times (C_c - C_r)] \times 100 \quad (10.6)$$

where E = solids capture, %;
M_c = mass of solids within the cake, kg (lb);
M_f = mass of solids within the feed sludge, kg (lb);
C_c = cake solids concentration, mg/L;
C_f = feed solids concentration, mg/L; and
C_r = filtrate/centrate/decant solids concentration, mg/L.

The capture rates of various dewatering processes typically range from 90% to 99%. Once the target cake solids content and solids capture rate have been determined, design engineers can use Equation 10.6 to determine the solids concentration in the liquid recycle stream returned to the main treatment process.

3.2.1.3 Cake Bulk Density

Cake bulk density is the weight of the cake divided by its volume. Alternatively, it can be calculated by multiplying the specific gravity of the bulk solids by the specific gravity of the weight of water (1000 kg/m^3 or 62.4 lb/cu ft). The bulk density takes into account air voids that are present as the cake solids concentration increases and the product takes on soil-like or granular properties. The bulk density will change as the cake is handled and conveyed. For example, a cake's bulk density as it leaves a dewatering device will be different than the bulk density when the cake is being trucked off site, as some cake compaction and water release may occur as the cake is conveyed or loaded onto trucks. Bulk density is useful when determining what volume of dewatered solids will be generated in a plate and frame press, sizing cake conveyance systems, and determining the number of trucks needed to transport dewatered cake offsite.

Example 10.3

A gravity belt thickener receives 1600 kg/d (3520 lb/d) of solids from an activated sludge system that has been clarified to 0.8% solids. The thickener is able to capture 92% solids in the process and produce a sludge with 4% solids. Ignoring the polymer and wash water used in the system, determine the volume of thickened sludge and the volume of filtrate that is produced from the thickening process.

Solution

A good way to approach this problem is to draw a diagram showing a mass balance of the system. Keep in mind that 1% solids is equivalent to 10 000 mg/L if the specific gravity is assumed to be one (valid when solids are less than 10%).

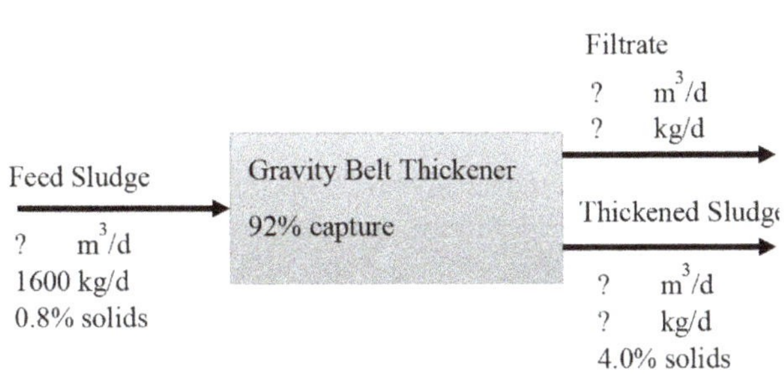

Determine the volume of the feed sludge (Equation 10.5):

$$M_s = VS_w\gamma_w p_s = 1600 \text{ kg/d} = (V)(1.0)(1000 \text{ kg/m}^3)(0.008)$$

$$\text{Volume} = 200 \text{ m}^3/\text{d}$$

Determine the mass of solids in the thickened sludge (Equation 10.6):

$$E = \left(\frac{M_c}{M_f}\right)(100) = 92 = \left(\frac{M_c}{1600 \text{ kg/d}}\right)(100)$$

$$\text{Thickened Sludge} = 1472 \text{ kg/d}$$

Determine the volume of the thickened sludge (Equation 10. 5):

$$M_s = VS_w\gamma_w p_s = 1472 \text{ kg/d} = (V)(1.0)(1000 \text{ kg/m}^3)(0.04)$$

$$\text{Volume of thickened sludge} = 36.8 \text{ m}^3/\text{d}$$

Determine the volume of filtrate:

$$\text{Volume of filtrate} = 200 \text{ m}^3/\text{d} - 36.8 \text{ m}^3/\text{d} = 163.2 \text{ m}^3/\text{d}$$

Note that thickening, blending, and dewatering processes all follow this same format with a constant application of the mass–volume relationship with the mass distribution based on the solids capture percentage.

4.0 SOLIDS STORAGE

The storage needs for liquid residuals depend on where and why they are being stored. Storage can be used to equalize flows and provide operational flexibility. For example, if dewatering equipment is operated only periodically, solids need to be stored between operating periods. The volume of storage required is process-specific. If biosolids are stored between stabilization and land application, the storage volume required will depend on agricultural schedules and climatic issues. If sludge or biosolids will be land applied or surface disposed, the reclamation facility must have enough storage available to allow for periods when the material cannot be used. A 120 to 180 day storage capacity is recommended, but this can be reduced if the facility has an available alternate disposal option. Regulations may also affect process storage systems. Some regulations require storage volumes ranging from 3 to 60 days of biosolids storage (Florida Department of Environmental Protection, 2008). Such storage may simply be excess capacity in the digesters or other process tanks, or it may be separate storage tanks.

The storage and transport of thickened residuals dewatered cake and dried biosolids is an energy-intensive process, potentially involving such equipment as screw pumps, piston pumps, diaphragm pumps, conveyors, and augers. Design engineers are encouraged to perform a life-cycle power consumption analysis, including a cost review, as part of any storage and transport system design. This analysis will give the owner and operator the information they need to make an informed decision as to how power consumption will affect their biosolids processing facility over the life of the proposed improvement.

In addition, many facilities are adding sustainability practices into their operating guidelines. Leadership in Energy and Environmental Design (LEED) and Envision evaluations are becoming popular in the industry, and energy consumption is a key component of these evaluations. Refer to *Sustainability and Energy Management for Water Resource Recovery Facilities* (Water Environment Federation [WEF] et al., 2018) for detailed information on this emerging field.

4.1 STORAGE TANKS

Although more costly per unit volume than earthen basins, storage tanks can be a good choice when residuals volumes are small, land costs are high, or other restrictions make earthen basins infeasible. These tanks typically are cylindrical with either a flat or sloped bottom (see Figure 10.4) (U.S. Environmental Protection Agency [U.S. EPA], 1979). The U.S. Army Corps of Engineers recommends a 4:1 floor slope and a minimum depth of 4.5 m (15 ft) (U.S. Army Corps, 1984). Cylindrical tanks are preferred because they do not have corners, which may become "dead spots." The tanks can be constructed of either concrete or steel. Steel tanks are susceptible to corrosion, however, which design engineers should consider when designing a sludge storage system.

Storage tanks should be mixed to ensure that the discharged solids are homogeneous (Spinosa & Vesilind, 2001). Mixer manufacturers recommend mixing energies ranging from 10 to 12 kW/ML (40–50 hp/mg) (S. Lottman of Siemens, personal communication, 2008). The key is to keep solids suspended without inducing excessive air into the sludge.

The tank(s) may also require aeration, especially if the solids are unstabilized or aerobically digested (U.S. EPA, 1979). If so, the tank's oxygen requirements should be similar to those for aerobic digesters. If the material was stabilized before storage, however, the tank's oxygen requirements to maintain aerobic conditions will be significantly less. Maintaining a minimum dissolved oxygen level of about 0.5 mg/L

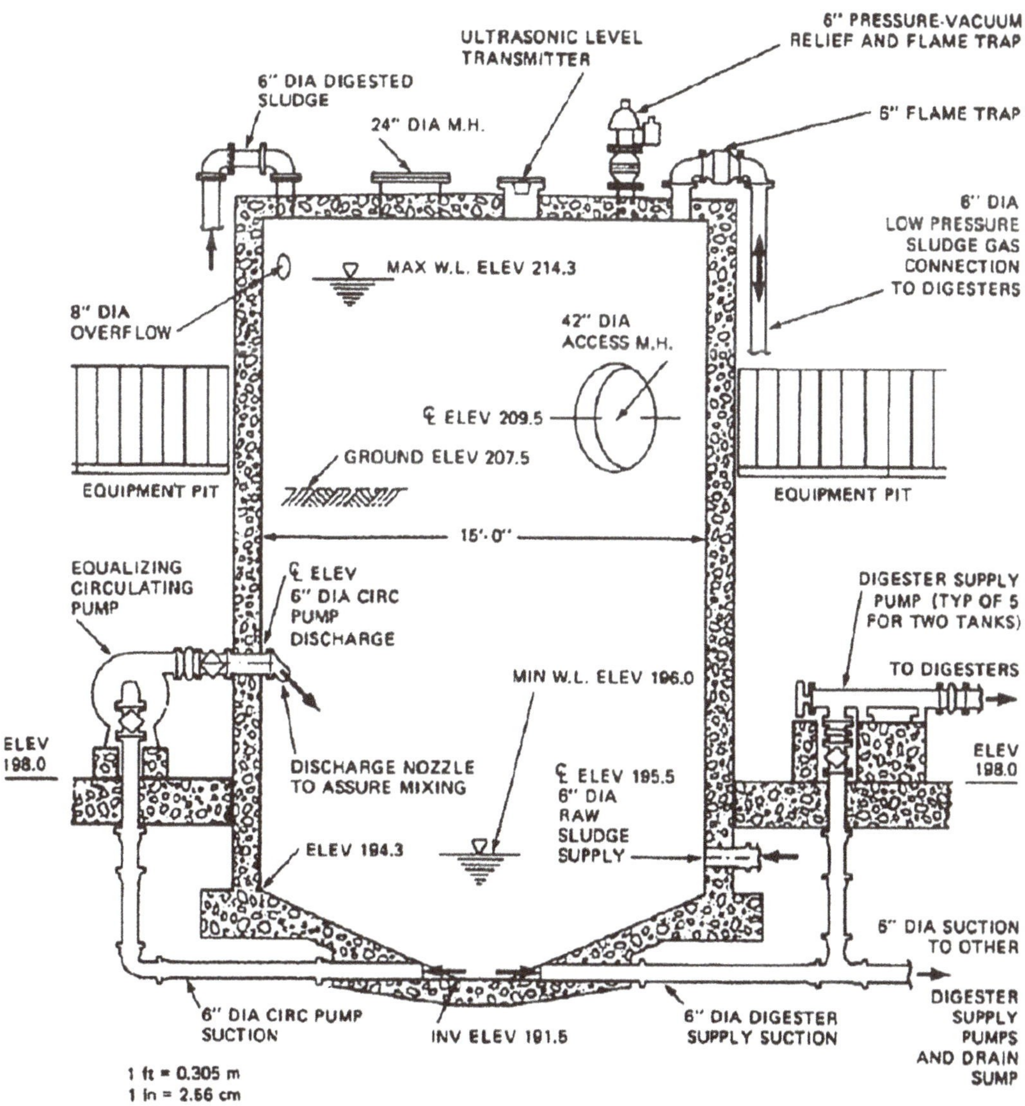

Figure 10.4 A 98 m³ (26 000 gal) Solids Equalization Tank (U.S. EPA, 1979)

should prevent anaerobic activity as long as the basin has adequate mixing. Otherwise, nuisance odors may be generated and more odor control may be required.

Odors may be an issue, depending on the type of residuals and the storage time. Well-stabilized biosolids may be stored for several days without odor control. However, residuals characteristics may change over time, so it is preferable to cover storage tanks to contain and minimize off-site odors. Covers provide the additional benefit of improved control over external environmental elements, such as excessive rainfall and high winds that could affect tank operation. The air in the tank can be vented to an odor control system. The air will be adversely affected by such compounds as dimethyl sulfide, dimethyl disulfide, and longer chain mercaptans. Odorous volatile fatty acid (VFA) compounds may also increase, yielding a sour odor (WEF, 2004).

4.2 STORAGE LAGOONS

A number of small water reclamation facilities use lagoons to store and stabilize solids. There are three types of solids storage basins: aerobic, facultative, and anaerobic (Lue-Hing et al., 1998).

4.2.1 AEROBIC STORAGE LAGOONS

Aerobic storage basins are designed to provide aeration and maintain a minimum dissolved oxygen concentration throughout the basin. Their aeration requirements should be similar to those for aerobic digesters, except that they take into account prior solids stabilization. Because of their size and earthen construction, a piped aeration system is typically not used; subsurface or surface free-standing mechanical aerators are most often used.

4.2.2 FACULTATIVE STORAGE LAGOONS

Facultative basins are unmixed and typically consist of three layers: an aerobic surface layer 0.3 to 1.0 m (1 to 3.3 ft) deep, a deeper anaerobic zone, and a sludge storage zone at the bottom. Both the aerobic and anaerobic zones are biologically active; anaerobic stabilization substantially reduces solids volume. The aerobic zone receives oxygen via surface transfer from the atmosphere, algal photosynthesis, and (if provided) surface mix aerators, which are operated periodically to break up scum on the pond surface and optimize oxygen transfer. Most of the satisfactory installations use brush-type surface mixers (U.S. EPA, 1979). The oxygenation rate is low, however, so U.S. EPA recommends that these basins only be used for anaerobically digested solids.

These basins are typically designed based on a volatile solids loading rate of 0.097 kg/m^2·d (0.023 lb/sq ft/d) (Lue-Hing et al., 1998). They are typically 5 m (16 ft) deep to provide enough space for aerobic and anaerobic layers. Because facultative solids lagoons' "sour," odor can be a major issue. When designing such storage basins, engineers should consider prevailing wind patterns, and minimize odor potential via proper loading and surface mixing.

4.2.3 ANAEROBIC BASINS

Anaerobic storage basins are similar to earthen basins but differ in that oxygen transfer from the surface is not considered in design. Therefore, earthen basins can be deeper than facultative ponds. Because maintaining an aerobic zone is not a key parameter, solids loading to earthen basins can be higher than with facultative ponds. Anaerobic ponds have essentially the same advantages and disadvantages as facultative ponds (Lue-Hing et al., 1998).

4.3 DEWATERED SOLIDS

Dewatered biosolids cake (biosolids) typically are stored somewhere before receiving more treatment (e.g., heat drying) or being transported off-site for use or disposal. For short-term disposal operations (further off-site treatment or direct to landfill hauling, for example), biosolids will only be held for a few days or weeks. In this case, storage alternatives include large roll-off containers, 18-wheel dump trailers, concrete bunkers with push walls, or bins with augurs.

If the biosolids will be land applied or surface disposed, however, long-term storage may be required. Weather-related application effects, regulatory limits on land-applied nutrient loading, and landfill restrictions are three instances that may dictate long-term storage need. In these cases, biosolids are often stockpiled on concrete slabs or other impervious pads. When designing long-term storage facilities, engineers need to consider facility buffering requirements (minimum distances between the facility and the nearest house or industry), whether odor control is required, and accessibility of machines to distribute the biosolids during storage or remove biosolids when removal is an available option. Designers also need to determine whether the storage facility should be open or covered, as rewetting of dewatered biosolids is often undesirable for land application and can cause biological reactivation of pathogens. For further guidance, see U.S. EPA's *Guide to Field Storage of Biosolids* (U.S. EPA, 2000).

Odor control can be an issue with dewatered biosolids, especially when larger quantities are stored or the solids storage area is relatively close to neighboring businesses or residences. Storing the cake for 20 to 30 days at 25 °C significantly cuts odor generation (Novak et al., 2004; Water Environment Research Foundation [WERF], 2008). Adding alum after digestion also reduced storage-related odors (Novak et al., 2004). However, the best way to minimize dewatered biosolids odors is to optimize the solids treatment processes before dewatering, according to the WERF study (2008).

4.4 DRIED BIOSOLIDS

Dried biosolids typically are stored either on site or at a land application site before disposal or beneficial use. They may be stored in stockpiles or silos. Because dried biosolids contain a significant amount of combustible organic material that can be released as dust, temperature control is important. If silos are used, engineers should design them to promote cooling and maximize heat dissipation, with fire safety. Tall, narrow silos are more effective at heat dissipation than wide ones. Narrow silos also make fires easier to control. However, if the silo is too narrow, it will make relief venting problematic. If multiple silos are used, there should be procedures to ensure that they are emptied cyclically to avoid exceeding safe residence times.

It is critical that dried biosolids be stored safely. Dried biosolids can self-ignite and have done so on several occasions. Explosions have also occurred in dried biosolids storage silos. If the storage area's temperature rises above a critical point, dried biosolids can begin to self-heat. Design engineers can calculate the critical temperature using isothermal basket tests and considering the effects of storage volume and residence time. In the United Kingdom, water recovery professionals performed a series of tests on 1 m^3 (35 cu ft) of biosolids and found that the self-heating temperature was typically above 60 °C (Health and Safety Executive, 2005). The Health and Safety Executive recommends that biosolids should be cooled to no more than 40 °C before being sent to a silo. (If the silo is particularly large, the critical temperature may be even lower.) Controlling biosolids temperature not only prevents silo fires but also avoids self-heating further along in the biosolids use or disposal process. Biosolids that are hotter than the critical temperature must not be removed from the WRRF until they are so cool that there is no risk of a fire either in transit or at the use or disposal site. There are cooling devices manufactured specifically for biosolids, and design engineers may want to consider using them in larger installations.

5.0 SLUDGE TRANSPORT

When designing solids transportation systems, engineers should consider service life and favor equipment that is flexible enough to remain useful despite changing technology, regulations, economics, and solids characteristics. They should also investigate full-scale working systems whenever possible to determine actual operating conditions and costs, and then make allowances for uncertainties that are specific to the project.

5.1 FLOW CHARACTERISTICS

The flow characteristics (rheology) of sludge and biosolids (residuals) cannot be simply defined. They vary widely from process to process and from facility to facility (Borrowman, 1985; Carthew et al., 1983; Levine, 1986; Mulbarger et al., 1981; U.S. EPA, 1979; Wagner, 1990). As a result, the characteristics and resulting headloss are difficult to predict. Solids content is an important rheological parameter. Generally, the higher a fluid's solids concentration, the higher its shear stress, density, and viscosity. Viscosity increases exponentially as solids concentration increases (Brar et al., 2005, and references therein).

A residual's rheological characteristics are strongly affected by the kind of treatment the material has undergone (Brar et al., 2005; Guibaud et al., 2004). For example, a raw, fresh, nonhydrolyzed solids stream at 3% solids has a higher apparent viscosity than a 4% material that has been thermally alkaline hydrolyzed (Brar et al., 2005). Similarly, digested biosolids can be pumped more easily than raw or undigested sludge

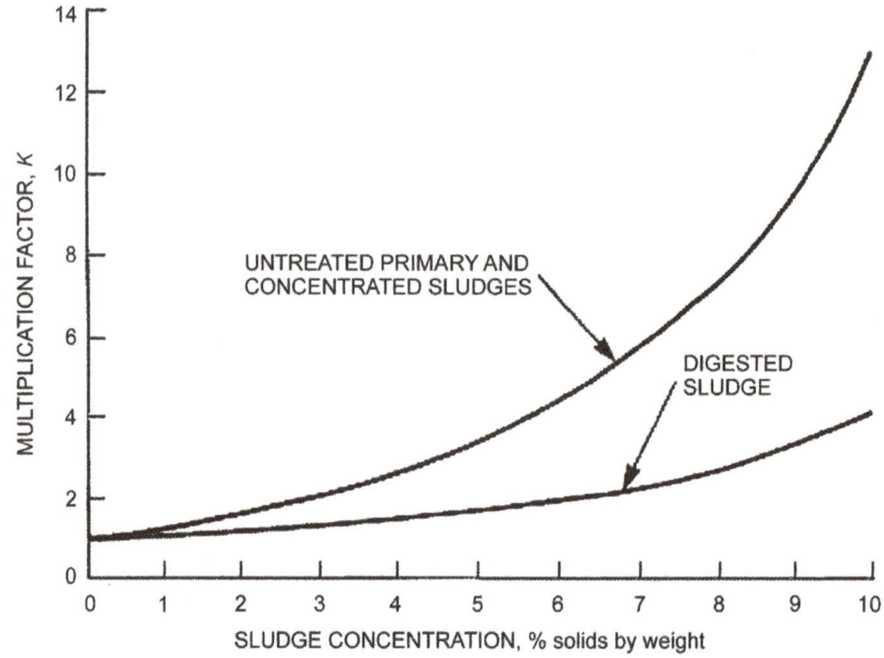

Figure 10.5 Approximate Multiplication Factors (Based on Solids Concentration) to be Applied to Headlosses Calculated for Water in Laminar Flow (U.S. EPA, 1979)

with the same moisture content (see Figure 10.5) (U.S. EPA, 1979). To derive approximate headlosses for residuals, design engineers should calculate the headloss for water and then multiply the result by the factor in Figure 10.5 corresponding to the residual's solids concentration. This will provide a rough estimate when velocities are between 0.76 and 2.4 m/s (2.5 and 8 ft/sec) and no **thixotropic** behavior or serious obstructions, such as grease, are anticipated.

Thixotropic refers to the time-dependent ability of some emulsions and solids to change viscosity when left at rest.

For design purposes, residuals can be divided into several distinct categories based on solids concentration as presented in Table 10.2. For more detailed information on solids concentrations and residuals behavior, see *Conveyance of Residuals from Water and Wastewater Treatment* (American Society of Civil Engineers [ASCE], 2000). Dilute residuals contain less than 5% solids. Such residuals typically include waste activated sludge (WAS), which typically contain less than 2% solids, and primary sludge, which contains less than 5% solids.

Table 10.2 Classification of Water and Wastewater Thixotropic Residuals by Type and Solids Content

			Thixotropic Solids		Granular-Compactable	
Solids Type[a]	Temperature (°C)	Thickened Solids, % total solids	Low-Medium Viscosity, % total solids	Medium-high Viscosity, % total solids	Wet Solids, % total solids	Dry Solids
RPS	10–20	5–10	20–26	24–40	35–65	65+
RWAS	10–25	3–7	10–16	14–15	22–65	65+
R (PS + WAS)[b]	10–25	4–8	13–20	18–31	28–65	65+
DPS	20–30	5–10	20–28	24–40	35–65	65+
SWAS	20–30	3–7	10–16	14–25	22–65	65+
D (PS + WAS)[b]	20–30	4–8	13–20	18–31	28–65	65+
Alum $[Al(OH)^3]^{(2)}$	5–25	3–7	8–15	13–30	25–60	65
Iron $[Fe(OH_3)]^{(2)}$	5–25	3–7	8–15	15–35	30–60	65
Lime $(CaCO_3)$	5–25	15–30	8–15	25–50	70–80	80+

[a]R = raw, D = digested, PS = primary sludge, and WAS = waste activated sludge.
[b]50 : 50 mixture of PS and WAS.

Thickened residuals, which are typically produced via a mechanical thickening process, have higher solids concentrations than gravity settled solids. Such residuals range from WAS with a 3% total solids content to primary sludge with up to 10% total solids content. Thickened residuals typically have a much higher viscosity and cannot be handled reliably by centrifugal pumps.

Dewatering processes typically used at water resource reclamation facilities tend to make residuals thixotropic, and their rheology is dependent on both time and applied stress. Although engineers can design pumping systems that handle thixotropic materials, the limited availability of related data makes site-specific studies important. Further dewatering, such as processes used to make pelletized fertilizer, makes residuals granular. Such residuals cannot be pumped; instead, they must be transported via conveyors or similar devices.

When designing solids transport systems with kinetic pumps (e.g., centrifugal), engineers need to be accurate rather than conservative. Such systems are most efficient when the system curve matches the pump curve. However, engineers should be conservative when designing transport systems for thicker residuals, either liquid or dewatered, that rely on positive displacement pumps.

5.2 LIQUID SOLIDS

There are two basic methods for transporting liquid solids at water reclamation facilities: trucking and pumping. Residuals are typically trucked off site and pumped on site. Liquid solids are defined as any solids that can be pumped and are typically those that have not been dried or dewatered.

5.2.1 TRUCKING

Although not typically part of the design process, trucking is often used to transport liquid solids, especially to land application sites. Liquid sludge and solids may also be trucked to another site for further treatment. Trucking liquids can be larger challenge than transporting dewatered or dried biosolids, and there is some basic information designers should know. Liquid sludge and biosolids are typically transported via tanker trailer trucks. Such trucks typically have nominal capacities ranging from 22 680 to 34 020 L (6000 to 9000 gal). However, depending on weight restrictions, a tanker trailer may not be filled completely. A 5850 kg (12 900 lb), 26 460 L (7000 gal) trailer typically is 13.1 m (40 ft) long and 2.4 m (8 ft) wide. Some are equipped with baffles to control movement and resulting surges of the liquid.

5.2.2 PUMPING

Pumping systems are an intrinsic part of solids management at water reclamation facilities. They typically transport solids from primary and secondary clarifiers to thickening, conditioning, or digestion systems. They are also used to transport solids from biological processes to further treatment units. Finally, pumping systems are used to transport solids from gritting facilities to temporary storage areas. Although specifying only one type of pump for all of a facility's solids transport systems might seem advantageous, the wide range of conditions involved typically exceeds the capabilities of any given pump. Fortunately, many types of pumps are available (see Table 10.3).

When designing pumping systems, design engineers should begin by asking: What sort of residuals will be pumped? Kinetic pumps, especially recessed impeller pumps, can handle some types of residuals, but other types may require positive displacement pumps. Kinetic pumps have lower capital costs (especially in large sizes), lower maintenance costs, and smaller footprints. They are also available in submersible form (although conventional dry well pumps are preferred for most applications). Positive displacement pumps have better process control because the pumping rate is less affected by fluid viscosity. They function better over the entire head range from zero to maximum without damaging the pump or motor, or changing drive speed. They work better under high pressure and at low flows. They are also less sensitive to non-ideal suction conditions (e.g., entrained air and gas) and less likely to disrupt fragile floc particles in flocculated solids.

The traditional approach to designing residuals transport systems is to minimize the pumping distance and apply a conservative multiplier to headlosses calculated for equivalent flows of water. However, this

Table 10.3 Sludge Pump Applications by Principle

Principle	Common Types	Typical Applications
Kinetic (rotodynamic) pumps	Nonclog mixed flow pump[a]	Grit slurry,[b] incinerator ash slurry[c]
	Recessed impeller pump (vortex pump, torque flow pump)	Unthickened primary sludge[b,c]
	Screw centrifugal pump	Return activated sludge[c,d]
	Grinder pump	Waste activated sludges from attached growth biological processes[c]
		Circulation of anaeronic digester[b]
		Drainage, filtrate, and centrate
		Dredges on sludge lagoons
Positive displacement pumps	Plunger pumps	Waste activated sludge
	Progressing cavity pump	Thickened sludges (all types)
	Air operated diaphragm pump	Unthickened primary sludge
	Rotary lobe pump	Feed to dewatering machines
	Pneumatic ejector	Unthickened secondary sludges
	Peristaltic pump	Dewatered cakes[e]
	Reciprocating piston	
Other	Air lift pump	Return activated sludge
	Archimedes screw pump	

[a]Limited solids capability; useful in larger sizes for return activated sludges. In most other applications, recessed impeller pumps are more popular.
[b]Abrasion is moderate to severe. Abrasion resisting alloy cast iron is usually specified.
[c]May contain precipitates from aluminum or iron salts added for phosphorus removal.
[d]Particular need for reliable flow meters, for process control, in this application.
[e]Reciprocating piston pumps and progressing cavity pumps only.

approach can be inaccurate, especially at higher solids concentrations with non-Newtonian characteristics. Such inaccuracies may not matter for short pumping distances, but they can be problematic for longer distances or critical applications.

The need to pump residuals long distances has increased in the last 20 years, so researchers have been developing methods to predict site-specific friction losses in pumping systems more accurately (Carthew et al., 1983; Honey & Pretorius, 2000; Mulbarger et al., 1981; Murakami et al., 2001; Wagner, 1990). Study results have shown that, once rheological properties have been determined, the Bingham plastic model (Carthew et al., 1983; Mulbarger et al., 1981) or the pseudoplastic model (Honey & Pretorius, 2000; Murakami et al., 2001) for **non-Newtonian** fluids may describe how reclaimed water residuals flow. They can also predict the critical velocity at which laminar flow changes to turbulent flow. In the turbulent flow range, dilute residuals obey conventional flow relationships for Newtonian fluids (Mulbarger et al., 1981).

Non-Newtonian fluids are fluids for which the shearing stress is not linearly related to the rate of shearing strain. The viscosity of these fluids change when under force.

In addition to friction-related headloss, design engineers should calculate static head and "minor" headlosses from valves and fittings. The sum of all these headlosses is the total dynamic head that the pump must provide. Design engineers also need to ensure that the available net positive suction head is sufficiently more than is needed, and should consider changes in thickened solids characteristics and evaluate multiple duty points, if necessary.

5.2.2.1 Dilute Sludge

Clarifiers often produce a relatively dilute settled solids (maximum concentrations of 1.2% to 1.5% are typical for activated sludge). At velocities greater than 0.3 to 0.6 m/s (1 to 2 ft/sec), such solids are in the turbulent flow regime and have a headloss essentially equal to that of water (Mulbarger, 1997). At lower velocities, the flow becomes laminar, and headlosses increase sharply. Therefore engineers should design dilute solids pumping systems to maintain a minimum velocity of 0.6 to 0.75 m/s (2 to 2.5 ft/sec) whenever possible to ensure turbulent flow.

5.2.2.2 Thickened Residuals

The concentration at which sludge and liquid biosolids can be defined as "thickened" depends on the type of solids and the preceding treatment processes. When designing pumping systems for thickened solids, design engineers can use the Darcy–Weisbach and Manning equations for water to determine headloss—regardless of the solids' flow regime (laminar, transition, or turbulent)—and apply a solids correction factor to the final calculation. The correction factor for residuals with up to 10% solids may be found in Figures 10.6 and 10.7.

5.2.2.3 Bingham Plastic Model

Another design approach is based on the assumption that the flow of thickened sludge and biosolids follows the **Bingham plastic** model. To use this model, design engineers need to know a solids' yield stress (S_y) and coefficient of rigidity (R_c), which may be determined experimentally. If solids-specific data are not available, then designers can use Figures 10.8 and 10.9 (ASCE, 2000) to estimate these values. Similar graphs have been created by Battistoni (1997); Guibaud et al. (2004); Laera et al. (2007); and Mori et al. (2008).

Bingham plastic is a fluid that can withstand some stress and thus can be either a fluid or a solid depending on the shear stress on the fluid.

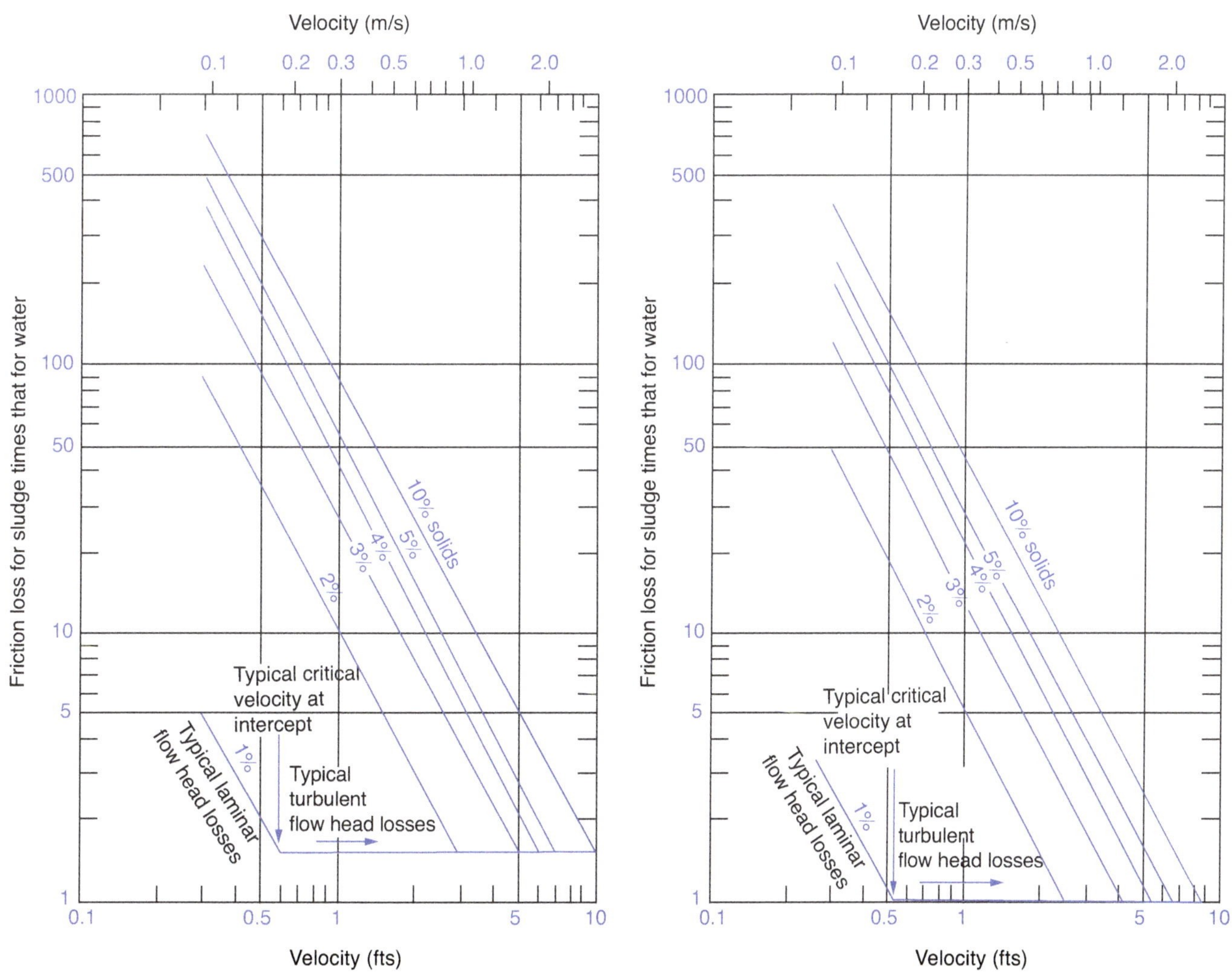

Figure 10.6 Multiplication Factor for Residuals Headloss (a) Routine Design and (b) Worst-Case Design (Sanks et al., 2008)

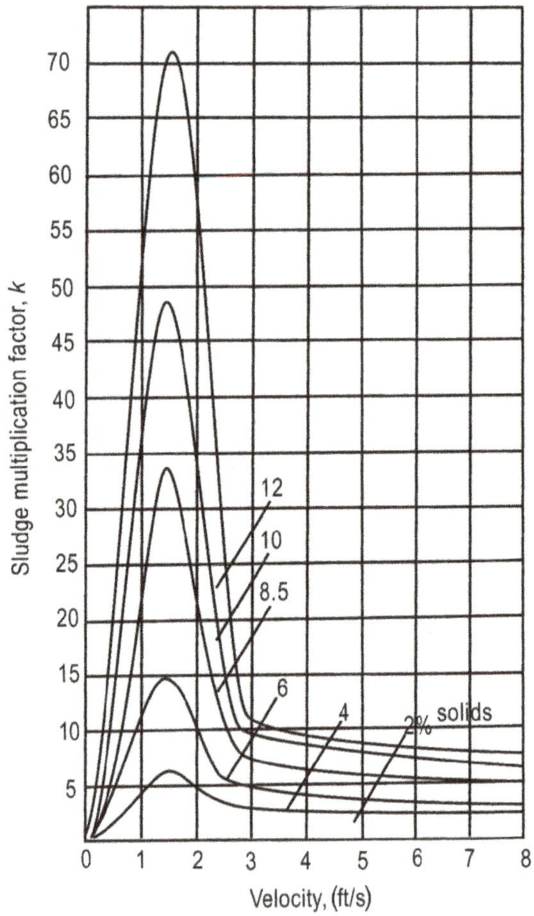

Figure 10.7 Multiplication Factor for Residuals Headloss (From Metcalf & Eddy, Inc./AECOM, 2014, with permission from the McGraw-Hill Companies)

The Reynolds number is a measure of the ratio of inertia forces on an element of fluid to the viscous forces on an element.

Once the rigidity coefficient and yield stress are known, designers can calculate the **Reynolds number** to determine the flow regime as follows:

$$Re = \rho VD/R_c \tag{10.7}$$

where ρ = fluid density (kg/m^3),
V = velocity (m/s),
D = pipe diameter (m), and
R_c = rigidity coefficient (Pa-s or N-s/m^2).

If $Re < 2000$, the flow is laminar, and if $Re > 3000$, the flow is turbulent.

For laminar flow $Re < 2000$ designers can calculate headloss using the Buckingham equation:

$$\frac{H}{L} = 32\left(\frac{S_y}{6\rho gD} + R_c\frac{V}{\rho gD^2}\right) \tag{10.8}$$

where H/L = headloss per unit length (m/m),
S_y = yield stress (Pa or N/m^2), and
g = gravitational acceleration (m/s^2).

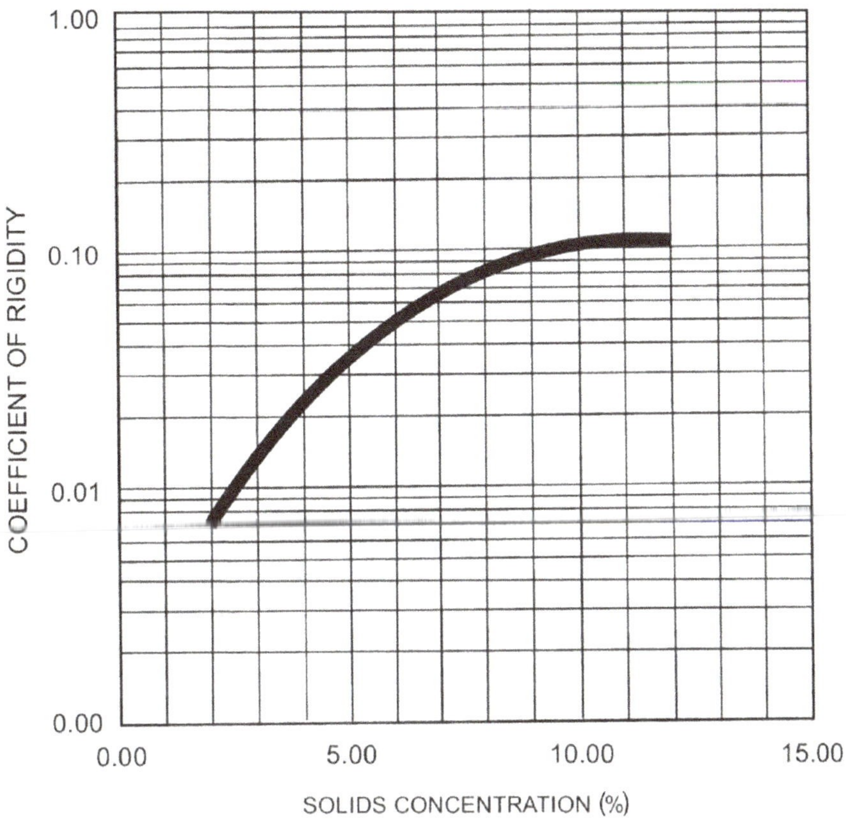

Figure 10.8 Coefficient of Rigidity Versus Solids Concentration (ASCE, 2000)

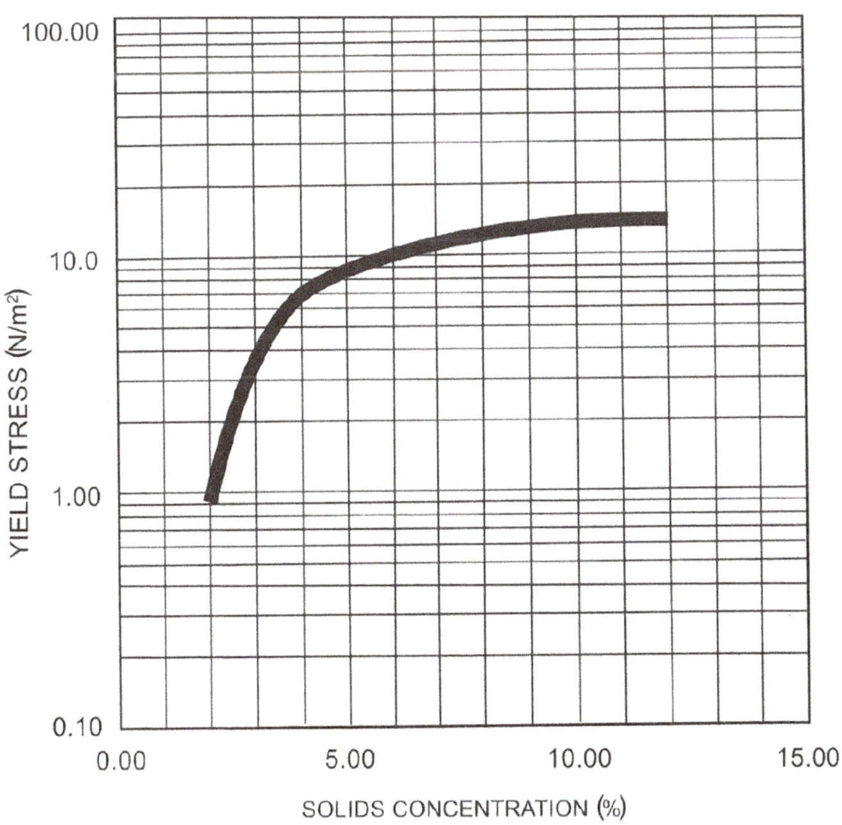

Figure 10.9 Yield Stress Versus Solids Concentration (ASCE, 2000)

For turbulent flow where velocities are greater than the upper critical velocity, or when $Re > 3000$, engineers can solve the Hazen–Williams equation for turbulent water and apply the solids correction factor (see Figures 10.6 and 10.7) to the result. When using this equation, design engineers should assume that C equals 140 under normal conditions and 112.4 under worst-case design conditions.

Designers can also use the Reynolds and Hedstrom numbers to calculate headlosses. To find the Reynolds number, see Equation 10.7. The Hedstrom number is calculated as follows:

$$He = D^2 \rho S_y / R_c^2 \tag{10.9}$$

After calculating the Hedstrom and Reynolds numbers, the friction factor (f) is then calculated using Figure 10.10. Design engineers then should use the Darcy–Weisbach equation to calculate headloss:

$$\Delta P = 2 f \rho L V^2 / D \tag{10.10}$$

Where ΔP = pressure headloss (Pa).

Example 10.4

Design engineers plan to pump a 3% sludge through a 152.4 mm (6 in.) diameter pipe for a distance of 3048 m (10 000 ft) at a velocity of 0.61 m/s (2 ft/sec) under normal conditions. The material is assumed to behave like a Bingham plastic. Determine the headloss due to friction in this system.

Solution

Use Figures 10.8 and 10.9 to determine the solids' yield stress (S_y) and coefficient of rigidity (R_c). From the figures, R_c = 0.015 N-s/m² and S_y = 3.4 N/m².

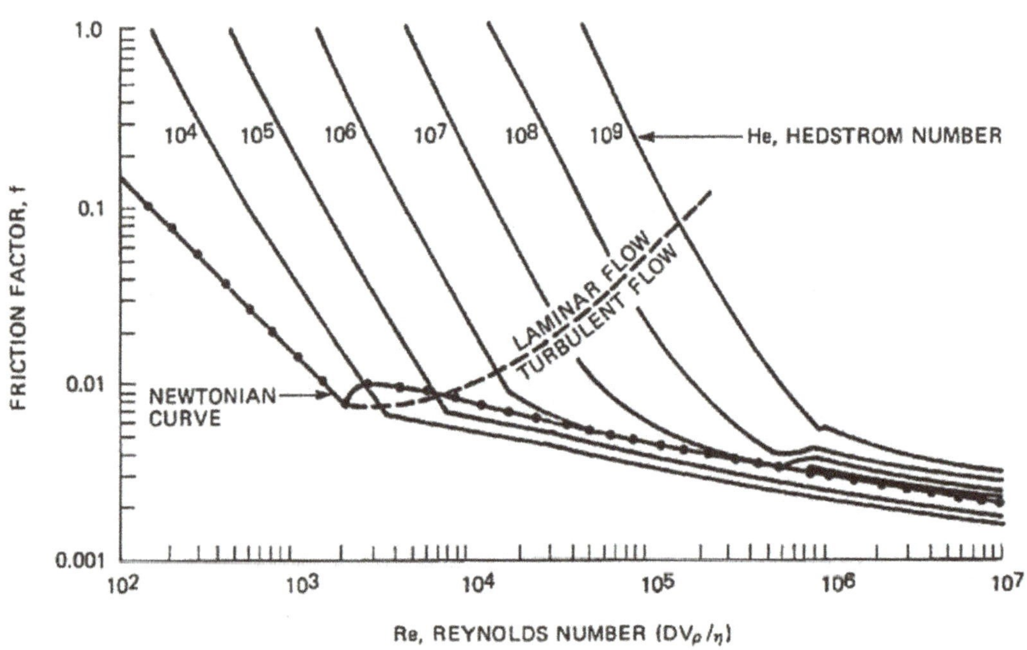

Figure 10.10 Friction Factor (f) for Solids, Assuming Bingham Plastic Behavior (U.S. EPA, 1979)

Calculate the Reynolds to determine the flow regime for the system (Equation 10.10).

$$Re = \rho V D/R_e = \frac{(1000\ \text{kg/m}^3)(0.61\ \text{m/s})(0.152\ \text{m})}{0.015\ \text{N} \cdot \text{s/m}}$$

$$R_e = 6181\ \text{(turbulent flow)}$$

There are two choices at this point: use the Hazen-Williams equation or use the Darcy-Weisbach equation. We will use the Hazen-Williams equation with the correction factor of 7.5 from Figure 10.7.

Find the flow in the pipe:

$$Q = VA = (0.61\ \text{m/s})\ \frac{\pi(0.152)^2}{4} = 0.011\ \text{m}^3/\text{s}$$

In SI units, the Hazen-Williams equation is (use C = 140 per text for normal conditions):

$$\frac{h_L}{L} = \frac{10.7\, Q^{1.85}}{C^{1.85} D^{4.87}} = \frac{h_L}{3048\, m} = \frac{(10.7)(0.011\, m^3/s)^{1.85}}{(140)^{1.85}(0.152\, m)^{4.87}}$$

$$\text{headloss} = 8.0\, m$$

Total headloss due to friction = (8.0 m)(7.5) = 60 m.

Notice the difference (60 m vs. 8 m) in the headloss for a 3% solids material versus pure water. It is important that designers take solids into account when determine the friction headlosses for pumping. If the flow would have been laminar, the headloss would have been determined from Equation 10.8.

5.3 DEWATERED SOLIDS

Modern dewatering operations can produce biosolids containing 15% to 40% total solids or more, depending on the conditioning chemicals and dewatering equipment used (see Chapter 11). The consistency of such biosolids ranges from pudding to damp cardboard, so they will not exit the dewatering equipment by flowing via gravity into a pipe or channel. Instead, they must be transported via mechanical devices, such as positive displacement pumps or belt or screw conveyors. Where there are receiving containers directly below dewatering equipment, the biosolids can be dropped by gravity into the containers.

5.3.1 PUMPING/CONVEYORS

Both progressing cavity pumps and hydraulically driven reciprocating piston pumps can handle dewatered biosolids. Compared to belt or screw conveyors, these pumps better control odors (because the biosolids travel in an enclosed pipe), eliminate spills, and have fewer maintenance requirements. The pumps also have much smaller footprints and, therefore, are suitable in buildings with space constraints. They can even reduce noise levels in some cases. However, pumps often need more electricity than conveyors to move a given volume of biosolids.

Choosing which pump to use depends on the application. Progressing cavity pumps provide a steady flow, while hydraulically driven reciprocating piston pumps pulsate (List et al., 1998). Progressing cavity pumps typically are preferred in applications where the biosolids are thinner and transport distances are short. Hydraulically driven reciprocating piston pumps are more expensive, but may handle greater pressures and thicker cake. Biosolids process discharge piping can be a high-pressure environment.

Engineers should avoid constrictions (e.g., smaller than line-size valves, valves that do not have full port openings, and short radius bends) and minimize any type of fitting when designing discharge piping; the focus should be on maximizing straight run of pipe to minimize headloss. The piping should be large enough that theoretical biosolids flow velocities never exceed 0.15 m/s (0.5 ft/sec), although maximum velocities of 0.08 m/s (0.26 ft/sec) are preferred, especially if dewatered biosolids concentrations exceed 30%. The pipe should be designed to allow flushing and pigging.

Unlike liquid solids, where pumping design equations are available for solids concentrations up to 12%, design approaches for pumping dewatered biosolids are more site-specific. Field testing is highly recommended, especially in relatively long-distance applications. A pipe pumping biosolids containing more than 30% total solids should not be more than 150 m (500 ft) long, and line lubrication using high-pressure fluid delivery systems is highly recommended.

Comparing actual field data, researchers developed a simple headloss equation for biosolids with 20% total solids concentration based on the pseudoplastic model, warning that it might only apply to the specific case (Bassett et al., 1991). Equation 10.11 may be valid for all pipe sizes from 100 to 300 mm (4 to 12 in.) with velocities ranging from 0.015 to 0.43 m/s (0.05 to 1.4 ft/sec):

$$\Delta P_d = \frac{3218.5}{D^{1.28}} + \frac{56636Q}{D^2} \tag{10.11}$$

where ΔP_d = pressure drop (N/m²·m),
D = inside diameter of the pipe (m), and
Q = biosolids flow (m³/s).

Table 10.4 illustrates the effect of pipe size on pipeline headloss (based on Equation 10.11). It shows that headloss increases almost 50% for each decrease in nominal pipe size. It also shows that pumping thicker biosolids (from 20% to 28% total solids) causes a minimal increase in headloss (approximately 10%) when a large enough pipe (e.g., 300 mm or 12 in.) is used.

Table 10.5 is a compilation of data from a progressing cavity pump manufacturer on transporting biosolids with different solids percentages (Bourke, 1997). Headlosses ranged from 6 to 68 KPa/m (0.25 to 3.0 psi per foot) of straight 150 mm (6 in.) pipe (Bourke, 1997). The table shows that centrifugally dewatered biosolids is pumped more easily than rotary drum dewatered cake. Centrifugally dewatered biosolids is also pumped more easily than belt pressed dewatered biosolids.

5.3.2 LINE LUBRICATION FOR LONG-DISTANCE PUMPING

A method for reducing pipe friction losses (mainly for long-distance pumping applications), called boundary layer injection, involves injecting a liquid lubricant into the discharge pipe via an annular ring, which distributes the liquid equally around the pipe's perimeter to create a "boundary layer." The lubricant could be water, a polymer, or oil-based. The designer should research the impact of adding this fluid on the end use of the dewatered biosolids as part of the design process. Field tests indicate that such lubrication can cut discharge pressure up to 80% (List et al., 1998). For a detailed description of this process, see *Conveyance of Wastewater Treatment Plant Residuals* (ASCE, 2000).

5.3.3 PROGRESSING CAVITY PUMPS

Progressing cavity pumps have low capital costs and can consistently transport thinner dewatered biosolids over short distances, typically 20% total solids or less. To ensure effective operations, pump manufacturers

Table 10.4 Headloss Data for Pumping Cake at Near-Maximum Recommended Fluid Velocities

Pipe Size	Percent Solids Concentration	Velocity (m/s)	H_f (m of water/10 m)	H_f (psi/ft)	Change in H_f From One Size Smaller
100 mm (4 in.)	20 (Bassett, 1991)	0.08	64.0	2.77	57%
150 mm (6 in.)	20 (Bassett, 1991)	0.08	41.0	1.77	46%
200 mm (8 in.)	20 (Bassett, 1991)	0.08	28.0	1.21	55%
300 mm (12 in.)	20 (Bassett, 1991)	0.08	18.0	0.78	Baseline
300 mm (12 in.)	28 (List et al., 1998)	0.10	19.5	0.86	10%

Table 10.5 Headloss Data (in m of water/10 m [psi/ft]) as Reported in Bourke (1997) for Fluid Velocities Less Than 0.06 m/s (2 ft/sec)

Method	Percent solids (%)	100 mm (4 in.)	150 mm (6 in.)	200 mm (8 in.)
Rotary drum	20–30	Slightly more than double of 150 mm	7.62–11.5 (0.33–0.50)	About half of 150 mm
Centrifuge	20–30	Slightly more than double of 150 mm	50% less than rotary drum	About half of 150 mm
Rotary drum and then heat treated	45	Slightly more than double of 150 mm	23.0–35.0 (1.00–1.50)	About half of 150 mm
Filter press	65	Slightly more than double of 150 mm	35.0–46.0 (1.50–2.00)	About half of 150 mm

and WRRF personnel recommend that design engineers use large diameter pipe to reduce friction to suitable levels. Design engineers should also limit the pumping distance to 50 m (164 ft) or less. Further distances (up to 90 m [295 ft]) can be researched by looking at other facilities and working with the manufacturers.

5.3.4 HYDRAULICALLY DRIVEN RECIPROCATING PISTON PUMPS

These pumps can handle both screenings and biosolids containing 5% to more than 40% dry solids that were dewatered via belt presses, centrifuges, plate and frame presses, screw presses, or rotary presses. The pump's principal advantage is that it can move high solids materials. Hydraulically driven reciprocating piston pumps have higher capital costs than other pumps, but can move very thick dewatered biosolids that other types of pumps cannot.

5.3.5 CONVEYORS

Conveyors typically move wet or dry solids (e.g., primary grit, screenings, and dewatered biosolids) that are not easily pumped. Municipal recovered water facilities typically use either belt or screw conveyors.

5.4 DRIED SOLIDS

5.4.1 BELT CONVEYORS

Belt conveyors are one of the most widely used and efficient means of transporting bulk materials. They typically transport relatively dry material (15% or more total solids). A chute deposits material onto the top of the belt at one end, and the belt transports it to the other end, where the material is discharged into another chute. Belt conveyors range from 0.35 to 1.5 m (1.15 to 5 ft) wide and can accommodate a broad range of capacities, speeds, and distances. The smallest belt conveyors can handle up to 76 kg/s (30 ton/hr) and operate at speeds up to 100 m/min (325 ft/min). The largest ones can handle up to 1134 kg/s (4500 ton/hr) and operate at speeds up to 200 m/min (650 ft/min). They work well transporting material 6 to 200 m (20 to 660 ft); if the material needs to move less than 6 m (20 ft), other conveying devices are preferred.

Belt conveyors are appropriate for transporting dried biosolids because they are gentle, efficient, and durable. They also prevent material degradation. When installing a belt conveyor, the belt should be angled to form a trough to prevent material from rolling off the belt. Belts used to transport dried biosolids typically are no more than 1 m (3.3 ft) wide. Although belt conveyors are preferred when transporting dried biosolids horizontally, other conveyors should be considered when moving them vertically. Typically, a belt conveyor should not be inclined more than 20 degrees when transporting dried biosolids, especially dried biosolids pellets, which tend to be free flowing. If dried materials must be moved at steeper angles, consider using sidewalls, cleats, and/or cover belts.

5.4.2 SCREW CONVEYORS

Screw conveyors are one of the most economical options for moving dried biosolids. Ranging from 150 to 600 mm (6 to 24 in.) in diameter, these conveyors use a shaft-mounted spiral helix or a self-supporting helix to move material in a covered trough. They typically are installed horizontally or at inclines less than 45 degrees. Shafted screw conveyors typically are available in 3 to 4 m (10 to 13 ft) long sections that are bolted together. Shaftless screw spirals typically are furnished in one piece (either fabricated or welded) that can be up to 50 m (165 ft) long, depending on the ability to ship to the site.

5.4.3 DRAG CONVEYORS

Drag conveyors have a slow-moving chain and flight assembly that typically pushes material along a steel pan or trough, which may be rectangular or U-shaped. Most drag conveyors are designed to transport material horizontally. Special designs, however, can move material vertically or in "Z" lifts. A "Z" lift conveyor system is designed to transport material horizontally, then vertically, and then horizontally again. Drag conveyors are used in many recovered water facility processes, such as primary clarification, grit removal, and solids treatment. They may also be used as live bottom feeders in hoppers. Their lengths typically are only limited by the chain's strength and the weight of the material being transported. In practical terms, drag conveyors typically are not designed to move material more than about 40 m (130 ft) horizontally or 10 to 15 m (33 to 50 ft) vertically.

5.4.4 BUCKET ELEVATORS

A bucket elevator is a simple, dependable device for vertically transporting dry materials. It consists of a series of buckets mounted on a belt or chain within a housing. The buckets are filled with material at the base of the unit and discharge it at the top. They are available in a wide range of capacities (10 to 350 kg/s [4 to 140 ton/hr]) and are totally enclosed to prevent dust and odors from escaping. At WRRFs, bucket elevators are used to transport dried materials vertically within solids processing units and to load dried biosolids products into trucks or railcars. For most dried materials, centrifugal discharge elevators are generally acceptable and the most cost-effective option. However, if the dried biosolids are intended for beneficial reuse, then dust content and degradation are concerns, so designers should consider continuous bucket elevators. There are three types of bucket elevators: centrifugal discharge, positive discharge, and continuous.

5.4.5 PNEUMATIC CONVEYORS

Pneumatic conveyors use air to move material through a pipeline. There are two types of these conveyors: dilute phase and dense phase. Dilute phase conveyors have low material-to-air ratios (less than 5:1 [5 kg material/kg air]) and use a large volume of air to move a small amount of material. The air velocity typically ranges from 20 to 40 m/s (65 to 130 ft/sec), which is high enough to suspend the material in the air stream. The systems then use either positive or negative pressure to push or pull the material through the pipeline. Dense phase conveyors have high material-to-air ratios (up to 100 kg material/kg air). A pressure tank and a high-pressure air compressor can provide 350 to 700 kPa (50 to 100 psig) of air, which typically moves at velocities less than 2.5 m/sec (8 ft/sec). In these systems, material enters the pressure tank or transporter via gravity and settles in the pipeline.

Pneumatic conveyors have been used to transport dried biosolids and grit in continuous processes, load railcars or trucks, and help collect and remove fugitive dust. Dilute phase conveyors may be used to transport lime, sawdust, and other chemicals typically used at WRRFs. They are ideal for transporting nonabrasive, powdered materials over short distances. Although successfully used to convey a wide variety of bulk solids, dilute phase conveyors may not be suitable for dried biosolids meant for beneficial reuse because they can be abrasive and degrade when exposed to high-velocity air. Dilute phase conveyors also have low capital costs, but their energy costs can be quite high because of the large air requirements.

5.5 KINETIC PUMPS

Kinetic (dynamic) pumps continuously add energy to the pumped fluid to make the velocity in the pump higher than the velocity at the discharge point and, therefore, increase pressure. Several types of these pumps and their common applications are discussed in the following subsections.

5.5.1 SOLIDS HANDLING CENTRIFUGAL PUMPS

A wide variety of centrifugal pumps are available. Except for special designs (e.g., recessed impeller), however, these pumps should only be used with relatively dilute (less than 1% solids), trash-free residuals. They typically are used to transport return activated sludge (RAS) because of the pump's high volumetric flow rate and excellent efficiency. The minimal debris in RAS typically does not clog the pumps.

Centrifugal pumps are not recommended for primary solids, primary scum, or thickened residuals for two reasons. First, there is no means to ensure that the pump's suction will positively draw thickened material to the pump impeller. Second, the system head curve depends on a solids concentration that is often inconsistent, leading to significant variations in liquid flow rate and pump power requirements. The GLUMRB (2014) recommends that, if these pumps are used for these applications, positive displacement pumps be provided in parallel to be used when the solids concentration increases past the capabilities of the centrifugal pumps.

5.5.2 RECESSED IMPELLER PUMPS

Recessed impeller pumps (also called a torque flow, vortex, or shear lift pumps) work well on untreated sludge containing no more than 2.5% solids or on digested biosolids with total solids as high as 4%. Although they can pump thicker residuals, varying friction losses cause erratic flow rates and heavy radial thrusts on the pump shaft. Positive displacement pumps perform better in such applications. If design engineers use recessed impeller pumps to transport thickened solids or biosolids, they should provide flowmeters and variable-speed drives to maintain a relatively constant flow. The primary drawback of recessed impeller pumps (compared to other non-clog centrifugal units) is their significantly lower efficiency. A recessed impeller pump's efficiency typically is between 5% and 20% lower than that of other non-clog centrifugal pumps.

5.5.3 SCREW/COMBINATION CENTRIFUGAL PUMPS

Screw/combination centrifugal pumps combine a screw-type impeller with a normal centrifugal impeller. They typically have a relatively high efficiency and relatively low net positive suction head requirements. In addition, the corkscrew action of screw impellers may provide more positive feed to the suction, so the pump better handles thicker sludge. These pumps are commonly used for primary sludge pumping and are capable of handling higher total solids concentrations up to 8%.

5.5.4 DISC PUMPS

Disc pumps operate on the principles of boundary layer and viscous drag. Their impellers are basically parallel discs installed at a certain distance apart. Fluid flows through the gap between the rotating discs, which transfer energy to the fluid and generate velocity and pressure gradients that force the fluid to flow through the pump. Pump wear is minimized because the fluid moves parallel to the discs and does not touch other pump parts.

Disc pumps traditionally are specified for residuals with total solids concentrations up 6% including slurries, viscous materials, and solids with high entrained air content (e.g., from dissolved air flotation units). They can run dry and handle abrasive materials, which makes them excellent candidates for pumping grit.

5.5.5 GRINDER/CHOPPER PUMPS

Special combination centrifugal pump grinders are also available. These pumps combine a hardened steel cutting bar with a relatively typical centrifugal vortex–type pump. They can be used as digester recirculation pumps and are gaining popularity for use in pumping thickened residuals to dewatering processes. These pumps are beneficial for existing WRRFs that do not have proper screening facilities to remove rags and stringy substances. These substances find their way into the thickening process and form rag balls that can clog other pump types. These pumps will act as a grinder to cut the rag balls and allow the residuals to pass to the dewatering process in a form that is typically not detrimental to the process. However, operating experience indicates that such pumps require as much maintenance as grinders.

5.6 POSITIVE DISPLACEMENT PUMPS

There are several types of positive displacement pumps that can be used to transport sludges.

5.6.1 PLUNGER PUMPS

Plunger pumps use pistons driven by either an exposed eccentric crank shaft or a walking beam to pump residuals. Plunger pumps have an output of 150 to 225 L/min (40 to 60 gpm) per plunger and can develop up to 70 m (230 ft) of discharge head. These pumps typically are designed for an efficiency of 40% to 50%, which leaves a power reserve to overcome changes in pumping head (Sanks et al., 2008).

Plunger pumps have several advantages over other pumps. They can transport residuals containing up to 15% total solids if the equipment is designed for load conditions. Plunger pumps are available in cost-effective options up to 30 L/s (500 gpm) and 60 m (200 ft) of discharge head, contain large port openings, and can operate at low pumping rates. Unlike pumps such as progressive cavity pumps, plunger pumps can operate under "no-flow" conditions without any damage.

5.6.2 PROGRESSING CAVITY PUMPS

A progressing cavity pump uses a worm-shaped metal rotor that turns eccentrically inside a pliable elastomeric stator. Cavities progress axially between these lines, moving residuals from the suction end of the pump to the discharge end. Compared to plunger pumps, progressing cavity pumps provide a more consistent flowrate, even with changes in discharge head. Design engineers should guard against pump operation in no-flow conditions because this can quickly damage the stator.

One advantage of a progressing cavity pump is that the stator acts as a check valve, preventing backflow under most conditions. An actual check valve or anti-reverse ratchet is only required if the pump's static backpressure is more than 50 m (165 ft) or stator wear is expected due to significant grit concentrations. However, design engineers should always include isolation valves on both the suction and discharge sides so the pump can be removed from service for routine maintenance.

Progressing cavity pumps can be used to transport biosolids containing about 20% total solids or less and pumping distance is limited to 50 m (164 ft) or less. Other design factors include restricting pump rotational speeds to 200 rpm or less (Bourke, 1997), limiting the pressure per stage to 52 kPa (7.5 psi), and to use either adjustable frequency or adjustable hydraulic motor drives.

5.6.3 DIAPHRAGM PUMPS

Diaphragm pumps typically transport solids from primary sedimentation tanks and gravity thickeners. These pumps are a relatively simple means of pumping thickened solids and can handle grit with minimum wear. An air-operated diaphragm pump typically consists of a single-chambered spring-return diaphragm, an air pressure regulator, a solenoid valve, a gauge, a muffler, and suction and discharge isolation valves (see Figure 10.11). Compressed air flexes a membrane that is pushed or pulled to contract or enlarge an enclosed cavity. Unless the WRRF already uses compressed air, providing this service can significantly increase pumping costs. The air exhausted from the pump valves also creates significant noise.

5.6.4 ROTARY LOBE PUMPS

A rotary lobe pump uses multi-lobed, intermeshed rotating impellers to transfer residuals containing up to 10% total solids. Like progressing cavity pumps, rotary lobe pumps offer a relatively smooth flow and do not require check valves in many applications with low- to moderate-discharge static heads. However, both suction and discharge isolation valves are needed so the pump can be removed from service for maintenance.

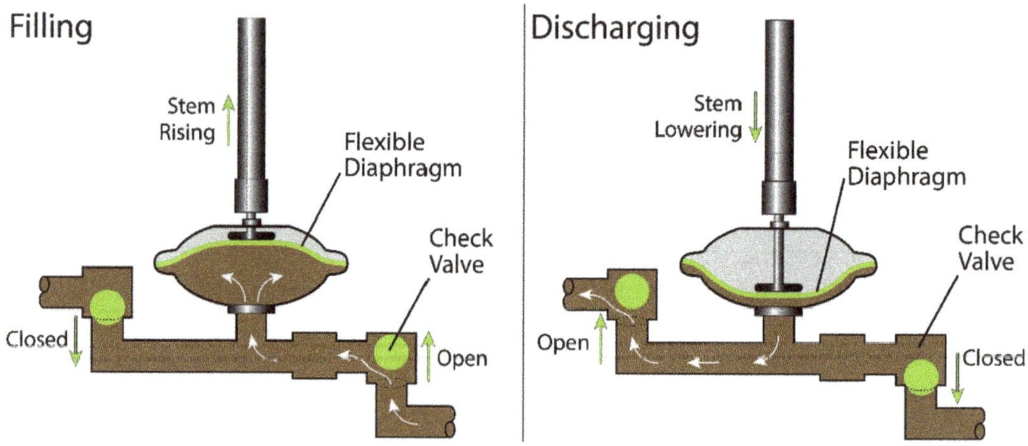

Figure 10.11 Diaphragm Pump Suction and Discharge Sequence

Pumping efficiency depends on maintaining relatively close tolerances between rotating lobes, so all large or abrasive material should be removed from residuals before they enter the pumps. Rotary lobe pumps are more suitable for applications with efficient grit removal and should not be used to transport grit. Engineers should also design the pumps to rotate at the lowest possible speed to minimize abrasion. Both rotary lobe and progressing cavity pumps react similarly to abrasion, so engineers can apply many of a progressing cavity's design considerations to a rotary lobe pump. Designers should also select appropriate lobe material for the residuals to be transported; otherwise, the lobes may fail prematurely. An advantage of these pumps over progressing cavity pumps is their ability to handle short periods of no flow without significant damage.

5.6.5 RECIPROCATING PISTON PUMPS

Reciprocating piston pumps are useful and cost-effective when dewatered biosolids must be transported to cake storage or loading facilities. They typically are not used ahead of the dewatering process. However, because these pumps can achieve discharge pressures up to 1.5×10^4 kPa (2200 psi) at flows up to 1500 L/min (400 gpm), they are the primary choice for pumping thickened solids over long distances. Given these pumps' high potential discharge pressures, however, engineers must design downstream piping systems properly. Designers should consider minimizing line pressure in the piping system when pumping material containing more than 25% total solids (either through minimizing fittings, lining the pipe, using boundary layer fluid injection, or a combination of any or all these methods).

5.7 STANDBY CAPACITY

Design engineers should consider the facility size, system's function, arrangement of units, anticipated service period, and time required for repair to determine the level of standby capacity to provide. For example, standby capacity for RAS pumping is important because a service interruption could quickly impair effluent quality. Primary and secondary solids pumping are also critical functions, so designers typically either provide dual units or use units that can perform dual duty. For example, primary sludge pumps also typically serve as standbys for scum pumps.

The designer should always propose standby capacity in all residuals pumping systems. If an owner must make the decision to remove standby capacity from the design, the pumps used should be heavy duty, have readily available spare parts, and be easy to repair quickly (preferably in place). Design engineers should ensure that the pump comes with adequate spare parts.

PRACTICE PROBLEMS

1. **What is the specific gravity of a sludge from a dewatering process that produces a sludge with 32% solids with a volatile solids content of 76%?**
2. **A centrifuge dewaters 1600 kg/d (3520 lb/d) of thickened sludge that enters the centrifuge with 4% solids. The capture rate of the solids is 93% in the centrifuge, and the dewatered solids exit the centrifuge at 31% solids. The dewatered solids also have a volatile solids content of 76%. Ignoring the polymer and wash water flows, determine the volume of solids leaving the centrifuge and the volume of centrate that will be returned to the head of the facility.**
3. **Waste activated sludge is to be pumped with a centrifugal pump from the bottom of the final clarifier to the digester. The solids concentration is 1% solids and the desired flowrate is 0.05 m³/s (793 gpm) against a static head of 16 m (52.5 ft). Design the pipe to handle this flow.**
4. **Raw primary sludge at a solids concentration of 2% is to be pumped for 6 km (3.46 mi) against a static head of 26 m (79.2 ft) at a minimum velocity of 0.9 m/s (3 ft/sec). Specify the required head and pipe diameter (use Figure 10.7 for pumping headloss and a C value of 140). Ignore minor losses other than friction.**
5. **Using resources on the internet, prepare a photo album of the different types of pumps used for pumping wastewater sludges.**

6. **A WRRF is pumping 5000 kg (11 000 lb) of digested sludge with a solids concentration of 3.5% to a sludge lagoon that is located 200 m (610 ft) from the facility each day. The pumps at the WRRF run 5 minutes each hour, and the line is 152 mm (6 in.) in diameter with a C value of 140.**
   a) **Determine the pumping rate when the pump is running.**
   b) **Determine the velocity in the line when the pump is running.**
   c) **Find the friction headloss and compare using information from the following methods:**
      - **Figure 10.7**
      - **Bingham Plastic Model (use Hazen-Williams method and Figure 10.6 if flow is turbulent)**

6.0 REFERENCES

American Society of Civil Engineers. (2000). *Conveyance of wastewater treatment plant residuals.*

Bassett, D. J., Howell, R. D., & Haug, R. T. (1991, October 7–10). Hydraulic properties evaluation for sludge cake pumping. *Proceedings of the 64th Annual Water Environment Federation Technical Exposition and Conference,* Toronto, Ontario, Canada.

Battistoni, P. (1997). Pretreatment, measurement execution procedure and waste characteristics in the rheology of sewage sludges and the digested organic fraction of municipal solid wastes. *Water Science and Technology, 36,* 33.

Borrowman, D. (1985). *Wastewater sludge characteristics and pumping application guide.* WEMCO Pump Co.

Bourke, J. D. (1997). *Handling high solids content non-Newtonian fluids.* Moyno Industrial Products.

Brar, S. K., Verma, M., Tyagi, R. D., Valéro, J. R., & Surampalli, R. Y. (2005). Sludge based *Bacillus huringiensis* biopesticides: Viscosity impacts. *Water Research, 39,* 3001.

Carthew, G. A., Goehring, C. A., & Van Teylingen, J. E. (1983). Development of dynamic head loss criteria of raw sludge pumping. *Journal of the Water Pollution Control Federation, 55,* 472.

Florida Department of Environmental Protection. (2008). *Chapter 62-640: Biosolids* (Draft Version).

Great Lakes–Upper Mississippi River Board of State and Provincial Public Health and Environmental Managers. (2014). *Recommended standards for wastewater facilities.* Health Research Inc.; Health Education Services Division.

Guibaud, G., Dollet, P., Tixier, N., Dagot, C., & Baudu, M. (2004). Characterisation of the evolution of activated sludges using rheological measurements. *Process Biochemistry, 39,* 1803.

Health and Safety Executive. (2005). *Control of health and safety risks at sewage sludge drying plants* (HSE 847/9).

Honey, H. C., & Pretorius, W. A. (2000). Laminar flow pipe hydraulics of pseudoplastic-thixotropic sewage sludges. *Water SA, 26,* 19.

Laera, G., Giordano, C., Pollice, A., Saturno, D., & Mininni, G. (2007). Membrane bioreactor sludge rheology at different solid retention times. *Water Research, 41,* 4197.

Levine, L. (1986). *Coming to grips with rheology.* Viscous Products.

List, E. J., Hannoun, I. A., & Chiang, W.-L. (1998). Simulation of sludge pumping. *Water Environment Research, 70,* 197.

Lue-Hing, C., Zenz, D., Tata, P., Kuchenrither, R., Malina, J., & Sawyer, B. (1998). *Municipal sewage sludge management: A reference text on processing, utilization and disposal.* Technomic Publishing Co., Inc.

Metcalf & Eddy, Inc./AECOM. (2014). *Wastewater engineering: Treatment and reuse* (5th ed.). McGraw-Hill.

Mori, M., Isaac, J., Seyssiecq, I., & Roche, N. (2008). Effect of measuring geometries and of exocellular polymeric substances on the rheological behaviour of sewage sludge. *Chemical Engineering Research and Design*, *86*, 554.

Mulbarger, M. C. (1997). Selected notions about sludges in motion, and movers. Paper presented at Central States Water Environment Association Education Seminar, Madison, Wisconsin, United States.

Mulbarger, M. C., Copas, S. R., Kordic, J. R., & Cash, F. M. (1981). Pipeline friction losses for wastewater sludges. *Journal of the Water Pollution Control Federation*, *53*, 1303.

Murakami, H., Katayama, H., & Matsuura, H. (2001). Pipe friction head loss in transportation of high-concentration sludge for centralized solids treatment. *Water Environment Research*, *73*, 558.

Novak, J., Adams, G., Chen, Y.-C., ; Erdal, Z., Forbes, R. H., Jr., Glindemann, D., Hargreaves, J. R., Hentz, L., Higgins, M. J., Murthy, S. N., Witherspoon, J., & Card, T. (2004, April 18–24). Odor generation patterns from anaerobically digested biosolids. *Proceedings of the Joint WEF/A&WMA Odors and Air Emissions Conference*, Bellevue, Washington, United States.

Sanks, R. L., Tchobanoglous, G., Bosserman, B. E., II, & Jones, G. M. (Eds.). (2008). *Pumping station design* (3rd ed.). Butterworth-Heinemann.

Spinosa, L., & Vesilind, P. A. (2001, reprinted 2007). *Sludge into biosolids—Processing, disposal, utilization.* IWA Publishing.

U.S. Army Corps of Engineers. (1984). *Engineering and design—Domestic wastewater treatment mobilization construction* (EM-1110-3-172).

U.S. Environmental Protection Agency. (1979). *Process design manual: Sludge treatment and disposal* (EPA-625/1-79-011).

U.S. Environmental Protection Agency. (2000). *Guide to field storage of biosolids and other organic by-products used in agriculture and for soil resource management* (EPA/832-B-00-007).

Wagner, R. L. (1990). *Sludge digester heating.* Alfa-Laval Thermal Co.

Water Environment Federation. (2004). *Control of odors and emissions from wastewater treatment plants* (Manual of Practice No. 25). McGraw-Hill.

Water Environment Federation, American Society of Civil Engineers, & Environmental and Water Resources Institute. (2018). *Sustainability and energy management for water resource recovery facilities* (WEF Manual of Practice No. 38; ASCE Manuals and Reports on Engineering Practice No. 137). Water Environment Federation and ASCE.

Water Environment Research Foundation. (2008). *Identifying and controlling odor in municipal wastewater environment Phase 3: Biosolids processing modifications for cake odor reduction.*

Chapter 11 | Solids Thickening and Dewatering

LEARNING OBJECTIVES

At the end of this chapter, students should be able to:

- Distinguish between thickening and dewatering.
- Describe the operation and design the following thickening processes:
  - Gravity thickener
  - Dissolved air flotation thickener
  - Gravity belt thickener
  - Rotary drum thickener
  - Disc thickener
  - Volute/screw thickener
  - Membrane thickener
- Describe the operation and design the following dewatering processes:
  - Belt filter press
  - Centrifuge dewatering
  - Screw press
  - Sand drying beds
  - Lagoons
  - Pressure filter press
  - Thermal drying
  - Rotary press
- Describe the role of chemical conditioning in thickening and dewatering processes.

1.0 INTRODUCTION

Water resource recovery facilities (WRRFs) typically use solids thickening processes to concentrate primary solids, secondary solids, or a combination of primary and secondary solids (typically waste activated solids). The reason for solids thickening is generally to increase the solids concentration of waste solids and reduce the hydraulic loading (total volume) to subsequent solids treatment processes such as anaerobic or aerobic digestion.

There are a number of widely accepted solids thickening technologies including: gravity thickeners, centrifuges, gravity belt thickeners, and rotary drum thickeners. These processes differ significantly in process configuration; size/footprint, degree of thickening provided; and chemical, energy, and labor requirements. Newer technologies such as disc thickeners, volute thickeners, membrane thickening systems, and recuperative thickening process are also gaining more attention as their use increases in the United States and abroad.

Sludge dewatering systems convey mixed and/or thickened liquid sludge, stabilized biosolids, and/or scum streams through a process that reduces the amount of excess water in the liquid/solids stream, thereby reducing the volume of sludge and/or biosolids for disposal. Dewatering produces a material suitable for further processing, beneficial use, or disposal. Additional processing may be required after dewatering to create a marketable product (such as drying and making fertilizer pellets) and/or prepare for other ultimate disposal methods (landfilling, land application, incineration, etc.). Dewatering differs from thickening in that thickening processes generate a product that behaves like a liquid (2% to 10% solids), while dewatering processes produce a denser material (i.e., cake) that behaves like a semisolid or solid material (12% to 50% solids).

2.0 THICKENING

Water resource recovery facilities typically use solids thickening processes to concentrate primary solids, secondary solids, or a combination of primary and secondary solids (typically waste activated solids). The reason for solids thickening is generally to increase the solids concentration of waste solids and reduce the hydraulic loading (total volume) to subsequent solids treatment processes, such as anaerobic or aerobic digestion. Additionally, thickening is used to reduce energy needs (i.e., heat) for processes such as anaerobic digestion.

There are several widely accepted solids thickening technologies, including gravity thickeners, dissolved air flotation thickeners (DAFTs), centrifuges, gravity belt thickeners, and rotary drum thickeners. These processes differ significantly in process configuration; size/footprint, degree of thickening provided; and chemical, energy, and labor requirements. Newer technologies such as disc thickeners, volute thickeners, membrane thickening systems, and recuperative thickening process are also gaining more attention as their use increases in the United States and abroad. Table 11.1 gives a general listing of the advantages and disadvantages of various thickening technologies.

2.1 GRAVITY THICKENER

Gravity thickeners function much like settling tanks: solids settle via gravity and compact on the bottom, while water or supernatant flows up over weirs. They also provide some solids equalization and storage, which may be beneficial to downstream processes. Gravity thickeners work best on primary and lime-conditioned solids but are also effective on primary solids combined with trickling filter secondary solids, primary and activated sludge, anaerobically digested solids and, to a lesser degree, waste activated sludge (WAS) and chemically enhanced primary solids. They are typically able to capture 80% to 90% of influent solids.

A basic gravity thickener typically has the following main components: center cage and column, center feed well, rake arm with squeegees and scraper blades, pickets, drive unit, weir plate, and overflow launder (Figure 11.1). Each gravity thickener has unique physical and mechanical features that can affect throughput, capture efficiency, polymer dose, and solids concentration. Some gravity thickeners have pickets attached to the top of the rake arm that are intended to help release water from the settling solids and improve the thickening process.

2.1.1 DESIGN CRITERIA

The two main parameters in gravity thickener tank design are depth and area. Engineers can calculate depth based on solids volume and storage requirements; it is not controlled by the type of sludge being thickened. Tank area, on the other hand, depends greatly on solids type and is typically determined via one of four methods: existing data, batch settling tests, bench-scale testing, or pilot-scale testing, with existing data being the most popular choice for municipal applications.

Table 11.2 presents typical surface area design criteria for various types of solids based on existing data. The mass loading rate is the quantity of solids allowable per unit area of gravity thickener per unit time ($kg/m^2{\cdot}h$) (lb/sq ft/hr) to achieve the indicated underflow solids concentration. The overflow rate is the quantity of liquid allowable per unit area of gravity thickener per unit time ($m^3/m^2{\cdot}d$) (gpd/sq ft). The gravity thickener area is determined by dividing the actual solids loading rate (SLR) and hydraulic loading rate (HLR) by the mass and hydraulic loading rate associated with the type of solids and desired underflow concentration. The larger of the areas is generally used for design.

Table 11.1 The Advantages and Disadvantages of Various Thickening Technologies

Method	Advantages	Disadvantages
Gravity	Simple Low operating cost Ideal for dense, rapidly settling sludges such as primary and lime Conditioning chemicals not typically required	Odor potential Erratic for WAS Thickened solids concentration limited for WAS High space requirements for WAS floating solids
Dissolved Air Flotation	Effective for WAS Will work without conditioning chemicals at reduced loadings Relatively simple equipment components	Relatively high power consumption Thickening solids concentration limited Odor potential Space requirements compared to other mechanical methods Moderate operator attention requirements
Centrifuge	Control capability for process performance Contained process minimizes housekeeping and odor considerations Will work without conditioning chemicals High thickened concentrations available	Relatively high capital cost and power consumption Sophisticated maintenance requirements Best suited for continuous operation Moderate operator attention requirements Vibration and noise
Gravity Belt Thickener	Control capability for process performance Relatively low capital cost Relatively lower power consumption High solids capture with minimum polymer High thickened concentrations available	Housekeeping Polymer dependent Moderate operator attention requirements Odor potential Building corrosion potential, if enclosed
Rotary Drum Thickeners	Low capital cost Relatively low power consumption High solids capture Can be easily enclosed	Polymer dependent Moderate operator attention requirements Odor potential if not enclosed

Note. WAS = waste activated sludge

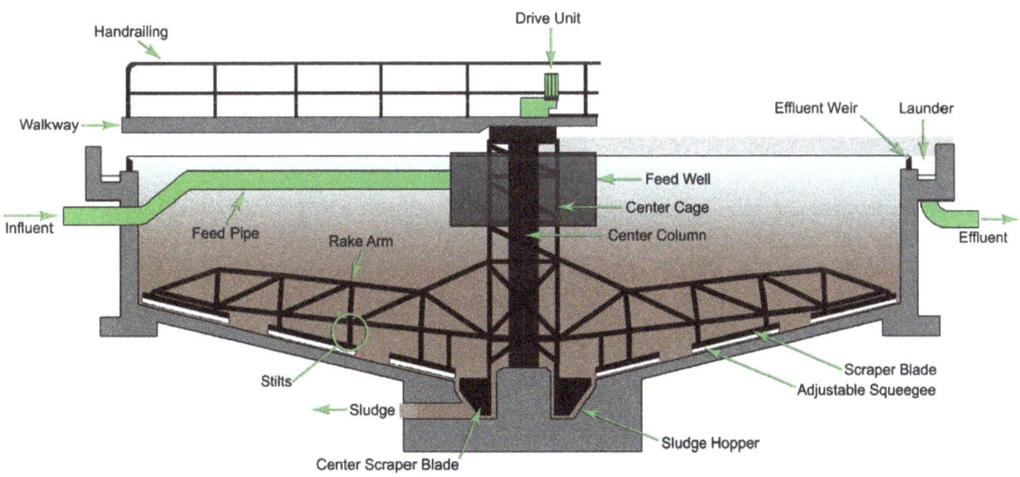

Figure 11.1 Cross-Sectional View of Typical Circular Gravity Thickener (Adapted from U.S. Environmental Protection Agency [U.S. EPA], 1979)

Table 11.2 Typical Surface Area Design Criteria for Gravity Thickeners (Adapted from U.S. EPA & American Society of Civil Engineers [ASCE], 1979)

Type of Solids	Influent Solids Concentration (% solids)	Expected Under Concentration Flow (% solids)	Mass Loading Rate (kg/m²·h)
Separate solids			
Primary	2–7	5–10	4–6
Trickling filter	1–4	3–6	1.5–2
Waste activated sludge (WAS)			
WAS–air	0.5–1.5	2–3	0.5–1.5
WAS–extended aeration	0.2–1.0	2–3	1.0–1.5
Thermally conditioned solids			
Primary only	3–6	12–15	8–10.5
Primary +WAS	3–6	8–15	6–9
WAS only	0.5–1.5	6–10	5–6
Mixed solids			
Primary + WAS	2.5–4.0	4–7	1.5–3.5
Primary + trickling filter	2–6	5–9	2.5–4
Primary + iron	2	4	1.5
Primary + lime	5 – 7.5	7 - 12	4 - 5
WAS + trickling filter	0.5–2.5	2–4	0.5–1.5
Anaerobically digested primary + WAS	4	8	3

The most common gravity thickener configuration is a circular tank with a side water depth designed at 3 to 4 m (10 to 13 ft). Such tanks typically range from to 21 to 24 m (70 to 80 ft) in diameter. Larger diameter tanks increase solids retention time (SRT), which can cause anoxic and anaerobic activity, leading to problems with gasification and solids flotation. The tank floor typically has a slope between 2:12 and 3:12, steeper than that of a standard settling tank.

2.1.2 DESIGN CONSIDERATIONS

When designing gravity thickeners, there are many important factors to consider for good operation. These factors are described in this section and include overflow rate, underflow piping solids source, and polymer usage.

2.1.2.1 Overflow Rate

The maximum overflow rate for primary solids is typically 15.5 to 31.0 m³/m²·d (380 to 760 gpd/sq ft); the maximum for secondary solids is typically 4 to 8 m³/m²·d (100 to 200 gpd/sq ft). If the hydraulic loading is too high, solids carryover can be excessive. If hydraulic loading is too low, detention times lengthen and anaerobic (septic) conditions (floating solids and odors) can occur. When thickening primary solids, design engineers often select a feed pumping rate that will maintain a desired overflow rate. They also add a dilution water supply (e.g., facility effluent), also called elutriation, to maintain aerobic conditions and may add chlorine, potassium permanganate, or hydrogen peroxide (typically via the dilution water supply) to control odor and septicity.

2.1.2.2 Underflow Piping

Underflow piping is a critical design element for gravity thickeners. Headlosses increase as solids concentrations increase, so the thickened solids discharge pipe should be designed as short as possible with a diameter large enough (minimum diameter of 15 cm [6 in.]) to avoid clogging, keep the headloss low, and maintain a minimum velocity of 0.6 m/s (2 ft/sec). For operation and maintenance (O&M) purposes, the underflow pump should be located close to the gravity thickener and below the gravity thickener's water surface level to provide a flooded suction condition.

Gravity thickener operations are most effective if underflow is withdrawn continuously. If intermittent withdrawal is necessary, a time-controlled system will allow operators to achieve efficient thickening performance. Pumping should be frequent and brief rather than extending for longer periods such as once or twice per shift. A certain minimum depth (typically about 1 to 2 m [3 to 6 ft]) is required to achieve a desired thickened solids underflow.

2.1.2.3 Feed Solids Source and Characteristics

The source and characteristics of feed solids greatly influence gravity thickener design and applicability. Depending on temperature, primary solids can be retained in the gravity thickener for 2 to 4 days before upset conditions develop. However, SRT of 1 to 2 days is best.

Waste activated sludge settles slowly and resists compaction, significantly reducing mass loading rates. To promote good operation, the SRT should be less than 18 hours to reduce the undesirable effects of continued biological activity, the gravity thickener diameter should be 10.7 to 13.7 m (35 to 45 ft) or less, and solids should be wasted directly from the aeration basin to the thickener (although solids are commonly wasted from the secondary clarifier).

2.1.2.4 Polymer

A polymer can be added to gravity thickeners to improve solids capture. Synthetic polyelectrolytes work better than inorganic coagulants (e.g., alum and ferric chloride) in this application because these do not yield metal hydroxides that add to the solids volume. Although adding chemicals improves the overflow concentrations, it may not lead to higher underflow concentrations but rather increases the solids inventory in the gravity thickener. Jar testing or pilot testing of polymers is recommended before design or implementation to determine removal efficiency and cost effectiveness.

Using about 2 to 4.5 g of **active polymer**/kg of dry solids (4 to 9 lb/ton) can produce a SLR of about 22 to 34 kg/m^2 d (4.5 to 7.0 lb/d/sq ft) when thickening primary solids. Adding about 4.5 to 6.0 g/kg (9.5 to 12.5 lb/ton) of active polymer can increase the thickener's loading rate to 12 to 16 kg/m^2 d (2.4 to 3.2 lb/d/sq ft) when treating WAS (Ettlich et al., 1978; U.S. EPA, 1978, 1979).

An active polymer refers to the percent of the molecular weight of the polymer that is able to react with the flocculate solid particles. Polymers are expressed in this method to be compared equivalently by mass.

2.2 DISSOLVED AIR FLOTATION THICKENER

A DAFT is typically used to thicken WAS, aerobically digested solids, and contact-stabilized, modified-activated, or extended-aeration solids without primary settling. It is typically not used for primary solids because gravity thickening is more economical. However, it can effectively **co-thicken** primary sludge with WAS or trickling filter solids. In contrast to gravity thickeners, the solids are floated in the DAFT process by using air bubbles to alter their specific gravity.

Co-thicken refers to the thickening of both primary and secondary solids.

The main components of a DAFT are the pressurization system and DAFT tank (Figure 11.2). The pressurization system has a recycle pressurization pump, an air compressor, an air saturation tank, and

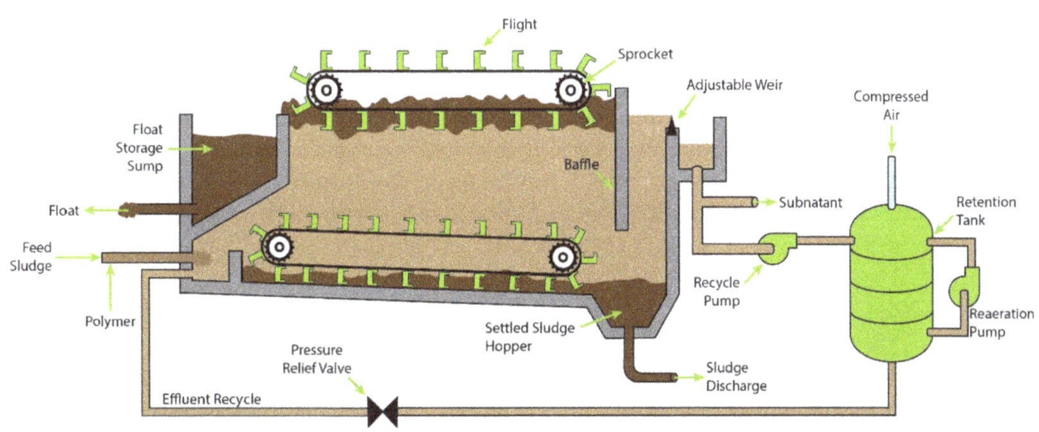

Figure 11.2 Diagram of a Dissolved Air Flotation Thickener (Reprinted with permission from Indigo Water Group)

a pressure release valve. The DAFT tank has a surface skimmer for float blanket depth control and float removal from the DAFT tank to a float collection box. The settled heavier solids are removed by a bottom collector. Most DAFT tanks are baffled and equipped with an overflow weir. Clarified effluent passes under an end baffle (rectangular units) or peripheral baffle (circular units) and then flows over the weir to an effluent launder. The weir controls the liquid level within the DAFT tank with respect to the float collection box and helps regulate the capacity and performance of the DAFT.

The DAFT tank serves to separate the solid phase from the liquid phase. The pressurization system dissolves air into the liquid stream, typically recycled subnatant, under pressure. As the pressure saturated recycled subnatant is introduced into the DAFT tank, its pressure is reduced, causing the air to precipitate out of solution in the form of very small bubbles. The precipitated bubbles are blended with the DAFT feed and become attached to the feed solids, forming bubble particle agglomerates with a density lower than water. The buoyant bubble particle agglomerates rise to the liquid surface and accumulate as a float while the heavier particles settle out as bottom sludge. The float and bottom sludge are removed from the DAFT tank by a surface skimmer and a bottom collector, respectively.

2.2.1 DESIGN CRITERIA

2.2.1.1 DAFT Tanks

Dissolved air flotation thickener tanks can be rectangular or circular. Both rectangular and circular units have been used in WRRFs ranging from 38 to more than 380 000 ML/d (1 to more than 100 mgd). The typically surface area for rectangular DAFTs varies from 9 to 167 m^2 (100 to 1800 sq ft). Length-to-width ratios typically are between 3 : 1 and 4 : 1. The typical surface area for circular DAFTs varies from 29 to 130 m^2 (300 to 1400 sq ft). They are often used when land availability is not a constraint.

2.2.2 PRESSURIZATION SYSTEM

Historically, three methods have been used to provide gas bubbles for a DAFT system: total, partial, and recycle pressurization flow schemes. The total pressurization flow scheme pressurizes the entire waste stream entering the DAFT; it is only practical for small flowrates, oily liquids, or other situations in which turbulence in the pressurization systems will not degrade solids enough to impair DAFT performance. This approach should not be used when the influent contains flocculated solids because the turbulence in the tank and pressure relief valve would destroy flocs. Partial pressurization systems pressurize a fraction of influent; how much depends on the air-to-solids ratio needed for optimal performance. This flow scheme is typically only practical for small rates of nonflocculated oily wastewaters.

Most DAFT units thickening WAS use recycle pressurization systems, in which some of the subnatant is pressurized. Influent solids do not pass through the pressurization system but are mixed with the pressurized recycle stream before entering the DAFT unit.

2.2.2.1 Hydraulic Loading Rate

The HLR is the sum of the feed and recycle flowrates divided by the net available flotation area. Engineers typically design DAFTs for HLRs of 30 to 120 m^3/m^2·d (0.5 to 2 gpm/sq ft), with a suggested maximum daily hydraulic loading of 120 m^3/m^2·d (2 gpm/sq ft) if no conditioning chemicals are used. If the hourly HLR exceeds 5 m^3/m^2·h (2 gpm/sq ft), the added turbulence may prevent a stable float blanket from forming and reduce the attainable floating solids concentration. Fewer solids also may be captured because increased turbulence forces the flow regime to convert from plug flow to mixed flow. A polymer flotation aid is typically required to maintain satisfactory performance when hourly HLRs are greater than 5 m^3/m^2·h (2 gpm/sq ft).

2.2.2.2 Solids Loading Rate

The SLR for a DAFT is denoted in terms of solids weight per hour per effective flotation area (see Table 11.3). Without chemical conditioning, the loading rates for DAFT processes thickening WAS range from about 2 to 5 kg/m^2·h (0.4 to 1 lb/hr/sq ft); this produces a thickened underflow of 3% to 5% total solids (Ashman, 1976; Burfitt, 1975; Jones, 1968; Mulbarger & Huffman, 1970; Reay & Ratcliff, 1975; U.S. EPA, 1979; Walzer, 1978). With polymer, the SLR can typically be increased from 50% to 100%, producing a thickened underflow that contains an additional 0.5% to 1% more solids.

Table 11.3 Percent Suspended Solids Captured When Using Dissolved Air Flotation to Thicken Waste Activated Sludge (Komline, 1976; U.S. EPA, 1979)

Type of Solids	Solids Loading Rate ($kg/m^2 \cdot h$)	
	No Chemical Addition	Optimal with Chemical Addition
Primary only	4–6	up to 12
Waste activated sludge (WAS)		
Air	2	up to 10
Oxygen	3–4	up to 11
Trickling filter	3–4	up to 10
Primary + WAS (air)	3–6	up to 10
Primary + trickling filter	4–6	up to 12

2.2.2.3 Air-to-Solids Ratio

The air-to-solids ratio, the ratio by weight of air available for flotation to the **floatable solids** in the feed stream, is the most important factor affecting DAFT performance. Reported ratios range from 0.01 : 1 to 0.4 : 1 (U.S. EPA & ASCE, 1979) at most municipal WRRFs, with adequate flotation occuring at ratios of 0.015 : 1 to 0.03 : 1. Up to a point, solids blankets increase as air-to-solids ratios increase; then, further increases in air-to-solids ratios result in little or no increase in floating solids (Gehr & Henry, 1978; Gulas et al., 1978; Maddock, 1976; Mulbarger & Huffman, 1970; Turner, 1975). The solids blanket is typically maximized when air-to-solids ratios are in the range of 0.1 : 1. The air-to-solids ratio for a system where the flow is fully pressurized is given as:

Floatable solids are materials that do not settle; instead, they float at the surface of a discharge.

$$\frac{A}{S} = \frac{1.3 s_a (fP - 1) Q_R}{C_s Q} \tag{11.1}$$

Where A/S = air-to-solids ratio;
s_a = solubility of air at the required temperature, mL/L;
f = fraction of air dissolved at pressure P, usually 0.5;
P = pressure in atmospheres, atm;
Q_R = pressurized recycle flow, m^3/d (mgd);
C_s = solids in incoming sludges, mg/L; and
Q = sludge flow to thickener unit, m^3/d (mgd).

The 1.3 factor in the equation is the weight of air in mg/mL, which makes the top part of the equation the mass of air that is applied to the DAFT. The bottom half of the equation is the mass of solids that are applied to the DAFT. The solubility of air (s_a) varies with temperature, with values of 29.2, 22.8, 18.7, and 15.7 mL/L at temperatures of 0, 10, 20, 30 °C, respectively.

2.2.3 *DESIGN CONSIDERATIONS*

2.2.3.1 Type of Solids

Dissolved air flotation thickeners can thicken a variety of solids, including conventional WAS, solids from extended aeration and aerobic digestion, pure oxygen activated sludge, and solids from dual biological processes (trickling filter plus activated sludge). The performance characteristics of each type of solids are difficult to document because site-specific conditions (e.g., type of process, SRT, and sludge volume index in the aeration basin) affect DAFT performance more than flotation equipment adjustments (e.g., air-to-solids ratio). Initially, the density of skimmed solids is about 700 kg/m^3 (6 lb/gal). After they are held for a few hours, the air escapes and solids return to normal densities.

2.2.3.2 Chemical Conditioning

Chemical conditioning can enhance DAFT performance. Conditioning agents can improve clarification or increase the floating solids concentration. Without polymer addition, a properly sized DAFT unit will typically recover more than 90% of solids. High loadings or adverse solids conditions can cut solids recovery to 75% to 90%. Polymer-aided recovery can exceed 95%. Design engineers should determine the amount of conditioning agent required, the point of addition (in the feed stream or recycle stream), and the intermixing method for each installation. Bench- or pilot-scale tests are the most effective methods of determining the optimal chemical conditioning scheme for a particular installation. Typical polymer doses range from 2 to 5 g dry active polymer/kg dry feed solids (4 to 10 lb/ton) as active content. Adding polymer typically affects solids capture more than floating solids content. For example, adding dry polymer at a dose of 2 to 5 g active polymer/kg dry feed solids typically increases floating solids content by no more than 1%.

2.2.3.3 Co-thickening Primary and Secondary Solids

In co-thickening processes, settled solids from primary and secondary clarifiers are mixed, then thickened. Co-thickening used to be rare; however, pilot testing and full-scale operations have indicated benefits of co-thickening over thickening primary and secondary solids separately (Butler et al., 1997).

A float total solids content of 6% approximates a typical minimum for DAFTs handling a 50:50 mixture of secondary and primary sludge compared to 4% for DAFTs handling solids without primary solids. Under optimum conditions, minimum solids contents ranging from 6% to 8% can be expected from thickening such mixtures (Bratby et al., 2004). Co-thickening can be used at small or large WRRFs but must include polymer to improve clarification and thickening. Design engineers should ensure that primary solids and WAS are mixed thoroughly because variable concentrations of mixed solids can cause operating problems and result in poor thickening performance.

Example 11.1

A municipal WRRF plans to use DAFTs to thicken WAS. The facility has complete-mix aeration basins. The following information is given to you as a designer, and you are asked to determine the number and size of circular DAFTs and the pressurization systems components.

WAS at 0.5% solids
Peak flow = 2400 kg/h (5292 lb/hr)
Operation: 24 hours per day, 7 days per week
Average DAFT loading rate = 6 kg/m²·h (1.23 lb/hr/sq ft)
Maximum DAFT loading rate = 10 kg/m²·h (2.0 lb/hr/sq ft)
No polymer will be used for this application
Pressurization systems = 448 kPa (65 psig)
Air-to-solids ratio = 0.03

Solution

Determine the area required for all DAFTs:

$$\text{Loading} = 6\ \text{kg/m}^2\cdot\text{h}\frac{(2400)\ \text{kg/h}}{\text{Area}} \qquad \text{Area} = 400\ \text{m}^2\ (4305\ \text{sq ft})$$

Use three DAFT tanks and find diameter of each tank"

$$A = \frac{\pi d^2}{4}\frac{400\ \text{m}^2}{3\ \text{DAFTs}} \qquad \text{Diameter} = 13\ \text{m}\ (42.8\ \text{ft})$$

Determine SLR with one unit out of service (not to exceed 10 kg/m²·h [2.0 lb/hr/sq ft]:

$$\text{Loading} = \frac{(2400)\ \text{kg/h}}{\left(\frac{\pi(13)^2}{4}\right)(2\ \text{DAFTs})} \qquad \text{Solids loading} = 9.0\ \text{kg/m}^2\cdot\text{h}\ (1.8\ \text{lb/hr/sq ft})$$

Determine the volume of solids at peak flow:

$$M_s = VS_w\gamma_w P_s = (2400\ \text{kg/h})(24\ \text{h/d}) = (V)(1.0)(1000\ \text{kg/m}^3)(0.005)$$

$$\text{Volume} = 11\,520\ \text{m}^3\text{/d}\ (2113\ \text{gpm})$$

Determine the pressure (P) in atmospheres of the pressured recycle:

$$P = \frac{448 \text{ kPa} + 101.35 \text{ kPa}}{101.35 \text{ kPa}} = 5.42 \text{ atm}$$

Determine the recycle flow required in each unit to maintain an air-to-solids ratio of 0.03 at a temperature of 20 °C (Equation 11.1) with one unit out of service:

$$\frac{A}{S} = \frac{1.3 s_a (fp - 1) Q_R}{C_s Q} = \frac{(1.3)(18.7 \text{ mL/L}[(0.5)(5.42 \text{ atm}) - 1](Q_R)}{(5000 \text{ mg/L})\left(\dfrac{11\,520 \text{ m}^3/\text{d}}{2 \text{ DAFTs}}\right)} = 0.03$$

Recycle flow = 20 784 m³/d (3812 gpm) per unit

Determine the HLR with one unit out of service:

$$\text{Hydraulic loading} = \frac{11\,520 \text{ m}^3/\text{d}}{\left(\dfrac{\pi(13)^2}{4}\right)(2 \text{ DAFTs})} = 0.03$$

Hydraulic loading = 43.4 m³/m²·d (0.72 gpm/sq ft)

This is within the range of typical hydraulic loadings.

2.3 CENTRIFUGE THICKENER

Centrifugal thickening is analogous to gravity thickening, except that centrifuges can apply a force 500 to 3000 times that of gravity (g-force). A centrifuge's main components are the bowl and the scroll. The bowl is mounted horizontally and turns rapidly to create the centrifugal force. The scroll is mounted horizontally inside the bowl and conveys solids from one end of the bowl to the other. Other important features of a centrifuge are identified in Figure 11.3.

G-force is a measurement of the acceleration of a free-falling body caused by the force of the Earth's gravity; equal to 9.81 m (32.2 ft) per second squared.

To thicken solids, the bowl and scroll typically operate at more than 1500 revolutions per minute (rpm); the scroll rotates just a few rpm faster (or slower) than the bowl to create a differential speed. Feed solids are injected into the scroll's hollow shaft and discharged into the spinning bowl. A conditioning agent, typically polymer, is added either at the centrifuge feed nozzle or further upstream within the feed solids piping, depending on solids characteristics and need for mixing. The bowl's centrifugal force causes solids to migrate toward the bowl wall. The scroll's screw conveyor moves the solids up the conical section of

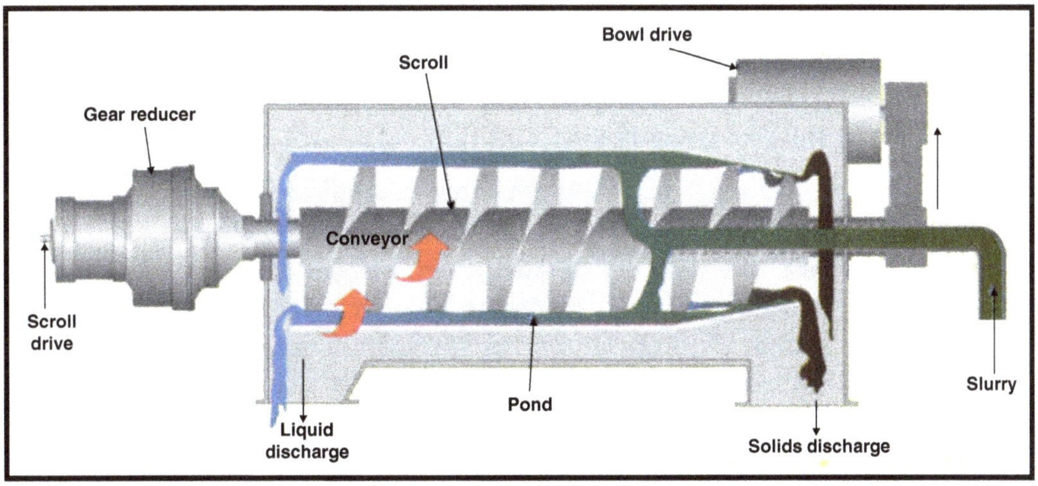

Figure 11.3 Centrifuge Components (Ireland & Balchunas, 1998)

the bowl and discharges them. Meanwhile, the liquid is discharged at the opposite end of the bowl via openings in the end plate.

2.3.1 DESIGN CRITERIA

Centrifuge manufacturers offer designs with substantially different features. Table 11.4 lists the major design and operating variables that influence the operation of a horizontal solid bowl centrifuge. These variables are discussed at length in the literature (U.S. EPA, 1979; Water Environment Federation [WEF], 2023). A desirable characteristic of the centrifuge is that its performance—as measured by thickened solids, polymer consumption, power demand, and solids capture—can be adjusted to desired values by modifying control variables such as feed flowrate, bowl and conveyor differential speed, conditioning chemical (polymer) dose, and pond depth.

Centrifuges have independent hydraulic loading and solids loading capacities. For typical WAS thickening applications, hydraulic loading capacity governs over solids loading. Hydraulic loading to the centrifuge controls the liquid phase residence time, which typically ranges from 30 to 60 seconds in thickening applications. For situations in which thicker feed solids are present, solids loads must be considered. Solids concentration determines the specific solids load applied to the centrifuge (kg [lb] of dry solids per day) and the thickened solids output volume (m^3/d [gal/d]). These measurements help centrifuge designers determine the machine's parameters. and also help operators adjust machine and process variables (e.g., conveyor differential and feed rates) to balance load demands.

2.3.1.1 Chemical Conditioning

Centrifugal thickening typically does not require polymer addition when treating biological and aerobically stabilized solids. Well-digested solids, however, have little natural flocculating tendency, and require polymer additions to achieve acceptable solids recovery levels. Engineers should make provisions for polymer addition in the initial design, even if chemical conditioning is not planned. The design should be flexible enough to allow conditioning chemical to be added at one of several points in the influent piping.

Dry or liquid high molecular weight cationic polymers are effective thickeners. When dry polymers are used, a 0.05% to 0.1% feed solution is used, while liquid polymers can range in concentration up to 0.5% on an active basis. It is important that a solids capture of at least 95% is obtained to prevent recycling filamentous bacteria and fines to the wastewater treatment process. Waste activated sludge produces a weak floc that tends to shear inside the centrifuge; a dose of up to 4 g active polymer/kg (8 lb/ton) of dry feed solids can be used to formulate a tougher floc. Aerobically and anaerobically digested solids have little natural floc and, therefore, require about 4 to 8 g active polymer/kg (8 to 16 lb/ton) of dry feed solids.

Table 11.4 Factors Affecting Centrifugal Thickening

Basic Machine Design Parameters	Adjustable Machine and Operational Features	Solids Characteristics
Flow geometry	Bowl speed	Particle and floc size
Countercurrent	Bowl and conveyor differential speed	Particle density
Concurrent		Consistency
Internal baffling	Pool depth and volume	Viscosity
Bowl/Conveyor geometry	Feed rate	Temperature
Diameter	Hydraulic loading	Sludge volume index
Length	Solids loading	Volatile solids
Conical angle	Flocculant use	Solids retention time
Pitch and lead		Septicity
Maximum pool depth		Floc deterioration
Solids and flocculant feed points		
Maximum operating speed		

2.3.2 DESIGN CONSIDERATIONS

2.3.2.1 Vibration

The high-speed revolution of a centrifuge creates significant vibration that must be addressed. To dampen vibrations transmitted to the foundation or piping, the centrifuge base should be supported on vibration isolators, and directly connected piping and conduits should have flexible connectors. Designers can also consider an isolated foundation for the centrifuges that is independent of the surrounding structure.

2.3.2.2 Particle Characteristics

The density of the feed material's solids and liquid fractions is important, and operators have limited control over this characteristic. Activated sludge and mixed liquor typically have similar low floc densities. Operators frequently need to add chemical conditioners to increase the effective density of the aggregate floc, thereby increasing sedimentation or centrifugal settling rates.

2.3.2.3 Effects of Different Feed Sludge Types

Primary solids settle rapidly and require lower bowl speed than WAS to thicken. It also requires more wear protection and screening/grinding due to the grit and debris typically present in primary sludge. Primary solids feed can be highly variable due to diurnal flows and primary sludge pump cycling. Substantial variation in feed solids can result in wide shifts in thickened sludge concentration and suboptimal polymer and power consumption. The variability in concentration can be accommodated through upstream sludge blending tanks or automated pond depth adjustment systems. If provided, solids blending tanks can be used to mix primary solids with WAS for co-thickening. These tanks should be mixed to maintain homogeneity of centrifuge feed. Some of the differences in thickening primary solids, WAS, or co-thickening both are indicated in Table 11.5.

Table 11.5 Differences in Thickening Primary Solids, Waste Activated Sludge, or Co-thickening with Centrifuges

	Primary Sludge	WAS	Co-Thickening Primary Sludge and WAS	Comments
Feed Concentration	0.2% to 4% TSS	0.4%–1.5% TSS	0.2%–5% TSS	Thin feed should be measured in TSS due to effect of dissolved solids; thick feed (>1.5% total solids) can be measured in total solids
Target Thickened Sludge Solids Concentration	5% to 6% total solids, but can easily increase to 12% total solids with excessive polymer	5% to 6% total solids	5% to 6% total solids, but can easily increase to 12% total solids with excessive polymer	Cake will overthicken if bowl speed and polymer dose are too high
Polymer Dose	0–2.2 kg active/tonne (0–5 active lb/dry ton)	0–2.7 kg active/tonne (0–6 active lb/dry ton)	0–4.5 kg active/tonne (0–10 active lb/dry ton)	For co-thickening, polymer use may increase due to ineffectiveness with either the primary sludge or the WAS
Target Centrate Quality	Less than 500 mg/L TSS	Less than 500 mg/L TSS	Less than 500 mg/L TSS	
Abrasion Protection	Robust abrasion projection (scroll tiles) required due to grit	100% WAS should have very little grit; tiles are beneficial but other types of abrasion protection may suffice	Robust abrasion projection (scroll tiles) required due to grit	
Torque	0–5, but may increase if slugs of sand/rags enter bowl	Typically consistent and near 0	0–5 with variation due to blend ratios	Gearboxes can be smaller than for dewatering, but must still have enough torque to clear a bowl if a plug should occur
Differential Speed Variation	10–20 rpm	1–15 rpm	1–15 rpm	For systems with automated pond depth adjustment, differential speed is not typically a control variable

Note. TSS = total suspended solids; WAS = waste activated sludge

2.4 GRAVITY BELT THICKENER

A gravity belt thickener (GBT) (Figure 11.4) is a solids separation device that typically relies on flocculation followed by a belt to physically separate the solids from the liquid. A GBT has a gravity drainage zone much like the upper gravity drainage zone of a belt filter press. The conditioned solids are evenly applied to the gravity feed belt via a distribution system, which is typically provided by the manufacturer. The continuous, porous belt provides a large surface area through which gravity drains free water. Filtrate from the gravity zone is collected and piped to a drainage system.

A GBT consists of a flocculation/retention tank with a slow speed mixer, a discharge chute, a gravity belt, chicanes or plows, a ramp, a spray wash system, and a variable- or constant-speed belt drive. A conditioning agent (typically polymer) is injected into the feed sludge and the mixture is sent to a flocculation/retention tank. The gravity belt is driven by a variable- or constant-speed belt drive and as the sludge moves along the belt, the plows spread the sludge and facilitate the separation of free water from the solids. As the free water releases from the sludge, it passes through the porous gravity belt to a filtrate drain.

Gravity belt thickeners are currently used to treat aerobically or anaerobically digested solids, alum and lime solids, primary solids, WAS, and blended solids that initially contain between 0.4% and 8% solids. However, in most applications the feed solids concentrations are less than 1.5%. Table 11.6 describes some of the performance differences in thickening primary sludge, WAS, or co-thickening both.

Gravity belt thickeners are often selected because of their relatively small footprint, low power use, and moderate capital costs. Experience has shown that GBTs work well with many types of wastewater solids and are less affected by facility operating problems than other thickening processes. They typically handle even difficult-to-thicken solids with minor modifications to polymer dose, HLRs, and SLRs.

The most commonly sized GBT has a belt width of 2 to 3 m. A GBT typically captures more than 95% of solids when a polymer dose of 1.5 to 5 g active polymer/kg dry weight of solids (3 to 10 lb/ton) is used to concentrate the material and avoid excessive solids losses. When treating municipal WAS and biosolids, GBTs can thicken the feed solids up to 6% to 8% solids.

Figure 11.4 Gravity Belt Thickener

Table 11.6 Performance Differences in Thickening Primary Sludge, Waste Activated Sludge, or Co-thickening With Gravity Belt Thickeners

	Primary Sludge	Waste Activated Sludge (WAS)	Co-thickening Primary Sludge and WAS
Typical Feed Solids Concentration	1%–4% total solids	0.5%–1% total solids	1.5%–3% total solids
Typical Thickened Sludge Solids Concentration	7%–10% total solids	4%–8% total solids	5%–9% total solids
Typical Polymer Dose	2.2–4.4 kg/tonne (5–10 active lbs/dry ton)	2.2–4.4 kg/tonne (5–10 active lbs/dry ton)	2.2–4.4 kg/tonne (5–10 active lbs/dry ton)
Typical Percent Capture	95%–99%	95%–99%	95%–99%

2.4.1 DESIGN CRITERIA

Gravity belt thickeners can be designed based on pilot test results and manufacturers' hydraulic and solids loading capacity criteria (Tables 11.6 and 11.7). Some testing is generally recommended to verify that a given solids can be thickened at typical polymer doses. Gravity belt thickeners are sized based on belt width, with widths ranging from 1 to 3 m in 0.5-m increments.

In lieu of pilot test data, engineers could use a conservative design value of 800 L/m·min (200 gpm/m) for the HLR. Typical HLRs range from 400 to 950 L/m·min (106 to 251 gpm/m). As feed flowrates increase, more operator attention is needed to maintain stable operations.

Operating experience demonstrates that some GBTs can treat WAS with 0.6% to 1.5% solids at a HLR of up to 1500 L/m·min (400 gpm/m) and SLR of up to 500 kg/h·m (1100 lb/hr·m). They also can treat digested solids with 2% to 4% solids at a HLR of up to 1100 L/m·min (300 gpm/m) and SLR of up to 770 kg/h·m (1700 lb/hr·m). In both cases, they can produce a thickened material containing 4% to 7% solids.

Polymer dosages range from between 1.2 and 4.4 g active polymer/kg dry solids (3 and 6 lb/ton) for raw primary solids, up to between 4 and 6 g/kg (8 and 12 lb/ton) for anaerobically stabilized solids. In all cases, solids capture should remain above 95%.

2.4.2 DESIGN CONSIDERATIONS

Gravity belt thickeners are less commonly used for thickening primary sludge. Two reasons for this are odor control and the possibility of belt blinding due to the amounts of grease, oils, and other debris in primary sludge. If design engineers intend to use a GBT as part of a co-thickening process, they should ensure that primary solids and WAS are mixed properly before the material is loaded onto the belts. This will help to control belt blinding caused by grease and debris in the primary solids.

When determining the size and number of units required for an application, the designer must consider operating conditions, **redundancy** requirements, and capacity of downstream processes to receive a

Redundancy is a design technique that allows for a unit(s) to be taken out of service while still retaining full design capacity.

Table 11.7 Typical Hydraulic Loading Ranges for Gravity Belt Thickeners (MacConnell, 1989)

Belt Size (Effective Width) (m)	Hydraulic Loading Range* (L/min)
1.0	400–950
1.5	570–1420
2.0	760–1900
3.0	1100–2800

*Assumes 0.5% to 1.0% feed solids for municipal sludges. Variations in sludge density, belt porosity, polymer reaction rate, filtration length, and belt speed will act to increase or decrease the rates of flow for any given belt size.

variable solids stream. Units that operate continuously (24 h/d; 7 d/wk) will be smaller (1 m vs. 2 to 3 m GBTs) than equipment required to operate only during staffed hours (e.g., 8 h/d; 7 d/wk). Fewer or larger units can be selected provided it is understood that loss of a redundant unit may decrease performance. For these reasons, the designer needs to consider if the added reliability of a larger number of smaller units is more important than the potential cost savings of fewer larger units.

2.4.2.1 Chemical Conditioning

Adding polymer to the feed solids is essential to successful thickening when using a GBT. It promotes flocculation of solids and release of free water. Without polymer, the belt would blind (because of fine solids filling belt pores) and flood (because of poor water release). Design engineers should test various polymers to determine which is the most effective. Cationic (positively charged) polymers typically are chosen because wastewater solids often are negatively charged. However, if the solids contain significant amounts of aluminum or ferric salts (which impart a positive charge), then an anionic (negatively charged) polymer may be better.

The mixing of polymer and solids to a homogenous blend is an important part of gravity belt thickening. The polymer mixer should be immediately downstream of the injection point. Design engineers should select adjustable mixers that allow mixing optimization both during initial operations and whenever solids feed rates or characteristics change.

Example 11.2

Determine the number of GBTs and holding tank size required for a WRRF that plans to install GBTs to treat a mix of primary solids and WAS. Design engineers have calculated that the facility produces 1150 L/min (304.2 gpm) of combined solids on a continuous basis. The following operating criteria apply:

- Thickening facility operates 8 h/d, 5 d/wk
- WAS solids concentrations = 0.5% to 1.0% solids
- Gravity belt thickeners have an effective thickening width of 2 m (6.6 ft)
- Design should allow for one standby unit

Solution

Using the criteria in Table 11.7 to select the HLR based on one unit out of service and one unit undergoing maintenance, choose a rate of 800 L/m·min.

Determine the actual daily flow rate to the GBTs during operational hours:

$$Q = \frac{(1150\ \text{L/min})(1440\ \text{min/d})(7\ \text{d/wk})(1\ \text{hr/60 min})}{(8\ \text{h/d})(5\ \text{d/wk})} = \mathbf{4830\ L/min}$$

Determine the number of GBTs to use if there is no standby:

$$\text{GBTs} = \frac{4830\ \text{L/min}}{(8000\ \text{L/min}\cdot\text{m})(2\ \text{m/thickener})} = \mathbf{3.0\ GBTs}$$

Design for a total of four GBTs using one as a standby:

Determine the holding tank volume required assuming that GBT operation is Monday through Friday for 8 hours. This means that solids need to be stored from approximately 5 pm on Friday through 9 am on Monday (40 hours).

$$V = (\text{HRT})(Q) = (40\ \text{h})(1150\ \text{L/min})(60\ \text{min/h})(1\ \text{m}^3/1000\ \text{L})$$

$$\text{Volume} = 2760\ \text{m}^3\ (0.73\ \text{Mgal})$$

2.5 ROTARY DRUM THICKENER

A rotary drum thickener (RDT; also called a rotary screen thickener) consists of a flocculation tank with slow speed mixer, an internally fed rotary drum, an integral internal screw or conveying flights, spray wash system, filtrate collection tank, and a variable- or constant-speed drum drive (Figure 11.5). A conditioning agent (typically polymer) is injected into the feed sludge and the mixture is sent to a flocculation tank. The tank is mixed slowly to allow sludge to flocculate, and the conditioned sludge flows into the drum. As the drum rotates, the sludge moves up the side of the drum and falls to the bottom. This movement releases free water, which flows through the drum covering material into the filtrate tank. Both GBTs and RDTs allow free water to drain through a moving, porous media while retaining flocculated solids. In RDTs, the rotating drum can consist of a filter cloth stretched over a frame or can be constructed as a metal wedge wire, wire mesh, or perforated plate screen itself. An enclosure at the end of the drum directs the thickened solids out of the drum. Advantages of RDTs include relatively low power use and ease of enclosure, which improves housekeeping and odor control.

Rotary drum thickeners can be used to thicken a variety of sludge types. Most municipal WRRFs use them to thicken WAS. However, the units can also be used to thicken primary sludge, and to co-thicken primary sludge and WAS. The type of sludge to be thickened impacts some of the equipment design and performance parameters because of sludge characteristics. Some of the performance differences in thickening primary sludge or WAS, or co-thickening both, are indicated in Table 11.8.

2.5.1 DESIGN CRITERIA

The most established RDT manufacturers to date produce units with a nominal maximum capacity of 90 m^3/h (400 gpm) (for a 1% solids WAS feed). However, as the technology becomes more popular, other manufacturers have started producing units with larger capacities up to 180 m^3/h (800 gpm). These larger capacity units often include two drums configured in parallel within one "unit."

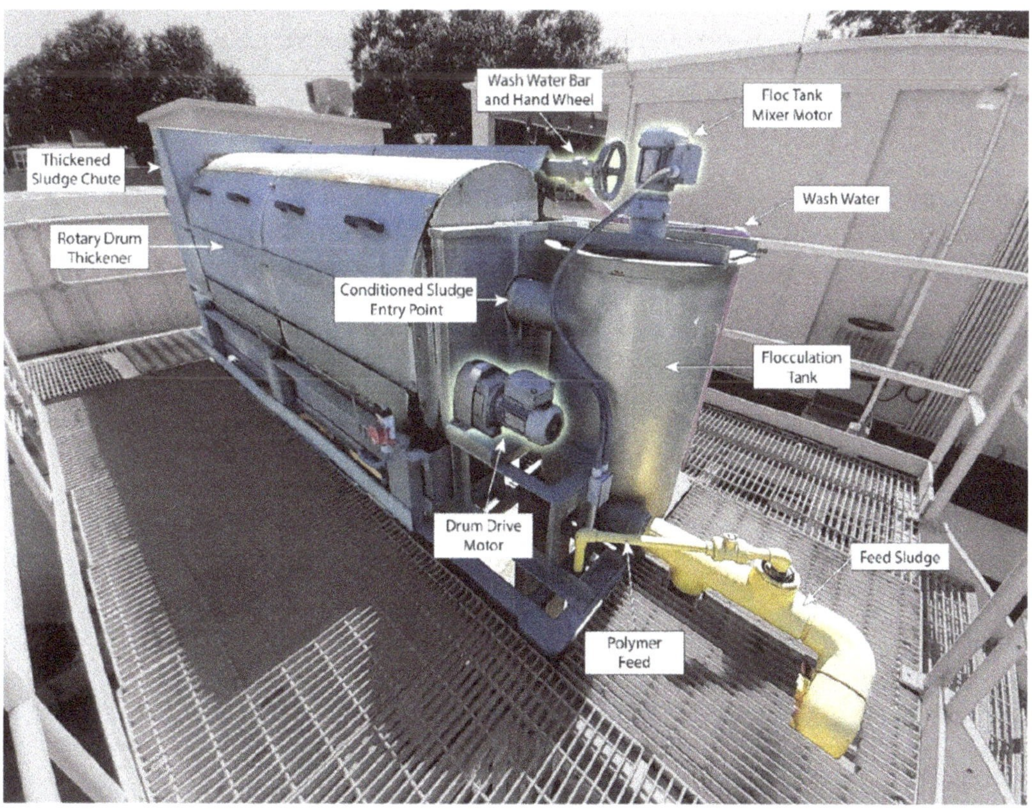

Figure 11.5 Rotary Drum Thickener (Reprinted with permission from the City of Dunedin Wastewater Division, Dunedin, Florida)

Table 11.8 Performance Differences in Thickening Primary Sludge, Waste Activated Sludge, or Co-thickening With RDTs

	Primary Sludge	Waste Activated Sludge (WAS)	Co-thickening Primary Sludge and WAS
Typical Feed Solids Concentration	1%–4% total solids	0.5%–1% total solids	1.5%–3% total solids
Typical Thickened Sludge Solids Concentration	7%–12% total solids	4%–7% total solids	5%–9% total solids
Typical Polymer Dose	2.2–6.7 kg/tonne (5–15 active lbs/dry ton)	2.2–6.7 kg/tonne 5–15 active lbs/dry ton)	2.2–6.7 kg/tonne (5–15 active lbs/dry ton)
Typical Percent Capture	93%–98%	93%–98%	93%–98%

Polymer requirements are about 10% to 20% greater than those associated with GBTs. Residuals typically can be thickened to 5% to 7% total solids (in some cases, more than 10% total solids) with up to 99% capture of feed solids at polymer dosages ranging from 4 to 6 g active polymer/kg dry solids (8 to 12 lb/ton).

2.5.2 DESIGN CONSIDERATIONS

Primary sludge thickens readily and requires different types of polymer, lower polymer doses, and lower polymer solution concentrations than WAS. Grease often found in primary sludge can blind drum screens, necessitating hot-water spray systems or the ability to periodically wash the screens with hot water. The presence of primary sludge in a co-thickening application has similar impacts. Primary sludge usually has higher solids content than biological sludge and therefore requires longer residence time in the drum, which is provided through lower drum speeds. The solids content also affects unit capacity, so this must be discussed with the manufacturers during system sizing.

Rotary drum thickeners are relatively simple machines, but require consideration of several items during design. When determining the required size and number of units required for an application, the designer must consider operating conditions and redundancy requirements. Units that operate continuously (24 h/d; 7 d/wk) will be smaller than equipment required to operate only during staffed hours (e.g., 8 h/d; 7 d/wk).

2.5.2.1 Chemical Conditioning

The polymer systems and associated considerations are similar to those described for GBTs and thickening centrifuges. The polymer feed system must be sized to handle both maximum and minimum feed rates, and it must be flexible enough to allow operators to vary the dose to meet flocculation requirements. Both underdosing and overdosing can degrade thickener performance. It is good practice to provide several locations for polymer injection upstream of the flocculation tank.

2.6 EMERGING TECHNOLOGIES

Several thickening technologies other than those described previously have been used in the United States or abroad for a number of years. The number of installations for these technologies is limited, so they are described here as "emerging." As more facilities use them, performance and design criteria for these systems will become better defined. They are described briefly to provide a fuller picture of thickening options.

2.6.1 DISC THICKENERS

The disc thickener shown in Figure 11.6 is an inclined, slowly rotating filter disc construction. The filter bottom consists of a perforated carrier disc covered with a microfilter that has a constant mesh size. The filter disc is installed into a closed stainless steel tank and divides the tank into a thickening zone (working area) and a filtrate collection zone.

As with GBTs and RDTs, polymer is injected into the sludge upstream of the thickener. The preflocculated sludge flows from the reaction tank onto the filter disc surface. The flocculated sludge settles on the filter

Figure 11.6 Disc Thickener Plan View (Courtesy of Huber Technology, Inc.)

surface while the filtrate flows to the collection zone and leaves the tank through the outlet. The residual solids on the microfilter thicken statically and are transported by disc rotation from the inlet section to a discharge opening. Cleaning of the filter disc is continuous by means of a spray nozzle bar, which is installed under the filter disc. The units are enclosed and can be connected to an odor control system if required.

2.6.1.1 Design Criteria

Disc thickeners can treat WAS with 0.5% to 1.5% solids at a HLR of up to 680 L/min (180 gpm) per unit and a SLR of up to 400 kg/h (880 lb/hr) per unit. Disc thickeners may also be used to thicken primary sludge. For a typical primary sludge with 2% to 3% solids, a disc thickener can process a HLR of up to 355 L/min (94 gpm) and a SLR of up to 465 kg/hr (1025 lb/hr). For thickening of both WAS and primary sludge, a thickened product material containing 5% to 7% solids is typical and may be optimized by the angle of disc inclination. Solids capture typically exceeds 95% in municipal applications.

2.6.2 VOLUTE/SCREW THICKENERS

The volute thickener is a mechanical screw thickener. Following flocculation, the sludge enters the thickener that consists of a cylindrical casing with openings, and a screw or auger that rotates. As the material is moved through the cylindrical casing by the rotating auger, the free water drains through the openings in the casing until the thickened product is discharged at the end of the cylinder. The volute thickener differs from a regular screw thickener in that the casing is made of a series of rings with fine gaps between them rather than a fixed material such as perforated plate or wedge wire. These rings continually move and prevent the fine gaps from plugging with the solids being thickened. The thickening mechanism can be completely enclosed to allow connection to odor control. The volute thickener is illustrated in Figure 11.7.

2.6.2.1 Design Criteria

Key design criteria are the hydraulic flow to the thickener and the desired thickened sludge concentration. Because there is a fixed volume, the only factor that affects the throughput is the screw speed. Higher

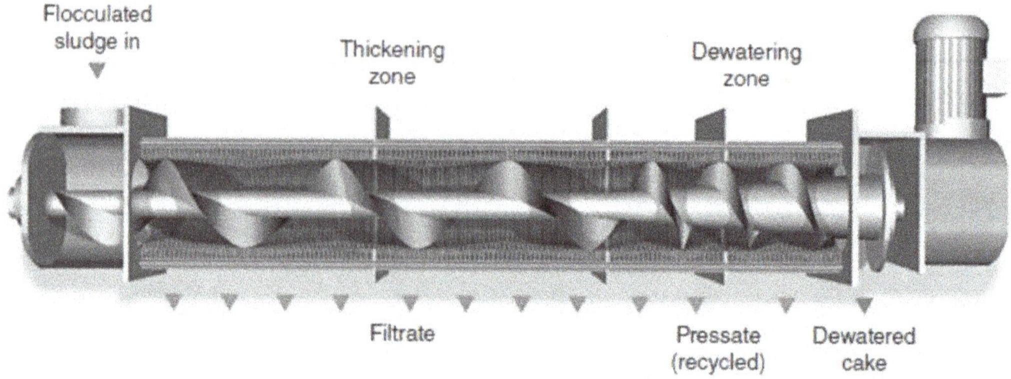

Figure 11.7 Volute Dewatering Schematic (Reprinted with permission from Process Wastewater Technologies, LLC [PWTech.us] *Volute is registered with the U.S. Patent and Trademark Office as a registered trademark of AMCON, Inc., Yokohama, Japan)

speeds will convey more material through the unit but allow less time for water to drain. Along with screw speed, thickened sludge concentration is also a factor of the nature of the sludge and how well the flocculated sludge releases water.

Most commonly used for thickening WAS, the volute thickener will achieve output solids concentrations between 4% and 8% when starting with 0.3% to 0.7% solids. The presence of primary sludge can significantly increase those numbers. Thickening only primary sludge can yield concentrations up to 16% solids for the thickened sludge.

2.6.3 MEMBRANE THICKENERS

Membrane thickeners thicken WAS, often upstream of aerobic digestion systems. Although membrane thickening does not produce as concentrated a product as other thickening processes, the thickened sludge concentration is suitable for feed into aerobic digesters. Polymer is not required for the process, but air is necessary for mixing and membrane scour. Either flat-sheet or hollow-fiber membranes used for membrane bioreactors can also be used for membrane thickening.

The process starts with the collection of WAS, which is initially held in a holding tank. When there is sufficient volume of WAS to thicken, the pumps or valves on the thickener membrane tanks are used to allow WAS to flow into the membrane thickener tank. The process pump starts and pulls the permeate through the membranes while leaving the thickened WAS in the thickener membrane tank. Once the target concentration is reached, the system then drains the thickened sludge to either a holding tank or further dewatering process.

It is important to note that a solids thickener membrane process uses aeration to keep the tank mixed and to minimize solids accumulation within the membrane fibers. Anytime there is mixed liquor within the system, the aeration system is called on to provide air for scouring.

2.6.3.1 Design Criteria

Membrane flux rates depend on the solids concentration in the membrane tank. Flat-sheet membranes are limited to a flux rate of 18.9 to 37.8 L/d·ft^2 (5 to 10 gpd/sq ft) for membrane thickening applications. Hollow-fiber membranes start thickening at a flux rate of 37.8 L/d·ft^2 (10 gpd/sq ft) and the rate gradually decreases to approximately 7.6 L/d·ft^2 (2 gpd/sq ft) as the sludge is thickened to 4% solids.

Membrane thickening can be used for WAS from conventional activated sludge facilities and membrane bioreactors. Typical solids concentrations for WAS from a conventional activated sludge facility range from 0.4% to 1%, whereas WAS from an membrane bioreactor can have concentrations exceeding 1%. For both sludge types, membrane thickening can produce thickened sludge with 4% solids.

3.0 SOLIDS DEWATERING

Solids dewatering systems convey mixed and/or thickened liquid sludge, stabilized biosolids, and/or scum streams through a process that reduces the amount of excess water in the liquid/solids stream, thereby reducing the volume of sludge and/or biosolids for disposal. Dewatering produces a material suitable for further processing, beneficial use, or disposal. Additional processing may be required after dewatering to create a marketable product (e.g., drying and making fertilizer pellets) and/or prepare for other ultimate disposal methods (landfilling, land application, incineration, etc.). Dewatering differs from thickening in that thickening processes generate a product that behaves like a liquid (2% to 10% solids), while dewatering processes produce a denser material (i.e., cake) that behaves like a semisolid or solid material (12% to 50% solids). Dewatering devices are typically placed after stabilization processes, whereas thickening typically occurs prior to stabilization processes. Some processes (e.g., screw press) are able to both thicken and dewater. These processes are usually placed where dewatering processes would normally go (after stabilization).

Dewatering systems typically require a large capital investment and a substantial share of a facility's annual budget for O&M. To design cost-effective dewatering facilities, engineers need to evaluate a wide array of dewatering options, solids characteristics, and site-specific variables (e.g., other treatment processes and sidestreams, disposal costs, energy costs, operator labor costs, and availability).

3.1 BELT FILTER PRESS

Belt filter press (BFP) machines evolved from papermaking applications to dewatering municipal wastewater solids. The BFP was introduced to North America in the 1970s as a lower-energy alternative to centrifuges and vacuum filter equipment. Compared to other mechanical dewatering devices, belt presses still have the lowest energy consumption per volume of solids dewatered. However, energy consumption is not the only factor to consider when selecting, sizing, and designing a dewatering system. Table 11.9 summarizes the advantages and disadvantages of BFPs.

The main components of a BFP system include feed equipment and piping, frame, belts, belt tracking and tensioning systems, belt wash system, rollers and bearings, cake discharge blades, chutes, cake conveyance, drive system, belt speed control, and chemical conditioning and flocculation. Some of these systems (for a two-belt press setup) can be seen in Figure 11.8.

Presses are available in widths ranging from about 0.5 to 3.5 m. Most municipal presses use 1 to 2 m belt widths. Care should be taken when specifying presses exceeding the 2.5 m width because deflection in the rollers could reduce machine performance (lower % dry solids).

Belt presses dewater solids in three stages: chemical conditioning, gravity drainage, and compaction in a pressure and shear zone. Dewatering operations begin when polymer flocculated solids enter the gravity drainage zone. The gravity drainage zone operates like a GBT. After gravity thickening, solids move into the pressure zones. Typically, a low-pressure zone is followed by a high-pressure zone. The low-pressure zone is the area where the two belts first come together with the gravity drained solids between them. This is a "wedge zone" where the solids are sandwiched between the two belts. The low-pressure zone provides enough dewatering to form a cake that can withstand the additional pressure and shear in the high-

Table 11.9 Advantages and Disadvantages of Belt Press Dewatering (Adapted from Metcalf & Eddy, Inc./AECOM, 2014)

Advantages	Disadvantages
Low energy requirements	High odor potential
Relatively low capital and operating costs	Requires sludge grinder in feed stream
Less complex mechanically and easier to maintain	Very sensitive to incoming sludge feed characteristics
High-pressure machines are capable of producing very dry cake	Automatic operation generally not advised
Minimal effort required for system shutdown	

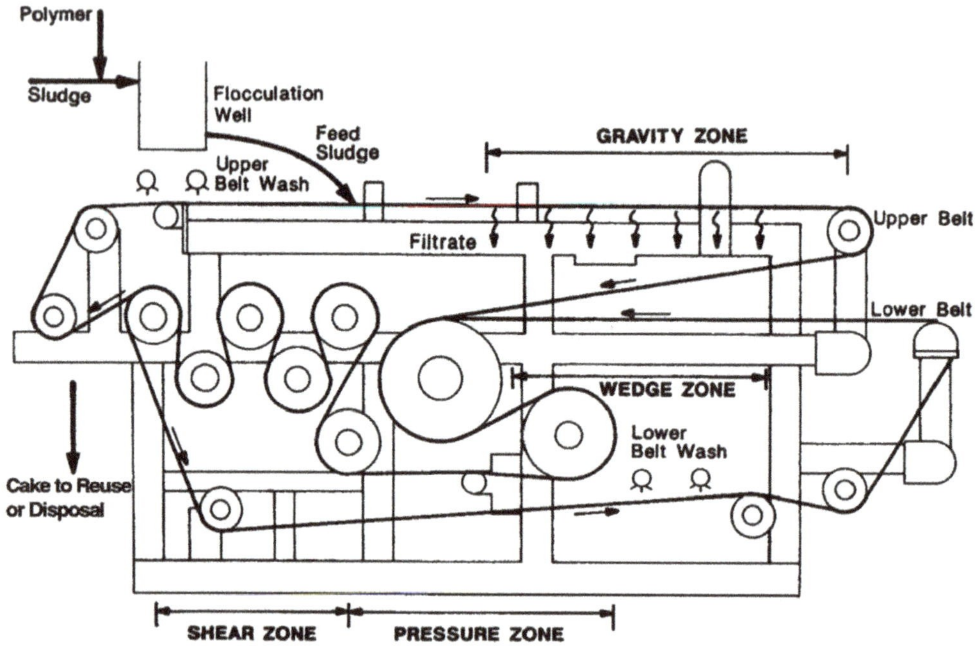

Figure 11.8 Schematic of a Belt Filter Press

pressure zone without extruding out the edges of the belts. A common operational mistake is applying too much pressure before enough water has been removed from the flocs.

In the high-pressure zone, forces are exerted on the solids by the movement of the upper and lower belts relative to each other as they move over and under rollers with decreasing diameters. Some machines have an extended roller section that provides higher pressures and sustains those pressures for a longer time.

3.1.1 DESIGN CRITERIA

The best method for evaluating BFP performance on a specific material is to dewater the residuals using a pilot unit. Most pilot units are small production machines that can perform comparably to larger models. As an alternative to a pilot test, bench testing can be used to predict belt press performance. A sample of the material to be dewatered can be sent to the manufacturer, who will provide design criteria and expected performance based on their tests. These tests have been shown to provide adequate projections of BFP performance.

The throughput capacity of a belt press is the primary design criterion when sizing a belt press system. Throughput capacity typically is considered to be either hydraulically limited (< 2.5% dry solids) or solids limited (> 2.5% dry solids), depending on feed solids concentration. Belt presses have a maximum hydraulic or solids loading capability for a given unit of belt width that can be attained only when solids are conditioned correctly. Typical design HLRs for a belt press range from 3 to 12.5 L/s·m of belt width (15 to 61 gpm/ft of belt width or 50 to 200 gpm/m of belt width). Typical SLRs range from 135 to 1136 kg/h·m (92 to 760 lb/hr/ft dry solids basis). However, these typical loading rates vary significantly depending on the source of the residuals to be dewatered and must be taken into account when designing the belt press system. Recommended loading limits also vary per machine manufacturer and should be verified when performing sizing calculations and comparing units.

The performance of BFPs depends on proper conditioning, and organic polymers traditionally are used. A properly conditioned product has a 95% to 98% solids recovery rate. However, the quantity of polymer required for proper conditioning varies widely; it depends on solids type, solids concentration, and ash content (typically, less polymer is needed when the ash content is high). For example, primary solids require a polymer dose of 3.5 to 5 g active polymer/kg dry solids (7 to 10 lb/ton); anaerobically digested solids require a dose of 7 g active polymer/kg dry solids (14 lb/ton); and autothermal thermophilic aerobically digested solids require a dose between 18 and 23 g active polymer/kg dry solids (37 to 47 lb/ton).

Insufficient conditioning causes inadequate dewatering in the initial sections of the press, which, in turn, can cause solids to extrude from the press section, overflow in the drainage section, or blind the belt. Over-conditioning can cause belt blinding and over-flocculation, which causes solids to drain too fast and mound on the belt, resulting in poor dewatering. The goal is to remove as much water as possible in the gravity section of the press.

3.1.2 DESIGN CONSIDERATIONS

Solids characteristics, origin, and degree of stabilization all significantly affect belt press loading and obtainable dewatering performance. Dilute solids require more gravity drainage, more polymer, and a longer dewatering time than more concentrated feed material. The type of process used for stabilization has a direct bearing on the maximum solids content achievable. Although manufacturers' estimates differ regarding the amount of dewatering that can be achieved, they typically agree that anaerobically digested residuals are easier to dewater than aerobically digested residuals. Typically, digested solids with lower volatile solids contents produce thicker dewatered cakes. In addition, as the primary to secondary solids ratio increases, dewatering becomes easier and the cake solids concentration increases.

The source of sludge has an effect on the loading and performance of dewatering equipment, including BFPs. Although the belt press typically can produce a dewatered cake containing anywhere from 12% to 35% solids when treating a typical combination of primary and secondary solids, these values vary significantly based on the type of sludge (primary, WAS) and treatment (aerobically digested, autothermal thermophilic aerobic digestion, etc.).

Capture efficiency is affected by the HLR, the SLR, the nature of the solids being dewatered, the mesh size of the belt, and chemical conditioning. It is important to test polymers to determine the best type of polymer, as well as the relationship between polymer doses, solids capture efficiency, and cake solids. Capture efficiency will decline if the solids are not properly conditioned. The solids capture rate for BFPs (total solids recovery, including wash water solids) is between 92% and 98%. However, most manufacturers guarantee solids capture rates exceeding 95% when the system is operated as designed. Table 11.10 includes the expected performance characteristics of a BFP for a number of municipal sludges.

A reasonably clean wash water supply is needed to ensure adequate belt cleaning, especially when dewatering secondary WAS and scum, which tends to rapidly clog the belt. This water supply, which amounts to 25% to 50% of the flowrate to the machine, typically is pressurized up to 700 kPa (100 psi).

Table 11.10 Typical Data for Various Types of Sludges Dewatered on Belt Filter Presses (U.S. EPA, 1987)

Type of Wastewater Sludge	Total Feed Solids (%)	Polymer (g/kg)	Total Cake Solids (%)
Raw Primary	3–10	1–5	28–44
Raw WAS	0.5–4	1–10	20–35
Raw Primary + WAS	3–6	1–10	20–35
Anaerobically Digested Primary	3–10	1–5	25–36
Anaerobically Digested WAS	3–4	2–10	12–22
Anaerobically Digested Primary + WAS	3–9	2–8	18–44
Aerobically Digested Primary + WAS	1–3	2–8	12–20
Oxygen Activated WAS	1–3	4–10	15–23
Thermally Conditioned Primary + WAS	4–8	0	25–50

Booster pumps are needed if the source water pressure is below 85 psi (these pressures may vary slightly with each manufacturer). Belt wash water can be potable water, secondary effluent, or even recycled filtrate water, although a clean supply is preferable. A good rule to determine the amount of wash water required is 1.3 L/s (20 gpm) per meter of belt per belt. Therefore, a 3 m, 3 belt unit would require about 11.4 L/s (180 gpm), while a 3 m, 2 belt unit would require 7.6 L/s (120 gpm).

Example 11.3

A belt filter press system is used to dewater anaerobically digested solids and operates with the following conditions:

- Two belt presses with an effective belt width of 2 m (6.6 ft); one is available as a standby;
- Operation is 5 d/wk, 7 h/d;
- Peak weekly solids production is 110 m^3/d (29059 gpd);
- Total solids in the BFP feed is 3.5%;
- Specific gravity of the solids is 1.03;
- Polymer solution is 0.2% and is added to the BFP at a rate of 25 L/min (6.6 gpm).

Determine if the BFP is operating within accepted standards and the amount of polymer required.

Solution

Determine the peak flow of solids that must be processed:

$$Q_{peak} = \frac{(110\ \text{m}^3/\text{d})(7\ \text{d/wk})}{(5\ \text{d/wk})(7\ \text{h/d})} = \mathbf{22.0\ m^3/h = 776\ cu\ ft/hr}$$

Determine the mass flow rate of solids to be processed:

$$M_s = VS_w\gamma_w p_s = (22.0\ \text{m}^3/\text{h})(24\ \text{h/d})(1000\ \text{kg/m}^3)(0.035) = \mathbf{793\ kg/h} = \mathbf{1745\ lb/hr}$$

Determine SLR to 1 BFP (1 BFP out of service):

$$\text{Solids Loading} = \frac{793\ \text{kg/h}}{2\ \text{m}} = \mathbf{396.5\ kg/m{\cdot}h} \qquad \text{Solids loading acceptable (Table 11.10)}$$

Determine the HLR and compare to values in Table 11.10:

$$\text{Hydraulic Loading} = \frac{(22.0\ \text{m}^3/\text{h})(1000\ \text{L/m}^3)}{(2\ \text{m})(3600\ \text{s/h})} = \mathbf{3.06\ L/s{\cdot}m}$$

Hydraulic loading is acceptable for anaerobically digested solids (1.89 – 6.31 L/s·m).

Determine the polymer dosage:

$$\text{Polymer dosage} = \frac{(25\ \text{L/min})(60\ \text{min/h})(0.002)(1\ \text{kg/L})}{(793\ \text{kg/h})(1\ \text{tonne}/1000\ \text{kg})} = \mathbf{3.78\ kg/\ tonne}$$

3.2 CENTRIFUGE DEWATERING

Centrifuges are used for both thickening and dewatering solids, although the internal design dimensions of thickening centrifuges may vary from dewatering centrifuges. Table 11.11 presents the advantages and disadvantages of using dewatering centrifuges. Like thickening centrifuges, dewatering centrifuges are comprised of the following major components: *bowl*, *scroll*, *main drive*, and *back drive* (see Figure 11.3). Because of their similarities, thickening and dewatering centrifuges behave similarly and thickening centrifuges were described in Section 1.3.

Table 11.11 Advantages and Disadvantages of Centrifuge Dewatering (Adapted from Metcalf & Eddy, Inc./AECOM, 2014)

Advantages	Disadvantages
Clean appearance, good odor containment, fast startup and shutdown capabilities	Scroll wear potentially a high maintenance problem
Produces a relatively dry sludge cake	Requires grit removal and possibly a sludge grinder in the feed stream
Low capital cost-to-capacity ratio	Skilled maintenance personnel required
High installed capacity-to-building area ratio	Moderately high suspended solids and nutrient content in centrate
	High energy requirements

3.2.1 DESIGN CRITERIA

Dewatering centrifuges are sized based on sludge HLRs or SLRs (kg/d or lb/d of dry solids) to the centrifuge and are typically designed and specified to meet process goals, such as cake dryness (often related to disposal costs), solids capture/centrate quality, polymer usage, and power usage. Consideration should be given to the facility's solids generation/influent, sludge/solids storage availability, preferred operating schedule (e.g., full-time versus part-time dewatering), footprint/space availability, solids disposal method (e.g., incineration, landfill), required equipment redundancy, and other facility-specific considerations when designing and specifying centrifuge equipment.

As is common with thickening centrifuges, greater **pond depths** provide more force to compress solids. Centrifuges operate under two different methods: (1) positive pond (i.e., liquid level is "below" the solids discharge radius, and (2) negative pond (i.e., liquid level is "above" the solids discharge radius). Centrifuges have negative ponds when starting up, resulting in sludge feed (startup liquid/semisolid "slop") discharging from the solids discharge port until a "seal" is established (typically within 5 to 15 minutes). Handling of startup "slop" discharged from the solids discharge port should be taken into account during design of a centrifuge system.

Pond depth is the radial dimension of the water that has been separated from thickened solids in a centrifuge. Also called pool depth.

Cake solids content should be determined by the solids disposal method being used (e.g., incineration, landfilling). Solids recovery or capture efficiency, as determined by Equation 10.6 in the previous chapter, should target approximately 95% to minimize recycle. Dewatering centrifuges can achieve up to and greater than 99% capture, but can result in excessive polymer and energy usage costs with little process benefit. Table 11.12 presents typical cake solids and polymer dosage for various types of dewatering feed sludge.

Table 11.12 Range of Expected Centrifuge Performance (U.S. EPA, 2000)

Type of Wastewater Solids	Feed %TS	Polymer lb/DTS	Cake %TS
Primary undigested	4–8	5–30	25–40
WAS undigested	1–4	15–30	16–25
Primary + WAS undigested	2–4	5–16	25–35
Primary + WAS aerobic digested	1.5–3	15–30	16–25
Primary + WAS anaerobic digested	2–4	15–30	22–32
Primary anaerobic digested	2–4	8–12	25–35
WAS aerobic digested	1–4	20	18–21
Hi-temp aerobic	4–6	20–40	20–25
Hi-temp anaerobic	3–6	20–30	22–28
Lime stabilized	4–6	15–25	20–28

Source: Various centrifuge manufacturers; Henderson & Schultz, 1999; Ireland & Balchunas, 1998; Leber & Garvey, 2000.

All dewatering centrifuges require polymer. Typically, a moderate- to high-charge, high molecular weight cationic polymer is used. Designs should include facilities for feeding both dry and liquid polymers. Polymer use typically increases solids capture; however, too much polymer can lead to a wetter cake because more fines are captured. Therefore, the relationship between recycled solids and cake dryness determines the dosage of polymer to be used. The type of polymer used is largely based on capital and long-term costs. Various polymer types and polymer solutions should be pilot tested prior to the design of polymer and dewatering systems.

3.2.2 DESIGN CONSIDERATIONS

Dewatered solids can be conveyed by four methods. The first method is gravity in which dewatered solids "fall" out of the centrifuge discharge chute. The second method of pumping generally uses positive displacement piston-style or progressing cavity pumps to convey dewatered solids. Centrifuges can discharge directly to a cake pump inlet hopper or into a separate "bin," where solids can be fed into a cake pump with other conveyance equipment (e.g., twin auger screw feeder, screw conveyor). The third method involves the use of belt conveyors; centrifuges that discharge directly to a conveyor belt should have adequate vertical distance between the conveyor and the solids discharge chute outlet to prevent "bridging." The fourth method uses screw conveyors, which are the most common conveyance devices used with centrifuges. Screw conveyors are less expensive than cake pumps, are enclosed systems (less messy and odorous), and can be used in an inclined layout to handle centrifuge startup/shutdown "slop."

Design of dewatered solids conveyance systems can be challenging. Elevating centrifuges, whether on the upper floor of a building or on a platform, is recommended so that solids can be transported horizontally or downward, thus avoiding having to convey solids upward.

3.3 SCREW PRESS

Although screw presses have been in existence since the 1960s, the technology originally was used solely in agricultural (e.g., dairy manure) and industrial applications (e.g., pulp and paper mills and food processing plants). In the U.S. municipal waste market, the application is relatively new, with most facilities installed after 2000. The increasing number of installations is still relatively small compared to other solids dewatering technologies.

The screw press is a simple, slow-moving mechanical device that gradually compresses conditioned, thickened solids as they move through the unit. Dewatering is continuous; it begins with gravity drainage at the inlet end of the screw and then continues as a result of increasing pressure at the end of the unit (Figure 11.9). Proper screw design is critical because different solids require different polymer dosages, screw speeds, and configurations to maintain a desired dewatered cake concentration and solids capture rate. The released water is allowed to escape through perforated plate or wedge-wire screens surrounding the screw while the solids are retained inside the press. The liquid forced out through the screens is collected and conveyed from the press, and the dewatered solids are dropped through the screw's discharge outlet at the end of the press. Screw speed and configuration, as well as screen size and orientation, can be tailored for each dewatering application.

The major elements of a screw press dewatering system are a solids feed pump, polymer makeup and feed system, polymer injection and mixing device, flocculation vessel with mixer (if applicable), solids inlet headbox or pipe, screw drive mechanism, shafted screw enclosed within a screen, rectangular or circular cross-section enclosure compartment, and outlet for dewatered cake (see Figure 11.10).

There are presently two major types of screw presses used in municipal dewatering applications: horizontal and inclined. Inclined screw presses are at angles 10° to 20° from the horizontal. Other areas of difference pertain to solids inlet configuration, screen basket design (perforated plate or wedge wire), basket cleaning from the inside and outside (brushes and rotating wash system), and filtrate water collection. Advantages and disadvantages of screw press systems are summarized in Table 11.13.

Based on a survey of existing installations, solids concentrations in screw press cake vary widely depending on polymer use, solids characteristics, and dewatering application. The higher solids concentrations typically are achieved in applications with primary solids or with the combined dewatering pasteurization

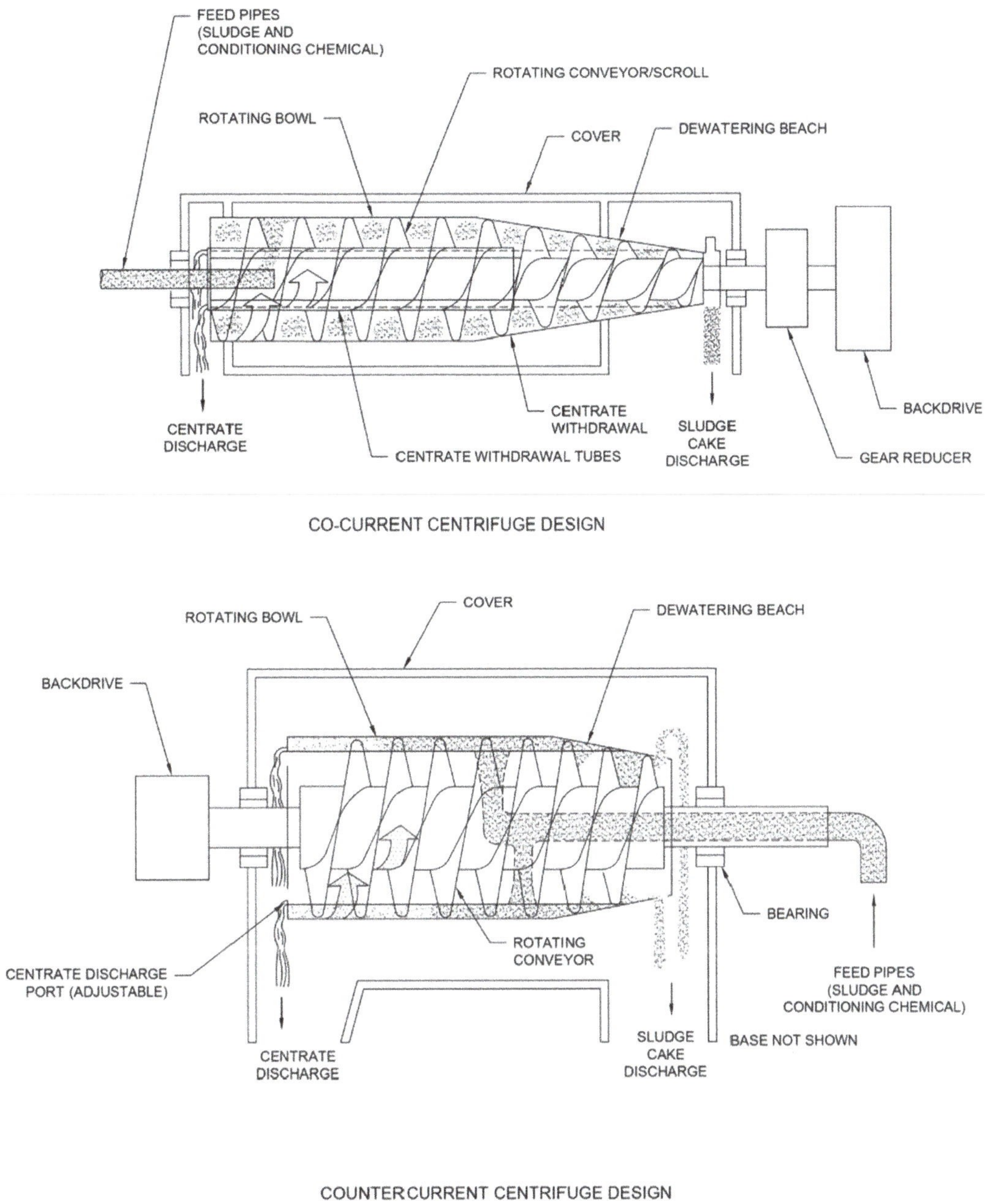

Figure 11.9 Co-current Versus Countercurrent Centrifuge

process. Typical solids concentrations are 15% to 28% for secondary WAS, and 13% to 40% for a blend of primary and secondary solids. Digested solids typically result in higher concentrations than undigested solids, but require a larger polymer dosage. Cake dewatering performance for solids with poor dewatering properties (e.g., in treatment facilities without primary clarifiers) can be significantly lower compared to other dewatering processes (Kabouris et al., 2005).

Screw presses run continuously at low speeds and do not require close operator supervision; therefore, they are easy to maintain and have low power consumption. The manual cleaning schedule ranges from once per week to once every 30 days.

3.3.1 DESIGN CRITERIA

Screw press designs are determined by solids characteristics, proper chemical conditioning, and hydraulic and solids loads, which affect the detention time (speed of the screw). It is essential to select the proper

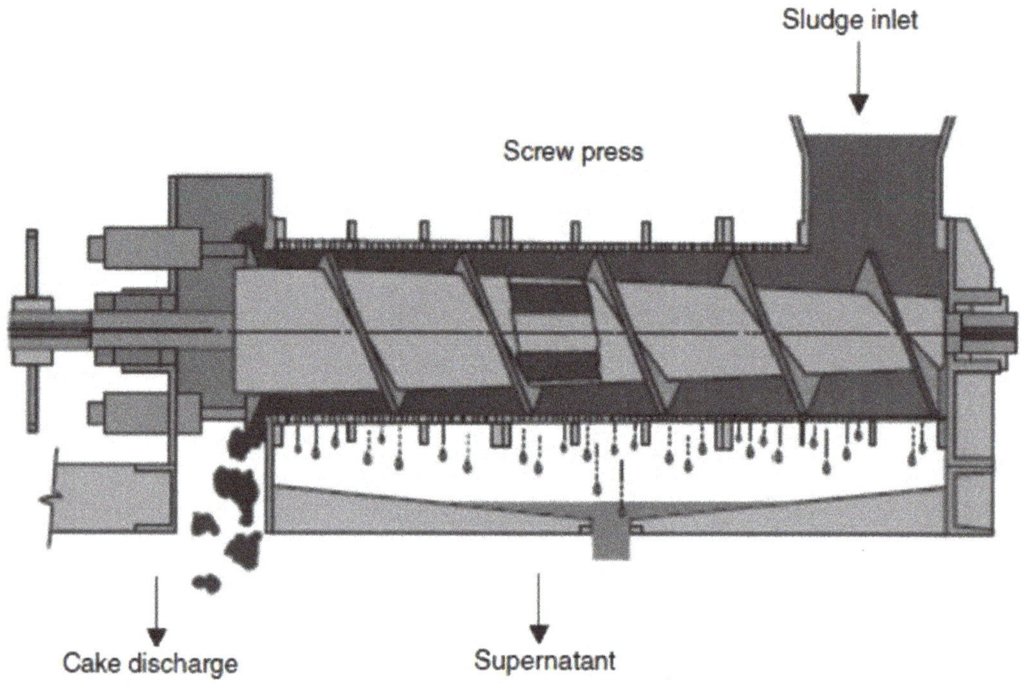

Figure 11.10 Cutaway View of a Horizontal Screw Press (Reprinted with permission from FKC Co., Ltd.)

conditioning agent (polymer); flocculation should agglomerate solids into large, strong flocs that release as much water as possible. The released water drains via gravity in the thickening section of the screw press. Screw press performance improves when less water must be removed in the dewatering zone.

The SLR for screw press dewatering varies, depending on solids characteristics, screw size, and rotational speed. The SLR capacity for a horizontal screw press ranges from 0.91 to 703 kg/h (2 to 1550 lb/hr), depending on screw press model. The typical SLR for an inclined screw press is 22.7 to 454 kg/h (50 to 1000 lb/hr), depending on screw press model. The inclined screw press design is mainly controlled by solids loading, but hydraulic loading is an important factor if the residuals have a low solids content.

In addition to solids loading, solids type, and desired discharge dryness, the HLR is a factor that should be considered in screw press sizing. Higher HLRs typically require larger diameter presses or a coarser screen. An important factor affecting the HLR is solids conditioning, because the screw press can only attain maximum capacity with the optimal polymer type and application. Optimal operation occurs when flocculated solids have sufficient time to fully gravity drain before being conveyed beyond the inlet end of the screw press. Typical HLRs for a horizontal screw press range from 3.8 to 2081 L/min (1 to 550 gpm),

Table 11.13 Advantages and Disadvantages of Screw Presses

Advantages	Disadvantages
Low rotational speed results in low maintenance and noise	Lower cake concentrations, especially when there are no primary clarifiers at the facility
Low energy consumption	Large footprint
Odors are contained	Requires wash water
Low operator attention required	Lower solids captures than other dewatering processes
Lower wash water demand and pressure than belt filter presses	Lower capacity per unit, requiring more units than other technologies for larger facilities

depending on the screw press model. Typical HLRs for an inclined screw press are between 18.9 and 454 L/min (5 and 120 gpm), depending on the model and solids concentration.

Polymer addition promotes particle flocculation and increases the dewatering and solids capture rates to produce a cake containing between 12% and 25% dry solids and a solids capture rate of 90% to 95%. Jar testing and pilot testing can be used to estimate the type and quantity of polymer necessary for each application, because it may vary significantly depending on solids characteristics. Polymer doses for screw press systems can range from 3 to 17.5 g of active polymer per kilogram of dry solids (6 to 35 lb of active polymer per dry ton of solids), with a typical range of 6 to 10 g/kg (12 to 20 lb/dry ton). In general, an increase in polymer dose increases cake solids concentration, although other factors (e.g., screw speed) can affect performance as well.

3.3.2 DESIGN CONSIDERATIONS

The inclined screw design (flight thickness, pitch, shaft diameter) is determined by the machine size and screen configuration. The diameter of the screw is 28 to 80 cm (11 to 31.5 in.), and the pitch of the flights typically is 15 to 25 cm (5.9 to 9.8 in.), although up to 40 cm (15.7 in.) is possible. Inclination varies from 10° to 20°. The inclination of the screw press allows the cake conveyor to fit under the discharge so that the screw press does not need another support/pedestal. The inclination also facilitates a consistent cake during startup, because the discharge is above the inlet elevation.

Screw press screen configurations also vary significantly among manufacturers. Screens can be either perforated stainless steel sheet or wedge wire. The holes used in perforated plate screens range from 1.0 to 3.0 mm in diameter, depending on solids type and press size. The open area of perforated plate screens ranges from 2% to 48%, depending on solids type, inlet consistency, and location on the screw press.

Screw press performance depends on a consistent inflow of solids. Therefore, positive displacement or progressing cavity pumps are highly recommended because their performance is not affected by solids consistency or changing water levels in the storage tanks. The progressing cavity pumps must be designed properly to minimize the wear and tear caused by grit and other abrasive material.

3.4 DRYING BEDS

Drying beds use a combination of drainage, evaporation, and time to dewater solids. Drying beds require considerable area compared to other dewatering methods. If well designed and properly operated, drying beds are typically less sensitive to influent solids concentration and can produce a drier product than most mechanical devices (see Table 11.14). Particularly suited to small facilities in arid locations, they can be used successfully in WRRFs of all sizes and in widely varying climates.

Sand drying beds are the oldest, most widely used drying bed dewatering method. The beds typically consist of a layer of sand underlain by a gravel layer and perforated drain piping. The beds are contained by concrete walls around the perimeter. To prevent the free liquid in the sludge from percolating into the groundwater, sand drying beds can be lined. Drying beds also may be enclosed to prevent rainwater from rewetting dewatered cake and to control odors (Figure 11.11).

Table 11.14 Advantages and Disadvantages of Using Drying Beds (Adapted from Wang et al., 2007)

Advantages	Disadvantages
Low capital cost if elaborate lining and leachate is not needed and land is available	Lack of rational design approach for sound economic analysis
Low operator attention and skill level	Large land requirement
Low electric power consumption	Stabilized sludge requirement
Low sensitivity to sludge variability	Climate significantly impacts design and performance
Low chemical consumption	Labor intensive sludge removal
High dry cake solids concentrations	Fuel and equipment costs for bed cleaning systems
	Real or perceived odor and visual nuisances

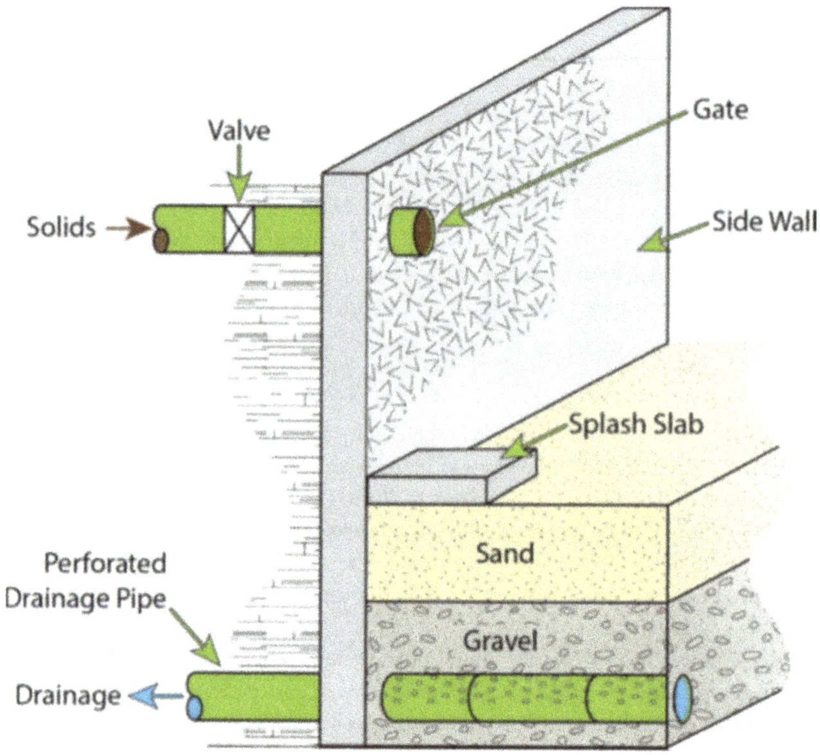

Figure 11.11 Schematic of a Typical Sand Drying Bed

Solids on sand beds are dewatered via drainage and evaporation. Initially, free water drains into the sand layer and is collected and removed via the underdrains. This step, which lasts a few days, continues until the sand becomes clogged with fine particles or all free water has drained. Once a supernatant layer forms, decanting can remove surface water. Decanting is important for removing rainwater, which can slow the drying process if allowed to accumulate on the surface. Decanting is also useful for removing free water released by chemical treatment. Solids drying in beds also can be enhanced by using auger mixing vehicles in paved beds. Water remaining after initial drainage and decanting is removed via evaporation.

The solids concentration of removed cake ranges from 44.5% to 95.5% dry solids, with higher cake moisture contents typically corresponding to higher initial solids concentrations. A cake with 40% to 45% solids may be achieved in approximately 2 to 6 weeks in good weather with well-digested sludges (Wang et al., 2007).

3.4.1 DESIGN CRITERIA

Per capita area criteria initially used to size sand drying beds are based largely on empirical studies of primary solids conducted in the early 1900s by Imhoff and Fair (1940), who recommended a range from 0.1 to 0.3 m^2/capita (1.0 to 3.0 sq ft/capita), depending on the type and solids concentration of the residuals applied to the bed. Other factors, such as applied solids depth and number of yearly applications, were also considered in this work.

Sizing sand drying beds based on solids loading is a better approach than sizing based on per capita area. Typical solids loading and per capita rates for sand drying beds vary (see Table 11.15). Furthermore, because the total quantity of solids produced daily by the overall treatment process can generally be predicted accurately, the risk of error is minimal. The best criteria to take into consideration are climatic conditions (e.g., temperature, wind velocity, humidity, and precipitation).

The total drying time required depends on the desired final moisture content, the solids removal method, and final use of the dewatered biosolids. The time required to achieve a liftable cake depends more on

Table 11.15 Summary of Recognized, Published Sand Bed Criteria for Anaerobically Digested, Unconditioned Solids

Initial Sludge Source	Uncovered Beds		Covered Bed Area (m^2/cap)
	Area (m^2/cap)	Solids Loading (kg/m^2·yr)	
Primary			
Imhoff & Fair (1940)	0.09	134	—
Rolan (1980)	0.09–0.14	—	0.07–0.09
Primary plus chemicals			
Imhoff & Fair (1940)	0.2	110	—
Rolan (1980)	0.18–0.21	—	0.09–0.12
Primary plus low-rate trickling filter sludge			
Quon & Johnson (1966)	0.15	110	—
Rolan (1980)	0.12–0.16	—	0.09–0.12
Primary plus waste activated sludge			
Quon & Johnson (1966)	0.28	73	—
Rolan (1980)	0.16–0.23	—	0.12–0.14

the initial solids content and percentage of total water drained than on the initial drainage rate. This is particularly significant from a dewatering standpoint because the time required for evaporation is considerably longer than for drainage. Therefore, the total time that solids must remain on the bed is controlled by the amount of water to be removed by evaporation, which, in turn, is determined by the amount removed by drainage and decanting. The percentage of drainable water strongly depends on initial solids concentration.

Quon and Johnson (1966) demonstrated that the percentage of drainable water in aerobically digested activated solids often considerably exceeds that reported for anaerobically digested solids. They also reported that the depth of applied solids affects the drainage rate and that the depth should not exceed 200 mm (8 in.). Haseltine (1951) reported an optimum depth of 230 mm (9 in.), depending on drying time and solids removal method; his suggested application depths ranged from 200 to 400 mm (8 to 16 in.). The applied depth should result in an optimum loading of 10 to 15 kg/m^2 (2 to 3 lb/sq ft). Even today, most design guidance documents recommend a maximum depth of applied sludge of 200 to 300 mm (8 to 12 in.).

Construction above the sand surface should include an embankment (vertical wall) with above-sand freeboard of 0.5 to 0.9 m (20 to 36 in.). Walls can be constructed of earth sodded with grass, wooden planks (preferably treated to prevent rotting), concrete planks, or reinforced concrete or concrete blocks that are set on an edge around the extremities of the sand surface and extended to the underdrain gravel to help prevent weed and grass encroachment.

3.4.2 DESIGN CONSIDERATIONS

Regional climatic conditions greatly affect dewatering performance in drying beds. In general, drying time is shorter in regions with abundant sunshine, low rainfall, and low humidity. Natural freezing in northern climates has been reported to improve dewaterability, but can also deactivate a bed for the winter. The prevalence and velocity of wind affects evaporation rates. Therefore, climatic conditions may warrant some modifications of design criteria. For example, storage of liquid residuals should be included if drying beds may be unavailable for extended periods because of climatic conditions.

For manual removal of dried solids in smaller facilities, the drying bed is typically divided into sections approximately 7.5 m (25 ft) wide. The width is designed to accommodate the removal method used (e.g., multiples of loader bucket width and span of vacuum removal system). Beds as long as 30 to 60 m

(100 to 200 ft) have been constructed. If polymer use is anticipated, however, the bed length should not exceed 15 to 25 m (50 to 75 ft) to avoid solids distribution problems. Provisions for flooding the bed with facility water before introducing solids can aid distribution at some facilities. In this approach, the bed drain valves are closed, the bed is flooded with water, liquid residuals are applied, and the drain valves are opened. Preflooding increases the rate of initial water removal because it adds to the hydraulic head by creating a vacuum under the bed when the drain valves are opened. This vacuum holds until air begins to leak through the bed into the underdrain system.

Supernatant can be removed continuously and/or intermittently from the drying bed using decanters provided on the perimeter of the bed (see Figure 11.12). Decanters are particularly useful for relatively dilute secondary and polymer-treated solids and for removing rainwater. Properly performed, decanting can reduce drying time significantly.

Drying beds can be fully covered on all sides or only on the top with glass or polyester glass fiber roofs. The roof-only configuration covers the top of the drying bed but leaves the sides open to the atmosphere, protecting the drying cake from precipitation but providing little temperature control. Completely enclosed drying beds, on the other hand, permit more cake withdrawals per year in most climates because of better temperature control. As a result, enclosed beds typically require less area than open beds. Favorable weather conditions, however, allow open beds to evaporate cake moisture faster than enclosed ones. Consequently, a combination of open and enclosed beds can achieve the most effective use of bed drying facilities.

Conditioning solids before sending them to drying beds is not widely practiced. In fact, the "Ten State Standards" (Great Lakes–Upper Mississippi River Board of State and Provincial Public Health and Environmental Managers, 2014) and other design guidelines do not consider using conditioning chemicals in this application, even though they may significantly reduce drying time and, therefore, the bed area required. Relatively small polymer doses (as little as 50 mg/L) can considerably improve the

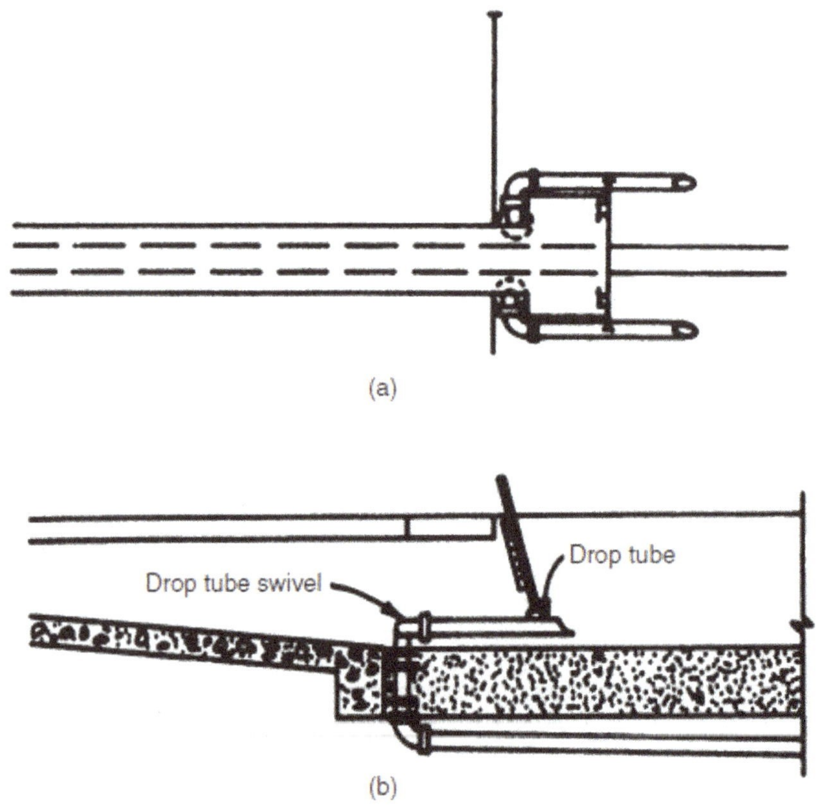

Figure 11.12 Typical Decant Piping: (a) Plan and (b) Elevation

drainage capabilities of properly digested solids by flocculating smaller particles. Flocculation speeds up the drainage period of the dewatering cycle and maintains a porous cake that is more readily susceptible to evaporation.

3.4.3 DRYING BED ALTERNATIVES

3.4.3.1 Polymer-Assisted Drying Bed

A polymer-assisted bed consists of a sand layer over specialized drainage panels. Physically, the filter beds look similar to conventional beds, but have a much quicker turnover time. The key to polymer-assisted filter beds is an upstream polymer activation system designed to properly mix polymer with feed solids and then properly distribute the mixture across the surface of the filter bed.

Sizing a polymer-assisted filter bed system is similar to sizing conventional solids drying beds, but should be confirmed with the manufacturer. Solids are applied to the system much as they would be on a conventional drying bed, but are first conditioned with polymer. The underdrain system provides a siphoning effect that gives vacuum assistance to the drying process. The plant water system is used to apply plant water to the underdrain system, which is key to the siphoning effect. Together, polymer conditioning and the siphoning effect allow solids to dewater more rapidly than in a conventional bed. A specially designed articulated vehicle is used to remove dewatered solids from the polymer-assisted filter bed.

3.4.3.2 Paved Drying Beds

Paved drying beds are constructed with concrete, asphalt, or soil cement liners to help front end loaders more easily remove cake and mix solids to speed up drying. A series of tests conducted by Randall and Koch (1969) indicated that drying beds with sand bottoms perform better than beds with impervious bottoms, although they may require more area than traditional sand drying beds for a given amount of sludge.

3.4.3.3 Wedge-Wire Drying Beds

Wedge-wire drying bed systems have been used in the United States since the early 1970s. In a wedge-wire drying bed, slurry is spread onto horizontal, relatively open drainage media in a way that yields a clean filtrate and provides a reasonable drainage rate. The cake typically is removed relatively wet (8% to 12% dry solids), which may complicate use or disposal. Wedge-wire drying beds may be better at drying difficult to dewater sludge.

3.4.3.4 Vacuum-Assisted Drying Beds

The principal components of vacuum-assisted drying beds include a bottom ground slab with reinforced concrete; a several-millimeters-thick layer of stabilized aggregate that supports the rigid multimedia filter top (this space is also the vacuum chamber and is connected to a vacuum pump); and a rigid multimedia filter top, which is placed on the aggregate.

Liquid residuals are spread onto the filter surface by gravity flow at a rate of 570 min (150 gpm) and to a depth of 300 to 750 mm (12 to 30 in.). Polymer is injected into the solids in the inlet line. Filtrate drains through the multimedia filter into the aggregate layer and then to a sump. From the sump, a level-actuated submersible pump returns filtrate back to the facility. After solids are applied and allowed to gravity drain for about 1 hour, the vacuum system is started and maintains a vacuum of 34 to 84 kPa (10 to 25 in. Hg) in the sump and under the media plates. Under favorable weather conditions, this system can dewater a dilute, aerobically digested solids feed to 14% solids concentration in 24 hours. The dewatered solids can be lifted from the bed by mechanical equipment. It will further dewater to an about 18% solids concentration in 48 hours.

3.4.3.5 Reed Beds

Reed beds to treat stabilized solids from secondary WRRFs have been used in a handful of U.S. locations as well as in many developing countries. The system combines the action of conventional drying beds with the effects of aquatic plants on water-bearing substrates. Although conventional drying beds drain more than 50% of the water content from solids, the resulting cake must be hauled away for further treatment or disposal at designated sites. When the drying beds are planted with reeds of the genus *Phragmites communis*, further desiccation results from the demand for water by these plants. To satisfy this demand, the plants continually extend their root system into the solids deposits. This extended root system establishes a rich population of microflora that feed on the organic content of the solids. This microflora is also partly kept aerobic by the action of the plants. Degradation by the microflora is so effective that eventually up to 97% of solids are converted into carbon dioxide and water, with a corresponding volume reduction. The beneficial result is that these planted drying beds can reportedly be operated for up to 10 years before the accumulated residues have to be removed. This represents a considerable monetary savings.

These systems are most applicable in climates in which winter temperatures ensure at least one prolonged frost. The planted reeds are harvested in the fall once they have become dormant. The use of reed bed systems is a reasonable application for WRRFs with less than 5.7 to 7.6 ML/d (1.5 to 2.0 mgd) capacity. The primary benefits of this system are the reduced O&M requirements and the frequency of disposal handling, which is once every 10 years.

Example 11.4

Use Table 11.15 to determine the number and area required for uncovered sand drying beds based on the SLR for a primary sludge and waste activated sludge that is produced at a rate of 250 kg/d (550 lb/d).

Solution

Determine the area required using Table 11.15 to find the SLR (73 kg/m²·yr).

$$73 = \frac{(250\ \text{kg/d})(365\ \text{d/yr})}{\text{Area}} \qquad \textbf{Area} = \mathbf{1250\ m^2\ (14\,455\ sq\ ft)}$$

Use a width of 7.5 m (25 ft) to determine the total length of the beds:

$$1250\ \text{m}^2 = (7.5)(\text{length}) \qquad \textbf{Length} = \mathbf{167\ m\,(547\ ft)}$$

Bed length is limited to 60 m (200 ft), so choose three beds. Each bed will be 55 m (180 ft) long. Provide three extra beds to allow drying of sludge for a total of 6 beds each 55 m (180 ft).

3.5 LAGOONS

Lagoons, which are natural or artificial earth basins, can be used for both solids drying and storage. However, the use of lagoons typically starts as a temporary expedient to handle solids volumes in excess of the original facility design. Climatic conditions have a decided effect on the functioning of a lagoon, with warmer, arid climates producing the best results. Lagoon operations typically involves a cycle of pumping stabilized liquid solids to the lagoon for several months, during which continuous or intermittent decanting of the supernatant from the lagoon surface is pumped back to the facility followed by removing the dewatered material with some type of mechanical removal equipment.

3.5.1 DESIGN CRITERIA

Storage lagoons can be 1.5 m (5 ft) deep or more and are primarily designed for storage rather than drying. To allow for cleaning, the dikes should be constructed to be about 3 m (10 ft) wide across the top to accommodate trucks and other mechanical equipment used for solids removal or for maintenance. The side slopes of the dikes should be a maximum of 3 : 1 (horizontal to vertical) to provide a slope surface that can be mowed by mechanical equipment. Residuals in storage lagoons typically do not dry or condense enough to permit removal by anything but a dragline, so the maximum width of lagoons must be less than twice the length of a dragline boom or other equipment to be used.

The design volume of lagoons depends on intended use and length of storage. Generally, solids lagoons are designed to provide for the emptying of one or more digesters and can be sized in reference to digester volume and frequency of emptying. Local climate and the proposed disposal of supernatant liquor and rainwater draw-offs influence the length of time between the initial placing of solids and their eventual concentration to a degree that permits mechanical removal.

Properly designed draw-off piping should be provided for the removal of supernatant liquid and rainwater from storage lagoons. The draw-off piping should be arranged to discharge these liquids to the WRRF's headworks. Removing these liquids prolongs the life of the lagoon, helps prevent insect breeding, and hastens dewatering.

3.5.2 DESIGN CONSIDERATIONS

The location of natural or artificial dewatering lagoons must be carefully considered before selecting lagoons as a treatment method. The proposed site should be sufficiently removed from dwellings and other areas where odors would produce problems. Because they go through a series of wet and dry conditions, large lagoons can produce nuisance odors.

The possibility of polluting groundwater or nearby surface waters should be investigated thoroughly. If the subsurface soil is permeable, the potential for groundwater contamination exists. Clay and/or membrane liners and underdrain systems can minimize this potential. Finally, all lagoon areas should be fenced to keep animals and other trespassers out and to prevent vandalism and potential liability problems.

3.6 PRESSURE FILTER PRESS

Pressure filter presses (commonly called recessed plate filter presses or plate and frame filter presses) have been in widespread use since the 1970s, evolving from a labor-intensive batch process to a partially automated operation in the presses of today. The reduction in overall labor requirements has made the filter press a more practical solids dewatering option for many facilities. Two types of filter presses typically are used to dewater wastewater solids. The most common is the fixed-volume, recessed chamber filter press. The other is the variable-volume, recessed chamber filter press (also called the diaphragm filter press).

The fixed-volume, recessed chamber filter press consists of a number of plates rigidly held in a frame to ensure alignment. These plates are pressed together either hydraulically or electromechanically between a fixed end and a moving end. The plates contain drainage surfaces, drainage ports for discharging filtrate, and a large centralized port for solids feed. A filter cloth covers the drainage surface of each plate and provides a filter medium. Once the filter press feed pump is stopped, the individual plates are shifted, and the sludge cakes are discharged.

The variable-volume press incorporates a flexible membrane across the face of the recessed plate. As with fixed-volume presses, the initial stage involves feeding sludge to the press. However, once the plate chamber is filled and the filter cake formation has started, the membrane is pressurized (between 600 and 1000 kPa [85 and 150 psi]) with compressed air or water, compressing the filter cake within the plate chamber. The physical compression of the filter cake increases the dewatering rate and shortens the cycle time. The results are higher production rates and more flexibility in achieving desired cake solids concentrations. The variable-volume press significantly differs from the fixed-volume press in that the volumetric capacity is typically less, cakes are much thinner, and the press is more highly automated.

Recessed plate filter presses typically can achieve capture efficiencies from 95% to more than 99%, depending on the nature of the solids. These capture efficiencies do not include the water used to wash the plates at the end of a filtration recycle, which also contributes solids to the recycle stream from the dewatering operation. Table 11.16 summarizes typical performance data of recessed plate filter presses for different types of solids. Advantages and disadvantages of recessed plate filter presses are shown in Table 11.17.

In terms of both capital and O&M costs, the pressure filter press system typically remains more expensive than other dewatering alternatives; however, when disposal requirements dictate drier cakes, pressure filter presses often have been proven cost-effective. Moreover, many landfills have adopted more stringent criteria for the moisture content of solids cakes; often, dewatered cake is required to contain more than 35% solids before it can be landfilled. Other dewatering devices cannot reliably and routinely meet this requirement. Pressure filter presses also are cost-effective when the dewatered cake must be incinerated. Often, the drier filter press cake (which increases the ratio of volatile matter to water content) enables autogenous combustion in incinerators, thus reducing the need for other fossil fuels (e.g., natural gas or fuel oil).

Table 11.16 Performance Characteristics of Recessed-Plate Filter Presses (Adapted from WEF, 2023)

Type of Sludge	Cycle Time[c] (min)	Conditioning Chemicals: Lime[d] (%)	Conditioning Chemicals: Ferric Chloride (%)	Feed Solids Concentration (%)	Cake Solids (%)
Raw primary	120	10	5	5 to 10	45
Primary/WAS (50:50)	150	10	5	3 to 6	40 to 45
Primary + Tricking filter sludge	120	20	6	5 to 6	38
Primary + Ferric chloride[a]	90	10	—	4	40
Primary + WAS[b]	180	10	5	8	45
WAS	150	15	7.5	5	45
Digested primary	120	30	6	8	40
Digested primary + WAS	120	10	5	6 to 8	45
Digested primary + WAS[b]	180	10	5	6 to 8	40
Digested primary + 50% WAS	120	10	5	6 to 10	45
Digested primary + 50% WAS	150	15	7.5	1 to 5	45

[a]Ferric chloride added at primary settling tank as a coagulant.
[b]Ferric chloride used as coagulant aid in the secondary treatment process.
[c]Cycle time is time from initiation of feed-to-feed pump discharge; time excludes cake discharge time.
[d]As CaO.

Pressure filtration is affected by several factors (e.g., particle size, specific gravity, and particle concentration). Wastewater solids consist of a wide variety of particle sizes and shapes that, under certain conditions, help keep an open matrix of particles, thus promoting free filtration. Most wastewater solids must be conditioned before entering the filter press. Creating an open matrix of biological solids is difficult; biological solids are gelatinous, leaving relatively small void spaces for filtrate even when they are well conditioned. If the feed solids consist of a large fraction of biological solids, lime may be needed to ensure that the solids matrix will have adequate flow channels.

3.6.1 DESIGN CRITERIA

The total cycle time is governed by solids characteristics, desired cake solids concentration, and the relationship between feed rate and pressure. Generally, longer cycle times result in drier cake, but after a while, the amount of water removed from the cake is negligible. Figure 11.13 illustrates the typical relationship between feed rate (which is equal to the water removal rate) and pressure in a filter press cycle.

Filter press plates can be round, square, or rectangular; range in length and width from 0.5 to 2.6 m (1.6 to 8.5 ft); and are available in several types of constructions, dimensions, and materials. Recessed plates are used almost exclusively in solids dewatering applications.

Table 11.17 Advantages and Disadvantages of Recessed Plate Filter Press Dewatering

Advantages	Disadvantages
High cake solids achievable	Batch operation
High capture rates	High equipment cost
More cost-effective than other technologies when drier solids are required	Higher operating costs
Can adapt to wide range of solids characteristics	Special support structure requirements
	Large footprint
	High labor requirements

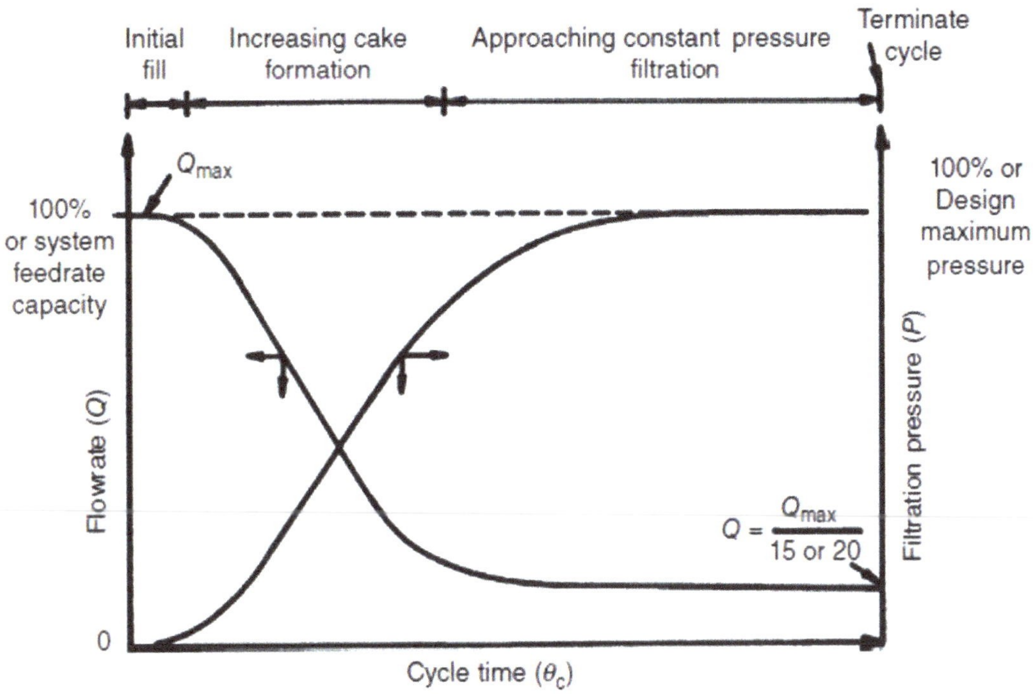

Figure 11.13 Filter Press Cycle Relationships

The choice of filter media is the most important equipment variable affecting cake quality and release, filtrate quality, and filter yield. Tightly woven cloths improve initial filtrate quality, but may extend filter cycles and result in difficulties in cake release. Open-weave media typically facilitate cake release during cake discharge but extend the initial cake formation time and reduce initial filtrate quality. Moreover, although multifilament fabric media typically improve filtrate quality by entrapping solids, they are susceptible to blinding and may result in poor cake release during discharge. Monofilament-type fabric construction facilitates cake releases during discharge and typically is easier to clean and to maintain.

3.6.2 DESIGN CONSIDERATIONS

Pressure is the driving force for filtration in these systems. Filter press systems typically are designed to operate at 700, 1600, or 2100 kPa (100, 225, or 300 psi). Researchers cite successful applications at all operating pressures; however, increases in filtration pressure can simultaneously increase cake resistance if the solids are compressible. Excessively high pressures can inhibit the process by tightly packing the solids, thereby reducing cake porosity and increasing resistance. Higher cake resistance reduces filtration flowrates. Increasing pressure on highly compressible solids often can decrease the rate of compression.

The number of plates in a filter press affects the overall efficiency of recessed plate filter presses. When well-conditioned solids are filtered in a press with a large number of plates, solids distribution throughout the filter chambers may be poor. In this situation, chambers nearest the entry points begin filling and start filtering, while chambers toward the center or end of the press have not yet received solids. As a result, unequal pressures develop throughout the length of the press during the filter cycle, producing cakes of randomly poor quality that often do not meet design criteria. The unbalanced forces created by poorly formed cakes can warp and eventually break plates made of weaker materials.

3.7 THERMAL DRYING

Thermal drying involves heating biosolids to evaporate water and thereby reduce the moisture content further than conventional mechanical dewatering methods can achieve. Thermally dried biosolids cost less to transport because of the mass and volume reduction that occur. Thermal drying meets the requirements to be a Class A process to further reduce pathogens (PFRP). There are

Class A is a designation for dewatered and heated sewage sludge that meets U.S. EPA guidelines for land application with no restrictions.

criteria associated with this PFRP, so all drying technologies do not automatically meet the Class A requirements. While most do, the system must be designed to meet the PFRP requirements. A Class A product that also meets certain vector attraction reduction requirements is more marketable than dewatered cake that does not. However, thermal drying processes are more complex and expensive to operate than conventional dewatering methods. Thermally dried biosolids can be used as a fertilizer, **soil conditioner**, or biofuel. The use depends on the source of the solids, the pretreatment and drying systems used, and local conditions.

Soil conditioner is a material that helps to improve soil structure by increasing aeration, water holding capacity, or nutrients. Also known as a soil amendment.

Thermal drying has been practiced since the 1920s when the Milwaukee (Wisconsin) Metropolitan Sewerage District began producing and selling Milorganite®, its thermally dried biosolids fertilizer. The popularity of thermal drying has grown since the early 1990s because of technology improvements, increasingly stringent regulations, and social factors affecting land application and landfilling. As of 2016, there were more than 105 thermal drying facilities operating or under construction in the United States and more than 375 worldwide. Their throughputs range from less than 1 to more than 100 dry metric ton/d (110 dry ton/d).

The drying rate for solids depends on both the solids flow and evaporation rates. During drying, a temperature gradient develops from the heated surface inward, causing moisture to migrate from inside wet solids to the surface via diffusion, capillary flow, and internal pressures generated as the solids dry and shrink. As heat transfers from the machine to the solids, the solids' temperature rises and water evaporates from the surface. Other conditions that affect the process include temperature, humidity, rate and direction of gas flow, exposed surface area, physical form of solids, agitation, detention time, and the support method used. Design engineers need to understand these conditions and their effects when investigating solids' drying characteristics, choosing the correct dryer, and determining the optimal operating conditions.

There are three stages of thermal drying: warmup, constant rate, and falling rate (see Figure 11.14). During the warmup stage, both the solids temperature and the drying rate increase until they reach steady-state conditions. This stage typically is short and results in little drying. During the constant rate stage, the surface of the solids remains saturated as interior moisture moves outward to replace the moisture that has evaporated. The drying rate depends on how much heat is transferred to the solids' surfaces (the heat or mass transfer coefficient). It also depends on how much area is exposed to the drying medium, as well

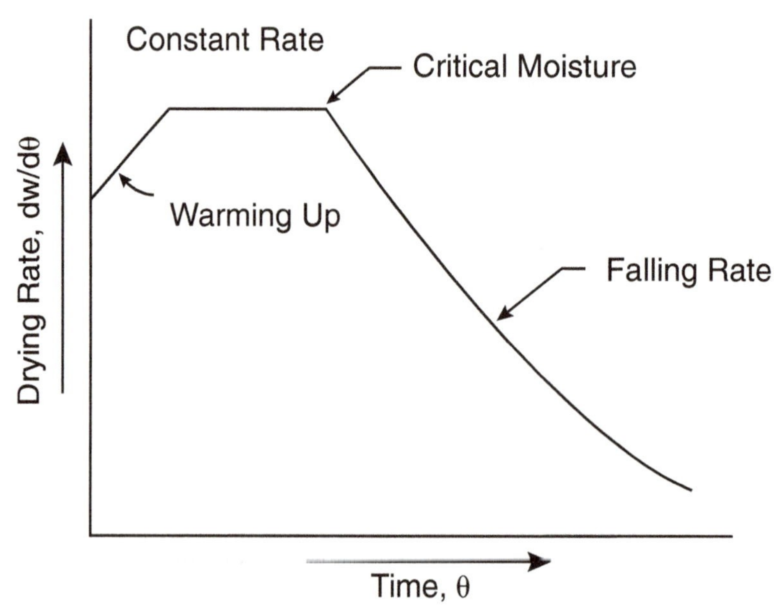

Figure 11.14 Three Stages of Drying

as temperature and humidity differences between the drying medium and the solids' surfaces. This stage typically is the longest and is when most drying occurs. During the falling rate stage, moisture evaporates faster from the solids' surfaces than it can be replaced by the moisture within the solids. Because the exposed surface is no longer saturated, solids do not transfer latent heat as rapidly as they receive sensible heat from the drying medium, and the temperature of the solids increases. This stage is when the drying rate decreases.

3.7.1 HEAT TRANSFER METHODS

Dryers are classified based on the method (convection or conduction) predominantly used to transfer heat to wet solids. Although most systems actually use multiple heat transfer methods, one method typically is more dominant than the others.

In convection (direct drying) systems, the heat transfer medium (e.g., hot gas) contacts the wet solids directly. Convection heat transfer is expressed mathematically as follows (U.S. EPA, 1979):

$$q_{conv} = h_c A(t_g - t_s) \tag{11.2}$$

Where q_{conv} = convective heat transfer (kJ/h [Btu/hr]),
h_c = convective heat transfer coefficient (kJ/h·m²·°C [Btu/hr·sq ft·°F]),
A = area of wetted surface exposed to gas (m² [sq ft]),
t_g = gas temperature (°C [°F]), and
t_s = temperature at sludge gas interface (°C [°F]).

Manufacturers have developed convective heat transfer coefficients for their proprietary systems that they use to size thermal drying equipment.

Conduction (indirect drying) systems transfer heat from the heat transfer medium (e.g., hot oil or steam) to the solids via a material (e.g., a steel plate) that separates the solids from the medium. The mathematical expression for conduction heat transfer is as follows (U.S. EPA, 1979):

$$q_{cond} = h_{cond} A(t_m - t_s) \tag{11.3}$$

Where q_{cond} = conductive heat transfer (kJ/h [Btu/hr]),
h_{cond} = conductive heat transfer coefficient (kJ/h·m²·°C [Btu/hr·sq ft·°F]),
A = area of heat transfer surface (m² [sq ft]),
t_m = temperature of heating medium (°C [°F]), and
t_s = temperature of solids at drying surface (°C [°F]).

The conductive heat transfer coefficient is a composite term that includes the effects of both the solids' and the medium's heat transfer surface films. Manufacturers have developed system specific coefficients that they use to size their proprietary systems.

3.7.2 CONVECTION/DIRECT DRYING SYSTEMS

Convection dryers that have been used successfully to dry municipal solids include the rotary drum dryer, fluidized bed dryer, and belt dryer.

3.7.2.1 Rotary Drum Dryer

Rotary drum dryers have been used to treat solids since Milwaukee first installed them in the 1920s. Although some system components are manufacturer-specific, all rotary drum dryers have long, cylindrical steel drums that rotate on bearings (see Figure 11.15). Some have one drum through which solids make a single pass, while others have concentric drums to allow solids to make multiple passes through the drum (e.g., triple-pass systems). Hot gases and dry solids and help move them through the system.

Actual influent moisture content requirements depend on the dryer supplier and solids characteristics. Feed solids and hot furnace gases (400 °C to 650 °C [750 °F to 1200 °F]) continuously enter the upper end of the drum dryer and are conveyed (typically concurrently) to the discharge end. As they travel through

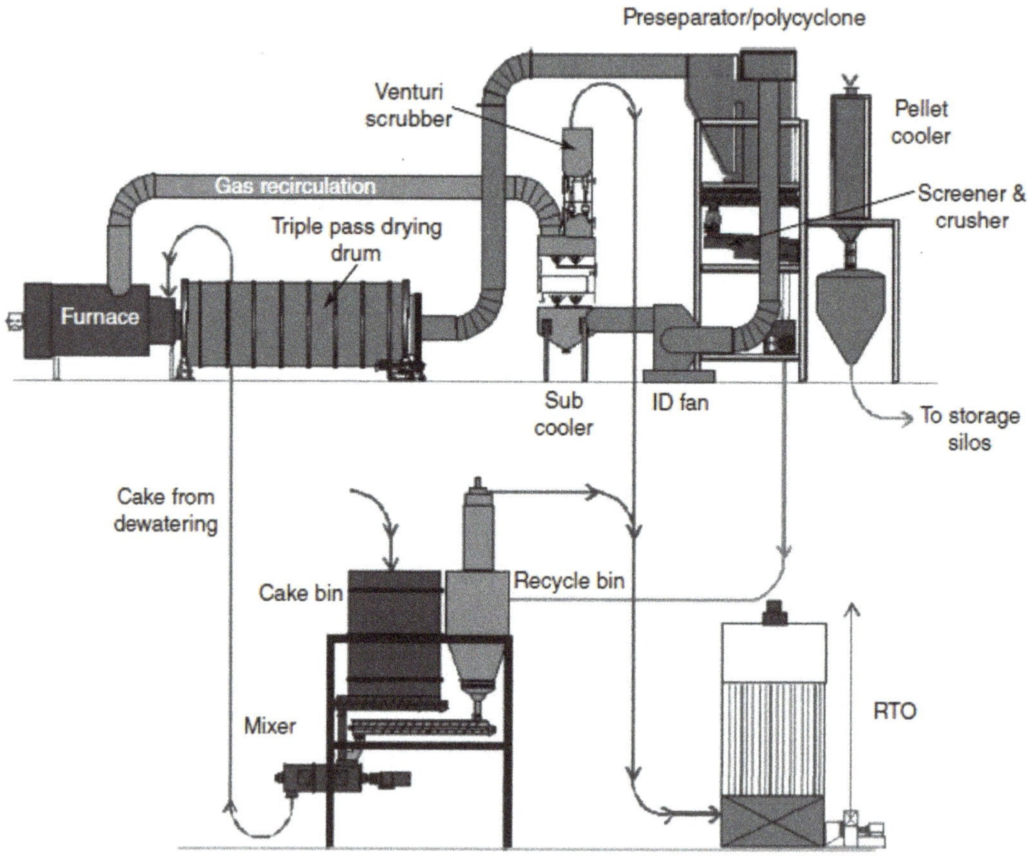

Figure 11.15 Schematic of Rotary Drum Dryer (Reprinted with permission from Andritz Separation, Inc.)

the drum, axial flights along the slowly rotating interior wall pick up and cascade solids through the dryer. This thin sheet of falling solids directly contacts the hot gases, which rapidly dry it. In most systems, about 80% of furnace gases are recycled to enhance the system's thermal efficiency and help maintain an inert atmosphere. Recycling gas also reduces the volume of exhaust gases to be treated. Exhaust gases exit the dryer at 66 °C to 105 °C (150 °F to 220 °F) and travel to emissions control equipment for particulate removal and odor control.

Rotary drum dryers have been successfully used to dry both digested and undigested solids. In Europe, where granules typically become a biofuel, dryers often process undigested solids. In the United States, where the granules primarily are used as a fertilizer or soil amendment, the dryers typically process digested solids. Digestion can break down much of the fiber in raw solids that can cause operating problems for dryers. It also reduces the putrescible compounds that can cause odor problems in granules in addition to reducing organic and nutrient content that reduces the soil conditioning value.

3.7.2.2 Fluidized Bed Dryer

The main component of a fluidized bed dryer is a stationary, vertical chamber segregated into three zones. The first zone is the wind box (gas plenum), which distributes hot fluidizing gases (typically air at about 85 °C [185 °F]) evenly through a bed of granules. The gases are forced or induced through the system by a blower arrangement in a closed-loop system. The second zone is a heat exchanger. A heat transfer medium (typically steam or hot oil) circulates through the heat exchanger tubes, while fluidizing gas keeps the granules in suspension, helps transfer heat from the tubes to the granules, and removes moisture from the system. The third zone is a hood, where fluidizing gases exit the unit and most solids are separated from the gas flow. Exhaust gas then is sent to emissions control equipment (e.g., cyclone separator, scrubber, and condenser), which removes particulates and moisture. As with rotary drum systems, fluidized bed dryers will reduce the solids moisture content to less than 10%.

The method for introducing dewatered solids to the dryer is manufacturer-specific. Some systems pump solids into the dryer, where a rotating device cuts dewatered cake into small, irregular pieces that fall into the fluidized bed. The fluidizing gas intimately contacts and highly intermixes the solids, resulting in high mass and heat transfer from gas to solids. Agitation in the fluidized bed removes rough edges, resulting in granules with a relatively uniform shape. Granules exit the dryer by overflowing from the chamber via an adjustable weir with a rotary airlock. All processing units are typically enclosed and maintained under slight negative pressure or sealed to capture dust and potential odors.

3.7.2.3 Belt Dryers

In belt dryers, solids are placed on a porous conveyor belt, and large volumes of heated gas are blown across or drawn through them. The belt speed and configuration control the SRT. The unit may have one or two belts, and its configuration depends on the manufacturer. The granules typically are less uniform in size than those from more complex systems because of the limited materials handling. Screens and conditioners can be used to improve uniformity.

Traditional belt dryers use a relatively low temperature gas (about 120 °C to 177 °C [250 °F to 350 °F]), so low-grade waste heat can be used as an energy source. Some very low temperature designs have evolved that can operate at temperatures as low as 65 °C to 80 °C (150 °F to 175 °F). These very low temperature dryers require increased effective belt area to achieve the same capacity as the higher temperature designs, but can use more energy from low-grade waste heat sources. This allows for a wide range of energy recovery options.

3.7.3 CONDUCTION/INDIRECT DRYING SYSTEMS

Over the last decade, conduction dryers have continued to become more popular as more small- and medium-size utilities have begun drying solids. The simple materials handling and emissions control systems associated with conduction dryers appeal to smaller utilities. Most facilities use the dryers to reduce the solids' moisture content to less than 10%, but some use these systems as scalping dryers to reduce the solids' moisture content to about 50% to 70%, depending on the requirements of downstream equipment or the ultimate market use. Conduction dryers that have been used successfully to dry solids include the hollow flight dryer, tray dryer, and pressure filter/vacuum dryer.

3.7.3.1 Hollow Flight Dryers

There are three types of hollow flight dryers: paddle, disc, and rotary chamber. Paddle dryers have been used in industrial applications for many decades but were first applied to drying biosolids in Japan in the late 1980s. Paddle and disc dryers have been used to treat biosolids in the United States since the mid-1990s.

Both paddle and disc dryers consist of a stationary horizontal vessel (trough) and a series of agitators (paddles, discs, or flights) mounted on a rotating shaft (rotor). In some systems, the trough has a jacketed shell through which a heat transfer medium (typically hot oil or steam) circulates. Rotary chamber dryers are similar, except that the trough's outer shell rotates and is heated by gas-fired burners. All these systems have a hollow rotor and agitators through which the heat transfer medium circulates. The surfaces of the trough, agitators, and rotors conduct heat to the biosolids.

Dewatered biosolids typically are fed directly to the dryer by a positive displacement pump or screw conveyor. (In these systems, feed solids typically are not mixed with recycled granules before treatment.) Paddle, disc, and rotary chamber dryers can be either continuous or batch fed. Granules exiting the dryer are irregularly shaped and vary in size because the agitation associated with paddles or discs can increase the concentration of fines in the product. Screens or conditioners can be used to produce a more uniformly sized product. The temperature of the heat transfer medium ranges from about 150 °C to 205 °C (300 °F to 400 °F). After transferring its available energy to solids, the heat transfer medium typically is recirculated through the heating system. A high-pressure condensate return system can be used for steam to reduce energy consumption.

3.7.3.2 Tray Dryers

Tray dryers are similar to disc dryers, but are configured vertically and mix recycled granules with dewatered solids before feeding them to the dryer. The blended solids are fed through a top inlet in the vertical multistage dryer. The dryer has a central shaft with attached rotating arms that move solids from one heated stationary tray to another in a rotating zigzag motion until they exit at the bottom as granules. The rotating arms have adjustable scrapers that move and tumble solids in 10 to 30 mm layers over the heated trays. After spiraling through the trays, solids exiting the dryer typically contain less than 10% moisture. These granules are screened and then either sent to a recycle bin for further processing or cooled to 30 °C (90 °F) and pneumatically conveyed to a storage silo.

3.7.3.3 Pressure Filter/Vacuum Dryer

A pressure filter/vacuum dryer is a variation on a standard pressure filter that combines dewatering and drying in one process. First, the process uses the standard pressure cycle to mechanically dewater solids, which typically have been conditioned with lime and ferric chloride before entering the unit. After the pressure cycle, hot water (80 °C [180 °F]) is circulated through the system to heat the chambers while a vacuum is drawn across the system to lower the boiling point of the water still in the solids. This is a batch process, which allows operators to control the level of dryness achieved, making the units suitable for partial or complete drying applications.

3.8 SOLAR DRYERS

Solar dryers represent a technology that has evolved from somewhat emerging to more established over the past decade. These systems started out in Europe, have been used there for more than 20 years, and have a more than 8-year history of operation in the United States. Solids have been dewatered naturally (e.g., sand drying beds) for decades, but such systems were susceptible to weather and typically produced solids containing about 70% moisture (for ease of disposal). In the 1990s, German researchers developed systems based on radiant (solar) energy and convective (air) drying theory that could produce solids containing no more than 10% moisture.

Solar dryers typically consist of a concrete pad with low walls that is surmounted by a greenhouse-type structure. Unthickened, thickened, or dewatered solids are trucked or pumped onto the pad. A program monitors climatic variables (e.g., humidity and temperature) both inside and outside the structure, and adjusts fans and louvers to provide enough ventilation for drying. An electrically powered, mobile mixer tills the solids to expose more surface for evaporation. Some tiller designs include the ability to convey granules to either end of the structure to facilitate product removal or automatic transport to the next process, to storage, or to vehicles for hauling. Solids may be spread in a relatively thin layer or arranged into windrows.

3.8.1 DESIGN CRITERIA

Dryers are evaporation devices, so a design engineer's first step is to calculate how much water needs to be evaporated during a given time. The key parameters when calculating the desired evaporative capacity are the solids concentration of both the feed solids and product granules, as well as operating time. Evaporative capacity typically is expressed in terms of kilograms (pounds) of water to be evaporated per hour. Designers should then consult with manufacturers, who have established their systems' evaporative capacity based on heat transfer calculations, lab tests, and demonstrated full-scale performance using various solids. A drying system may have multiple evaporative capacities based on the type of solids (e.g., undigested, digested, or combined primary and secondary), which affect heat transfer coefficients and materials handling characteristics.

Design engineers should keep in mind that the performance of the mechanical dewatering process substantially affects the size of a thermal dryer and its energy requirements. It typically is more cost-effective to remove water mechanically than thermally (to the extent possible).

3.8.2 DESIGN CONSIDERATIONS

Storage is an important but often overlooked element of a thermal drying system. It typically is needed both before and after the dryer. The system requires enough storage before the dryer to attenuate solids variations, to accommodate intermittent operating schedules, and to allow for system shutdowns. It also needs enough storage after the dryer to handle changes in hauling schedules and seasonal fluctuations in market demand.

Granules often are stored in silos that are sealed from the outside environment. Because of the cost of silos, design engineers should size them carefully. Alternative use or disposal methods (e.g., landfilling) may be more cost-effective than significant storage capacity. Small- and medium-size facilities can use other storage methods (e.g., bulk bags or covered pads) as long as dust control and other safety issues are addressed.

Design engineers should carefully consider the planned use for granules when selecting a thermal dryer. Some uses (e.g., high-end fertilizer) will require a higher quality product than others (e.g., soil amendment, bulk land application, or biofuel). Although thermal dryers typically produce similar granules in terms of solids/moisture content, physical characteristics of the product granules (size distribution, granule shape, etc.) vary significantly based on the dryer technology and associated processing.

To minimize energy consumption, WRRFs should try to operate with as high feed cake solids as is possible, consistent with maintaining acceptable product characteristics. Slight improvements in dewatering system performance can significantly reduce the overall thermal dryer system capacity requirements and energy consumption needs.

Post-drying systems (e.g., screening, cooling, and dust control) depend on dryer type and manufacturer. These systems can fulfill processing requirements, improve granule handling characteristics, and make the overall system safer. For example, screening makes the granules more uniform, and rejects can be recycled for blending with feed cake. Care should be taken for product storage to prevent fires and explosions. Granules must be cooled before storage to reduce the risk of auto-oxidation. Dust control systems (e.g., mineral oil spray and mixing systems) agglomerate dust particles to improve granule quality and enhance system safety. Minimizing the moisture content in the product, injecting nitrogen to create an inert atmosphere, and limiting the duration of storage are some additional considerations.

Example 11.5

Determine the daily heat energy requirements and volume of dried solids for an anaerobically digested and dewatered sludge. The sludge characteristics and manufacturers recommendations are as follows:

- Daily average sludge = 50 metric ton/d (55 dry ton/d) at 25% solids
- Heat energy consumed = 3.72 MJ/kg H_2O (1600 BTU/lb H_2O evaporated)
- Dried products bulk density = 729 kg/m^3

Solution

Determine the amount of water that is too be evaporated (conservatively assume all water will be evaporated)

$$H_2O = \frac{(50 \text{ metric ton/d})(0.75)}{0.25} = \mathbf{150\ metric\ ton\ H_2O/d}$$

Determine the daily heat energy required:

$$\text{Heat} = (150 \text{ metric ton } H_2O/d)(1000 \text{ kg/(metric ton)}(3.72 \text{ MJ/(kg } H_2O)$$

$$\mathbf{Heat\ required = 558\,000\ MJ/day\ (528\,872\ MBTU/day)}$$

Determine the daily volume of product:

$$\text{Volume} = \frac{(50 \text{ metric ton/d})(1000 \text{ kg/metric ton})}{729 \text{ kg/m}^3} = 68.6 \text{ m}^3 \text{ (2423 cu ft)}$$

3.9 ROTARY PRESS

Rotary and rotary fan presses can achieve cake solids and solid captures similar to belt presses and centrifuges relying on gravity, friction, and pressure differential to dewater solids (Crosswell et al., 2004). A typical rotary fan press is shown in Figure 11.16 with a schematic of the dewatering mechanims shown in Figure 11.17. Solids are dosed with polymer and fed into a channel bound by screens on each side. The channel curves with the circumference of the unit, making a 180° turn from inlet to outlet. Free water passes through the screens, which move in continuous, slow, concentric motions. The motion of the screens creates a "gripping" effect toward the end of the channel where cake accumulates against the outlet gate, and the motion of the screens squeezes out more water. The cake is continuously released through the pressure-controlled outlet. A key feature of both rotary press and rotary fan press dewatering technology is their slow rotational speed. Typical installations use speeds of 1 to 3 revolutions per minute. This provides low vibration, low shear, low noise, and low energy use.

The flow path through the units differs between the rotary press and the rotary fan press. Rotary presses are fed from the top and discharge on the bottom, while rotary fan presses are fed on the bottom and discharge from the top. Table 11.18 summarizes the advantages and disadvantages of rotary type presses.

3.9.1 DESIGN CRITERIA

The two primary design parameters for a rotary press are the HLR and SLR. The HLR is a function of the equipment's size and number of channels. The technology is modular, and the HLR of single-drive units ranges from 0.5 to 15 L/s (7 to 250 gpm), although a maximum HLR of 3 L/s (50 gpm) per channel is typical. Because solids capture is a function of the adjustable back pressure, the SLR varies with the HLR. At higher solids concentrations, residuals will accumulate in the outlet zone, form cake, and extrude more quickly. Rotary presses provide better performance on residuals with higher fiber content (e.g., primary solids). Capture rates depend on solids type and polymer use but can exceed 95% solids capture.

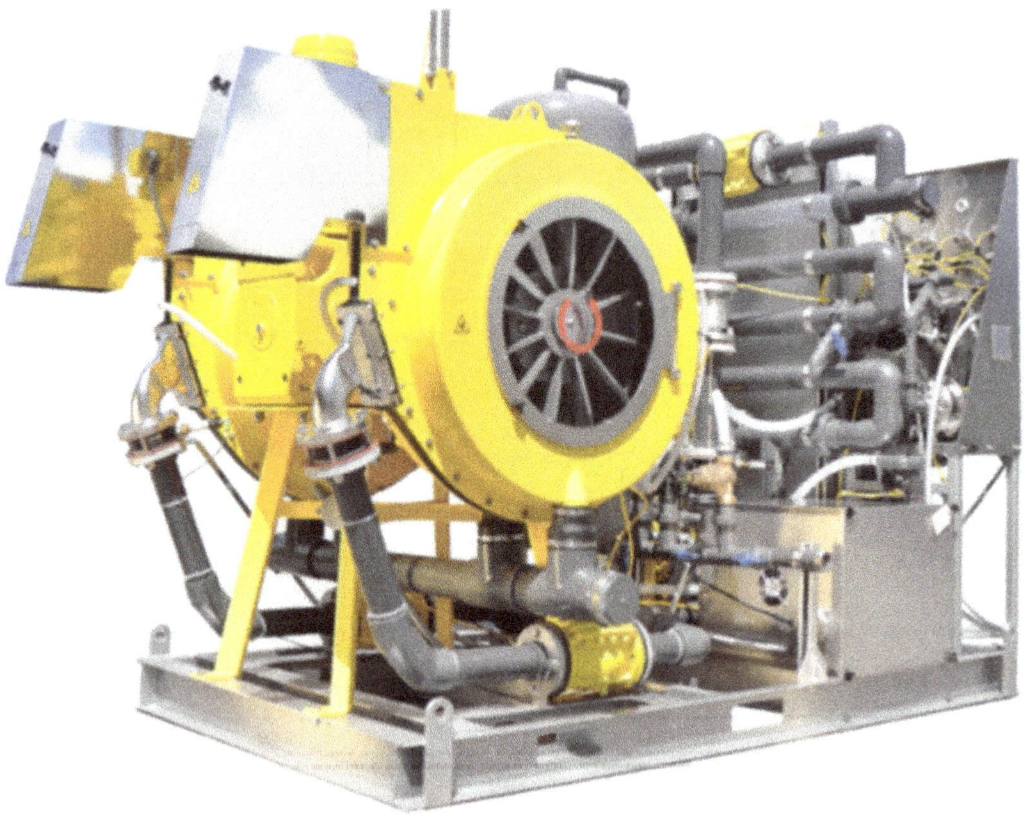

Figure 11.16 Two-Channel Rotary Fan Press System (Reprinted with permission from Prime Solution, Inc.)

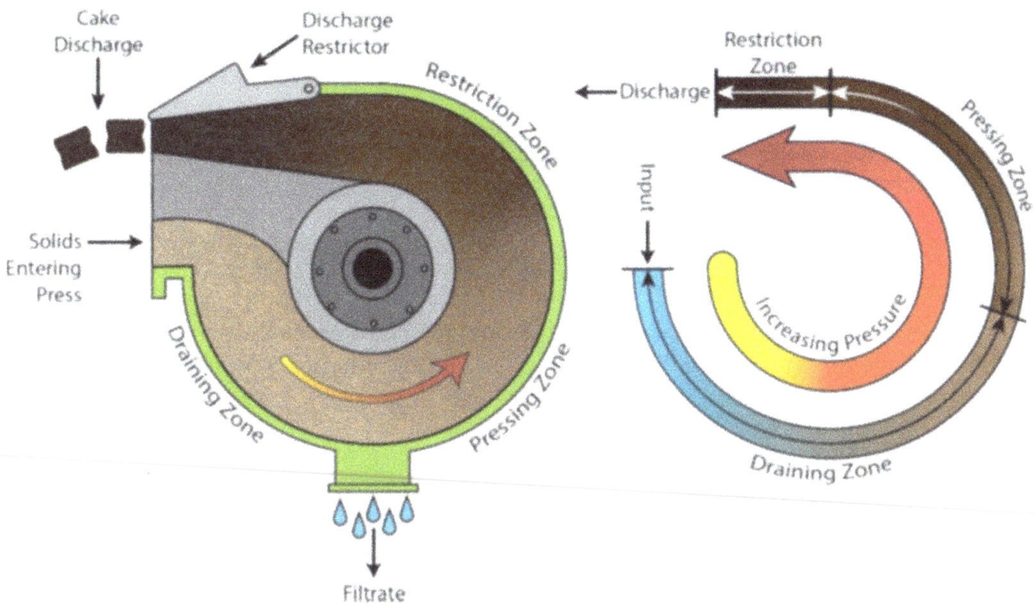

Figure 11.17 Cross-Sectional View of a Rotary Press Channel

Performance depends largely on solids consistency with discharge solids ranging from 12% to 28% (WEF, 2023).

3.9.2 DESIGN CONSIDERATIONS

Rotary presses and rotary fan presses are typically provided with control systems by the manufacturer. Optimum feed rates and discharge pressures are determined at startup and programmed into the controls. Chemical conditioning is required in rotary press or rotary fan press dewatering. Polymer feed systems can be supplied by the press manufacturer or can be procured independently. Rotary presses and rotary fan presses include a self-cleaning system that must run for 5 minutes per day at the end of use to flush all lines and equipment. Typically, the normal in-facility water source has sufficient pressure, but in some cases, high-pressure booster pumps may be required.

Table 11.18 Advantages and Disadvantages of Rotary Presses

Advantages	Disadvantages
Uses less energy than centrifuges or belt filter presses	May be more dependent on polymer performance than centrifuges or belt filter presses
Small footprint	Low throughput compared to other mechanical dewatering processes
Containment of odors	Screen clogging potential
Low shear	Need for heavy rated overhead crane to lift and maintain channels
Minimal moving parts	High capital cost
Minimal building requirement	
Minimal startup and shutdown time	
Uses less wash water than belt filter presses	
Low vibration	
Low noise	
Modular design	

PRACTICE PROBLEMS

1. Two gravity thickeners, each with a diameter of 10 m (32 ft), are fed primary and waste activated sludge with the characteristics listed below. For the thickeners, find the overflow rate and the SLR. Compare the SLR to the values in Table 11.2 and comment. Assume all solids have a specific gravity of 1.0.
   - Primary solids: 5.8 m^3/h (36 770 gal/d) at 4% solids
   - Waste activated sludge: 27.1 m^3/h (171 805 gal/d) at 1% solids

2. A single DAFT will be used to treat waste activated sludge. The design information is presented below. Determine the area required for the thickener (m^2 or sq ft) and the amount of pressured recycle required (m^3/hr or gpm).
   - WAS: 130 m^3/hr (5 729 gpm) at 0.8% solids
   - HLR: 120 $m^3/m^2 \cdot d$ (2 gpm/sq ft)
   - SLR: 9 $kg/m^2 \cdot h$ (1.8 lb/hr/sq ft)
   - Air-to-solids ratio: 0.02 : 1
   - Temperature: 20 °C
   - Recycle pressure (gage): 413.5 kPa (60 psig)

3. A WRRF is installing rotary drum thickeners to treat combined primary solids and WAS. The combined solids are generated at a rate of 2200 L/min (580 gpm) on a continuous basis. The following criteria are provided by the facility and the drum manufacturer:
   - Thickening facility will operate 8 hours per day, 5 days per week (2400 min/wk)
   - The rotary drum thickeners have an effective drum size of 1.5 m (60 in.) and can be operated at a rate of 1100 L/min/unit (290 gpm/unit).

   Determine the hydraulic loading that the thickeners will have to provide (L/min or gpm); and determine the number of thickeners required for this facility, assuming one will be always on standby and one on maintenance at all times.

4. Design engineers need to size circular gravity thickeners for a WRRF with a primary sludge with 22 680 kg/d (49 896 lb/d) at 5% solids. The criteria for the designer are as follows:
   - SLR not to exceed 6 $kg/m^2 \cdot hr$ (1.2 lb/sq ft/hr)
   - Hydraulic (overflow) loading rate not to exceed 31 $m^3/m^2 \cdot d$ (760 gpd/sq ft)
   - Depth between 3 to 4 m (10 to 13 ft)

   Determine the diameter and number of thickeners, assuming there are no standby units and they are operated continuously. Check detention time and comment.

5. The solids from Problem 4, 22 680 kg/d (49 896 lb/d), are thickened to 5% solids and sent to a belt filter press where they are dewatered to 18% solids (s.g. 1.04). There are three belt filter presses, each with a 2 m (6.6 ft) belt width. If the presses are operated 5 days a week for 8 hours per day, determine the following:

   a. The HLR on the belt filter presses during operation;

   b. The SLR on the belt filter presses during operation;

   c. The polymer required if the polymer dosage is 7.5 kg active polymer/1000 kg of sludge (15 active lb/ton of sludge);

   d. The mass and volume of solids produced if the capture rate is 96%;

   e. The required storage volume if solids are processed Monday through Friday.

6. A community is going to implement sand drying beds as an economical method for drying sludge. The community has 4000 people and produces 138.6 kg/d (305 lb/d) of digested solids. Determine the area required for uncovered sand drying beds by two methods. The first method uses a solids loading of 100 $kg/m^2 \cdot yr$ (20.5 lb/sq ft/yr). The second method uses 0.1 m^2/capita (1.0 sq ft/capita). Which would you choose and why?

4.0 REFERENCES

Ashman, P. S. (1976). Operational experiences of activated sludge thickening by dissolved air flotation at the Aycliffe Sewage Treatment Works. Paper presented at the Conference on Flotation in Water and Waste Treatment, Felixstowe, Suffolk, United Kingdom.

Bratby, J., Jones, G., & Uhte, W. (2004, October 2–6). State-of practice of DAFT technology–Is there still a place for it? Paper presented at the Water Environment Federation's 77th Annual Technical Exposition and Conference, New Orleans, Louisiana, United States.

Burfitt, M. L. (1975). The performance of full-scale sludge flotation plant. *Water Pollution Control (Great Britain), 74,* 474.

Butler, R. C., Finger, R. E., Pitts, J. F., & Strutynski, B. (1997). Advantages of cothickening primary and secondary sludges in dissolved air flotation thickeners. *Water Environment Research, 3,* 69.

Crosswell, S., Young, T., & Benner, K. (2004, October 2–6). Performance testing of rotary press dewatering unit under varying sludge feed conditions. *Proceedings of the 77th Annual Water Environment Federation Technical Exhibition and Conference* [CD-ROM], New Orleans, Louisiana, United States.

Ettlich, W. F., Hinrichs, D. J., & Lineck, T. S. (1978). *Operations manual: Sludge handling and conditioning* (EPA-68/01-4424).

Gehr, R., & Henry, J. G. (1978). Measuring and predicting flotation performance. *Journal of the Water Pollution Control Federation, 50,* 203.

Great Lakes–Upper Mississippi River Board of State and Provincial Public Health and Environmental Managers. (2014). *Recommended standards for wastewater facilities "Ten States Standards".* Health Research, Inc.

Gulas, V., Lindsey, R., Benefield, L., & Randall, C. (1978). Factors affecting the design of dissolved air flotation systems. *Journal of the Water Pollution Control Federation, 50,* 1835.

Haseltine, T. R. (1951). Measurement of sludge drying bed performance. *Sewage and Industrial Wastes, 23,* 1065.

Henderson, R. T., & Schultz, S. T. (1999). *Centrifuges versus belt presses in San Bernardino, California.* WEF/AWWA Joint Residuals and Biosolids Management Conference; Strategic Networking for the 21st Century. Charlotte, NC.

Imhoff, K., & Fair, G. M. (1940). *Sewage treatment.* Wiley & Sons.

Ireland, J. S., & Balchunas, B. M. (1998, October). High-speed, high-solids centrifuges: Sorting through mechanical features, manufacturers' claims, and owners' opinions. *Water Environment & Technology.*

Jones, W. H. (1968). Sizing and application of dissolved air flotation thickeners. *Water Sewage Works, 115,* R-177.

Kabouris J. C., Gillette, R. A., Jones, T. T., & Bates, B. R. (2005, October 29–November 2). Evaluation of belt filter presses, centrifuges, and screw presses for dewatering digested activated sludge at St. Petersburg's Water Reclamation Facilities. *Proceedings of the 78th Annual Water Environment Federation Technical Exhibition and Conference* [CD-ROM], Washington, DC, United States.

Komline, T. R. (1976). Sludge thickening by dissolved air flotation in the USA. Paper presented at the Conference on Flotation in Water and Waste Treatment, Felixstowe, Suffolk, United Kingdom.

Leber, R. S., & Garvey, D. (2000, March). *Centrifuge or belt press? Odor control may be a pivotal factor.* WEF 14th Annual Residuals and Biosolids Management Conference.

MacConnell, G. S. (1989, October 15–19). Full scale testing of centrifuges in comparison with DAF units for WAS thickening. Paper presented at the 62nd Annual Water Pollution Control Federation Technical Exposition and Conference, San Francisco, California, United States.

Maddock, J. E. L. (1976). Research experience in the thickening of activated sludge by dissolved air flotation. Paper presented at the Conference on Flotation in Water and Waste Treatment, Felixstowe, Suffolk, United Kingdom.

Metcalf & Eddy, Inc./AECOM. (2014). *Wastewater engineering: Treatment and reuse* (5th ed.). McGraw-Hill.

Mulbarger, M. C., & Huffman, D. D. (1970). Mixed liquor solids separation by flotation. *Journal of Sanitary Engineering, 96* (SA4), 861.

Quon, J. E., & Johnson, G. E. (1966). Drainage characteristics of digested sludge. *Journal of the Sanitary Engineering Division, American Society of Civil Engineers, 92*, 4762.

Randall, C. W., & Koch, C. T. (1969). Dewatering characteristics of aerobically digested sludge. *Journal of the Water Pollution Control Federation, 41*, R215.

Reay, D., & Ratcliff, G. A. (1975). Experimental testing on the hydrodynamic collision model of fine particle flotation. *Canadian Journal of Chemical Engineering, 53*, 481.

Rolan, A. T. (1980). Determination of design loading for sand drying beds. *Journal of the North Carolina Section, American Water Works Association and North Carolina Water Pollution Control Association, L5*, 25.

Turner, M. T. (1975). The use of dissolved air flotation for the thickening of waste activated sludge. *Effluent Water Treatment Journal (Great Britain), 15*(5), 243.

U.S. Environmental Protection Agency. (1978). *Sludge treatment and disposal: Sludge treatment, Vol. 1* (EPA-625/4-78-012).

U.S. Environmental Protection Agency. (1979). *Process design manual for sludge treatment and disposal* (EPA-625/1-79-011). U.S. Environmental Protection Agency, Municipal Environmental Research Laboratory, Office of Research and Development.

U.S. Environmental Protection Agency. (1987, September). *Design manual: Dewatering municipal wastewater sludges* (EPA/625/1-87/014). U.S. Environmental Protection Agency, Office of Research and Development.

U.S. Environmental Protection Agency. (2000, September). *Biosolids technology fact sheet: Centrifuge thickening and dewatering* (EPA 832-F-00-053) [Fact sheet].

U.S. Environmental Protection Agency & American Society of Civil Engineers. (1979). *Proceedings of the workshop towards developing an oxygen transfer standard*; (EPA-600/9-78-021). U.S. Environmental Protection Agency.

Walzer, J. G. (1978). Design criteria for dissolved air flotation. *Pollution Engineering, 10*, 46.

Wang, L. K., Li, Y., Shammas, N. K., & Sakellaropoulos, G. P. (2007). Chapter 13: Drying beds. In *Handbook of environmental engineering, Vol. 6: Biosolids treatment processes.* Humana Press.

Water Environment Federation. (2023). *Solids process design and management* (2nd ed.).

Chapter 12 Stabilization

LEARNING OBJECTIVES

At the end of this chapter, students should be able to:

- Describe how solids stabilization changes solids residuals generated at a water resource recovery facility (WRRF).
- Describe the purpose or goals of solids stabilization.
- Describe the benefits and drawbacks of different stabilization approaches.
- Calculate the size solids stabilization processes including tank volumes.
- Evaluate digestion efficiency and biogas production rates from anaerobic digestion.
- Determine digestion efficiency and aeration requirements for aerobic digesters.
- Estimate the lime dose requirement for lime stabilization.

1.0 INTRODUCTION

With growing focus on renewable energy production and biosolids utilization, stabilization processes have gained more attention in recent years. Stabilization treats the solids generated in the liquid treatment train, converting them to a stable product for beneficial use or disposal. Pathogens and odors are reduced, making the resulting biosolids appropriate for beneficial use. This chapter discusses four key stabilization processes, as identified by Part 40 of the *Code of Federal Regulations*, Part 503: Standards for the Use or Disposal of Sewage Sludge (40 CFR Part 503; U.S. Environmental Protection Agency [U.S. EPA], 1999), used in the United States today: anaerobic digestion, aerobic digestion, composting, and alkaline stabilization.

When designing a stabilization process, engineers should start by identifying the reasons for stabilization and all the stabilization options that could be integrated into the existing wastewater treatment scheme. Before selecting a process, the engineer must consider local demand and regulations for end use of the biosolids. Once a process is chosen, engineers should review all aspects (e.g., sidestream management, effectiveness in producing the desired biosolids quality, safety, ease of operation, and ancillary equipment needs) to ensure that the system is well designed. Design engineers should use a systematic approach that addresses both the economic and noneconomic ramifications of any proposed processing. This includes evaluating all parts of the system to ensure that they are flexible and reliable enough to consistently produce biosolids with the required characteristics. Furthermore, any potential effects on the liquid treatment processes (e.g., nutrient recycle loads) that may result from the selected system must be carefully evaluated and mitigated when necessary.

Not all water resource recovery facilities (WRRFs) stabilize their solids. Those that do, however, typically stabilize solids for one or more of the following reasons:

- Reduction of pathogens (typically to comply with 40 CFR Part 503);
- Vector-attraction reduction and/or volatile solids reduction (VSR) (typically to comply with 40 CFR Part 503);

Vector-attraction reduction is a U.S. EPA requirement reflecting stability of biosolids theoretically associated with a decrease in disease transmission potential.

- Mass and volume reduction;
- Biogas (renewable energy) production;
- Aesthetic reasons (e.g., product appearance and odor); and
- Product usefulness and marketability.

Facility planners should consider biosolids management when designing or upgrading WRRFs. They should start by deciding how the solids will be used or disposed of, as this will determine the type of stabilization required. For example, if solids will be landfilled with routine cover, federal and many state agencies may not require that the material be stabilized. Stabilization also may not be desirable if the solids will be thermally oxidized, because doing so will reduce the energy content. On the other hand, when treated biosolids are used in agriculture or silviculture, or distributed commercially, the material is stabilized to reduce pathogens, odor, and vector attraction in accordance with local and federal requirements.

The implementation of 40 CFR Part 503 and the public's growing concern for the environment have increased research into new technologies for beneficial use of biosolids. These factors have also prompted many engineers and municipalities to investigate or design more effective stabilization systems. Table 12.1 summarizes many of the advantages and disadvantages of the principal stabilization processes used today.

2.0 ANAEROBIC DIGESTION

Anaerobic digestion is a common stabilization process that is increasingly emphasized at facilities that generate high-energy primary sludge. It produces stable biosolids as well as biogas for fueling boilers, cogeneration systems (e.g., engine generators or microturbines with heat recovery), thermal hydrolysis conditioning systems, biosolids dryers, and injection into natural gas pipelines as **renewable natural gas** (RNG).

Renewable natural gas (RNG) is a biogas that has been upgraded to natural gas quality and can be substituted for natural gas in application.

Compared to other stabilization options, these systems may be expensive to build, require significant tankage (Figure 12.1) and equipment, produce a strong ammonia and phosphorus sidestream, require extra heat to maintain the desired temperature. Moreover, the process biology is slow growing and somewhat sensitive to environmental conditions. Disadvantages to anaerobic digestion are resolved via proper design, careful operations, and pretreatment programs.

2.1 PROCESS DEVELOPMENT

Anaerobic digestion is the most common and established technology used to stabilize wastewater solids and has been in use for nearly a century. The major objectives of the technology have historically been to stabilize primary and secondary/tertiary solids, reduce pathogens, reduce the mass of material (hauling cost savings), and produce usable biogas (typically for boilers).

Recent practice includes consideration of **co-digestion** of outside feedstocks or hauled waste in WRRF anaerobic digestion systems as part of a co-digestion process. By bringing in outside wastes, digester biogas production increases resulting in a more favorable WRRF energy balance. Some potential feedstocks include industrial residuals, such as food processing wastes and glycerin for soy biodiesel facilities, residential food wastes, and oil and grease from restaurant grease traps. The reader is referred to outside references for additional reading on anaerobic treatment of outside wastes (Jones et al., 2019; Speece, 1996).

Co-digestion is the anaerobic digestion of residuals at a WRRF with outside wastes.

As sustainable practices have developed in the United States, the role of anaerobic digesters at WRRFs has evolved to include generating a biosolids product with fertilizer value, recovering resources by co-digesting solids with other organic wastes, and developing power and energy via biogas use in cogeneration facilities.

Anaerobic digestion is a relatively complex process biochemically, but it is mechanically straightforward. It requires proper design that coordinates with the planned operations approach. When designing an anaerobic digestion process, the designer must consider that it involves handling explosive and corrosive

Table 12.1 Comparison of Stabilization Processes

Process	Advantages	Disadvantages
Anaerobic digestion	Good volatile suspended solids destruction (40% to 60%)	Requires skilled operators
		May experience foaming
	Net operational cost can be low if gas (methane) is used	Methane formers are slow growing; hence, "acid digester" sometimes occurs
	Broad applicability	Recovers slowly from upset
	Biosolids suitable for agricultural use	Supernatant strong in ammonia and phosphorus
	Reduces total sludge mass	
	Low net energy requirements	Cleaning is difficult (scum and grit)
		Can generate nuisance odors resulting from anaerobic nature of process
		High initial cost
		Potential for struvite (mineral deposit)
		Safety issues concerned with flammable gas
Advanced anaerobic digestion (many process options)	Excellent volatile solid destruction	Requires skilled operators
	Can produce Class A biosolids using time and temperature based batch operations	Can be maintenance intensive (see Anaerobic digestion for other disadvantages)
	Can increase gas production	
	Can reduce solid retention time	
Aerobic digestion	Low initial cost, particularly for small facilities	High energy cost
	Supernatant less objectionable than anaerobic	Generally lower volatile suspended solids destruction than anaerobic
	Simple operational control	
	Broad applicability	Reduced pH and alkalinity
	If properly designed, does not generate nuisance odors	Potential for pathogen spread through aerosol drift
	Reduces total sludge mass	Biosolids typically are difficult to dewater by mechanical means
		Cold temperatures adversely affect performance
		May experience foaming
Autothermal thermophilic aerobic digestion	Reduced hydraulic retention compared with conventional aerobic digestion	High energy costs
		Potential of foaming
	Volume reduction	Requires skilled operators
	Excess heat can be used for building heat	Potential for odors
	Pasteurization of the sludge pathogen reduction	Requires 18% to 30% dewatered solids
Composting	High-quality, potentially saleable product suitable for agricultural use	Requires bulking agent
	Can be combined with other processes	Requires either forced air (power) or turning (labor)
	Low initial cost (static pile and window)	Potential for pathogen spread through dust
		High operational cost: can be power, labor, or chemical intensive, or all three
		May require significant land area
		Potential odors

(Continued)

Table 12.1 Comparison of Stabilization Processes (*Continued*)

Process	Advantages	Disadvantages
Lime stabilization	Low capital cost	Biosolids not always appropriate for land application
	Easy operation	
	Good as interim or emergency stabilization method	Chemical intensive
		Overall cost very site specific related to product management costs
		Volume of biosolids to be managed is increased
		pH drop after treatment can lead to odors and biological growth in product odorous operations
Advanced alkaline stabilization	Produces a high-quality Class A product	Operator intensive
	Can be started quickly	Chemical intensive
	Excellent pathogen reduction	Potential for odors
		Volume of biosolids to be managed is increased
		May require significant land area

gases; results in recycling streams high in nitrogen and phosphorus; decreases dewaterability (whereby enhanced biological phosphorus removal liquid processes are used); and requires the use of completely closed tanks, which can make process monitoring challenging. When designing an anaerobic digestion system, it is important to plan for the operating conditions shown in Table 12.2.

2.2 PROCESS FUNDAMENTALS

An anaerobic digester's ecosystem is driven by environmental conditions created by a combination of factors including design, operation, and the composition of the feedstock. Feed solids composition, solids retention time (SRT), hydraulic retention time (HRT), temperature, and mixing all exert selective pressures

Figure 12.1 Anaerobic Digesters at Loudon Water in Loudon, Virginia (Reprinted with permission from CROM, LLC)

Table 12.2 Typical Operating Parameters for Mesophilic Anaerobic Digestion of Wastewater Solids

Parameter	Value	Units
Volatile solids destruction	45–55	%
pH	6.8–7.2	
Alkalinity	2500–5000	mg $CaCO_3$/L
Methane content	60–65	percent by volume
Carbon dioxide content	35–40	percent by volume
Volatile acids	50–300	mg VA/L
Vol. acid : alkalinity ratio	<0.3	mg $CaCO_3$/mg VA
Ammonia	800–2000	mg N/L

on the microbial populations in the digester. Selective pressure can affect overall digestion capacity due to the relative population of species, such as reported for *Methanosarcina* spp. and *Methanosaeta* spp. (Conklin et al., 2006).

The relationship between microorganism groups in an anaerobic digester is considered a syntrophic relationship. The metabolic activity of one population supports another. These relationships result in some distinct control points in the digestion process.

Feed solids characteristics affect which populations are dominant and can set up conditions where there is competition between a desired population and one that is less desirable. For example, acetate, the substrate from which about 75% of the methane in biogas is generated, is also the preferred substrate of sulfate-reducing bacteria. If the feed solids have high sulfate concentrations, conditions create direct competition between methanogens and sulfate-reducing bacteria. When this happens, the biogas methane content, and overall biogas production, may decrease.

2.2.1 MICROBIOLOGY AND BIOCHEMISTRY

Anaerobic digestion is driven by a series of **syntrophic** relationships that convert complex organic matter via a series of intermediate compounds to a variety of low-molecular-weight reduced compounds. The primary products of complete anaerobic metabolism are methane (CH_4), carbon dioxide (CO_2), biomass, and residual organic matter. Minor products include hydrogen (H_2), hydrogen sulfide (H_2S), ammonia (NH_3), and phosphate (PO_4^{-3}). Digestion can be viewed as a series of steps in which the waste products of one organism are the substrate for another. Solids destruction is the result of a balanced coupling of a variety of metabolisms.

Syntrophic refers to a phenomenon involving mutual dependency between groups of organisms.

Figure 12.2 is a simplified flow diagram of the major metabolic processes in anaerobic digestion for converting organic matter to methane and carbon dioxide. Each metabolic pathway represents a myriad of microorganisms, many of which have yet to be speciated. The microorganisms responsible for digestion are *Bacteria* and *Archaea*. Hydrolysis, acidogenesis, and methanogenesis are the three major metabolic steps in anaerobic digestion. Each step involves several biochemical reactions to convert complex organics

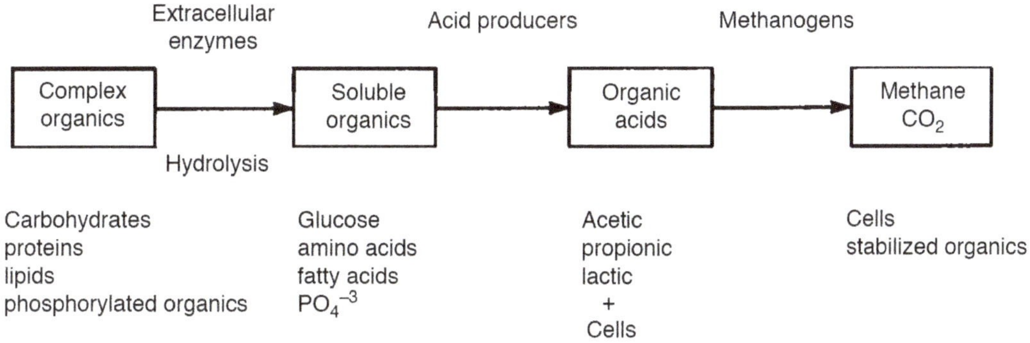

Figure 12.2 Major Metabolic Processes and Products of Anaerobic Digestion

to intermediates, such as short-chained organic acids, and final products, such as methane and carbon dioxide. Each group provides a unique and indispensable biotransformation that links one group to the other. If an imbalance occurs in the linkages, the overall process may be upset.

2.2.2 PROCESS RATES AND KINETICS

Process rates are affected by several external factors, including temperature, substrate, interspecies competition, and the presence of toxicants. While the environmental conditions affect the observed process rates, the fundamental limits of a process are regulated by kinetics. The kinetic limitations are often encapsulated as loading parameters and reaction time used for sizing anaerobic digestion, such as the volatile solids loading rate (VSLR) and SRT.

2.2.3 FEEDSTOCK CHARACTERISTICS

Key objectives of anaerobic digestion are to stabilize raw solids and reduce the mass of the material. Raw, primary, and secondary solids are primarily composed of proteins, polysaccharides, nucleic acids, fatty acids, and lipids. The relative concentrations of these compounds are a direct function of influent wastewater characteristics and the liquid treatment train used. Incorporation of outside feedstock into digesters is becoming increasingly common, and the impacts of outside wastes on the process must be considered carefully (Jones et al., 2019).

Different mixes of the undigested solids can result in different effects on treatment. A strong characterization of the raw solids can help fine-tune the overall digestion design. For example, waste mixtures with a higher protein content results in higher alkalinity for buffering, but it also results in a higher ammonia content and the potential for ammonia toxicity. On the other hand, a high lipid content can result in more acid production and less buffering leading to the potential for digester souring if a proper balance with methanogenesis is not maintained.

2.2.4 HYDRAULIC AND SOLIDS RESIDENCE TIME

Anaerobic digesters are sized to provide enough residence time in well-mixed reactors to meet regulatory requirements, allow significant VSR to occur, and prevent slower-growing microbial populations from washing out. Sizing criteria are defined as follows:

- Solids retention time (days) is equal to the mass of solids (often quantified as total solids in digestion) in the digester (kilograms) divided by the solids removed (kilograms per day).
- Hydraulic retention time (days) is equal to the working volume (liters) divided by the volume flow of solids removed (liters per day).

Typically, HRT is calculated based on either digester inflow or outflow rate. The SRT and HRT are equal in digestion systems without recycle or supernatant withdrawal. However, in older, existing systems in which supernatant is removed from the digester, SRT is calculated based on the solids volume removed.

The SRT (or HRT) and the extent of hydrolysis, acid formation, and methane formation during anaerobic digestion are directly related: an increase in SRT increases the extent of each reaction, while a decrease in SRT decreases the extent of each reaction (Figure 12.3). Each reaction has a minimum SRT (e.g., 4.2 d for wastewater sludge) (Lawrence, 1971), below which bacteria cannot grow rapidly enough to remain in the digester. Below the minimum SRT, the reaction mediated by the bacteria ceases, and the digestion process fails. Excessively long SRTs help prevent washout, but the extra equipment and infrastructure costs typically are not justified by the marginal increase in process performance.

2.2.5 ORGANIC LOADING RATE AND FREQUENCY

Volatile solids loading rate refers to the mass of volatile solids added to the digester each day divided by the digester's working volume (kg VS/m^3·d [lb VS/cu ft/d]). Loading criteria typically are based on sustained loading conditions (typically peak month or peak week solids production), with provisions for avoiding excessive loading during shorter periods. A typical design sustained-peak loading rate for **mesophilic** digesters is 1.9 to 2.5 kg VS/m^3·d (0.12 to 0.16 lb VS/cu ft/d). The upper limit of the VSLR typically is determined by the rate at which toxic or inhibitory substances accumulate, resulting in the washout of slow-growing microorganisms. A limiting value of 3.2 kg-VS/m^3·d (0.20 lb-VS/cu ft/d) is often used.

Mesophilic refers to the operation of treatment systems between the temperatures of 20 °C and 45 °C; optimally at 35 °C for anerobic digestion.

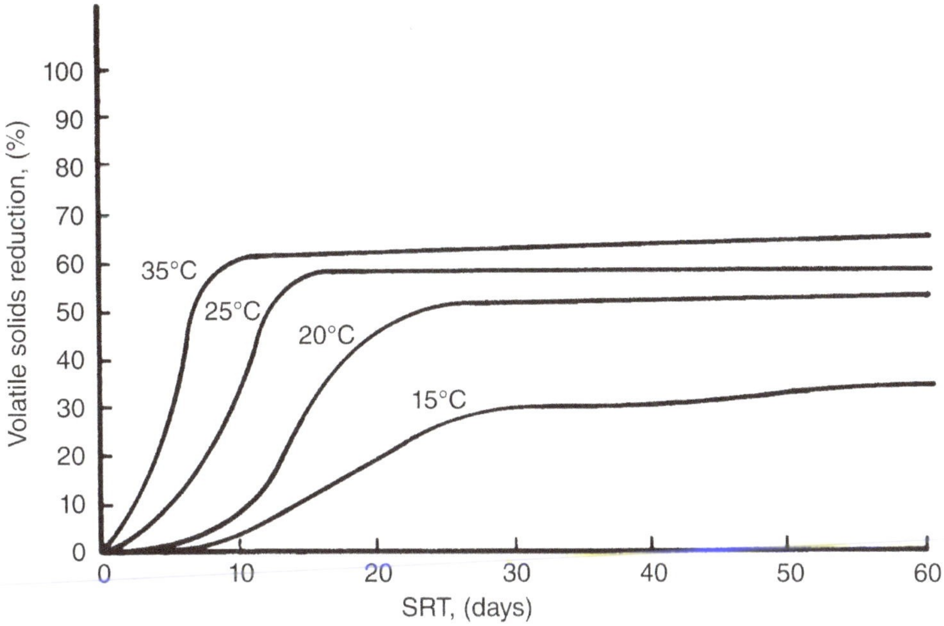

Figure 12.3 The Effect of Solids Retention Times (at Different Temperatures) on Chemical Oxygen Demand Removal and Methane Production in Anaerobic Digesters

Higher loading rates may be achieved with certain pretreatment technologies or other configurations. Some **thermophilic** systems have been able to attain higher VSLRs than mesophilic systems. With pre-dewatering to increase feed solids concentration and thermal or thermochemical pretreatment, some digesters may be loaded with 2.5 kg VS/m^3·d on a constant or consistent basis. When designing these highly loaded systems, engineers may want to pilot-test digester configurations to determine loading limits or perform a thorough review of comparable systems.

Thermophilic refers to the operation of treatment systems between the temperatures of 45 °C and 80 °C ; optimally at 55 °C for anerobic digestion.

Although design engineers should be conservative when setting an upper loading limit for a design, excessively low VSLRs (e.g., due to poor thickening) may result in designs that are expensive to construct (large tank volume) and operate (high heating demand of low percent solids feed). Thickening of solids before digestion is an efficient method for increasing a digester's HRT and SRT while decreasing the energy used to heat the process. Advances in thickening technology (e.g., thickening centrifuges) may allow digesters to be fed at 6% to 8% solids with a higher VSLR. When considering thickening improvements, pumps and mixing systems should be evaluated to ensure they will handle the increased viscosity, and care should be taken to avoid overloading the digester (excessive VSLR).

Continuous feed or frequent incremental feeding is ideal for supporting optimum design but may be challenging due to operational constraints. Slug or pulse feeding (semi-continuous feeding) is often used; this involves feeding and wasting solids at set intervals, typically at peak staffing times. This type of operation has been implemented at many utilities but has some process risks (e.g., overloading, upsets, rapid volume expansion, and foaming). However, some research has shown that pulse feeding may increase the diversity and resilience of microbial communities within an anaerobic digester (De Vrieze et al., 2013).

2.2.6 TEMPERATURE

An anaerobic digester's operating temperature significantly affects its observed performance and stability. The design operating temperature establishes the minimum SRT (or HRT) required to destroy a given amount of volatile solids with lower temperatures operating less efficiently (Figure 12.4). Currently, most anaerobic digesters are designed to operate at the idealized mesophilic temperature (about 35 °C [95 °F]). Some systems have been designed to operate at the idealized thermophilic temperature (about 55 °C [131 °F]). Many digesters are being designed so they can operate at both thermophilic and mesophilic temperatures, allowing future process flexibility.

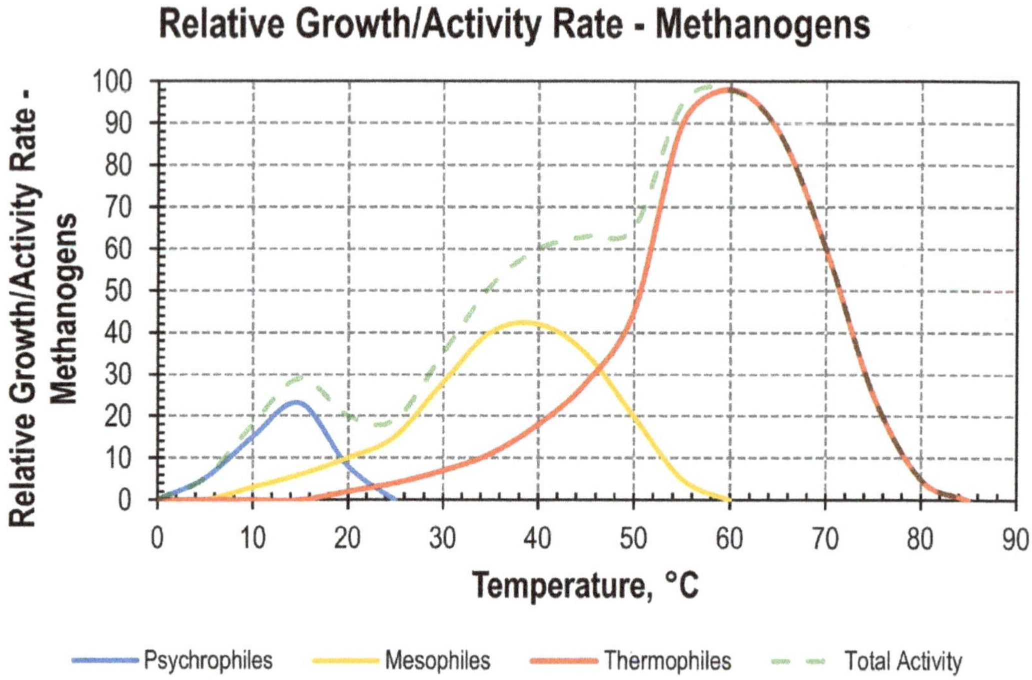

Figure 12.4 Temperature Effect on Anaerobic Organisms

Regardless of which temperature is selected, keeping it constant, or minimizing the rate of change, is of utmost importance. The microorganisms involved (particularly methanogenic populations) are sensitive to temperature changes; fluctuations in temperature can stress the organisms, thereby destabilizing the process. The organisms can, however, adapt to temperature changes over time. Therefore, it is desirable to limit temperature changes to 1 °C/d. A good design avoids temperature changes greater than 0.5 °C/d.

Temperature stability affects not only microbial stability, but also the classification of the resulting biosolids. Under 40 CFR 503 Alternative 1 (time and temperature), solids must be maintained at a specific temperature above 50 °C for a set period of time (depending on the temperature but a minimum of 20 minutes) to achieve Class A status; see Equation 12.1 (U.S. EPA, 1999). In this instance, time may be separate from HRT because every particle must be treated, requiring a batch held in an isolated tank, unless process equivalency has been granted. The effect of temperature on pathogen inactivation has made it one of the core mechanisms for achieving Class A biosolids, so it is important to design the system to maintain temperature (within close tolerances) under varying loads.

$$D = \frac{50\,070\,000}{10^{0.14T}} \tag{12.1}$$

Where D = Detention time (days) and
T = temperature (°C).

2.2.7 VOLATILE FATTY ACIDS CONCENTRATION AND COMPOSITION

When designing anaerobic digestion systems, engineers need to understand how volatile fatty acid (VFA) concentrations affect system design. Volatile fatty acids are the primary intermediates between complex organic matter and methanogenesis. The gross concentration of VFAs can indicate how complete digestion is, while the composition of the acids can indicate or cause a process upset or disturbance. Some acids have been reported to be variably inhibitory to anaerobic consortia (Franke-Whittle et al., 2014). Volatile fatty acid concentrations from 50 to 300 mg/L are considered normal for an anaerobic digester operating at mesophilic temperatures.

2.2.8 ALKALINITY AND PH

The production of hydrogen from fatty acid metabolism is not, thermodynamically, a highly favorable reaction. When hydrogen accumulates in the system at partial pressures above 5 mPa, there is a feedback inhibition of the acid oxidation process (McCarty & Smith, 1986). For the process to continue, as it does under stable digestion conditions, the hydrogenotrophic methanogenic population must be well established and respiring. What is evident from this one example is that a stress or toxin that affects one population may be enough upset the entire digestion process.

Anaerobic microorganisms, particularly methanogens, are sensitive to pH. Optimum methane production typically occurs when the pH is maintained between 6.8 and 7.2. Acid forms continuously during digestion, in the forms of VFAs and CO_2, and tends to lower pH. However, anaerobic digestion also produces alkalinity due to ammonia release, which buffers changes in pH. A reduction in pH promotes more acid formation and inhibits methane formation. As acid accumulation increases, the process is further stalled or inhibited, possibly leading to process failure. Mixing, heating, and feed-system designs are important in minimizing the potential for such upsets. Design engineers may include provisions for adding chemicals (e.g., lime, sodium bicarbonate, or sodium carbonate) to neutralize excess acid in an upset digester.

Useful alkalinity in anaerobic digesters is measured as bicarbonate alkalinity and ranges from 1500 to 5000 mg/L as calcium carbonate in anaerobic digesters. Both carbon dioxide and VFAs produced by the acid producers depress pH. Under stable conditions, VFA concentrations range from 50 to 150 mg/L as acetic acid (mg-HAc/L). A general guideline is to maintain a constant ratio of VFA to alkalinity that is less than 0.3. For most digesters, maintaining this ratio allows the system's buffering capacity to maintain a pH supportive of methanogens.

The bicarbonate alkalinity concentration can be calculated from the total alkalinity (which also includes the alkalinity of VFA [e.g., acetate]) as follows:

$$BA = TA - (0.71 \cdot VA) \tag{12.2}$$

Where BA = Bicarbonate alkalinity (mg $CaCO_3$/L),
TA = Total alkalinity (mg $CaCO_3$/L), and
VA = Volatile ccids or VFAs (mg HAc/L).

Sodium bicarbonate, lime, sodium carbonate, and ammonium hydroxide all have been used successfully to increase the alkalinity of anaerobic digesters, especially during startup or upset conditions. However, most well-designed and well-operated digester facilities do not require alkalinity addition as long as the wastewater has sufficient buffering capacity. Design engineers should evaluate the wastewater's alkalinity before determining whether an alkalinity feed system should be constructed.

2.2.9 TOXICITY IN DIGESTERS

If concentrations of certain substances (e.g., ammonia, heavy metals, light metal cations, and sulfide) increase sufficiently, they can create unstable conditions in an anaerobic digester resulting from process inhibition or toxicity (Tables 12.4 and 12.5). A shock load of such materials in facility influent or a sudden change in digester operation (e.g., overfeeding solids or adding excessive chemicals) can create toxic conditions in the digester.

Typically, excess concentrations of toxins or toxicants inhibit methane formation, which typically leads to VFA accumulation, pH depression, and digester upset. Depending on the concentration and type of toxin or toxicant, the effect can be acute (e.g., instant process failure) or chronic (e.g., depressed performance). Chemicals can be added to control the concentrations of dissolved forms of some toxicants (e.g., using iron salts to control sulfide). Sometimes, toxicity is a result of more than one factor (e.g., ammonia toxicity increases at higher pH) (Finger & Butler, 1996).

Table 12.3 Concentrations of Selected Inorganic Compounds That Inhibit Anaerobic Processes (Parkin & Owen, 1986)

Substance	Moderately Inhibitory Concentration (mg/L)	Strongly Inhibitory Concentration (mg/L)
Na+	3500–5500	8000
K+	2500–4500	12 000
Ca++	2500–4500	8000
Mg−+	1000–1500	3000
Ammonia-nitrogen	1500–3000	3000
Sulfide	200	200
Copper (Cu)	—	0.5 (soluble)
		50–70 (total)
Chromium VI (Cr)	—	3.0 (soluble)
		200–250 (total)
Chromium III	—	180–420 (total)
Nickel (Ni)	—	2.0 (soluble)
		30.0 (total)
Zinc (Zn)	—	1.0 (soluble)

2.2.10 BIOGAS PRODUCTION AND CHARACTERIZATION

The quality and quantity of digester biogas produced is specific to digester biological performance and chemistry. Biogas production is directly related biochemically to the mass of VSR; it often is expressed as volume of gas per unit mass of VSR. The gas production rate varies for organic substance in the digester (5). On a chemical oxygen demand (COD) basis, 3.5 standard cubic meters of methane is produced for

Table 12.4 Concentrations of Select Organic Chemicals That Reduce Anaerobic Digester Activity by 50% (Parkin & Owen, 1986)

Compound	Concentration Resulting in 50% Activity (mM)
1-Chloropropene	0.1
Nitrobenzene	0.1
Acrolein	0.2
1-Chloropropane	1.9
Formaldehyde	2.4
Lauric acid	2.6
Ethyl benzene	3.2
Acrylonitrile	4
3-Chlorol-1,2-propandiol	6
Crotonaldehyde	6.5
2-Chloropropionic acid	8
Vinyl acetate	8
Acetaldehyde	10
Ethyl acetate	11
Acrylic acid	12
Catechol	24
Phenol	26
Aniline	26
Resorcinol	29
Propanol	90

Table 12.5 Biogas Production Rates From Various Organic Substrates (Buswell & Neave, 1939)

	Specified Gas Production Per Unit Mass Destroyed	
Material	m^3/kg	Methane Content (%)
Fats	1.2–1.6	62–72
Scum	0.9–1.0	70–75
Grease	1.1	68
Crude fibers	0.8	45–50
Protein	0.7	73

every kilogram of COD metabolized; however, the overall biogas production is higher due to the CO_2 balance of the biogas (which typically varies from 20% to 40%). The overall biogas balance depends on the chemistry in the anaerobic digester.

When considering specific substrates, biogas production rates are higher with some feed sources and lower with other feed sources. The biogas production rate of fats ranges from about 1.2 to 1.6 m^3/kg (20 to 25 cu ft/lb) of VSR; the biogas production rate of proteins and carbohydrates is 0.7 m^3/kg (12 cu ft/lb) of VSR. The biogas production rate of a typical anaerobic digester treating a combination of primary solids and waste activated sludge (WAS) should be about 0.8 to 1.0 m^3/kg (13 to 18 cu ft/lb) of VSR. The amount of gas produced is a function of temperature, SRT, and VSLR.

In addition to the two main constituents of biogas, CH_4 and CO_2, biogas also contains trace amounts of nitrogen (N_2), H_2, and H_2S. Performance data from healthy municipal digesters show typical CH_4 concentrations ranging from 60% to 70% (by volume) and CO_2 concentrations ranging from 30% to 35% (by volume). Tortorici and Stahl (1977) have published data on typical digester gas characteristics. An increase in CO_2 levels (%) may indicate an upset digester. Excessive concentrations of H_2S can indicate unbalanced digestion, industrial waste sources, or saltwater infiltration. Hydrogen sulfide may be responsible for odor problems and excessive corrosion in the digester and adjacent piping. Heavy metals can precipitate as metallic sulfide, thereby reducing H_2S concentrations in biogas.

The digester gas collection and distribution system (Figure 12.5) must be maintained under positive pressure to avoid the possibility of explosion by allowing air into the system. A mixture of 5% to 15% methane to air is considered to be in the explosive range and is dangerous. The gas line from the digester to the biogas handling system should be located a minimum of 1.2 m above the maximum liquid level in the tank. A greater distance may be necessary to minimize the amount of solids and foam entering the piping system.

Table 12.6 A Survey of the Characteristics of Biogas From Anaerobic Digesters (Tortorici & Stahl, 1977)

Constituent	Values for Various Plants, % by Volume[a]							
Methane	42.5	61.0	62.0	67.0	70.0	73.7	75.0	73–75
Carbon dioxide	47.7	32.8	38.0	30.0	30.0	17.7	22.0	21–24
Hydrogen	1.7	3.3	[b]	—	—	2.1	0.2	1–2
Nitrogen	8.1	2.9	[b]	3.0	—	6.5	2.7	1–2
Hydrogen sulfide	—	—	0.15	—	0.01	0.06	0.1	1–1.5
Heat value, Btu/cu ft[c]	459	667	660	624	728	791	716	739–750
Specific gravity (air = 1)	1.04	0.87	0.92	0.86	0.85	0.74	0.78	0.70–0.80

[a]Except as noted
[b]Trace
[c]Btu/cu ft × 37.26 = kJ/m^3

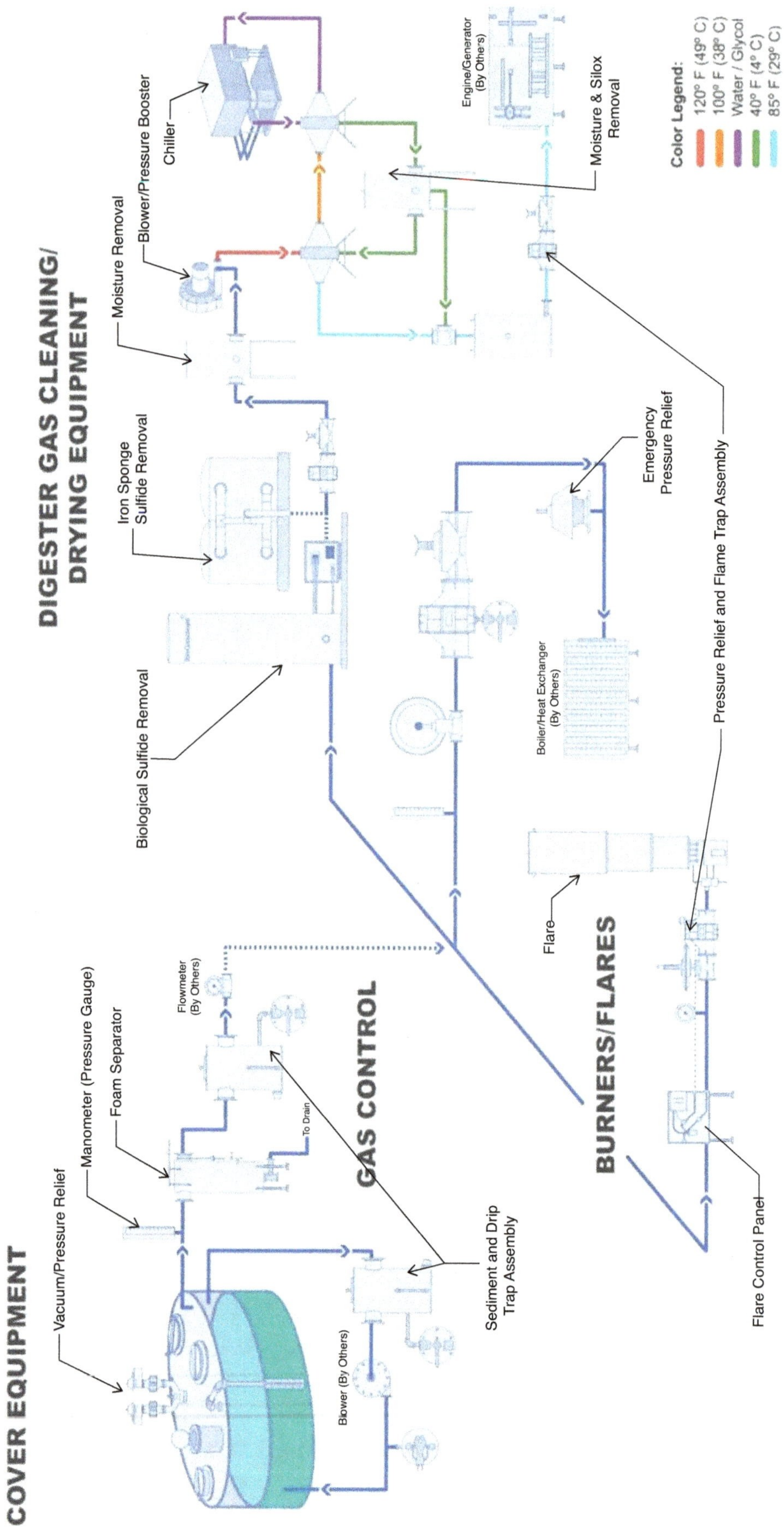

Figure 12.5 Diagram of a Gas Control System (Courtesy of Varec Biogas, a division of Ovivo USA, LLC. All rights reserved.)

Biogas and biogas combustion products can result in air pollution concerns. The traditional air pollutants of concern (criteria pollutants) from gas-combustion equipment are nitrogen oxides (NO_x), carbon monoxide (CO), sulfur oxides (SO_x), nonmethane hydrocarbons (NMHC), and particulate matter (PM). Nitrogen oxides (e.g., nitrogen dioxide, NO_2, and nitric oxide, NO) traditionally have been the most significant criteria pollutants. They are formed via combustion from nitrogen in the air. This group typically does not include nitrous oxide (N_2O). Carbon monoxide is formed via the partially complete combustion of methane (CH_4). Its emissions are controlled by combustion modifications. Sulfur oxides (e.g., sulfur dioxide) typically form when the hydrogen sulfide in digester gas combusts. It is controlled by eliminating hydrogen sulfide from the gas. Nonmethane hydrocarbons typically are not found in significant quantities in digester gas, but may need to be controlled for some applications; e.g. use of biogas as RNG. Particulate matter can include both PM_{10} and $PM_{2.5}$ (PM_{10} are particles less than 10 μm in diameter, while $PM_{2.5}$ are particles less than 2.5 μm in diameter).

One project consideration that has become a significant public concern is the reduction of greenhouse gases (climate change emissions). There are three greenhouse gases of concern in digester gas use evaluations: CO_2, CH_4, and nitrous oxide (N_2O).

Probably the best-known greenhouse gas, CO_2, is a relatively heavy gas. Digester gas can contain as much as 40% CO_2 by volume or about 60% CO_2 on a per-weight basis. For most of the digester biogas uses under consideration, the CO_2 initially in digester gas passes through unreacted and unchanged.

Carbon dioxide is formed from the complete combustion of any fuel that contains carbon (e.g., methane). Any boiler, flare, incinerator, or power-generation technology that combusts methane will produce a corresponding predictable amount of carbon dioxide. The basic chemical reaction is $CH_4 + 2\,O_2 \rightarrow CO_2 + 2\,H_2O$.

In other words, when 1 mole of methane is combusted completely, it forms 1 mole of CO_2. This conversion is essentially the same in a boiler, engine, gas turbine, or flare. Because CH_4 has a molecular weight of 16 and CO_2 has a molecular weight of 44, each completely combusted kilogram of methane produces 44/16 = 2.75 kg of carbon dioxide. Biogenic CO_2 is produced by life processes and is not included in greenhouse gas inventories.

Methane is the principal component of both digester gas and natural gas. It is a light gas with a specific gravity of less than 1.0. Methane is an exceptionally important greenhouse gas; its global warming potential is 21 to 23 times that of carbon dioxide. From a greenhouse gas perspective, completely combusting all methane without atmospheric release is vital. Beneficial reuse of methane should be strongly considered for heating, electricity generation, or generation of RNG for pipeline injection.

2.2.11 PROCESS STABILITY

Stable anaerobic digester operations can provide significant benefits (e.g., consistent solids destruction, pathogen reduction, and biogas generation). Stability is achieved through consistent loading, effective temperature control, and proper mixing. A well-functioning, stable anaerobic process exhibits specific digester and biosolids characteristics. Unstable digesters are more likely to foam, go acidic, and have microbial populations that are more susceptible to toxins.

A good design can help promote stable digestion operations. To ensure a constant temperature, boilers and heat exchangers should be sized to meet both the heat demands of raw solids and the shell losses that occur under the coldest influent and ambient conditions expected.

Influent blending tanks improve process stability by homogenizing raw solids and metering them more constantly to the digester. Minimizing fluctuations in solids strength and loading rate help the microorganisms in the digester maintain a constant metabolic state, which minimizes their stress. Though some research indicates that minor variations in feed may help promote more robust microbial populations, wide, frequent fluctuations in loading can stress the biomass, especially if loadings, substrates, and nutrients are insufficient.

Table 12.7 Typical Design Parameters for Low- and High-Rate Digesters (Burd, 1968)

Parameter	Low Rate	High Rate
Solids retention time, days	30–60	15–20
Volatile suspended solids loading, lb/cu ft/d (kg/m³·d)	0.04–0.1 (0.64–1.6)	0.12–0.16 (1.9–1.5)
Volume criteria, cu ft/cap (m³/cap)		
Primary sludge	2–3 (0.06–0.08)	1.3–2 (0.03–0.06)
Primary sludge + trickling filter sludge	4–5 (0.11–0.14)	2.6–3.3 (0.07–0.09)
Primary sludge + waste-activated sludge	4–6 (0.11–0.17)	2.6–4 (0.07–0.11)
Combined primary + waste biological sludge feed concentration, % solids-dry basis	2–4	4–6
Anticipated digester underflow concentration, % solids-dry basis	4–6	4–6

2.3 PROCESS OPTIONS

Process options for anaerobic digestion of wastewater solids have advanced significantly since the early 1990s. This section provides some historical context and discusses process options that are being considered and implemented more frequently in the 21st century (e.g., staged and phased systems, and mesophilic and thermophilic processes). The first option is high-rate versus low-rate digestion, which are compared in Tables 12.8 and 12.9.

Before the 1950s or 1960s, solids were anaerobically digested in "low-rate" systems characterized by intermittent feeding, low organic loading rates, little to no external heating or mixing, and detention times of 30 to 60 days. The tanks were large because grit and scum accumulated on the bottom and top, respectively, thereby decreasing the effective volume. Optimum digestion conditions were not maintained. Relatively few low-rate digestion systems are in service today because technology advancements have made them uneconomical and unattractive. Wastewater treatment professionals kept tinkering with the low-rate system, making improvements that eventually led to the "high-rate" anaerobic digestion system that was more common in the 1960s (Figure 12.6). This system is still widely used for mesophilic digestion.

High-rate anaerobic digestion was developed after research demonstrated the benefits of controlling environmental conditions in the digester. High-rate digestion is characterized by supplemental heating and mixing, relatively uniform feed rates, and thickening of solids (solids feedstock typically should contain 4% to 5% solids, although some recent improvements allow for thicker or thinner solids). These factors result in relatively uniform conditions throughout the reactor, leading to lower overall tank volume requirement and increased process stability.

For high-rate digesters, the digesters may be designed to operate as a mesophilic or thermophilic process. Table 12.9 compares typical range of the design basis for several parameters with these two processes side

Table 12.8 Estimated Volatile Solids Destruction

Digestion Type	Digestion Time (Days)	Volatile Solids Destruction (%)
High-rate (mesophilic range)	30	55
	20	50
	15	45
Low-rate	40	50
	30	45
	20	40

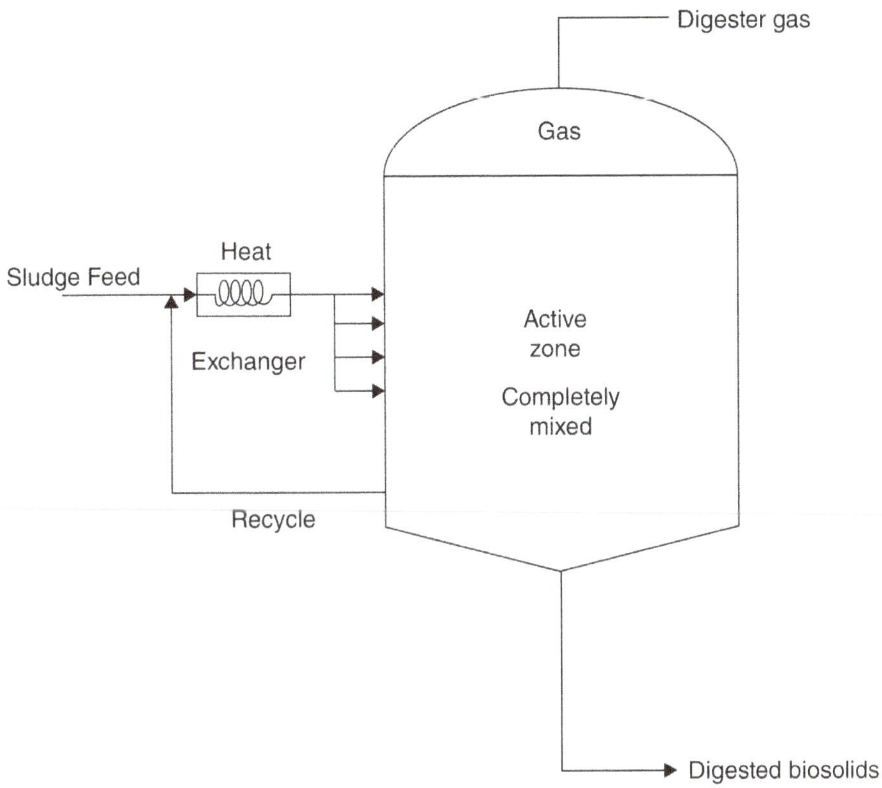

Figure 12.6 Simplified Flow Schematic of High-Rate Anaerobic Digestion

by side. The subsections that follow describe the mesophilic and thermophilic digestion, respectively, in more detail.

2.3.1 MESOPHILIC

Typically, high-rate systems achieve increased gas production, solids destruction, and overall process stability when compared to the increasingly rare low-rate systems. In the 1970s, the U.S. EPA extensively evaluated single-stage, high-rate anaerobic digesters operated at mesophilic temperatures with SRTs exceeding 15 days; regulators found that the process achieves significant pathogen reduction and solids stability. The agency defined it as a process to significantly reduce pathogens (PSRP) in its 1979 rule (40 CFR Part 257), and it essentially became a baseline for wastewater solids stabilization. In addition to SRT, other basic design criteria for mesophilic digesters are shown in Table12.9. The table shows typical operating ranges, which should be evaluated for both average and peaking conditions.

Table 12.9 Comparison of Mesophilic and Thermophilic Design Ranges

Loading Parameter	Units	Typical Mesophilic Range	Typical Thermophilic Range
Feed Solids (total solids)[1]	%	2–7	4–8
Solids Loading (VSLR)	kg-VS/m^3·d	1.2–2.5	2.2–2.8
	lb-VS/cu ft/d	0.08–0.16	0.14–0.18
Detention Time (SRT)	day	15–30	10–20
Temperature	°C	35–40	50–58
	(°F)	95–104	122–136
Mixing Intensity (Shear Rate)	s^{-1}	50–150	—
Volatile Solids Reduction (VSR)	%	38–60	45–70

[1]Higher feed concentrations possible with pretreatment; e.g., thermal hydrolysis process (THP).

Adequate mixing to ensure that the temperature is relatively consistent throughout the reactor (and to minimize bottom deposits and surface scum/debris, although this is only partially achieved in many high-rate digesters), and frequent or continuous solids feeding helps maintain steady-state conditions in the digester. Digesters are sensitive to changes in substrate levels; uniform feeding and multiple feed-point locations in the tank reduce instability.

2.3.2 THERMOPHILIC

Although most anaerobic digesters are operated at mesophilic temperatures (i.e., 35 °C [95 °F]), they also can be operated at thermophilic temperatures (typically between 50 °C and 57 °C [122 °F and 135 °F]). Thermophilic digestion has several advantages over mesophilic digestion. For example, solids from thermophilic digesters generally have better dewatering characteristics, so heating costs may be offset by reduced dewatering costs. Thermophilic digestion systems also destroy more volatile solids (Schafer et al., 2002) and produce biosolids containing fewer pathogens.

Thermophilic digesters have somewhat different design and performance criteria than those for mesophilic digestion. According to 40 CFR Part 503, a thermophilic digestion process typically meets Class A requirements by maintaining its temperature at or above 50 °C (more typically, 55 °C) for a specific period of time in a batch operation. The amount of time is calculated using the Equation 12.1. With the increased temperature, VSLRs can be higher and SRTs can be lower (Schafer et al., 2002). Thermophilic digesters also reduce more volatile solids and generate more biogas than identical-size mesophilic digesters, as governed by the Arrhenius relationship. Because the temperature is higher, however, more energy is needed to provide heat. To reduce heating costs, energy recovery heat exchangers may be used.

Early full-scale tests at Los Angeles indicated that thermophilic digesters were difficult to operate (Garber, 1982). However, thermophilic digester operations can be reliable when temperatures are constant and good mixing and feeding systems are used (Krugel et al., 1998). Inadequate mixing and temperature control systems may have contributed to challenges with early thermophilic digesters.

2.3.3 PRIMARY-SECONDARY DIGESTION

Anaerobic digesters are commonly configured with primary digesters providing treatment and secondary digesters receiving treated primary digesters. Historically, the primary tank was a typical mixed and heated anaerobic digester, and the secondary tank was a solid–liquid separator (supernatant withdrawn from the top). The secondary reactor traditionally did not have mixing or heating, but this configuration is not common anymore. In modern systems or facilities, secondary digesters have been upgraded with mixing and heating capabilities. As a result, the second tank may serve several functions including providing storage capacity, acting as a holding tank for dewatering, providing insurance against process short-circuiting, and acting as a standby primary digester.

2.3.4 STAGED DIGESTION

The concept of staged (phased) digestion has been used in various ways over the years (e.g., to stage metabolisms, operating temperatures, and redox conditions), but it is increasingly being recognized for its pathogen-control benefits. Primary–secondary digestion, for example, is basically a two-stage mesophilic process that produces well-stabilized solids because of the relatively long retention times and reduction of solids short-circuiting. Other staged mesophilic digestion options also are being recognized and used.

Two-stage mesophilic digestion is an extension of single-stage high-rate anaerobic digestion that uses two complete-mix digesters in series. The first stage must be designed to provide reliable mesophilic digestion (e.g., sufficient SRT and reasonable VSLR). Because much of the process considerations are met in the first reactor, the second stage can operate with a relatively shorter SRT. Both stages are heated and mixed. Placing the tanks in series causes the reaction kinetics to behave more like a plug-flow rather than a complete-mixed process, thus reducing short circuiting and improving process efficiency.

Schafer and Farrell (2000) and Chapman and Muller (2010) reported that, compared to single-stage digestion, two-stage mesophilic digestion can improve product stability (because more volatile solids are destroyed) and reduce short-circuiting of raw solids and pathogens.

Zahller et al. (2005) and McCarthy et al. (2004) reported that two-stage digestion provided better VSR and biogas composition (i.e., more methane content) than a single-stage digester with an equivalent SRT. However, the study also noted that the two-stage system seemed to have less capacity to absorb large variations in loadings than the single-stage system.

2.4 DESIGN CONSIDERATIONS

Historically, anaerobic digesters have been designed based on SRT, organic loading rate (evaluated as VSLR), and volume per capita. In the absence of operating data (or estimates of projected facility flows) and calibrated system modeling, volume figures per capita can be used to estimate influent volumes at WRRFs. Anaerobic digester design can also be evaluated as part of a modeling exercise using available simulation tools (Batstone et al., 2002; Jones et al., 2008).

2.4.1 PROCESS DESIGN

When sizing digesters, the key parameters are SRT and VSLR. For digestion systems without recycle, the SRT and HRT are equal. The VSLR is also used and is important in situations, such as co-digestion, whereby the volatile solids content in the feed may vary. The selection of the design SRT and VSLR should consider several factors (e.g., process microbiology, stability, biosolids regulatory requirements, biosolids stability, and industrial inputs). Generally, VSLR is limiting at higher total solids concentrations, and SRT is limiting at lower total solids concentrations.

The design minimum SRT typically is selected based on experience, general rules, and regulatory requirements. Design engineers should note that a range of solids production conditions must be considered when developing appropriate SRT design criteria. Typical SRTs are about 30 to 60 days for low-rate digestion and 15 to 30 days for high-rate digestion at mesophilic temperatures. The SRT is the ratio of the total mass of solids in the system to the quantity of solids withdrawn per day. In two-stage digestion, the typical SRT above refers to that of the first reactor in the system. For anaerobic digesters that are continuously mixed, the SRT equals the HRT. If settled solids are recycled, the SRT would be larger than the HRT. Recycling is another option with anaerobic digestion (two-phase digestion process or recuperative thickening) that may result in a different SRT and HRT.

Researchers are developing more quantifiable approaches to understanding digestion and its limits. The selection of a design SRT directly affects system kinetics and, consequently, process performance. Both complex models (e.g., the anaerobic digestion model no. 1 [ADM-1]; Batstone et al., 2002) and simplified models have been suggested as a means of estimating or predicting the limiting (minimum) digester SRT, although the data required for its application are limited. The Parkin and Owen (1986) approach is based on applying a safety factor to a limiting SRT to establish the design SRT. If the limiting SRT is based on a desired digestion efficiency and the digester approaches complete-mix conditions, the limiting SRT can be estimated as follows:

$$\mathrm{SRT}_{\mathrm{min}} = \frac{YkS_{\mathrm{eff}}}{K_c + S_{\mathrm{eff}}} - b^{-1} \tag{12.3}$$

Where $\mathrm{SRT}_{\mathrm{min}}$ = limiting SRT for required digester performance;
Y = yield of anaerobic organisms resulting from growth (g VSS/g COD destroyed);
K = maximum specific substrate use rate (g COD/g VSS·d);
S_{eff} = concentration of biodegradable substrate in digested solids (therefore, in digester) (g COD/L), which is equivalent to $S_o \cdot e$. S_o is the concentration of biodegradable substrate in the feed solids (g COD/L), and e is the digestion efficiency at removing S_o;
K_c = half saturation concentration of biodegradable substrate in feed solids (g COD/L); and
b = endogenous decay coefficient (d^{-1}).

Values for constants in Equation 12.3 have been proposed for typical municipal primary solids within a temperature range of 25 °C to 35 °C (77 °F to 95 °F). Table 12.10 gives proposed values for evaluation based on laboratory experiments (Parkin & Owen, 1986).

Table 12.10 Proposed Design Values

Parameter	Value
k	6.67 g COD/g VSS·d (1.035^{T-35})
K_c	1.8 g COD/L (1.112^{T-35})
b	0.03 d^{-1} (1.035^{T-35})
Y	0.04 g VSS/g COD removed
T	Temperature (°C)

Using the calculated value of the limiting SRT for required digester performance, the safety factor (*SF*) for anaerobic digestion then can be calculated as:

$$SF = \frac{SRT_{actual}}{SRT_{actual}} \tag{12.4}$$

For example, suppose engineers are designing a new anaerobic digestion system with the help of data from similar existing systems. The data indicate a median average SRT of about 20 days. Assuming the new system would have a digestion efficiency of 90%, a design temperature of 35 °C (95 °F), and feed solids with a biodegradable COD concentration of 19.6 g/L, the engineers used Equation 12.3 to calculate that the minimum SRT would be 9.2 days (Parkin & Owen, 1986). Then, using Equation 12.4 and the median SRT of the existing systems, engineers calculate that the new system's safety factor is 2.2. The safety factor obtained should be used to estimate whether short-term or dynamic increase in hydraulic loadings could be accommodated within the process.

Pilot testing may be used to determine a system's actual limits when typical and/or lumped kinetic parameters are uncertain due to the addition of new substrate. For example, if feedstocks contain significant amounts of materials (e.g., lipids) that are more difficult to degrade than typical municipal primary solids, then the constants given for Equation 12.3 do not apply. In these cases, engineers may need to use higher design SRT values than those listed in Table 12.9 or Table 12.10. If solids readily degrade (e.g., typical primary solids without biological solids), then engineers may be able to use slightly lower design SRT values.

Other limits (e.g., regulatory limits to meet PSRP requirements) need to be considered as well. Although a shorter design SRT may reduce tank and ancillary equipment size and cost, the ease of biosolids use or disposal also must be considered. Any initial cost savings may be lost if the biosolids are poor quality, do not meet regulatory standards, or incur significant operations and maintenance (O&M) costs to ensure compliance.

Furthermore, to minimize the likelihood of a digester upset, design engineers should select the design SRT based on a critical operating period (e.g., a high-solids-loading period when grit and scum have accumulated, or when a digester is out of service). The choice of critical operating period depends on facility size, anticipated solids production, and other site-specific factors.

A digester's loading rate and frequency significantly affects digester performance. Constant loading produces the most stable operations because the microorganisms reach and maintain steady-state conditions. Intermittent loading affects the process' overall stability because both over- and underloading can impair process performance. The design loading rate should be coupled with the process selected.

Furthermore, design engineers also should consider the potential effects of a specific feeding regime. Continuous feeding provides the best digestion efficiency and stability, but it may necessitate storage tanks or 24-hour thickening and dewatering operations.

In addition, design engineers should ensure that upstream processes are adequate to meet design loading. Thickening performance could prevent a design loading rate or SRT from being met consistently. Design engineers should take this into account when sizing tanks because it could lead to process limitations that could reduce the expected life of the system.

Anaerobic digestion system design requires an estimate of VSR efficiency and biogas production rates. Design engineers can estimate the expected VSR using previous data (40% to 70%) or equations relating VSR to detention time (Liu & Liptak, 1997). When reviewing VSR efficiency from previous data, a mass balance is performed to determine VSR based on mass of solids in versus mass of solids out. One key equation for evaluation of VSR-based data is derived from a mass balance and described by Van Kleek (Switzenbaum et al., 2003) as follows:

$$VSR = 100\left(\frac{VS_i - VS_o}{VS_i - (VS_i \cdot VS_o)}\right) \tag{12.5}$$

Where VS_i = Digester influent, volatile solids fraction; and
VS_o = Digester effluent, volatile solids fraction.

In the absence of data, the VSR for a standard-rate system may be estimated as follows:

$$V_d = 30 + \frac{t}{2} \tag{12.6}$$

Where t = time of digestion (days).

For a high-rate digestion system, digestion VSR is estimated as follows:

$$VSR = 13.7 \ln (\mathrm{SRT}) + 18.94 \tag{12.7}$$

An estimation of the volatile solids (kg/d) that would be fed into the second-stage digester of a two-stage system is given by the following equation:

$$VS_r = (VS/100 \cdot TS_r \cdot (1 - VSR)) \tag{12.8}$$

Where VS_r = volatile solids entering second-stage digester (kg/d [lb/d]),
TS_r = total solids input into primary digester (kg/d [lb/d]),
VS = volatile solids concentration (%), and
VSR = VSR of primary digester (%).

Once VSR is determined, design engineers can estimate the specific biogas production at WRRFs by using the relationship of approximately 0.8 to 1.1 m^3/kg (13 to 18 cu ft/lb) of VSR. Biogas production increases as the percentage of fats, oil, and grease (FOG) in the feedstock increases. The total gas volume produced is as follows:

$$G_T = (G_{sgp})V_d \tag{12.9}$$

Where G_T = volume of total biogas produced (m^3/d [cu ft/d]);
V_d = rate of VSR (kg/d [lb/d]); and
G_{sgp} = specific biogas production, taken as 0.8 to 1.1 m^3/kg VSR (13 to 18 cu ft/lb VSR).

When evaluating methane and biogas production from organic feedstock, the total amount of methane produced is estimated by looking at organic removal efficiency:

$$G_m = M_{sgp}[\Delta COD - 1.42(\Delta X)] \quad (12.10)$$

Where G_m = volume of methane produced (m^3/d [cu ft/d]),
M_{sgp} = specific methane production per mass of organic material (COD) removed (0.35 m^3/kg COD [cu ft/lb COD]),
ΔCOD = COD removed daily (kg COD/d [lb COD/d]), and
ΔX = biomass produced (kg VSS/d [lb VSS/d).

After calculating methane production using Equation 12.10, the total biogas production rate needs to be calculated next. Digester biogas is about two-thirds methane, which means the total digester gas produced is equal to the following:

$$G_T = \frac{G_m}{0.67} \quad (12.11)$$

With external organic feedstocks, methane concentrations can range from 45% to 75%; typical methane concentrations range from 60% to 75% (by volume), and carbon dioxide concentrations range from 25% to 40% (by volume). The expected heat value of digester gas depends on the composition of the biogas.

Example Problem 12.1

Design a heated primary mesophilic anaerobic digester for wastewater from 50 000 people. The digester feed sludge (sp gr = 1.01) contains 0.13 kg dry solids/cap × d at design average and 80% of the solids are volatile. The feed solids concentration is 6%, and the digester temperature is maintained at 35 °C. Digestion reduces the volatile solids by 50% (sp gr = 1.03). Design for an SRT safety factor of 2. Check VSLR to make sure it is within typical range. How much biogas would be expected from this digester and how much methane? Assume 65% methane content.

Summary input data:

Population	=	50 000 people
Solids contribution	=	0.13 kg dry solids/cap × d
Percent of sludge volatile	=	80%
Percent solids in sludge	=	6%
Specific gravity = sp gr	=	1.01
Target VSR in digester	=	50%
Percent of dry solids in digested sludge	=	8%
Specific gravity of digested sludge	=	1.03
Operating temperature	=	35 °C

Solution

Average solids entering the digester = (Population)(solids/cap × d) = 6500 kg/d
Volatile solids entering = (Solids entering)(Fraction volatile) = 5200 kg/d
Nondigestable solids = (Solids entering)(1 – Fraction volatile) = 1300 kg/d
Volatile solids removed in digester = (Volatile solids entering)(Fraction removed) = 2600 kg/d
Total solids in digested sludge = (nondigestable) + (Volatile solids in digested) = 2900 kg/d
Volume entering the digester = (Solids entering)/(Fraction)(sp gr) = 107.26 m^3/d

$VSR = 13.7 \ln(SRT) + 18.94$
$SRT_{min} = \exp((VSR - 18.94)/13.7) = \exp((50 - 18.94)/13.7) = 9.65$ d
$SRT = 2 \cdot 9.65$ d $= 19.3$ d
Digester Volume = 107.26 m^3/d * 19.3 d = 2070 m^3
$VSLR$ = 5200 kg/d / 2070 m^3 = 2.51 kg/(d·m^3)

Biogas production rate:

$$G_T = (G_{sgp})\,V_d$$
$$= 1.0\ m^3/\text{kg-VSR}\cdot 2900\ kg/d = 2900\ m^3/d$$
$$G_m = 2900\ m^3/d\cdot 0.65 = 1885\ m^3/d$$

2.4.2 TANK CONFIGURATION

The tank configuration (shape) significantly affects the operating characteristics of anaerobic digestion, as well as the cost of construction and O&M. Typically, digesters are available in three basic shapes: short cylinder ("pancake"), tall cylinder ("silo"), and egg-shaped.

Egg-shaped digesters (Figure 12.7) have been in service for more than 50 years in Europe, and since the early 1990s in the United States (Volpe et al., 2004). This shape typically is considered the optimal shape for a digester, providing excellent mixing characteristics, very few dead zones, and good grit suspension. Egg-shaped digesters do well at controlling foam when there is low to moderate foaming potential.

This digester's shape improves mixing. The tapered base, with centrally located mixing, is designed to promote the resuspension and removal of grit and other heavy materials (Figure 12.8). This minimizes the amount of material retained in the digester, increasing the active fraction and reducing the out-of-service time for cleaning. In a review of egg-shaped digesters, Volpe et al. (2004) reported that some have been in service for 20 years without needing to be cleaned.

The digester's small gas dome provides operational advantages and disadvantages. First, the degree of liquid agitation in the dome due to mixing and the foam-suppression system is typically high, preventing a scum layer from forming. Instead, it remains entrained in the liquid phase and can be withdrawn. Depending on tank configuration, however, withdrawal from the small gas dome can be problematic. This design may be susceptible to significant liquid-level variation during rapid volume expansion events, but

Figure 12.7 Egg-Shaped Digesters From Lincoln, Nebraska

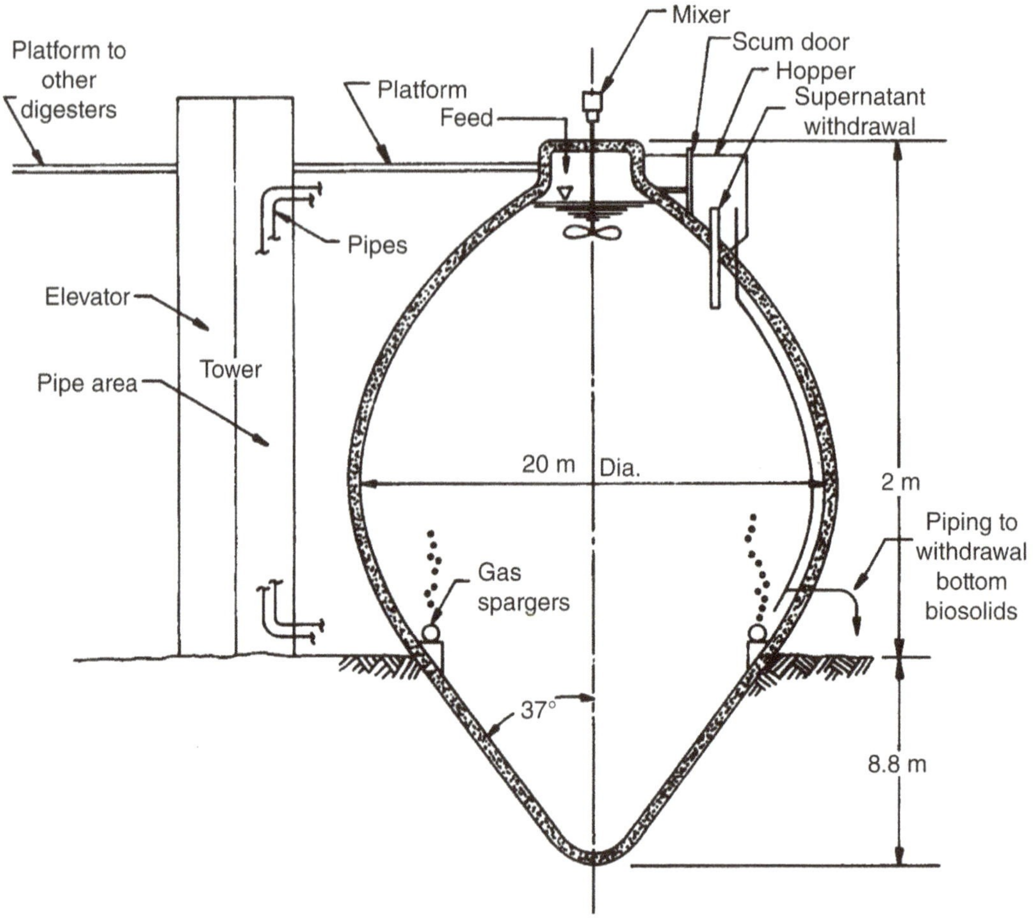

Figure 12.8 Schematic of a Typical Egg-Shaped Digester

this can be overcome through properly sized overflows. In order to provide maximum operational control, both surface and bottom withdrawals should be provided.

Volatile solids loading in these digesters ranges from 0.64 to 2.4 kg-VS/m^3·d (0.040 to 0.150 lb-VS/cu ft/d), with a maximum loading of 1.6 to 2.9 kg-VS/m^3·d (0.106 to 0.175 lb-VS/cu ft/d). SRT ranges from 15 to 45 days, and VSR ranges from 42% to 65%, depending on the intended use for the resulting biosolids. Gas production ranges from 0.68 to 1.08 m^3/kg volatile solids destroyed (11 to 17.5 cu ft/lb VSR). Though traditionally suited to mechanical mixers, gas, mechanical, and pump mixing systems can be used, and the mixing energy varies from 3.16 to 13.19 W/m^3 (0.12 to 0.5 hp/1000 cu ft). Heating systems primarily are heat exchangers, although a few use steam injection.

Silo digesters are a newer version of cylindrical (pancake) digesters, with a greater height-to-diameter ratio (Figure 12.9). This achieves some of the mixing advantages of egg-shaped digesters without the cost. For this reason, the DC Water Blue Plains facility, despite early plans for new egg-shaped digesters, ended up with silo digesters in 2014.

Silo digesters can resemble egg-shaped digesters when equipped with a submerged fixed cover. The small liquid–gas interface of the submerged cover is similar to the gas dome of the egg-shaped digester and provides many of the same advantages. Other cover options include gasholder (membrane or steel), fixed, and floating, allowing design engineers more flexibility to customize the tank for a specific use or even multiple uses.

Conventional digesters are by far the most common tank design for anaerobic digesters in the United States (Figure 12.10). Their cylindrical design makes them easier and less expensive to construct. However, this tank does not have the process benefits realized by egg-shaped and silo digesters.

Figure 12.9 Silo-Style Digesters Constructed in Green Bay, Wisconsin, for NEW Water, the brand of the Green Bay Metropolitan Sewerage District, Using the AWWA D110, Type III, Wire-Wound, Prestressed Concrete Tank Technology (Photos provided by DN Tanks)

Conventional digesters may be prone to dead zones and poor mixing regimens, resulting in lower VSR and more grit deposition. As a result, digester cleaning may be needed more often, increasing O&M costs. However, conventional digesters can be fitted with a variety of cover configurations, which allows for significant process flexibility and multiple roles (e.g., gas storage and variable liquid level).

Figure 12.10 Anaerobic Digesters (Conventional Configuration) at King County's South Wastewater Treatment Plant in Renton, Washington (Reprinted with permission from Brown and Caldwell and King County)

These tanks typically are made of reinforced concrete, with sidewall depths ranging from 6 to 14 m (20 to 45 ft) and diameters ranging from 8 to 40 m (25 to 125 ft) (U.S. EPA, 1979). Conical bottoms are preferred for cleaning purposes, with slopes varying between 1:3 and 1:6. Slopes greater than 1:3, although desirable for grit removal, are difficult to construct and difficult to stand on while cleaning. Conical bottoms also minimize grit accumulation, instead providing for relatively continuous grit removal. The floors can have either one central withdrawal pipe or be divided into wedges, each with its own withdrawal pipe (waffle-bottom digester). The latter are more expensive to construct but may reduce cleaning costs and frequency. Where necessary, cylindrical digesters have been insulated using brick veneer and an air space, earth fill, polystyrene plastic, fiber glass, or insulation board.

2.4.3 DIGESTER COVERS

Fixed digester covers can be constructed of either concrete or steel (Figure 12.11). In some cases, stainless steel has been used to eliminate the need for coatings and to speed installation. The fixed cover minimizes fugitive odors, which can be released from the annular space between any floating cover and the digester wall. Because it rests on the tank wall and not on the liquid, it is not vulnerable to tipping; however, provisions must be made to prevent overpressurization. To prevent recurring damage from foaming incidents, several facilities installed emergency pressure-relief manhole covers. These covers do not require the ballasting that floating or steel gasholder covers do; however, they must be anchored to the top of the digester to resist the uplift caused by the operating gas pressure.

A floating cover is traditionally used to allow for significant liquid-level variation because it floats on the digester liquid. Floating covers require ballasting to ensure that the cover floats at the proper level in the liquid. Weights may also need to be added to balance uneven loads on the cover because piping and equipment on the cover could make it float unevenly on the liquid surface. Design engineers should ensure that the operating liquid level is above the corbels to maintain the perimeter gas seal. Floating covers are inherently immune to hydraulic overloading because if the gas safety devices are blocked during an

Figure 12.11 Photograph of an Anaerobic Digester With a Fixed Cover at East Bay Municipal Utility District, Oakland, California (Reprinted with permission from WesTech Engineering, LLC)

overpressure event, the cover will effectively "burp" gas, typically at the lightest (or highest) radial location on the cover.

All connections on floating covers between the cover and the digester wall must be flexible enough to allow full movement of the cover without stressing any of the equipment (e.g., gas pipes, or stairways/walkways on and off the cover). Also, guides are needed to ensure that the cover moves up and down without getting bound and tipped. Original designs involving rollers and spring shoes have been largely replaced by slide guides, a series of cover-mounted plastic bars that slide vertically in a wall-mounted channel.

Like floating covers, steel gasholder covers are not fixed in place. The cover has a skirt that extends into the liquid, allowing the cover to float on a bubble of gas stored for off-peak use. As with floating covers, hydraulic overpressurization of the tank is difficult because solids will escape from the annular space.

Although gasholder covers provide extra gas storage, they have some distinct disadvantages. They have no surface withdrawal above the corbels, so foam and scum become entrapped and accumulate in the digester. The cover is effectively a large gas–liquid interface, which can lead to corrosion. Furthermore, mixing alternatives are limited because the cover moves and must be ballasted to ensure that it is balanced. Also, because the cover is riding on a gas bubble, the total volume of the reactor is not used for solids digestion (i.e., overall maximum hydraulic capacity is reduced).

Membrane digester covers (Figure 12.12) provide maximum gas storage capacity, ensure absolute gas containment, provide for the highest degree of liquid level variation, and can be economical. Though there are differences in operating these covers when compared to steel covers, proper design and operation can help ensure a successful outcome. With an increase in the need for gas storage for cogeneration and other uses, it is not surprising that several hundred are in successful operation at WRRFs in the United States.

2.4.4 DIGESTER HEATING

Anaerobic digesters must operate at suitable environmental conditions for optimal biological activity. Mesophilic digestion ideally operates between about 35 °C and 39 °C (95 °F and 102 °F), and thermophilic digestion is targeted between 50 °C and 56 °C (122 °F and 133 °F). The amount of heat needed varies seasonally, mainly in relation to the temperature of the raw solids and in relation to heat losses from the reactor to the environment. Therefore, to be effective, anaerobic digesters need a consistent, reliable heating system.

Figure 12.12 Des Moines WRA—Anaerobic Digester With Membrane Cover

The majority of a digester's total heating load is the energy needed to heat raw solids to the temperature needed for anaerobic digestion. This energy is calculated as follows:

$$q = m \cdot C_p \cdot T \tag{12.12}$$

Where q = heat load (J/h or MJ/h [Btu/hr or million Btu/hr]);
m = mass flowrate of the cake's sludge's liquid, treated as water (kg/h [lb/hr]);
C_p = solids' specific heat or heat capacity (J/kg·°C or MJ/kg·°C [Btu/lb·°F]); and
T = temperature difference between the cold, raw solids and desired heated solids temperature (°C [°F]).

To accurately compute the sheating needs of the solids, design engineers need to know the actual solids temperature, which may not be recorded. However, the WRRF's influent and effluent temperatures are typically known and are representative of the raw solids' temperature.

In all but the very hottest weather, digesters lose heat to the environment via their covers, walls, sides, and bottom. However, because digester temperatures are near ambient, radiant heat loss is small, and virtually all heat is lost via convection. The general formula for heat loss (heat transfer) from these areas is as follows:

$$Q = U \cdot A \cdot \Delta T \tag{12.13}$$

Where q = heat load (J/h or MJ/h [Btu/hr or million Btu/hr]),
U = overall heat transfer coefficient (J/h·m²·°C [Btu/hr·sq ft·°F]) (see Table 12.11),
A = surface area (m^2 [sq ft]), and
ΔT = temperature difference between the digester and the environment (°C [°F]).

2.4.5 DIGESTER MIXING

Auxiliary mixing of digester contents is beneficial for reducing thermal stratification, dispersing substrate for better contact with the active biomass, reducing scum buildup, diluting any inhibitory substances or adverse pH and temperature feed characteristics, increasing the reactor's effective volume, allowing biogas to separate and rise more easily, and keeping in suspension inorganic material that has a tendency to settle.

Table 12.11 Overall Heat Transfer Coefficients

Surface	Overall Heat Transfer Coefficient (W/(m²·°C))
Concrete Wall (above ground, not insulated)	4.7–5.1
Concreate Wall (above ground, insulated)	0.6–0.8
Concrete Floor (in contact with dry earth)	1.6–2.0
Concrete Cover (insulated)	1.2–1.6
Floating Cover (insulated)	0.9–1.0
Steel Plate Wall (insulated)	0.4–0.9

Three types of mixing methods typically have been used: mechanical, hydraulic recirculation, and gas recirculation. Mechanical or internal mixers use impellers, propellers, and turbine wheels. Although some technologies have proven very reliable, problems may arise because of the internal surfaces of the mixers. Shafts and impellers are subject to vibration (due to collected materials) and wear (due to grit and debris). The linear motion mixer is relatively new technology used in digesters over the past 10 years. Typically mounted in the center of the tank cover, this mixer does not spin but uses a cam-drive to operate a shaft to move a disc up and down. The oscillating motion of the disc in the fluid promotes mixing in the tank.

Selection of a mixing system is based on costs; maintenance requirements; process configuration; and the screenings, grit, and scum content of the feed. Suggested parameters for sizing digester mixing systems include unit power, velocity gradient, unit gas flow, and digester volume turnover time. Some newer technologies, such as the linear motion mixer, fall outside these conventional recommendations and, therefore, use less energy than other mixers.

Unit power is defined as delivered motor watts per cubic meter (horsepower per 1000 cu ft) of digester volume. Actual energy applied, viscosity, and digester configuration are not considered. Several values have been suggested for unit power selection, ranging from 5.2 to 40 W/m³ (0.2 to 1.5 hp/1000 cu ft) of reactor volume. Using laboratory data, Speece (1972) predicted a level of 40 W/m³ to be sufficient for a complete mix reactor.

The velocity gradient parameter as a measure of mixing intensity was presented by Camp and Stein (1943). It is expressed as the following equation:

$$G = (W/\mu)^{1/2} \qquad (12.14)$$

Where G = root-mean-square velocity gradient (s^{-1}),
W = power dissipated per unit volume (W/m³ or N/m²·s [lb·s/sq ft]), and
μ = dynamic viscosity (N·s/m² or Pa·s [lb·s/sq ft]) (for water, $7.2 \cdot 10^{-4}$ N·s/m² or Pa·s at 35 °C [$1.5/10^{-5}$ lb·s/sq ft at 95 °F]).

$$W = P/V \qquad (12.15)$$

Where P = power dissipated (Watts [ft·lb/s]), and
V = tank volume (m³ [cu ft]).

The power for gas injection can be determined from the following equation:

$$P = P_1(Q_g)(\ln P_2/P_1) \quad \text{(SI units)} \qquad (12.16)$$

or

$$P = 2.40P_1(Q_g)(\ln P_2/P_1) \quad \text{(U.S. customary units)} \qquad (12.17)$$

Where Q_g = gas flow (m^3/s [cu ft/min]),
P_1 = absolute pressure at surface of liquid (Pa [psi]),
P_2 = absolute pressure at depth of gas injection (Pa [psi]), and
2.40 = lumped conversion factor for U.S. customary units.

By rearranging the preceding equations, the unit gas flow relationship to the root-mean-square velocity gradient can be solved by the following equation:

$$\frac{Q_g}{V} = \frac{G^2}{P_1\left(\ln\frac{P_2}{P_1}\right)} \tag{12.18}$$

Suggested values of gas flow/tank volume (Q_g/V) for a free-lift system range from 76 to 83 mL/m^3·s (4.5 to 5.0 cu ft/min·1000 cu ft). For a draft tube system, the suggested values range from 80 to 120 mL/m^3·s (5 to 7 cu ft/min·1000 cu ft).

Turnover time is defined as digester volume divided by the flowrate through the draft tube. This concept is used only with draft tube gas and mechanically pumped recirculation systems. Typical digester turnover times range from 20 to 30 minutes.

3.0 AEROBIC DIGESTION

Aerobic digestion is typically used at smaller WRRFs (smaller than about 19 000 m^3/d [5 mgd]) and those that only produce biological solids or WAS. Compared to anaerobic digestion, aerobic digestion is a power-intensive process (because of the power needed for oxygen transfer), but it may be less expensive to construct and simpler to operate than anaerobic digestion.

Researchers have developed many methods for increasing VSR and destroying pathogens in aerobic digestion processes. Some use higher temperatures to destroy pathogens and increase VSR, while others use phasing or mechanical disruption to improve performance. Autothermal thermophilic aerobic digestion (ATAD) is an advanced, self-heating aerobic digestion process that operates at 50 °C to 65 °C (131 °F to 149 °F). Early ATAD experience in the United States established the process and allowed it to evolve with improvement of subsequent systems.

Sometimes aerobic digestion is not a separate process; many extended aeration facilities (e.g., oxidation ditches) have a long SRTs to provide at least partial digestion via endogenous respiration. However, Part 503 does not permit the VSR achieved in aeration tanks to be included as part of the 38% VSR required for biological solids stabilization.

Stabilization during aerobic digestion occurs from the destruction of degradable organic components and the reduction of pathogens by aerobic, biological mechanisms. Aerobic digestion is a suspended-growth biological treatment process based on biological theories similar to those of the extended aeration modification of the activated sludge process (see Chapters 5 and 6). The objectives of aerobic digestion, which can be compared to those of anaerobic digestion, include producing a stable biosolids product, reducing mass and volume, reducing pathogens, and conditioning solids for further processing.

Advantages of the aerobic process compared to anaerobic digestion are the production of an inoffensive, biologically stable product; lower capital costs; simpler operational control; safer operation with no potential for gas explosion and less potential for odor problems; and discharge of a supernatant with a 5-day carbonaceous BOD (cBOD$_5$) concentration typically less than that found in the anaerobic process. There is also some evidence that aerobic digestion is helpful in reducing some antibiotics (Burch et al., 2013). In addition, it is less prone to upsets and less susceptible to toxicity.

The primary disadvantage typically attributed to aerobic digestion is the higher power cost associated with oxygen transfer. Other disadvantages cited include reduced efficiency of the process during cold weather;

its inability to produce a useful byproduct, such as methane gas; and the mixed results achieved during mechanical dewatering of aerobically digested solids.

Aerobic digestion has been used successfully in facilities with capacities up to $1.89 \cdot 10^5$ m^3/d (50 mgd), but is most common in facilities with design capacities of less than 19 000 m^3/d (5 mgd). It has been used successfully in extended aeration and nutrient-removal activated sludge facilities, both with and without primary settling, and in many package treatment facilities. In larger facilities, mixed primary and biological solids are most often handled, and their oxygen requirements are greater than those of waste biological solids alone. Because of the high energy cost for aeration in such facilities, and the potential for resource recovery, it may be more economical to anaerobically digest primary solids separately while aerobically digesting biological solids.

3.1 PROCESS FUNDAMENTALS

The primary purpose of aerobic digestion is to produce biosolids that are stabilized and amenable to various disposal options. In this discussion, stabilized means that the biological organisms, particularly pathogens, have been reduced to a level at which the use or disposal of the biosolids does not result in a significant adverse environmental impact.

Aerobic digestion is based on the biological principle of endogenous respiration. Endogenous respiration occurs when the supply of available substrate (food) is depleted and microorganisms begin to consume their own protoplasm to obtain energy for cell-maintenance reactions.

Depending on feedstock characteristics, aerobic digestion can reduce volatile solids by 35% to 50%. Part 503 regulations require that a 38% VSR across the digester be met to attain the VSR requirements. This requirement can be achieved in facilities with shorter sludge ages and more volatile solids in the digester feed. However, if significant biodegradable solids reduction has already occurred in a secondary treatment system with a long SRT, it can be difficult to achieve this 38% reduction. Before the implementation of the 503 regulations, credit was sometimes given by regulators for the VSR that occurred in extended aeration facilities. Therefore, the overall VSR takes into account the reduction through both the activated sludge and the solids-handling processes. Another alternative to meet VSR requirements is by achieving a specific oxygen uptake rate (SOUR) of less than 1.5 mg of oxygen per gram per hour total solids at a temperature of 20 °C on the digested solids.

3.1.1 PROCESS REACTIONS

Aerobic digestion actually involves two steps: direct oxidation of biodegradable matter and subsequent oxidation of microbial cellular material by organisms. These processes are illustrated by the following formulas (U.S. EPA, 1979):

$$\text{Organic matter} + NH_4 + O_2 \xrightarrow{\text{bacteria}} \text{Cellular material} + CO_2 + H_2O \tag{12.19}$$

$$\text{Organic matter} + O_2 \xrightarrow{\text{bacteria}} \text{Digested biosolids} + CO_2 + H_2O + NO_3 \tag{12.20}$$

These equations describe the oxidation of organic matter to cellular material, which then is oxidized to digested biosolids. The processes represented by these equations are typical of endogenous respiration and is the predominant mechanism in aerobic digesters.

As shown in Equation 12.19, during digestion, cell tissue is oxidized aerobically. Because aerobic oxidation is exothermic, heat is released during the process. Although digestion should theoretically go to completion given an infinite SRT, in fact only 75% to 80% of cell tissue is oxidized. The remaining 20% to 25% is composed of refractory solids and organic compounds that are not biodegradable via aerobic digestion (Park et al., 2006). The material that remains after digestion exists at a low energy state that it is essentially biologically stable, rendering it suitable for a variety of disposal options.

To optimize aerobic digestion and maintain the process in the endogenous respiration phase, aerobic digestion typically is used to stabilize WAS, whereby the organic matter has already been converted to biomass. In contrast, primary solids contain a lower fraction of cellular material, and most of the organic and particulate material in primary solids is an external food source for the active biomass in biological solids. This means that longer retention times may be required to aerobically digest primary solids.

Using the formula $C_5H_7O_2N$ as representative of a microorganism's cell mass, the stoichiometry of aerobic digestion of the biomass fraction can be represented by either of the following equations:

$$C_5H_7O_2N + 5O_2 \rightarrow 5CO_2 + 2H_2O + NH_3 + \text{Energy} \tag{12.21}$$

$$C_5H_7NO_2N + 7O_2 \rightarrow 5CO_2 + 3H_2O + NO_3 + H^+ + \text{Energy} \tag{12.22}$$

Equation 12.21 represents a system designed to limit nitrification (because it is oxygen limited); thereby, nitrogen appears in the form of ammonia. The stoichiometry of a system in which nitrification occurs is represented by Equation 12.22, whereby nitrogen is oxidized to nitrate.

As indicated by Equation 12.22, nitrification during aerobic digestion increases the concentration of hydrogen ions and subsequently decreases pH if the solids have insufficient buffering capacity. As in the activated sludge process, about 7.1 kg of alkalinity is destroyed per kilogram of ammonia oxidized (7.1 lb/lb). The pH may drop as low as 5.5 during long aeration times, but aerobic digestion does not seem to be adversely affected. If excessive pH depression is a problem (as a result of alkalinity consumption by nitrification), it may be possible to control this problem through alkaline addition (e.g., lime) or periodic denitrification.

Equations 12.21 and 12.22 indicate that, theoretically, 1.5 kg of oxygen is required per kilogram of active cell mass (1.5 lb O_2/lb) in the non-nitrifying system, while 2 kg of oxygen per kilogram of active cell mass (2 lb O_2/lb) is required when nitrification occurs. Actual oxygen requirements for aerobic digestion depend on such factors as operating temperature, inclusion of primary solids, and the SRT in the activated sludge system.

This discussion on the design of aerobic digestion systems is directed toward conventional aerobic systems (i.e., systems operating at temperatures between 20 °C and 30 °C that use air as the oxygen source for biological activity). Thermophilic aerobic digestion systems are discussed later in this chapter. Cryophilic (low-temperature) aerobic digestion systems have also been used in some situations, but are covered in other references (Koers & Mavinic, 1977; Water Environment Federation [WEF], 2023a).

Factors that govern the design of aerobic digestion systems include feedstock characteristics, desired reduction in volatile solids, process operating temperature, oxygen-transfer and mixing requirements, tank volume/detention time, and method of system operation. However, since the U.S. EPA's 40 CFR 503 regulations went into effect, the overriding factor in the design of these systems has been meeting the requirements for vector-attraction and pathogen reduction.

3.1.2 FEEDSTOCK CHARACTERISTICS

As noted previously, aerobic digestion typically is used to stabilize biological solids (e.g., WAS). When the process is used to stabilize mixtures of primary and biological solids, the oxygenation requirements are increased by the additional demand of oxidizing the organic matter in primary solids to biomatter. In addition, longer detention times are required for these mixtures, as compared to a pure WAS feedstock. The mechanism of aerobic digestion being similar to that of the activated sludge process, the same concerns regarding variations in influent characteristics and levels of biologically toxic materials apply.

The influent concentration is important in the design and operation of an aerobic digestion process. Sludge digestion at concentrations of 1% to 2% allows the use of conventional equipment for mixing and aeration but requires a greater volume to achieve treatment objectives. Thickening solids to higher concentrations reduces oxygen-transfer efficiency, but results in smaller digester volume requirements and potentially higher VSR.

3.1.3 DETENTION TIME

Before the implementation of the Part 503 regulations, the required volume of an aerobic digester was typically governed by the detention time needed to achieve a desired reduction in volatile solids. In other cases, as in the "Ten States Standards" (Great Lakes–Upper Mississippi River Board of State and Provincial Public Health, 2014), the volume was based on a per capita loading.

The Part 503 regulations have two requirements, either of which may govern the required detention time and, thus, the aerobic digester volume. The regulations require compliance with both VSR criteria and pathogen reduction requirements. Typically, the pathogen reduction requirements will govern. To comply with U.S. EPA requirements for aerobic digestion as a PSRP requires a residence time of 60 days at 15 °C and 40 days at 20 °C.

Full-scale aerobic digestion studies have shown that a total aeration time (including time in the extended aeration process) of 35 to 50 days is required to consistently meet Part 503's VSR requirement for a SOUR of less than 1.5 mg oxygen/g-VS·h at WRRFs located at higher altitudes with colder climates (Maxwell, 1992). The total required aeration time depends significantly on operating temperature and WAS biodegradability. Although VSR will continue as detention time increases, the oxidation rate significantly decreases, and continuing digestion past the typical detention time is not economical.

The reduction in biodegradable solids during digestion typically is described by a first-order biochemical reaction at constant volume conditions, similar to the following:

$$dVS/dt = K_d \cdot VS_t \cdot V \tag{12.23}$$

Where dVS/dt = rate of change of biodegradable volatile solids per unit of time (kg/d),
K_d = reaction rate constant (d^{-1}), and
V = Active treatment volume (m^3).

Solving Equation 12.23 for time results in a solution for the SRT of an aerobic digester. Krishnamoorthy and Loehr (1989) found that first-order kinetics adequately represent digestion of WAS-only and WAS/primary sludge mixtures. On the other hand, a first-order equation may not accurately describe aerobic digestion of primary solids, but instead, use of Monod kinetics (or other kinetic models) may be more applicable.

Factors affecting the rate constant include the method of digester operation, solids concentration, operating temperature, and the SRT of the activated sludge system. Figure 12.13 depicts a graph of the change in reaction rate constant versus increasing operating temperature. Other studies have shown that the reaction rate constant declines as the digester's suspended solids level increases (D'Antonio, 1983) (Figure 12.14).

The product of temperature and SRT appears to correlate with the percentage of VSR that can be achieved during digestion. Figure 12.15 (Daigger et al., 1999) shows the combined effect of the temperature–SRT product on VSR. The selection of a desired percentage of reduction in volatile solids, coupled with an assumed operating temperature for the system, can be used to estimate the required digester SRT.

Benefield and Randall (1980) developed equations for determining required digester detention times. These equations resulted from an analysis of the kinetics associated with the digestion process, and the understanding that a portion of the volatile solids in the process are refractory and a portion of the nonvolatile solids is solubilized from microbial cells contained within the solids. The basic equation is as follows:

$$SRT = \frac{VSR \cdot V_{i,v}}{K_d \cdot [X_{i,v} - (1 - f_d) \cdot X_{i,v} - VSR \cdot X_{i,v}]} \tag{12.24}$$

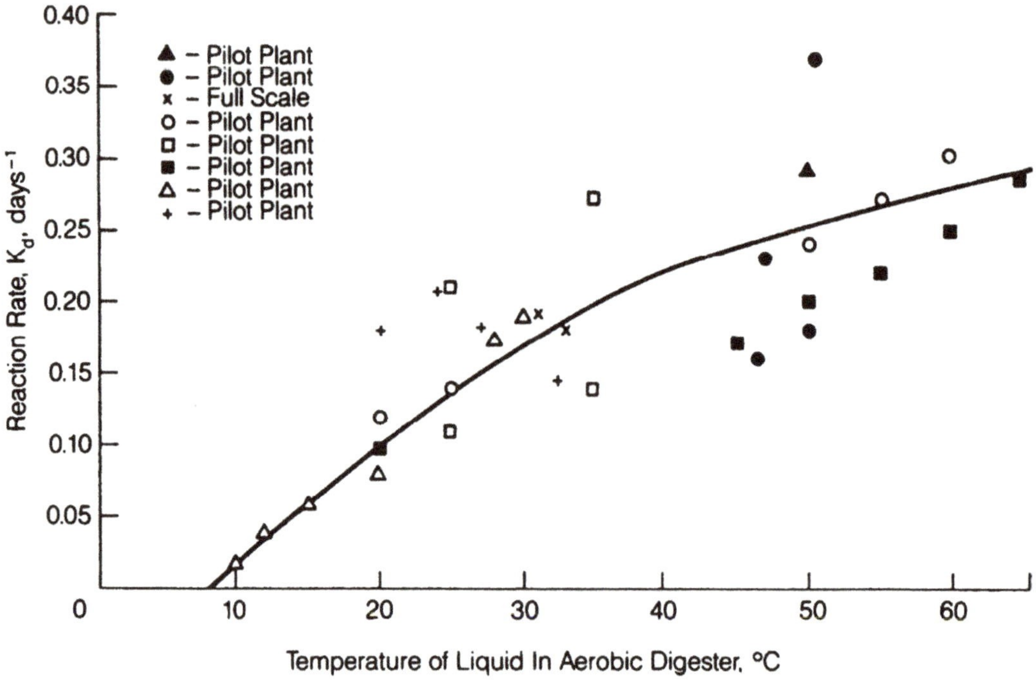

Figure 12.13 Experimentally Determined Reaction Rate (K_d) Versus Aerobic Digester Liquid Temperature. The value of K_d depends on solids characteristics and digester operating conditions (e.g., pH, TSS, and oxygen level) (U.S. EPA, 1978)

Where SRT = digester solids retention time (days),
$VSR = (X_{i,v} - X_{e,v})/X_{i,v}$ volatile solids reduction expressed as a fraction (dimensionless),
$X_{i,v}$ = VSS concentration in influent (mg/L),
$X_{e,v}$ = VSS concentration in effluent (mg/L),
K_d = reaction rate constant (d^{-1}), and
f_d = fraction of biodegradable active biomass (dimensionless).

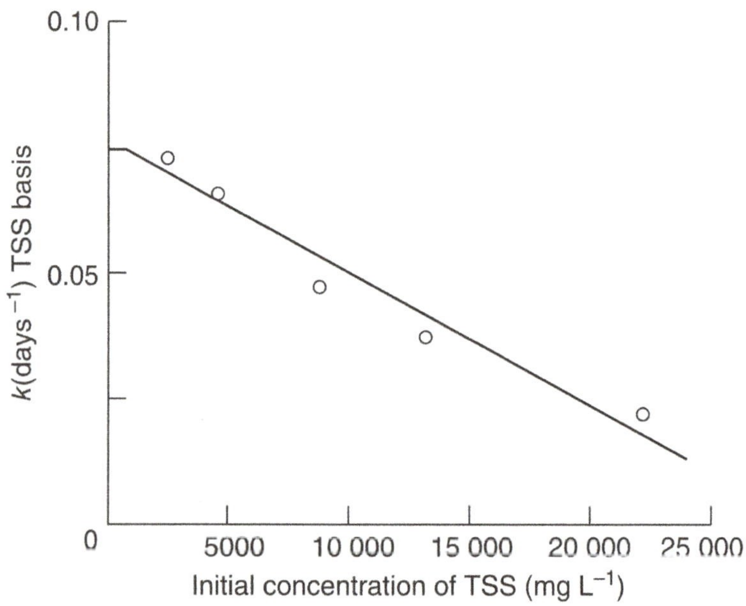

Figure 12.14 Effect of Solids Concentration on Reaction Rate (D'Antonio, 1983)

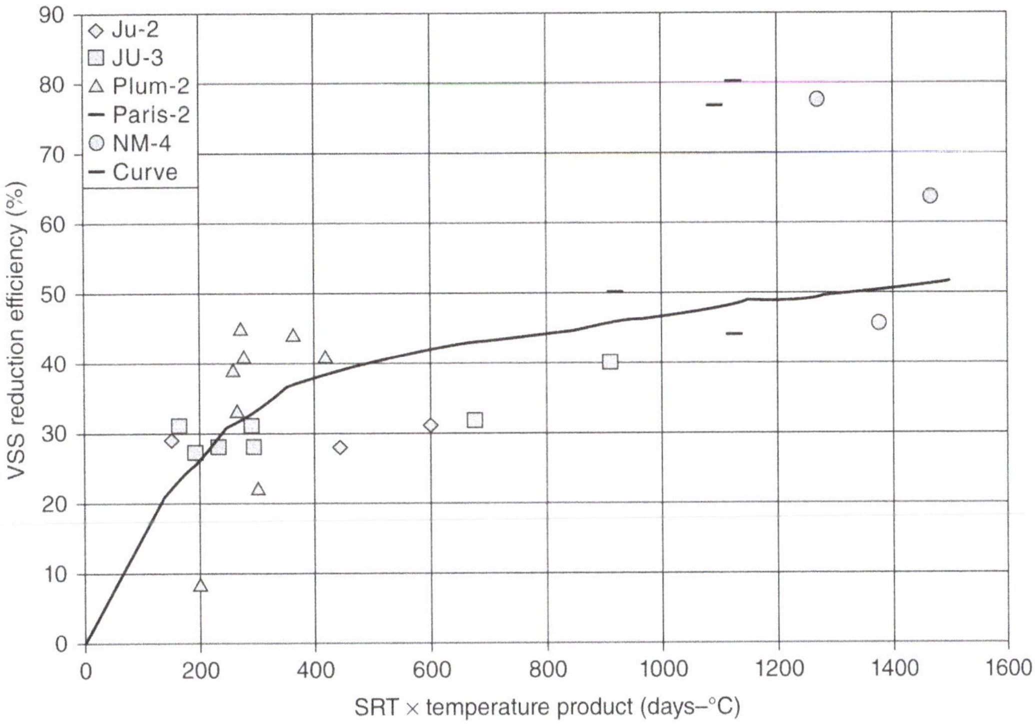

Figure 12.15 Volatile Solids Reduction as a Function of Digester Liquid Temperature and SRT (Daigger et al., 1999)

The relationships shown in Equation 12.24 are further defined and stated in the Activated Sludge Model 1 (ASM1) terminology by Grady et al. (2011). In cases of a mixture of primary solids and WAS, the equation should be modified with the inclusion of factors describing the primary solids component. Refer to Benefield and Randall (1980) or Grady et al. (2011) for more detail on this aerobic digestion design approach based on fundamental mechanisms.

Equation 12.25 supports the assumption that for equivalent solids reduction and constant solids loading, SRT must be increased as the active fraction of the influent biomass decreases. Actual operating experience indicates that for systems with a low fraction of active biomass in the feed, which is typical of extended aeration systems, this trend does not hold. For these systems, the detention times computed by Equation 12.24 may be reduced in proportion to the decrease in the active fraction of the biomass. After the SRT is determined, one method of determining the volume of a continuously operating aerobic digester is to apply the following formula:

$$V = Q_i(TSS_i + f_p\, S_i)/TSS(K_d P_v + 1/\text{SRT}) \quad (12.25)$$

Where V = volume of the aerobic digester (L [cu ft]),
Q_i = influent average flowrate (L/d [cu ft/d]),
TSS_i = total suspended solids concentration in influent (mg/L),
f_p = fraction of the digester influent cBOD_5 attributed to raw primary solids (−),
S_i = cBOD_5 in digester influent (mg/L),
TSS = (digester) total suspended solids (mg/L),
K_d = reaction rate constant (d^{-1}),
P_v = volatile fraction of digester suspended solids (%), and
SRT = solids retention time (days).

The term $f_p \cdot S_i$ in equation 12.25 can be disregarded if no primary solids are included in the load to the aerobic digester. Equation 12.25 should not be used to compute digester volumes in systems in which significant nitrification is intended.

3.1.4 TEMPERATURE

The aerobic digester's operating temperature is a critical parameter. Conventional aerobic digestion systems are operated in the mesophilic zone of bacterial action (between 10 °C and 40 °C). Operating temperatures are typically closely related to ambient temperatures because most aerobic digestion systems use open tanks and the temperature is not controlled.

A frequently cited disadvantage of the aerobic process is the variation in process efficiency that results from changes in operating temperature. Because aerobic digestion is a biological process, the effects of temperature can be estimated by applying a temperature correction factor (θ). Temperature coefficients ranging from 1.02 to 1.10 have been reported.

3.1.5 OXYGEN TRANSFER AND MIXING

The biological reaction that occurs during aerobic digestion requires oxygen for the respiration of cellular material in activated sludge and, in the case of mixtures with primary solids, the oxygen needed to convert organic matter to cellular material. In addition, proper system operations require adequate mixing of the contents to ensure contact of oxygen, cellular material, and organic matter (food source). Because introducing oxygen to maintain the biological process typically mixes the contents in the process, these parameters are interrelated.

In aerobic digestion systems that strictly treat biological solids with a total solids concentration in the range of 1% to 2%, the need for adequate mixing typically governs the sizing or power requirement of the oxygenation equipment. Systems treating primary and biological mixtures require more oxygen for the biological oxidation process and, in most cases, this requirement governs aeration equipment sizing. For designs with a higher total solids concentration (3% to 6%), process air requirements usually control aeration equipment sizing and design. Actual mixing requirements typically range from 10 to 100 W/m^3 (0.5 to 4.0 hp/1000 cu ft) of digester volume; however, this value will vary depending on tank geometry and type of mixing device.

As noted previously, aerobic digestion theoretically requires 1.5 to 2.0 parts of oxygen by mass per part of applied organic cell mass, depending on whether nitrification is supported. Design experience has shown that 2.0 parts of oxygen per part of organic cell mass destroyed is a standard minimum value for biological stabilization. Including primary solids in the digestion process requires another 1.6 to 1.9 parts of oxygen per part of volatile solids destroyed to convert the organic matter to cell tissue and satisfy the endogenous demand of the resulting cell mass. U.S. state regulators may stipulate in their design standards for WRRFs that the aeration system should account for another 0.45–0.91 kg (1–2 lb) of oxygen per kg of $cBOD_5$ applied by primary solids in the aerobic digestion design.

Oxygen requirements for aerobic digestion systems typically represent airflow rates of 0.25 to 0.33 L/m^3·s (15 to 20 cu ft/min·1000 cu ft) for WAS. Airflow rates increase to a range of 0.40 to 0.50 L/m^3·s (25 to 30 cu ft/min·1000 cu ft) for a mixture of primary solids and WAS (Benefield & Randall, 1980). These airflow rates are valid for solids concentrations between 1% and 2% and increase with increased solids content. Dissolved oxygen levels in the aerobic digester typically are maintained at about 2 mg/L; however, this level may be reduced if the oxygen uptake rate is less than 20 mg/L·h.

After the requirements for adequate mixing and oxygen transfer have been separately computed, the larger of the two requirements governs overall system design. If the mixing requirement exceeds the oxygen-transfer requirement, design engineers may provide supplemental mechanical mixing rather than overdesigning the oxygen-transfer system. The increased capital cost of supplemental mechanical mixers must be balanced against the power and maintenance costs of more aeration to determine the optimum configuration.

3.1.6 SUPERNATANT

As noted previously, one of the benefits of aerobic digestion is that the sidestream from the process typically has less effect on facility loadings compared to anaerobic digestion. On the other hand, depending on how the digesters are operated, there can be significant release of both nitrogen and phosphorus in the supernatant. Careful monitoring of solid–liquid separation in continuous and batch-feed digesters and proper operation both increases aerobic digester performance and reduces the recycle loadings to the facility influent. Table 12.12 lists "acceptable" supernatant values from aerobic digestion processes (Metcalf & Eddy, Inc./AECOM, 2014).

Although the Metcalf and Eddy, Inc./AECOM data are widely used, other data developed in additional studies provide some additional guidelines in managing sidestream loadings. Aerobic digestion recycle streams consist of high levels of nitrates (>200 mg/L), significant amounts of both ortho- and total phosphorus (>100 mg/L), and high COD (118 mg/L), but operating with intermittent anoxic cycles may reduce these concentrations (Jenkins & Mavinic, 1989).

3.1.7 REQUIREMENTS FOR LAND APPLICATION

The implementation of 40 CFR 503 has substantially changed the typical design parameters used for the standard aerobic digestion process. Although shorter detention times may meet VSR requirements, the U.S. EPA requirements for pathogen reduction and the specific time-temperature guidelines for aerobic digestion as a PSRP typically result in a design residence time of 40 days at a temperature of 20 °C and 60 days at 15 °C. For temperatures between these two extremes, the relationship that the U.S. EPA requires (U.S. EPA, 2003) is:

$$\frac{\text{Time @ } T\ (°\text{C})}{40 \text{ days}} = 1.08(20 - T(°\text{C})) \tag{12.26}$$

If design engineers use aerobic digestion for stabilization and do not meet the specified time-temperature requirements, then the facility operator must monitor fecal coliform and demonstrate compliance with the PSRP requirements through testing. In both cases, design engineers must demonstrate that the VSR criteria have been achieved by a 38% reduction in volatile solids, an oxygen uptake rate of less than 1.5 mg of oxygen per gram of total solids per hour, or a VSR of less than 15% on further testing after 30 days of additional aerobic digestion. Alternately, the VSR criteria may be met by managing land application practices.

An important deviation from this rule is obtained in the design of two-stage or batch operation. U.S. EPA's guidance manual on control of pathogens (U.S. EPA, 2003) recognizes the higher efficiency of two-stage or batch systems and allows a 30% reduction in the time required to obtain the pathogen reduction criteria specified by the regulations. Acceptance of this guidance varies with state regulatory agencies, but using this credit reduces the time required to achieve PSRP standards from 40 days to 28 days at 20 °C (68 °F), and from 60 days at 15 °C (59 °F) to 42 days at 15 °C (59 °F).

The Class B biosolids criteria that a conventional mesophilic aerobic digestion system typically is designed to meet can be summarized as follows: A system with one digester, or with multiple digesters operating in parallel, would be designed to meet the following criteria:

1. Meet one of the following pathogen-reduction requirements:
   - 60-day SRT at 15 °C or 40-day SRT at 20 °C, or
   - Fecal coliform density of less than 2 million most probable number (MPN)/g total dry solids.

Table 12.12 Acceptable Characteristics for Supernatant From Aerobic Digestion Systems (Data from Metcalf & Eddy, Inc./AECOM, 2014)

Parameter	Acceptable Range
pH	6.0–8.0
5-day BOD (mg/L)	30–2000
TSS (mg/L)	50–2000
Total Kjeldahl nitrogen (mg-N/L)	10–400
Nitrate-nitrogen (mg/L)	<30
Total phosphorus (mg/L)	20–200

2. Meet one of the following VSR requirements:
   - At least 38% VSR during biosolids treatment, or
   - A SOUR of less than 1.5 mg/g·h of total solids at 20 °C (68 °F), or
   - Less than 15% additional VSR after 30 days of further batch digestion at 20 °C (68 °F).

A system with multiple digesters in series or true batch configuration could be designed to meet the following criteria (U.S. EPA, 2003), taking advantage of the credit offered in the U.S. EPA document. However, the acceptability of this credit should be verified with the biosolids authority (some states have delegated authority):

1. Meet one of the following pathogen-reduction requirements:
   - Fecal coliform density of less than 2 million MPN/g total dry solids, or
   - 42-day SRT at 15 °C or 28-day SRT at 20 °C.
2. Meet one of the following VSR requirements:
   - At least 38% VSR during biosolids treatment, or
   - A SOUR of less than 1.5 mg/g·h of total solids at 20 °C (68 °F), or
   - Less than 15% additional VSR after 30 days of further batch digestion at 20 °C (68 °F).

3.2 DESIGN CONSIDERATIONS

Design of conventional aerobic digesters is similar in many ways to the design of activated sludge systems. Much of the equipment is similar and may even be shared. Several devices (e.g., diffused air, mechanical surface aeration, mechanical submerged turbines, jet aeration, and combined systems) have been used successfully to provide oxygenation and mixing in a number of applications. Control of the aerobic digestion process is different from activated sludge. Design considerations are discussed in detail within this section.

3.2.1 AERATION

The design of diffused-air systems for aerobic digesters is similar to the design of those used in standard activated sludge systems. Diffusers typically are located near the tank bottom. They can be placed along one side of the tank to produce a spiral or cross-roll pattern, or they may be installed as a floor-mounted grid system. Both fine-bubble and coarse-bubble diffusers have been used in aerobic digesters. Airflow rates of 0.33 to 0.67 L/m^3·s (20 to 40 cu ft/min·1000 cu ft) typically are required to ensure that mixing is adequate. The airflow rates needed to meet oxygen-transfer requirements depend on digester loading, diffuser type, system layout, and overall oxygen-transfer efficiency.

Diffused-air systems provide the following advantages: oxygen transfer is controlled by varying the air-supply rate; the introduction of compressed air to the digester typically adds heat to the system, which minimizes temperature loss during cold weather; and overall heat loss from the system is minimized because of the relatively small degree of surface turbulence.

Advantages of diffused-air systems may be outweighed by clogging problems that can occur in aerobic digesters, especially those for which operation includes periodic settling and supernatant removal. While the air is turned off, solids can enter the air piping and adhere to the inner walls of piping or diffusers. Nonclogging and porous media devices are more resistant to this type of plugging than large-bubble, orifice diffusers. However, surface fouling of porous diffusers can occur. If a diffused-air system is to be used, it is imperative that provisions be included for easy cleaning of the diffusers and air drop pipes.

For higher solids concentrations, there is a significant decrease in oxygen-transfer efficiency with conventional diffusers (Krampe & Krauth, 2003; WEF, 2023a). In addition, the clogging issues with conventional diffusers may be exacerbated. An alternative to the floor-covering diffuser systems is high-shear, nonclogging aeration equipment designed specifically for high solids concentrations (4% to 8% solids). This aeration system combines draft tubes or shear tubes with an adjustable, above-water orifice for airflow control and nonclogging point-source diffusers. The shear tube and draft tubes provide mixing and shearing to transfer oxygen and achieve volatile reduction with high solids. The limitation of this system is that it works better when the liquid is more than 6.1 m (20 ft) deep (Daigger et al., 1997).

Mechanical surface aerators typically are floating, pontoon-mounted devices of either low- or high-speed design. Low-speed aerators are more often used in aerobic digesters. Compared to diffused-air systems, mechanical surface-aeration systems typically are simpler and easier to maintain, and less prone to fouling. Disadvantages typically attributed to surface aeration include the lack of control of the oxygenation rate, performance deterioration if excessive foam is present, more potential for foaming because of high surface turbulence, increased heat loss from the system, and the potential for ice accumulation during winter in cold climates as a result of the device's splashing.

Mechanical submerged turbine aerators (and other combined mechanical mixing and diffused-air systems) provide several advantages and eliminate some disadvantages of the diffused-air and surface-aerator devices. Oxygenation rates can be controlled by varying the airflow rate to the submerged impeller. Because impellers are submerged, they are not as sensitive to foaming conditions as surface aerators and avoid the ice and heat dissipation problems associated with surface aerators. Additionally, the submerged unit can be operated as a mixer only, thereby promoting denitrification. Although the oxygen-transfer efficiency is similar to that of fine-bubble diffusers, the penalty resulting from two motors (turbine and blower) reduces the aeration efficiency (standard aeration efficiency) (Stenstrom & Rosso, 2008).

Jet aeration devices provide many of the advantages of submerged turbines. These devices typically are more easily installed and have a similar oxygen-transfer efficiency to submerged turbines. Problems with device plugging have occurred when liquid flow paths were not large enough to pass the stringy solids typically found in aerobic digesters. The use of jets may also promote floc shear and subsequent dewatering difficulties.

3.2.2 CONTROL OF THE AEROBIC DIGESTION PROCESS

Aerobic digestion typically is controlled manually. The operating variables that currently lend themselves to automatic control are dissolved oxygen, oxidation-reduction potential, and tank level. The dissolved oxygen signal can be used to control the aeration system so it maintains an optimum dissolved oxygen level (typically between 1 and 2 mg/L). This can conserve energy. Low- (and occasionally high-) dissolved oxygen conditions can trigger an alarm to allow operators to take corrective action. However, except for inadvertent digester overloads, dissolved oxygen changes in aerobic digesters typically are minimal, and maintaining dissolved oxygen monitors may be time-consuming.

A tank-level signal, useful in preventing overfilling, can be used for on–off control of digester feed pumps. Intermittent feeding of primary solids to digesters via timer-controlled pumps has been used. This controlled-feeding technique also has been used for waste biological solids feeding. Automatically controlled feeding is successful when care is taken to establish the proper time as facility operating conditions change. With manually controlled feed systems, a tank-level signal can be used to warn of a high-level condition.

Withdrawal of solids from aerobic digesters typically is manually controlled and done intermittently. Manual withdrawal allows rational reaction to variable solids production, solids concentration, digestion rates, and capacity of subsequent processing.

The important factors when controlling aerobic digestion operations are similar to those for other aerobic biological processes (Table 12.13) (Stege & Bailey, 2003). Operators should monitor the primary process indicators daily. The secondary indicators are useful in monitoring long-term performance and for troubleshooting problems associated with the primary indicators. While monitoring and controlling these parameters is important, the degree of control that can be exercised on each parameter varies. Analysis frequency should be increased during startup and whenever large changes are made to operating conditions, such as solids flowrate and large increases or decreases in the feedstock's solids concentration.

3.2.3 AEROBIC DIGESTION WITH THICKENED SOLIDS

Typically, waste activated solids from secondary clarifiers have solids concentrations in a range from 7500 to 20 000 mg/L. If WAS is being digested without primary sludge, it may be beneficial to prethicken the solids. Thickening is required for ATAD systems and can be beneficial for conventional mesophilic

Table 12.13 Monitoring Parameters for Aerobic Digestion Performance (Adapted from Stege & Bailey, 2003; WEF, 2023b)

	Operating Range			
Monitoring Parameter*	**Frequency**	**Minimum**	**Nominal**	**Maximum**
Temperature (°C)	Daily	15	20	37
pH	Daily	6.0	7.0	7.6
Dissolved oxygen (mg/L)	Daily	0.1	0.4 to 0.8	2.0
Alkalinity (mg/L as $CaCO_3$)	Weekly	100	>500	—
Ammonia-nitrogen (mg/L)	Weekly	—	<20	40
Nitrate (mg/L)	Weekly	—	<20	—
Nitrite (mg/L)	As required	—	<10	—
SOUR (mg oxygen/h/g of total solids)	As required	—	<1.5	—
Phosphorus (mg/L)	As required		<5	

digestion. The main advantages of this technique include increased SRT and VSR for a given volume; the higher solids concentrations also can yield auto-heating.

The oxidation of biodegradable organic matter elevates digester temperatures via its enthalpy release (about 3.6 kcal/g [6500 Btu/lb] of VSS destroyed) and accelerates digestion and pathogen destruction rates (Grady et al., 2011). If the heat generated by this process can be maintained in the digester, it can be used to increase the reactor's temperature. This could be beneficial in cold climates. Overheating is possible in warmer climates or seasonally. When the digester temperature exceeds the target, loading should be adjusted. If loading cannot be adjusted, venting or cooling may be implemented.

Thickened aerobic digestion may require an additional unit process, which may increase O&M needs. Solids can only be prethickened to a maximum solids concentration wherein oxygen can be successfully transferred in the solids using available aeration equipment, and to the maximum extent that solids rheology does not significantly affect mixing characteristics in the aerobic digester. It is also important to avoid excessive temperatures, especially in summer months.

Thickened aerobic digestion can be achieved using several different approaches based on the thickening treatment processes used to increase the feed cake's solids concentration and the position of thickening in the digestion process.

The first approach is to use batch operation. Batch operation involves the practice of manually removing digested solids from digester tanks. Originally, aerobic digestion was operated as a draw-and-fill process, a concept still used at many facilities. Solids are pumped directly from the clarifiers or sequencing batch reactors (SBRs) to the aerobic digester. The time required to fill the digester depends on the tank volume available and the volume of solids. When a diffused-air aeration system is used, the solids being digested are aerated continually during the filling operation. When the solids are removed from the digester, aeration is discontinued, and the biosolids are allowed to settle. The clarified supernatant is then decanted before removing solids and returned to the treatment process. The remaining biosolids can contain up to 2.5% solids, depending on the sludge and the time allowed for settling.

The second approach for thickening is to use sedimentation. This mode of thickening treatment process consists of a continuous-feed operation using sedimentation (e.g., a gravity thickener) after digestion. This is typically a continuous aerobic digestion process that closely resembles the activated sludge process. Solids are pumped directly from the clarifiers into the aerobic digester. The digester operates at a fixed level, with the overflow going to a solid–liquid separator. Thickened and stabilized solids are removed for further processing. Continuous operation typically produces biosolids with lower solids concentrations. This process has some advantages over the semi-batch operation because the aerobic digestion basin is operated at a fixed level and the aeration-transfer efficiency is optimized.

The third approach is to include a gravity thickener in the digestion loop. This process typically consists of two main phases (in-loop and isolation) and four main basins (two digesters, one premix basin, and a gravity thickener). During the in-loop phase, a premix basin, a digester, and a thickener operate in a loop, which reduces volatile solids, reduces ammonia, and increases solids concentration. The in-loop thickener has two main functions: thickening and denitrification. The in-loop digester acts as a stabilization step and reduces most volatile solids. The in-loop digester is fed in batches 8, 16, or 24 times per day for a period that typically lasts 10 to 20 days. The digester then enters the isolation phase. During the isolation phase, no solids are introduced to the digester, which completes the additional pathogen reduction needed to meet regulatory requirements. The process is considered a "modified batch process" because of the multiple feedings to the loop digester. This process can produce biosolids containing 2.5% to 3% solids.

The fourth approach is mechanical prethickening. In this process, a mechanical prethickening device (e.g., a gravity belt thickener, dissolved air flotation (DAF) mechanism, centrifuge, or drum thickener) is used before aerobic digestion. The designer can choose the ideal mechanical device (refer to Chapter 11 for more information) and desired operating solids concentration (e.g., 4%, 5%, or 6%) to minimize the aerobic digestion basins. This process gives flexibility to meet performance requirements in summer and winter by modifying the mechanical device's operating schedule as desired. For this process, two digesters in series are recommended as a minimum (series operation will be addressed later in this chapter). Because of the flexibility and reliability, and the capital and O&M cost savings, this process is preferred in WRRFs designed for average daily flows greater than 7600 m^3/d (2.0 mgd).

3.2.4 AEROBIC DIGESTERS IN SERIES OR BATCH OPERATION

Aerobic digesters traditionally have been designed with one basin, or with multiple basins operated in parallel. Multiple tanks in series or in isolation operating in a batch operation have proven to improve both pathogen destruction and compliance with VSR requirements. According to the U.S. EPA (2003), solids can be aerobically digested using a variety of process configurations, including continuous-flow or batch systems, in one or multiple stages. From a process basis, single-stage completely mixed reactors with continuous feed and withdrawal are the least effective option for bacterial and viral destruction, mainly because of the potential for short-circuiting.

Operation in series is simpler, as it reduces the complexity of controlling a batch system. Using two or more completely mixed digesters in series reduces the potential for short-circuiting and improves the process kinetics. Although calculations indicate that a two-stage system would theoretically allow a 50% reduction in volume and detention time, not all factors involved in the decay of microorganism densities are known. Therefore, to allow a factor of safety, for staged operation (using two stages with about equal volume) it is recommended that the required time be reduced to 70% of that needed for single-stage aerobic digestion in a continuously mixed reactor. The same reduction is recommended for true batch operation or for more than two stages in series. This is consistent with the credit allowed by U.S. EPA (2003) for the improved efficiency of both batch and multistage operation. The benefits of a two-stage reactor system (in series or in isolation) include improvement in pathogen destruction, improvement in VSR, smaller tank volumes required, and lower airflows required because of smaller tank volumes.

3.3 AUTOTHERMAL AEROBIC DIGESTION

Aerobic digestion is an exothermic reaction, generating heat. The ATAD process uses this heat to help achieve operating temperatures of 40 °C to 80 °C. It relies on having efficient oxygen-transfer systems (to provide oxygen without heat stripping), insulated vessels (to minimize heat loss), and higher concentrations of solids to provide enough heat generation to maintain the thermophilic temperatures. If sufficient insulation and adequate solids concentrations are provided, the process can be controlled at thermophilic temperatures to achieve greater than 38% VSR and meet Part 503's Class A pathogen requirements (U.S. EPA, 1990).

Autothermal thermophilic aerobic digestion (ATAD) has been studied since the 1960s. Much of the developmental work was done by Popel (1971a, 1971b), who, along with his coworkers, studied animal manure and wastewater residuals in Germany. They developed an aspirating aeration device that was key to the process success. The initial full-scale installations in the United States occurred during the 1990s. The use of ATAD digestion gained popularity because it can produce Class A biosolids. In 2003, there were

35 ATAD systems operating in North America. Second-generation systems are designed to address issues through changes in operating parameters, addition of appropriate odor control systems, and provisions of mesophilic stabilization (Smith et al., 2012).

The major advantages of ATAD include shorter retention times of approximately 5 to 6 days to achieve 30% to 50% VSR, lower oxygen requirements than nitrifying mesophilic aerobic digestion, and Class A biosolids yields. The major disadvantages of ATAD include the potential for poor dewatering characteristics, the potential for objectionable odors, reduced potential for nitrification and/or denitrification (Daigger et al., 1998), and foam management is needed to ensure effective oxygen transfer (Metcalf & Eddy, Inc./AECOM, 2014). Several of these disadvantages have been addressed with improvements in the design ATAD process.

Autothermal thermophilic aerobic digestion systems should include two or more enclosed, insulated reactors in series. Both continuous and batch processing are acceptable. To comply with the U.S. EPA definition of ATAD as a process to further reduce pathogens (PFRP), a batch process should be used. In this case, pumps should be designed to withdraw and feed the daily allotment of solids in 1 hour or less. A specific volume of solids is removed daily from the second-stage reactor (which is operating in a range of 55 °C to 65 °C). After the solids are removed, biomass from the first-stage reactor is transferred into the second-stage reactor. The second-stage reactor is then isolated for the remaining 23 hours each day, at a minimum temperature of 55 °C. After the biomass is transferred from the first-stage reactor to the second stage, raw feed is then introduced to the first stage to make up the volume removed.

The VSR achieved by the ATAD process depends on the feedstock(s), SRT, operating temperature, and reactor loading. The Bowling Green, Ohio, WRRF, which is a second-generation facility, reported a VSR of 75% (Scisson, 2006).

3.4 NUTRIENT CONTROL

The conventional aerobic digestion process typically operates in a nitrifying mode. Supernatant from these digesters typically has high concentrations of nitrate; however, at longer detention times, the supernatant can also contain significant concentrations of nitrite and ammonia (Jenkins & Mavinic, 1989). When conventional digesters are operated in the aerobic-anoxic mode, nitrification and denitrification occurs with a reduction in total nitrogen. Typically, nitrate levels are reduced to less than 5 mg/L (Jenkins & Mavinic, 1989). Because nitrification is inhibited in the ATAD process, the supernatant from the ATAD process has low nitrates but very high ammonia concentrations.

From an overall mass-balance perspective, phosphorus entering the digester ends up either in waste solids or in effluent. If phosphorus is not partitioned into the solids, then it is recycled back to the head of the facility. Phosphorus is released into the liquid phase in both anaerobic and aerobic digestion; however, the release is lower in aerobic processes than anaerobic processes.

The amount of phosphorus released in aerobic digestion is dependent on the liquid treatment process (conventional vs. biological phosphorus removal), the aerobic digestion time, and the digestion temperature, as well as the operational mode of the digester (Jenkins & Mavinic, 1989). Operating the system in an aerobic-anoxic mode or under low-dissolved-oxygen conditions (which provides simultaneous nitrification and denitrification) reduces the phosphorus release (Daigger et al., 2001). Based on research (Jenkins & Mavinic, 1989), the release in a continuously aerated digester is two to three times the release in intermittently aerated systems.

Another factor that affects phosphorus release is pH. Several studies show that a pH less than 6.0 should be avoided, because it encourages inorganic metal phosphates to dissolve (Daigger et al., 2001; Jenkins & Mavinic, 1989). Alternatively, feeding lime to the digester to raise the pH significantly lowers the phosphorus release.

Dealing with phosphorus in the digesters depends on both the overall nutrient requirements for the facility and any phosphorus restrictions on land application. Depending on the controlling parameters, there are options that can be used to manage this phosphorus release.

When controlling phosphorus in the recycle stream is the primary concern, there are several potential options for reducing phosphorus loads in the sidestreams. One approach is maintaining the solids being fed to the prethickening device in an aerobic state (to minimize anaerobic phosphorus release). To prevent anaerobic conditions, solids should be wasted directly from the liquid sidestream to the prethickening device; or if storage is required before prethickening, the detention time should be minimized. If the solids can be prethickened while still in an aerobic state, and no further decanting or dewatering occurs, the phosphorus is wasted as part of the biosolids.

When thickening or dewatering occurs after digestion, there is greater potential for return of phosphorus to the influent of the facility. In this case, operating an aerobic/anoxic system and maintaining a neutral pH helps reduce phosphorus release. In a multistage system, the first stage can be operated with limited aeration to control nitrification and promote the formation of struvite (Daigger & Bailey, 2000), which can be separated and recovered. Remaining phosphorus in the supernatant or filtrate from the postdigestion dewatering can be chemically bound, passing through the secondary process and allowing the phosphorus to be removed as part of the final biosolids.

The final alternative occurs when there is a limit on phosphorus loading for land application sites. As in the previous alternative, limiting nitrification in the first state promotes the formation of struvite. The supernatant or filtrate from postdigestion dewatering is again chemically treated, but is sent to an inclined plate separator (Daigger & Bailey, 2000). This allows the bound phosphorus to be removed for separate disposal, while the final biosolids have a lower phosphorus concentration.

4.0 COMPOSTING

Composting often is used to convert solids into a soil amendment or conditioner. The feedstock can be either raw solids or biosolids but should contain at least 40% solids. A bulking agent is added to increase solids content, provide carbon for the process, improve the material's structural properties, and promote adequate air circulation. Composting typically is a labor-intensive process (e.g., adding bulking agent, turning the material, and recovering the bulking agent). It can also emit odors, especially if the site is poorly designed or operated. In addition, the process may increase the mass of biosolids to be used or disposed and could spread pathogens via dust from the material.

Composting is a biological process in which organic matter is decomposed under controlled, aerobic conditions to produce humus. Operators can accelerate the process by using the proper blend of materials (organics or solids and bulking agents) and controlling the temperature, moisture content, and oxygen supply. The resulting compost is stable and can be safely used in many landscaping, horticulture, or agriculture applications.

Composting can be used to treat both unstabilized solids and partially stabilized biosolids. Odors would generally be reduced in the latter case due to a lower oxygen demand that allows aerobic conditions to be more easily maintained. In both solids and biosolids composting, operators control several essential process variables to optimize the material's decomposition/stabilization rate including solids content, carbon-to-nitrogen ratio (C:N), aerobic conditions, and temperature. With the right tools, owners can induce the composting mass to achieve thermophilic temperatures, which destroy pathogens. Well-stabilized compost can then be stored indefinitely and has minimal odor, even if rewetted.

4.1 PROCESS FUNDAMENTALS

Compositing relies on three major categories of organisms: bacteria, actinomycetes, and fungi. Bacteria are responsible for decomposing a major portion of organic matter. Actinomycetes are microorganisms common to soil environments; they metabolize a wide variety of organic compounds (e.g., sugars, starches, lignin, proteins, organic acids, and polypeptides). Fungi are present at both mesophilic and thermophilic temperatures. Chang (1967) indicated that mesophilic fungi metabolize cellulose and other complex carbon sources. Their activity is similar to that of actinomycetes; both typically are found in the exterior portions of compost piles.

Microbial activity during composting occurs in three stages: mesophilic, when temperatures in the pile range from ambient to 40 °C (104 °F); thermophilic, when temperatures range from 40 °C to 70 °C (104 °F to 158 °F); and a cooling period associated with a reduction in microbial activity and the completion of composting. The optimum temperature in the thermophilic range seems to be between 55 °C and 60 °C (131 °F and 140 °F) where the maximum rate of VSR occurs.

When compositing, the initial solids content depends on how much amendment or bulking agent is mixed with dewatered cake. For good process performance, the dewatered cake should contain between 14% and 30% solids. It then is blended with the bulking agents, which are drier materials, to achieve a solids content of about 38% to 45%. The target solids content depends on the composting technology used. Solids content is controlled throughout the process via aeration, material agitation, or both.

The amount of carbon and nitrogen used by microorganisms depends on the composition of the microbial biomass. Ideally, the ratio of available carbon to nitrogen is between 25 : 1 and 35 : 1. If the ratio is less than 25 : 1, excess nitrogen is released as ammonia, reducing the compost's nutrient value and emitting odor. Calculating the carbon-to-nitrogen ratio is complicated, because some of the carbon becomes available more slowly than the nitrogen (Kayhanian & Tchobanoglous, 1992). If wood chips are the bulking agent, for example, only a thin surface layer of the wood provides available carbon. The carbon in sawdust, on the other hand, is more readily available to biodegradation.

Composting is an aerobic process. Microbial oxygen demand during composting can reduce the available oxygen in air to as low as 3% to 5% in as little as 15 minutes. Aerobic conditions are maintained via forced or convective aeration, material agitation, or both, depending on the composting technology used.

The temperature varies during compositing. At first, the challenge is to heat the material up to the thermophilic range as quickly as possible. Then, the challenge is removing excess heat to maintain the process in the thermophilic range. It is also difficult to achieve uniform temperatures throughout the pile. Covers may help in this regard. Near the end of composting, the goal is to dry the material without removing too much heat. All of these are achieved using aeration, agitation, or both.

Heat is generated when organic carbon converts to carbon dioxide and water vapor. The fuel is provided by rapidly degraded volatile solids. Heat primarily is removed by the evaporative cooling promoted by aeration and agitation. Some heat also is lost at the pile surface. The process temperature does not rise if heat is lost faster than it is generated.

4.2 PROCESS AND DESIGN CONSIDERATIONS

The primary objective of composting is to produce a nutrient-rich soil amendment that complies with federal, state, and local requirements for the beneficial use of biosolids. The compost must meet both environmental and public health requirements and be attractive for use. This primary objective is met via the following process objectives: pathogen reduction, maturation, and drying.

The general process approach to composting is shown in Figure 12.16. The compost process can be achieved with aerated static pile composting, windrow composting, and in-vessel composting. Table 12.14

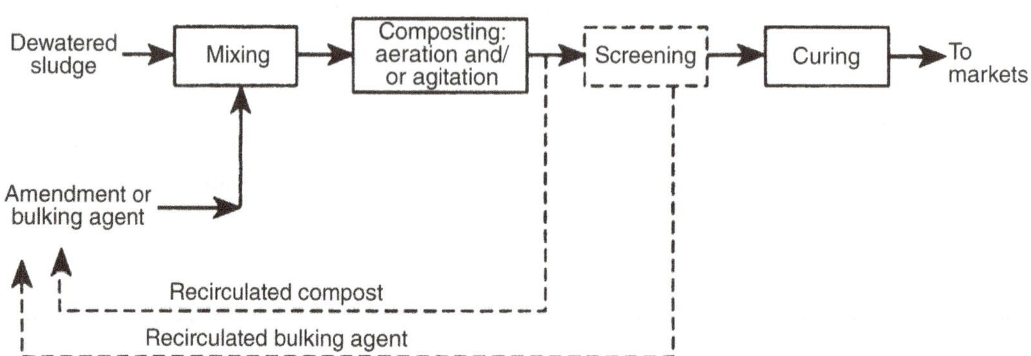

Figure 12.16 Generalized Composting Flow Chart (Dashed Lines Indicate Optional Steps; Screening May Follow Curing; Drying Step May Precede Screening)

Table 12.14 Key Advantages and Disadvantages of Composting Systems

Composting Technology	Advantages	Disadvantages
Aerated static pile	• Adaptability to various bulking agents • Flexibility to handle changing feed conditions and peak loads (volume not fixed) • Relatively simple mechanical equipment	• Relatively labor intensive • Relatively large area required • Operators exposed to composting piles • Potentially dusty working environment
Windrow	• Adaptability to various bulking agents. • Flexibility to handle changing feed conditions and peak loads (volume not fixed) • Relatively simple mechanical equipment • Requires no fixed mechanical equipment	• Very large area required • Relatively labor intensive • Operators exposed to composting piles • Dusty working conditions
Vertical plug flow	• Completely enclosed reactors in some systems improve ability to control odors • Relatively small area required • Operators not exposed to composting material	• Single outfeed device per reactor (large reactors), potential bottleneck • Potential inability to maintain uniform aerobic conditions throughout reactor • Relatively maintenance intensive. • Limited flexibility to handle changing conditions • Materials-handling system may limit choice of bulking agents
Horizontal plug flow (tunnel)	• Completely enclosed reactors improve ability to control odors • Relatively smaller area required (composting mix compacted) • Operators not exposed to composting material	• Fixed-volume reactors (no flexibility) • Limited ability to handle changing conditions. • Relatively maintenance intensive • Materials-handling system may limit choice of bulking agents
Agitated bin	• Mixing enhances aeration and uniformity of compost mixtures • Ability to mix compost (advantage in handling some bulking agents) • Adaptability to various bulking agents	• Fixed-volume reactors (no flexibility) • Relatively large area required • Potentially dusty working environment • Operators exposed to composting piles • Relatively maintenance intensive

lists the advantages and disadvantages of five composting technologies based on physical facilities, processing aspects, and O&M. No one process is appropriate for every situation. The choice depends on many factors (e.g., climate, siting considerations, operational concerns, and sensitivity to odors).

4.2.1 TEMPERATURE

Heat, or high temperature, is one of the most effective methods for destroying pathogens. Table 12.15 summarizes time-and-temperature relationships for inactivating pathogens in actual composting operations. Note that temperatures measured in a composting pile or vessel may not be uniform because of variations in heat loss, solids-mixture characteristics, and airflow.

Table 12.15 Temperature Exposure Required for Pathogen Destruction in Compost (Knoll, 1964; Morgan & MacDonald, 1969; Shell & Boyd, 1969; Wiley & Westerberg, 1969)

	Exposure Time for Destruction at Various Temperatures Hours		
Microorganisms	45–55 °C	60 °C	65 °C
Salmonella Newport		25	
Salmonella	168	116	
Poliovirus type 1		1.0	
Candida albicans		72	
Ascaris lumbricoides		4.0	1.0
Mycobacterium tuberculosis			336

Composting systems must control temperatures to optimize decomposition and provide pathogen reduction. While higher temperatures are ideal for pathogen reduction, it is also desirable to prevent material temperatures from climbing too high. Pile temperatures in excess of 70 °C inhibit the biological decomposition process. If high temperatures persist for periods longer than several weeks, the potential of spontaneous combustion can occur in very dry material (>75% solids).

4.2.2 AERATION

Compost typically can be aerated either by forcing air up through the material (positive aeration) or by pulling it down through the material (negative aeration) (Figure 12.17). Positive aeration typically requires less energy than negative aeration to move the same volume of air. In positive aeration, the air is cooler, drier and, therefore, has less volume.

Negative aeration is better in enclosed and worker-occupied operations because it captures most of the material's odors and moisture, preventing them from entering the air above the pile (where greater airflows are required to capture and treat such emissions). However, condensation accumulates in the ductwork and blowers, so ample drainage must be provided.

Most in-vessel systems use positive aeration for system-specific reasons. For example, in-tunnel systems have little headspace above the piles and are not occupied by workers during active composting, so there is no advantage to negative aeration. Negative aeration is popular in aerated static-pile operations because it directly captures odors and moisture. Many aerated static-pile operations are configured to allow for both negative and positive aeration. During decomposition, negative aeration captures odors and moisture. Afterward, positive aeration can provide more air to accelerate drying before the compost is screened.

Higgins et al. (2006) reported that an aeration rate of 34 $m^3/Mg \cdot h$ (1100 cu ft/hr·dry ton) provided adequate drying and high-enough temperatures for pathogen destruction. Early in the composting process, higher aeration rates may be needed to prevent excessive pile temperatures.

To maintain temperatures less than 60 °C during peak activity, aeration rates may need to approch 300 $m^3/Mg \cdot h$ (10 000 cu ft/hr·dry ton). Such aeration capacity may be impractical in large systems. Practical aeration capacities are in the range of 90 to 160 $m^3/Mg \cdot h$ (3000 to 5000 cu ft/hr·dry ton) of wastewater solids. Aeration in this range helps control temperatures throughout most of the composting period and provide adequate moisture-removal capacity.

4.2.3 DETENTION TIME

The time required to stabilize organic material typically is divided between an active composting stage and a curing stage (see Figure 12.16). When the U.S. Department of Agriculture (USDA) developed the aerated static-pile process, researchers found that 21 days of aerated composting followed by 30 days of unaerated curing would adequately stabilize a raw feed with wood chips as the bulking agent. However, to create fully stabilized compost suitable for any use, another 20 days or more of detention time is recommended. This detention time criterion has been codified in a number of state regulations and incorporated into some design standards. Most horizontal agitated-bed systems are designed for 21 days of aerated composting

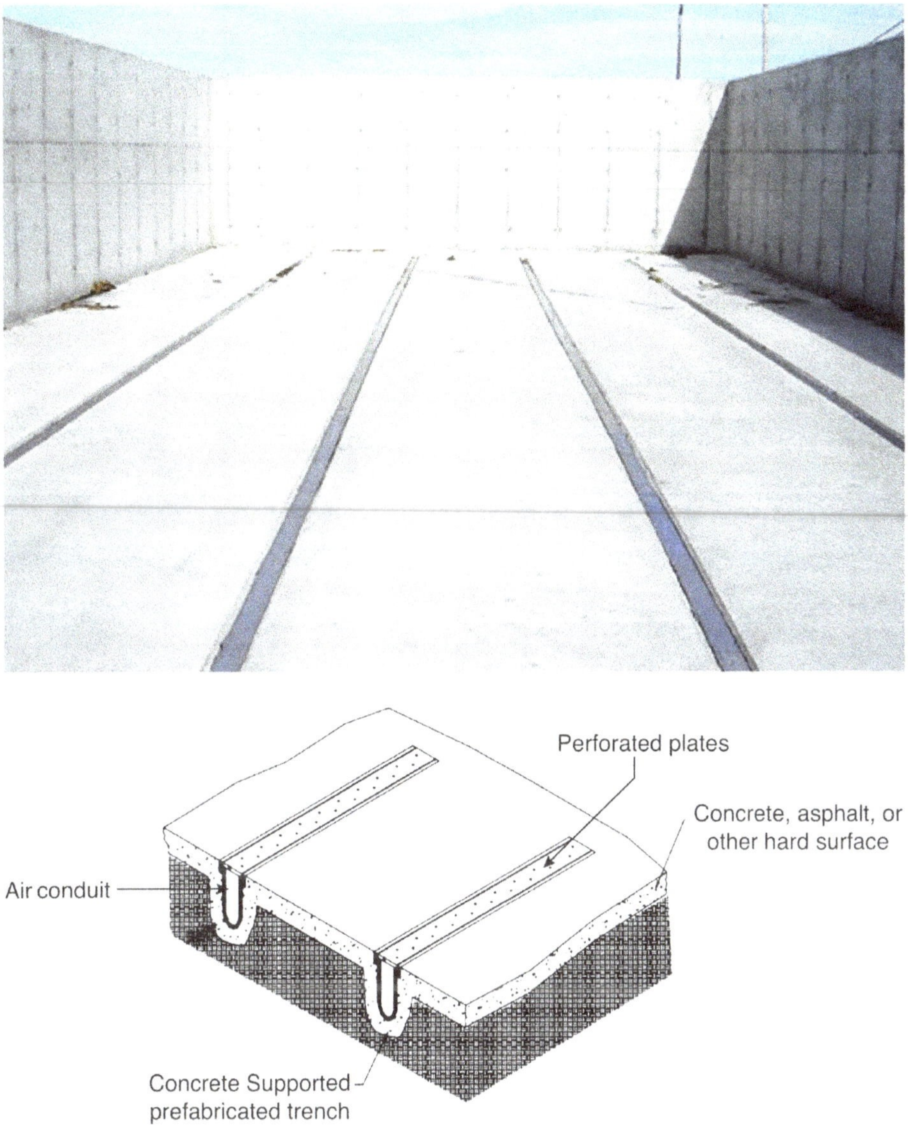

Figure 12.17 Example of Compost Aeration System–Aeration Trenches

followed by curing. However, other in-vessel systems use shorter active composting times (often 14 days) to minimize the system's capital costs.

4.2.4 BULKING AGENT

All composting technologies require mixing enough bulking agent with dewatered solids to adjust the initial solids content and provide porosity. Bulking agents also provide supplemental carbon to adjust the carbon-to-nitrogen ratio and energy balance. Unfortunately, these bulking agents also increase the quantity of solids that must be handled. Table 12.16 lists some typically used bulking agents and their characteristics.

Materials balance calculations track the weight and volume of each material through each stage of the composting process. Table 12.17 shows a typical materials balance for 1 dry ton of biosolids (20% solids). In this example, solids were mixed with yard waste, stacked over a layer of yard waste to provide air distribution, and covered with a layer of unscreened compost. The entire pile (except for the volume reserved for the cover layer) was screened after composting, and the oversized particles were recycled as a bulking agent. Screening typically recovers between 50% and 80% of the bulking agent (by volume), so it must be supplemented with makeup bulking agent. The recovery rate depends on the compost's moisture content (stickiness), the bulking agent's particle size, and the screen's loading rate.

Table 12.16 Types and Characteristics of Bulking Agents

Bulking Agent	Characteristics
Wood chips (1 to 2 in.*)	• Must typically be purchased • High recovery rate in screening (60% to 80%) • Good source of supplemental carbon
Chipped yard or land-clearing debris	• May be available as waste material • Low recovery rate in screening (40% to 60%) because of higher percentage of fines • Green waste fraction adds nitrogen; more may be needed for C:N ratio • Good source of supplemental carbon
Ground waste lumber	• May be available as waste material • May be poor source of supplemental carbon if old and extremely dry because more volatile forms of carbon will be missing
Leaves	• Insufficient porosity to be used alone • Rapidly available source of supplemental carbon • Available as waste material • Not recovered by screening and adds to compost volume
Sawdust	• Insufficient porosity to be used alone • Rapidly available source of supplemental carbon • Must be purchased and typically is expensive • Not recovered by screening and adds to compost volume
Shredded paper	• Insufficient porosity to be used alone • Rapidly available source of supplemental carbon • Available as waste material • Not recovered by screening and adds to compost volume

*in. × 25.4 = mm.

4.2.5 STORAGE AND MATURATION

Composting basically has two phases: a rapid decomposition period (14 to 21 days) followed by a longer, slower one with significantly lower oxygen and moisture-removal demands. This second phase (called curing or maturation) is typically about 30 days long and is needed to produce a stable, usable product. Curing sometimes consists of merely stockpiling the material, but this can prolong curing time and increase the danger of fires. Low-rate aeration better controls curing time and product stability. The curing material should be covered to control its moisture content and thereby prevent the material from compacting and going anaerobic.

Maturation refers to the conversion of a solids-amendment mixture's rapidly biodegradable components into substances similar to soil humus, which decomposes slowly. Stability refers to the reduction in the microbial-degradation rate of the mixture's biodegradable components. Stabilization is achieved by maintaining optimal conditions for a sufficient period of time.

To dry compost, operators provide enough aeration or agitation to facilitate the removal of water vapor. This increases the solids content from about 40% to 55% or higher. Drying is critical in processes that include screening, because screens do not perform well if the compost contains less than 50% to 55% solids.

4.3 COMPOSTING ALTERNATIVES

4.3.1 AERATED STATIC PILE COMPOSTING

Aerated static-pile composting is also called the Beltsville method, because it was developed in Beltsville, Maryland, in the 1970s by the USDA. As the name suggests, it involves aerating piled feedstock (Figure 12.18). This flexible method is popular in the United States.

Table 12.17 Materials Balance for 1 Dry Ton of Biosolids in Aerated Static-Pile Composting

Material	Volume (cu yd)	Total Weight (Ton)	Dry Weight (Ton)	Volatile Solids (Ton)	Bulk Density (lb/cu yd)	Solids Content	Volatile Solids
Biosolids	6.3	5.0	1.0	0.5	1600	20.0%	53.0%
Yard waste (processed)	10.9	3.3	1.8	1.3	600	55.0%	70.0%
Wood waste	0.0	0.0	0.0	0.0	500	60.0%	95.0%
Screened recyled bulking agent	9.9	3.5	1.9	1.8	695	55.0%	93.0%
Unscreened recycle	0.0	0.0	0.0	0.0	780	55.0%	88.6%
Mixture	25.7	11.7	4.7	3.6	911	40.1%	75.7%
Base (recycled bulking agent)	1.4	0.5	0.3	0.3	695	55.0%	93.0%
Cover (unscreened)	2.9	1.1	0.6	0.5	780	55.0%	88.6%
Composting losses			9.4			0.4	
Cover (unscreened)	2.9	1.1	0.6	0.5	780	55.0%	88.6%
Screen feed	21.5	8.4	4.6	3.5	780	55.0%	74.8%
Recycled bulking agent	11.4	4.0	2.2	2.0	695	55.0%	93.0%
Curing	9.9	4.4	2.4	1.4	900	55.0%	58.6%
Curing losses		0.2		0.1		0.1	
Compost to storage	9.5	4.3	2.3	1.3	900	55.0%	56.9%

Assumptions:
Recovery by screening
- Yard waste 50% by volume
- Wood waste 70% by volume
- Recycled bulking agent 50% by volume
- Pile base 95% by volume

Processing losses
- Losses during composting 10% of volatile solids
- Losses during curing 5% of volatile solids

Note. cu yd × 0.7646 = m^3; lb/cu yd × 0.5933 = kg/m^3; ton × 0.9072 = Mg.

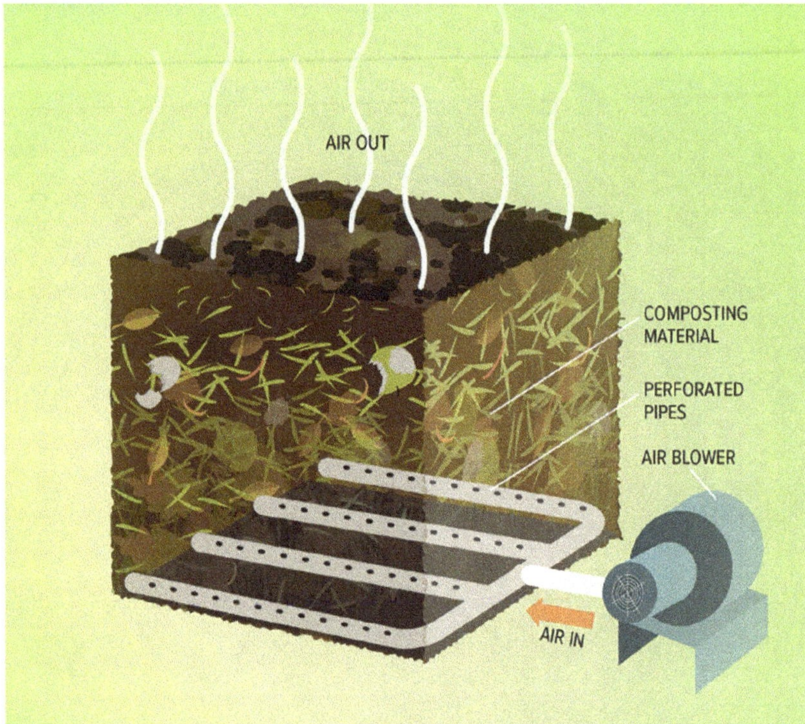

(a)

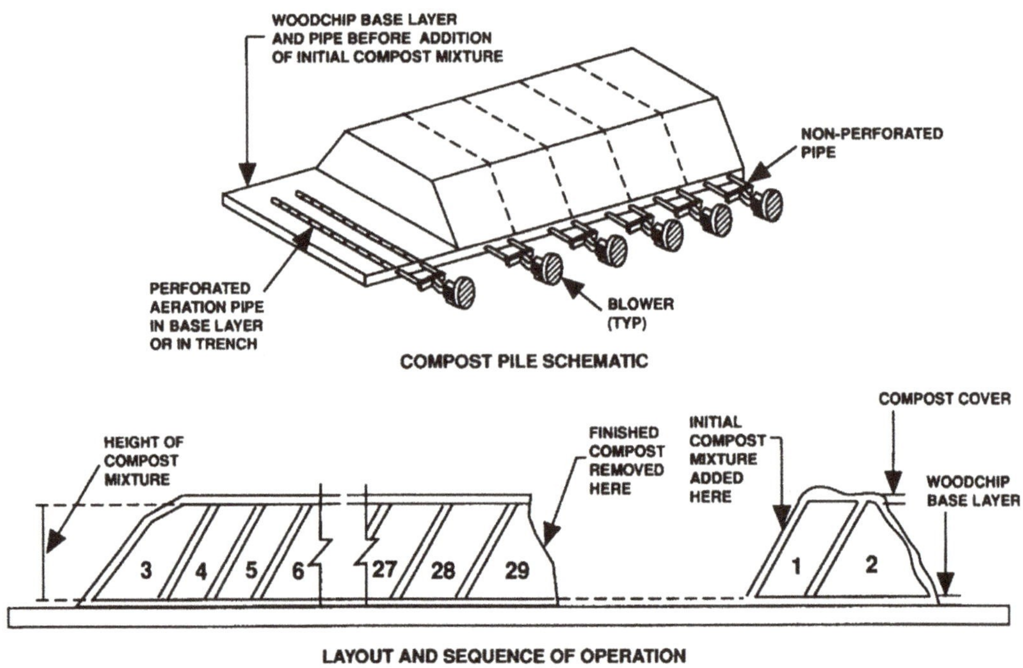

(b)

Figure 12.18 Illustration (a) (Reprinted with permission from Juan Moreira) and Schematic (b) of an Aerated Static-Pile Composting System

In this method, the solids-amendment mixture is constructed into a 2 to 4 m (6 to 12 ft) deep pile over an aeration floor (plenum) and then covered with a 150 to 300 mm (6 to 12 in.) deep insulating blanket of wood chips or unscreened finished compost to ensure that all of the mixture meets the temperature standards for pathogen destruction and vector-attraction reduction. Small operations may construct individual piles, while large ones may divide a continuous pile into sections representing each day's contribution. The mixture typically remains in the pile for 21 to 28 days while the plenum forces air through the material to provide an aerobic composting environment. Then the piles are broken down, and the material is either moved directly to a curing area or screened and then moved to the curing area. In some facilities, an intensive drying step (with a higher aeration rate than active composting) is needed before screening. Compost typically is cured for at least 30 days to further stabilize the material. Some facilities screen the compost after curing (rather than before curing).

Aerated static-pile composting originally was developed for outdoor sites, but many systems are either partially or fully enclosed to control odors or facilitate operations during unfavorable environmental conditions (e.g., temperature or rainfall extremes).

4.3.2 WINDROW COMPOSTING

In windrow composting, the solids-amendment mixture is formed into long parallel windrows whose cross-sections are either trapezoidal or triangular (Figure 12.19). The material then is turned periodically by a front-end loader or a dedicated windrow-turning machine to release moisture, expose more particles to the air, and loosen (fluff) the material to facilitate air movement through the windrow. Windrow composting occurs at open outdoor sites or covered sites. This system needs more space than other composting technologies because of pile geometry and the room needed to maneuver a windrow-turning machine.

In the aerated windrow method, windrows are constructed over air channels to protect aeration piping from the turning equipment. Air can either be forced up through the windrow or be pulled down through the windrow into the channel. The windrows are turned periodically to expose more particles to air. Aeration and turning optimize the composting rate and release of moisture.

4.3.3 IN-VESSEL COMPOSTING

In-vessel systems typically combine aeration with some type of automated material movement in a reactor. A wide variety of such systems has been developed over the years, but only a few have been installed in more than one or two sites.

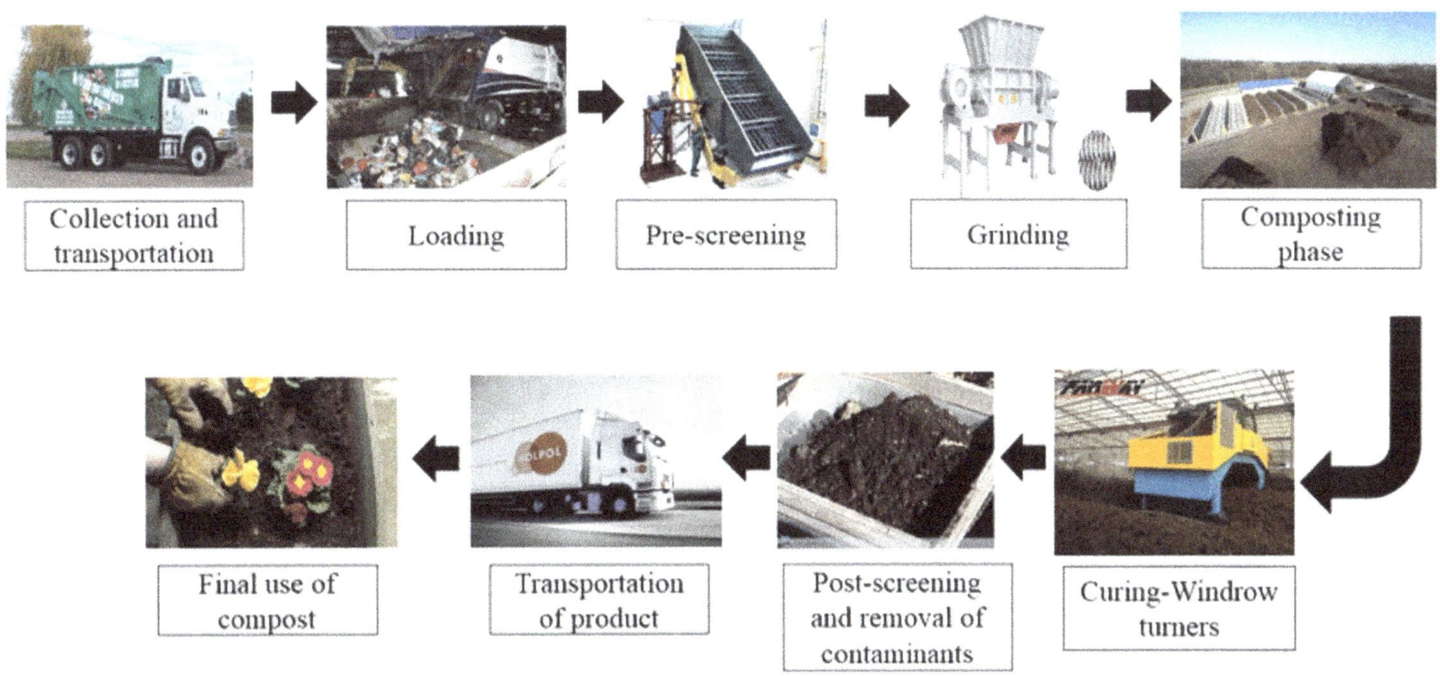

Figure 12.19 Process Flow Diagram for Windrow Composting (Adapted from Saer et al., 2013)

The detention time ranges from about 10 to 21 days, depending on system-supplier recommendations, regulatory requirements, and costs. It also should be based on desired product characteristics, especially stability, and take into account the overall solids residence time in the entire composting operation (all process phases). Once discharged from the reactor, the composted biosolids typically must be further stabilized for 30 to 60 days to achieve the desired product stability.

There are basically three types of in-vessel composting systems: vertical plug-flow reactors, horizontal plug-flow reactors, and agitated-bay systems. Vertical plug-flow reactors are made of steel, concrete, and/or reinforced fiber-glass panels (Figure 12.20). A mix of dewatered cake, amendment, and recycled solids is loaded in the top of the reactor, where it is aerated but not agitated (mixed). It moves as a plug to the bottom of the reactor, where it is removed via a traveling auger.

Horizontal plug-flow reactors are similar to vertical ones, except that the solids-amendment mixture is moved laterally through the reactor by a hydraulic ram (Figure 12.21).

Agitated-bay reactors are open-topped bays with blowers and piping systems that supply air from the bottom (Figure 12.22). Unlike plug-flow reactors, they also have mechanical devices that periodically agitate the mixture during its stay in the reactor. These systems are designed to function much like aerated windrows. A variety of methods are used to transfer compost from the reactors.

The most commonly used in-vessel system is the horizontal agitated-bed reactor. These reactors are rectangular, aerated from the bottom with independently programmable aeration zones, and enclosed in a building. A loader places the solids-amendment mixture into the front end. The agitation device is completely automatic, operates only in agitation mode, and typically makes one pass through the reactor each day. The composting material is dug out and redeposited about 4 m (11 ft) behind the machine until it has moved through the entire length of the reactor.

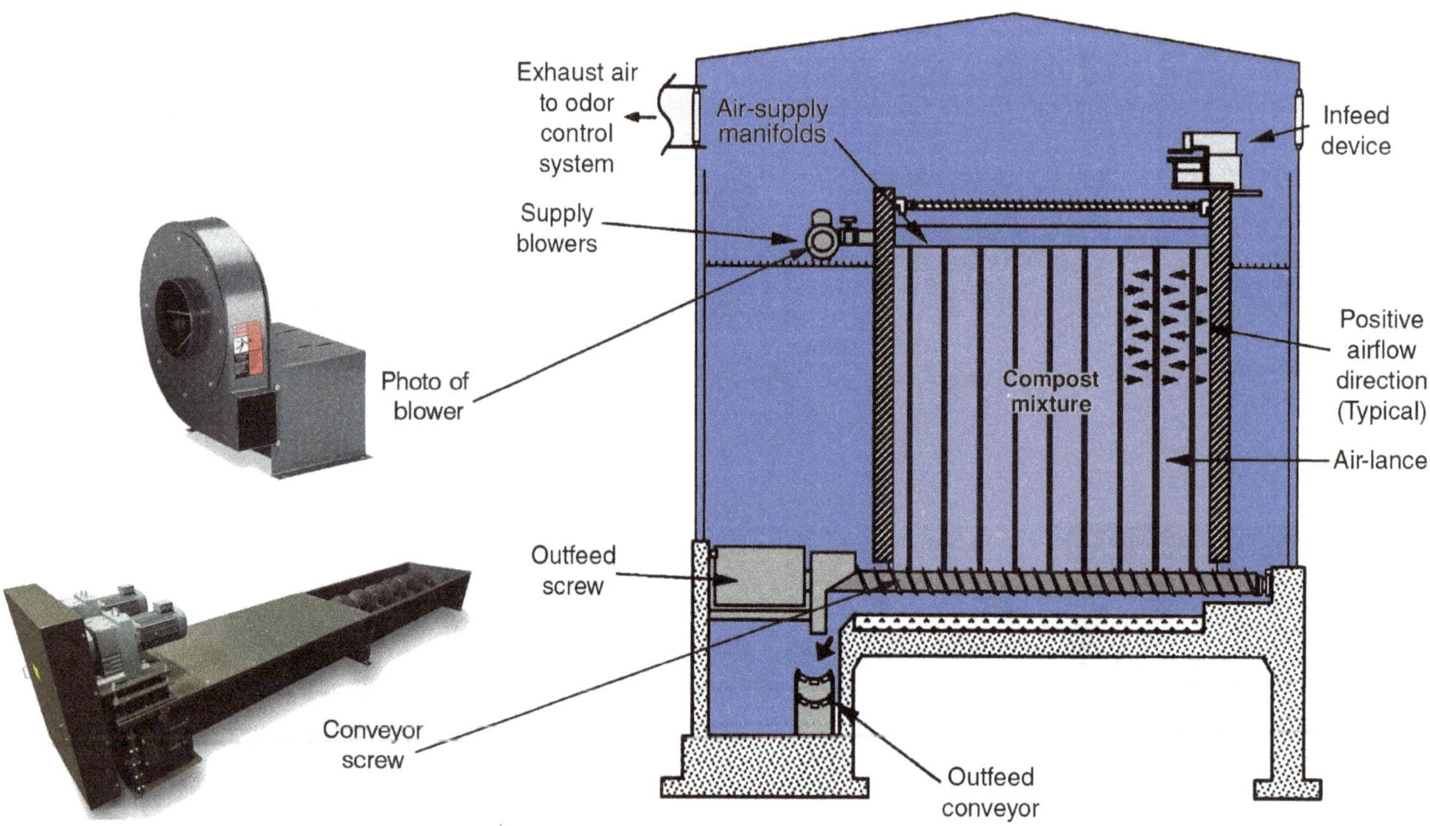

Figure 12.20 Cross-Section of a Vertical Plug-Flow Reactor (Rectangular Design, Made of Steel)

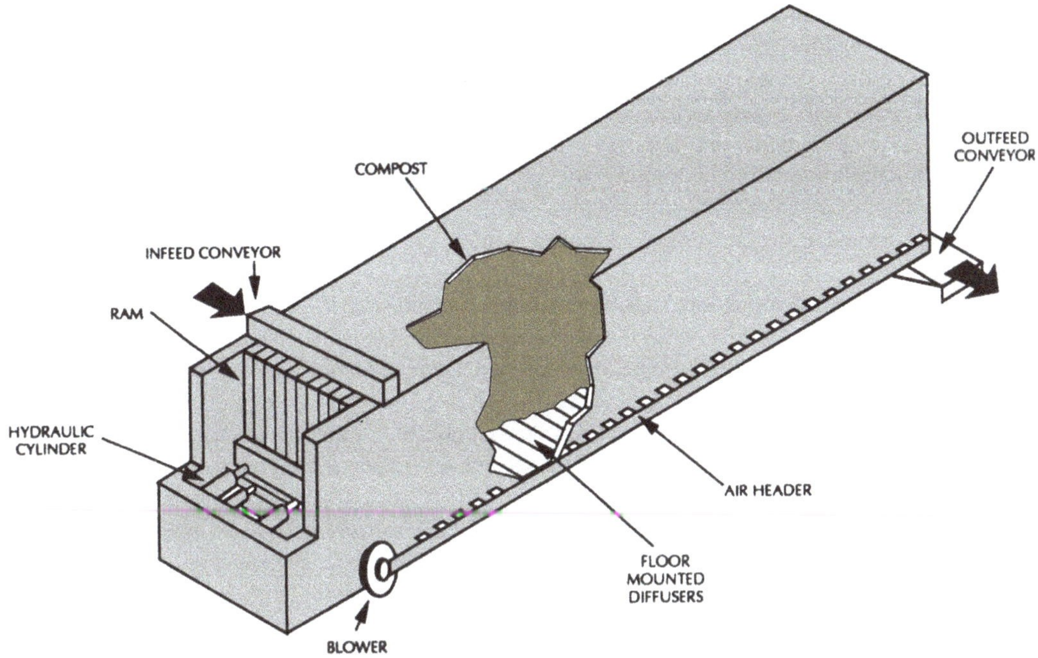

Figure 12.21 Schematic of a Horizontal Plug-Flow Reactor

5.0 ALKALINE STABILIZATION

Lime or alkaline stabilization frequently is used to meet the 40 CFR Part 503 requirements for Class B biosolids. In some cases, this process can produce a soil amendment or conditioner that meets Class A requirements. Alkaline stabilization typically is simpler to operate than digestion or composting. However, the resulting biosolids can become unstable if the pH drops after treatment and organisms regrow. The lime or alkaline agent often is costly and can significantly increase the mass and cost of solids to be used or disposed of. Some facilities have experienced odors and undesirable working conditions.

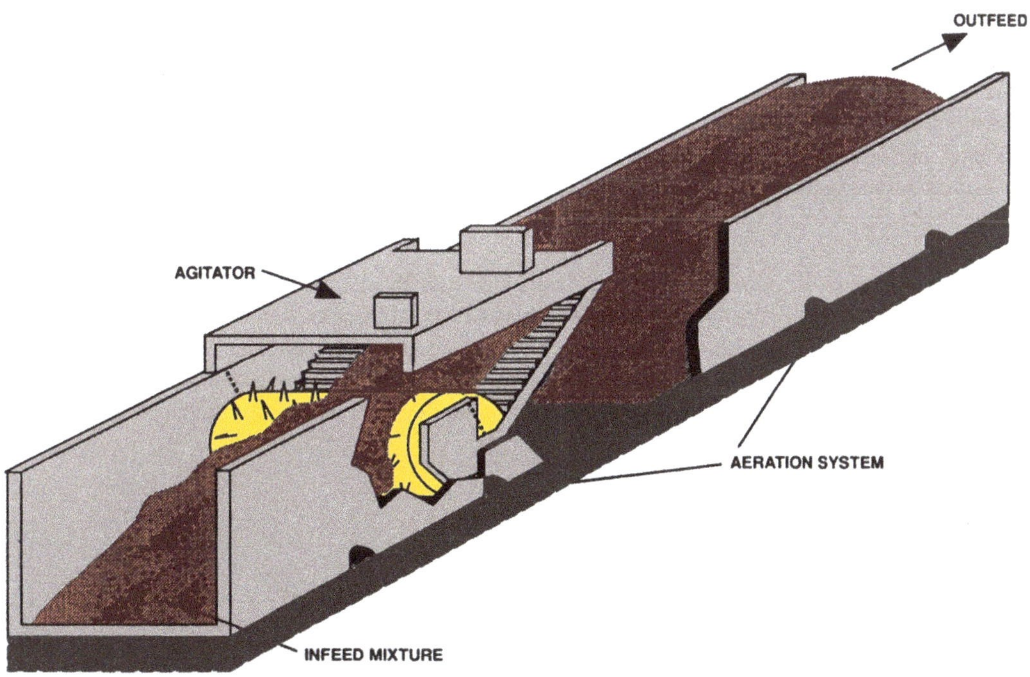

Figure 12.22 Schematic of a Horizontal Agitated-Bed Reactor

A number of advanced alkaline-stabilization technologies now used in the wastewater treatment field include chemical additives in addition to, or instead of, lime. Alkaline-stabilization processes produce a rich, soil-like product containing a few pathogens. The biosolids also have a higher pH, which is desirable at farms with acidic soils. However, alkaline stabilization increases the mass of biosolids to be managed and generates strong ammonia and amine odors that may need to be treated. One of the more important parameters for alkaline stabilization is mixing efficiency, which depends on the raw materials used in the process (rheology of dewatered cake and gradation of lime). Improper mixing results in variable biosolids characteristics and odors during storage and land application.

Adding alkaline chemicals to solids is a reliable stabilization method that WRRFs have practiced since the 1890s. The chemicals traditionally used are quicklime and hydrated lime. In recent years, a number of advanced alkaline-stabilization technologies have emerged. These technologies, which use new chemical additives, special equipment, or special processing steps, all claim advantages over traditional lime stabilization (e.g., enhanced pathogen control and a more publicly acceptable product). They also produce a biosolids that sometimes is called artificial soil because it has been successfully used as a soil substitute.

Lime is the most widely used and one of the least expensive alkaline materials available in the wastewater industry. It has been used to reduce odors in privies, increase pH in stressed digesters, remove phosphorus in advanced wastewater treatment processes, treat septage, and condition solids before and after mechanical dewatering. It is also the principal stabilizing chemical at WRRFs with capacities ranging from 379 m^3/d to approximately 1.13 million m^3/d (0.1 to 300 mgd) (U.S. EPA, 1979).

Alkaline-stabilized biosolids can be beneficially used in many ways, depending on the particular quality requirements and associated standards. Traditional lime stabilization is classified as a Class B process (PSRP) (U.S. EPA, 2007). Many of the advanced alkaline-stabilization technologies meet the U.S. EPA's definition of a Class A process (PFRP). Many of the beneficial use and disposal options for alkaline-stabilized biosolids are further discussed in Oerke (1999).

5.1 PROCESS FUNDAMENTALS

The purposes of alkaline stabilization are to substantially reduce the number and prevent the regrowth of pathogenic and odor-producing organisms, thereby preventing biosolids-related health hazards; to create a stable product that can be stored; and to reduce the short-term leaching of metals from biosolids not incorporated with natural soil.

Both liquid- and dry-lime stabilization achieve significant pathogen reduction, provided that a sufficiently high pH or temperature is maintained for an adequate period of time (Bitton et al., 1980; Christensen, 1982). Table 12.18 lists bacteria measurements during full-scale studies at the Lebanon, Ohio, WRRF; it shows that liquid-lime stabilization at pH 12.5 and a 25% dose (dry-weight) reduced total coliform, fecal coliform, and fecal streptococci concentrations by more than 99.9%. Table 12.19 shows that pathogen concentrations in liquid-lime-stabilized biosolids ranged from 10 to 1000 times less than those in anaerobically digested biosolids from the same WRRF.

Effective lime stabilization depends on raising the pH high enough and maintaining it at that level long enough to halt or substantially retard the microbial reactions that otherwise could lead to odor production and vector attraction. The process also can inactivate viruses, bacteria, and other microorganisms. Lime stabilization involves a variety of chemical reactions that alter the chemical composition of solids. The following equations (simplified for illustrative purposes) show the types of reactions that may occur:

Reactions with inorganic constituents:

Water: $$CaO + H_2O \rightarrow Ca(OH)_2 \quad (12.27)$$

Calcium: $$Ca^{2+} + 2HCO_3^- + CaO \rightarrow 2CaCO_3 + H_2O \quad (12.28)$$

Phosphorus: $$2PO_4^{3-} + 6H^+ + 3CaO \rightarrow Ca_3(PO_4)_2 + 3H_2O \quad (12.29)$$

Carbon dioxide: $$CO_2 + CaO \rightarrow CCO_3 \quad (12.30)$$

Table 12.18 Bacteria Reduction Via Liquid-Lime Stabilization at Lebanon, Ohio (U.S. EPA, 1979)

Type of Solids	Bacterial Density, Number/100 mL				
	Total Coliforms[a]	Fecal Coliforms[a]	Fecal Streptococci	*Salmonella*[b]	*Ps. aeruginosa*
Raw sludge					
Primary	2.9×10^9	8.2×10^8	3.9×10^7	62	195
Waste activated	8.3×10^8	2.7×10^7	2.7×10^7	6	5.5×10^3
Anaerobically digested biosolids					
Mixed primary and waste activated	2.8×10^7	1.5×10^5	2.7×10^5	6	42
Lime stabilized biosolids[c]					
Primary	1.2×10^5	5.9×10^3	1.6×10^4	<3	<3
Waste activated	5.2×10^5	1.6×10^4	6.8×10^3	<3	13
Anaerobically digested	18	18	8.6×10^3	<3	<3

[a]Millipore filter technique used for waste activated sludge. Most probable number technique used for other sludges.
[b]Detention limit = 3
[c]To pH equal to or greater than 12.0

Reactions with organic constituents:

Acids: $$RCOOH + CAO \rightarrow RCOOCAOH \tag{12.31}$$

Fats: $$\text{"Fat"} + CaO \rightarrow \text{Fatty Acids} \tag{12.32}$$

Lime initially raises the pH of solids. Then, reactions occur (e.g., those in the equations above) that lower the pH unless excess lime is added. The amount of excess lime needed depends on the length of time that a high pH must be maintained (e.g., during extended storage).

5.1.1 PH

To meet Class B stabilization requirements, the pH of the feed cake–chemical mixture must be elevated to more than pH 12.0 for 2 hours and then maintained above pH 11.5 for another 22 hours to meet VSR requirements. To meet Class A stabilization requirements, the elevated pH is combined with elevated temperatures (70 °C for 30 minutes or other U.S. EPA–approved time and temperature combinations listed in U.S. EPA, 1999). As long as the pH remains above 10 to 10.5, microbial activity and the associated odorous gases are greatly reduced or eliminated (U.S. EPA, 1979). However, other odorous gases (e.g., ammonia and trimethyl amine) may be produced under high pH and high temperature conditions.

5.1.2 TEMPERATURE

If quicklime (or any compound with high quicklime concentrations) is added to solids, it initially reacts with the water in solids to form hydrated lime. This exothermic reaction releases about 15 300 cal/g·mol ($2.75 \cdot 10^4$ BTU/lb·mol) (U.S. EPA, 1982). The reaction between quicklime and carbon dioxide is also exothermic, releasing about $4.33 \cdot 10^4$ cal/g·mol ($7.8 \cdot 10^4$ Btu/lb·mol).

Both reactions can raise the temperature substantially, particularly in solids cake with a low moisture content. For example, focusing on the first reaction, adding 250 g (0.55 lb) of quicklime per kg (2.2 lb) of total solids to a cake containing 15% total solids can result in a temperature of 12 °C (21.6 °F). This means a mix of 250 g CaO with 1 kg solids and 5.67 kg H_2O (1 kg solids/6.67 kg TS = 15%). The temperature rise is calculated as follows:

$$\begin{aligned} &250\ \text{g CaO}/5.67\ \text{kg-H}_2\text{O} \cdot (1\ \text{mol}/56\ \text{g})(64\ \text{kJ/mol}) = 50.4\ \text{kJ/kg H}_2\text{O} \\ &50.4\ \text{kJ/kg-H}_2\text{O} \cdot (1/4.18\ \text{kJ/kg °C}) = 12\ \text{°C} \end{aligned} \tag{12.33}$$

$$(0.55 \text{ lb-CaO}/12.5 \text{ lb-H}_2\text{O})(1 \text{ lb-mol}/56 \text{ lb})(27{,}500 \text{ BTU/lb-mol}) = 21.6 \text{ BTU/lb-H}_2\text{O} \quad (12.34)$$

$$(21.6 \text{ BTU/lb-H}_2\text{O})(1\ ^\circ\text{F/lb H}_2\text{O/Btu}) = 21.6\ ^\circ\text{F} \quad (12.35)$$

The example above assumes the specific heat of the solids is negligible. In reality, the energy input increases the temperature of the solids mixture, which has an overall specific heat less than the specific heat of water. This correction can be applied by using an overall specific heat instead of the specific heat of water.

5.1.3 FEEDSTOCK CHARACTERISTICS

The amount, sources, and composition of the feed cake determine the overall size of the alkaline-stabilization system. Variable thickening or dewatering performance is an important consideration because poor performance significantly increases the size of the stabilization system. The dewatered cake's solids concentration affects both chemical dose and system size. Equipment capacities must be able to accommodate the volume of feed cake to be processed. The system needs larger equipment and more alkaline chemical to process a "wet" cake (10% to 15% solids) than a drier one (20% to 25% solids).

The feed cake's nutrient content affects the biosolids characteristics. The agronomic benefit of an alkaline-stabilized biosolids depends on the amount of nutrients it contains and the need for a liming agent at the application site. Alkaline stabilization may be advantageous for untreated solids with relatively high metal concentrations because alkaline additives dilute metals (on a dry-weight basis) and immobilize some trace metals.

5.1.4 CHEMICAL AND DOSAGE

The types and doses of alkaline chemicals are important design criteria. The quality of the chemicals (e.g., lime, cement plant kiln dust, Portland cement, and lime plant kiln dust) should be consistent. Lime is available from numerous sources, ranging from a high-calcium lime in oyster or clam shells to a relatively low-calcium dolomitic lime. Major considerations when selecting a chemical include economics, availability, desired mixing, and desired product characteristics.

The two predominant types of lime are quicklime (calcium oxide) and calcium hydroxide. Slaked lime in a liquid slurry (carbide lime) is also available, and the principles of application are similar to other lime sources. Design engineers should select the type of lime based on economics and material-handling characteristics (e.g., alkaline-material particle size). Calcium hydroxide costs about 30% more to produce and transport than quicklime, but it requires less equipment on site because it already has been hydrated (slaked). Calcium hydroxide typically is economical for use at small facilities, but if more than 9000 to 13 000 m^3/d (3 to 4 ton/d) is needed, quicklime should be considered.

Quicklime typically requires on-site slaking equipment. Dry-lime stabilization (i.e., adding quicklime directly to dewatered cake) does not require that the chemical be slaked first, but additional handling precautions must be addressed because of the exothermic reaction of quicklime and water. Dry-lime stabilization also eliminates lime sidestreams and the related abrasion and scaling of piping and mechanical equipment.

The required doses of specific chemicals will depend on the type of feed solids (e.g., primary, WAS, trickling filter, or septage), its quality and chemical composition (including organic content), its solids concentration, the desired final product characteristics, and the type and quality of the alkaline material.

Table 12.19 shows the range of liquid lime doses required to maintain pH 12 for 30 minutes (U.S. EPA, 1979). Numerous researchers have confirmed these doses (Ramirez & Malina, 1980).

The chemical dose is affected by the feed cake's chemical composition, which depends on the type of solids and the treatment process used (e.g., chemical coagulation). Another factor that affects chemical dose is solids concentration (Figure 12.23) (U.S. EPA, 1979). Table 12.20 shows a wide range of lime doses (from 10% to 60% on a dry-weight basis). As the solids concentration increases, the required dose typically increases. The required dose per unit mass of solids tends to be somewhat higher for dilute feeds (less than

Table 12.19 Lime Dose Required for Liquid Lime Stabilization at Lebanon, Ohio (U.S. EPA, 1979)

Type of Sludge	Average Solids Concentration, %	Average Lime Dosage, lb Calcium Hydroxide/lb[a] Dry Solids	Average pH	
			Initial	Final
Primary sludge[b]	4.3	0.12	6.7	12.7
Waste activated sludge	1.3	0.30	7.1	12.6
Anaerobically digested combined	5.5	0.19	7.2	12.4

Note. Dose required to maintain pH 12 for 30 minutes.
[a]lb/lb × 1000 = g/kg.
[b]Includes waste activated sludge.

2.0% solids) because more lime is required to raise the pH of water. However, liquid lime requirements are more closely related to the feed cake's total mass than to its volume when its solids concentration ranges from 0.5% to 4.5% (U.S. EPA, 1979). Thickening solids to reduce the volume may have little or no effect on lime requirements, because the mass is not significantly changed.

Minimum doses of 13% to 40% (on a dry-weight basis as calcium hydroxide) typically are required for effective dry lime stabilization. Figure 12.24 shows the theoretical dry lime stabilization dose for both Class B and Class A stabilization. The lower line shows maximum pH requirements, and the upper line shows Class A temperature requirements. Figure 12.24 is based on a quicklime dose requirement of 25% (dry-weight basis).

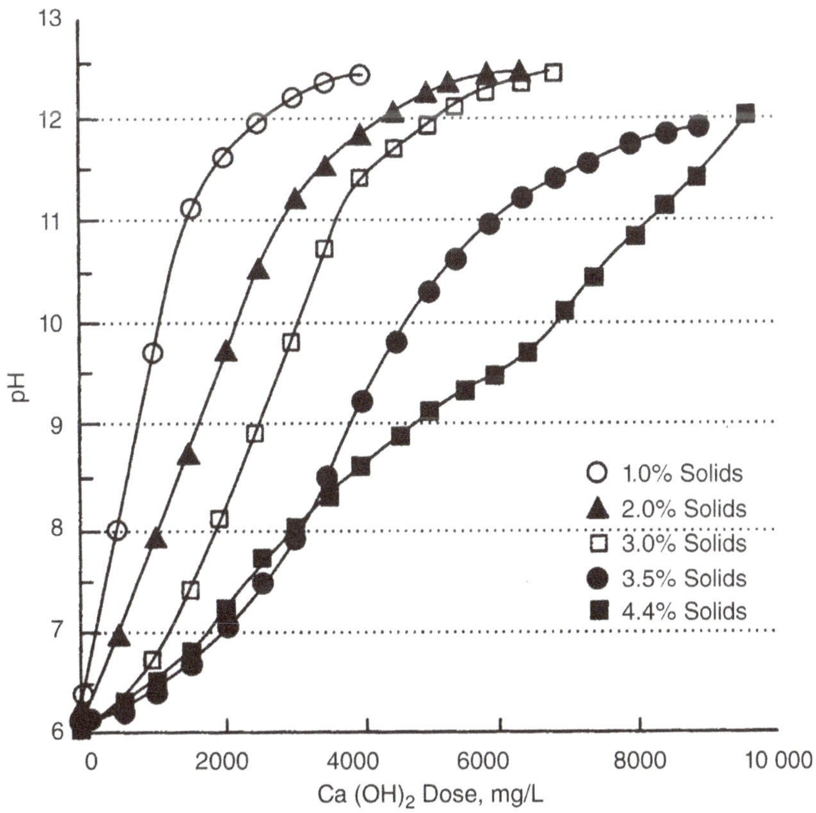

Figure 12.23 Dose of Liquid Lime Required to Raise the pH in a Stabilization System Feedstock (Primary Solids and Trickling Filter Humus) With Various Solids Concentrations (U.S. EPA, 1979)

Table 12.20 Liquid Lime Stabilization Doses Required to Keep pH Above 11.0 for at Least 14 Days (Farrell et al., 1974)

Type of Raw Sludge	Lime Dose, lb Calcium Hydroxide/lb Suspended Solids[a]
Primary sludge	0.10–0.15
Activated sludge	0.30–0.50
Septage	0.10–0.30
Alum sludge[b]	0.40–0.60
Alum sludge[b] plus primary sludge[c]	0.25–0.40
Iron sludge[b]	0.35–0.60

[a]lb/lb × 1000 = g/kg
[b]Precipitation of primary treated effluent
[c]Dry-weight basis

Design engineers can use the data in Table 12.21 and Table 12.22 for preliminary design of liquid and dry lime stabilization facilities; however, the required dose should be determined on a case-by-case basis because of the many factors involved (Farrell et al., 1974). To prevent pH decay and the associated regrowth of organisms, the lime dose may have to be higher than necessary for stabilization (Ramirez & Malina, 1980).

5.1.5 STORAGE

The system's solids storage facilities should be tailored to the facility's actual needs. Both intermediate and final storage of solids should be provided, and chemical storage is needed. Intermediate storage is useful to properly size the process and to ensure that less chemical is needed. Intermediate storage units can include insulated steel, live-bottom hoppers, concrete bunkers, or an uninsulated stockpile in an open concrete pad.

Biosolids storage is another important design consideration. Facilities need adequate storage capacity if the biosolids markets are seasonal or have not been established. The amount of storage needed depends

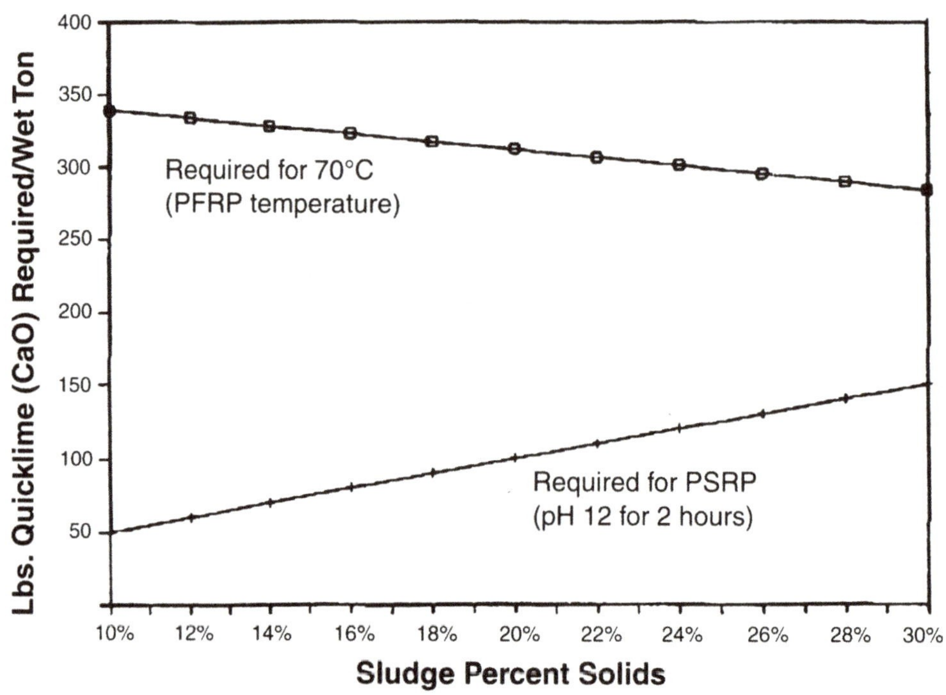

Figure 12.24 Theoretical Dry Lime Requirements to Stabilize Cake With Various Solids Concentrations So It Meets Class B or Class A Standards (PFRP = process to further reduce pathogens; PSRP = process to significantly reduce pathogens) (Lue-Hing et al., 1992)

Table 12.21 Typical Advanced Alkaline Stabilization Design Criteria (Fergen, 1991)

Item	Description or Equipment	Parameter	Units	Range of Value		Selected Design Value
				Minimum	Maximum	
Materials	Sludge	Solids	Percent	20	30	25
		Density	lb/cu ft[a]	45	55	50
	Alkaline bulking chemical	Solids	Percent	90	98	95
		Density	lb/cu ft[a]	50	65	65
	Lime	Solids	Percent	90	96	95
		Density	lb/cu ft[a]	55	60	60
	Stabilized product	Solids	Percent	55	65	60
		Density	lb/cu ft[a]	65	75	75
Curing	Technology	Windrow				
	Detention time	Average	Days	3	7	6
		Peak	Days	3	7	4
	Temperatures		°C	—	—	52 to 12 ℩r
	Pile dimensions	Bottom width	ft[b]	6	14	10
		Mix height	ft[b]	2	3	3
		Top width	ft[b]	4	8	6
		Area/unit length	sq ft/ft[c]	10	33	24
		Pile spacing	ft[b]	—	—	5
	Pile turning		lb/d[d]	—	—	1 (typical)
Odor control	Building air	Number of stages	Number	1	3	1
		Air changes	Number/hr	6	15	12
	Product storage	Number of stages	Number	1	3	1
Storage	Sludge	Days of storage	Days	0	1	1
	Chemicals	Days of storage	Days	5	30	5
	Product	Days of storage	Days	80	180	60

[a]lb/cu ft × 16.02 = kg/m^3
[b]ft × 0.3048 = m
[c]sq ft/ft × 0.3048 = m^2/m
[d]lb/d × 0.4536 = kg/d

Table 12.22 Typical Materials Balance for Advanced Stabilization Facilities (Fergen, 1991)

		Solids Balance				
Process	Item	Solids Content (%)	Volume (cu ft)[a]	Total Weight (Ton)	Dry Weight (Ton)	Bulk Density (lb/cu ft)[b]
Mixing	Sludge cake	25.0	6400	160.0	40.0	50.0
	Chemicals	95.0	3186	103.5	98.4	65.0
	Initial mix	52.5	9586	263.5	138.4	55.0
Windrow	Initial mix	52.5	9586	263.5	138.4	55.0
	Evaporation loss	—	3437	32.9	—	—
	Product	60.0	6149	230.6	138.4	75.0

Notes.
Chemical dose:
Wet-weight basis 65%
Dry-weight basis 245%
For peak conditions multiply all the values by peaking factor (except density and percent solids).
[a]cu ft × 28.32 = L
[b]lb/cu ft × 16.02 = kg/m^3

on both the type of biosolids and the distribution and marketing methods involved. At least 30 to 90 days' worth of storage should be provided if biosolids curing is required; storage also is needed to accommodate road and weather conditions, as well as fluctuations in the biosolids marketing and distribution schedule.

Alkaline stabilization requires special chemical storage and feeding equipment. Traditionally, an alkaline chemical storage system should be able to meet at least 7 days' worth of demand (although a 2- to 3-week supply is preferred). Calcium hydroxide can be stored up to 1 year. Quicklime deteriorates more rapidly; it should not be stored longer than 3 to 6 months.

5.2 PROCESS AND DESIGN CONSIDERATIONS

Product quality and process design are interdependent, and engineers should evaluate a number of design criteria before implementing an alkaline-stabilization process. Although they vary from site to site, typical design criteria include sources and characteristics of feed cake (e.g., quantity, type, quality, and solids content); contact time, pH, and temperature; alkaline chemical types and doses; solids concentration of the feed cake–chemical mixture; energy and storage requirements; pilot-scale test results; and product storage.

5.2.1 CONTACT TIME, PH, TEMPERATURE

Contact time and pH are directly related because the pH must be maintained at the required level for enough time to destroy pathogens. The treatment chemical must have enough residual alkalinity to maintain a high pH in the biosolids until they are used or disposed of. The high pH prevents odorants and pathogenic organisms from growing or reactivating.

A drop in pH (pH decay) occurs when biosolids absorb atmospheric carbon dioxide or acid rain (which forms a weak acid when dissolved in water), which gradually consumes the residual alkalinity. The pH gradually decreases, eventually dropping below a pH of 11. Bacterial action then resumes, and the renewed production of organic acids causes the pH to continue decaying (similar to the reactions in anaerobic digestion).

The pH typically drops during stabilization, so it should be raised to and maintained at more than pH 12. Biosolids do not have to be inside a contact vessel as long as the pH can be monitored to ensure that it remains at the desired value for the desired time.

5.2.2 SOLIDS CONTENT

The solids concentration of the feed cake–chemical mixture is an important design consideration for material-handling purposes. Regulations may require a minimum solids concentration (e.g., for landfilling or extended storage). The final product solids concentration (dryness) and granularity also affect the type of biosolids trucks and application/disposal equipment needed.

The solids concentration of the initial feed cake–chemical mixture also affects any supplemental drying step in advanced alkaline-stabilization processes. The alkaline additive causes chemical reactions to occur that increase the mix's apparent solids content. This increase in solids is caused by the addition of solids (treatment chemical), chemical binding, and evaporation of water from the feed cake. The alkaline material—particularly quicklime—produces a fast reaction that increases temperature in a matter of minutes. Thorough mixing of feed cake and alkaline material is important to achieve the target solids content and pathogen destruction, and to reduce residual odors (e.g., ammonia) in biosolids.

5.3 LIME STABILIZATION ALTERNATIVES

5.3.1 LIQUID-LIME (PRE-LIME) STABILIZATION

In liquid-lime (pre-lime) stabilization, a lime slurry is added to feed solids to meet Class B stabilization requirements (Figure 12.25). The lime typically is added to thickened solids at WRRFs that land-apply

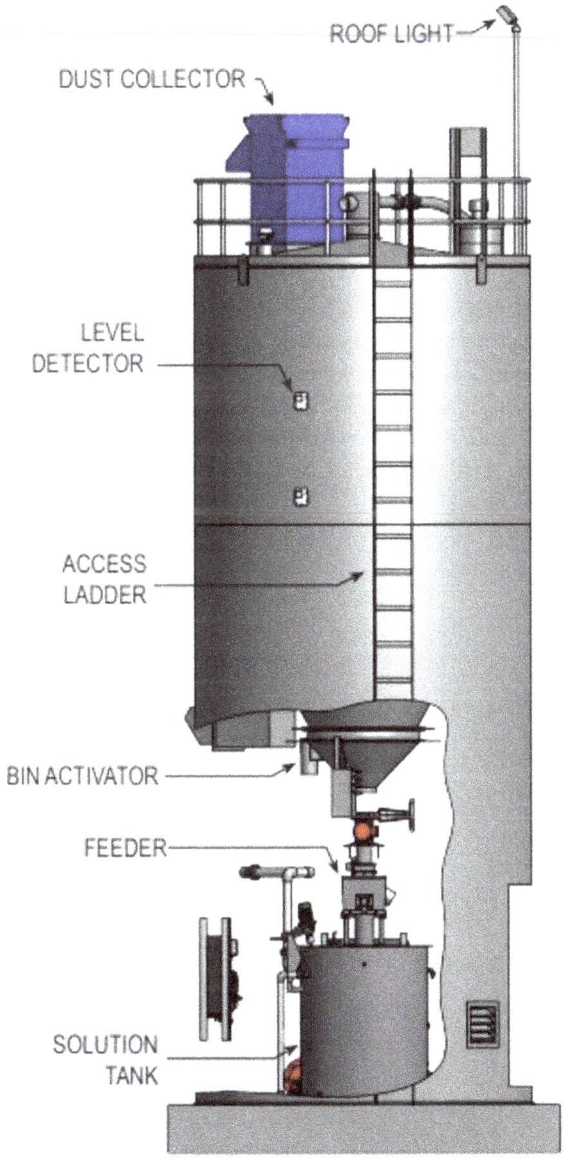

Figure 12.25 Integrity Municipal Systems Storage Silo (Reprinted with permission from Integrity Municipal Systems)

liquid biosolids (e.g., subsurface injection on agricultural land). This practice typically has been limited to smaller WRRFs or those with nearby land application or use sites.

Another liquid-lime stabilization method involves conditioning solids or septage with lime before dewatering. The lime typically is combined with other conditioners (e.g., aluminum or iron salts) to improve solids dewatering. This method primarily has been used with vacuum filters and recessed-plate filter presses; in such cases, the lime dose needed to condition solids typically exceeds that required to stabilize them.

5.3.2 DRY LIME (POST-LIME) STABILIZATION

In dry-lime (post-lime) stabilization, dry quicklime or hydrated lime is added to dewatered cake. This process has been practiced at WRRFs since the 1960s (Stone et al., 1992). The lime typically is mixed with the cake via a pug mill, plow blender, paddle mixer, ribbon blender, screw conveyor, or similar device. Figure 12.26 is a process schematic for a typical dry-lime stabilization system with a pneumatic lime-conveyance system.

Quicklime, hydrated lime, or other dry alkaline materials can be used in this process, although the use of hydrated lime typically is limited to smaller installations. Quicklime is less expensive and easier to handle than hydrated lime, and the heat of hydrolysis released when quicklime is added to dewatered cake can enhance pathogen destruction. If enough dry alkaline material is added to feed solids, the resulting biosolids can meet either Class B or Class A requirements.

5.3.3 ADVANCED ALKALINE-STABILIZATION TECHNOLOGIES

In the last 30 years, alkaline-stabilization methods have been developed that use materials other than lime; these methods are being used by a number of municipalities. Most of those that rely on additives (e.g., cement kiln dust, lime kiln dust, Portland cement, or fly ash) are modifications of traditional dry lime stabilization. The most common modifications include the use of other chemicals, a higher dose (depending on the chemical), and supplemental drying. These processes alter the feed material's characteristics and, depending on the process, increase stability, decrease odor potential, reduce pathogens, and otherwise enhance the resulting biosolids.

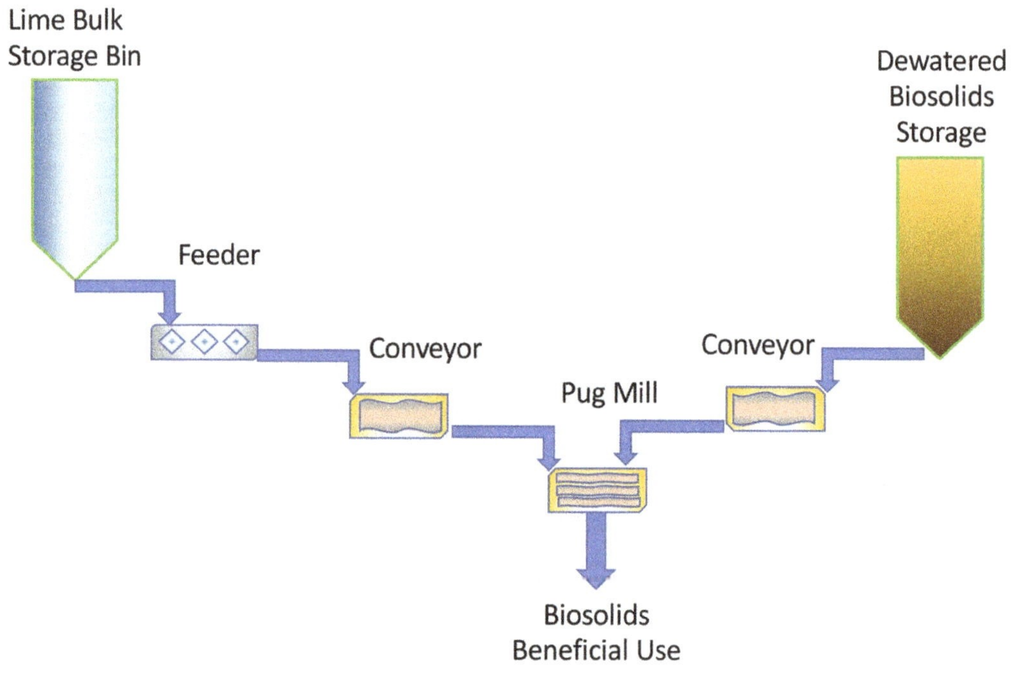

Figure 12.26 Process Schematic of a Typical Dry Lime Stabilization System

Pasteurization processes use the exothermic reaction of quicklime with water to raise process temperatures above 70 °C. This temperature is then maintained for more than 30 minutes as required by federal regulations for add-on pasteurization to meet Class A criteria. This pasteurization reaction must occur under carefully controlled and monitored mixing and temperature conditions to ensure that all of the solids' particles are uniformly treated and pathogens are inactivated by the heat generated during the reaction.

The process produces a soil-like material that is nonviscous and, therefore, not subject to liquefaction under mechanical stress. Varying the process additives and mixing ratios results in a range of biosolids-derived materials suitable for use as daily, intermediate, and final landfill cover or in land reclamation (Sloan, 1992). Figure 12.27 is a process schematic for a typical pasteurization process. In a variation of this process, pasteurization occurs in a heated and insulated vessel reactor, whereby temperatures are maintained at 70 °C or higher for at least 30 minutes.

One proprietary process (the N-Viro process) combines advanced alkaline stabilization with accelerated drying (Figure 12.28). U.S. EPA has approved two versions of this technology as systems that produce Class A biosolids. Both versions involve adding quicklime, cement plant kiln dust, lime plant kiln dust, alkaline fly ash, or other alkaline admixtures and further processing the solids to stress pathogens via pH, temperature, ammonia, salts, and dryness (Burnham et al., 1992). In one version, chemical addition is followed by raising the material's temperature to between 52 °C and 62 °C for at least 12 hours so the heat generated by the chemical reaction can further reduce pathogens. The second version uses chemical addition to raise the solids' pH above 12, and then mechanically dries the material in windrows or a rotary-drum dryer to produce biosolids containing 50% to 60% solids. The biosolids primarily are used as an agricultural liming agent, soil conditioner, landfill cover, or component of blended topsoil.

A second proprietary process (RDP EnVessel pasteurization) uses electrically generated heat to supplement the heat generated by quicklime, which purportedly reduces lime consumption. An electrically heated screw auger transfers the solids-lime mix to an enclosed reactor, where the material is held for 30 minutes at 70 °C to achieve pasteurization.

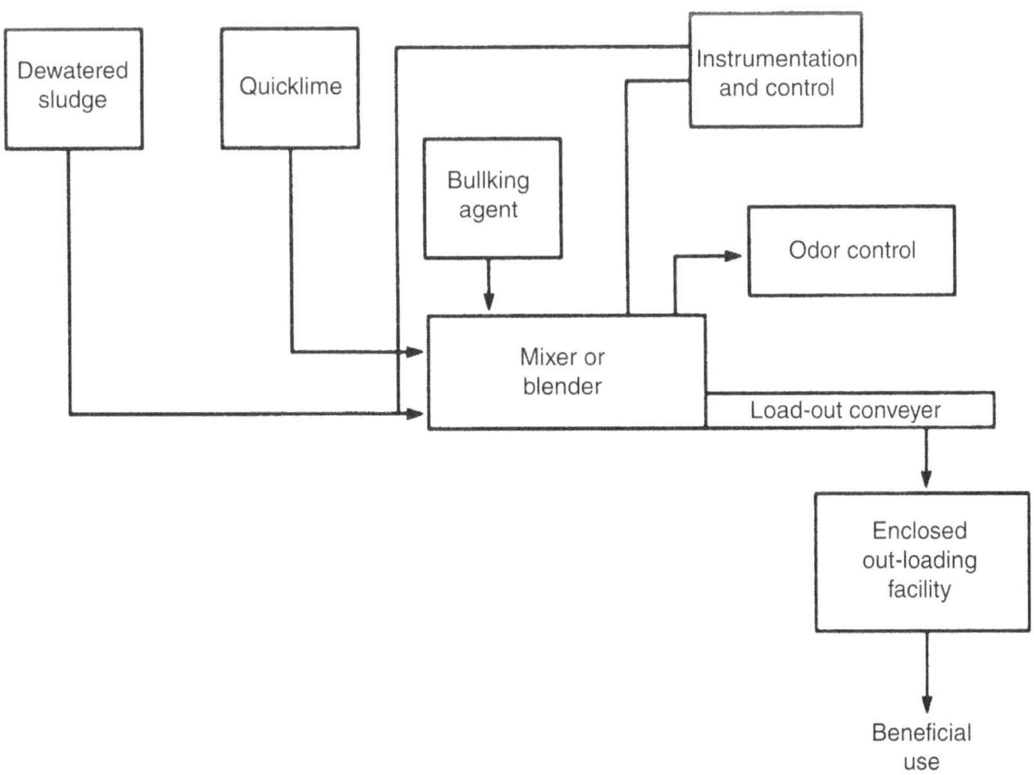

Figure 12.27 Process Schematic of a Typical Pasteurization System

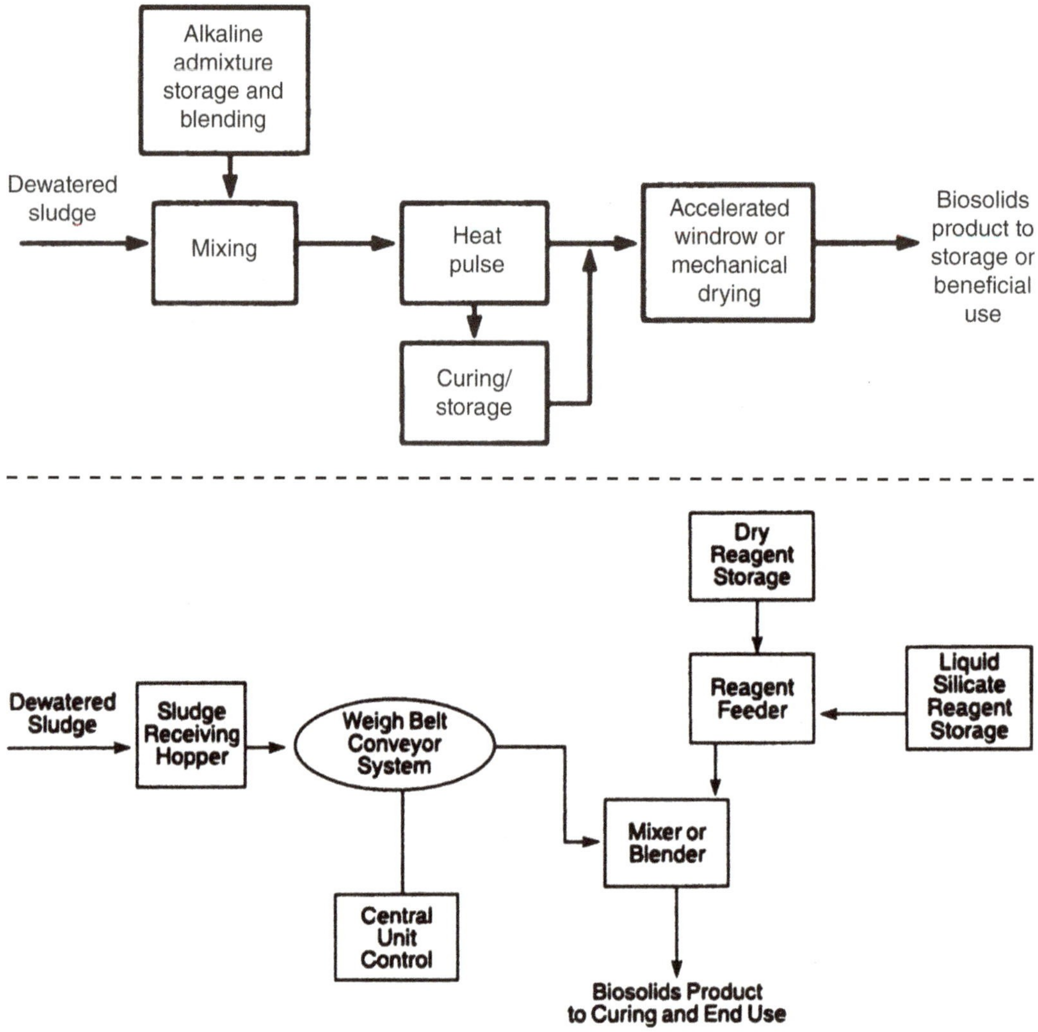

Figure 12.28 Process Schematic of a Typical Alkaline Stabilization System With a Subsequent Drying Process

A third proprietary process (Bioset) uses sulfamic acid to supplement the heat produced by quicklime (both lime and water, and lime and sulfamic acid, react exothermically). The process occurs in a pressurized vessel to achieve pasteurization conditions. Bioset has been recommended by the U.S. EPA Pathogen Equivalency Committee for certification of its process as a Class A biosolids technology.

PRACTICE PROBLEMS

1. What are three reasons for applying solids stabilization processes?
2. What general thresholds do federal requirements specify for land application of biosolids?
3. Why is the development of a mass balance important for evaluating solids stabilization?
4. Identify four solids stabilization processes used to make wastewater solids amenable to land application.
5. What are three benefits and three drawbacks of anaerobic digestion?

6. Before biosolids can be land applied, they must meet the requirements of which important regulation?
   a. 40 CFR Part 258
   b. Resource Conservation and Recovery Act (RCRA)
   c. 40 CFR Part 503
   d. Occupational Safety and Health Act (OSHA)
7. Why would aerobic digestion be preferred over anaerobic digestion and vice versa?
8. What three categories of organisms are critical to composting?
9. Calculate the VSR efficiency for a WRRF with anaerobic digesters that has a digester influent volatile solids fraction of 0.80 and a digester effluent volatile solids fraction of 0.64 (hint: van Kleek equation).
10. What is the minimum SRT needed to achieve a VSR of 48% in a mesophilic anaerobic digester?
11. What volume of digester is needed if the design volatile solids production rate is 26 000 kg/d and the volatile solids loading rate should be 1.3 kg/(m³·d) or less?
12. What is the target pH, and how long does the target pH need to be maintained for alkaline stabilization?
13. What are three different types of biosolids composting alternatives, and what distinguishes them from each other?
14. How do biosolids characteristics affect alkaline stabilization (in particular, solids concentration; e.g., lime dosing, contact time)?
15. How do advanced alkaline stabilization technologies differ conventional lime stabilization?
16. How does thickening affect sizing or dosing requirements for anaerobic digestion, aerobic digestion, composting, and alkaline stabilization?
17. Design a mesophilic aerobic digester processing solids from a nonprimary solids secondary biological treatment process. The following conditions are applicable to the design:
    - Secondary solids concentration = 0.8%
    - Total solids = 1144 kg/d (2522 lb/d)
    - Volatile solids = 894 kg/d (1971 lb/d) or 78.15%
    - Decanted solids concentration = 1.5%
    - Minimum liquid temperature (winter) = 15 °C (59 °F)
    - Maximum liquid temperature (summer) = 30 °C (86 °F)
    - The biosolids must meet Class B conditions in a single-tank configuration.
18. Design an aerated static-pile system based on the following design criteria:
    - 20 dry ton/d of cake containing 20% solids
    - Operations occur 7 days per week
    - The bulking agents are yard waste supplemented by ground wood waste
    - All storage, mixing, active composting, and screening operations are fully enclosed
    - Enough covered storage space for 30 days' worth of new bulking agent
    - Enough storage space for 1 day's worth of feedstock biosolid.
    - Enough covered storage space for 7 days' worth of recycled bulking agent
    - 21-day minimum SRT in active composting area
    - 28-day minimum SRT in aerated curing area
    - Enough outdoor storage for 90 days' worth of finished product

    Determine the mass balance, the volume of solids and bulking agent, and the areas required, including pile areas, storage areas, and areas for front end loader and truck loading/unloading. Consider sketching the layout to capture the elements needed.

19. **Design a solids stabilization system that meets Class B pathogen requirements using quicklime and a lime mixer. The WRRF's peak production is 20 dry tonne/d of solids. Solids are dewatered by a belt filter press that can produce a cake containing 18% solids. The facility initially will produce biosolids 24 hours a day, 7 days a week, loading directly into trucks that each hold 21 wet tonne. Assume the chosen mixer has a rated capacity of 100 kg/min, and provide about 15% to 25% extra mixing capacity (beyond the manufacturer-suggested rated capacity) in mixer and standby mixer equipment. Assume a lime dose of 15% on a dry-weight basis.**

6.0 REFERENCES

Batstone, D. J., Keller, J., Angelidaki, I., Kalyuzhnyi, S. V., Pavlostathis, S. G., Rozzi, A., Sanders, W. T. M., Siegrist, H., & Vavilin, V. A. (2002). The IWA anaerobic digestion model no. 1 (ADM-1). *Water Science and Technology, 45*(10), 65–73.

Benefield, L. D., & Randall, C. W. (1980). *Biological process design for wastewater treatment.* Prentice-Hall.

Bitton, G., Damron, B. L., Edds, G. T., & Davidson, J. M. (1980). *Sludge health risks of land application.* Ann Arbor Science Publishers.

Burch, T. R., Sadowsky, M. J., & LaPara, T. M. (2013). Aerobic digestion reduces the quantity of antibiotic resistance genes in residual municipal wastewater solids. *Frontiers in Microbiology, 4,* 12.

Burd, R. S. (1968). *A study of sludge handling and disposal* (Publication No. WP-20-4). U.S. Department of the Interior, Federal Water Pollution Control Administration.

Burnham, J. C., Hatfield, N., Bennett, G. F., & Logan, T. J. (1992) *Use of kiln dust with quicklime for effective municipal sludge treatment with pasteurization and stabilization with the N-Viro soil process* (Stand. Tech. Publication 1135). American Society for Testing and Materials.

Buswell, A. M., & Neave, S. L. (1939). Laboratory studies of sludge digestion. *Illinois State Water Survey Bulletin, 30.* Institute for Natural Resource Sustainability, University of Illinois.

Camp, T. R., & Stein, P. C. (1943). Velocity gradients and internal work in fluid motion. *Journal of the Boston Society of Civil Engineers, 30,* 219–237.

Chang, Y. (1967). The fungi of wheat straw compost: Part II—biochemical and physiological studies. *Transactions of the British Mycological Society, 50,* 667.

Chapman, T., & Muller, C. (2010, May 23–26). Impact of series digestion on process stability and performance. *Proceedings of the 24th Annual Water Environment Federation Residuals and Biosolids Conference,* Savannah, Georgia, United States.

Christensen, G. L. (1982). Dealing with the never-ending sludge output. *Water Engineering and Management, 129,* 25.

Conklin, A., Stensel, H. D., & Ferguson, J. (2006). Growth kinetics and competition between methanosarcina and methanosaeta in mesophilic anaerobic digestion. *Water Environment Research, 78,* 486–496.

D'Antonio, G. (1983). Aerobic digestion of thickened activated sludge—Reaction rate constant determination and process performance. *Water Research, 17*(11), 1525–1531.

Daigger, G. T., & Bailey, E. (2000). Improving aerobic digestion by prethickening, staged operation, and aerobic–anoxic operation: Four full-scale demonstrations. *Water Environment Research, 72,* 260–270.

Daigger, G. T., Ju, L. K., Stensel, D., Bailey, E., & Porteous, J. (2001, October 14–19). Can 3% SS digestion meet new challenges? *Proceeding of the Aerobic Digestion Workshop* (Vol. 5; featured presentation sponsored by Enviroquip, Inc.), Water Environment Federation 74th Annual Exposition and Conference, Atlanta, Georgia, United States.

Daigger, G. T., Novak, J., Malina, J., Stover, E., Scisson, J., & Bailey, E. (1998, October 3). Panel of experts. *Proceedings of the Aerobic Digestion Workshop* (Vol. 2), Orlando, Florida, United States. Enviroquip.

Daigger, G. T., Scisson, J., Stover, E., Malina, J., Bailey, E., & Farrell, J. (1999, October 10). Fine tuning the controlled aerobic digestion process. *Proceedings of the Aerobic Digestion Workshop,* (Vol. 3), New Orleans, Louisiana, United States. Enviroquip.

Daigger, G. T., Yates, R., Scisson, J., Grotheer, T., Hervol, H., & Bailey, E. (1997, October 18). The challenge of meeting Class B while digesting thicker sludges. *Proceedings of the Aerobic Digestion Workshop,* (Vol. 1), Chicago, Illinois, United States. Enviroquip.

De Vrieze, J., Verstraete, W., & Boon, N. (2013). Repeated pulse feeding induces functional stability in anaerobic digestion. *Microbial Biotechnology, 6*(4), 414–424.

Farrell, J. B., Smith, J., Hathaway, S., & Dean, R. (1974). Lime stabilization of primary sludge. *Journal of the Water Pollution Control Federation, 46,* 113.

Fergen, R. E. (1991, August 11–14). Stabilization and disinfection of dewatered municipal wastewater sludge with alkaline addition. *Proceedings of the 5th Annual American Water Works Association/Water Pollution Control Federation Joint Residuals Management Conference*; Durham, North Carolina, United States. Water Pollution Control Federation, Alexandria, Virginia.

Finger, R. E., & Butler, R. C. (1996). *The effect of sampling procedures on digester pH measurement.* Paper presented at the 69th Annual Water Environment Federation Technical Exhibition and Conference, Dallas, Texas, United States.

Franke-Whittle, I., Waltera, A., Ebnerb, C., & Insama, H. (2014). Investigation into the effect of high concentrations of volatile fatty acids in anaerobic digestion on methanogenic communities. *Waste Management, 34*(11), 2080–2089.

Garber, W. F. (1982). Operating experience with thermophilic anaerobic digestion. *Journal of the Water Pollution Control Federation, 54,* 1170.

Grady, C. P. L., Jr., Daigger, G. T., Love, N. G., & Filipe, C. D. M. (2011). *Biological wastewater treatment* (3rd ed.). CRC Press.

Great Lakes–Upper Mississippi River Board of State and Provincial Public Health and Environmental Managers. (2014). *Recommended standards for wastewater facilities.* Health Education Services, Inc.

Higgins, M. J., Yarosz, D. P., Chen, D. P., Murthy, S. N., Maas, N., Cooney, J., Glindemann, D., & Novak, J. T. (2006). Cycling of volatile organic sulfur compounds in anaerobically digested biosolids and its implications for odors. *Water Environment Research, 78,* 243–252.

Jenkins, C. J., & Mavinic, D. S. (1989). Anoxic-aerobic digestion of waste activated sludge: Part II–Supernatant characteristics, ORP monitoring results and overall rating system. *Environmental Technology Letters, 10*(4), 371–384.

Jones, C. A., Coker, C., Kirk, K., & Reynolds, L. (2019). *Food waste co-digestion at water resource recovery facilities: Business case analysis.* Water Research Foundation Project ENER19C17/4792.

Jones, R., Parker, W., Khan, Z., Murthy, S., & Rupke, M. (2008). Characterization of sludges for predicting anaerobic digester performance. *Water Science and Technology., 57,* 721–726.

Kayhanian, M., & Tchobanoglous, G. (1992). Computation and importance of carbon to nitrogen (C/N) ratios for various organic fractions of municipal solid waste. *BioCycle, 33,* 58–60.

Knoll, K. H. (1964). *Information Bulletin No. 13-20.* International Research Group on Refuse Disposal, U.S. Public Health Service.

Koers, D. A., & Mavinic, D. S. (1977). Aerobic digestion of waste activated sludge at low temperature. *Journal of the Water Pollution Control Federation, 49*(3), 460–468.

Krampe, J., & Krauth, K. (2003). Oxygen transfer into activated sludge with high MLSS concentration. *Water Science and Technology, 47*(11), 297–303.

Krishnamoorthy, R., & Loehr, R. C. (1989). Aerobic sludge stabilization–Factors affecting kinetics. *Journal of Environmental Engineering, 115,* 283–301.

Krugel, S., Nemeth, L., & Peddie, C. (1998). Extending thermophilic anaerobic digestion for producing Class-A biosolids at the Greater Vancouver Regional Districts Annacis Island Wastewater Treatment Plant. *Water Science and Technology, 38*(8), 409–416.

Lawrence, A. W. (1971). Application of process kinetics to design of anaerobic processes. In Pohland, F. G. (Ed.), *Anaerobic biological treatment processes*, Advances in Chemistry series. American Chemical Society, 105.

Liu, D. H. F., & Liptak, B. G. (Eds.). (1997). *Environmental engineers' handbook* (2nd ed.). CRC Press.

Lue-Hing, C., Zenz, D. R., & Kuchenrither, R. (Eds.). (1992). *Municipal sewage sludge management: Processing, utilization and disposal.* Technomic Publishing Co.

Maxwell, M. J. (1992, September 20–24). Impact of new sludge regulations on aerobic digester sizing and cost-effectiveness. *Proceedings of the 65th Annual Water Environment Federation Exposition and Conference*, New Orleans, Louisiana, United States.

McCarthy, W. C., Nelson, C. J., Vandenburgh S. J., & Butler, R. C. (2004). Improved digester performance without the acroynms: Full-scale evaluations of series mesophilic digestion at King County's South Treatment Plant. *Proceedings of the Water Environment Federation, Residuals and Biosolids Management, 28*, 42–69.

McCarty, P. L., & Smith, D. P. (1986). Anaerobic wastewater treatment: Fourth of a six-part series on wastewater treatment processes. *Environmental Science and Technology, 20*(12), 1200–1206.

Metcalf & Eddy, Inc./AECOM (2014). *Wastewater engineering: Treatment and resource recovery* (5th ed.). McGraw-Hill.

Morgan, M. T., & MacDonald, F. W. (1969). Tests show MB tuberculosis doesn't survive composting. *Journal of Environmental Health, 32*, 101.

Oerke, D. W. (1999, March). Alkaline stabilization of biosolids can save money, space. *Water World*, 14–16.

Park, C., Abu-Orf, M. M., & Novak, J. T. (2006). The digestibility of waste activated sludges. *Water Environment Research, 78*, 59.

Parkin, G. F., & Owen, W. F. (1986). Fundamentals of anaerobic digestion of wastewater sludge. *Journal of Environmental Engineering, 112*, 5.

Popel, F. (1971a). *Die theoretischen und praktischen grunlagen der flussig-kompostierung hockkonzentrierten substrate.* Ausgearbeitet fur die Badische Anilin -und Sodafabrik.

Popel, F. (1971b). Energieerzeugung beim biologischen abbau organischer stoffe. *Gewaesser Abwaesser* (Ger.), 112.

Ramirez, A., & Malina, J. (1980). Chemicals disinfect sludge. *Water Sewage Works, 127*(4), 52.

Saer, A., Lansing, S., Davitt, N., & Graves, R. E. (2013, August). Life cycle assessment of a food waste composting system: Environmental impact hotspots. *Journal of Cleaner Production, 52*, 234–244.

Schafer, P. L., & Farrell, J. (2000, October 17). Performance comparisons for staged and high-temperature anaerobic digestion systems. *Proceedings of the Annual Water Environment Federation Technical Exposition and Conference* [CD-ROM], Anaheim, California, United States.

Schafer, P. L., Farrell, J., Newman, G., & Vandenburgh, S. (2002, September 28–October 2). Advanced anaerobic digestion performance comparisons. *Proceedings of the 75th Annual Water Environment Federation Technical Exposition and Conference* [CD-ROM], Chicago, Illinois, United States.

Scisson, J. P. (2006, March 12–15). ATAD, the far country: Improvements to the 3rd, 2nd generation ATAD yield better than expected returns. *Proceedings of the Water Environment Federation Residuals and Biosolids Management Specialty Conference*, Cincinnati, Ohio, United States.

Shell, G. L., & Boyd, J. L. (1969). *Composting dewatered sewage sludge* (SW-12c). U.S. Department of Health, Education, and Welfare.

Sloan, D. (1992). *Design and process considerations for advanced alkaline stabilization with subsequent accelerated drying facilities.* Paper presented at the 5th Annual International Conference on Alkaline Pasteurization Stabilization, Somerset, New Jersey, United States.

Smith, J. F., Bizier, P., & Sobrodos-Bernardos, L. (2012, April 25–28). Global development of the ATAD process and its significant achievements in energy recovery and utilization. *Proceedings of the Residuals and Biosolids Conference 2012*, Raleigh, North Carolina, United States.

Speece, R. E. (1972). Anaerobic treatment. In E. L. Thackston and W. W. Eckenfelder (Eds.), *Process design in water quality engineering.* Jenkins Publishing Company.

Speece, R. E. (1996). *Anaerobic Biotechnology for Industrial Wastewater.* Archae Press, Nashville.

Stege, K., & Bailey, E. (2003) *Aerobic digestion operation of Stockbridge Wastewater Treatment Plant, GA.* Georgia Water Pollution Control Association.

Stenstrom, M. K., & Rosso, D. (2008). Aeration and mixing. In M. Henze, M. C. M. van Loosdrecht, & G. A. Ekama (Eds.), *Biological wastewater treatment: Principles, modeling and design*, Chapter 9. International Water Association.

Stone, L. A., et al. (1992). *The historical development of alkaline stabilization.* Paper presented at the Water Environment Federation Specialty Conference on Future Directions in Municipal Sludge (Biosolids) Management: Where We Are and Where We're Going, Portland, Oregon, United States.

Switzenbaum, M. S., Farrell, J. B., & Pincince, A. B. (2003). Relationship between the Van Kleeck and mass-balance calculation of volatile solids loss. *Water Environment Research, 75*(4), 377–380. http://www.jstor.org/stable/25045709

Tortorici, L., & Stahl, J. F. (1977). Waste activated sludge research. *Proceedings of the sludge management, disposal, and utilization conference*, Rockville, Maryland, United States. Information Transfer.

U.S. Environmental Protection Agency. (1978). *Sludge treatment and disposal—Volume I* (EPA-625/4-78-012). Technology Transfer; U.S. Environmental Protection Agency.

U.S. Environmental Protection Agency. (1979). *Process design manual for sludge treatment and disposal* (EPA-625/1-79-011).

U.S. Environmental Protection Agency. (1982). *Guide to the disposal of chemically stabilized and solidified waste* (SW-872). Office of Solid Waste Emergency Response, U.S. Environmental Protection Agency.

U.S. Environmental Protection Agency. (1990). *Autothermal thermophilic aerobic digestion of municipal wastewater sludge* (EPA-625/10-90-007).

U.S. Environmental Protection Agency. (1999). Standards for the use or disposal of sewage sludge. *Code of Federal Regulations*, Part 503, Title 40.

U.S. Environmental Protection Agency. (2003). *Control of pathogens and vector attraction in sewage sludge* (EPA-625/R-92013).

U.S. Environmental Protection Agency. (2007). *Alkaline treatment of municipal wastewater treatment plant sludge technical guide.* National Risk Management Research Laboratory, Office of Research and Development, U.S. Environmental Protection Agency.

Volpe, G., Keaney, J., Schlegel, P., Tyler, C., Carr, J., & Nagel, J. (2004, October 2–6). Large egg-shaped digesters: Issues and improvements. *Proceedings of the 77th Annual Water Environment Federation Technical Exhibition and Conference* [CD-ROM], New Orleans, Louisiana, United States.

Water Environment Federation. (2023a). *Solids process design and management* (2nd ed.).

Water Environment Federation. (2023b). *Standards for operation of water resource recovery facilities* (Manual of Practice No. 11).

Wiley, J. S., & Westerberg, S. C. (1969). Survival of human pathogens in composted sewage. *Applied Microbiology, 18*, 944.

Zahller, J. D., Bucher, R. H., Ferguson J. F., & Stensel, H. D. (2005, October 29–November 2). Performance and stability of two-stage anaerobic digestion. *Proceedings of the 78th Annual Water Environment Federation Technical Exhibition and Conference* [CD-ROM], Washington, DC, United States.

7.0 SUGGESTED READINGS

Abu-Orf, M. M., Griffin, P. P., & Dentel, S. K. (2001). Chemical and physical pretreatment of ATAD biosolids for dewatering. *Water Science and Technology*, 44(10), 309–314.

Agarwal, S., Abu-Orf, M., & Novak, J. T. (2005). Sequential polymer dosing for effective dewatering of ATAD sludges. *Water Research, 39*, 1301–1310.

Al-Ghusain, I., Hamoda, M. F., & El-Ghany, M. A. (2004). Performance characteristics of aerobic/anoxic sludge digestion at elevated temperatures. *Environmental Technology, 25*, 501–511.

American Society of Heating, Refrigerating, and Air Conditioning Engineers. (2013). *ASHRAE handbook fundamentals, inch-pound edition.*

Avallone, E. A., & Baumeister, T. (1996). *Marks' standard handbook for mechanical engineers* (10th ed.). McGraw-Hill.

Brinkman, D. G., & Voss, D. (1997). *Egg-shaped digesters: Are they all they're cracked up to be? Operational survey*. Black and Veatch.

Bird, E. B., Stewart, W. E., & Lightfoot. E. N. (1960). *Transport phenomena.* Wiley & Sons.

Chen, Y., Higgins, M. J., Maas, N. A., Murthy, S. N., Toffey, W. E., & Foster, D. J. (2005). Roles of methanogens on volatile organic sulfur compound production in anaerobically digested wastewater biosolids. *Water Science and Technology*, *52*, 67–72.

Christensen, G. L. (1987). Lime stabilization of wastewater sludge. In K. A. Gutschick (Ed.). *Lime for environmental uses.* American Society for Testing and Materials.

De la Rubia, M.A., Riau, V., Raponso, F., & Borja, R. (2013). Thermophilic anaerobic digestion of sewage sludge: focus on the influence of the start-up. A review. *Critical Review of Biotechnology, 33*(4), 2013.

Gebreeyessus, G. D., & Jenicek, P. (2016, June 18). Thermophilic versus mesophilic anaerobic digestion of sewage sludge: A comparative review. *Bioengineering*. www.mdpi.com/2306-5354/3/2/15/pdf

Golueke, C. G. (1977). *Biological reclamation of solid waste.* Rodale Press.

Green, D. W., & Perry, R. H. (2007). *Perry's chemical engineers' handbook* (8th ed.). McGraw-Hill.

Griffin, M. E., McMahon, K. D., Mackie, R. I., & Raskin, L. (2000). Methanogenic population dynamics during start-up of anaerobic digesters treating municipal solid waste and biosolids. *Biotechnology and Bioengineering, 57*, 342–355.

Hartman, R. B., Smith, D. G., Bennett, E. R., & Linstedt, K. D. (1979). Sludge stabilization through aerobic digestion. *Journal of the Water Pollution Control Federation, 51*, 2353.

Haug, R. T., LeBrun, T. I., & Tortoriei, L. D. (1983). Thermal pretreatment of sludges—A field demonstration. *Journal of the Water Pollution Control Federation, 55*, 23–34.

Iacoboni, M. D., Leburn, T. J., Lingston, J. (1980). Deep windrow composting of dewatered sewage sludge. *Proceedings of the National Conference of the Municipal and Industrial Sludge Composting Hazardous Materials Control Research Institute*, Silver Spring, Maryland, United States, 88–108.

Jewell, W. J., & Kabrick, R. M. (1980). Autoheated aerobic thermophilic digestion with air aeration. *Journal of the Water Pollution Control Federation, 52*, 512.

Jones, R., Parker, W., Zhu, H., Houweling, D., & Murthy, S. (2008, October 18–22). Predicting the degradability of waste activated sludge. *Proceedings of the 81st Annual Water Environment Federation Technical Exhibition and Conference* [CD-ROM], Chicago, Illinois, United States.

Kabouris, J., et al. (2013, October 5–9). Positive business case for codigestion and resource recovery facilities in Australia without pre-existing anaerobic digesters. *Proceedings of the 86th Annual Water Environment Federation Technical Conference and Exhibition*, Chicago, Illinois, United States.

Kelly, H. G., Mavinic, D. S., Trueblood, B., Zhou, J., Hystad, B., Frese, H., & Cheshuk, J. (2003, October 11–15). Autothermal thermophilic aerobic digestion research application and operational experience. *Proceedings of the 76th Annual Water Environment Federation Technical Exhibition and Conference*, Workshop W104—Thermophilic Digestion: Hot Update!, Los Angeles, California, United States.

Lawrence, A. W., & McCarty, P. L. (1969). Kinetics of methane fermentation in anaerobic treatment. *Journal of the Water Pollution Control Federation, 41*, 1–17.

Matsch, L. C., & Drnevich, R. F. (1977). Autothermal aerobic digestion. *Journal of the Water Pollution Control Federation, 49*, 296.

McHugh, S., Carton, M., Mahony, T., & O'Flaherty, V. (2006). Methanogenic population structure in a variety of anaerobic bioreactors. *FEMS Microbiology Letters, 219*, 2297–2304.

Muller, C. D., & Novak, J. T. (2007, April 15–18). The influence of anaerobic digestion on centrifugally dewatered biosolids. *Proceedings of the 21st Annual Water Environment Federation Residuals and Biosolids Conference*, Denver, Colorado, United States.

Murthy, S. N., Novak, J. T., & Holbrook, R. D. (2000). Optimizing dewatering of biosolids from autothermal thermophilic aerobic digesters (ATAD) using inorganic conditioners. *Water Environment Research, 72*, 714–721.

Murthy, S. N., Novak, J. T., Holbrook, R. D., & Surovik, F. (2000). Mesophilic aeration of autothermal thermophilic aerobically digested biosolids to improve plant operations. *Water Environment Research, 72*, 476–483.

Nielsen, H. B., & Angelidaki, I. (2008). Strategies for optimizing recovery of the biogas process following ammonia inhibition. *Bioresource Technology, 99*, 7995–8001.

Pavlostathis, S. G., & Gossett, J. M. (1988). Preliminary conversion mechanisms in anaerobic digestion of biological sludges. *Journal of Environmental Engineering, 114*, 575–592.

Pavlostathis, S. G., & Gossett, J. M. (2004). A kinetic model for anaerobic digestion of biological sludge. *Biotechnology and Bioengineering, 28*, 1519–1530.

Poincelot, R. P. (1975). *The biochemistry and methodology of composting, Bulletin 754*. Connecticut Agricultural Experiment Station.

Reimers, R. S., Little, M. D., Englande, A. J., Leftwich, D. B., Bowman, D. D., & Wilkinson, R. F. (1981). *Parasites in southern sludge and disinfection by standard sludge treatment* (EPA-600/2-81-166). U.S. Environmental Protection Agency.

Rittmann, B. E., & McCarty, P. L. (2001). *Environmental biotechnology: Principles and applications*. McGraw-Hill.

Salsali, H. R., & Parker, W. J. (2007). An evaluation of 3 stage anaerobic digestion of municipal wastewater treatment plant sludges. *Water Practice, 1*, 1–12.

Sanders, W. T. M., Geerink, M., Zeeman, G., & Lettinga, G. (2000). Anaerobic hydrolysis kinetics of particulate substrates. *Water Science and Technology, 41*, 17–24.

Scisson, J. P. (2009, May 3-6). "As good as the hype": An overview of the second generation ATAD performance. *Proceedings of the Residuals and Biosolids Management Conference 2009*, Portland, Oregon, United States. Water Environment Federation.

Stensel, H. D., & Coleman, T. E. (2000). Assessment of innovative technologies for wastewater treatment: Autothermal aerobic digestion (ATAD), (Preliminary report, Project 96-CTS-1). U.S. Environmental Protection Agency.

Stutzenberger, F. J. (1971). Cellulase production by *thermomonospora curvata* isolated from municipal solid waste compost. *Applied Microbiology, 22*(2), 147.

Van Lier, M. J. B., Martin, M. J. L. S., & Lettinga, M. G. (1996). Effect of temperature on the anaerobic thermophilic conversion of volatile fatty acids by dispersed and granular sludge. *Water Research, 30*, 199–207.

Water Environment Federation. (2013). *Wastewater treatment process modeling* (2nd ed.; Manual of Practice No. 31).

Waksman, S. A., & Cordon, T. C. (1939). Thermophilic decomposition of plant residues in composts by pure and mixed cultures of microorganisms. *Soil Science, 47*, 217.

Westphal, A., & Christensen, G. L. (1983). Lime stabilization: Effectiveness of two process modifications. *Journal of the Water Pollution Control Federation, 55*, 1381.

Wilson, C. A., Fang, Y., Novak, J. T., & Murthy, S. N. (2008). The effect of temperature on the performance and stability of thermophilic anaerobic digestion. *Water Science and Technology, 57*, 297–304.

Wilson, C. A., Murthy, S. N., & Novak, J. T. (2008, March 30–April 2). Laboratory digestibility study of wastewater sludge treated by thermal hydrolysis. *Proceedings of the 22nd Annual Water Environment Federation Residuals and Biosolids Conference: Traditions, Trends & Technologies*, Philadelphia, Pennsylvania, United States.

Zhou, J., Mavinic, D. S., Kelly, H. G., & Ramey, W. D. (2002). Effects of temperature and extracellular proteins on dewaterability of thermophilically digested biosolids. *Journal of Environmental Engineering and Science, 1*, 409–415.

Chapter 13 | Chemical Conditioning and Thermal Processing

LEARNING OBJECTIVES

At the end of this chapter, students should be able to:

- Describe what solids conditioning is, what it does, and why it is used.
- Describe how conditioning changes residuals being processed.
- Describe the components of conditioning systems.
- Determine what kind of chemical conditioning is needed for solids streams.
- Determine an appropriate dose for chemical conditioning for different solids thickening and dewatering processes.
- Size a conditioning system's process components based on solids type and production rates.

1.0 INTRODUCTION

The importance of municipal wastewater sludge processing and disposal has grown since the establishment of secondary treatment standards by the Clean Water Act of 1972 and subsequent construction of water resource recovery facilities (WRRFs) that generate increasingly larger volumes of sludge. At the same time, increasingly strict regulations governing wastewater solids disposal combined with the decreasing availability of disposal sites have limited disposal options and increased disposal costs. In general, the goals of solids handling include reducing the volume, reducing the mass, reducing the percentage of organic matter, inactivating or reducing the number of pathogens, and meeting the requirements of the 40 CFR Part 503 or Part 258 regulations.

It is preferable to maximize the dry solids content of solids before stabilization and then before trucking and final disposal (Figure 13.1). Therefore, conditioning of solids is used to facilitate processing by reducing the water balance in the solids-water matrix (Figure 13.2).

Conditioning alters the physical properties of solids to facilitate the release of water during thickening and dewatering. Mechanical thickening and dewatering techniques are often uneconomical for a utility without chemical conditioning upstream. This chapter presents two approaches to conditioning. The first approach is chemical conditioning. The second approach is thermal conditioning.

Conditioning is a process whereby WRRF solids are treated with chemicals or various other means (e.g., thermal processing) to prepare the sludge for thickening or dewatering.

Conditioning is a treatment used to improve the efficiency of downstream processes (e.g., thickening or dewatering). Chemical conditioning processes use inorganic chemicals, organic polymers, or both to improve solids' thickening and dewatering characteristics. Physical conditioning techniques (e.g., thermal conditioning) use heat to condition and stabilize solids.

This chapter presents theory and design considerations. Some thickening and most dewatering methods, particularly those containing solids from biological treatment processes (e.g., fixed-film and suspended-growth activated solids treatment systems), are not practical without some type of conditioning.

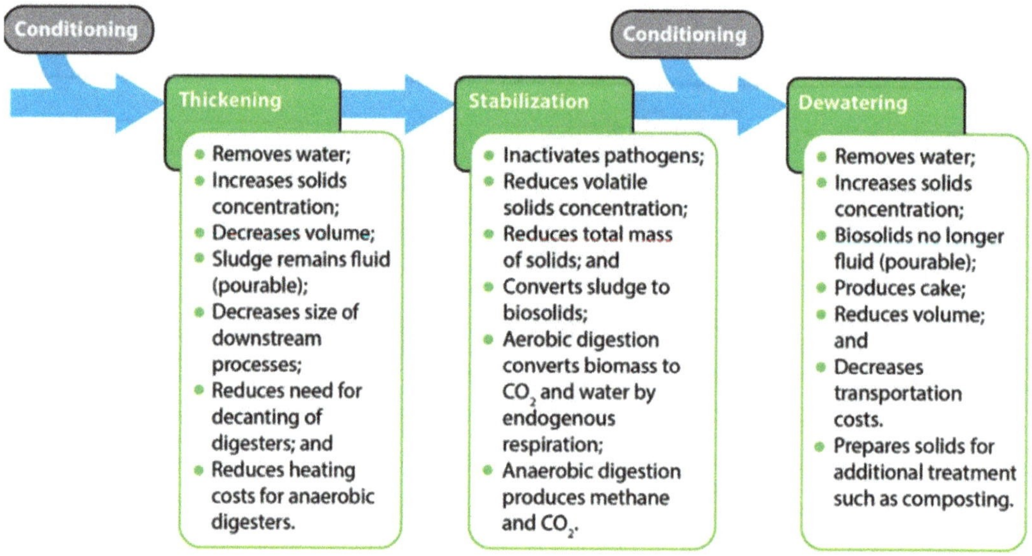

Figure 13.1 Typical Steps in Solids Processing

Conditioning can significantly alter the characteristics of the incoming solids, and the effectiveness of thickening or dewatering processes. Typically, chemical conditioning paired with the appropriate thickening and dewatering process can increase the total solids content from as low as 0.4% before thickening to between 18% and 40% after dewatering.

In the United States, most WRRFs use polymers for thickening and dewatering optimization, and no longer use inorganic chemicals like lime or metal salts (e.g., ferric chloride, alum). These chemicals are now more frequently used for other types of treatment such as stabilization or phosphorus treatment.

The use of organic polyelectrolytes, or polymers, in municipal WRRFs was introduced during the 1960s and rapidly adopted for both thickening and dewatering processes. The primary advantage of polymers is that they do not significantly increase solids production; in contrast, every kilogram of inorganic chemicals (lime, ferric, alum) added during conditioning creates an additional kilogram or more of solids that must be managed.

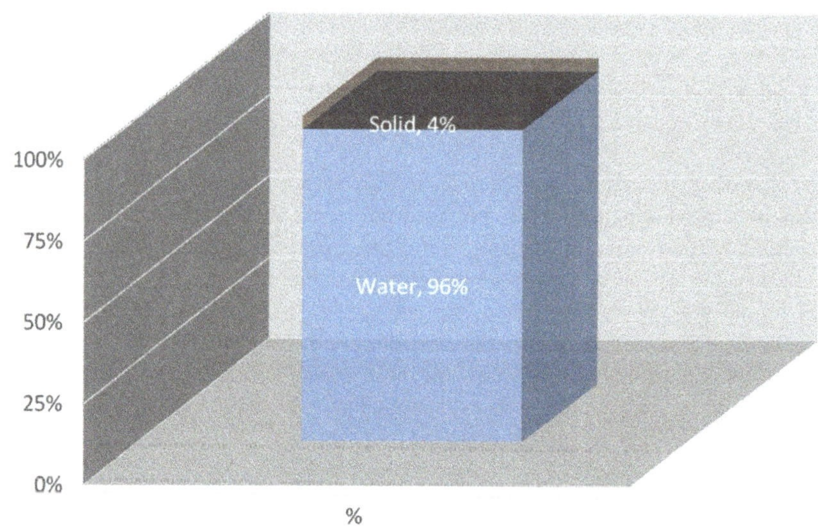

Figure 13.2 Solids-Water Matrix

Polymers are intended to increase the capture rate and reduce the total solids in the sidestreams, or reject water streams such as filtrate, centrate, or supernatant sent back to the liquid process. Additionally, polymers have less of an effect on the water chemistry than do lime or metal salts. Metal salts and inorganic chemicals such as lime have a significant effect on the overall alkalinity, pH, and hardness of the liquid fraction of the solids.

The conditioning method must be compatible with the proposed thickening or dewatering method. For example, centrifuges use centrifugal force to compact the solids. On the other hand, belt filter presses permit the water to pass through the void spaces utilizing gravity and pressure. Due to the different physical approach to water separation, a single type of conditioning agent is not expected to be useful for all applications (Water Environment Federation [WEF], 2014). Furthermore, varying types of organic and synthetic polymers are developed for use with solids of varying characteristics and water chemistry. Cross-linked polymers and varying charges and molecular weights of cationic and anionic polymers exist to suit the needs of the unique characteristics of the residuals generated and the thickening or dewatering process used.

For final disposal of solids after dewatering, inorganic chemicals such as lime can be added to stabilize solids and increase the final product total solids to 20% to 30% (dry solids). Fly ash (as permitted) or other bulking agents can also be added increase the final product total solids to 50% to 100% (dry solids) and create a product more suitable for certain end uses.

Interest continues in thermal processing methods as a means of producing a more marketable product. **Thermal processing** acts as a complement to chemical conditioning and it produces an improved product by increasing the stability of the residuals and reducing the mass and volume of biosolids generated. Economic, environmental, and sociopolitical analyses provide the best basis for deciding whether to use thermal processing methods. Such an analysis considers all available reuse and disposal options and their costs and revenues. Costs to consider include process costs, energy costs, and the cost of further handling and managing solids or residuals (hauling, tipping fees, and so on). Revenues include those generated by production of a marketable product (fertilizer, electricity, fuel, etc.).

Thermal processing is a conditioning process that uses high temperatures to improve the dewaterability of WRRF residuals or biosolids; typically applied as part of the stabilization process.

The selection of a thermal processing system should recognize new, rapid advancements in dewatering technology. Some of these advances may eliminate the future need for, or economic feasibility of, some thermal processing alternatives. Further, the selection process must weigh the likely effect of present and future regulations on the thermal processing system, associated air pollution control equipment, current and future energy prices, and the reuse or disposal option. The process should also assess whether the options are flexible enough to accommodate likely regulation changes.

A sound analysis of thermal processing options requires realistic expectations for advanced technology. Some existing sludge-processing systems were planned with high expectations for advanced technology that have often not been realized. As a result, the best reuse or disposal option available for these facilities may have been overlooked.

The design and procurement of a thermal processing system often depends on proprietary vendor information, which may not be readily accessible. As a result, a designer should avoid specifying equipment or processes in a manner that may unnecessarily restrict some vendors from bidding on the project. In some cases, however, restrictive specifications reflect good engineering judgment. An engineer must set basic requirements such as solids and liquid loading capacities to the unit, types of support equipment, and minimum standards for materials.

2.0 CONDITIONING FACTORS

The type and dosage of conditioning agent needed depends on the characteristics of the residuals, and the solids handling and processing before and after conditioning. In addition, it is important to provide the right type of mixing process after adding the chemical conditioning agent in order to ensure complete mixing of the agent through the combined mixture.

Solids residuals' characteristics that affect conditioning requirements (adapted from U.S. Environmental Protection Agency [U.S. EPA], 1979b) include the following:

- Source of the residuals,
- Concentration,
- Alkalinity and pH,
- Biocolloids and biopolymer content,
- Particle size distribution,
- Degree of hydration,
- Particle surface charge,
- Volatile suspended solids (VSS) content, and
- Phosphate content.

Several of these characteristics are discussed individually; however, the combined effect is not always directly additive.

2.1 RESIDUALS SOURCE

The biosolids matrix is composed of solids, water, and gas when dewatered. It can be viewed in different ways (Figure 13.3), but it is fundamentally composed of volumes and masses associated with the solids, water, and gas. The water and gas fill voids between solids. The goal of conditioning is to support removal of as much water as possible while reducing voids.

To some extent, the conditioning method depends on the type of solids that must be treated (Figure 13.4). Particle size distribution affects dewaterability due to the surface area and space between particles, and a review of particle size distribution may provide insight into dewaterability and conditioning requirements (Figure 13.5). For example, cationic polymers typically work best on primary, secondary, and digested solids, while anionic polymers may be effective with inorganic solids or negatively charged solids particles (Figure 13.6).

An examination of published data for a variety of thickening and dewatering devices suggests that primary solids require lower doses than secondary solids do, and that fixed-film secondary solids require lower doses than suspended-growth secondary solids (U.S. EPA, 1979b). Depending on the thickening or dewatering method used, aerobically and anaerobically digested solids typically require conditioning doses comparable to those for secondary solids. Similarly, combined solids (primary and secondary solids)

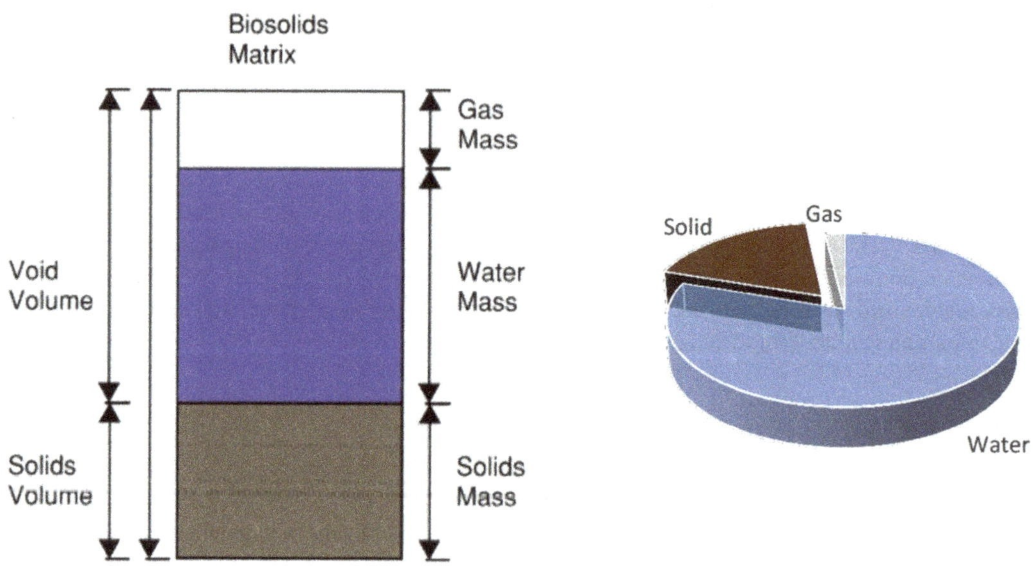

Figure 13.3 Biosolids Matrix

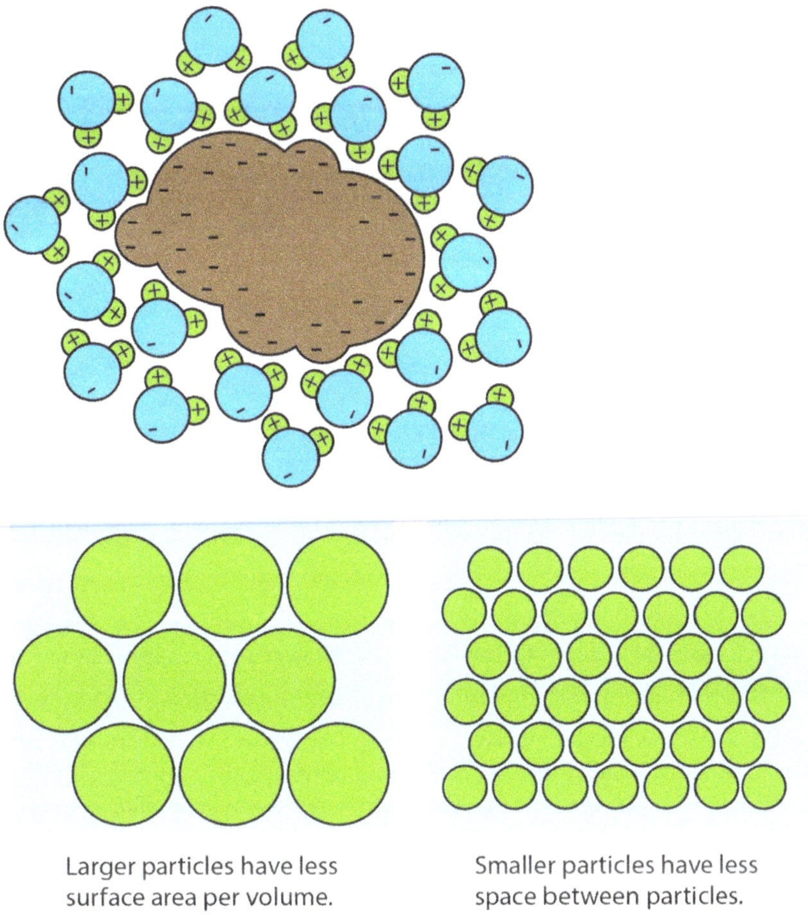

Figure 13.4 Particle Charge and Size Affect Surface Area and Space Between Particles (Reprinted with permission from Indigo Water Group)

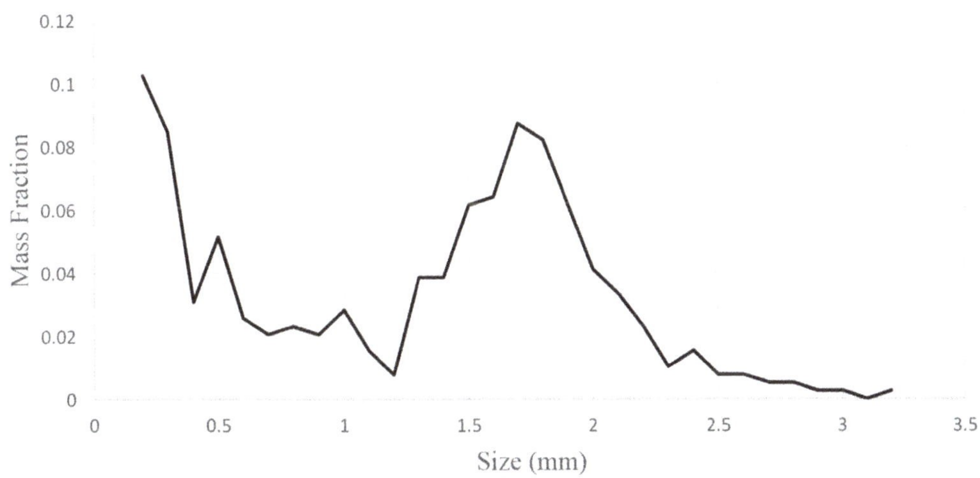

Figure 13.5 Example Particle Size Distribution Plot (Larger Than 0.1 mm)

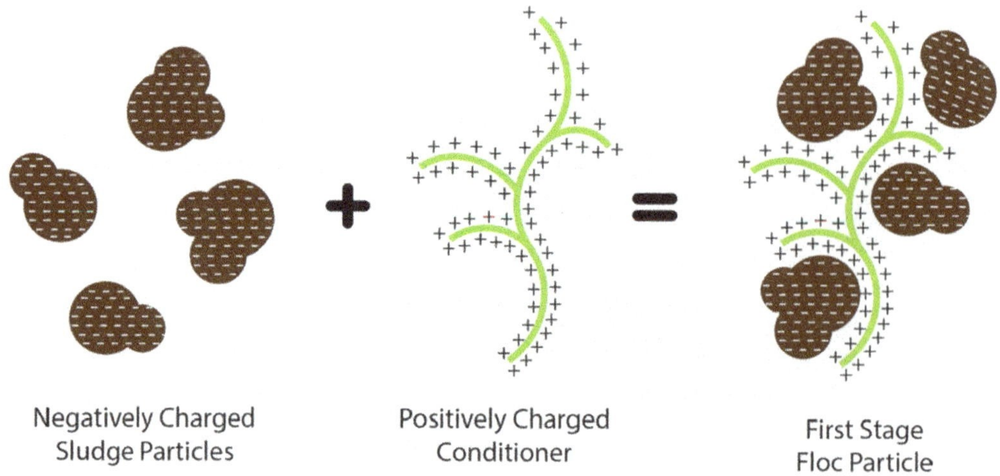

Figure 13.6 Sludge Conditioning (Drawn by Indigo Water Group based on figure from Jennett & Santry, 1969)

have properties that are closer to those of secondary solids, although they are affected by the respective composition of each type.

Equally important is that characteristics of solids vary from facility to facility and can also vary seasonally. The conditioner dose depends on the specific conditioning agent used, the process applied (thickening or dewatering), and the goal (solids content and capture rate). Constituents like sodium or phosphate can also affect the dewaterability of sludge, such that adding more polymer or different types of polymers has negligible benefit. Also, some sludges cannot be dewatered beyond a certain point due to the structure of the solids. For example, typical waste activated sludge is limited to 20% total solids due to bacteria structures' inherent ability to bind water at 4 to 5 parts water to 1 part solids (mass basis).

Chemical solids are solids that have been mixed with an inorganic conditioning agent (e.g., the addition of aluminum, iron salts, or lime). These types of solids can be more variable than traditional primary and secondary solids and, as a result, are hard to categorize with respect to dose. The conditioning requirements of chemical solids are often qualitatively different from those for primary and secondary solids. For example, adding lime to mixed liquor suspended solids (MLSS) before secondary clarification may improve suspended solids removal (in the clarifier); however, the resulting solids may require an anionic polymer instead of the cationic polymer typically used for primary or secondary solids. This change in polymer requirement results from the positively charged calcium neutralizing part of the negative surface charge from the biological process. Although charge neutralization is a fundamental part of the process, an equally significant piece is the interconnection of solids particles via the polymer chain.

Active refers to the concentration of polymer added to biosolids condition based strictly on the polymer content; i.e., excluding the water in the as-received polymer solution.

2.2 SOLIDS CONCENTRATION

In many applications, conditioning neutralizes the surface charge in the solids with oppositely charged organic polymers or inorganic complexes. Therefore, the residuals' solids concentration affects the dosage and dispersal of the conditioning agent. For a given particle size distribution, increasing the suspended solids concentration increases the required coagulant dose for effective surface coverage (on a volumetric basis). For this reason, polymer dosing and performance is typically discussed on a dry-weight basis; for example, 5 kg of active polymer per 1000 dry kg of solids (10 lb per dry ton). Note that the terms "active" and "neat" are used when referring to polymer concentrations. The active polymer content of a polymer solution refers to the quantity of pure polymer (or useful, reactive polymer) in the solution. A neat solution, on the other hand, is the as-received polymer solution before accounting for the active content of the polymer. Concentrations provided in this chapter are all presented as active polymer content.

Neat refers to the concentration of polymer solution added to biosolids conditioning system; i.e., including water in the as-received polymer solution.

The suspended solids concentration of the residuals also affects two additional aspects of conditioning agents. First, the process is less susceptible to overdosing at higher solids concentrations. Second, the solids and the conditioning agent are more difficult to mix at higher solids concentrations.

2.3 ALKALINITY AND PH

When inorganic conditioning agents are used, alkalinity and pH are the most important chemical parameters affecting conditioning. Coagulation occurs when the coagulant interacts with the surface of colloids in the solids. It is important to recognize that the surface chemistry of the solids and the charge of the polymer are both pH-dependent. When inorganic conditioners are used, the solids' pH determines which chemical species are present.

Iron and aluminum salts behave like acids when added to water (i.e., these conditioners reduce pH), so the pH of the conditioning process depends on the solids' alkalinity and the dosage of iron or aluminum salts. The pH, in turn, determines the predominant coagulant species and the nature of the charged colloidal surface. The high alkalinity typically associated with anaerobically digested solids is one reason for the higher coagulant doses required. Low-molecular-weight coagulants tend to be more effective over a broader pH range than inorganic conditioning agents.

2.4 BIOCOLLOIDS AND BIOPOLYMERS

Although more commonly measured by researchers than by design engineers, the concentrations of biocolloids and biopolymers represents fundamental parameters that provide insight into the conditioning process. The biopolymers in activated solids flocs affect the physico-chemical properties of flocs (e.g., floc density, floc particle size, specific surface area, charge density, bound water content, and hydrophobicity).

Studies have shown that cations can affect bioflocculation and change the settling and dewatering properties of activated sludge flocs (Bruus et al., 1992; Eriksson & Alm, 1991; Higgins & Novak, 1997a, 1997b). Divalent cations bridge across negatively charged biopolymers to form a dense, compact floc structure. Monovalent cations tend to prevent proper flocculation by forming a much weaker structure. As a result, divalent cations promote bioflocculation and produce subsequent improvements in settling and dewatering properties, while monovalent cations tend to degrade settling and dewatering properties. Settling and dewatering properties are further improved when the two divalent cations are added to the feed rather than superficially added to the settling tank (Higgins & Novak, 1997a).

2.5 PARTICLE SIZE AND DISTRIBUTION

Particle size distributions affect the total particle surface area and the porosity of cakes formed from these particles. These properties affect required coagulant doses and dewaterability. Several researchers (Karr & Keinath, 1978; Novak et al., 1988; Sorensen & Sorensen, 1997) studied the effect of particle size on dewaterability and concluded that particle size was one of the most important parameters determining dewaterability. Smaller particles (colloidal and supracolloidal) can blind filters and deter the release of water from the solids cake. Also, another study has suggested that an increase in floc density improves dewatering properties via a decrease in bound water associated with the flocs (Kolda, 1995). These studies concluded that dewaterability improvements often associated with other factors (e.g., pH, mixing, biological degradation, and conditioning) all could be explained by the effects of these factors on particle size distributions.

2.6 DEGREE OF HYDRATION

Excessive bound water has been suggested as the cause of dewatering difficulties. The percent of bound water associated with the floc also indicates the maximum dryness that can be achieved in the solids cake by mechanical means (Robinson, 1989). Additionally, Vesilind (1979) reviewed the work of several investigators on water distribution in activated sludge. This water was described as free water, interstitial (floc) water, vicinal (capillary) water, and bound water (water of hydration). These categories are defined based on the amount of centrifugal acceleration required to release a given portion of water. Vesilind (1979) suggested that the water distribution in solids could determine the applicability of a specific thickening or dewatering operation.

2.7 PARTICLE SURFACE CHARGE

Solids particles (e.g., subcolloidal and macromolecular constituents) typically have a negative surface charge, so they tend to repulse each other. The resulting spaces between these constituents are occupied by cations and water. If the charge can be eliminated, thickening or dewatering improves. This is why chemical conditioners are positively charged, or become positively charged when added to water.

In most cases, polymer conditioning is optimal when the charge is neutralized, so measuring charge can be useful in laboratory comparisons of polymers and doses. A streaming current detector (zeta meter) can be used to measure this charge. It also can be used for real-time monitoring or control of polymer dosing in dewatering processes (Dentel et al., 1995).

Shear during mixing or dewatering tends to open up new negative surfaces in the biocolloids, thereby undoing the charge effects of cationic polymers. Therefore increases in mixing shear can increase the required polymer dose (Dentel, 2001).

2.8 WASTEWATER CATIONS

Numerous studies have suggested that cations interact with the negatively charged biopolymers in activated solids to change the structure of the floc (Bruus et al., 1992; Eriksson & Alm, 1991; Higgins, 1995; Novak & Haugan, 1979; Tezuka, 1969). One study indicated that monovalent cations tend to deteriorate settling and dewatering characteristics, whereas divalent cations tend to improve them (Higgins, 1995). The effects of charge density on activated sludge properties could decrease the polymer dose needed to condition secondary solids.

Researchers have studied influent concentrations of cations (e.g., aluminum, ammonium, calcium, iron, magnesium, potassium, and sodium) extensively. They postulated that cations play a critical role in bioflocculation. Cations have been found to influence the thickening and dewatering characteristics of biological solids. For example, high concentrations of sodium typically resulted in poor dewatering; however, if the floc contained enough aluminum and iron concentrations, it typically offset the deleterious effects of sodium (Park et al., 2006). The data associated with aluminum further revealed that waste activated sludge (WAS) with low aluminum levels contained high concentrations of soluble and colloidal biopolymer (proteins and polysaccharides), resulting in a high effluent chemical oxygen demand concentrations, the need for larger doses of conditioning chemical, and poor solids dewatering properties. Studies have shown that iron may contribute to floc strength, and it seems that the reduction and solubilization of iron during anaerobic digestion may be a reason why digested solids dewater poorly. Also, the presence of proteins in solution contributes to poor dewatering and larger conditioning chemical doses (Novak et al., 2001).

2.9 RHEOLOGY

Many studies have focused on the rheological properties of wastewater solids in an attempt to correlate solids properties with chemical conditioning requirements (Ormeci et al., 2004). For example, yield strength and viscosity have been used to optimize chemical conditioning. Another study demonstrated that mixing considerably affected the rheological characteristics of conditioned solids (Abu-Orf & Dentel, 1999). Bache and Dentel (2000) showed that solids conditioning could be improved by monitoring centrate or filtrate viscosity. Both laboratory- and full-scale testing showed that the network strength of the sludge could be used to optimize chemical conditioning and achieve drier solids (Abu-Orf & Ormeci, 2005). In general, the type of conditioner (e.g., polymers or fly ash) and wastewater solids (e.g., chemical, WAS, or biosolids) involved will determine which rheological parameters affect conditioning needs.

3.0 HANDLING AND PROCESSING SOLIDS

The efficiency of any conditioning process depends to a large degree on chemical and physical characteristics (e.g., origin, solids concentration, inorganic content, chemistry, storage time, and mixing) of the solids before conditioning. The solids' physical characteristics are a function of the physical stresses they were exposed to before conditioning. For example, any process that damages the flocculant nature of solids particles typically either increases chemical conditioning demand or reduces performance in the final treatment stage. The extent of mixing and shear stress before and after conditioning can significantly affect conditioning efficiency and, ultimately, solids treatment performance.

3.1 STORAGE

There are two types of storage for liquid residuals: long-term and short-term. Long-term storage may occur in stabilization processes with long detention times or in specially designed tanks. Long-term storage is on the order of weeks to months. Short-term storage may occur in wastewater treatment process (e.g., increasing solids inventory) or in smaller, specially designed tanks. Short-term storage may be as short as a couple of hours, or as long as a few days. Storage helps smooth out fluctuations in solids production and make the solids feed rate more uniform. It also provides a place to keep solids during equipment downtime. Most important in terms of conditioning, long-term storage has been reported to negatively affect the dewaterability of solids.

Unstabilized solids that have been stored for long periods typically require more conditioning chemicals than fresh solids because the degree of hydration and percentage of fine particles has increased. Also, storing activated sludge increases the sludge's specific resistance to filtration and, subsequently, conditioning requirements (Karr & Keinath, 1978). Storing aerobically or anaerobically stabilized solids for extended periods typically will lower temperature significantly and can change pH and alkalinity. Temperature drops typically increase conditioning requirements. However, if the temperature decrease is small, the negative effect on conditioning may be more than offset by an increase in solids concentration.

3.2 PUMPING

Pumping subjects solids to shear forces; the level of shear depends on the type of pump and the flowrate. Solids particles are fragile, and pumping typically causes some of them to fragment. Researchers have shown that the major demand for chemical conditioning is associated with the fraction of particles in the colloidal and supracolloidal range (Karr & Keinath, 1978; Roberts & Olsson, 1975), so any process that reduces particle size will increase conditioning chemical requirements.

Conditioned solids should not be pumped, because pumping introduces shear forces that tend to break down flocs. If required, however, then the pump should be designed to minimize shear.

3.3 MIXING

During conditioning, the solids and added chemicals must be mixed enough to ensure that the chemical is evenly dispersed throughout the solids. However, the mixer must not break the floc once it has formed. Design engineers should optimize the mixing time with these two goals in mind. Mixing requirements depend on the thickening or dewatering method used. In-line mixers typically are used with most modern thickening and dewatering units. A separate mixing and flocculation tank is provided with some older thickening and dewatering devices.

For many municipal solids, conditioning requires mixing intensity and time to optimize binding of polymer to the solids. A common approach is to evaluate conditioning based on identifying an optimum shear rate (velocity gradient) multiplied by mixing time; that is, the G t value. Shear rates may range from 50 to 500 s^{-1} with mixing times as short as 15 seconds and as long as several minutes (Lynch & Novak, 1991). The G t value can vary over a range from 10 000 to 100 000. Optimization of the G t value is dependent on solids characteristics and the stabilization technology applied. Bench- and/or pilot-testing may be appropriate to find the optimum value. Solids are further flocculated in the pipeline or flocculation tank before entering the thickening or dewatering equipment.

3.4 SOLIDS CONCENTRATION

The solids concentration of biosolids can significantly affect conditioning system performance and cost. In most cases, as the influent solids concentration increases, the conditioning cost decreases to a certain level. However, it becomes increasingly difficult to evenly mix coagulant in residuals containing 4% solids (or more) before further dewatering.

3.5 STABILIZED VERSUS UNSTABILIZED SOLIDS

Polymer requirements also are affected by the type of solids to be conditioned. In fact, this may have the most effect on the quantity of chemical needed. Solids that are difficult to dewater require the largest doses of chemicals, typically yield a wetter cake, and result in poorer quality sidestreams (filtrate, centrate, etc.). The following types of solids are listed in increasing order (approximately) of conditioning chemical requirements (Metcalf & Eddy, Inc./AECOM, 2014):

- Untreated (raw) primary solids,
- Untreated mixed primary and trickling filter solids,

- Untreated mixed primary solids and WAS,
- Anaerobically digested primary solids,
- Anaerobically digested mixed primary solids and WAS,
- Untreated WAS, and
- Aerobically digested primary solids.

Digestion changes the chemical and physical characteristics of solids, increasing alkalinity while reducing mass. However, stabilized solids typically are more difficult to dewater than unstabilized solids. Anaerobically digested solids contain considerably more colloidal and supracolloidal solids than do primary solids or activated sludge (Karr & Keinath, 1978). Aerobic digestion detains solids for 30 days or more, greatly reducing the dewatering characteristics of the resulting biosolids. Digested solids typically have higher specific resistance values (Karr & Keinath, 1978), which, in turn, mean higher chemical conditioner doses to achieve a specified dewatered solids concentration.

Solids with inorganic contents in the range of 15% to 35% (e.g., biological solids) typically have cationic charge-neutralization requirements. Digestion also produces solids with cationic charge-neutralization requirements, although the inorganic solids content may increase to between 30% and 50%. Occasionally, lime-stabilized or chemically treated solids contain higher levels of inorganic solids, which respond better to an anionic or non-ionic polymer. As a general rule, residuals containing less than 50% inorganic solids have a cationic charge demand, while those containing more than 50% inorganic solids have an anionic or non-ionic charge demand (regardless of whether they are accompanied by a pH shift).

Whenever possible, raw, undigested, or unprocessed solids should remain separate from biological or chemically treated solids until just before dewatering. This is especially true if biological solids are generated at a facility that practices biological nutrient removal. Septic conditions cause bacteria to release their bound phosphorus to the filtrate, which then is recycled back to the influent. If such biological solids will be dewatered with primary solids, they should not be mixed until just before they enter the thickening and dewatering device.

4.0 CHEMICAL CONDITIONING

The fundamental objective of chemical conditioning is to cause fine solids to aggregate via coagulation with inorganic or organic coagulants, flocculation with organic polymer, or both (Institute of Water Pollution Control, 1981). It should improve the efficiency of thickening, dewatering, and other subsequent treatment processes. Also, conditioning is a significant item in a solids-management operation and maintenance budget, so it is desirable to select the most cost-effective method that produces acceptable liquid and solid output streams.

To be effective, the conditioning method must be compatible with the proposed methods of solids thickening, dewatering, and ultimate use or disposal. For example, belt filter presses, gravity belt thickeners, and rotary drum thickeners perform better when the solids are a uniform floc size that increases the voids between particles, thus allowing free water to filter more rapidly through the porous belt or drum. Polymer conditioning is the easiest way to produce such a floc. Other dewatering methods (e.g., pressure filtration and sand bed filtration) perform well when the solids are conditioned via the addition of chemical solids (organic and inorganic) or bulking materials.

Design engineers should take all subsequent solids treatment processes into account. For example, if the dewatered solids will be sent to a composting system or thermal dryer, then the conditioning and dewatering systems must produce a cake with maximum solids content. However, in concept, the conditioning and dewatering systems should not reduce the fraction of volatile solids; use an exotic, expensive polymer; or add inorganic chemicals that dramatically increase the volume of material to be dried or composted.

Solids can be conditioned via a number of methods (e.g., chemical, heat, pressure, electrolysis, freeze–thaw). Historically, the most popular has been chemical conditioning (e.g., polymers, inorganic chemicals, or both). Other conditioning methods have been used and have grown in popularity for a variety of purposes (e.g., thermal hydrolysis paired with digestion), discussed in the next section.

In the mid-20th century, wastewater treatment professionals used various inorganic chemicals and natural organics to condition solids. The most common inorganic chemicals used were lime and metal salts. When synthetic organic polymers were introduced in the late 1960s, they were quickly adopted for solids conditioning because they did not significantly increase the amount of solids to be thickened and dewatered, and were able to achieve better capture with less side effects (e.g., fluctuations in alkalinity and pH).

4.1 INORGANIC CHEMICALS

Inorganic chemical conditioning is principally associated with recessed-chamber plate and frame filter presses, although they have also been used for belt filter presses. Lime and liquid ferric chloride are the two most widely used inorganic conditioning agents. These conditioning agents are readily available and can condition a wide range of solids. In addition, the resulting biosolids are suitable for land application or composting. Compared to polymers, larger doses of inorganic chemicals are required to condition solids, and this affects the volume of solids to be managed. For example, adding iron salts and lime can increase the solids mass (and volume) by as much as 20% to 40% (WEF, 2014).

Less commonly used inorganic coagulants include liquid ferrous sulfate, anhydrous ferric chloride, aluminum sulfate, and aluminum chloride. Other inorganic materials (e.g., fly ash, power plant ash, cement kiln dust, pulverized coal, diatomaceous earth, bentonite clay, and sawdust) have been used to improve dewatering, increase cake solids and, in some cases, reduce the required dosage of other conditioning agents.

In addition to increasing the volume of solids to be managed, inorganic chemical conditioners reduce the solids' heat value. However, cake combustibility depends on the ratio of water to dry volatile solids, not the level of chemical precipitates in the cake.

4.1.2 LIME

Lime is often used to control pH and improve settling in wastewater treatment processes, as well as condition and stabilize solids. Lime is commercially available in two main dry forms:

- Pebble quicklime (CaO)
- Powdered hydrated lime [$Ca(OH)_2$]

In either form, lime is caustic, tends to produce dust, and tends to precipitate when slurried, forming a calcium carbonate scale on conveyance equipment.

As a conditioner, lime is typically used to raise the pH, which is lowered by ferric chloride addition. It also forms calcium carbonate and calcium hydroxide precipitates, which improve dewatering by acting like a bulking agent, increasing porosity while resisting compression. Some dissolved calcium hydroxide also is available at high pH levels.

Quicklime is typically 85% to 95% pure; it is typically called calcined lime because it is manufactured by burning crushed limestone (calcium carbonate) in high-temperature kilns to drive off carbon dioxide, leaving calcium oxide (quicklime). It is typically purchased in pebble form to minimize dust problems during handling. However, it rarely is applied in dry form, except to stabilize solids. Instead, it is typically mixed with water and converted to the more reactive hydrated form (calcium hydroxide) before application. This hydration reaction (typically called slaking) emits heat as part of the reaction:

$$CaO + H_2O \rightarrow Ca(OH)_2 + \text{Heat} \tag{13.1}$$

The quicklime pebbles rupture during slaking, splitting into microparticles of hydrated lime, which have a large total surface area and are highly reactive. A high-grade quicklime produces a quick-slaking, highly reactive, calcium hydroxide slurry, while a low-grade quicklime produces a slow-slaking, less reactive slurry (Table 13.1). Low-grade quicklime requires critical water control to maximize slaking efficiency and minimize calcium hydroxide particle size.

Table 13.1 Relative Slaking Ability of Quicklimes (National Lime Association, 1982)

CaO Content (%)	Degree of Burn	Slaking Ability
High	Soft	Very quick
	Normal	Medium
	Over to hard	Medium to slow
Medium	Soft	Quick to medium
	Normal	Medium
	Over to hard	Slow
Low	Soft	Quick to medium
	Normal	Medium
	Over to hard	Slow to very slow

Quicklime must be stored under controlled conditions, because prolonged contact with carbon dioxide in moist air causes quicklime to air slake, cake, and become less reactive. Likewise, air or excessively hard water (alkalinity more than 180 mg/L as calcium carbonate) in a hydrate slurry encourages carbonate scaling, which can eventually lead to plugging problems in conveyance pumps and piping.

Hydrated lime is a powdered form of calcium hydroxide; its composition and characteristics depend on the quality of its parent quicklime (Table 13.2). Hydrated lime typically costs 30% more than quicklime with the same calcium oxide content because of its higher production and transportation costs. However, at small facilities where daily requirements for lime are intermittent or minimal, hydrated lime often is preferred because it does not require slaking. The storage and mixing operations are relatively simple (e.g., typically a dedicated storage area and minimal manual labor). Hydrated lime is more stable than quicklime, so storage precautions are satisfied more easily. However, because of its dusting characteristics, handling is more difficult.

Table 13.2 Characteristics of Quicklime and Hydrated Lime (Wang et al., 2007)

Material	Available Forms	Containers and Requirements	Appearances and Properties	Weight	Commercial Strength	Solubility in Water
Quicklime	Pebble, 6–19 mm	80–100-lb moisture-proof bags, barrels, and container cars Store dry, maximum 60 days in tight container and 3 months in moisture-proof bag	White (light grey to tan) Lumps to powder Unstable caustic irritant slakes to hydroxide slurry evolving heat Saturated volume pH approximately 12.5	3.4–4.7 kg/m³ sp gr: 3.2–3.4	70%–96% CaO	Reacts to form $Ca(OH)_2$; 1 lb of quicklime will form 1.16 to 1.32 lb of $Ca(OH)_2$ with 2%–12% grit, depending on the purity
Hydrated lime	Powder, <200 mesh	50-lb bags, 100-lb barrels, and container cars Store dry, maximum 1 year	White Powder free of lumps. Caustic dust irritant absorbs H_2O and CO_2 to form $Ca(HCO_3)_2$ Saturated volume pH approximately 12.4	1.6–2.5 kg/m³ sp gr: 2.3–2.4	82%–98% $Ca(OH)_2$ 62%–74% CaO	10 lb/1000 gal at 70 °F 5.6 lb/1000 gal at 175 °F

Note. gal × 3.785 = L and lb 0.4536 = kg

4.1.3 FERRIC SALTS

Both ferric chloride and ferric sulfate react with the bicarbonate alkalinity in solids to form ferric hydroxide precipitates. The precipitate can lead to both charge neutralization and floc aggregation. The chemical reaction may be written as follows:

$$Fe^{3+} + 3H_2O \rightarrow Fe(OH)_3 + 3H^+ \tag{13.2}$$

$$2FeCl_3 + 3Ca(HCO_3)_2 \rightarrow 2Fe(OH)_3 + 3CaCl_2 + 6CO_2 \tag{13.3}$$

Ferric-chloride coagulation is pH-sensitive; it works best above pH 6. Below pH 6, floc formation is weak and dewaterability is sometimes poor. Therefore, lime is used to adjust the pH to optimize ferric chloride use and solids dewatering. The acid formed during the reaction causes the pH to drop to 6.0. Adding lime raises the pH as high as 8.5, thus allowing the ferric chloride reaction to be more efficient in forming hydroxides.

Depending on the type of solids involved, the ferric chloride dosage ranges from 2% to 10% (dry solids basis), and lime dosages range from 5% to 40% (dry solids basis). Activated sludge requires high ferric chloride dosages, anaerobically digested solids require mid-range dosages, and fresh raw primary solids require low dosages (Table 13.3).

Typically used to flocculate solids, ferric chloride is sold as an orange-brown liquid—containing between 30% and 40% (by weight) ferric chloride. At 30 °C (86 °F) and a specific gravity of 1.39, a 30% ferric

Table 13.3 Typical Dosages of Ferric Chloride and Lime for Dewatering Wastewater Solids (U.S. EPA, 1979a)

Application	Sludge Type	Ferric Chloride (g/kg)[a]	Lime (g/kg)[a]
Vacuum filter	Raw primary	20–40	70–90
	Raw WAS	60–90	0–140
	Raw (primary + TF[b])	20–40	80–110
	Raw (primary + WAS)	22–60	80–140
	Raw (primary + WAS + septic)	25–40	110–140
	Raw (primary + WAS + lime)	15–25	None
	Anaerobically digested primary	30–45	90–120
	Anaerobically digested (primary + TF)	40–60	110–160
	Anaerobically digested (primary + WAS)	30–60	140–190
Recessed-chamber filter press	Raw primary	40–60	100–130
	Raw WAS	60–90	180–230
	Anaerobically digested (primary + WAS)	40–90	100–270
	WAS + TF	40–60	270–360
	Anaerobically digested WAS	70	360
	Raw primary + TF + WAS	75	180

[a]All values shown are mass of either $FeCl_3$ or CaO per unit mass of dry solids pumped to the dewatering unit.
[b]Trickling filter.

chloride solution typically contains 1.46 kg (3.24 lb) of ferric chloride. Liquid ferric chloride is corrosive, so it must be handled and stored properly. In colder climates, for example, the shipping strength is reduced to prevent a crystalline hydrate from forming on cold rail cars.

Liquid ferric sulfate typically is sold as a reddish-brown liquid—water containing 50% to 60% of ferric sulfate. It is a cationic coagulant and flocculant typically used with another conditioning agent (e.g., lime or polymers). It has been reported that using ferric sulfate before solids thickening or dewatering will reduce the amount of polymer needed and improve the filtrate or centrate quality. However, its use as a solids conditioner is limited; it is primarily used in water and wastewater treatment to remove turbidity, color, suspended solids, and phosphorus.

Ferrous sulfate ($FeSO_4 \cdot H_2O$, also called copperas) is similar to ferric chloride in terms of handling, storage, and stoichiometry. Ferrous sulfate is available in granular form in bags, barrels, and in bulk. The product has a bulk density of about 1000 to 1100 kg/m^3 (62 to 66 lb/cu ft). Dry ferrous sulfate begins to cake when stored at temperatures above 20 °C (68 °F) and further oxidizes and hydrates in moist and humid conditions. Ferrous sulfate should be stored in a dry area, and care should be taken to control dust, which can stain and also irritate skin, eyes, and the respiratory tract. Ferrous sulfate forms an acidic solution, so manufacturer precautions should be followed when storing, feeding, and transporting the material. Ferrous sulfate in granular (dry) form may be fed using gravimetric or volumetric feeding equipment; it also may be fed as a solution.

The effectiveness of ferric coagulation depends on pH and alkalinity. A lower pH favors the formation of positively charged hydroxoiron (III) complexes; a higher pH favors the solid species $Fe(OH)_3(s)$ (Figure 13.7). Because hydroxoiron (III) complexes are effective coagulants, a lower pH should produce better results. In a study conducted by Tenney et al. (1970), ferric iron was most effective between pH 5 and 8, which is near the pH of maximum precipitation (Figure 13.7).

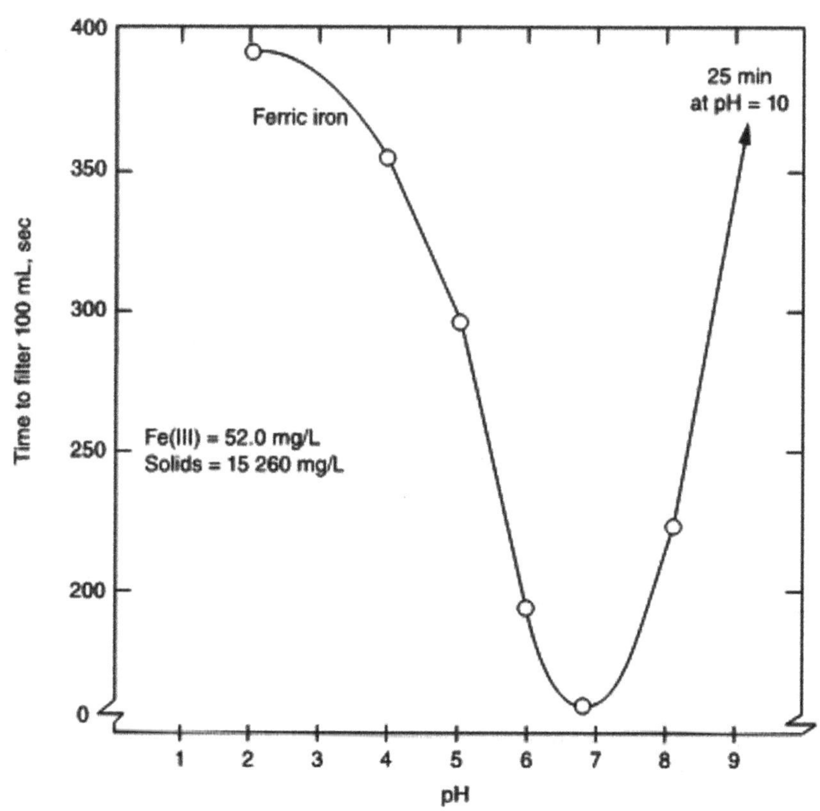

Figure 13.7 Effectiveness of Ferric Iron as a Function of pH (Tenney et al., 1970)

Alkalinity is important in ferric solids conditioning because it controls the solids' pH during conditioning. The ferric ion functions as an acid, lowering pH, while alkalinity maintains the existing pH. For a given solid, the pH decreases as ferric doses increase (Figure 13.8).

4.1.4 FERRIC SALT WITH LIME

Precipitation of $Fe(OH)_3$ can neutralize charge and lead to effective aggregation and filtration in the pH range 6 to 8. Practically, however, the majority of wastewater solids cannot be adequately conditioned unless lime is added after the ferric salt (Christensen & Stulc, 1979). The iron neutralizes and precipitates organic constituents, but the lime creates a much more rigid framework of calcium carbonate, which provides a rigid shell around the organic material. Full-scale filtration involves much more pressure than that typically applied in laboratory tests, which is why lime is required at full-scale tests. The key ingredients are a pH of 11 to 12, a high calcium ion concentration (1022 M), and the presence of solid ferric species (Christensen & Stulc, 1979). Conditioning with ferric salts and lime is not practiced in centrifugation because the solids cannot resist the imposed shear stresses but corrode and abrade metallic surfaces. Ferric should be added before lime at a separate addition point because adding ferric and lime to thickened solids together in the same tank (or in close proximity) adversely affects ferric conditioning (Christensen & Stulc, 1979; Webb, 1974). When conditioning with both ferric chloride and lime, it typically takes two to four times more quicklime than ferric chloride to reach a pH between 11 and 12.

The choice of ferric salt for conditioning is more significant when the ferric salt is followed by lime (Table 13.4). Ferric sulfate followed by lime deteriorates more rapidly and produces a poorer result than ferric chloride followed by lime. The deterioration of solids dewaterability seems to be associated with the formation of insoluble calcium sulfate.

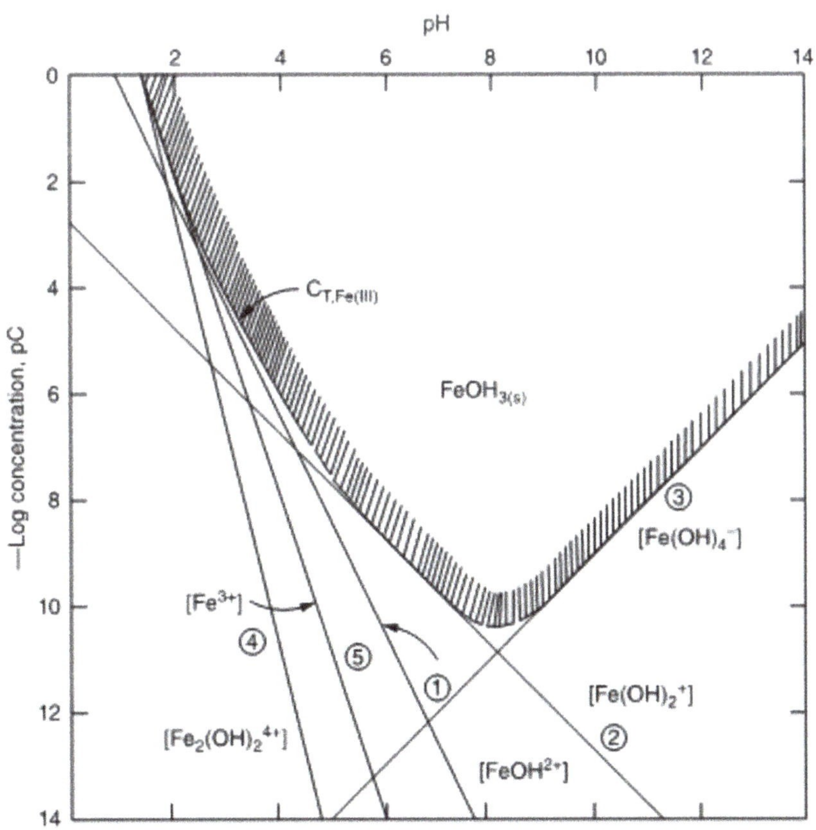

Figure 13.8 Equilibrium Concentrations of Hydroxoiron (III) Complexes in a Solution in Contact With Freshly Precipitated $Fe(OH)_3(s)$ at 25 °C (Snoeyink & Jenkins, 1980; reprinted with permission from Wiley & Sons, Inc.).

Table 13.4 Comparison of Iron Conditioners Used With and Without Lime (Lewis & Gutschick, 1988; reprinted with permission from the National Lime Association)

Sludge[a] Total Solids (%)	Iron Conditioner	Iron Dose (% Fe)	CST[b] After Iron (Seconds)	Lime Dose (% CaO)	Specific Resistance After Line (Tm/kg[c])
5.5	$FeSO_4 \cdot 7H_2O$	1.72	208	15	1.40
5.5	$FeCl_2 \cdot 4H_2O$	1.72	157	15	0.79
5.5	$Fe2(SO_4)_3 \cdot 6H_2O$	1.72	41	15	0.50
5.5	$FeCl_3 \cdot 6H_2O$	1.72	26	15	0.26
5.5	$FeSO_4 \cdot 7H_2O$	3.44	180	30	0.60
5.5	$FeCl_2 \cdot 4H_2O$	3.44	139	30	0.29
5.5	$Fe_2Cl_4 \cdot 6H_2O$	3.44	27	30	0.23
5.5	$FeCl_3 \cdot 6H_2O$	3.44	19	30	0.12
7.0	$FeSO_4 \cdot 7H_2O$	3.44	480	20	1.1
7.0	$FeCl_2 \cdot 4H_2O$	3.44	—	20	0.56
7.0	$Fe_2(SO_4)_3 \cdot 6H_2O$	3.44	117	20	0.53
7.0	$FeCl_3 \cdot 6H_2O$	3.44	58	20	0.18

[a]Mixture of raw primary sludge and WAS
[b]Capillary suction time
[c]Terameters/kg (Tm/kg) = 10_{12} m/kg

4.1.5 ALUMINUM SALTS

Aluminum salts have been used at some facilities with limited degrees of success. Coagulants such as polymerized aluminum chloride and aluminum chlorohydrate that are widely used in the water treatment industry are also being used in the wastewater treatment industry for phosphorus removal and, to a limited degree, solids conditioning. Whereas aluminum salts are not widely used as a solids conditioner in the United States, aluminum chlorohydrates have been a popular conditioner in the United Kingdom for some time.

The primary differences between aluminum and ferric chemistry are the relative solubility of aluminum above pH 7 and the relative insolubility of ferric above pH 7. The practical significance is that ferric hydroxide is relatively insoluble at the highest pH values used in ferric and lime conditioning (i.e., pH 12 to 12.5), whereas aluminum hydroxide is quite soluble above pH 10. Aluminum salts, therefore, are unlikely to be effective with the same lime doses often used with iron salts.

4.2 ORGANIC CHEMICALS: POLYMERS

The organic chemicals used to condition solids are primarily long-chain, water-soluble, synthetic organic polymers. Polyacrylamide, the most widely used polymer, is formed by the polymerization of a monomer acrylamide. Polyacrylamide is non-ionic. To carry a negative or positive electrical charge in aqueous solution, the polyacrylamide must be combined with anionic or cationic monomers. Because most solids carry a negative charge, cationic polyacrylamide copolymers typically are the polymers most used to condition biological solids. Polymers are further categorized by the following characteristics: molecular weight (varies from 0.5 to 18 million), charge density (varies from 0% to 100%), active solids levels (varies from 2 to 100), and form (e.g., dry, liquid or solution, emulsion, gel, or mannich).

High-molecular-weight, long-chain polymers are highly viscous in liquid form, extremely fragile, and difficult to mix into aqueous solution. Unmixed polymers in a diluted solution look like fish eyes. As the polymer's molecular weight increases, so does the difficulty in mixing and diluting it.

Unlike the inorganic chemicals discussed earlier, polymers have become attractive because they do not appreciably add to the volume of solids to be used or disposed of. Nor do they lower the fuel value of thickened or dewatered solids. Also, polymers are safer and easier to handle and result in easier maintenance than inorganic chemicals, which require frequent cleaning of equipment, typically via acid baths. However, polymers are not completely stable, can plasticize at high temperatures, and are slippery when spilled on floors.

4.2.1 PROPERTIES OF ORGANIC POLYMERS

Polymers are classified by the polymer compound's charge (e.g., anionic, non-ionic, or cationic), molecular weight, and form (when received). A combination of the polymer molecule's charge and molecular weight is useful in product identification.

Conditioners typically can be categorized as low, intermediate, and high molecular weight. A polymer's molecular weight is a rough indication of the length of polymer chain that holds the charged sites apart. It also affects other product attributes (e.g., solubility, viscosity, and charge density in aqueous solution).

Although there are exceptions, lower-molecular-weight products tend to be more soluble, less viscous, and have higher charge density in water. Low-molecular-weight polymers are often called primary coagulants, a term typically reserved for products ranging from $2.0 \cdot 10^4$ to $1.0 \cdot 10^5$ (Kemmer & McCallion, 1979). These water-soluble products typically are marketed in concentrations of 30% to 50%. They have low viscosities (close to the viscosity of water) and can be easily diluted and mixed with water at the application point. These polymers are useful for clarification applications where there are many small, dispersed particles to be destabilized and settled. They are typically in oily waste and biological waste treatment applications where low concentrations of solids are being treated. These polymers are sometimes also used as the first part of a two-polymer program in which high charge density is required to break the suspension.

Intermediate-molecular-weight products are available as solutions and in dry and liquid emulsion forms. It is difficult to generalize about the entire class of intermediate-molecular-weight products; however, most require wetting (e.g., mixing activation to disperse the polymer) and aging to develop full product activity in application. Solutions of intermediate-molecular-weight products are typically more viscous than low-molecular-weight products. In fact, product handling of feeding characteristics typically limits commercial solutions of these products to 1% (dry solids basis) or less. Consequently, supplemental dilution water is typically needed to improve polymer disbursement in the solids being conditioned.

Intermediate-molecular-weight products are common in thickening and dewatering systems treating wastewater solids, especially those with high concentrations of secondary solids. Virtually all charge variations are available in the intermediate-molecular-weight range.

High-molecular-weight polymers can be cationic, anionic, and non-ionic, and are available as liquid viscous solutions, emulsions, or dry powder. Their molecular weights vary from $2 \cdot 10^6$ to more than $12 \cdot 10^6$. Solubility and viscosity considerations typically dictate the solution concentrations available. Product solutions are made up at 0.25% to 1.0% solids concentration and allowed to age for several hours before further dilution at the application point.

Besides polymer weight, a relatively recent development in polymer formulation is the use of controlled degrees of cross-linkage. Such polymers can be highly branched rather than linear and are called structured polymers. Larger doses of structured polymers may be needed to reach an optimum performance, but the resulting floc is stronger (Dentel, 2001). High-shear dewatering applications (e.g., centrifuges and some recessed-plate filter presses) can benefit from such flocculants. Some suppliers use such terms as XL, FS, FL, and FLX to indicate the cross-linked forms (Dentel et al., 2000).

4.2.2 POLYMER FORMS AND STRUCTURE

Polymers are available in two physical forms: dry and liquid. Dry polymers can be delivered in a microbead or gel powder form, while liquid polymers can be delivered as a solution or emulsion. All dry and liquid polymers can be prepared with three charge types—cationic, anionic, and non-ionic—and can be purchased in a wide array of molecular weights, charge densities, and active solids levels. The form, charge, and activity level of the polymer can greatly affect their reactivity with solids.

A polymer's "activity" relates to the percent of the molecular weight that is available to react with and flocculate solids particles; it can greatly vary with the form of the polymer. The polymer dosing criteria are stated in grams of active polymer per kilogram of dry solids. This method allows polymer types with different activity levels to be compared on an equivalent basis. For example, a polymer with an activity of 9% will require 10 times more grams of bulk polymer than a similar polymer with an activity of 90%.

Dry polymers can have an active solids level as high as 94% to 100%. The shelf life of dry polymers is typically 2 years. Storage areas that are susceptible to wet and humid conditions should be avoided, because dry polymers will tend to cake and deteriorate.

Dry polymers may be difficult to dissolve. To make up a working solution, an eductor is used as a pre-wetting device to disperse polymers in water (Figure 13.9). The solution is slowly mixed in a mixing tank until the dry polymer particles are dissolved, and then aged in accordance with manufacturer recommendations. Aging time typically ranges from 30 minutes to 2 hours. Aging allows polymer particles to "unfold" into long chains. However, once the dry polymer is diluted and converted into a solution, it is only stable for about 24 hours.

The quality of the water used to dissolve dry polymer particles is important. Hard water (greater than 120 mg/L as calcium carbonate) or water containing more than 0.5 mg/L of free chlorine can cause the solution to deteriorate within a few hours.

Emulsions are dispersions of polymer particles in a hydrocarbon oil or light mineral oil. Surface active agents typically are applied to prevent the polymer–oil phase from separating from the water phase. These polymers can then be fed using a chemical feed system (Figure 13.10). Provisions must be made to mix the bulk storage tank regularly to prevent the oil and water from separating. With emulsions, it is possible to achieve a high molecular weight and maintain an active solids level of 30% to 50% without producing a solution that has a high viscosity. The approximate viscosity of emulsion polymer in its as-delivered state ranges from 300 to 5000 centipoise (cP). The shelf life of emulsion polymers is typically 6 months to 1 year. The initial breaking of the emulsion and aging are critical for optimum performance, which can be accomplished with a static mixer, high-speed mixer, or wet dispersal unit.

Emulsion polymers can have higher molecular weights and higher charges than dry polymers without the operating problems. The primary disadvantages of emulsion polymers are the potential for oil

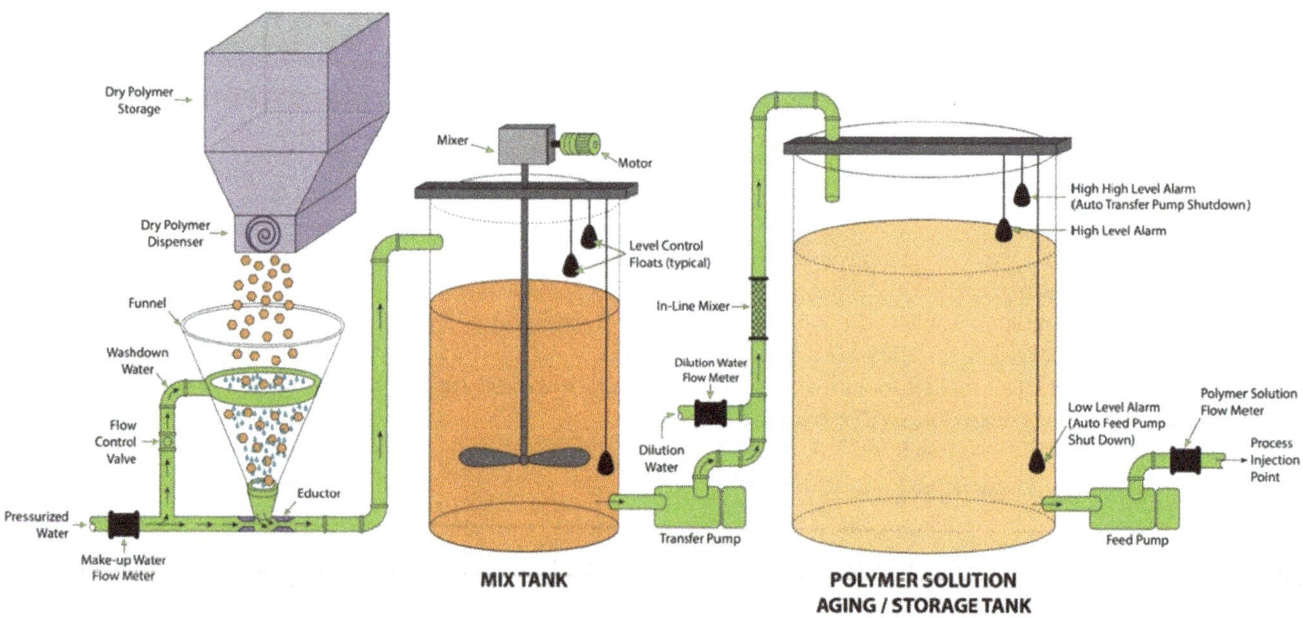

Figure 13.9 Schematic of Dry Polymer Makeup System (Reprinted with permission from PASSARO ENGINEERING, LLC, and Indigo Water Group)

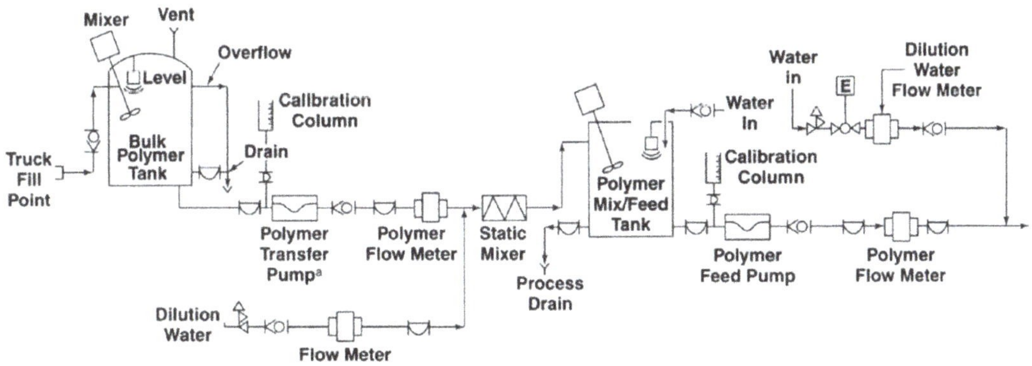

Figure 13.10 Typical Liquid Polymer Batch System

and water separation and the higher cost per volume of active material. A concern about emulsion polymers is the adverse environmental effects of the surfactants used in them. Such surfactants include alkylphenoethoxylates, which decompose to nonylphenol, a known endocrine disruptor. Recent developments in emulsion polymer manufacturing have been to abandon the use of mineral oils and surfactants for a new class of water-soluble emulsions. The process essentially involves dissolving the polymers in an aqueous salt of ammonium sulfate. A low-molecular-weight dispersant polymer is added to prevent aggregation of polymer chains.

Additionally, some of the typically used copolymers are susceptible to chemical hydrolysis at high pH values. If dewatered solids will be stabilized using an alkaline chemical (e.g., kiln dust, lime), odor problems could occur from the generation of trimethylamine, which has a "fishy" odor (Chang et al., 2005).

A Mannich polymer typically contains 3% to 8% active polymer; it is produced by using a formaldehyde catalyst to promote the chemical reaction to create the organic compound. Because vapors from formaldehyde pose a safety hazard and can be carcinogenic, Mannich polymers should be stored carefully and only used in well-ventilated areas. Mannich polymers are viscous (from 50 000 to more than 150 000 cP), difficult to pump, and have a relatively short shelf life. However, they can be effective and economical for large WRRFs, depending on the shipment cost.

4.2.3 POLYMER DOSAGE

Various polymers can enhance the performance of thickening or dewatering processes. The dose needed depends on the specific process used and the solids or biosolids to be thickened (Table 13.5).

Table 13.5 Typical Dosages of Polymer for Thickening Wastewater Solids (U.S. EPA, 1979a)

Application	Sludge Type	Polymer Dosage (g/kg)
Gravity thickening	Raw primary	2–4
	Raw (primary + WAS + TF *)	0.8
	Raw WAS	4.3–5.6
Dissolved air flotation	WAS (oxygen)	5.4
	WAS	0–14
	P + TF	0–3
	P + WAS	0–14
Solid-bowl centrifuge	Raw WAS	0–3.6
	Anaerobically digested WAS	2–7.2
Rotary drum	WAS	6.8
Gravity belt	Digested secondary	5

*Trickling filter

Most dewatering processes (except recessed-chamber filter presses) require polymer addition (Table 13.6). (Recessed-chamber filter presses typically use ferric chloride and lime as solids conditioners, either alone or with fly ash or polymers, and typically produce a slightly thinner cake with polymer conditioning than with ferric chloride and lime conditioning.) Centrifuges and belt filter presses cannot achieve optimum dewatering performance without polymer addition. Both applications require polymers with high positive charge and high molecular weight to produce a strong and durable floc.

Example Problem 13.1

For a neat polymer dose of 200 kg/d (440 lb/d) using a 25% active polymer, 10 000 kg/d (22 000 lb/d) of dry solids are dewatered. Calculate the active polymer dose normalized to dry solids processed.

Solution

Active polymer content = 200 kg/d · 0.25 = 50 kg/d

Normalized dose = 50 kg/d/10 000 kg/d · 1000 kg/1 metric tonne = 5 kg/metric tonne

4.2.4 APPLICATION OF POLYMER

Because of the wide range of polymers now available, the performance of almost any conditioning or dewatering process can be enhanced by their use. Depending on the application, polymers may improve unit throughput, solids capture, filtrate quality, thickened or dewatered solids, or a combination of these parameters.

As with inorganic chemical conditioning, proper organic chemical conditioning centers on four basic requirements:

- Correct dosage of polymer;
- Proper wetting, mixing, and aging of raw polymer;

Table 13.6 Typical Dosages of Polymer for Dewatering Solids (Dague, 1983; U.S. EPA, 1979a)

Application	Sludge Type	Polymer Dosage (g/kg)
Belt filter press	Raw (primary + WAS)	2–5
	WAS	4–9
	Anaerobically digested (primary + WAS)	6–10
	Anaerobically digested primary	4–7
	Raw primary	2–3
	Raw (primary + TF*)	3–6
Solid-bowl centrifuge	Raw primary	0.5–2.3
	Anaerobically digested primary	2.7–5
	Raw WAS	5–10
	Anaerobically digested WAS	1.4–2.7
	Raw (primary + WAS)	2–7
	Anaerobically digested (primary + WAS + TF)	5.4–6.8
Vacuum filter	Raw primary	1–5
	Raw WAS	6.8–14
	Raw (primary + WAS)	5–8.6
	Anaerobically digested primary	6–13
	Anaerobically digested (primary + WAS)	1.4–7.7
Recessed chamber filter press	Raw (primary + WAS)	2–2.7

*Trickling filter

- Correct solids and polymer mixing procedures; and
- Continuous observation of results and response to those observations.

Adhering to these requirements is more critical to conditioning performance when using polymers than when using inorganic chemicals. The polymers perform under a narrower range of operating conditions than inorganic agents do, so they are more sensitive to dosage and mixing. Although this sensitivity requires more operator attention, it can promote efficiency, because the dewatering process does not work if conditioning is not closely controlled.

The correct chemical dosage is critical to proper operation. Chemical conditioning tests (e.g., Buchner funnel or capillary suction time) should be conducted frequently to determine conditioning requirements. Maintaining the correct dosage requires knowledge and control of the solids stream and chemical feed(s). Continuous metering equipment should be used and the solids content monitored to determine the mass flow. Chemical mass feed rates also should be continuously monitored and controlled to maintain desired dosages. Flow-measuring devices should be used to determine the chemical volume being fed. The consistency of chemical concentration also should be monitored and maintained throughout the process. Although feed-solution concentrations can be measured using total solids or viscosity measurements, an accurate flow-metering system on the feed solution makeup system is the preferred option.

Polymers should be used at specific solution strengths based on the manufacturer's recommendation and the results of any laboratory tests performed. Dilute solutions may be required because of the polymer's chemical activity or to allow good contact of relatively small chemical quantities with large solids volumes. In some cases, poor quality dilution water (e.g., secondary effluent) affects polymer solution activity. Although this is not a common concern, high-quality water typically should be used to prepare feed solutions.

Proper mixing is critical to conditioning. It has two primary components: intensity and duration. Getting a viscous material thoroughly dispersed in the solids is of utmost importance when using polymers. Supplemental dilution water is used to reduce the polymer's viscosity, and high-intensity, short-duration mixing is needed to disperse the polymer. This high-intensity mixing is often accomplished with a polymer-injection ring and an adjustable check-valve device that delivers a high shearing action. Some facilities have also had success injecting polymer immediately upstream of the sludge feed pumps to achieve high-energy mixing; however, this can result in other complications if not done properly. The flocculation phase requires 15 to 60 seconds of gentle agitation to allow the chemical reaction between the polymer and solids to occur.

Flocculation mixing duration typically is on the order of 15 to 60 seconds and can be accomplished a number of ways (e.g., in the pipeline to the equipment or in a separate flocculation tank). For example, multiple addition points can be located at mixing times of 15, 30, and 45 seconds from the inlet to the thickening or dewatering equipment. At each addition point in the solids feed pipe, a flexible coupling and a polymer pipe drawoff point should be provided to allow the insertion of a polymer-injection ring and an in-line, high-intensity mixing unit (Figure 13.11). This arrangement provides the most flexibility in allowing operators to fine-tune the process feed rate.

Another method of providing flocculating mixing is to provide a flocculating tank directly upstream of the dewatering unit. This tank provides 15 to 30 seconds of flocculating time based on the design loading rate. A disadvantage of the flocculant tank is that it can create dead zones, which could result in improper conditioning.

To ensure that thickening or dewatering performance is optimal, both processes should be monitored and frequently tested. The thickened or dewatered cake's total solids and released water should be analyzed at least once per shift to monitor solids loading and polymer performance. Although one sample per shift is sufficient, taking composite samples each shift would allow an operator to detect any operational changes and potential equipment problems.

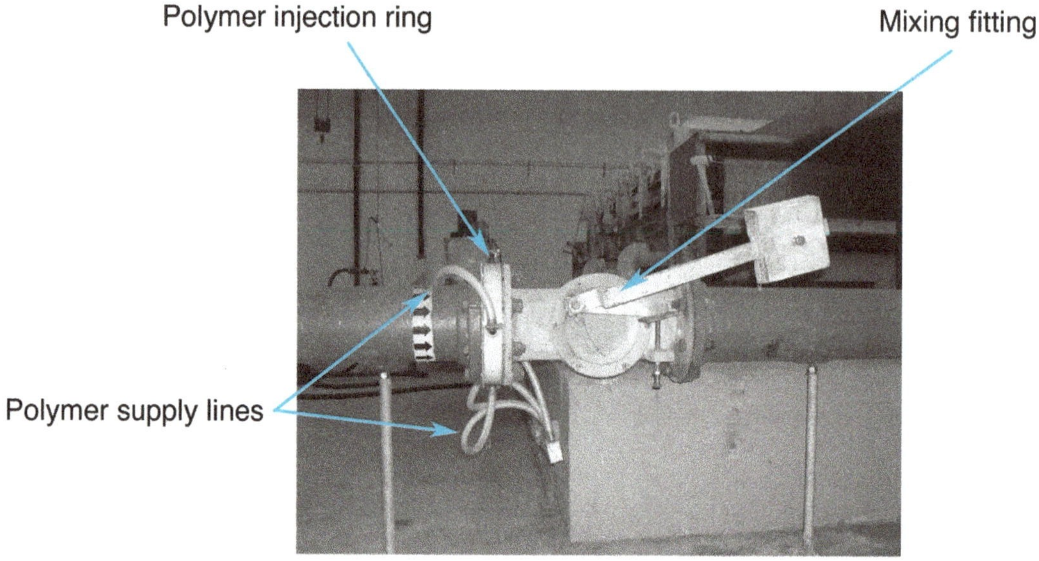

Figure 13.11 Polymer-Injection Ring and In-Line High-Intensity Mixing Unit (Courtesy of the City of Orlando, Florida)

Polymer monitoring includes performance and quality control checks. Polymer use should be recorded during each shift via drawdown readings in bulk or solution-storage tanks, timers on polymer-transfer pumps, or flowmeters on polymer feed lines. To confirm solution strengths, total solids tests should be conducted regularly. From these data, product performance or process variations may be detected.

There has been much technological development on a variety of automated process sensors, controllers, and related software for managing thickening and dewatering operations to optimize performance and polymer use. These systems typically include a solids probe in the drain line from the dewatering unit's filtrate line and a controller on the solids feed pump and polymer feed equipment. Typically, when the solids probe detects a sudden increase in filtrate solids concentration, the controller first tries to increase the polymer flowrate. If this does not clear the filtrate after a specific time period, the controller sends a signal to decrease the solids feed rate. These alternate steps are repeated until the filtrate clears. Other control systems monitor the viscosity or particle charge in filtrate. The particle charge measurement informs the control system whether to increase or decrease the polymer dose (Dentel et al., 2000). The primary benefit of such automated systems is that they tend to smooth out variations in unit operations and decrease polymer use by 20% to 50%. Systems that control dewatering based on filtrate or centrate properties typically will not provide optimal cake solids (Abu-Orf & Dentel, 1999). Therefore, if cake dryness significantly affects handling costs, a control system should not be based solely on polymer savings.

5.0 PROCESS DESIGN CONSIDERATIONS

Summaries of design considerations for several types of thickening and dewatering processes are discussed in this section. Typical polymer dosages for thickening and dewatering applications are provided in Tables 13.7 and 13.8, respectively.

5.1 GRAVITY THICKENING

Conventional gravity thickening typically does not require the use of organic polymers. That said, using these chemicals increases solids and hydraulic loading rates by a factor of 2 to 4 and improves solids capture. However, they have minimal effect on the resultant underflow solids concentration. Also, using polymer increases the overall cost of gravity thickening, so it only should be used to prevent operating problems caused by solids carryover.

Bench tests and other laboratory or field investigations should be performed to test the relative effectiveness of flocculating aids (either alone or in combinations). Also, care should be taken during feeding and mixing to prevent the overfeeding or poor mixing that causes "islands" to form.

Table 13.7 Polymer Dosages Associated With Various Solids Thickening Processes

Facility Name/Location	Type of Sludge	Type of Stabilization	Influent Feed Solids (%)	Method of Dewatering	Polymer Dosage (g/kg)	Dewatered Cake Solids (%)
Water Reclamation Facility Bend, Oregon	Primary (55%) Secondary (45%)	Anaerobic digestion	1.8–2.1	Belt filter press	3.75–6	12–15
Irwin Creek WRRF Charlotte, North Carolina	Primary (60%) Secondary (40%)	Anaerobic digestion	1.4	Belt press	3.5–4	18.75
City of Greeley WPCF Greeley, Colorado	Primary (60%) Secondary (40%)	Anaerobic digestion	1.5–2.0	Centrifuge	5–8	19–22
Buckman Street WRRF, Jacksonville, Florida	Primary (60%) Secondary (40%)	Anaerobic digestion	2–4	Centrifuge	3.75–6.25	19–22
Greenfield Road WRP Mesa, Gilbert and Queen Creek, Arizona	Primary and Secondary	Anaerobic digestion	2.75–3.0	Centrifuge	5.75	22–23
Water Conserv II Orlando, Florida	Secondary	Anaerobic digestion	2.0–2.5	Belt filter presses	3.5–4	12
Eastern Water Reclamation Facility Orange County Utilities Orlando, Florida	Secondary (100%)	Contract lime stabilization	<1	Belt filter presses (three belt)	3–3.5	16–17
Iron Bridge Water Reclamation Facility Orlando, Florida	Secondary	Lime stabilization	2.0–3.0	Belt filter presses	3.5–4	17
Iron Bridge Water Reclamation Facility Orlando, Florida	Secondary	Lime stabilization	0.4–0.6	Belt filter presses (three belt)	3.5–4	17
Northwest Water Reclamation Facility Orange County Utilities Orlando, Florida	Secondary (100%)	Contract lime stabilization	1–2	Belt filter presses	3.75–6.25	14–16
Water Conserv II, Orlando, Florida	Secondary (100%)	0.4–0.8	Gravity belt thickening	10–12	2–4	Anaerobic digestion
Orange County Utilities, South Water Reclamation Facility, Orlando, Florida	Secondary (100%)	0.4–1.25	Gravity belt thickening	2.4–6.5	3.1–6.4	Anaerobic digestion
91st Avenue Wastewater, Treatment Plant, Phoenix, Arizona	Secondary (100%)	1.0–1.5	Centrifuge	2.0–3.0	5.7–6.2	Anaerobic digestion
McAlpine Creek WRRF, Pineville, North Carolina	Primary (50%) Secondary (50%)	1.0	Conventional gravity thickening	0	3–5	Anaerobic digestion
			Centrifuge	3–5	0.9–1	
Roger Road Wastewater Reclamation Facility	Secondary	0.2–0.4	Gravity belt thickening	9–10	5–6	Anaerobic digestion
Pima County, Tucson, Arizona	Primary	0.4–0.6	Conventional gravity thickening	0	4–5	Anaerobic digestion
Columbia Boulevard WRRF, Portland, Oregon	Secondary (100%)	0.5–1.0	Gravity belt thickening	5–9	4–6	Anaerobic digestion

Table 13.8 Polymer Dosages Associated With Various Solids Dewatering Processes

South Water Reclamation Facility Orange County Utilities Orlando, Florida	Secondary	Anaerobic digestion	2.3–3.7	Belt filter presses	0.5–1.1	9.3–19.7
McAlpine Creek WRRF Pineville, North Carolina	ORC reported a 1:1 ratio of primary to secondary solids	Anaerobic digestion	2.4	Centrifuge	3.75	20
Columbia Boulevard WRRF Portland, Oregon	Freshly digested (65%–75%) thickened WAS (20%) Primary (80%) and previously digested, lagoon stabilized thickened WAS (25%–35%)	Anaerobic digestion	1.5–2.0	Belt filter presses	3.75–5	19–22
Thomas P. Smith Water Reclamation Facility Tallahassee, Florida	Secondary (100%)	Anaerobic digestion	2.81–4.39	Screw presses	3.75–6.75	13–20

Notes.
WAS = waste activated sludge;
WPCF = water pollution control facility;
WRP = water reclamation plant;
WRRF = water resource recovery facility.

Using about 2 to 4.5 g of active polymer/kg of dry solids (4 to 9 lb/ton) can produce a solids loading rate of about 22 to 34 $kg/m^2{\cdot}d$ (4.5 to 7.0 lb/d/sq ft) when thickening primary solids. Adding about 4.5 to 6.0 g/kg (9.5 to 12.5 lb/ton) of polymer can increase the thickener's loading rate to 12 to 16 $kg/m^2{\cdot}d$ (2.4 to 3.2 lb/d/sq ft) when treating WAS (Ettlich et al., 1978; U.S. EPA, 1978, 1979b).

5.2 DISSOLVED AIR FLOTATION THICKENING

Chemical conditioning is unnecessary for dissolved air flotation (DAF) thickening if low hydraulic and solids loading rates are used. However, if high loading rates are required, or compaction is poor and the sludge volume index is high, chemical conditioning improves solids capture and can increase the float solids concentration. Although the increase in float solids is typically small (in the range of 0.5%), polymers may be required for WAS if a 4% float solids concentration is to be achieved. Float solids can routinely be 6% or higher for co-thickened mixtures of primary solids and WAS. Unless problems exist, solids capture without polymers is typically about 95%. With polymers, solids capture can increase to 97% or 98%, thereby improving subnatant quality and lessening the effect of recycle solids on facility performance. Also, with polymer addition, it is possible to as much as double the solids loading rate (a typical rate is 10 $kg/m^2{\cdot}h$ [2 lb/sq ft/hr]).

Typically, a cationic polymer with a moderate charge and high molecular weight is used; however, lower-charge cationic polymers are starting to show better performance. Typical dosages range from 2 to 5 g/kg (4 to 10 lb/ton) up to 7.5 g/kg (15 lb/ton) of dry solids.

A common problem when conditioning solids for DAF thickening is improper mixing of conditioner and solids. To mitigate this problem, a more dilute polymer solution (0.25% to 0.5%) should be used, or the flocculant should be mixed with pressurized recycle before contacting the solids (Ettlich et al., 1978; U.S. EPA, 1978).

5.3 CENTRIFUGAL THICKENING

A solid-bowl conveyor centrifuge has been used to thicken a wide variety of solids. Centrifugal thickening typically does not require polymer addition when treating biological and aerobically stabilized solids. Well-digested solids, however, have little natural flocculating tendency, and require polymer additions to achieve acceptable solids recovery levels. Engineers, therefore, should make provisions for polymer addition in the initial design, even if chemical conditioning is not planned. The design should be flexible enough to allow conditioning chemical to be added at one of several points in the influent piping.

Dry or liquid high-molecular-weight cationic polymers are effective thickeners. When dry polymers are used, a 0.05% to 0.1% feed solution is used, whereas liquid polymers can range in concentrations up to 0.5% on an active basis. It is important that a solids capture of at least 95% is obtained to prevent recycling filamentous bacteria and fines to the wastewater treatment process. Waste activated sludge produces a weak floc that tends to shear inside the centrifuge; a dose of up to 4 g/kg (8 lb/ton) of polymer can be used to formulate a tougher floc. Aerobically and anaerobically digested solids have little natural floc and, therefore, require about 4 to 8 g/kg (8 to 16 lb/ton) of polymer.

5.4 GRAVITY BELT THICKENING

Gravity belt thickening works well with many types of solids. Difficult-to-thicken solids only require minor modifications of polymer dosages and solids loading rates to maintain high effluent solids concentration and percent solids capture. Gravity belt thickeners have been used to treat solids containing as little as 0.4% solids or as much as 10% solids with polymer addition. Polymer dosages range from between 1.5 and 3 g/kg (3 and 6 lb/ton) (dry-weight basis) for raw primary solids up to between 4 and 6 g/kg (8 and 12 lb/ton) for anaerobically stabilized solids. In all cases, solids capture remains above 95%.

5.5 ROTARY DRUM THICKENING

Rotary drum thickeners work much like gravity belt thickeners: in both systems, a moving, porous media retains conditioned solids while allowing free water to drain through. A polymer is injected into the feed line and mixed with incoming solids before entering the flocculation tank.

Drum speed, mixer speed, and spray water cycling is adjustable to ensure maximum performance with minimal polymer and water use. Polymer requirements are about 10% to 20% greater than those associated with gravity belt thickeners. Rotary drum thickeners are suited for high-fiber solids, as well as raw and digested solids with a significant fraction of primary solids. Their success with municipal WAS is variable and depends on solids characteristics. Residuals typically can be thickened to 5% to 7% total solids (in some cases, more than 10% total solids) with up to 99% capture of feed solids at polymer dosages ranging from 4 to 6 g/kg (8 to 12 lb/ton) (dry-weight basis).

5.6 CENTRIFUGAL DEWATERING

Polymers have been used with solid-bowl conveyor centrifuges to increase machine throughput without lowering cake dryness, to improve solids recovery, or both. Typically, a moderate- to high-charge, high-molecular-weight cationic polymer is used. Pilot studies are needed to determine the correct conditioning agent and dosage. Designs should include facilities for feeding both dry and liquid polymers. Polymer use typically increases solids capture; however, too much polymer can lead to a wetter cake because more fines are captured. Therefore, the relationship between recycled solids and cake dryness determines the dosage of polymer to be used.

5.7 BELT FILTER PRESS DEWATERING

The performance of belt filter presses depends on proper conditioning, and organic polymers traditionally are used. A properly conditioned product has a 95% to 98% solids recovery rate. However, the quantity of polymer required for proper conditioning varies widely; it depends on solids type, solids concentration, and ash content (typically, less polymer is needed when the ash content is high). For example, primary solids require a polymer dose of 3.5 to 5 g/kg (7 to 10 lb/ton), anaerobically digested solids require a dose of 7 g/kg (14 lb/ton on a dry-weight basis), and autothermal thermophilic aerobically digested solids require a dose between 18 and 23 g/kg (37 and 47 lb/ton).

Insufficient conditioning causes inadequate dewatering in the initial sections of the press, which, in turn, can cause solids to extrude from the press section, overflow in the drainage section, or blind the belt. Overconditioning can cause belt blinding and overflocculation, which causes solids to drain too fast and mound on the belt, resulting in poor dewatering. The goal is to remove as much water as possible in the gravity section of the press. Overflocculation may be mitigated by using deflection plates to even out the mounds before pressing begins, or by selecting a belt filter press with an extended gravity table.

Because of the shearing action between belts, Novak and Haugan (1980) have suggested using turbulent mixing when adding polymers for conditioning before dewatering. The best dosage and overall system performance depend on solids concentration, mixing intensity, and mixing time.

5.8 SCREW PRESS DEWATERING

The screw press is a simple, slow-moving mechanical device that gradually compresses conditioned, thickened solids as they move through the unit. Dewatering is continuous; it begins with gravity drainage at the inlet end of the screw and then dewaters the result of increasing pressure at the end of the unit. Proper screw design is critical, because different solids require different polymer dosages, screw speeds, and configurations to maintain a desired dewatered cake concentration and solids capture rate.

Proper solids conditioning is essential to produce a consistent dewatered cake. Slower operation produces a dryer cake but also reduces solids throughput. Therefore, it is important that a relationship between polymer dosage, solids throughput, and cake dryness be established. Depending on the influent solids characteristics, the polymer dosage may range from 8 to 12 g/kg (16 to 24 lb/ton) to produce a cake containing between 12% and 25% dry solids and a solids capture rate of 90% to 95%.

5.9 ROTARY PRESS DEWATERING

The polymer dosage for a rotary press depends on the type of solids to be dewatered. Work performed at the Daniels and Plum Island facilities in Charleston, South Carolina, indicated that 4.5 to 6 g/kg (9 to 12 lb/ton) of polymer was needed for a mixture of raw primary and secondary solids, resulting in an average dewatered cake concentration of 25%. However, in St. Petersburg, Florida, polymer dosages ranging from 15 to 18 g/kg (31–37 lb/ton) were required to dewater aerobically digested secondary solids to an average of 15% solids. In both cases, solids capture was more than 95%.

5.10 DRYING BEDS

Conditioning solids before sending them to drying beds is not widely practiced. In fact, the "Ten States Standards" (Great Lakes–Upper Mississippi River Board of State and Provincial Public Health, 2014) and other design guidelines do not consider using conditioning chemicals in this application, even though they may significantly reduce drying time and, therefore, the bed area required. Relatively small polymer doses (as little as 50 mg/L) can considerably improve the drainage capabilities of properly digested solids by flocculating smaller particles. Flocculation speeds up the drainage period of the dewatering cycle and maintains a porous cake that is more readily susceptible to evaporation.

Studies have indicated that conditioning significantly increases the loading rate for digested primary solids and WAS; the unconditioned solids loading rate was 73 kg/m^2·yr (15 lb/sq ft/yr), while the conditioned solids loading rate was 270 kg/m^2·yr (55 lb/sq ft/yr). A well-conditioned product dries in about one-third the time (approximately 10 to 15 days) required for unconditioned solids. Such performance improvements typically result from a dosage of about 15 to 23 g/kg (31 to 46 lb/ton) of a cationic polymer with a moderately high or high charge and a high molecular weight.

The conditioning system must be designed to avoid rupturing the conditioned floc during transport to the drying beds. Rupturing typically occurs during pumping; it can be overcome by locating the flocculation tank close to the drying beds, and allowing conditioned solids to flow by gravity from the flocculating chamber. Excessive holding increases the percentage of fines, which impairs flocculation and dewatering.

Both wedge-wire and vacuum-assisted drying beds use polymers to coagulate fines and promote rapid cake formation. The polymer is injected into the solids in the inlet line or in a flocculation tank next to the bed. Typical doses are between 1.5 and 3 g/kg (3 and 6 lb/ton).

6.0 THERMAL CONDITIONING AND PRETREATMENT

Thermal conditioning is the simultaneous application of heat and pressure to solids to enhance digestibility and dewaterability without adding conditioning chemicals. During thermal conditioning, heat lyses the cell walls of microorganisms in biological solids, releasing bound water from the particles. This process further hydrolyzes and solubilizes hydrated particles in biological solids and, to a limited degree, organic compounds in primary solids. Conventional mechanical dewatering devices can then readily separate released water from particles as long as the solids have enough fibrous solids for cake structure.

There are two main thermal pretreatment processes in use today: thermal dewatering and thermal hydrolysis. Thermal hydrolysis is a process used to enhance the digestibility of sludge in a digestion system. Thermal dewatering is used as a pretreatment process upstream of a thermal oxidization system to reduce or eliminate auxiliary fuel usage. Thermal dewatering is essentially a partial thermal drying process, sometimes called "scalping," that involves removing sufficient water content from the thermal oxidation system feed material to make the combustion process self-supporting or "autogenous" to the extent practical. This is done using waste heat recovered from the thermal oxidation system as the energy source for the dewatering process. Thermal dewatering is a useful way to both minimize auxiliary fuel usage and take advantage of more of the recoverable energy available in the thermal oxidation system exhaust gases, provided that the thermal oxidation system can handle the higher feed solids concentration. The utility of thermal dewatering is greater for fluidized-bed systems than it is for multiple-hearth systems, as the latter are not as amenable to operation with higher feed solids concentrations.

Figure 13.12 provides a schematic of a thermal dewatering system in conjunction with a fluidized-bed thermal oxidation system. Fluidized-bed systems normally use a portion of the recoverable energy in the system exhaust gases to preheat combustion air, which also serves to reduce auxiliary fuel usage. Depending on the solids content from the mechanical dewatering system and the thermodynamic characteristics of the feed material, air preheating alone may not be sufficient to enable the designer to achieve autogenous operation. Generally speaking, air preheating only requires 40% to 60% of the recoverable energy in the system exhaust gases, so there is still a significant amount of energy remaining downstream of the air preheater to support thermal dewatering. Technically, this energy can be recovered to produce either a separate heated air stream, steam, or a heated thermal fluid stream. The designer should evaluate these options in conjunction with the requirements of the thermal dewatering equipment, along with other planned uses for the supplemental recovered energy, such as to meet building or other process heating loads. Generally speaking, indirect drying systems, such as disc or paddle dryers, are most appropriate for thermal dewatering applications because the solids content of their product can be more closely controlled. Indirect drying systems can utilize energy in the form of steam or heated thermal fluids.

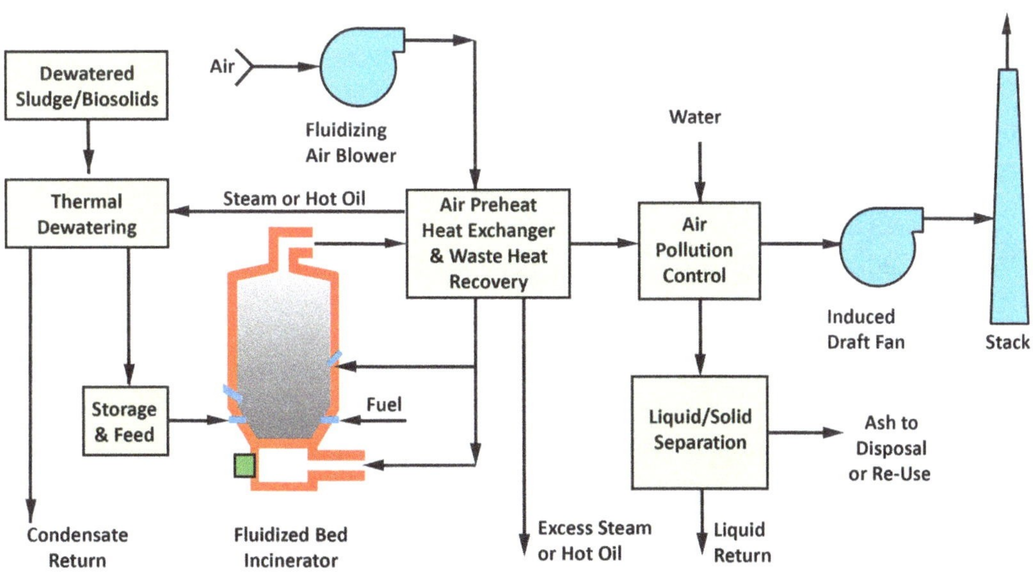

Figure 13.12 Thermal Dewatering System Schematic

PRACTICE PROBLEMS

1. Sludge conditioning adds chemicals to sludge or biosolids to improve separation of water.
   a. True
   b. False
2. Diluted polymer is not stable and may become ineffective if allowed to age in the mix tank too long.
   a. True
   b. False
3. Polymer spills should be cleaned up immediately to prevent slippery, hazardous conditions.
   a. True
   b. False
4. The conditioning process is used to prepare sludge or biosolids for more efficient:
   a. Thickening and disinfection
   b. Stabilization and dilution
   c. Thickening and dewatering
   d. Neutralization and dewatering
5. Organic chemicals that are commonly used for conditioning include
   a. Alum
   b. Lime
   c. Polymers
   d. Ferric chloride
6. Lime is sometimes added to sludge along with ferric chloride for this reason:
   a. To fully activate the ferric chloride
   b. Prevent an unacceptable drop in pH
   c. Decrease sludge production
   d. Add ballast for improved settling
7. Polymers are preferred over inorganic chemicals for sludge dewatering because
   a. Inorganic chemicals increase sludge mass
   b. Polymers are more soluble in oily sludge
   c. Inorganic chemicals decrease handling costs
   d. Polymers require prewetting and aging
8. Jar testing has determined that the optimum dose for 600 mL of sludge is 5 mL of a 20 000 mg/L polymer solution. What is the concentration of polymer in the jar test?
   a. 122 mg/L
   b. 167 mg/L
   c. 184 mg/L
   d. 221 mg/L

9. Polymer is fed at a dose of 75 mg/L. Find the polymer feed rate in kg/d (lb/d) if the sludge feed rate is 5 L/s (79.3 gpm).
   a. 32.4 kg/d (71.4 lb/d)
   b. 48.6 kg/d (107 lb/d)
   c. 64.8 kg/d (143 lb/d)
   d. 81.0 kg/d (179 lb/d)
10. For a polymer dose of 50 kg/d (110 lb/d), 3500 kg/d (7700 lb/d) of dry solids are dewatered. Calculate the active polymer dose normalized to dry solids processed.
11. Example calculation for polymer feed to a gravity belt thickener: Waste activated sludge is thickened using a gravity belt thickener before anaerobic digestion. Emulsion polymer is added upstream of the thickener to condition the WAS. The WAS is fed at 20.5 L/s (325 gpm) and has an average total solids concentration of 1.4%.
   a. If the target polymer feed rate is 2.1 kg of active polymer per metric ton of dry solids (4.2 1b of active polymer per short ton of dry solids), how many kilograms (pounds) per day of polymer should be fed?
   b. If feeding 0.25% active polymer solution, what polymer feed rate in L/h (gph) should be added to the gravity belt thickener? The specific gravity of the polymer solution is 1.0.

7.0 REFERENCES

Abu-Orf, M. M., & Dentel, S. K. (1999). Rheology as tool for polymer dose assessment and control. *Journal of Environmental Engineering, 125*(12), 1133–1141.

Abu-Orf, M. M., & Ormeci, B. (2005). Measuring sludge network strength using rheology and relation to dewaterability, filtration, and thickening—31 laboratory and full-scale experiments. *Journal of Environmental Engineering, 131*(8), 1139–1146.

Bache, D. H., & Dentel, S. K. (2000). Viscous behaviour of sludge centrate in response to chemical conditioning. *Water Research, 34*(1), 354–358.

Bruus, J. H., Nielsen, P. H., & Keiding K. (1992). On the stability of activated sludge flocs with implication to dewatering. *Water Research, 26*(37), 1597–1604.

Chang, J. S., Abu-Orf, M. M., & Dentel, S. K. (2005). Alkylamine odors from degradation of flocculant polymers in sludges. *Water Research, 39*, 3369–3375.

Christensen, G. L., & Stulc, D. A. (1979). Chemical reactions affecting filterability in iron–lime sludge conditioning. *Journal of the Water Pollution Control Federation, 51*, 2499.

Dague, R. R. (1983). *A modified approach to activated sludge design.* Paper presented at 56th Annual Water Pollution Control Federation Conference, Atlanta, Georgia, United States.

Dentel, S. K. (2001). Conditioning. In L. Spinosa & P. A. Vesilind (Eds.), *Sludge into biosolids.* IWA Publishing.

Dentel, S. K., Abu-Orf, M. M., & Griskowitz, N. J. (1995). *Polymer characterization and control in biosolids management* (Publication 11 D43007). Water Environment Research Foundation.

Dentel, S. K., Gucciardi, B. M., Griskowitz, N. J., Chang, L., Raudenbush, D. L., & Arican, B. (2000). Chemistry, function, and fate of 14 acrylamide-based polymers. In H. H. Hahn, H. Odegaard, & E. Hoffmann (Eds.), *Chemical water and wastewater treatment VI*, pp. 35–44. Springer Verlag.

Ettlich, W. F., Hinrichs, D. J., & Lineck, T. S. (1978). *Operations manual: Sludge handling and conditioning* (EPA-68/01-4424).

Eriksson, L., & Alm, B. (1991). Study of bioflocculation mechanisms by observing effects of a complexing agent on activated sludge properties. *Water Science and Technology, 24*, 21–28.

Great Lakes–Upper Mississippi River Board of State and Provincial Public Health and Environmental Managers. (2014). *Recommended standards for wastewater facilities*. Health Research, Health Education Division.

Higgins, M. J. (1995). The roles and interactions of metal salts, proteins, and polysaccharides in the settling and dewatering of activated sludge [Doctoral dissertation, Virginia Polytechnic Institute and State University].

Higgins, M. J., & Novak J. T. (1997a). The effect of cations on the settling and dewatering of activated sludge: Laboratory results. *Journal of Water Environment Research, 69*, 215–224.

Higgins, M. J., & Novak J. T. (1997b). Dewatering and settling of activated sludges: The case for using cation analysis. *Journal of Water Environment Research, 69*, 225–232.

Institute of Water Pollution Control. (1981). *Sewage sludge II: Conditioning, dewatering and thermal drying* (Manual of British Practice in Water Pollution Control).

Jennett, J. C., & Santry, I. W., Jr. (1969, October). Characteristics of sludge drying. *Journal of the Sanitary Engineering Division, 95*(5).

Karr, P. R., & Keinath, T. M. (1978). Influence of particle size on sludge dewaterability. *Journal of the Water Pollution Control Federation, 50*, 1911.

Kemmer, F. N., & McCallion, J. (1979). *The NALCO water handbook*. McGraw-Hill.

Kolda, B. C. (1995). Impact of polymer type, dosage, and mixing regime and sludge type on sludge floc properties [Master's thesis, Virginia Polytechnic Institute and State University].

Lewis, C. J., & Gutschick, K. A. (1988). *Lime in municipal sludge processing*. National Lime Association.

Lynch, D. P., & Novak, J. T. (1991). Mixing intensity and polymer dosing in filter press dewatering. *Journal of the Water Pollution Control Federation, 63*(2), 160.

Metcalf & Eddy, Inc./AECOM. (2014). *Wastewater engineering: Treatment and resource recovery* (5th ed.). McGraw-Hill.

National Lime Association. (1982). *Lime handling, application, and storage in treatment processes* (4th ed., Bulletin 213).

Novak, J. T., & Haugan, B. E. (1979). Chemical conditioning of activated sludge. *Journal of Environmental Engineering, 105*(EE5), 993.

Novak, J. T., & Haugan, B. E. (1980). Mechanisms and methods for polymer conditioning of activated sludge. *Journal of the Water Pollution Control Federation, 52*, 2571.

Novak J. T., Goodman G. L., Pariroo, A., & Huang, J. C. (1988). The blinding of sludges during filtration. *Journal of the Water Pollution Control Federation, 60*, 206–214.

Novak, J. T., Miller, C. D., & Murthy, S. N. (2001). Floc structure and the role of cations. *Water Science and Technology, 44*(10), 209–213.

Ormeci, B., Cho, K., & Abu-Orf, M. M. (2004). Development of a laboratory protocol to measure network strength of sludges using torque rheometry. *Journal of Residuals Science and Technology, 1*(1), 35–44.

Park, C., Muller, C. D., Abu-Orf, M. M., & Novak, J. T. (2006). The effect of wastewater cations on activated sludge characteristics: Effects of aluminum and iron in floc. *Water Environment Research, 78*, 31–40.

Roberts, K., & Olsson, O. (1975). The influence of colloidal particles on the dewatering of activated sludge with polyelectrolyte. *Environmental Science and Technology, 9*, 945.

Robinson, J. K. (1989). *The role of bound water content in designing sludge dewatering characteristics* [Master's thesis, Virginia Polytechnic Institute and State University].

Snoeyink, V. L., & Jenkins, D. (1980). *Water chemistry*. Wiley and Sons.

Sorensen, B. L., & Sorensen, P. B. (1997). Applying cake filtration theory to membrane filtration data. *Water Research, 31*(3), 665–670.

Tenney, M. W., Echelberger, W. F., Jr., Coffey, J. J., & McAloon, T. J. (1970). Chemical conditioning of biological sludges for vacuum filtration. *Journal of the Water Pollution Control Federation, 42*, R1.

Tezuka, Y. (1969). Cation-dependent flocculation in flavobacterium species predominant in activated sludge. *Applied Microbiology, 17*, 222.

U.S. Environmental Protection Agency. (1978). Sludge treatment and disposal. In *Sludge treatment, Vol. 1* (EPA-625/4-78-012).

U.S. Environmental Protection Agency. (1979a). *Chemical aids manual for wastewater treatment facilities* (EPA-430/9-79-018).

U.S. Environmental Protection Agency. (1979b). *Process design manual for sludge treatment and disposal* (EPA-625/1-79-011). U.S. Environmental Protection Agency, Municipal Environmental Research Laboratory, Office of Research and Development.

Vesilind, P. A. (1979). *Treatment and disposal of wastewater sludges*. Ann Arbor Science Publishers.

Wang, L. K., Pereira, N. C., & Hung, Y. T. (2007). *Handbook of environmental engineering biosolids treatment processes* (6th ed.). Humana Press.

Water Environment Federation. (2014). *Introduction to water resource recovery facility design* (2nd ed.).

Webb, L. J. (1974). A study of conditioning sewage sludges with lime. *Journal of the Water Pollution Control Federation, 73*, 192.

8.0 SUGGESTED READINGS

Dentel, S. K., Abu-Orf, M. M., & Walker, C. A. (2000). Optimization of slurry flocculation and dewatering based on electrokinetic and rheological phenomena. *Chemical Engineering Journal, 80*(1–3), 65–72.

Gillette, R. A., & Scott, J. D. (2001). Dewatering system automation: Dream or reality? *Water Environment Technology, 13*(5), 44–50.

Novak, J. T., Sadler, M. E., & Murthy, S. N. (2003). Mechanisms of floc destruction during anaerobic and aerobic digestion and the effect on conditioning and dewatering of biosolids. *Water Research, 37*, 3236.

Pramanik, A., LaMontagne, P., & Brady, P. (2002). Automation improvements: Installing an integrated control system can improve sludge dewatering performance and cut costs. *Water Environment Technology, 14*(10), 46–50.

U.S. Environmental Protection Agency. (1978). *Innovative and alternative technology assessment manual* (EPA-430/9-78-009). U.S. Environmental Protection Agency, Office of Water Program Operations.

U.S. Environmental Protection Agency. (1978). *Operations manual for sludge handling and conditioning* (EPA-430/9-78-002).

U.S. Environmental Protection Agency. (1979). *Chemical primary sludge thickening and dewatering* (EPA-600/20-79-055). U.S. Environmental Protection Agency, Municipal Environmental Research Laboratory, Office of Research and Development.

U.S. Environmental Protection Agency. (1979). *Evaluation of dewatering devices for producing high-solids sludge cake* (EPA-600/2-79-123). U.S. Environmental Protection Agency, Water Resources Management Administration, Municipal Environmental Research Laboratory.

U.S. Environmental Protection Agency. (1979). *Review of techniques for treatment and disposal of phosphorus-laden chemical sludges* (EPA-600/2-6 79-083). U.S. Environmental Protection Agency, Municipal Environmental Research Laboratory, Office of Research and Development.

Chapter 14 | Ultimate Disposal

LEARNING OBJECTIVES

At the end of this chapter, students should be able to:

- Describe the differences between multiple hearth and fluidized bed incinerators regarding how they function and what emissions are a concern from each.
- Explain the different types of landfills that accept biosolids.
- Detail how the land application of biosolids can be practiced in a responsible and sustainable manner to provide nutrient reuse.
- Describe specific considerations for finding a suitable site for biosolids land application and how public perception affects decision making.
- Determine the land required for land application of biosolids, considering nutrient and metal loadings.

1.0 INCINERATION

1.1 PROCESS FUNDAMENTALS

Thermal oxidation systems, including incinerators, totally or partially convert organic solids into oxidized products (primarily carbon dioxide and water) or else partially oxidize and volatilize organic solids via starved air combustion into products with a residual caloric value. Thermal oxidation in the presence of excess air is generally referred to as incineration. Starved air combustion is thermal oxidation that restricts airflow (oxygen) to produce three potentially energy-rich products: gas, oil or tar, and a char. Pyrolysis, wet air oxidation, and gasification are examples of processes using starved air combustion.

Although wastewater solids typically are organic, they only will sustain combustion without auxiliary fuel if enough water has been removed. Hence, dewatering should always precede thermal oxidation. Dewatered cake with 20% to 35% total solids can be burned with or without auxiliary fuels, depending on the volatile solids-to-moisture ratio, heating value of volatile solids, amount of excess air, and the temperature of the combustion air. Dewatered cake with a water content of 70% to 74% (total solids of 26% to 30%) and volatile solids content of 75% to 80% have supported autogenous combustion conditions in both **fluidized bed incinerators** (FBIs) and **multiple hearth furnace** (MHF)–style thermal oxidizers.

A fluidized bed incinerator is an incinerator with a bed of inert material that is maintained in suspension by introducing the inlet air in the bottom of the incinerator.

Combustion (burning) is the rapid combination of oxygen with fuel, resulting in the release of heat. The typically combustible elements of wastewater solids—carbon, hydrogen, and sulfur—chemically combine in organic solids as grease, carbohydrates, and protein. The combustible portion of wastewater solids typically has a heating value about equal to that of lignite coal. Adding air provides oxygen to support the combustion. The major products of complete combustion are carbon dioxide, water vapor, sulfur dioxide, and inert ash. Good combustion requires proper proportioning and thorough mixing of fuel and air, and the initial and sustained ignition of the mixture. To achieve complete combustion, three fundamental principles must be followed: temperature, time, and turbulence.

A multiple hearth furnace is an incinerator consisting of numerous hearths that is used to combust organic solids. Solids travel down through the hearths by the action of a rotating arm.

The gaseous products of thermal oxidation include dry combustion gases, excess air, and water vapor from the moisture in the solids and the oxidation of hydrogen. The heat content of dry combustion gases (which is predominantly CO_2 and N_2 with trace amounts of SO_2) is determined by multiplying the individual weight of each gas by its specific heat content at the exit temperature. The weight of moisture in the exit gases (resulting from the combustion of hydrogen) typically is combined with the weight of the moisture in the wet feed to calculate the total heat content of the total water vapor exiting in combustion gases. If supplemental fuel is used, the volumes and heat contents of stack gases resulting from fuel combustion and solids combustion must be calculated separately.

The main objective of thermal oxidation is to reduce the quantity of solids requiring disposal. This process reduces the volume and weight of feed solids by about 80% to 95%, oxidizes or reduces toxics, can operate without consuming fossil fuels (autogenous combustion), and can yield exhaust heat that can be recovered for use in beneficial heating applications (process heat or building heat) or used to produce electrical power or mechanical energy. Although the capital costs for thermal oxidation may be higher than those for other solids use and disposal options, the overall economics have shown to be favorable due to reduction of final solids product; bioenergy potentials; and the flexibility, in some cases, for economic revenue from contract burning. Incorporation of energy recovery and utilization systems can greatly mitigate operation and maintenance costs and often make thermal oxidation more sustainable and competitive with other options.

1.2 SOLIDS CALORIFIC VALUES

The composition and quantity of fuels to be burned are fundamental inputs for the design of any combustion system. The composition determines the fuel's calorific or heating value, and the fuel's quantity determines the required size of the unit.

A solid's heating value is the amount of heat that can be released per unit mass of the solid. Its gross heating value—the prime indicator of combustion potential—depends on how much carbon, hydrogen, and sulfur it contains. Carbon burned to carbon dioxide has a gross heating value of 3.4×10^4 kJ/kg (1.46×10^4 Btu/lb). Hydrogen has a gross heating value of 1.44×10^5 kJ/kg (6.23×10^4 Btu/lb). Sulfur has a gross heating value of 1.0×10^4 kJ/kg (4500 Btu/lb). So, any changes in the solid's carbon, hydrogen, or sulfur content will raise or lower its heating value.

Each fuel has a characteristic ("effective") heating value and a gross heating value. Effective heat (available heat) is the amount of heat released in a combustion chamber minus both the dry flue-gas loss and the moisture loss. It depends on the solids' characteristics and the combustion chamber's design and operating characteristics; it cannot be determined via a bomb calorimeter alone. Incinerators that need large amounts of excess air offer less available heat because of the energy needed to heat the excess air (see Figure 14.1) (Los Angeles County Sanitation Districts, 1988; North American Manufacturing Co., 1965).

The typical heating values of wastewater solids obtained via bomb calorimetry are shown in Table 14.1 (U.S. Environmental Protection Agency [U.S. EPA], 1979). Generally, the higher heating value of the combustible fraction of dry primary and waste activated sludge (WAS) solids will range from 2.2×10^4 to 2.6×10^4 kJ/kg (9500 to 11 200 Btu/lb). The literature contains several methods for calculating the heating values of organic materials using proximate and ultimate analyses, but none of these, including the Dulong formula, accurately approximate the heating value of wastewater solids. Laboratory determination of the higher heating value of the wastewater solids is recommended when considering a thermal oxidation system.

1.3 FLUIDIZED BED INCINERATOR

The principle of using a high-temperature gas to fluidize solid materials was first commercially developed by the oil refining industry in a technology called fluidized bed catalytic crackers. Although FBIs only vaguely resemble catalytic crackers, they function similarly. In the FBI, combustion air is passed through an enclosed space (fluidized bed zone) in a way that sets all the particles in that zone in a homogeneous, boiling motion (i.e., fluidizes them) (Figure 14.2). In this state, particles are separated from each other by an envelope of fluidizing gas (combustion air), thereby presenting more surface for a gas-to-solid reaction (e.g., air to carbon). The maximized surface area is what gives most FBIs their high thermal efficiency.

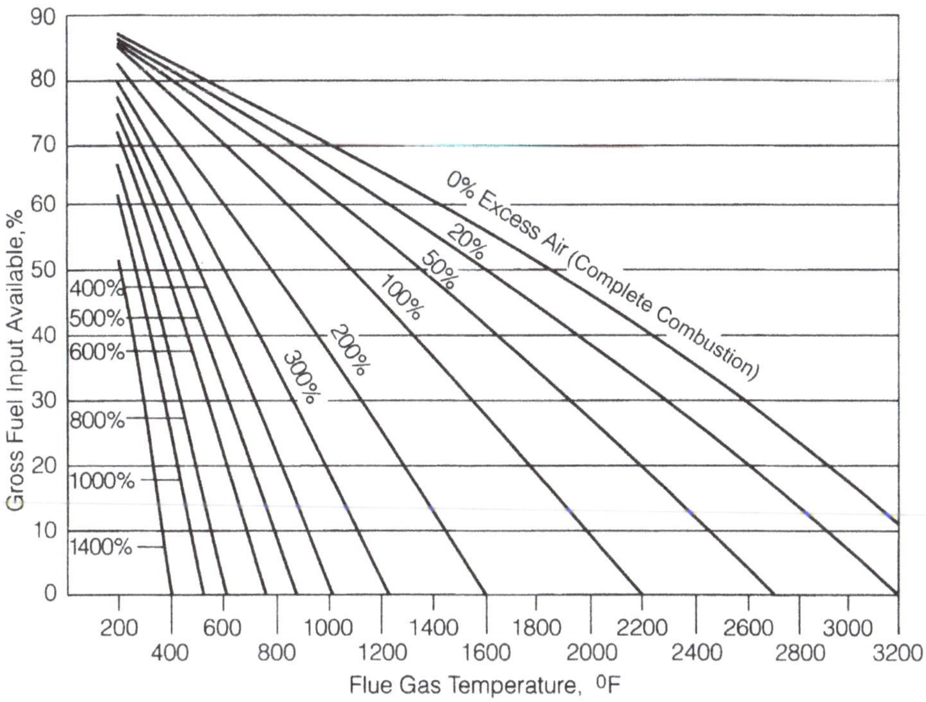

Figure 14.1 Generalized Available Heat Chart

Treatment capacity, therefore, is a function of total reactor bed volume (although typically expressed as either fluidized bed surface area or freeboard area) and the maximum gas velocity permitted because of particle entrainment.

At combustion equilibrium, the fluidized bed resembles a boiling liquid and obeys most hydraulic laws. A specially designed orifice plate or series of nozzles (tuyeres) are used to disperse fluidizing gas (air) throughout the fluidized bed zone to ensure complete mixing. The orifice or nozzles are designed to have sufficient pressure drop to provide even distribution throughout the cross-section of the bed. When the bed is properly fluidized, temperature variations between any two spots in the fluidized bed typically will not exceed 6 °C to 8 °C (10 °F to 15 °F).

Fluidized bed incinerators have thermally oxidized various wastewater solids successfully. They have several advantages:

- The smooth, liquid-like flow of particles allows continuous, automatically controlled operations with ease of handling.
- They are well suited to large-scale operations.
- Heat- and mass-transfer rates between gas and particles are high.

Table 14.1 Representative Heating Values of Some Solids (Reprinted with permission from Suez)

Material	Combustible Solids (%)	High Heating Value (kJ/kg [Btu/lb] of Dry Solids)
Grease and scum	88	38 816 (16 700)
Raw wastewater solids	74	17 678 (7600)
Fine screenings	75	15 700 (6750)
Ground garbage	75	13 095 (5630)
Biosolids	60	12 793 (5500)
Grit	30	9304 (4000)

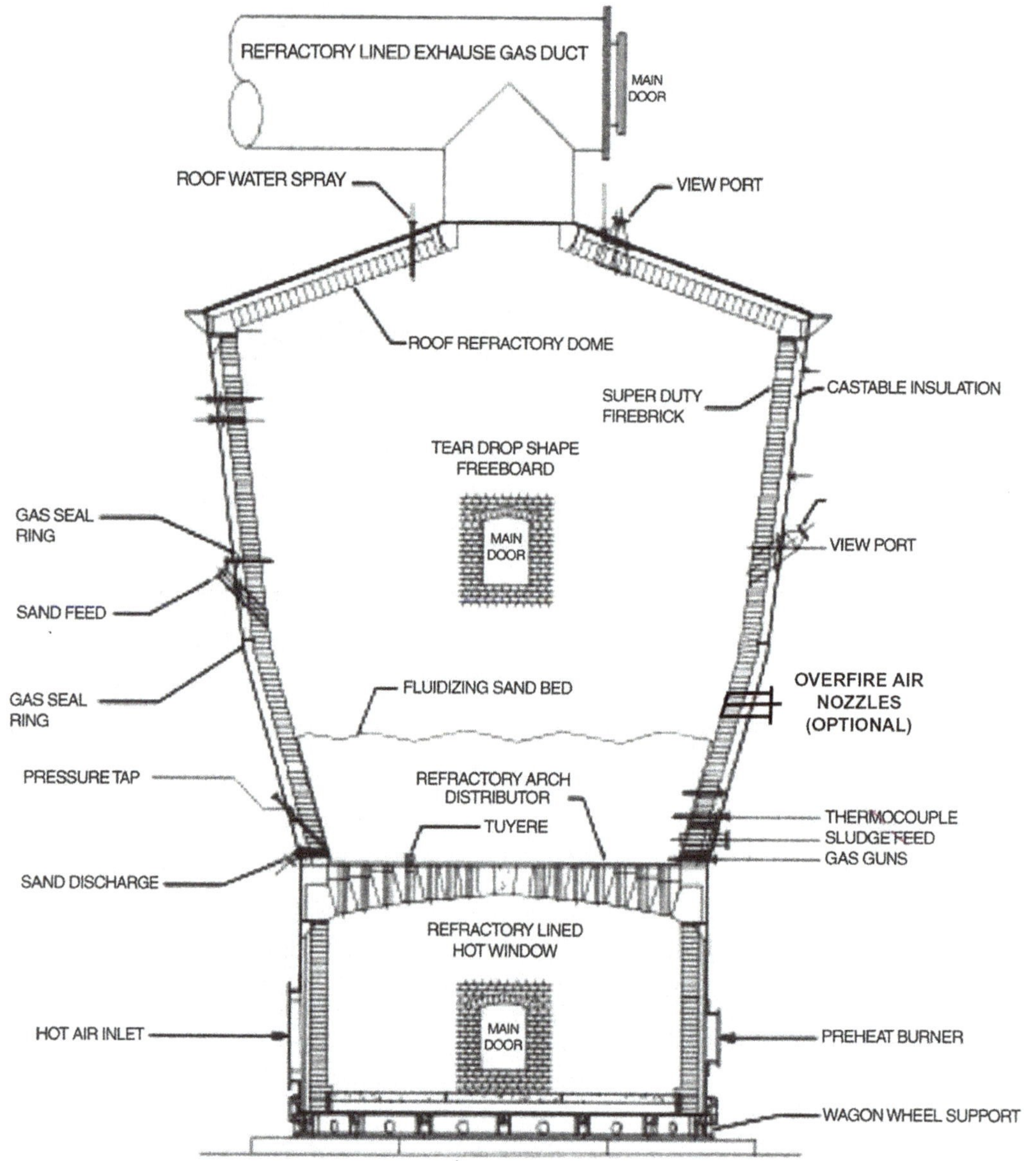

Figure 14.2 Typical Section of a Fluidized Bed Furnace (Reprinted with permission from Suez)

- Heat exchangers inside FBIs require relatively small surface areas because the heat-transfer rate between the incinerator and an object immersed in the bed zone is high.
- The incinerator can be shut down and started up daily without a preheating period because the bed cools slowly (3 to 8 °C/h [6 to 15 °F/hr]).

There are two types of FBIs: hot wind-box and cold (warm) wind-box. The hot wind-box systems incinerate wastewater solids with typically low heat values that require intensive air preheating to minimize auxiliary fuel consumption. The cold wind-box system typically is used to burn heat-treated solids, dried solids, primary solids, scum, grease, and other materials (e.g., wood chips or sawdust) that can burn autogenously without heat recovery or with moderate heat recovery.

1.4 MULTIPLE HEARTH FURNACE

An MHF consists of a vertical, refractory-lined steel cylinder containing a series of stacked horizontal refractory shelves (hearths) (Figure 14.3) (U.S. EPA, 1979, 1983b). A hollow cast-iron rotating shaft runs through the center of the hearths. Rabble arms are attached to this shaft above each hearth, and cooling air may flow through them. Metal blades (rabble teeth) on each arm may be angled in the direction of rotation (forward rabbling) or may be reversed (back rabbling). Back rabbling increases the cake's

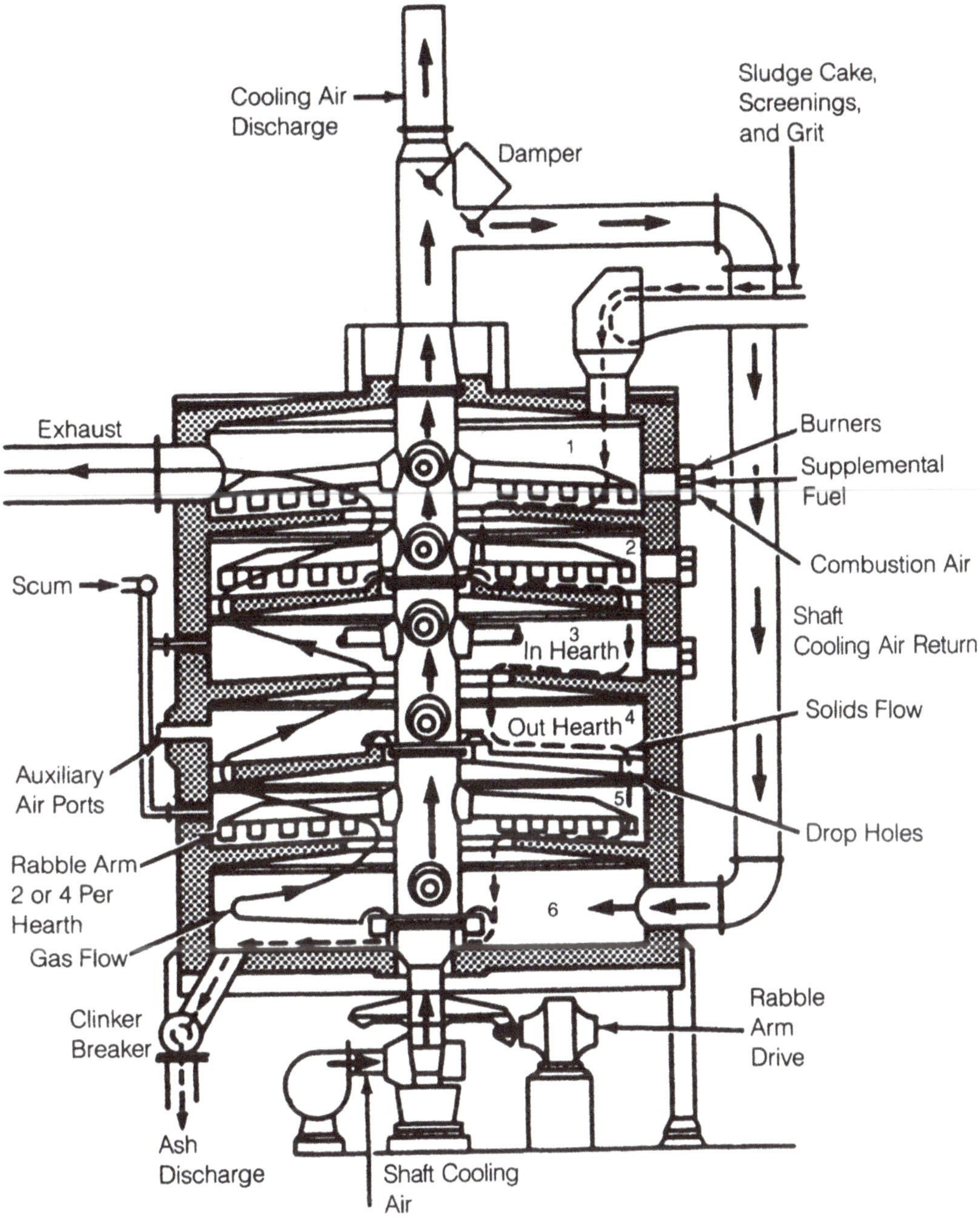

Figure 14.3 Typical Cross-Section of a Multiple Hearth Furnace

detention time and improves drying. Combustion air enters at the bottom of the hearth and circulates upward through drop holes in the hearths, countercurrent to solids flow. It exhausts through the top hearth to the waste heat exchanger (if present) and then to the air pollution control equipment.

Solids flow down through the furnace in a countercurrent pattern to the combustion gas and airflow, motivated by the rabble teeth, or plows, mounted on rabble arms that are attached to the rotating central shaft. Solids move outward and inward on successive hearths, dropping through either peripheral holes in a hearth ("out-hearth") or an annular opening at the center ("in-hearth"). The rabble arms and teeth sweep solids forward and radially several times on each hearth, resting after each plowing action. In addition to moving solids, the rabble teeth cut, furrow, turn, and open them, thereby exposing new surfaces for drying. Therefore the effective drying surface is estimated to be 130% of the hearth's actual area (Niessen, 1988; U.S. EPA, 1979).

Multiple hearth furnaces have three zones: drying, combustion, and char burning and cooling. Most of the water in solids evaporates in the upper hearths (drying zone). The heat-transfer mechanisms in this zone are about 15% convection and 85% radiation (Lewis & Lundberg, 1988). The temperature in the drying zone ranges from 315 °C to 480 °C (600 °F to 900 °F) as solids move from hearth to hearth. In the central hearths (combustion zone), the solids' combustibles are burned at 760 °C to 925 °C (1400 °F to 1700 °F). Volatile gases and solids burn in the upper hearths in the combustion zone, while fixed carbon burns in the lower hearths in the combustion zone. Incoming combustion air cools ash to between 95 °C and 205 °C (200 °F and 400 °F) in the lowest hearths (ash cooling zone). Ash exits through the bottom of the furnace (Water Environment Federation [WEF], 2009).

The MHFs have successfully treated various solids for more than 100 years. They have minimum space requirements and are reliable, easy to operate, and effective; reduce solids volume; and produce a sterile ash (U.S. EPA, 1983a). However, MHF-based systems have relatively high capital costs.

1.5 EMISSIONS CONTROL

Odors, the most noticeable form of air pollution, are the reason for about half of all citizen complaints to local air pollution control agencies. Thermal processing odors are particularly offensive, so design engineers must make every effort to minimize them.

The quantity and quality of combustion emissions from thermal processes depend on the thermal processing method used and the composition of the solids and auxiliary fuel. Completely combusted solids produce carbon dioxide, sulfur dioxide, water vapor, hydrogen chloride, and hydrogen fluoride. The exhaust gases will contain nitrogen, excess oxygen, and an inert ash that contains heavy metals, some of which (e.g., mercury) are volatile and present in both solid and gaseous forms. Exhaust gases could also contain ammonia or other compounds evaporated from the biosolids or formed during processing. In real life, however, combustion is never complete, so thermal processes release both nitrogen oxides and products of incomplete combustion. Such products include carbon monoxide, volatile organic compounds (VOCs), and polycyclic organic matter.

1.5.1 CARBON MONOXIDE

Carbon monoxide is a product of incomplete combustion that occurs when carbon in the feed cake partially oxidizes with oxygen in the combustion air. It is the result of one or more of the following combustion-system deficiencies: inadequate temperatures, inadequate residence time for combustion gases, or inadequate mixing or turbulence (needed to bring combustion gases in dynamic contact with oxygen in the air supply).

In general, multiple hearth and fluidized bed furnaces have markedly different combustion environments. In an MHF, the feed cake dries on the upper hearths before being combusted on the middle hearths. Although this arrangement is efficient (using the heat from combusting dried cake to dry incoming wet feed cake), it releases partially oxidized combustion gases and products of incomplete combustion from the upper hearths of the furnace when dried feed cake begins burning. The slow, stratified flow of combustion gases (i.e., lack of turbulence) in this part of the furnace leads to emission of high levels of carbon monoxide and products of incomplete combustion.

Carbon monoxide emissions from a MHF without an afterburner can range from 900 to 2700 mg/Nm^{3}dv (1000 to 3000 ppmdv); the typical carbon monoxide mass emission rate is 15.5 g/kg (31 lb/dry ton) of solids incinerated. High carbon monoxide and VOC emissions are one reason why the use of MHFs has declined steadily in recent decades. A high-temperature afterburner can control such emissions. Multiple hearth furnaces that have been retrofitted with top-hearth ("zero-hearth") afterburners have significantly lower carbon monoxide and VOC emissions. However, afterburners also increase fuel requirements. Post-scrubber afterburners have been shown to reduce carbon monoxide and VOC emissions with a fraction of the fuel demand required by zero-hearth afterburners (Chilson, 1994).

A fluidized bed furnace is a completely mixed, highly turbulent system in which drying and combustion take place concurrently and in a matter of seconds. Turbulence provides complete, intimate contact of feed cake, volatilized gases, and oxygen in the fluidizing air. As hot combustion gases rise from the bed, they enter the freeboard, which provides a long residence time that allows carbon monoxide and other volatilized organics to fully burn out. Carbon monoxide emissions from a fluidized bed furnace are invariably less than 45 mg/Nm^{3}dv (50 ppmdv) and often less than 9 mg/Nm^{3}dv (10 ppmdv). Mass emission rates are typically less than 0.5 g/kg (1.0 lb/dry ton) of solids incinerated.

State regulators typically require a new facility to meet a carbon monoxide emission limit of 90 mg/Nm^{3}dv (100 ppmdv). A fluidized bed furnace can easily meet this limit. An MHF would require an afterburner operating at a minimum of 816 °C (1500 °F) to meet this standard.

1.5.2 VOLATILE ORGANIC COMPOUNDS

The VOCs are products of incomplete combustion that occur when organic matter in feed cake vaporizes and volatilized compounds partially oxidize. Incomplete combustion happens when the temperature, residence time, oxygen content, and/or mixing in the thermal oxidizer is insufficient.

Chemically, a wide variety of compounds (e.g., straight and branched-chain aliphatic hydrocarbons [methane, ethane, acetylene, etc.]; oxygenated hydrocarbons [acids, aldehydes, ketones, etc.]; chlorinated hydrocarbons [perchloroethylene, trichloroethane, etc.]; and saturated and unsaturated ring compounds [benzene, toluene, phenols, etc.]) are volatile and organic. Such compounds are regulated under 40 CFR Part 503 (U.S. EPA, 1993) and 40 CFR, Parts 60 and 62 (U.S. EPA, 2011, 2016), which require that thermal oxidizers limit concentrations of total nonmethane hydrocarbons to less than 100 ppm as propane on a dry volume basis corrected to 7% oxygen (i.e., 100 ppmdv [140 mg/Nm^{3}dv]).

Because multiple hearth and fluidized bed furnaces have different combustion conditions, the VOC emissions from each are quite different. The upper drying hearths of an MHF typically will be hot enough to volatilize organic compounds but not fully oxidize them. Also, VOC emissions from an MHF depend on the cake's feed rate and combustion characteristics (percent solids, percent volatile solids, and heating value), as well as the furnace's operating conditions (hearth temperatures, excess air, and burner firing rates on different hearth levels). Some MHFs can meet the 140 mg/Nm^{3}dv (100 ppmdv) standard without an afterburner. Maintaining a top hearth temperature of at least 595 °C (1100 °F) typically is important for achieving the standard (Baturay, 1990;Waltz, 1990).

In a fluidized bed furnace, temperatures and turbulence are high, and total hydrocarbon (THC) emissions are low, typically less than 14 mg/Nm^{3}dv (10 ppmdv) as propane.

1.5.3 POLYCYCLIC ORGANIC MATTER

Polycyclic organic matter (e.g., polychlorinated biphenols [PCBs], polychlorinated dibenzo-p-dioxin [PCDD], and polychlorinated dibenzo furan [PCDF]) constitutes a subset of VOCs that is of particular concern to regulators because of its potentially high health risk effects. The U.S. EPA does not have specific emission limits for PCBs; however, total PCDD and PCDF emissions are regulated under the sewage sludge incineration maximum achievable control technology rule (U.S. EPA, 2011). In addition, some state regulatory agencies have included emission criteria for a few of these compounds in permits for new thermal oxidizers. The U.S. EPA's Compilation of Emission Factors (AP-42) indicates that both multiple hearth and fluidized bed furnaces emit low levels of these pollutants, in the range from 0.5×10^{-7} to 0.5×10^{-9} g/kg (U.S. EPA, 1998). To minimize emissions of polycyclic organic matter, system design should maximize the thermal oxidizer's combustion efficiency. The MHFs may need high-temperature afterburners if more control of polycyclic organic matter is necessary. This likely would not be necessary for a fluidized bed furnace.

1.5.4 NITROGEN OXIDES

Nitrogen oxides (e.g., nitrogen dioxide, nitrous oxide, and nitric oxide) are formed by the oxidation of nitrogen during combustion. The two types of nitrogen oxides are thermal nitrogen and fuel nitrogen. Thermal nitrogen is formed by atmospheric nitrogen in the combustion air. Fuel nitrogen is formed by the chemical oxidation of fuel- or solids-bound nitrogen. Emissions of NO_x attributable to furnace-burner firing can be significant and should not be overlooked in the design of a thermal oxidation system. Low NO_x burners should be used where feasible, particularly for multiple hearth systems where burner operation may be continuous in the furnace and an afterburner, if provided.

1.6 EMISSION REGULATIONS

Federal, state, and local governments all promulgate air pollution regulations, and they require emission sources to have permits to operate. The Clean Air Act of 1970 gave the U.S. EPA the responsibility and authority to establish a nationwide program to abate air pollution and enhance air quality, and the 1990 amendments strengthened that authority. State governments have primary responsibility for implementing the program under federal supervision, and each state has developed its own plan for achieving and maintaining the National Ambient Air Quality Standards (NAAQS). State governments often delegate the responsibility and authority for implementing the plan to local pollution control agencies.

The U.S. EPA promulgated Standards for the Use or Disposal of Sewage Sludge (40 CFR Part 503) in 1993. This rule, which falls under the Clean Water Act, includes emission standards and management practices for solids incinerators.

There are two types of air pollution regulations: limiting and administrative. Limiting regulations establish specific limits for specific air pollutants. Such regulations include the New Source Performance Standards, National Emission Standards for Hazardous Air Pollutants, local prohibitory rules, and source-specific standards. Administrative regulations establish requirements for gathering data (i.e., source testing and ambient monitoring), controlling the growth of emission sources, and using pollution control technologies. Such regulations include prevention of significant deterioration regulations for attainment areas (areas that meet NAAQS) and new source review regulations for nonattainment areas (areas that do not meet NAAQS).

1.7 ASH USE AND DISPOSAL

Ash is the product of solids incineration, essentially consisting of the noncombustible portions of the feed material. There are numerous methods and equipment available for handling ash and various uses for ash. Handling methods and the final destination or use of material is often site-specific. Ash handling equipment—especially conveyance—can be the most troublesome subsystem associated with incinerators.

Multiple hearth furnaces and fluidized bed incinerators use either wet or dry systems for ash conveyance. Ash is abrasive and sometimes non-uniform, making it difficult to convey as a bulk material, and conveyance systems must be designed accordingly. The design of such systems is not covered in this chapter, but is detailed in the WEF Manual of Practice titled *Wastewater Solids Incineration Systems* (WEF, 2009). That document also covers another critical aspect of ash handling: storing wet and dry ash.

1.7.1 REGULATORY CONSIDERATIONS

Regulations vary from state to state. Some states do not regulate ash; others treat it as a waste. Design engineers should check local, state, and federal regulations against biosolids characteristics before determining whether to use or dispose of incinerator ash. Each facility should conduct its own research on local, state, and federal regulations concerning ash disposal and reuse. Some landfills will require a toxicity characteristic leaching potential test or a pH test before accepting ash.

1.7.2 USE AND DISPOSAL OPTIONS

Biosolids incinerator ash is similar in physical and chemical characteristics to coal ash and can be used in applications where coal ash is permitted. Incinerator ash may be acidic, neutral, or alkaline, depending on the solids characteristics. In addition, incinerator ash will contain more phosphorus than coal ash. Although ash has historically been landfilled, interest in beneficial use of this material is increasing (Dominak et al., 2005). Beneficial use options for ash include:

- Landfill: Ash can be used as a landfill cover or blended with soil and used as a cover.
- Fill material: Ash can be used as fill material for excavations. One utility, for example, has a contractor using the material to fill old solids lagoons. The material also can be used as a flowable fill.
- Soil amendment: In some areas (e.g., those with high clay soils), incinerator ash may be used as a soil additive that produces a soil that handles more easily, allows better drainage and airflow, and includes some valuable minerals.
- Concrete fly ash: Ash has been used as a fly ash substitute in concrete mixes.

- Asphalt additive: The ash has been used as a mineral filler and fine aggregate in asphalt mixes. New Jersey's Department of Solid and Hazardous Waste has permitted the use of incinerator ash for this purpose.

Use options for incinerator ash tend to be site-specific. Utilities should pursue all avenues available for using ash. State departments of transportation should be contacted to determine requirements for using incinerator ash as a fly ash substitute in concrete or as a mineral filler or fine aggregate substitute in asphalt mixes. If the incinerator ash is approved for use in mix designs by the state department of transportation, then a considerable market can be opened for incinerator ash.

2.0 LANDFILLING

Landfilling is an option for disposing of residuals from wastewater treatment. Considerations for adopting landfilling as an ultimate disposal method include planning, designing, constructing, monitoring, and closing such landfills. Important aspects include:

- Landfill siting and capacity needs;
- Liner and leachate-collection system design;
- Surface water control;
- Landfilling methods;
- Daily, intermediate, and final covers;
- Monitoring requirements;
- Gas migration control, collection, and reuse;
- Covers and cap systems; and
- Landfill closure and reuse.

2.1 LANDFILL TYPES AND REGULATORY CONSIDERATIONS

There are two types of landfill sites that typically accept solids. One is a monofill—a landfill that only accepts stabilized or unstabilized municipal wastewater solids. The other is a co-disposal landfill, which accepts both residuals and municipal solid waste (MSW).

2.2 MONOFILLS

The U.S. EPA regulates monofill design and operation under Subpart C of 40 CFR Part 503, which governs surface disposal of wastewater solids. Part 503 standards include general requirements; pollutant limits; management practices; pathogen-reduction and vector-attraction reduction (VAR) alternatives; and monitoring, recordkeeping, and reporting requirements.

Subpart C addresses two types of monofill operations. Unstabilized solids must adhere to VAR Alternative 11 and must be covered with soil or another material at the end of each operating day. Stabilized solids, on the other hand, must meet one of the other VAR alternatives and do not have to be covered each day. Monofills for stabilized solids operate more like land application sites with exceptionally high application rates.

2.3 CO-DISPOSAL LANDFILLS

The U.S. EPA regulates co-disposal landfill design and operation under 40 CFR 258, which governs the disposal of MSW (e.g., household wastes) (U.S. EPA, 1991). Because wastewater solids typically are a small percentage of the waste at such sites, Part 258 regulations are not referenced in detail in this section.

The disposal rate partially depends on the residuals' solids content. At co-disposal landfills, solids typically are spread in the active area (the part of the landfill accepting waste) and mixed with incoming solid waste to ensure that the material has acceptable handling characteristics. For example, a cake containing 20% solids might be mixed with solid waste 4:1 (i.e., 4 metric ton of solid waste to 1 metric ton of solids [wet ton basis]). Residuals with lower solids contents would need to be mixed with more solid waste. The mixing process (also called bulking operation) typically depends on the type and quantities of wastes delivered to the landfill.

Solids delivered to a co-disposal site must not contain any free liquids, as defined via the paint filter liquids test (U.S. EPA, 1995b). Dewatered cakes containing 20% solids typically pass this test.

3.0 LAND APPLICATION

Land application is the practice of adding biosolids to land for beneficial purposes (e.g., to promote crop growth, to promote forest growth, to improve soil quality, and to reclaim former mining sites and other disturbed land). In these applications, both plants and soil benefit from the nutrients and organic matter in biosolids.

The popularity of land application has grown dramatically over the last 30 years. Rising disposal costs and encouragement from the U.S. EPA have substantially increased the number of facilities undertaking land application programs. According to a 2018 survey conducted by the North East Biosolids and Residuals Association (NEBRA, 2018), "the percent of beneficial use of biosolids increased, from 49% in 2004 (not including stored biosolids) to 53% in 2018" (NEBRA, 2018). The NEBRA survey also noted that the term "beneficial use" historically referred to "biosolids that are applied to soils to take advantage of the nutrients and organic matter they contain." In the future, it noted, this definition may be too narrow, as biosolids are used in ways that provide other benefits (e.g., energy)(NEBRA, 2018).

3.1 REGULATORY CONSIDERATIONS

Federal and state regulations establish controls for the land application of biosolids. Some activities (e.g., the selection and management of land application sites) also can be governed on a local level. The U.S. EPA has delegated responsibility for compliance with federal regulations to a limited number of states.

3.1.1 FEDERAL REQUIREMENTS

At the federal level, biosolids are regulated under Section 405(d) of the Clean Water Act. Specific criteria are set forth in 40 CFR Part 503 (U.S. EPA, 1993), which is the baseline for beneficial use across the country. States and local jurisdictions may add more stringent or site-specific requirements.

3.1.2 STATE REQUIREMENTS

Some states have enacted regulations more stringent than Part 503. Wastewater treatment professionals should consult state regulatory requirements when assessing the feasibility of land application or planning a land application program.

State requirements vary. However, an increasing number of states require nutrient management planning and some states (e.g., Virginia) may require conservation planning (Evanylo, 1999). In addition, a number of states require permits for land application sites. More recently, the issue of per- and polyfluoroalkyl substances, collectively referred to as PFAS, has received national attention. The state of Maine, for instance, has passed legislation banning land application of biosolids containing PFAS. Other states are considering similar measures. In 2021, U.S. EPA published a strategic roadmap on the monitoring, measurement, and regulation of PFAS, including biosolids, and when the agency expects action to be taken (U.S. EPA, 2021).

3.1.3 NUTRIENT MANAGEMENT PLANNING

Land-applied biosolids supply nutrients (e.g., nitrogen, phosphorus, and micronutrients) to plants. However, too much of any nutrients can lead to water quality issues. Excess nitrogen is associated with groundwater concerns, for example, and excess phosphorus is associated primarily with surface water concerns. (For more information on nutrients and biosolids, see *Comparing the Characteristics, Risks and Benefits of Soil Amendments and Fertilizers Used in Agriculture* [Moss et al., 2002] and *National Manual of Good Practice for Biosolids* [National Biosolids Partnership, 2005].)

Part 503 requires that biosolids be applied at the agronomic rate for nitrogen (i.e., the rate needed to meet a given crop's nitrogen requirements), and many state regulations share this requirement. However, increasingly more states are requiring that land appliers manage phosphorus as well. Phosphorus-based management can quadruple the area needed for land application programs. Some of the states that require phosphorus-based management rely on the Nutrient Management Standard (Code 590) issued by the U.S. Department of Agriculture's (USDA's) Natural Resources Conservation Service (NRCS) (USDA,

1999). Where used, this national standard (called the P index) typically is modified by states to account for local conditions.

3.1.4 SITE PERMITS

Site-permitting programs typically are administered by state agencies, although local entities can be involved in some states. The permits contain the issuing agency's specific requirements for land application, which are designed to ensure that the biosolids are being beneficially used, not disposed of. Permit requirements also are typically designed to protect public health, surface water, and groundwater, as well as address the aesthetic concerns (e.g., odor) of the land application sites' neighbors.

For example, most state programs impose separation distances (setbacks) between the areas receiving biosolids and adjacent site features (e.g., developments, dwellings, surface water, wells, roads, and rights of way). Setbacks are based primarily on the potential for surface runoff, leaching, and aesthetic concerns and will vary depending on the environmental resource being protected (see Table 14.2). Many states will reduce the setback requirements if the biosolids are immediately incorporated into the soil, applied via subsurface injection, or if the neighboring property owner(s) waive the setbacks (Forste, 1996).

3.2 BIOSOLIDS CHARACTERISTICS AND SUITABILITY

When biosolids are used in agriculture, regulations and crop needs typically determine which biosolids characteristics (e.g., pathogen content; vector attraction; and nutrient content—typically nitrogen, phosphorus, and/or calcium) will help to drive the development of a land application program (see Table 14.3). Site conditions and economics (e.g., distance to land application sites, application equipment typically in use) also may determine whether the biosolids should be applied in liquid or cake form.

3.3 PATHOGENS AND VECTORS

Pathogen content of wastewater treatment solids depends on the level of treatment they have received. Raw primary solids contain the most pathogens due to lack of treatment. In contrast, biosolids that meet Part 503's Class A standards have very low pathogen concentrations, which can, in some cases, even be below detection limits. Concentrations of specific pathogens (e.g., helminths) depend on the sources in a sewershed, as well as the climate.

Sewershed is the area that drains into a sewerage or collection system.

Before they can be land applied, biosolids must meet Class A or B pathogen-reduction standards (as well as relevant VAR and pollutant concentrations). Land application sites that receive Class B biosolids have additional management restrictions (e.g., waiting periods before specific crops can be grown) (U.S. EPA, 1994).

Part 503 requires that Class B biosolids contain less than 2 million colony-forming units (CFU) (most probable number [MPN]) of fecal coliforms per gram of dry biosolids. Class A biosolids must contain less than 1000 MPN/g of fecal coliforms or less than 3 MPN/4 g of *Salmonella* bacteria, and meet one of six alternatives. One alternative includes testing for enteric viruses and helminth ova (U.S. EPA, 2003). Larger water resource recovery facilities (WRRFs) may have the personnel, equipment, and accreditation to handle bacterial testing on site, whereas smaller WRRFs may send samples for bacterial testing to commercial laboratories with appropriate accreditation. Virus and helminth ova testing is nearly always performed by commercial laboratories with appropriate accreditation.

Table 14.2 Representative Setback Requirements for Land Application

	Setback Distance	
Feature	**ft**	**m**
Public road	0–50	0–15
House	20–500	6–152
Well	100–500	30–152
Surface water	25–300	8–91
Property line	None listed–100	None listed–30
Intermittent stream	10–200	3–61

Table 14.3 Typical Biosolids Physical Characteristics (Logan, 2008; Lue-Hing et al., 1998)

Property	Class A Alkaline Stabilized	Compost	Liquid	Cake	Heat-Dried
Total solids (%)	30–65	58	4	22	95
Volatile solids (% of total solids)	12–20	60	61	60	67
Bulk density (kg/m³ [lb/cu ft])	1040–1200 (65–75)	720 (45)	960 (60)	880 (55)	530–740 (33–46)
Organic content (%)	12–20	50–75	60–75	60–75	78
pH	—	—	—	—	6.4–8.0

Land-applied biosolids also must meet VAR requirements via thermal drying, volatile solids reduction (VSR), oxygen uptake rate, time and temperature, or pH increases (U.S. EPA, 2003). Eight VAR options are based on process parameters or testing; two are based on land management (e.g., soil incorporation or injection). In order to qualify for exceptional quality, the generator must meet one of the first eight VAR options.

For details on pathogen and VAR requirements, methodologies, monitoring, and testing, see *Environmental Regulations and Technology: Control of Pathogens and Vector Attraction in Sewage Sludge* (also called the "White House document") (U.S. EPA, 2003).

3.4 METALS

Biosolids contain a wide range of chemical elements, many of which are trace elements, also called heavy metals. This is a misnomer because the density of some trace elements (e.g., boron and arsenic) is too low to be considered a heavy metal. Likewise, not all trace elements behave as metals in the environment. The ones with metallic chemical behavior are those that are cations in the environment (e.g., copper [Cu^{2+}], cadmium [Cd^{2+}], and lead [Pb^{2+}]). Other trace elements are anions in the environment (e.g., arsenic [AsO_4^{3-}], molybdenum [MoO_4^{2-}], and selenium [SeO_4^{2-}]. This distinction is important because cationic trace elements are more bioavailable when soil pH is low, while anionic trace elements are more bioavailable when soil pH is high.

When developing 40 CFR Part 503, the U.S. EPA conducted a risk assessment on a number of trace elements in biosolids and developed a hazard index, which considered various pathways of exposure (e.g., direct ingestion of biosolids, plant uptake, and uptake by soil organisms) from trace elements in land-applied biosolids (U.S. EPA, 1995a). The agency eliminated two trace elements (fluorine and iron) based on the hazard index. Then, regulators developed acceptable concentrations for the remaining 10 elements based on a pathway analysis. The ceiling concentration is the maximum safe concentration for biosolids land-applied at agronomic rates. A lower concentration also was established for what was later called exceptional quality (EQ) biosolids (this term refers to biosolids that meet Class A pathogen-reduction requirements, meet VAR requirements [503.33(a)(1) through (8)] at the same time or after pathogen reduction is met, and have low concentrations of regulated pollutants [503.13, Table 3]). The intent of the concentrations for EQ biosolids is that these levels would be safe no matter how much biosolids were applied to land. It is important to note that while biosolids may meet the metals levels associated with EQ biosolids, they cannot be considered EQ biosolids without meeting the Class A pathogen requirements.

Part 503 originally regulated 10 trace elements: arsenic, cadmium, chromium, copper, lead, mercury, molybdenum, nickel, selenium, and zinc. Lawsuits filed against the U.S. EPA resulted in the removal of chromium as a regulated metal and the removal of the EQ limit for molybdenum.

When Part 503 was promulgated, there were a number of municipal WRRFs whose biosolids failed to meet the EQ limits for one or more elements. Most, however, met the ceiling concentrations so their biosolids could be land-applied, but there was a limit on the cumulative loading to a given site (U.S. EPA, 1995a). By the late 1990s, industrial pretreatment was so successful that most WRRFs could meet EQ limits. Today, trace elements in biosolids are no longer considered an important limitation in land application programs.

3.5 NUTRIENTS

Biosolids contain various concentrations of macro- and micronutrients. As shown in Table 14.4, the nitrogen content in biosolids ranges from 1.0% (dry-weight) for Class A alkaline-stabilized biosolids to up to 6.0% (dry-weight) for heat-dried biosolids. Primary and waste activated solids mostly contain organic nitrogen (in the form of protein), while up to a third of total nitrogen in aerobically and anaerobically digested biosolids is ammonia. Biosolids contain very little, if any, nitrate.

The total phosphorus content in biosolids typically ranges from 0.9% to 3.1% (dry-weight). This includes both organic phosphorus and various forms of inorganic phosphorus. Municipal WRRFs that use iron and aluminum salts for phosphorus removal will have solids that contain inorganic phosphorus in the form of iron and aluminum phosphates. The WRRFs that use lime for dewatering or disinfection will have solids that contain inorganic phosphorus in the form of calcium phosphates. Iron, aluminum, and calcium phosphates are relatively insoluble, which will limit the availability of phosphorus in soils receiving repeated applications of biosolids.

Biosolids also contain small amounts of potassium—typically 0.1% to 0.3% (dry-weight)—and are not a significant source of potassium for crops. Most of the potassium in biosolids is in a water-soluble form. In addition, biosolids contain various concentrations of other macro- and micronutrients (e.g., calcium, magnesium, sulfur, iron, manganese, as well as the regulated elements [copper, molybdenum, selenium, zinc]). Some plant species also require nickel for growth. The plant availability of these nutrients depends on the major chemical phases in the biosolids. Elements like copper and zinc are known to be complexed with the biosolids organic matter, and most sulfur in biosolids is in the form of protein. On the other hand, iron and manganese exist as relatively insoluble oxides.

3.6 OTHER CONSTITUENTS

Biosolids contain myriad organic and inorganic chemical compounds. Some are discharged into the collection system as household chemicals (e.g., surfactants and conditioning agents used in detergents) and very low levels of medicines (e.g., birth control chemicals) and concentrate in biosolids. Biosolids also contain natural materials (e.g., silica and aluminosilicate clays) that enter the collection system as sediments. For quantitative estimates, the U.S. EPA has conducted three national sewage sludge surveys with the most recent in 2009. These survey results present national estimates of the concentrations of more than 500 pollutants in sewage sludge, including metals; dioxins and dioxin-like compounds; inorganic ions; certain organics (e.g., polycyclic aromatic hydrocarbons, semivolatiles); polybrominated diphenyl ethers (flame retardants); and pharmaceuticals, steroids, and hormones in sewage sludge managed by land application (U.S. EPA, 2009).

In the case of alkaline-stabilized biosolids, lime and lime-containing chemicals (e.g., alkaline coal ash, cement kiln dust, and lime kiln dust) are added to biosolids for disinfection. The lime reacts with water in biosolids to form calcium hydroxide, which has a pH of 12 to 12.5 and adds soil liming value to the biosolids. Alkaline-stabilized biosolids often are used as a substitute for agricultural limestone.

3.7 SITE SUITABILITY

After establishing that the biosolids are suitable for land application, program staff must determine if suitable land is available. The objective of a site evaluation is to select application sites that not only meet the technical requirements of land application but also balance prevailing economic and social constraints. When determining how much land is needed for a land application program, it is important to note that adequate acreage must be available during all but the most inclement weather conditions. Site availability

Table 14.4 Primary Nutrient Concentrations in Biosolids (Adapted from Moss et al., 2002)

Nutrient Content (as % of Dry Weight)	Class A Alkaline Stabilized	Compost	Liquid	Cake	Heat-Dried
Nitrogen (N)	1.0	2.8	5.3	4.1	6.0
Total phosphorus (P)	0.4	1.7	2.2	2.2	3.1
Potassium (K)	0.3	0.3	0.3	0.2	0.3

is determined by local farming practices, and there are periods of the year when certain types of cropland are unavailable for the application of biosolids. Additionally, a number of states have seasonal restrictions, such as prohibitions on applications on frozen or snow-covered ground. Therefore, to ensure land availability, it is advisable to maintain two to three times the acreage actually needed in any given year to accommodate all of the biosolids produced. Typically, storage capacity at the WRRF or the beneficial use site will also be necessary at some time during the year.

3.7.1 RESOURCES

Soil surveys are a useful source of information for initially evaluating a site's suitability for land application. The surveys provide delineated, broad-scale soil maps on a photographic background; a soil description by series and mapping unit; data on the drainage and agronomic properties of soils; and interpretive tables. Soil surveys are prepared by the USDA's NRCS in cooperation with agricultural experiment stations and local government units. They are available from local NRCS offices or online (http://websoilsurvey.nrcs.usda.gov). The soil surveys are used extensively to identify potential sites that meet the regulatory and agronomic requirements for land application.

3.7.2 SITE EVALUATION CRITERIA

Although selection criteria for land application sites vary from state to state, the *National Manual of Good Practice for Biosolids* discusses a variety of factors that should be considered in any land application program (NBP, 2005). Soil surveys contain essential information for determining whether a field has appropriate soil characteristics for biosolids application (e.g., soil texture, erodibility, soil drainage characteristics, and slopes). Favorable soil characteristics are discussed in Appendix C of the *National Manual of Good Practice for Biosolids* (NBP, 2005). U.S. Geological Survey quadrangle maps are also helpful during preliminary planning and screening to estimate slope, topography, depressions or wet areas, rock outcrops, drainage patterns, and water table elevations. The following topographical and soil characteristics are unfavorable for biosolids application:

- Steep areas with sharp relief;
- Undesirable soil conditions (sandy, shallow, highly erodible, or poorly drained);
- Environmentally sensitive areas (e.g., intermittent streams, ponds);
- Rocky, non-arable land;
- Wetlands and marshes; and
- Areas bordered by surface water bodies without appropriate setback areas.

In addition to topography and soil characteristics, any preliminary site evaluation should assess water quality-based requirements at each farm (e.g., buffer zones, conservation planning, and nutrient management). Part 503 requires a buffer of at least 10 m (33 ft) to surface waters, and states often have additional buffer requirements to surface water bodies, water supply wells, property lines, outcroppings, or public roads. Because the buffer requirements limit the actual land available for application, factoring in these limitations may make some sites unsuitable for land application.

Some states require that farms implement conservation plans to control soil erosion and nutrient management plans to limit excess nutrient transportation to water resources. Certain conservation measures (e.g., no-till or residue management) and nutrient management issues (e.g., fields with a long history of land-applied manure or biosolids) could limit a site's suitability for land application.

Site criteria important to the economic viability of a land application program include the ease of access to the farm (e.g., road restrictions or traffic limitations) and the hauling distance or travel time (and therefore cost).

3.7.3 PUBLIC PERCEPTION

Assessing the public acceptance of and attitudes toward land application is an important part of the site evaluation process. The attitudes of stakeholders (e.g., neighbors) can be critical to the success of a land-application program, and discussions with farmers or landowners beforehand often can provide key insights in this regard.

Farmer and landowner attitudes are critical to program success as well. According to the *National Manual of Good Practice for Biosolids* (NBP, 2005), the landowner or farmer "must be confident about the benefits of biosolids and be willing to accept and comply with all of the…regulatory requirements, as well as meet biosolids program needs." To maximize program success, publicly owned treatment facilities s should provide outreach and education to famers and to all stakeholders during the evaluation process.

3.8 ODOR MANAGEMENT

Odors are frequently cited as the basis for opposing land-application programs. One approach to minimizing the potential for off-site odors is maximizing the buffers between land application fields and the public. As many agencies have found, however, development can encroach on previously remote sites, reducing buffers and increasing the potential for odor complaints. Therefore, odor-management plans cannot be based on setbacks and buffers alone. The *National Manual of Good Practice for Biosolids* (NBP, 2005) recommends a comprehensive odor-management approach that addresses liquid and solids treatment at the WRRF, transportation, site storage, field operations, and community relations. Addressing potential odors (and their effects) at each "critical control point" is essential to effectively managing odors at land-application sites. In all cases, generating a stable, low-odor product is essential to minimize the potential for odor-related challenges.

Example Problem 14.1

A farm growing cotton is considering the application of anaerobically digested biosolids. Metals limits must be considered, and the design will be based on nitrogen application. The characteristics of the biosolids are as follows:

Sludge production:	3 tonne/d
Solids content:	7%
Total nitrogen:	3%
Ammonia-nitrogen:	2%
Nitrate:	0%
Arsenic:	50 mg/kg
Cadmium:	18 mg/kg
Chromium:	1000 mg/kg
Copper:	400 mg/kg
Lead:	430 mg/kg
Mercury:	20 mg/kg
Molybdenum:	15 mg/kg
Nickel:	0 mg/kg
Selenium:	50 mg/kg
Zinc:	900 mg/kg
Volatilization factor, *K*:	0.5
Annual uptake by cotton:	100 kg/hay

Solution

Assume surface application of biosolids with volatilization losses. Calculate the plant available nitrogen using Equation 14.1:

$$N_{PA} = 1000(NH_4^+ \cdot K + NO_3^- + N_O \cdot f) \tag{14.1}$$

Where N_{PA} = plant available nitrogen during application year (kg/metric tonne dry solids)
NH_4^+ = fraction of ammonia in biosolids
NO_3^- = fraction of nitrate in biosolids
N_O = fraction of organic nitrogen in biosolids
f = mineralization factor

Organic nitrogen = total N − $(NH_4^+ + NO_3^-)$ = 1%

Mineralization factor for Year 1 = $f_1 = 0.2$

$N_{PA} = 12$ kg/tonne dry solids

Residual N in biosolids in year 2, $N_{O,2} = N_{O,1} - f_1 \cdot N_O = 0.008$

$f_2 = 0.1$

Year 2 mineralization = $f_2 \cdot N_{O,2} = 0.0008$

Residual N in biosolids in Year 3 = residual in Year 2 − Year 2 mineralization = 0.0072, and so forth

$f_3 = 0.05$

Mineralization in Year 3 = $f_3 \cdot N_{O,3} = 0.0072$

$f_4 = 0.03$

Mineralization in Year 4 = 0.0002

$f_5 = 0.03$

Mineralization in Year 5 = 0.0002

Total available nitrogen in Year 2 is the Year 2 contribution plus the residual from Year 1

$N_{PA,2} = N_{PA,1} + 1000 \cdot f_1 \cdot N_{O,2} = 12.8$ kg/tonne dry solids

$N_{PA,3} = N_{PA,1} + 1000 \cdot f_1 \cdot N_{O,2} + f_3 \cdot N_{O,3} = 13.2$ kg/tonne dry solids

$N_{PA,4} = N_{PA,1} + 1000 \cdot f_1 \cdot N_{O,2} + f_3 \cdot N_{O,3} + f_4 \cdot N_{O,4} = 13.4$ kg/tonne dry solids

$N_{PA,4} = N_{PA,1} + 1000 \cdot f_1 \cdot N_{O,2} + f_3 \cdot N_{O,3} + f_4 \cdot N_{O,4} + f_5 \cdot N_{O,5} = 13.6$ kg/tonne dry solids

Future N_{PA} will remain 13.6 assuming the biosolids do not change.

Calculating the nitrogen limiting sludge loading:

$$R_N = \frac{U_N}{N_{PA} + N_{PM}}$$

Where R_N = annual biosolids loading for year of concern (tonne/ha)
U_N = annual crop uptake (kg/ha)
N_{PA} = plant available nitrogen for year of concern (kg/tonne)
N_{PM} = plant available nitrogen from mineralization of all previous years' applications (kg/tonne)
R_N = 7.35 tonne/ha

Calculate lifetime metal loading:

Lifetime biosolids application rate = cumulative pollutant loading/(metal limit 0.001)

Metals limits are provided in Table 14.4.

Arsenic (limit 41 kg/ha):	820 tonne/ha
Cadmium (39):	2167
Chromium (3000):	3000
Copper (1500):	3750
Lead (300):	698
Mercury (17):	850
Molybdenum (18):	1200
Nickel (420):	5250
Selenium (100):	2000
Zinc (2800):	3111

The lowest biosolids loading is controlled by the application of lead. The useful lifetime of the site is limited by the application of 698 tonne/ha. If all concentrations in the biosolids are below the limits shown above, heavy metals would not constrain the biosolids land application.

The useful life of the site is limited by the lead concentration; therefore:

Useful life = application limit for lead/application rate for nitrogen = 94.9 yr

An application rate of 698 tonne/ha and an annual production of 1095 tonne/y requires 1.6 ha of land to be reclaimed each year.

PRACTICE PROBLEMS

1. Why is the pH of biosolids and soil important for land application?
2. The limiting factor for land application is often the nitrogen content of biosolids. Why is this? Isn't more nitrogen better for crop production, and if so, why is it limiting?
3. Why are liquid biosolids not able to be landfilled? Is this a reasonable prohibition?
4. Suppose you have a process to produce Class A biosolids. What would be involved in having this process added to the list of accepted treatment practices?
5. Rework the example problem (14.1) given the following for land application of biosolids on a farm growing corn for bioethanol production:

Sludge production:	50 tonne/d
Solids content:	22%
Total nitrogen:	3.5%
Ammonia-nitrogen:	4%
Nitrate:	0%
Arsenic:	60 mg/kg
Cadmium:	20 mg/kg
Chromium:	1100 mg/kg
Copper:	500 mg/kg
Lead:	300 mg/kg
Mercury:	30 mg/kg
Molybdenum:	10 mg/kg
Nickel:	100 mg/kg
Selenium:	45 mg/kg
Zinc:	1000 mg/kg
Volatilization factor, K:	0.5
Annual uptake by corn:	172 kg/ha·yr

4.0 REFERENCES

Baturay, A. (1990). *Case studies of total hydrocarbon emissions (THC) from multiple hearth sewage sludge incinerators and THC reduction strategies.* Prepared for the Association of Metropolitan Sewerage Agencies Incinerator Workgroup Meetings, New Orleans, Louisiana, United States.

Chilson, S. J. (1994, October). Hatfield targets 20% of EPA 503 regulations at 85% fuel savings. *Proceedings of the 67th Annual Water Environment Federation Exposition and Conference*, Chicago, Illinois, United States.

Dominak, R. P. (2005). *Long term residuals management plan for the Northeast Ohio Regional Sewer District.* Northeast Ohio Regional Sewer District.

Evanylo, G. K. (1999). *Agricultural land application of biosolids in Virginia* (Publication No. 452-303). Virginia Cooperative Extension.

Forste, J. (1996). Land application. In M. Girovich, (Ed.), *Biosolids treatment and management.* Marcel Dekker.

Lewis, F. M., & Lundberg, L. A. (1988, June). Modifying existing multiple-hearth incinerators to reduce emissions. *Proceedings of the National Conference on Municipal Sewage Treatment Plant Sludge Management*, Palm Beach, Florida, United States. Hazardous Materials Research Institute.

Logan, T. (2008, July). Personal communication.

Lue-Hing, C., Tata, P., Granato, T. A., Pietz, R., Johnson, R., & Sustich, R. (1998). *Sewage sludge survey.* Association of Metropolitan Sewerage Agencies.

Los Angeles County Sanitation Districts. (1988). *1988 survey of thermal conditioning and wet air oxidation facilities.*

Moss, L., Epstein, E., & Logan, T. (2002). *Comparing the characteristics, risks and benefits of soil amendments and fertilizers used in agriculture* (Report 99-PUM-1). Water Environment Research Foundation.

National Biosolids Partnership. (2005). *National manual of good practice for biosolids.*

Niessen, W. R. (1988, June). Thermal processing of wastewater treatment plant sludges. *Proceedings of the National Conference on Municipal Sewage Treatment Plant Sludge Management*, Palm Beach, Florida, United States. Hazardous Materials Research Institute.

North American Manufacturing Co. (1965). *North American combustion handbook.*

North East Biosolids and Residuals Association. (2018). *A National Biosolids Regulation, Quantity, End Use and Disposal Survey.*

U.S. Department of Agriculture, Natural Resources Conservation Service. (1999). *Nutrient Management Code 590* (Conservation Practice Standard Publication).

U.S. Environmental Protection Agency. (1979). *Process design manual for sludge treatment and disposal* (EPA-625/1-79-011).

U.S. Environmental Protection Agency. (1983a). *Assessment of sludge processing problems at selected POTWs.* U.S. Environmental Protection Agency, Office of Water Management.

U.S. Environmental Protection Agency. (1983b). *Municipal wastewater sludge combustion technology* (EPA-625/4-85/015).

U.S. Environmental Protection Agency. (1991). Solid waste disposal facility criteria (Final Rule, Part II). *Code of Federal Regulations*, Parts 257 and 258, Title 40.

U.S. Environmental Protection Agency. (1993). Standards for the use or disposal of sewage sludge. *Code of Federal Regulations*, 40 CFR, Part 503, Subpart E.

U.S. Environmental Protection Agency. (1994). *Plain English guide to the EPA Part 503 Rule* (EPA 832R93003).

U.S. Environmental Protection Agency. (1995a). *A guide to the biosolids risk assessment for the EPA Part 503 Rule* (EPA832B93005).

U.S. Environmental Protection Agency. (1995b). *Process design manual: Surface disposal of sewage sludge and domestic septage* (EPA 625R95002).

U.S. Environmental Protection Agency. (1998). *Compilation of air pollutant emission factors, Volume I: Stationary point and area sources* (AP-42; Section 2.2). U.S. Environmental Protection Agency, Office of Air Quality Planning and Standards.

U.S. Environmental Protection Agency. (2003). *Environmental regulations and technology: Control of pathogens and vector attraction in sewage sludge* (EPA 625R92013).

U.S. Environmental Protection Agency. (2009). *Targeted national sewage sludge survey sampling and analysis technical report* (EPA-822-R-08-016).

U.S. Environmental Protection Agency. (2011). *Standards of performance for new stationary sources and emission guidelines for existing sources: Sewage sludge incineration units* (Final Rule). *Code of Federal Regulations*, 40 CFR, Part 60.

U.S. Environmental Protection Agency. (2016). Federal plan requirements for sewage sludge incineration units constructed on or before October 14, 2010 (Final Rule). *Code of Federal Regulations*, 40 CFR, Part 62.

U.S. Environmental Protection. (2021). *EPA's PFAS strategic roadmap: A year of progress.*

Waltz, E. W. (1990). *Technical discussion of proposed EPA hydrocarbon regulation for sludge incinerators—Charts and Graphs*. Prepared for the Association of Metropolitan Sewerage Agencies Incinerator Workgroup Meetings, New Orleans, Louisiana, United States.

Water Environment Federation. (2009). *Wastewater solids incineration systems* (Manual of Practice No. 30).

Chapter 15 | Tertiary Treatment for Sustainable Water Reuse

LEARNING OBJECTIVES

At the end of this chapter, students should be able to:

- Provide examples of two common chemical disinfectants and describe what is involved with each with respect to design and operational considerations.
- Describe the main aspects of disinfection with ultraviolet light, including its advantages and disadvantages when compared to chemical disinfection.
- Explain the main operational and design considerations for granular media filtration.
- Provide examples of two non-granular media alternatives for filtration and explain how they function.
- Describe the principles behind membrane filtration and give examples for the spectrum of membrane processes.

1.0 DISINFECTION

Disinfection is the most critical component of wastewater treatment for the protection of public health. Improperly disinfected water and wastewater have been responsible for major disease outbreaks in both the developing and developed worlds. Readers are encouraged to consult other literature, such as *Ultraviolet Disinfection for Wastewater* (Water Environment Federation [WEF], 2007) and *Disinfection of Wastewater Influent* (Water Environment Research Foundation [WERF], 2015) to obtain additional information on the importance of the disinfection practice on public health.

Disinfection is the selective destruction of disease-causing microbes via the application of chemicals or energy (e.g., heat or light).

A key concept in disinfection design is the indicator organism or target organism—an organism whose concentration serves as a conservative indicator of the presence of other pathogens. The reduction of such organisms indicates a reduction of a broader group of pathogens and thus the success of disinfection or relative safety of the water. The target pathogens, regulatory standards applied, and treatment technologies used have evolved in recent years as more is understood about the risks associated with the various pathogens. In the United States, the Federal Water Pollution Control Act Amendments of 1972 provided a unified basis for determining secondary treatment standards, including disinfection standards mandated through the establishment of fecal coliform criteria. Increasingly, however, other indicator organisms have been considered to be more conservative indicators of disinfection and thus are increasingly used as a basis of regulatory limits and disinfection design, including *Escherichia coli* (*E. coli*), Enterococci, total coliform, viruses, and protozoa. Tables 15.1 and 15.2 provide background information on density and removal rates of various coliform, virus, and protozoa in wastewater.

Permitting agencies may limit the maximum allowable concentration of bacterial and viral indicator organisms based on daily, monthly, or seasonal requirements. Currently, regulators are using many methods to monitor and control the disinfection processes at municipal water resource recovery facilities (WRRFs) and maintain adherence to design standards. Historically, standards have been directed toward the control of chlorine-based disinfection systems. In addition, compliance with specific water-quality discharge or receiving water limitations commonly is required in WRRF discharge permits. Such limitations vary, but often consist of fecal coliform concentrations of 200 or 400 most probable number (MPN) fecal coliforms (FC)/100 mL, 240 MPN total coliforms (TC)/100 mL, or 2.2 MPN TC/100 mL.

Table 15.1 Secondary Effluent Ranges for Pathogenic and Indicator Organisms Before Disinfection (U.S. EPA, 1986)

Organism	Minimum (MPN/100 mL)	Maximum (MPN/100 mL)
Total coliform	45 000	2 020 000
Fecal coliform	11 000	1580 000
Fecal streptococci*	2000	146 000
Viruses	0.05	1000
Salmonella sp.	12	570

*Assuming removal efficiencies for fecal streptococci similar to fecal coliform removal efficiencies.

More recently, the U.S. Environmental Protection Agency (U.S. EPA) has encouraged states to adopt either *E. coli* or *Enterococcus* for fresh waters, and *Enterococcus* for marine waters for National Pollutant Discharge Elimination System (NPDES) permitting. Limitations on effluent chlorine residuals and toxicity (via dechlorination, for chlorine disinfection practices) are also common in many areas of the United States.

With the heightened awareness of sustainability, there is a growing interest in improving wastewater disinfection practices, while continuing to develop effective and affordable low-impact disinfection alternatives. Applying sustainability principles to the disinfection system requires the careful consideration of technologies that not only meet regulatory requirements, but also achieve the economic, social, and environmental goals of the project.

1.1 CHLORINATION

Chlorine is the most widely used chemical for the disinfection of wastewater. It can be applied either as gaseous chlorine (elemental chlorine, Cl_2), a hypochlorite compound, or as chlorine dioxide. Chlorine dioxide doses may be limited in water reuse applications due to concerns about the potential health effects of chlorite and chlorate ions (World Health Organization, 2016).

1.1.1 CHEMISTRY OF CHLORINE AS A DISINFECTANT

When chlorine is added to water or wastewater, hydrolysis occurs, and a mixture of hypochlorous acid and hydrochloric acid is formed. The reaction is pH and temperature dependent, completed within milliseconds, and reversible (see Equation 15.1). The equilibrium approaches 100% dissociation ($H^+ + OCl^-$) when the pH exceeds 8.5 and approaches 100% hypochlorous acid when the pH is less than 6.5 (pKa = 7.53).

$$HOCl \rightleftarrows H^+ + OCl^- \tag{15.1}$$

For water and wastewater, the hypochlorous acid and hypochlorite ions are the oxidizing agents that provide disinfection. Hypochlorite solutions of sodium hypochlorite and calcium hypochlorite also dissociate in water to form hypochlorite ions yielding, in essence, the same chemistry as chlorine gas. The addition of chlorine gas will decrease the pH, because it forms both hypochlorous acid and hydrochloric

Table 15.2 Typical Levels of Coliform Bacteria in Domestic Wastewater After Various Wastewater Treatment Steps

Wastewater	Total Coliforms (no./100 mL)	Fecal Coliforms (no./100 mL)
Raw	10^7–10^8	10^6–10^7
Primary effluent	10^7–10^8	10^6–10^7
Secondary	10^5–10^6	10^4–10^5
Filtered secondary	10^4–10^5	10^3–10^5
Nitrified	10^4–10^5	10^3–10^5
Filtered nitrified	10^4–10^5	10^3–10^5

acid, while sodium and calcium hypochlorite solutions have a minimal effect on pH. Gaseous chlorine also reduces alkalinity by as much as 2.8 parts chlorine to 1 part calcium carbonate ($CaCO_3$) for the same reason. For low-alkalinity wastewaters, the drop in pH may be a concern in meeting effluent permit requirements and in providing a stable effluent for water reuse applications. The chlorine present as both hypochlorous acid and hypochlorite ions is defined as **free available chlorine**, or simply free chlorine. In terms of disinfecting effectiveness, HOCl is approximately 10 times as powerful as OCl^-, resulting in a greater disinfecting ability at a lower pH.

Free available chlorine refers to the portion of the total residual chlorine in water or wastewater after a specified contact period that will react chemically and biologically as hypochlorous acid or hypochlorite ion.

1.1.2 INORGANIC REACTIONS

In addition to oxidizing and disinfecting pathogens, chlorine compounds will react with and oxidize many inorganic compounds present in wastewater, including hydrogen sulfide (H_2S), nitrite (NO_2^-), ferrous iron (Fe^{+2}), and manganous manganese (Mn^{+2}). These reducing agents exert a chlorine demand and react rapidly with the free chlorine.

When chlorine is added to water containing hydrogen sulfide or sulfite compounds, these compounds are oxidized to elemental sulfur or sulfate, depending on the conditions of the reaction. Nitrite is oxidized to nitrate (NO_3^-), ferrous iron to ferric (Fe^{+3}), and manganous manganese to manganic (Mn^{+4}) ions. The chlorine product in the reaction is the chloride ion. Chloramines are products of the reactions between chlorine and ammonia-nitrogen found in wastewater. Ammonia-nitrogen may be present in appreciable amounts, typically 10 to 40 mg/L, in wastewater as either dissolved ammonia gas (NH_3) or the ammonium ion (NH_4^+) depending on pH.

Three chloramines are formed—monochloramine (NH_2Cl), dichloramine ($NHCl_2$), and trichloramine or nitrogen trichloride (NCl_3)—in a stepwise process, beginning with the reaction of hypochlorous acid and ammonia in dilute aqueous solutions. Chlorine and ammonia can also react to produce nitrogen gas, although this reaction should not be considered a representation of the reaction mechanism.

How far these stepwise reactions proceed and how much of each compound is formed depends on the pH, temperature, time of contact, and ammonium ion and hypochlorous acid concentrations. In general, low pH levels and high chlorine-to-ammonia ratios favor dichloramine ($NHCl_2$) formation. At a pH of greater than approximately 8.5, monochloramine (NH_2Cl) exists almost exclusively. At a pH between 8.5 and 5.5, monochloramine and dichloramine exist simultaneously; between a pH of 5.5 and 4.5, dichloramine exists almost exclusively. At a pH of less than 4.4, nitrogen trichloride will be produced (Jafvert & Valentine, 1992).

Chloramines have disinfecting properties and are part of the combined available chlorine (CAC) measurement, also called *total chlorine*. Chloramines are a weaker oxidant compared to chlorine, and are more specific. Their application for disinfection generally is limited to providing drinking water distribution system disinfectant residuals, because they are not effective as a primary disinfectant.

1.1.3 THE BREAKPOINT PHENOMENON

Breakpoint chlorination is the process of using chlorine's oxidative capacity to oxidize ammonia. Although breakpoint chlorination is a practice used in water treatment to obtain a free chlorine residual in the presence of ammonia-nitrogen, it is not typically used in conventional wastewater treatment because of the large quantities of chlorine required (approximately a 10:1 mass ratio of chlorine to ammonia-nitrogen). An exception to this occurs when a WRRF knowingly or unknowingly nitrifies most of its ammonia-nitrogen, leaving little or no ammonia present to form chloramines. In such a case, the addition of chlorine can produce the breakpoint reaction. In breakpoint chlorination, ammonia is progressively oxidized until a point is reached, beyond which combined chlorine or chloramines and ammonia react with chlorine to produce nitrogen gas and excess chlorine is present as free chlorine. In this case, ammonia is no longer present as ammonia gas or ammonium ion, and only a minimum is present in combined chlorine forms.

Breakpoint chlorination is a treatment approach in which chlorine is added to the disinfection process until all chlorine demand (including ammonia) has been oxidized. Further addition will result in a chlorine residual.

Breakpoint reactions proceed at a rate that is highly pH-sensitive. At pH 7 to 8, and when the mass ratio of chlorine to ammonia-nitrogen is 5:1 or less, all free chlorine is converted to monochloramine. At a lower pH, dichloramine is formed. When the applied chlorine-to-ammonia-nitrogen ratio exceeds 5:1, the total chlorine residual decreases and the breakpoint is approached, with nitrogen (N_2) as a product. When the breakpoint is reached (a chlorine-to-ammonia mass ratio from 8 : 1 to 10 : 1), ammonia-nitrogen is oxidized and free chlorine begins to appear in small amounts in the resultant residual. Figure 15.1 shows a typical breakpoint curve for effluent with ammonia.

Organic chloramines are formed when chlorine reacts with amino acids, proteinaceous material, and other organic nitrogen forms. Organic chloramines are not effective disinfectants, and the presence of organic nitrogen can exert considerable chlorine demand, in addition to interfering with differentiation between free and combined forms. Chlorine existing in chemical combination with ammonia or organic nitrogen chloramines is termed **combined available chlorine** (CAC).

Combined available chlorine is the total chlorine residual including free chlorine and chloramines.

1.1.4 ORGANIC REACTIONS

Chlorinated organic compounds, or chloro-organics, are formed by the action of chlorine on organic carbon in wastewater, just as organic chloramines are products of the chlorination of the organic nitrogen fraction. Organic nitrogen compounds contain nitrogen atoms, in addition to carbon atoms, and include urea, amino acids, and proteins. Most of the reactions between chlorine and organic compounds produce compounds that no longer have oxidizing potential and do not contribute to the CAC.

Trihalomethanes are some of the disinfection byproducts (DBPs) of complex chemical reactions of chlorine with a group of organic acids known as **humic acids**. The trihalomethanes (THMs) are single carbon molecules containing three halogen atoms present in varying combinations. There is concern regarding the formation of THMs and other DBPs because of their effects on the environment and on human health. Chloroform, a well-known THM, is a documented animal carcinogen, and all haloforms are believed to act in a similar manner. U.S. EPA has issued regulations for the drinking water industry via the Safe Drinking Water Act targeted at minimizing public exposure to this class of compounds as along with state and national water quality standards for the discharge of these types of compounds. In areas where water plants are having difficulty meeting DBP regulatory limits, WRRFs may be required to meet individual THM and/or DBP concentrations in their effluent.

Humid acids are natural high molecular weight organic acids that result from the decay of organic matter.

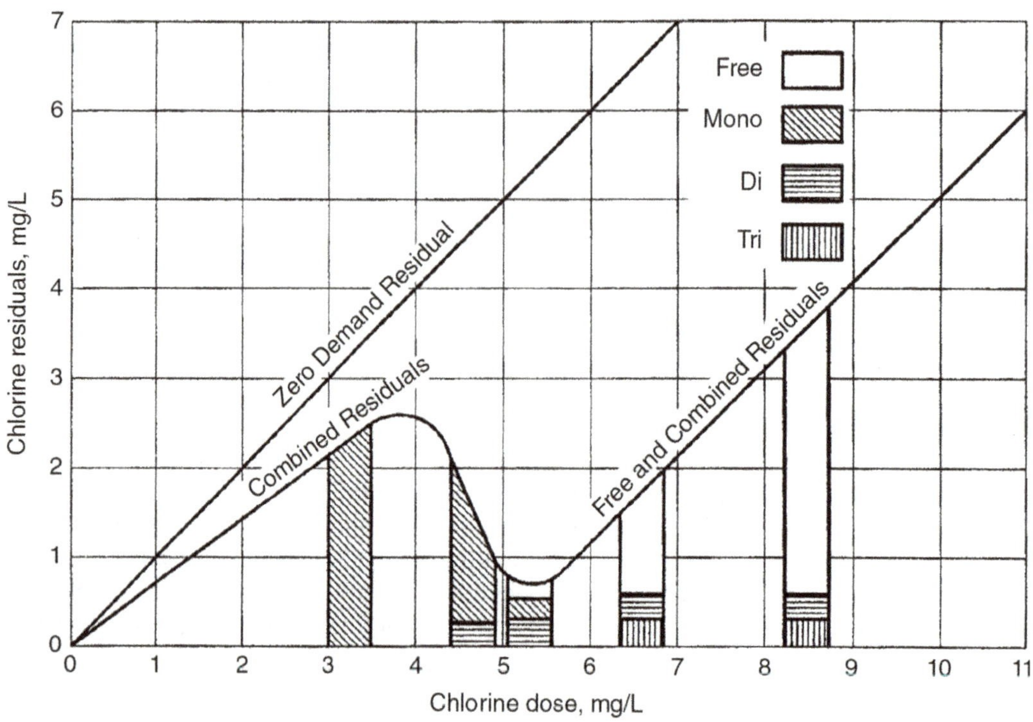

Figure 15.1 Chlorine Residual With Ammonia-Nitrogen (Black & Veatch Corporation, 2010)

1.1.5 BYPRODUCT TOXICITY

Chlorine is an extremely reactive element, rapidly undergoing chemical reactions with inorganic and organic substrates. When the organic substrate is part of a living organism, the reaction can have a toxic effect on the organism. This toxicity may affect the organism's ability to reproduce or metabolize, cause genetic dysfunctions (mutations), or ultimately kill the organism. The active agents are hypochlorous acid, hypochlorite ion, monochloramine, and dichloramine. Hypochlorous acid reacts with organic nitrogen compounds and may alter the chemical structure of the organism's organic materials and change the genetic information. Undesirable effects of chlorination on receiving water biota also have been documented.

1.1.6 AFTERGROWTH

In the receiving stream, aftergrowth of some organisms occur after discharge of chlorinated effluents. The extent of these effects is governed primarily by the amount of biologically oxidizable material in the effluent. Aftergrowth observed in waters receiving chlorinated effluent are presumed to be a result of the destruction of large numbers of protozoa by chlorination. This permits subsequent multiplication of surviving bacteria unhampered by predatory protozoa, such as ciliates and flagellates. When effluent is chlorinated, the greater the initial reduction in bacterial population, the longer the lag time will be before the multiplication of surviving organisms becomes apparent.

1.1.7 SAFETY AND HEALTH

Perhaps the most substantial drawback associated with the use of chlorine gas is the safety risk of a gas leak. Safety issues related to chlorine are documented in the Uniform Fire Code (UFC), (International Fire Code Institute, 1994) and in regulations by the Occupational Safety and Health Administration (OSHA, 1996).

Chlorine gas has a detectable odor at low levels of concentration, and has a greenish-yellow color at higher levels of concentration. At volumes of less than 0.1 ppm in air, chlorine gas is undetectable, except by instruments. The maximum contaminant level established for chlorine gas by OSHA is a 1-ppm, time-weighted average over 8 hours. The harmful effects of chlorine gas exposure begin to become evident at approximately 5 ppm and higher. Between 5 and 10 ppm, however, these effects (choking, coughing, watery eyes, mild skin irritation, and lung irritation) are often temporary. At higher concentrations, the effects become more long lasting and can result in serious health consequences or death.

High-capacity storage and transportation facilities should be located in isolated areas and equipped with scrubbers. The designer should consult the latest edition of *The Chlorine Manual* (Chlorine Institute, 1986); Compressed Gas Association data; the appropriate U.S. Department of Transportation and OSHA regulations; and any applicable fire codes, including local codes and the Uniform Fire Code, Standard Fire Prevention Code, and National Fire Code. U.S. Department of Transportation regulations govern the use of tank cars, trucks, and 900-kg (1-ton) containers. Additional regulations pursuant to the Clean Air Act requires an air permit and risk-management plan for chlorination facilities depending upon the amount of chlorine gas stored at the WRRF.

1.1.8 CHLORINATION OF RECLAIMED WATER

Title 22 of the California Code of Regulations (2006) governs recycled water treatment in California and provides a model for recycled water regulation nationwide. Title 22 requires that recycled water disinfection practices result in a minimum of a 5-log (99.999%) reduction in viruses, with poliovirus being the standard. Title 22 requirements are based on a chlorination concentration time (combination of chlorine residual and modal contact time) of 450 mg·min/L providing 4 logs of poliovirus kill without media filtration and 5 logs of poliovirus kill with media filtration. Thus, Title 22 requires approved media filtration of clarified secondary effluent followed by chlorination (free or combined chlorine) at a concentration time of 450 mg·min/L; or an alternative filtration/disinfection process that results in 5-log inactivation of MS2 (a non-pathogenic indicator virus used in seeding studies) or poliovirus.

Example Problem 15.1

Calculate the required chlorine contact tank volume to meet California's Title 22 concentration time requirement of 450 mg·min/L given the following:

- Peak hourly flowrate, $Q = 37\,850\ m^3/d$
- Free chlorine residual concentration = 2.0 mg/L

Solution

The first step is to calculate the required detention time for this *T* value:

$$detention\ time = \frac{CT}{C} \tag{15.2}$$

$$detention\ time = \frac{450}{2.0} = 225\ \text{min}$$

If we design a basin with an inlet distribution baffle, serpentine flow pattern, and an outlet weir, standard practice suggests this will provide a baffling factor of 70%. The required volume would be: (calculated detention time/baffling factor) × flow rate:

$$volume = 225\ \text{min} \cdot \frac{1}{0.7} \cdot 37\,850 \frac{\text{m}^3}{\text{d}} \cdot \frac{\text{d}}{24\ \text{h}} \cdot \frac{\text{h}}{60\ \text{min}} = 4140\ \text{m}^3$$

Therefore, a total basin volume of 4140 m³ would be required to satisfy the Title 22 requirement. If we divide this into two basins with common wall construction, each basin will have a volume of 2070 m³. This represents over a 5-hour detention time at peak flow and will likely be one of the largest basins at the facility. If the WRRF employs filtration, the contact basin can serve as a clearwell to provide water for filter backwash. It should be noted that direct potable reuse (DPR) and indirect potable reuse (IPR) requirements will vary. In addition, requirements will vary from state to state. Alternative disinfection practices (e.g., ultraviolet light) may also be allowed as a disinfection alternative.

1.1.9 DECHLORINATION

Much attention has been focused on the toxic effects of chlorinated effluent. Both free chlorine and chloramine residuals are toxic to fish and other aquatic organisms, even at concentrations less than 0.02 mg/L. Although fish are repelled by low levels of chlorine and frequently escape harm, other aquatic organisms in the food chain may be killed by chlorine discharges. Dechlorination, the removal of remaining chlorine, is required in most states. Stream standards have been established in most parts of the country to limit residual chlorine in the effluent. However, carcinogenic chlorination byproducts will not be reduced by dechlorination.

Free and combined chlorine residuals can be effectively reduced by sulfur dioxide and sulfite salts. The sulfite ion reacts rapidly with free and combined chlorine. The sulfite ion is the active agent when sulfur dioxide or sulfite salts are dissolved in water. Their dechlorination reactions are identical. Sulfite reacts instantaneously with free and combined chlorine, yielding small amounts of acidity, which is neutralized by the alkalinity of the wastewater. The stoichiometric amount of sulfur dioxide required per part chlorine is 0.9, but typical actual practice calls for a ratio of 1 : 1.

It should be noted that previous research (Helz & Nweke, 1995) has shown that complete dechlorination may not be achieved or may be delayed in the presence of monochloramine.

Granular and powdered carbon may be used to dechlorinate free, and some combined, chlorine residuals. Carbon requirements for dechlorination typically are determined by on-site pilot testing. Parameters of significance include mean particle diameter of the carbon (pressure drop within a contactor) and influent quality (pH, organics, and colloids). For typical municipal effluent, costs are high. Doses in the range of 30 to 40 mg/L have been reported.

Sulfur dioxide is more soluble than chlorine in water. Sodium sulfite and bisulfite, typically provided in solution form, are reducing agents like sulfur dioxide. Detailed information on safe handling procedures for all chemicals is available from various chemical suppliers. Training of all personnel in the handling and use of sulfur dioxide is available from the Compressed Gas Association (1988) and individual manufacturers.

1.2 ULTRAVIOLET DISINFECTION

Ultraviolet (UV) disinfection acts primarily by damaging the genome of viruses, bacteria and parasites. Ultraviolet light is typically divided into three bands depending on wavelength (UVA: 310–400 nm, UVB: 280–315 nm, and UVC: 100–280 nm), with UVC being the most potent as a disinfectant. The primary effect of the absorbed UVC photons is the production of pyrimidine (thymine and cytosine) dimers in the DNA or RNA of the irradiated organism. A sufficient number and location of dimers will prevent replication of the organism, rendering it harmless, unable to infect or infest. While there are other effects of UV radiation, these dimers are the factors primarily responsible for disinfection (Jagger, 1967). As a consequence, the UV wavelengths that are strongly absorbed by nucleic acids are most effective at disinfection. The extent to which the organism absorbs the light and is damaged determines the sensitivity of the organism to UV disinfection. A number of references (Haji Malayeri et al., 2017; U.S. EPA, 1986; WEF, 2015) provide information on the kinetics of UV disinfection.

Ultraviolet systems offer chemical-free disinfection and proven protection against waterborne illnesses. These systems have been successfully installed in waters of all qualities and temperatures ranging from challenging primary effluent and combined sewer overflows (CSO) to secondary and tertiary effluent. Installation, operation, and maintenance of the UV systems is well-established, and the latest generation technologies bring greater efficiency, fewer lamps, and better automation and control. There have been many advances in UV technology since it was first introduced. Lamp efficiency and UV intensity monitoring are some of the key features that contributed to the growing popularity of UV systems (WEF, 2015; WERF 2007). In the mid-1980s, U.S. EPA named UV disinfection as a "best available technology" for wastewater disinfection (Jeyanayagam & Cotton, 2002).

The majority of UV disinfection systems for wastewater disinfection currently use an open-channel modular design, though pressurized systems are often used for water reuse applications following membrane treatment (in order to not break "head"). Three principal lamp geometries have been adopted—horizontal, uniform arrays with flow directed parallel to lamp axes; vertical, staggered arrays with flow directed perpendicular to lamp axes; and inclined, staggered arrays with flow directed at an angle to lamp axes.

A fundamental variable that strongly affects the amount of UV equipment that needs to be installed in any application is the UV transmittance (UVT, units %/cm). This measure is an indication of how much light the water lets through. The higher the number, the further the UV light is able to penetrate. Waters with low UVT absorb more UV light and, therefore, require higher energy for a desired inactivation; hence, the capital and operating and maintenance costs are generally higher for lower UVT installations. For high UVT applications, the UV reactor efficiency can be improved by increasing the lamp spacing, which would result in lower headloss across the UV system. Such system design changes can only be made by or in consultation with system suppliers who fully understand the operational range (UVT, headloss, disinfection target dose/fluence) of their system designs. Arbitrary changing of lamp spacing is not recommended.

The UVT of the wastewater to be disinfected is one of the most important parameters in determining the cost of UV disinfection, with reduced UVT having an exponential (not linear) effect on equipment and power to maintain a target dose. Therefore, frequent (ideally online) monitoring of UVT is critical to optimize UV system operation, and maintaining higher effluent quality is important to maintain efficient UV operation. However, the assumption of a conservative UVT can be made to allow systems to run very safely even without constant online UVT monitoring, where concerns around energy optimization are not paramount, such as in small systems.

1.2.1 ULTRAVIOLET LAMPS

Practical UV disinfection requires an efficient and low-cost source of UVC photons. The mercury-vapor lamp is typically used for disinfection, because these lamps are very efficient and the physics and manufacturing were well developed in the 1960s for fluorescent lamps that rely on the same basic process, the ionization of mercury. (Fluorescent lamps use a fluorescent coating, which is absent in disinfection lamps.) There are three main types of mercury lamp used in disinfection applications: low-pressure, amalgam, and medium-pressure lamps. These lamps are used in both open-channel and pressurized (closed vessel) applications. The lamps are enclosed within UV transparent quartz sleeves to protect the lamps from exposure to water and to control lamp temperature.

Mercury-vapor lamps require power supplies that are suited to the unique characteristics of the mercury vapor gas discharge. These power supplies are sometimes referred to as ballasts, and older designs rely primarily on large inductors and capacitors. Modern electronic power supply designs incorporate sophisticated and efficient solid-state circuit designs able to precisely control lamp power.

Two standard lamp lengths are typically used in conventional disinfection systems: 0.9 m (0.7 m arc length) (36 in. [30 in. arc length]) and 1.6 m (1.5 m arc length) (64 in. [58-in. arc length]). This lamp configuration is used in pressurized closed-vessel systems and in open-channel systems. Although the low-pressure lamp is efficient at producing effective germicidal radiation, its output intensity is relatively low.

Amalgam lamps decrease and control the vapor pressure of the mercury by combining it with other metals to create a solid amalgam. As a result, the mercury vapor can be maintained near its optimum value over a broader range of higher temperatures. Accordingly, amalgam lamps (also known as "low-pressure, high-output") have become increasingly popular. The amalgam lamps generally operate at higher electrical current, higher temperature, and higher total internal pressure than low-pressure lamps. The actual operating pressure is as much as 40% higher than that of its low-pressure counterpart. Amalgam lamps are used in enclosed pressurized reactors, and in open-channel reactors in vertical, horizontal, or inclined orientations.

Medium-pressure lamps do not use amalgams, but operate at higher power density and temperature, so that the mercury is completely vaporized and many additional energy levels and emission transitions are populated. As a result, these lamps have a complex polychromatic broad emission of wavelengths from vacuum UV through to the infrared region. Medium-pressure lamps have a rated life of 4000 hours or more, although experience has shown an expected life exceeding 8000 hours. Actual lamp life depends on lamp operating power. A higher operating power results in higher lamp temperatures and typically lower lamp life. Medium-pressure lamps are generally more expensive than low-pressure or amalgam lamps, but this is offset by their greater power, which requires fewer lamps for a given output. Medium-pressure systems typically include automatic in-place cleaning systems (wipers). Medium-pressure lamps are available in either closed-vessel or open-channel systems. Figure 15.2 provides an example of a typical UV installation.

1.2.2 ULTRAVIOLET SYSTEM DESIGN

The hydraulic behavior of a UV disinfection system is a critical element to ensure adequate UV disinfection performance. The hydraulic behavior influences the retention time of the various fluid elements passing through the complex UV intensity profile within the system and, thus, directly affects UV dose. Inadequate hydraulic conditions (especially those caused by incorrect installation of the UV system in the field, e.g., lack of attention to the manufacturer's installation specifications) are the most common cause of UV disinfection system failures. Important hydraulic influences include longitudinal and radial mixing and turbulence, the inlet structure, outlet structure, and headloss through the system. Consultation with equipment manufacturers is critical during the design of specific UV systems for a particular WRRF.

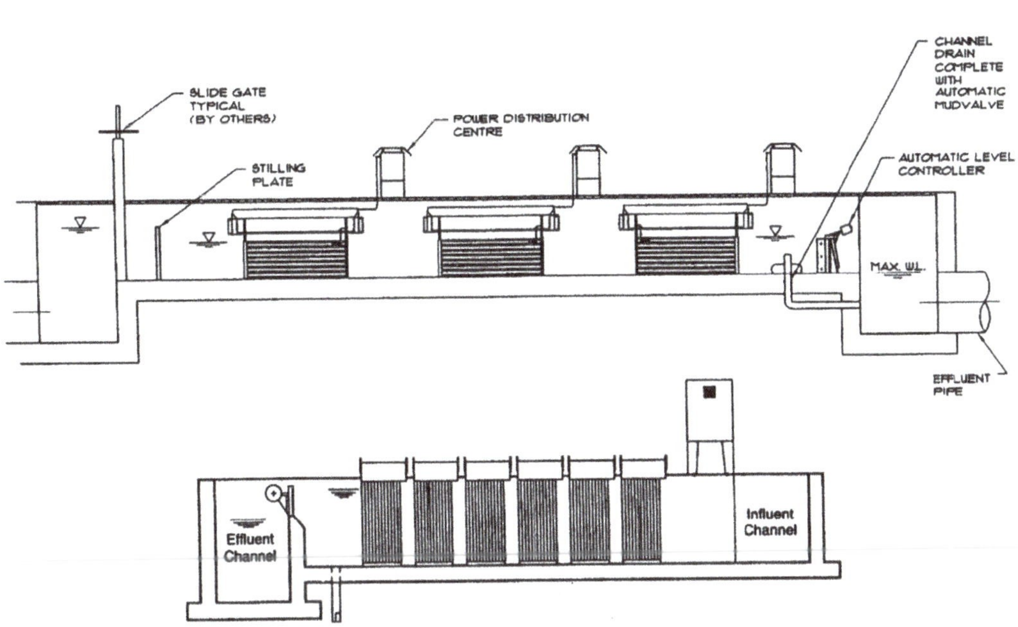

Figure 15.2 · Typical Ultraviolet Installation: Top— schematic illustration of open-channel ultraviolet disinfection system with horizontal lamp configuration; Bottom—schematic illustration of open-channel ultraviolet disinfection system with vertical lamp configuration

1.2.3 ULTRAVIOLET INACTIVATION KINETICS

Ultraviolet inactivation of microorganisms is described by U.S. EPA (1986, 2006) and within other literature (WEF, 2015). A very comprehensive, curated compilation of the sensitivity of many microorganisms from many classes can be found in the literature (Haji Malayeri et al., 2017.) To date, all microorganisms that have been studied appear to respond to UV treatment and do so over a narrower dose range (from most sensitive to most resistant organism) compared to other disinfectants.

Organisms respond to UV irradiation by first-order kinetics, with sensitivities that range from ~<2 mJ/cm^2 per log of reduction for UV-sensitive organisms such as *E. coli* or *Cryptosporidium parvum* (depending on the type) to >40 mJ/cm^2 per log of reduction for certain types of UV-resistant organisms such as adenoviruses; however, most pathogens tested so far are shown to be quite sensitive to UV disinfection, and the adenoviruses are generally considered outliers. Designers of UV systems and regulators need to consider the target design pathogen of concern and its kinetics when determining the design dose of a system.

1.2.4 PHOTOREACTIVATION AND DARK REPAIR

Microorganisms have effective systems for repairing damage caused by hostile environmental conditions, such as exposure to disinfectants. Repair and recovery from inadequate damage is known to occur following all disinfection operations. Designers and operators of disinfection processes should understand these repair processes and their potential consequences When the inflicted damage is small due to the application of a very low UV dose relative to the target design dose, the photochemical damage to an organism's nucleic acids caused by UV irradiation can be repaired. These repair mechanisms can allow UV-treated microorganisms to regain viability (ability to reproduce) following the disinfection process. Two principal repair mechanisms have significance relative to UV disinfection—photorepair, which occurs in the presence of visible light; and dark repair, which can occur even in the absence of light.

Photorepair can occur in some organisms when only minor damage has occurred, and as such is an indication of underdosing. At higher doses, the same organisms will not be able to photorepair. The design dose must be carefully selected to ensure that adequate damage to the target pathogen's nucleic acids is provided. Historic evidence suggests that at least 4 logs inactivation or higher are required to minimize the effect of photorepair on final counts of pathogens that can photorepair (Lindenauer & Darby, 1994). In the case of protozoan parasites and *Legionella*, and other UV-sensitive bacteria, this could mean a UV dose of at least 10 mJ/cm^2, but when attempting to provide redundancy for chlorination by targeting the majority of the photorepairable pathogens, this would mean a UV dose of closer to 25 mJ/cm^2. *Cryptosporidium* has been shown not to photorepair, and laboratory lines of *Giardia* have been shown to have no manifestations of repair at UV doses of 16 and 40 mJ/cm^2 (Linden et al., 2002); that is, at UV doses used respectively to polish municipal potable water through point-of-use/point-of-entry UV technologies and to disinfect municipal potable water, respectively. The historical target dose for drinking water of 40 mJ/cm^2 results from the European target for the pathogen rotovirus, and both the original assessment of the target design dose and subsequent epidemiological assessments through years of use in municipal applications suggests that photorepair is adequately controlled at this dose.

A consensus does not exist within the engineering, scientific, or regulatory communities regarding the inclusion of repair in UV-disinfection system design, although the work of Lindenauer and Darby (1994) suggests that targeting 4 logs of inactivation will eliminate concerns regarding photorepair because the damage inflicted will be too severe for organisms to withstand. Although many operating WRRFs have been designed and are operating successfully with and without consideration of repair, photorepair can only occur with very low doses.

1.2.5 EFFECTS OF UPSTREAM PROCESSES

Upstream treatment processes have a significant effect on the disinfection of secondary effluent. Because of "shading," particle-associated coliforms (PACs) are often difficult to inactivate to levels required for reuse (e.g., 2.2 total coliforms/100 mL). Additionally, certain upstream processes, such as fixed-film biological treatment, yield recalcitrant effluent with a high concentration of PAC in large particles. Loge et al. (1999) reported that PACs were found to decline exponentially with increasing mean cell residence times (MCRTs). The factors influencing the formation of PACs included the concentration of particles, concentration of dispersed (non-particle-associated) coliform bacteria, and the MCRT. The concentration of dispersed coliform bacteria was found to decline with increasing MCRTs. The rate of decline was greater than the typical half-life attributed to endogenous decay, suggesting that other factors (e.g., predation by protozoa) influence the concentration of dispersed coliform bacteria and subsequently the formation of PACs (Loge et al., 1999).

A number of industries have been implicated as discharging wastewater with high dissolved UV absorbance (low filtered UVT) as a result of the presence of organic compounds not readily degraded (Swift et al., 2007), including sunblock, coffee, pharmaceutical, and chemical and photochemical manufacturers; centralized WRRFs; and printed-circuit-board manufacturers. Several WRRFs have been documented as having potentially violated effluent disinfection standards as a result of the presence of refractory organic compounds passing through the WRRF and lowering effluent UVT. Compounds in the wastewater responsible for lowering effluent UVT can be considered to be "pollutants of concern" for source-control programs and thus appropriate for regulatory control through the use of local limits established through an evaluation of maximum allowable headworks loading using methods analogous to those used for other pollutants (Swift et al., 2007). The use of ferric chloride as a coagulant to lower phosphorus concentrations has been shown to interfere with UV disinfection (Gere & Wright, 1998).

1.3 OZONE DISINFECTION

Although ozone and chlorine were both developed as disinfection methods at roughly the same time in the late 19th and early 20th centuries (von Sonntag & von Gunten, 2012), chlorine became the dominant wastewater disinfection technology due to its substantially lower capital cost. Moreover, many early ozone disinfection systems were abandoned because of maintenance and operational problems (Leong et al., 2008). However, concerns about the residual oxidizing effects of chlorine and the production of chlorinated DBPs have caused WRRF owners to look again at alternative treatment technologies. Improvements in ozone-production technologies, combined with the capacity of ozone to oxidize micropollutants and color- and odor-causing compounds, has brought ozone back into consideration as a disinfection method, especially in applications involving water reuse.

Leong et al. (2008) reported that, as of 2007, there were seven major WRRFs in the United States using ozone with a median design flow of 37 900 m^3/d (10 mgd) and a range of 11 400 to 129 000 m^3/d

(3–34 mgd). As of 2016, ozone disinfection systems are in use or in construction at approximately 20 U.S. WRRFs, many of which cite color removal or the destruction of trace organics as deliberate supplementary goals. Although not free of DBP formation, ozone's strong oxidizing abilities will lead to its more frequent adoption in IPR and DPR applications.

1.3.1 GENERAL DESCRIPTION

Ozone acts as an effective disinfection agent through the following four mechanisms: cell lysis by oxidation of cell walls; breakdown of purines and pyrimidines, the building blocks of nucleic acids; breakdown of carbon-nitrogen bonds, leading to depolymerization of organic molecules; and production of hydroxyl radicals in water, which are powerful oxidants.

The effectiveness of ozone for disinfection depends on several factors, including target pathogen, the pathogen's exposure to ozone, particulate shielding, and temperature. In municipal wastewater effluent, ozone exposure, defined as the integral of ozone concentration over time, is influenced by the transferred ozone dose and the concentration and type of dissolved organic matter present.

The disinfection process can be summarized as follows. Ozone, an unstable gas, must be generated on site and then transferred into the treated water. Once transferred, a large portion of the initial ozone concentration rapidly decays within the first few seconds ("instantaneous demand"); the remaining concentration ("residual") decays more slowly, with a typical half-life of less than 30 seconds in municipal wastewater effluent (Metcalf & Eddy, Inc./AECOM, 2014). However, because ozone decays so quickly, and much of the actual concentration time is accounted for before measurement is practical, treatment trains should be designed using site-specific bench and pilot studies.

1.3.2 OZONE KINETICS

For a given pathogen, inactivation performance is partially a function of the concentration–time relationship, which is in turn determined by ozone kinetics. Ozone is unstable in aqueous solution and will proceed along one of the two reaction pathways: the direct pathway or the indirect pathway. Relatively few substrates react directly with ozone, so the direct pathway is the slower and more selective of the two. However, in some cases ozone will form hydroxide radicals (•OH) or other radicals in its reaction products. These radicals react quickly and non-selectively with neighboring molecules, sometimes forming new radicals and perpetuating the reaction cycle. This indirect pathway greatly increases the ozone reaction rate and range of potential substrates. The following section is a summary of the mechanisms behind aqueous ozone kinetics. For a more detailed discussion, the reader is referred to von Sonntag and von Gunten (2012).

The major factors influencing ozone reaction kinetics are dissolved organic matter, carbonaceous alkalinity, ozone concentration, pH, and temperature. By far the most significant factor affecting ozone kinetics is dissolved organic matter, which competes with pathogens for ozone exposure through both the direct and indirect pathways. There is increasing evidence that the ozone demand in municipal wastewater effluent scales linearly with dissolved organic carbon, which can be taken as a reasonable proxy for dissolved organic matter (von Sonntag & von Gunten, 2012).

Inorganic carbon is a major inhibitor to the indirect pathway because the radical forms of carbonate and bicarbonate are not reactive with most organic compounds. Hence, carbonate alkalinity generally increases the stability of dissolved ozone (von Sonntag & von Gunten, 2012). An increase in the concentration of ozone in the bulk liquid will cause an increase in the reaction rate constants for direct (Bellamy et al., 1991; Prados et al., 1995) and indirect (Prados et al., 1995) reactions. This increase will result in an overall increase in the ozonation reaction rate constant. Because hydroxide ions are the primary initiators of the indirect pathway, increasing pH promotes the indirect pathway. The direct pathway dominates for acidic conditions ($pH < 4$), while the indirect pathway dominates above pH 10 (Staehelin & Hoigne, 1985). However, because pH is typically within a narrow range in wastewater effluent, it is not a factor of concern. Based on the Arrhenius expression, increasing the temperature will increase the ozone reaction rate but will decrease the solubility of ozone in the liquid phase, which decreases gas-to-liquid transfer efficiency.

1.3.3 THE ROLE OF THE HYDROXYL RADICAL

Although •OH is a much stronger oxidant than ozone and is the primary mechanism for the destruction of most micropollutants, its effectiveness as a disinfectant is debated. Several studies have shown that increasing the ratio of •OH:O_3 in solution correlates with lower inactivation rates against *E. coli* and other heterotrophic bacteria (Hunt & Mariñas, 1999; von Gunten, 2003; Wolfe, 1990). Other studies have shown improved inactivation rates against *C. perfringens* (a spore-forming bacterium) and *B. subtilis* spores (Cho et al., 2003; Leong et al., 2008). It may be that bacterial spores are particularly resistant to ozone and require attacks via the indirect pathway. Nevertheless, ozone's strong oxidizing capabilities make it attractive for intentional •OH production in advanced oxidation systems for the destruction of contaminants of emerging concern. These systems are often followed by biologically active filters (Zhang et al., 2020).

1.3.4 NON-KINETIC FACTORS IN OZONE DISINFECTION EFFICIENCY

Suspended solids can inhibit the performance of ozone disinfection by shielding pathogens (Gottschaulk et al., 2010). Xu et al. (2002) observed a 1-log improvement in inactivation for the same concentration × time values after wastewater effluent had been filtered at 10 μm. Generally, decreasing temperatures will require a higher concentration time to achieve the same inactivation rates (Gottschaulk et al., 2010), although *B. subtilis* spores appear to exhibit a local maximum inactivation rate at 10 °C, with decreasing efficiency at temperatures above and below (von Gunten, 2003).

1.3.5 OZONATED EFFLUENT CONCERNS

Environmental issues associated with ozone include the formation of DBPs, including bromate (formed from bromide), and aldehydes. There may also be a slight increase in the biochemical oxygen demand (BOD) of the effluent resulting from the breakdown of recalcitrant compounds associated with natural organic matter and the formation of more readily biodegradable smaller molecules. For this reason, ozonation is often followed by a tertiary biological treatment step (such as biological aerated filtration), which may, in turn, be followed by additional downstream "post-ozonation." Ozonation will often result in high concentrations (10–20 mg/L) of effluent dissolved oxygen, typically obviating the need for post-aeration filtration. The high dissolved oxygen combined with readily degradable organics creates an ideal environment for regrowth.

2.0 FILTRATION

Several wastewater filtration technologies now exist to remove suspended solids remaining in the secondary clarifier effluent. The need for filtration is often related to the need to meet effluent total phosphorus (TP) limits. Enhanced biological phosphorus removal (EBPR) systems encourage the growth of phosphate-accumulating organisms as discussed in Chapter 8. Consequently, suspended solids leaving an EBPR process will have a higher phosphorus content than solids leaving non-EBPR systems, and it becomes imperative to limit effluent solids to meet low total phosphorus permit requirements (i.e., 0.5 mg/L).

Alternative filtration technologies involve use of various filtering media and configurations. Filtration technologies use granular or synthetic nongranular-type filter media such as compressible media, disc filters, and membrane filters. There are significant design and/or operational differences between different types of filtration technologies. The basic principle of filtration is the same, however, regardless of the technology used. Particle removal is achieved by passing wastewater through a porous filtering medium. As the suspension flows through the filtering medium, particles are transported to the surface of the collecting medium where they are removed by various mechanisms. Membrane filtration is considered more advanced compared to compressible- and disc-filtration technologies because of higher effluent quality. The capital and operational costs are also higher, but membrane filtration may be the feasible filtration alternative for certain applications such as pretreatment for desalination.

2.1 GRANULAR MEDIA FILTRATION

Granular media filtration (GMF) was the sole filtration technology used for wastewater treatment until the 1990s and has numerous different types and configurations. All GMF technologies are designed and operated based on depth filtration principles.

The filtration rate for GMFs varies significantly depending on the type of the technology. In slow sand filtration, the rate is approximately 1 to 5 L/m^2·min (0.025 to 0.125 gpm/sq ft). Filtration rates typically range from 80 to 240 L/m^2·min (2 to 6 gpm/sq ft) for rapid sand filtration. Higher filtration rates up to 400 L/m^2·min (10 gpm/sq ft) have been reported for GMF, but nominal maximum application rates typically range from 200 to 300 L/m^2·min (5 to 7.5 gpm/sq ft) in high-rate applications. Gravity filters involving the use of granular media have been shown to perform successfully at higher filtration rates (e.g., 300 L/m^2·min [7.5 gpm/sq ft]) especially with optimized chemical addition (Holden et al., 2006; Williams et al., 2007).

Use of two or more layers of media, each with a different specific gravity, prevents stratification and promotes filtration with inherently longer filter runs. Typical combinations of media used in dual-media beds include

- Anthracite and sand,
- Activated carbon and sand,
- Resin beds and sand,
- Resin beds and anthracite, and
- Garnet or ilmenite (often used in multimedia beds).

Table 15.3 presents typical characteristics of some granular filter media. Table 15.4 presents a design example for a depth filter involving the use of granular media.

In slow sand filtration, the filtration rate is approximately 1 to 5 L/m^2·min (0.025 to 0.12 gpm/sq ft). Slow sand filters typically consist of a sand layer 150 to 380 mm (6 to 15 in.) in depth placed on an equivalent depth of coarser material that rests on an underdrain system. Slow sand filters are regenerated by remaining idle for a period and then being cleaned manually. These filters typically can achieve 60% suspended solids removal from a trickling-filter effluent containing 20 to 90 mg/L total suspended solids (TSS). Although simple to construct, slow sand filters have high maintenance and space requirements. They also are temperature-sensitive and subject to rapid clogging. Because these filters are not considered competitive with other techniques for TSS removal, they are seldom used in urban areas. In the rural areas, however, they may be an attractive alternative.

Deep-bed, coarse-medium filters typically involve single-medium bed depths of 1.2 to 1.8 m (4 to 6 ft). The proper size for sand medium depends on secondary effluent characteristics and required filtered effluent quality. A diameter of 1.5 mm or larger is typical for secondary effluent filtration. A highly uniform sand (e.g., uniformity coefficient of 1.1) is best for allowing full use of the bed.

Early problems with effective bed cleaning and economical volumes of backwash water seem to have been solved by using an air backwash at 0.015 to 0.025 m^3/m^2·s (3 to 5 scfm/sq ft), with a water rinse at 240 to 320 L/m^2·min (6 to 8 gpm/sq ft). Total rinse-water consumption per unit of filter surface per wash varies between 4000 and 8000 L/m^2 (100 to 200 gal/sq ft) and is typically approximately 6000 L/m^2 (150 gal/sq ft). Coarser media decrease headloss development and provide a higher solids storage capacity.

Dual-media and multimedia filter beds are commonly used in WRRFs. These systems consist of two or more media, such as anthracite and sand, in layers (Figure 15.3). The anthracite layer may be distinct or mixed with sand. The specific gravity of the anthracite is typically 1.6; that of silica sand is 2.65. Garnet sand, used with the multimedia system, has a specific gravity of approximately 4.2.

2.1.1 HEADLOSS DEVELOPMENT

Headloss developent in a GMF depends on hydraulics and solids retention capacity. The total headloss at any time is the sum of the hydraulic headloss for a clean bed and additional headloss caused by the deposition of wastewater solids. Both are influenced largely by media size.

Hydraulic headloss encompasses losses associated with piping, valving, metering, bends, constrictions, underdrains, collection systems, and media. Of these, the media bed composes the predominant headloss.

Table 15.3 Media Characteristics for Dual-Media and Multimedia Filters

Characteristics	Values	
	Range	Typical
Dual media		
Anthracite		
Depth, mm[a]	300–600	450
Effective size, mm	0.8–2.0	1.2
Uniformity coefficient	1.3–1.8	1.6
Sand		
Depth, mm	150–300	300
Effective size, mm	0.4–0.8	0.55
Uniformity coefficient	1.2–1.6	1.5
Filtration rate, m/h[b]	5–24	12
Multimedia		
Anthracite (top layer of quad-media filter)		
Depth, mm	200–400	200
Effective size, mm	1.3–2.0	1.6
Uniformity coefficient	1.5–1.8	1.6
Anthracite (second layer of quad-media filter)		
Depth, mm	100–400	200
Effective size, mm	1.0–1.6	1.2
Uniformity coefficient	1.5–1.8	1.6
Anthracite (top layer of trimedia filter)		
Depth, mm	200–500	400
Effective size, mm	1.0–2.0	1.4
Uniformity coefficient	1.4–1.8	1.6
Sand		
Depth, mm	200–400	250
Effective size, mm	0.4–0.8	0.5
Uniformity coefficient	1.3–1.8	1.6
Garnet or ilmenite		
Depth, mm	50–150	100
Effective size, mm	0.2–0.6	0.3
Uniformity coefficient	1.5–1.8	1.6
Filtration rate, m/h	5–24	12

[a]mm × 0.039 37 = in.
[b]m/h × 0.409 2 = gpm/sq ft.

For the laminar flow conditions that occur for normal operating conditions of media filtration, the Reynolds number based on superficial velocity is less than 3 and headloss through the filter media varies directly with flow velocity (American Water Works Association [AWWA], 1990). For any media filtration system, a headloss development curve for clean water as a function of filtration rate can be obtained either from pilot-plant study or directly from the manufacturer.

Clean water headloss curves for a given hydraulic load are influenced by media type, size, uniformity, and depth. As the filtration rate increases, clean water headloss increases, and the available headloss development for the filtration run decreases. The filtration run length is directly related to the difference between the clean filter headloss and the terminal headloss value.

Table 15.4 Secondary Effluent Filtration Design Example

Input Parameters	Units	Design Values/ Considerations
Annual average flow	m^3/d	19 000
Peak-day flow	m^3/d	38 000
Peak-hour flow	m^3/d	57 000
Basis of design flow*		Peak-day flow
Influent turbidity range	NTU	2–10
Average influent turbidity	NTU	5
Influent TSS range	mg/L	5–25
Average influent TSS	mg/L	12
Maximum design filtration rate	$L/m^2 \times min$	200
Average water production ratio	Percentage	90
Backwash water reject ratio range	Percentage	5–15
Average backwash water reject ratio	Percentage	10
Filter effluent turbidity limit to start chemical addition	NTU	1.8
Redundancy requirement	Provide filtration capacity with largest filter unit out of service	
Design output		
General		
Filtration type		Depth
Flow direction during filtration cycle		Downflow
Flow direction during backwash cycle		Upflow
Unit sizing		
Minimum total filtration surface area required	m^2	130
Number of filter units	#	8
Filtration surface area, width (each filter unit)	m	4.25
Filtration surface area, length (each filter unit)	m	4.25
Filtration surface area for each filter unit	m^2	18
Total filtration surface area provided	m^2	145
Filtration rate		
Annual average flow	$L/m^2{\cdot}min$	90.0
Peak-day flow	$L/m^2{\cdot}min$	180
Annual average flow, one filter out of service	$L/m^2{\cdot}min$	105
Peak-day flow, one filter out of service	$L/m^2{\cdot}min$	210
Peak-day flow, one filter out of service and one filter in backwash	$L/m^2{\cdot}min$	240
Medium properties		
Medium type	Mono medium–sand, unstratified	
Medium depth	mm	1200
Effective size	mm	2.00
Uniformity coefficient	Unitless	1.20
Porosity	Percentage	40
Backwash requirements		
Backwash type	Combined air-water backwash	
Design backwash average rate, water	$L/m^2{\cdot}min$	400

(continued)

Table 15.4 Secondary Effluent Filtration Design Example (*Continued*)

Input Parameters	Units	Design Values/ Considerations
Design backwash rate range, water	L/m²·min	200–600
Auxiliary air scour rate	L/m²·min	2000
Backwash period	min	20
Assumed backwash frequency, average loading	h	18
Assumed backwash frequency, peak loading	h	6
Total daily backwash water required, average loading	m³/d	1500
Total daily backwash water required, peak loading	m³/d	4600
Backwash reject ratio, average conditions	Percentage	8%
Backwash reject ratio, peak conditions	Percentage	12%
Clearwell volume (sufficient volume for two backwashes)	m³	290
Backwash reject water TSS concentration range	mg/L	50–250
Backwash reject water typical TSS concentration	mg/L	150
Backwash reject water storage basin volume (sufficient volume for six filter backwashes)	m³/d	900
TSS/turbidity removal		
Effluent turbidity, average conditions	NTU	<1.6
Effluent turbidity, cannot exceed more than 5% of the time	NTU	4
Effluent turbidity, cannot exceed anytime	NTU	8
Effluent TSS, average conditions	mg/L	<4
Effluent TSS, maximum daily	mg/L	15
Headloss development		
Clean medium headloss at maximum design filtration rate	m of water	1
Terminal headloss	m of water	3

*In this design example, it is assumed that the flows greater than peak day are equalized upstream of filtration facilities.
TSS = total suspended solids.

Example Problem 15.2

Calculate the area required, loading rate, and clean bed headloss for a rapid sand filter given the following:

Peak hourly flowrate, $Q = 37\,850\ \text{m}^3/\text{d}$
Effective size $(ES) = 0.5$ mm
Uniformity coefficient $(UC) = 1.4$
Shape factor (sphericity, or Φ) = 0.8
Bed porosity $(\varepsilon) = 0.4$
Temperature = 15 °C

Using the Rose equation for headloss calculation:

$$h_L = \frac{1.067(v_a)^2 D}{\Phi \cdot g \cdot \epsilon^4} \Sigma \frac{C_D \cdot f}{d} \quad (15.2)$$

Where h_L = clean bed headloss, m;
v_a = approach velocity (loading rate), m/s;

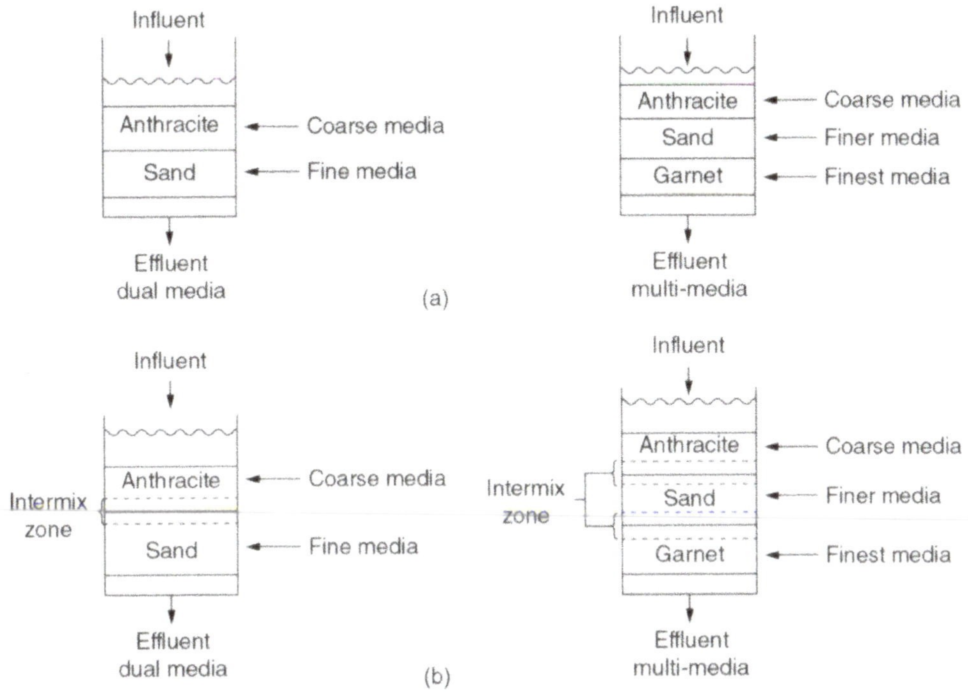

Figure 15.3 Dual-Media and Multimedia Filter Configurations: (a) Nonintermixed and (b) Intermixed Media

D = media depth, m;
Φ = shape factor;
g = acceleration due to gravity, m/s^2;
ϵ = bed porosity;
C_D = drag coefficient;
f = mass fraction of media with size = d; and
d = media diameter, m.

Analyzing the headloss requires dividing the media into layers according to size, because smaller sized particles will contribute more to headloss than larger sized particles. When plotted on log-probability paper, the size distribution of media will produce a straight line. Using this relationship allows us to determine various layers within the bed according to size. Figure 15.4 shows the size distribution for our media ($ES = d_{10} = 0.5$ mm, because $UC = d_{60}/d_{10}$, $d_{60} = 0.7$ mm). If we divide the media into five equal layers (0%–20%, 20%–40%, 40%–60%, 60%–80%, and 80%–100%), we can use the average size in each layer (i.e., d_{10}, d_{30}, d_{50}, d_{70}, d_{90}) for our calculation. Now that we have the media characteristics, we can begin the headloss analysis. At this point we will want to assume a loading rate. A reasonable rate might be in the range of 12 m^3/m^2·h (0.00333 m/s). The next step is to calculate the Reynolds number for each layer in our filter so that we can calculate the C_D for each layer to use in Equation 15.2.

$$R = \frac{\Phi \cdot d \cdot v_a}{v} \tag{15.3}$$

Where R = Reynold's number and
v = kinematic viscosity, m^2/s.

For the 0%–20% layer, the average size is d_{10} and the resulting R is as follows ($v = 1.14 \cdot 10^{-6}$ at 15 °C):

$$R = \frac{0.8 \cdot 0.0005\ \text{m} \cdot 0.00333\frac{m}{s}}{1.14 \cdot 10^{-6}} = 1.170$$

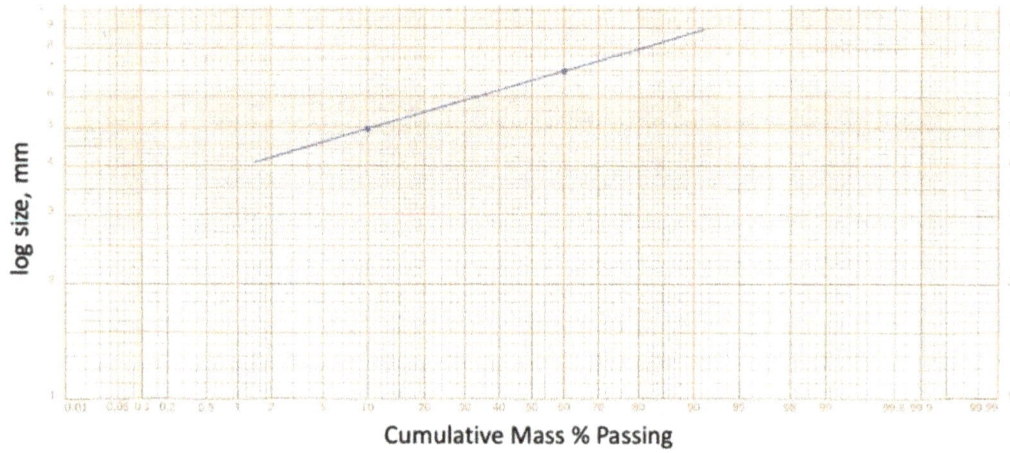

Figure 15.4 Size Distribution of Media with ES = 0.5 mm and UC = 1.4

With the Reynolds number, we can calculate the C_D from Equation 15.4:

$$C_D = \frac{24}{R} + \frac{3}{R^{0.5}} + 0.34 \tag{15.4}$$

$$C_D = \frac{24}{1.17} + \frac{3}{(1.17)^{0.5}} + 0.34 = 23.63$$

At this point, we can go ahead and calculate R and C_D for the remainder of the layers as summarized in Table 15.5.

The only thing left is to assume a filter depth. A reasonable depth for a single media filter would be 0.45. With this, we can now calculate the total headloss for this loading rate:

$$h_L = \frac{1.067(0.0033)^2 \cdot 0.45}{0.8 \cdot 9.8 \cdot 0.4^4} \cdot 29\,650 = 0.79 \text{ m}$$

Considering this is calculated at the peak hourly flow rate, this is an acceptable headloss (<1.0 m). Accepting this headloss and loading rate, we can calculate the total filter area (with one filter out of service for backwashing):

Area = peak hourly flow rate/loading rate = (37 850 m^3/d)/288 $m^3/m^2 \cdot d$ = 131.4 m^2

Table 15.5 Calculated Properties of Media for Headloss Calculation

Media layer	d, m	R	C_D	f	$\frac{C_D \cdot f}{d}$
0%–20%, d_{10}	0.0005	1.1696	23.63	0.2	9453.59
20%–40%, d_{30}	0.00059	1.3801	20.28	0.2	6875.76
40%–60%, d_{50}	0.00066	1.5439	18.3	0.2	5545.42
60%–80%, d_{70}	0.00073	1.7076	16.69	0.2	4572.76
80%–100%, d_{90}	0.00088	2.0585	14.09	0.2	3202.29
					Total = 29 650

If we decide that six filters for a facility of this size is appropriate, we divide the required surface area by 5 (to meet the design loading rate with one filter out of service for backwashing) giving an individual filter area of 26.3 m^2 (total filter area of 157.7, including a sixth filter).

The solids capture efficiency and solids storage capacity of the filter determine its effective run time for a given headloss. A substantial solids breakthrough typically does not occur when filtering secondary effluent if the medium grain size is smaller than 2 mm (0.08 in.). All filters containing a medium of this size or smaller are likely capable of producing water with approximately 5 mg/L TSS, especially if the system includes upstream coagulant addition. The solids storage capacity of a filter will vary with the nature of the solids applied. Headloss development should be associated with accumulated solids in addition to other hydraulic considerations.

Figure 15.5 data show that anthracite has a clear advantage in terms of solids storage capacity. It also reveals a diminishing return (exaggerated as the medium size increases) as higher filter headloss, resulting from solids accumulation, is achieved.

2.1.2 BACKWASH AND CLEANING

Effective medium cleaning is not ensured by expanding the bed. Fluidizing a bed is intended to allow trapped particles to be washed away from the medium. A minor expansion will promote cleaning by allowing particles to work against each other, thereby abrading more tenacious solids. Figure 15.6 shows backwash rate requirements for 10% bed expansion as a function of medium size, type, and water temperature (Cleasby & Baumann, 1974). Typically, it is desirable to provide to 10% expansion of the 60% finer-by-weight particle size to have 30% to 50% expansion of the entire bed. The bottom of the wash trough is used as a guide, corresponding to the approximate rise anticipated for the effective size.

A filter-bed expansion of 30% to 50% overall without the use of supplemental air wash will ensure that even the largest diameter media is fluidized (and cleaned). There are substantial differences in backwash water flowrates required to achieve the same bed expansion at differing water temperatures. Therefore, the controlling backwash water flowrate must provide the appropriate degree of bed expansion at the highest backwash-water temperature anticipated, and the backwash trough should be placed higher than the height of the expanded media determined with coldest water temperature.

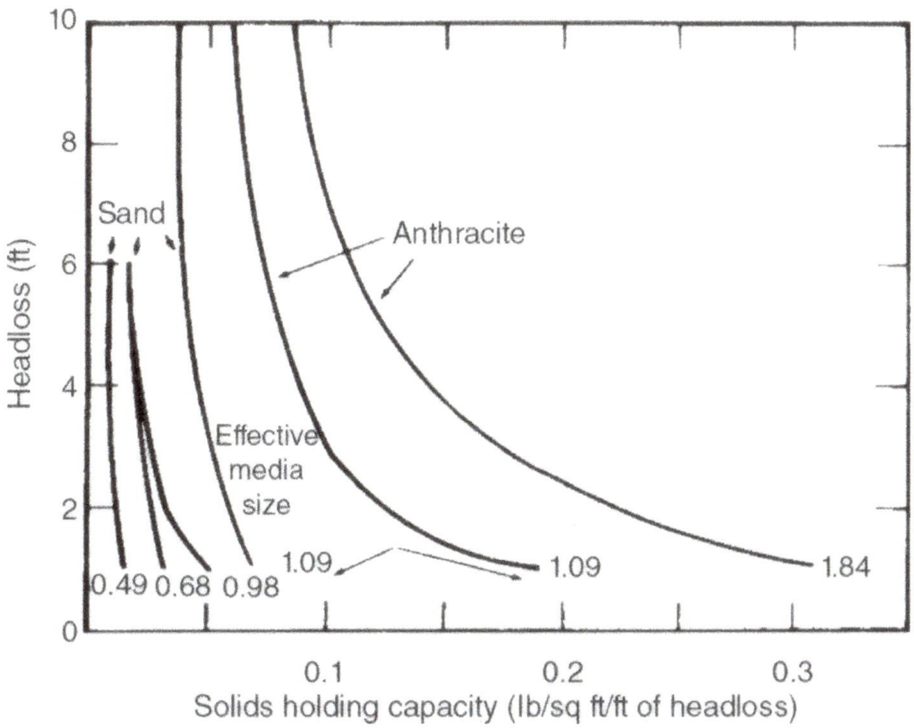

Figure 15.5 Solids Storage Capabilities of Filter Media (ft × 0.3048 = m; lb/sq ft/ft × 16.02 = kg/m^2·m)

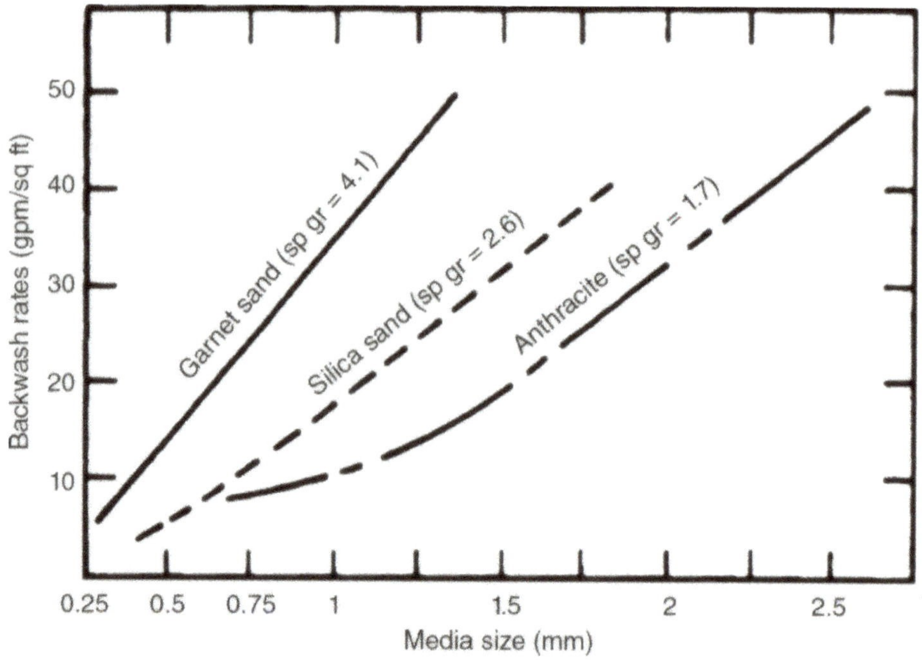

Figure 15.6 Minimum Backwash Rates for a Bed Expansion of 10% at 25 °C (77 °F) (gpm/sq ft × 2.444 = m/h)

Some researchers recommend selecting the maximum backwash rate from the 90% finer medium size (Cleasby & Baumann, 1974). Backwash pump sizing should account for the warmest water temperature anticipated. The total backwash requirement will typically be approximately 4000 to 10 000 L/m^2 (100 to 250 gal/sq ft) and will be largely independent of the backwash rate (Cleasby & Baumann, 1974). This observation is typical for conventional U.S. wash-trough spacing with edges approximately 0.9 m (3 ft) above the filter media. This requirement should not be applied to proprietary systems without rigorous review of the proposed system. Selection of relative grain sizes for components of dual-media and multimedia filters typically should allow equal fluidization among bed components. Failure to provide equal fluidization may result in unplanned intermixing or, in the worst extreme, bed turnover.

Relying solely on an upward flow of water to clean granular filters is inadequate because abrasive action among particles is the most effective cleaning mechanism during backwash (Babbitt et al., 1962; Fair & Geyer, 1954). Fluidized media are separated by the flow of water, thus reducing the frequency of grain collisions (Amirtharajah, 1979). Also, biological solids in secondary effluent are strongly attached to media because of their "sticky" nature. Consequently, it is not surprising that fluidization alone cannot effectively clean the bed (Amirtharajah, 1979; Cleasby & Lorence, 1978; Cleasby et al., 1975, 1977). Many investigators have reported and cited benefits associated with air scour before and during backwash with both subcritical and fluidizing backwash rates (Huang, 1979). Surface and subsurface washing or agitating equipment are commonly used to assist cleaning.

Air scour of approximately 0.015 to 0.025 m^3/m^2·s (3 to 5 scfm/sq ft) is advisable as a minimum for conventional systems; deeper beds may require more and shallow beds may require less. Although most of the beneficial action of air scour will occur during the first 2 minutes, sufficient operational flexibility should be provided for 10 or more minutes of scour with some control of the applied rate. Designers also should consider concurrent washing and air scouring. This requires the capability of draining the filter to just above its surface, followed by a brief simultaneous period of air-and-water backwash until the water reaches within 150 to 200 mm (6 to 8 in.) of the wash trough. Then, either the air or the water wash is stopped. The design of the backwash reservoir should provide for some flexibility by affording between a 3-minute and a 15-minute capacity at the maximum backwash rate. The simultaneous backwash filtration process is sustained by simultaneous air-and-water backwash and specially designed baffles developed by equipment manufacturers.

Another benefit is that simultaneous backwash leaves the filter bed unstratified. Conventional water-wash or air-scour systems leave filter beds stratified. Stratified beds have small particle sizes at the surface of the bed and large particles at the bottom of the bed. This causes rapid accumulation of solids on the surface of the bed, shortening filter run lengths. An unstratified filter bed has uniform media grain sizes throughout the filter bed allowing deep-bed penetration of TSS and increasing filter run lengths significantly. It also increases the net production of filtered water.

Filter effluent is typically the source of backwash water. Chlorine contact basins can be sized to provide a convenient reservoir with sufficient capacity for backwash requirements. Clearwells that are used for chlorination contact should not be located below filtering equipment because of possible corrosion. They also should not receive any bypassed, untreated flow from the process sequence preceding the filter. The clearwell should have adequate capacity for at least two consecutive backwashes and, in small facilities, the capability to receive supplementary clean water from a source such as a high-volume fire hydrant. Filter backwash should not return directly to the facility if the instantaneous recycle rate could adversely affect the treatment processes. If a backwash recovery water (BRW) storage capability is provided, then the reservoir should be designed for maximum filtration with the longest duration of complete, consecutive filter washes.

Grease that is not effectively removed prior to filtration may remain in suspension as an emulsion that coats the filter media. Chlorination before the filter allows some of the grease in the influent stream or on the media to pass through the filter and into the facility's final effluent. All filtering units should be installed with provisions and operating instructions for the removal of grease accumulated in the media. Grease removal may be accomplished by superdosing with chlorine, hypochlorite, or commercially available industrial cleaning solutions. This entails applying either the concentrated or the appropriately diluted solution to the bed, allowing it to stand, and then vigorously backwashing the bed. Weak acids have been used successfully to free bed encrustation associated with upstream use of metal salts or lime for phosphorus removal.

Example Problem 15.3

Continuing from Example Problem 15.2, we now need to calculate bed expansion during backwashing for a given backwash velocity. Using the equation from Fair and Geyer (1954):

$$D_e = (1 - \in) D \Sigma \frac{f}{(1 - \in_e)} \qquad (15.5)$$

Where D_e = expanded bed depth, m; and
$\in_e$ = expanded bed porosity,

The expanded bed porosity can be calculated from Equation 15.6:

$$\in_e = \left(\frac{v_b}{v_s} \right)^{0.22} \qquad (15.6)$$

Where v_b = backwash velocity, m/s; and
v_s = media settling velocity, m/s.

To calculate the settling velocity to use in Equation 15.6, we will use the following expression to represent turbulent flow particle settling:

$$v_s = \left[\frac{4g(\rho_s - \rho)d}{3C_D \rho} \right]^{\frac{1}{2}} \qquad (15.7)$$

Where ρ_s = particle density, kg/m³; and
ρ = water density, kg/m³.

The drag coefficient is calculated using Equation 15.4 as before, only the Reynolds number is calculated from Equation 15.8 with v_s in place of v_a in Equation 15.3:

$$R = \frac{\Phi \cdot d \cdot v_s}{\nu} \tag{15.8}$$

Where v_s = media settling velocity, m/s.

At this point, we recognize that the settling velocity must be known to calculate the Reynolds number for the calculation of C_D (Equation 15.3) and settling velocity (Equation 15.7). To get around this difficulty, one can use a trial-and-error process (assume a settling velocity to calculate R and C_D) and repeat the process to zero in on a calculated v_s. Alternatively, a tool like Solver in Excel can be used to calculate the v_s directly.

Using the same media and specific layers used in Example Problem 15.2 and an assumed backwash rate of 410 L/m²·min (v_b = 0.00683 m/s), we can calculate R as follows for the d_{10} layer assuming a settling velocity for this size particle of 0.07 m/s (followed by calculation of C_D):

$$R = \frac{0.8 \cdot 0.0005 \cdot 0.07}{1.14 \cdot 10^{-6}} = 24.56$$

$$C_D = \frac{24}{24.56} + \frac{3}{(24.56)^{0.5}} + 0.34 = 1.92$$

For the calculated v_s, assume a sand and water density of 2560 and 1000 kg/m³, respectively:

$$v_s = \left[\frac{4 \cdot 9.8(2560 - 1000)0.0005}{3 \cdot 1.92 \cdot 1000}\right]^{\frac{1}{2}} = 0.0749 \text{ m/s}$$

Iteration converges on a settling velocity for this sized particle of 0.0774 m/s (see Table 15.6 for summary of values).

$$D_e = (1 - 0.4)0.45 \cdot 2.2197 = 0.599 \text{ m}$$

This equates to an expansion of 33% that is acceptable, if not a little on the low end of the desired 30%–50% expansion. The total backwash flowrate for a filter area determined previously as 26.3 m = 0.41 m³/m²·min · 26.3 m = 10.78 m³/min. Backwash troughs should be placed at least 0.3 m above the top of the non-expanded media to allow for 0.15 m buffer from the top of the expanded media to the bottom of the wash trough: (0.6 m − 0.45 m) + 0.15 m = 0.3 m.

Table 15.6 Calculated Properties of Media for Backwash Expansion

Media layer	d, m	Assumed v_s, m/s	R	C_D	Calculated v_s, m/s	ϵ_e	$\frac{f}{1-\epsilon_e}$
0%–20%, d_{10}	0.0005	0.077	27.017	1.805	0.077 27	0.586	0.484
20%–40%, d_{30}	0.00059	0.094	38.919	1.438	0.094 07	0.562	0.456
40%–60%, d_{50}	0.00066	0.108	50.021	1.244	0.106 95	0.546	0.441
60%–80%, d_{70}	0.00073	0.12	61.474	1.113	0.118 91	0.533	0.429
80%–100%, d_{90}	0.00088	0.14	86.456	0.94	0.142 05	0.513	0.411
							Total = 2.2197

2.1.3 COLLECTION AND DISTRIBUTION SYSTEMS

In conventional downflow filters, the underdrain system collects the filtrate and distributes the backwash water. The latter function controls the design. The traditional system uses several layers of graded gravel under the filter bed and a lateral header positioned on the filter floor. This header, equipped with orifices, provides the preliminary distribution of backwash water. Final distribution occurs as water percolates upward through the gravel. Typical gravel-bed designs for a pipe underdrain system can be found in standard references, which also provide accepted rules for pipe lateral underdrain design (AWWA & American Society of Civil Engineers [ASCE], 2012; Eliassen & Bennett, 1967; Fair et al., 1968; Weber, 1972). A disadvantage of gravel-pipe lateral systems is the tendency for the gravel to intermix with filter media because of high local velocities during backwash, introduction of air to the backwash system, or excessive backwash flowrates. Gravel-pipe systems also require a large vertical space.

Other underdrain systems have been developed to eliminate the need for a gravel supporting bed. Potential benefits include ease of installation, smaller vertical footprint (allowing more depth for media), and improved distribution of air and water (for effective backwash). Plastic or steel nozzles of various sizes and shapes are available. Nozzles inserted into precast channels, poured-in-place concrete, or steel plates are designed with built-in orifices to evenly distribute backwash water and air. Nozzles selected for the air-water backwash system should ensure that air and water are evenly distributed and that the finest filter media do not pass through or lodge in the opening. The underdrain plenum below the strainers must be cleaned meticulously before use to prevent construction dirt or debris from clogging the strainers during backwash. At a minimum, a clogged nozzle may result in unequal backwash; at worst, excessive uplift pressures may cause structural failure of the underdrain plate or slab.

Each filter should have the capability to operate when isolated completely from other parallel service units. Typically, for proper operation a filter includes valves for influent, effluent, wash water supply, wash water drain, surface wash and, in some facilities, filter-to-waste lines. Filter influent is admitted to the bed so that the medium surface is not disturbed. This entails directing the influent stream to an initial gullet to dissipate the velocity head, using a throttling control on the influent valve, or using a series of distribution troughs and splash plates. A filter-to-waste phase is typically exercised at the end of backwash cycles (or after a maintenance/cleaning event) to ensure that backwash reject or cleaning water does not mix with the filtered effluent. The effluent wash water supply and filter-to-waste lines typically are manifolded for a common connection with the filter underdrain system.

Filter pipes and other conduits, valves, and gates ordinarily are designed for velocities and flows shown in Table 15.7 (AWWA & ASCE, 2012).

2.1.4 FLOW CONTROL METHODS

Design of a filtration system requires an understanding of system operation and requirements and common problems. The main methods used to control the rate of flow through downflow gravity filters may be classified as (1) constant-rate filtration with fixed head, (2) constant-rate filtration with variable head, and (3) variable, declining-rate filtration.

In constant-rate filtration with fixed head, the flow through each filter is the same. Constant-rate systems are either influent controlled or effluent controlled. Pumps or weirs are used for influent-control systems.

Table 15.7 Filter Design Velocities and Flow Volumes (AWWA & ASCE, 2012)

Flow Description	Velocity (ft/s)[a]	Maximum Flow per Unit of Filter Area (gpm/sq ft)[b]
Influent	1–4	3–8
Effluent	3–6	3–8
Wash water supply	5–10	15–25
Wash water drain	3–8	15–25
Filter to waste	6–12	1–6

[a]ft/s × 0.304 8 = m/s.
[b]gpm/sq ft × 2.444 = m/h.

A modulating valve that can be operated manually or mechanically is used for effluent-control systems. In effluent-control systems, at the beginning of the run, a large portion of the available head is dissipated at the valve (almost-closed position). The effluent valve is opened as the headloss builds up within the filter during the run. This method is not as common in WRRFs because the required control valves are expensive and have malfunctioned on a number of occasions. Disadvantages of this system include initial and operating costs of the rate control system, which may require frequent maintenance. Alternative methods of flowrate control involving pumps and weirs are more commonly used.

In variable, declining-rate filtration, flow through the filter decreases as the rate of headloss builds up with time. Declining-rate filtration systems are either influent controlled or effluent controlled. When flow decreases to the minimum design rate, the filter is removed from service and backwashed.

Typically, declining-rate filters provide the best mode of gravity filter operation unless the design terminal headloss exceeds 3 m (10 ft). In this situation, constant-level control of pressure filters may be a more economical choice. This prospective economy, however, must be weighed against a potential disadvantage resulting from terminal headloss conditions occurring simultaneously in all filters. This could cause a system failure under a declining-rate filtration mode unless backwash is consistently initiated before terminal conditions, an upstream reservoir is provided, or both. Figure 15.7 shows a typical filter-and-clearwell arrangement for both influent splitting, constant-rate, and variable, declining-rate filtration operations.

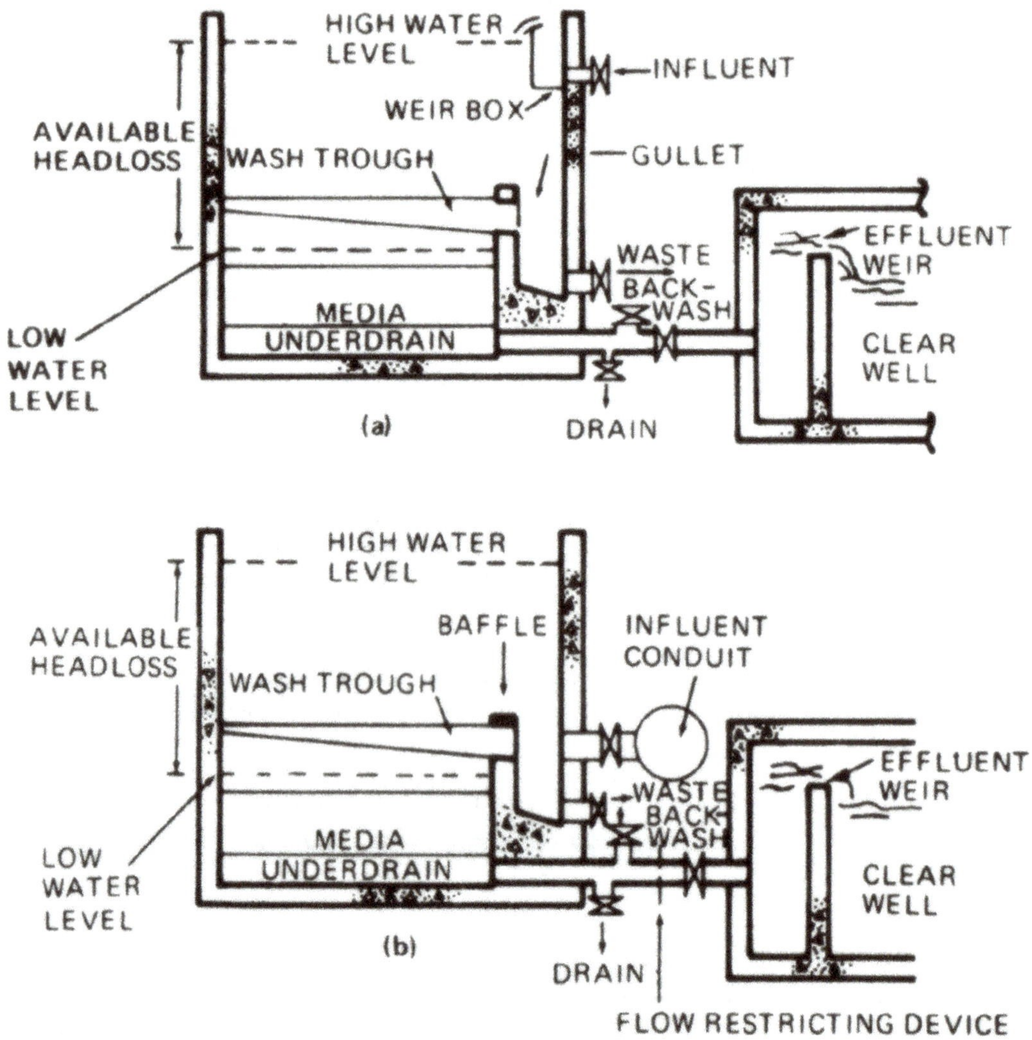

Figure 15.7 Typical Filter-and-Clearwell Arrangement Elevations for (a) Influent Splitting, Constant-Rate Filtration, and (b) Variable, Declining-Rate Filtration

2.1.5 AUTOMATIC BACKWASHING AND MOVING BED FILTERS

Several nongraded-media, continuous-operation filter configurations exist specifically for wastewater treatment operations. Most of these systems have some proprietary features, and several are briefly described in this section. The most common of these filtration processes are the automatic backwashing (ABW) filter and moving bed filter.

Moving bed filters provide a continuous supply of filtered water without the interruption of backwash cleaning cycles. In a typical downflow configuration, influent enters the top of the filter and flows down through layers of increasingly finer sand. Filtered water is collected in a central filtrate chamber before exiting. Solids captured in the filter bed are drawn downward with the sand into an air-lift pump. The turbulent, upward flow in the air lift provides scrubbing action that effectively separates sand and solids before discharging to the filter wash box. The wash box is a baffled chamber that allows gravity separation of cleaned sand from concentrated waste solids. From the wash box, regenerated sand returns to the top of the filter bed, and the solids and BRW are piped to a suitable disposal site. Filter media size is tailored to individual application requirements, and effective size ranges from 0.6 to 1.0 mm (0.02 to 0.04 in.).

The ABW filter—which is often referred to as a traveling bridge filter, continuous backwash filter, or low-head filter—typically is constructed in modules that are 4.9 m (16 ft) wide. The length of the unit and the number of units are then adjusted to provide the required filter surface area. The filter uses a relatively shallow depth of media, typically less than 300 mm (12 in.), and the filter is backwashed continuously. The change in backwash procedure eliminates the need for BRW storage basins and reduces capital costs.

Figure 15.8a shows a longitudinal section through an ABW filter. The filter underdrain system is divided into compartments, or cells, and each cell is individually backwashed as the traveling bridge and backwash hood are positioned over the cell to be cleaned. Figure 15.8b is a transverse section through an ABW filter. Water enters through the influent channel, flows into the filter box, and is collected in the filtrate channel. Headloss from the influent to the effluent channel is typically less than 1.5 m (4.9 ft), which is the source of the term "low-head filter." Typical configurations for the influent and filtrate channels are indicated.

2.2 NON-GRANULAR MEDIA FILTRATION

Several filtration technologies using non-granular media have been developed in the last 20 years to minimize the needs of the backwash operation and the BRW ratio and to provide different operational and design advantages. These filtration technologies can be grouped under two main categories: (1) compressible media filtration and (2) disc filtration.

2.2.1 COMPRESSIBLE MEDIA FILTRATION

One of the filtration technologies using non-granular (i.e., synthetic material) is the compressible media filter (CMF), which was first developed in Japan and has been installed in many facilities around the world since its development. The filter bed consists of a large number of compressible balls, which are formed by shrinking synthetic fibers into a quasi-spherical shape resulting in a complex woven structure. The CMF was first introduced in the United States about 20 years ago, and its application has increased in the last 10 to 15 years.

One of the CMF technologies, also known as the Fuzzy Filter™, involves the use of mechanical compression of the filter medium. The medium has some unusual properties, including being highly porous (approximately 90%) and compressible. The Fuzzy Filter is also a depth filter similar to GMFs. Important design parameters include medium depth, compression ratio, and filtration rate (Caliskaner et al., 1999). The filter medium is resistant to temperature, pH, and acid and has good memory characteristics. The density of the medium is slightly greater than that of water. Because of its low density, the filter medium is retained between two perforated plates (see Figure 15.9). The filter medium properties, such as effective collector size, porosity, and depth, can be adjusted in response to changing influent conditions because it is compressible. Properties of the filter medium are altered by adjusting the position of the upper moveable plate.

Typical uncompressed medium depth for secondary effluent filtration is 760 mm (30 in.). For example, to compress the medium 30% (average value with respect to depth), the compression plate is lowered

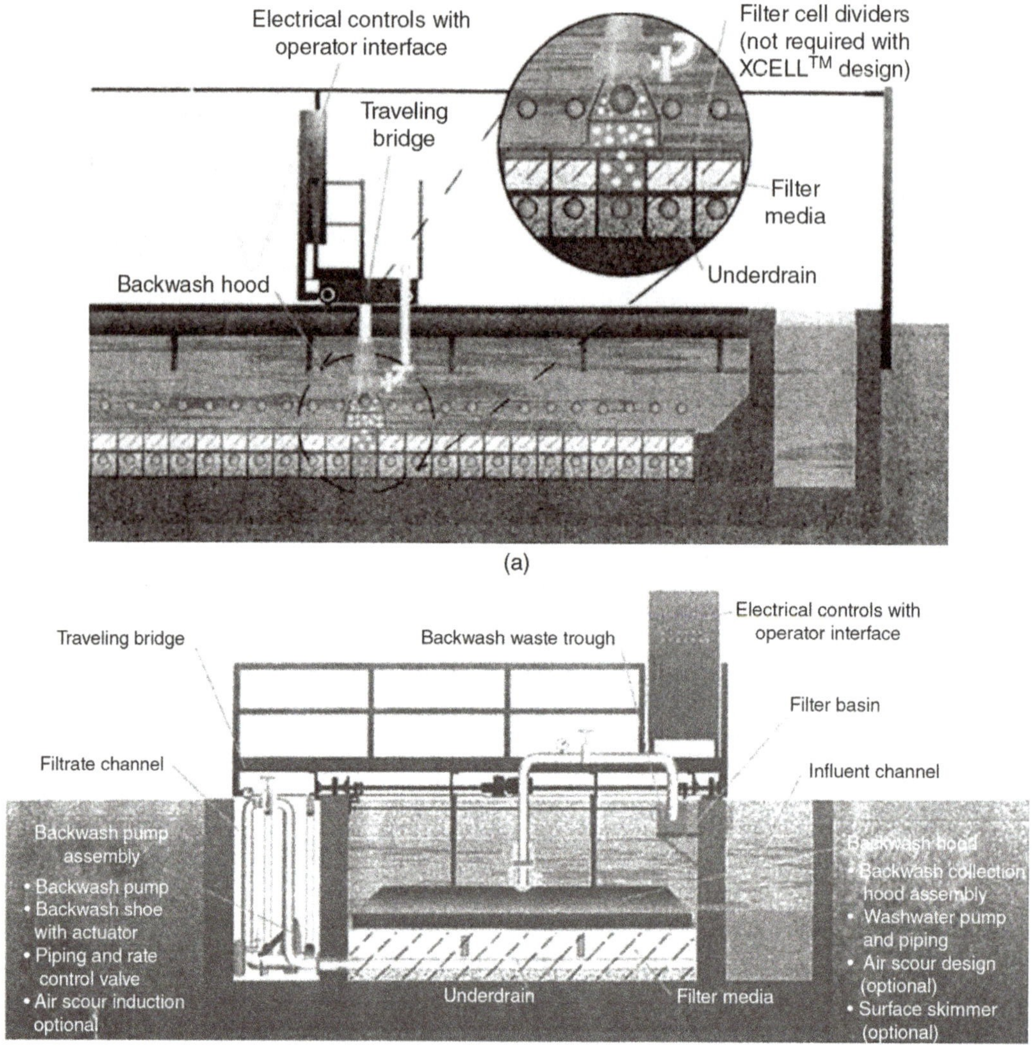

Figure 15.8 (a) Longitudinal Section Through Automatic Backwashing Filter, and (b) Transverse Section Through an Automatic Backwashing Filter

until the medium thickness is 530 mm (21 in.). Medium properties such as the porosity, collector size (i.e., pore/interstice size), and depth all decrease with the increase in the applied compression ratio. The corresponding porosity, medium depth, and collector size values of the medium are tabulated in Table 15.8 at different medium compression ratios.

The CMF is able to operate at filtration rates up to 1600 L/m^2·min (40 gpm/sq ft) because of reduced headloss development resulting from high porosity of the medium (Caliskaner et al., 2011b). Typical average design filtration rates are between 800 and 1200 L/m^2·min (20 to 30 gpm/sq ft). Footprint requirements are reduced significantly (approximately 75%) compared to the granular filters because of the high filtration rates. Removal performance can be increased approximately 50% to 70% with chemical addition. Chemical type, dose, and addition duration should be selected carefully to prevent medium blinding.

During the filtration cycle, secondary effluent is introduced in the bottom of the filter into a plenum. The influent wastewater flows upward through the filter medium, and is discharged from the top of the filter. On a regular operational basis, the filtration cycle is interrupted for backwashing when the terminal headloss or breakthrough (as measured in effluent turbidity) is reached. At the inception of the backwash

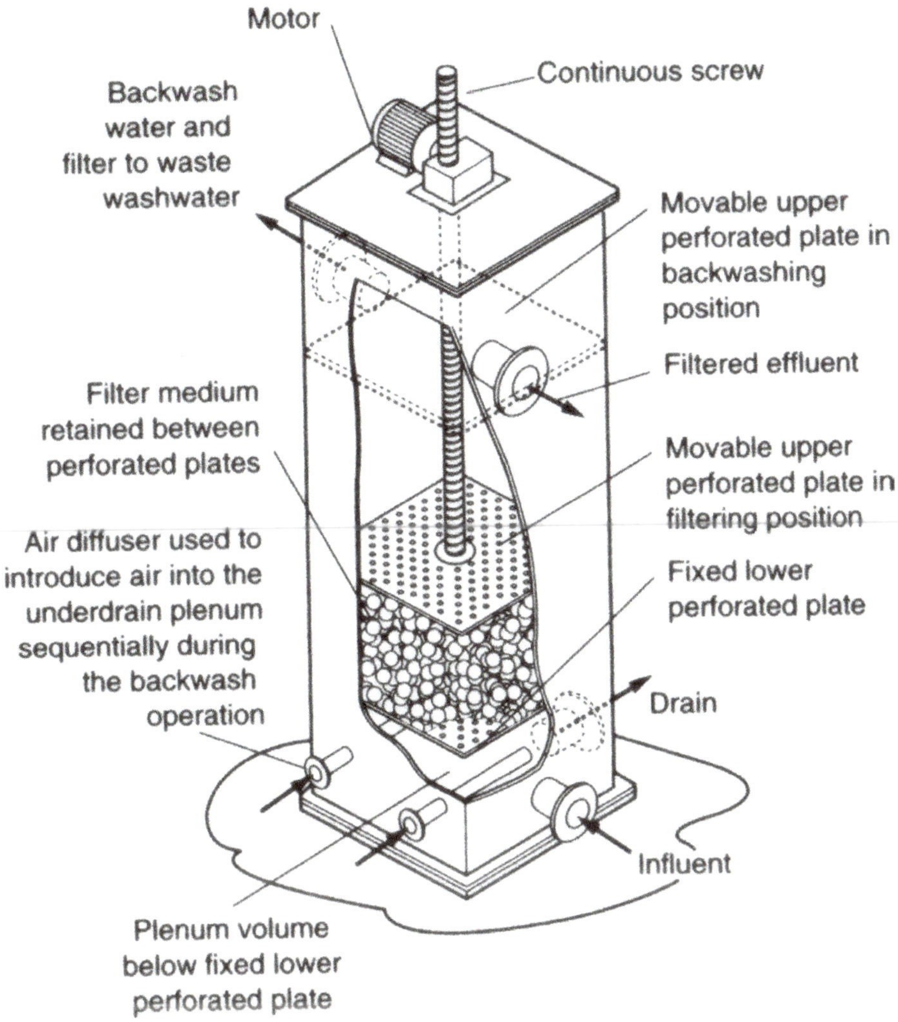

Figure 15.9 Schematic View of Filter With a Compressible Filter Medium. Porosity, Collector Size, and Depth of the Filter Bed Can Be Adjusted by Compressing the Filter Material (Caliskaner et al., 1999)

cycle, the flow is diverted to the backwash water line, and the upper plate is moved up mechanically to increase the bed volume for backwashing. Secondary effluent typically is used to backwash the filter, eliminating the need for a clean water tank. While secondary effluent continues to flow up through the filter, air is introduced below the lower perforated plate, first from the left side and then from the right side of the filter (through separate air pipes in both sides). The complete backwash cycle takes approximately 20 minutes. To start the next filtration cycle, the upper perforated plate is returned to its desired position to obtain the required level of medium compression. The typical backwash rate is 410 L/m^2·min (10 gpm/sq ft); therefore, the secondary influent flowrate is reduced at the beginning of the backwash cycle (e.g., from 1230 L/m^2·min [30 gpm/sq ft] to 410 L/m^2·min [10 gpm/sq ft]). The BRW ratio ranges between 1% and 5% depending on the specific filtration conditions, with average typical values around 2% to 3%. Underdrain systems are eliminated in the CMF system because a granular medium is not used.

Clean filter headloss increases linearly with filtration rate. Headloss also increases as porosity and collector size of the medium decrease as compression ratio is increased. The typical terminal headloss value is between 1.8 and 3.7 m (6 and 12 ft).

2.2.2 CLOTH DEPTH FILTRATION

Cloth depth filters (CDFs) use different types of cloth material (such as woven nylon or polyester construction) to create a dense fiber arrangement to separate the TSS remaining in the secondary effluent.

Table 15.8 Properties of Commercially Available Carbons

Properties and Specifications	ICI America Hydrodarco 3000	Calgon Filtrasorb 300M (8 to 30 mesh)	Westvaco Nuchar WV—L (8 to 30 mesh)	Witco 517 (12 to 30 mesh)
Surface area, m²/g	600–650	950–1050	1000	1050
Apparent density, g/cm³	0.43	0.48	0.48	0.48
Density, backwashed and drained, lb/cu ft[a]	22	26	26	30
Real density, g/cm³	2.0	2.1	2.1	2.1
Particle density, g/cm³	1.4–1.5	1.3–1.4	1.4	0.92
Effective size, mm	0.8–0.9	0.8–1.0	0.85–1.05	0.89
Uniformity coefficient	1.7	£2.1	£1.8	1.44
Pore volume, cm³/g	0.95	0.85	0.85	0.60
Mean particle diameter, mm	1.6	1.5–1.7	1.5–1.7	1.2
Sieve size (U.S. standard series)				
Larger than No. 8, maximum %	8	8	8	—
Larger than No. 12, maximum %	—[b]	—	—	5
Smaller than No. 30, maximum %	5	5	5	5
Smaller than No. 40, maximum %	—[b]	—	—	—
Iodine No.	650	900	950	1000
Abrasion No., minimum	Not available	78	70	85
Ash, %	Not available	8	7.5	0.5
Moisture as packed, maximum %	Not available	2	2	1

[a]lb/cu ft × 16.02 = kg/m³.
[b]Not applicable for this carbon size.

In its operational state, the cloth medium provides an approximate 5 mm (0.2 in.) thickness, which allows depth filtration to occur in addition to surface filtration. The pore size of the media is significantly smaller compared to the GMFs and CMFs, resulting in similar medium depth-to-media diameter (L/d) ratios (i.e., similar medium depth–to–medium size ratios observed for GMF technologies). The CDF is one of the most common disc filtration technologies used for secondary effluent filtration.

Different types of cloth media are used, depending on the manufacturer. Manufacturers may also offer different cloth media in order to address site-specific conditions (e.g., chemical resistance, different pore size characteristics). Cloth media technologies can vary in their construction with materials that can be woven, knitted, or produced by needle felting. Materials can be selected for the application and often include nylon, acrylic, or polyester material. The use of woven-pile cloth materials has emerged as the most common type of CDF due to improvements in backwash efficiency. The density and thickness of the filter cloth can also be selected according to the characteristics of the influent wastewater and desired effluent quality. The pore size of the cloth media is not absolute because of the random arrangement of the CDF fibers. Nominal pore size ranges between 5 and 10 mm for different type of cloth materials, but significant removals can be realized with smaller particle size ranges.

Typical maximum design filtration rates are between 240 to 280 L/m²·min (6 and 7 gpm/sq ft), although testing has shown that these filters can operate at hydraulic loading rates up to 800 L/m²·min (20 gpm/sq ft) for short periods. The maximum hydraulic loading rate can also be limited by the influent TSS when the solids loading rate exceeds the manufacturer's recommendation (typically 400 to 700 g/m²·h).

Turbidity and TSS removal performance of CDF is similar to GMF and CMF technologies (Caliskaner et al., 2011a). Cloth depth filter removal performance can be increased approximately 50% to 70% with chemical addition, but chemical type, dose, and addition duration should be selected carefully to prevent medium blinding. The footprint requirement is 70% to 75% less compared to GMFs (Caliskaner et al.,

2011a; Haecker & Healy, 2006). The vertical configuration of the filtration discs provides a relatively large filter surface area in a small footprint. An individual filter can contain 1 to 24 discs. The number of discs in a filter unit depends on the design flow and filter loading rates and the size of an individual disc. The diameter of an individual disc ranges from 0.8 m (2.8 ft) to 3.05 m (10 ft) depending on the disc manufacturer, model, and design flow and loading rates. Different CDF systems, such as the diamond or MegaDisk configurations that use the same type of cloth filter medium are available. The diamond configuration was manufactured to accommodate low-profile/shallow basin designs (Baumann et al., 2006). A typical CDF is shown in Figure 15.10.

3.0 MEMBRANE PROCESSES

Membrane processes can be pressure or vacuum driven, or depend on electrical potential gradients, concentration gradients, or other driving forces. Some facilities that are using the vacuum-driven membranes are applying the gravity-assisted mode of operation when membranes are operated under the gravity (static pressure) until the desired permeability is achieved.

Four main membrane categories, classified by the size of the separated particles or materials, are commercially used at the present time and can be expressed in microns (µm) and/or Daltons molecular weight cutoff (Da MWCO) (Frenkel, 2004). Microfiltration separates particles from 0.1 to 10.0 µm (>100 000 Da). Ultrafiltration rejects materials from 0.01 to 0.1 µm (2000 to 100 000 Da). Nanofiltration rejects materials from 0.001 to 0.01 µm (200 to 1000 Da), and reverse osmosis ranging in molecular size less than 0.001 µm (<200 Da). Microfiltration and ultrafiltration membranes are low pressure, and nanofiltration and reverse osmosis membranes are high pressure.

These four classes of membranes have been established primarily based on the size range of particles or ions each type will remove from water or wastewater. This characteristic is described in terms of nominal pore size for microfiltration and ultrafiltration membranes and by MWCO for nanofiltration and

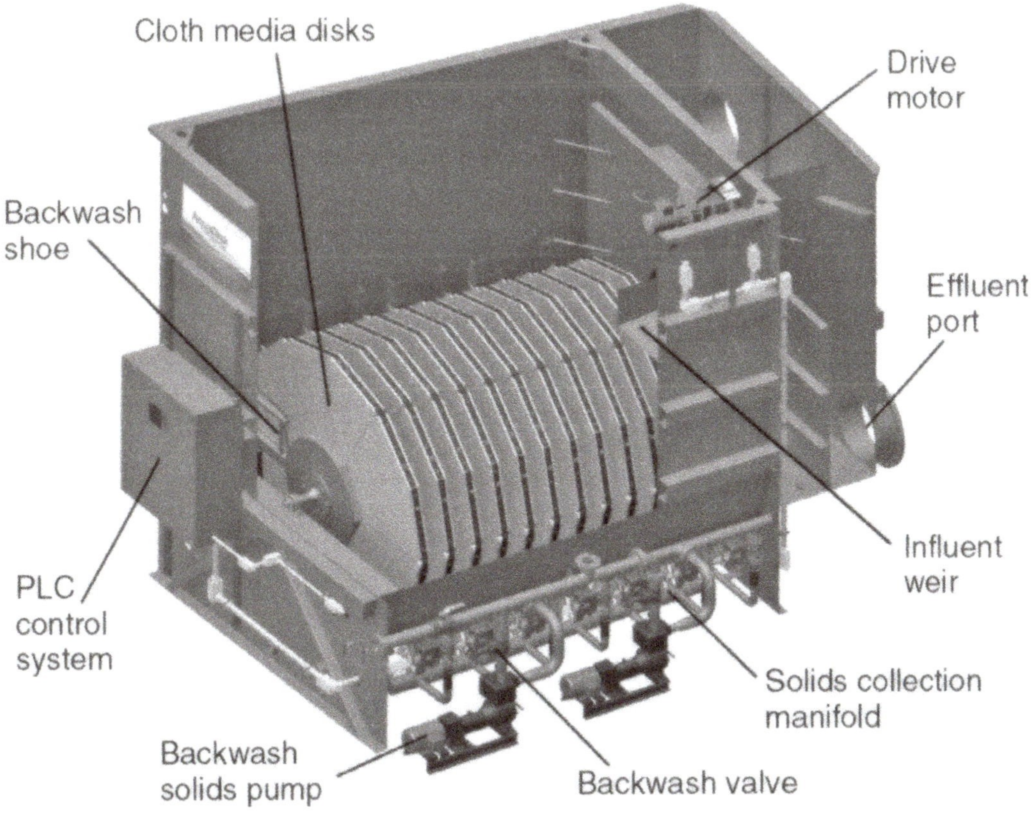

Figure 15.10 Typical Schematic of a Cloth Depth Filter System (Reprinted with permission from Aqua-Aerobic Systems, Inc.)

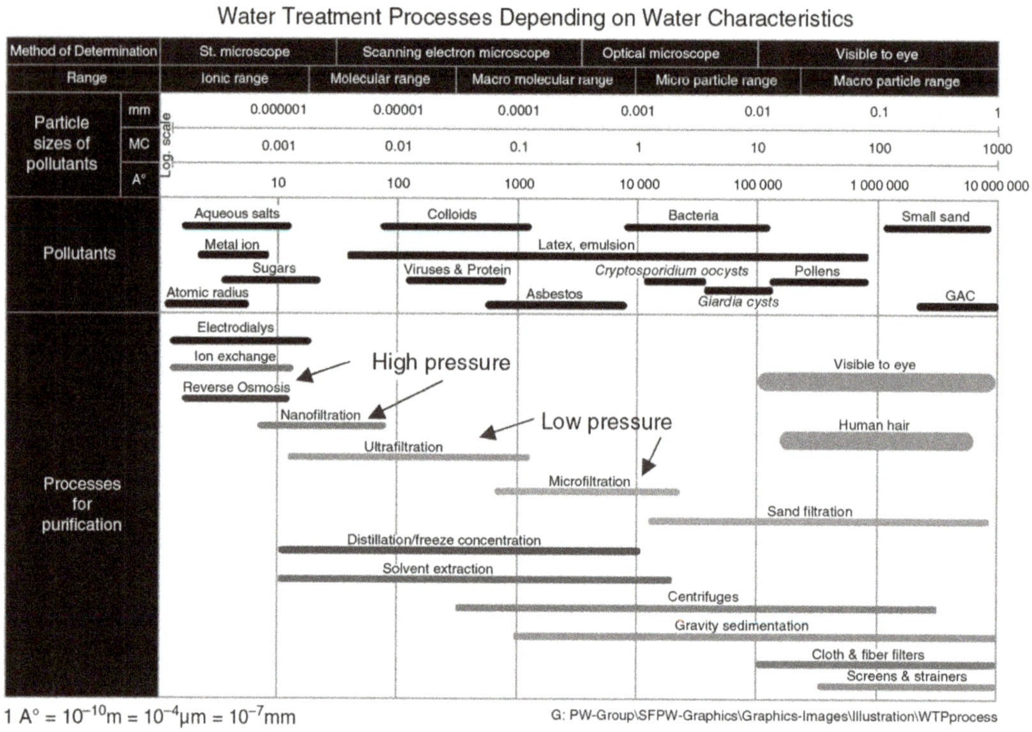

Figure 15.11 Size Ranges for Various Treatment Processes and Membrane Separation (Courtesy of Val S. Frenkel, PhD, PE, D.WRE)

reverse osmosis membranes. Microfiltration and ultrafiltration membranes typically are used as particle (suspended and colloidal matters) separation processes, whereas nanofiltration and reverse osmosis membranes can be classified as ion rejection processes.

Membrane processes originally were developed in the early 1960s as desalination or demineralization processes. Membranes initially were commercially applied in reverse osmosis, and then in the 1990s were developed for microfiltration and ultrafiltration. Membrane shapes include spiral wound, hollow fiber, flat sheet, and tubular.

Figure 15.11 illustrates the relative rejection properties of four types of membranes for a range of contaminants that may be found in secondary wastewater effluent. Although low-pressure membranes are designed to remove suspended and colloidal matter from water, high-pressure membranes are designed to remove dissolved constituents.

Compared to other membranes, microfiltration membranes have the largest pore size and are designed to remove relatively large suspended particles such as colloids, bacteria, cysts, and others. Ultrafiltration membranes typically have a one-order-of-magnitude smaller pore size than microfiltration membranes and can achieve greater logs of virus removal than the microfiltration membranes. Nanofiltration and reverse osmosis membranes are applied to remove dissolved constituents of different sizes and charges from the water.

PRACTICE PROBLEMS

1. **Provide examples of two common chemical disinfectants and describe what is involved with each with respect to design and operational considerations.**

2. **For Example Problem 15.1, how much smaller could the contact tank be if the free chlorine residual was increased to 4.5 mg/L?**

3. For a wastewater with a pH = 7, the concentration-time (CT) value for 3-log *Giardia* removal at 15 °C is reported to be 76 mg·min/L for a free chlorine residual concentration of 1.2 mg/L. Calculate the required contact basin volume for a flow of 50 000 m^3/d. How does this volume compare to that for 4-log virus removal (CT = 4 mg·min/L)?
4. Describe the main aspects of disinfection with ultraviolet light, including its advantages and disadvantages when compared to chemical disinfection.
5. Explain the main operational and design considerations for granular media filtration.
6. Design a rapid filter facility for a flow of 80 000 m^3/d with sand media, given the following:

   Peak hourly flowrate Q = 80 000 m^3/d
   Effective size (ES) = 0.55 mm
   Uniformity coefficient = 1.5
   Shape factor (sphericity, or Φ) = 0.84
   Bed porosity (ε) = 0.45
   Temperature = 10 °C

7. Determine a backwash velocity for the previous filter design to provide 30% to 50% expansion. What is the backwash flow rate for a single filter?
8. Provide examples of two non-granular media alternatives for filtration and explain how they function.
9. Describe the principles behind membrane filtration and give examples for the spectrum of membrane processes.

4.0 REFERENCES

American Water Works Association. (1990). *Water quality and treatment* (4th ed.). McGraw-Hill.

American Water Works Association & American Society of Civil Engineers. (2012). *Water treatment plant design* (5th ed.). McGraw-Hill.

Amirtharajah, A. (1979). Discussion of "effectiveness of backwashing for wastewater filters," by J. L. Cleasby and J. C. Lorence. *Journal of Environmental Engineering, 105*(EE2), 59.

Babbitt, H. E., Doland, J. J., & Cleasby, J. L. (1962). *Water supply engineering* (6th ed.). McGraw-Hill.

Baumann, P., Kieffer, C., Morrall, T., & Bauer, R. (2006, October 21–25). Fox Metro Water Reclamation District solves filtration problems with innovative Aquadiamond® cloth media filter. *Proceedings of the 79th Annual Water Environment Federation Technical Exposition and Conference* [CD-ROM], Dallas, Texas, United States.

Bellamy, W., Awad, J., Wei, J., & Gramith, J. (1991). In-line ozone dissolution demonstration-scale evaluation. *Ozone Science Engineering, 13*(5), 559–591.

Black & Veatch Corporation. (2010). *White's handbook of chlorination and alternative disinfectants* (5th ed.). John Wiley & Sons.

California Code of Regulations. (2006). California Wastewater Reclamation Criteria, Title 22, Division 4, Chapter 3, CCR 002E.

Caliskaner, O., Bennett, G. B., & Celeste, S. (2011a). Comparison of three filtration technologies to meet tertiary level discharge and unrestricted reuse requirements. *Proceedings of the Water Environment Federation Technical Exposition and Conference 2011*, Los Angeles, California, United States.

Caliskaner, O., Tchobanoglous, G., & Carolan, A. (1999). High-rate filtration of activated sludge effluent with a synthetic compressible media. *Water Environment Research, 71*(6), 1171.

Caliskaner, O., Tchobanoglous, G., Carolan, A., & Downey, M. (2011b). Evaluation of the new compressible media filter at filtration rates between 30 and 40 gpm/ft^2 for unrestricted water reuse. *Proceedings of the Water Environment Federation Technical Exposition and Conference 2011*, Los Angeles, California, United States.

Chlorine Institute. (1986). *The chlorine manual* (5th ed.).

Cho, M., Chung. H., & Yoon, J. (2003). Disinfection of water containing natural organic matter using ozone-initiated radical reactions. *Applied Environmental Microbiology*, *69*(4), 2284–2291.

Cleasby, J. L., Arboleda, J., Burns, D. E., Prendiville, P. W., & Savage, E. S. (1977). Backwashing of granular filters. *Journal of the American Water Works Association*, *69*(2), 115.

Cleasby, J. L., & Baumann, E. R. (1974). *Wastewater filtration design considerations* (U.S. Environmental Protection Agency Technology Transfer Seminar Publication). U.S. Environmental Protection Agency.

Cleasby, J. L., & Lorence, J. C. (1978). Effectiveness of backwashing wastewater filters. *Journal of Environmental Engineering*, *104*(EE4), 749.

Cleasby, J. L., Stangl, E. W., & Rice, G. A. (1975). Developments in the backwashing of granular filters. *Journal of Environmental Engineering*, *101*(5)(EE5), 713.

Compressed Gas Association. (1988). *Sulfur dioxide* (4th ed., Pamphlet G-3).

Eliassen, R., & Bennett, G. (1967). *Progress report: Reclamation of re-usable water from sewage (Federal Water Pollution Control Association Demonstration Grant WPD 21-05)*. U.S. Department of the Interior.

Fair, G. M., & Geyer, J. C. (1954). *Water supply and waste-water disposal.* John Wiley and Sons.

Fair, G. M., Geyer, J. C., & Okun, D. A. (1968). *Water and wastewater engineering* (Vol. 2). Wiley and Sons.

Frenkel, V. (2004, August 5–7). Membranes vs. conventional treatment in municipal and industrial applications. Paper presented at American Membrane Technology Biennial Conference, San Antonio, Texas, United States.

Gere, R., & Wright, H. (1998). UV disinfection of wastewater coagulated with ferric chloride: Recalcitrance and fouling problems. *Water Science and Technology*, *38*(3), 15–23.

Gottschaulk, C., Libra, J. A., & Saupe, A. (2010). *Ozonation of water and waste water: A practical guide to understanding ozone and its applications.* Wiley-VCH.

Haecker, S., & Healy, J. (2006, October 21–25). Innovative technology to implement a reuse water program. *Proceedings of the 79th Annual Water Environment Federation Technical Exposition and Conference* [CD-ROM], Dallas, Texas, United States.

Haji Malayeri, A., Mohseni, M., Cairns, B., & Bolton, J. R. (2017). Fluence (UV dose) required to achieve incremental log inactivation of bacteria, protozoa, viruses and algae. *IUVA News*, *18*(3), 4–6 + Supplemental Tables.

Helz, G. R., & Nweke, A. C. (1995). Incompleteness of wastewater dechlorination. *Environmental Science and Technology*, *29*(4), 1018–1022.

Holden, B., Nelson, K., Crook, J., Cooper, R. C., Williams, G., Kouretas, T., & Sheikh, B. (2006, October 21–25). Higher filter loading rates for greater water reuse capacity. *Proceedings of the 79th Annual Water Environment Federation Technical Exposition and Conference* [CD-ROM], Dallas, Texas, United States.

Huang, J. Y. C. (1979). Filter backwashing by ejector. *Journal of Environmental Engineering*, *105*(EE5), 915.

Hunt, N. K., & Mariñas, B. J. (1999). Kinetics of *Escherichia coli* inactivation with ozone: chemical and inactivation kinetics. *Water Research*, *33*, 2633–2641.

International Fire Code Institute. (1994). *Uniform fire code.*

Jafvert, C. T., & Valentine, R. L. (1992). Reaction scheme for the chlorination of ammoniacal water. *Environmental Science and Technology*, *26*, 577–586.

Jagger, J. (1967). *Introduction to research in ultra-violet photobiology*. Prentice-Hall.

Jeyanayagam, S., & Cotton, C. (2002). Practical considerations in the use of UV light for drinking water disinfection. *Proceedings of the CASE/ASCE Joint Conference: An International Perspective on Environmental Engineering.*

Leong, L. Y. C., Kuo, J., & Tang, C. (2008). *Disinfection of wastewater effluent—Comparison of alternative technologies.* Water Environment Research Foundation.

Linden, K.G., Shin Gwy-am, Faubert G., Cairns, W., & Sobsey, M. D. (2002). UV disinfection of *Giardia lamblia* cysts in water. *Environmental Science and Technology, 36*, 2519–2522.

Lindenauer, K. G., & Darby, J. L. (1994). Ultraviolet disinfection of wastewater: Effect of dose on subsequent photoreactivation. *Water Research, 28*, 805–817.

Loge, F., Emerick, R., Thompson, D., Nelson, D., & Darby, J. (1999). Factors influencing UV disinfection performance—Part 1: Light penetration into wastewater particles. *Water Environment Research, 71*, 377–381

Metcalf & Eddy, Inc./AECOM. (2014). *Wastewater engineering: Treatment and resource recovery* (5th ed.). McGraw-Hill.

Occupational Safety and Health Administration. (1996). *Process Safety Management of Highly Hazardous Chemicals* (Regulation 1910.119, 57 FR 23060, June 1, 1992; 61 FR 9227, March 7, 1996).

Prados, M., Paillard, H., & Roche, P. (1995). Hydroxyl radical oxidation processes for the removal of triazine from natural water. *Ozone Science Engineering, 17*(2), 183–194.

Staehelin, J., & Hoigne, J. (1985). Decomposition of ozone in water in the presence of organic solutes acting as promoters and inhibitors of radical chain reactions. *Environmental Science and Technology, 19*(12), 1206–1213.

Swift, J. L., Wilson, J. P., & Hunter, G. (2007, October 13–17). Implementing local limits for the control of WWTP effluent ultraviolet transmittance. *Proceedings of the 80th Annual Water Environment Federation Technical Exposition and Conference* [CD-ROM], Anaheim, California, United States.

U.S. Environmental Protection Agency. (1986). *Municipal wastewater disinfection design manual* (EPA-625/1-86-021).

U.S. Environmental Protection Agency. (2006). *Ultraviolet disinfection guidance manual for the final long-term 2 Enhanced Surface Water Treatment Rule* (EPA-815/R-06-007).

von Gunten, U. (2003). Ozonation of drinking water: Part 1. Ozonation kinetics and product formation. *Water Research, 37*(7), 1443–1467.

Von Sonntag, C., & Von Gunten, U. (2012). *Chemistry of ozone in water and wastewater treatment: From basic principles to applications.* IWA Publishing.

Water Environment Federation. (2015). *Ultraviolet disinfection for wastewater.*

Water Environment Research Foundation. (2007). *Disinfection of wastewater effluent—Comparison of alternative technologies* (Report No. 04-HHE-4).

Weber, W. J., Jr. (1972). *Physiochemical processes for water quality control.* Wiley-Interscience.

Williams, G., Sheikh, B., Holden, R., Kouretas, T., & Nelson, K. (2007). The impact of increased loading rate on granular media, rapid depth filtration of wastewater. *Water Research, 41*(19), 4535–4545.

Wolfe, R. L. (1990). Ultraviolet disinfection of potable water. *Environmental Science and Technology, 24*, 768–773.

World Health Organization. (2016). *Chlorine dioxide, chlorite and chlorate in drinking-water* (Background document for development of WHO Guidelines for Drinking-Water Quality). WHO Press.

Xu, P., Janex, M., Savoye, P., Cockx, A., & Lazarova, V. (2002). Wastewater disinfection by ozone: Main parameters for process design. *Water Research, 25*, 761–773.

Zhang, K., Zhang, Z. H., Wang, H., Wang, X. M., Zhang, X. H., & Xie, Y. F. F. (2020). Synergistic effects of combining ozonation, ceramic membrane filtration and biologically active carbon filtration for wastewater reclamation. *Journal of Hazardous Materials, 382*, 121091.

5.0 SUGGESTED READINGS

American Public Health Association, American Water Works Association, & Water Environment Federation. (2023). *Standards methods for the examination of water and wastewater* (24th ed.). American Public Health Association.

Culp, R. L., & Culp, G. L. (1978). *Handbook of advanced wastewater treatment* (3rd ed.). Van Nostrand Reinhold Co.

Whitby, G. E., & Palmateer, G. (1993, May 23–25). The effects of UV transmission, suspended solids, wastewater mixtures and photoreactivation in wastewater treated with UV light. *Proceedings of the Water Environment Federation Specialty Conference: Planning, Design, and Operation of Effluent Disinfection Systems,* Whippany, New Jersey, United States.

Chapter 16 Modeling and Mass Balances

LEARNING OBJECTIVES

At the end of this chapter, students should be able to:

- Describe the basics of process modeling and mass balances.
- Describe the underlying basis of model simulators from a high level.
- Use model simulations for process design including sizing aeration basins, sedimentation basins, and chemical dosing systems.
- Develop model simulations from existing tools that predict operational conditions for a water resource recovery facility [WRRF].
- Describe the limitations of process model simulations.
- Develop a summary of output for process model simulations.
- Generate mass balances from model output.

1.0 INTRODUCTION

Modeling is important to the engineering design of modern water resource recovery facilities (WRRFs) that are experiencing increasing demands on effluent quality with ever-tightening resources. A wastewater process model consists of a number of mathematical equations that balance process reactions and reaction rates of biological, chemical, and physical phenomena of various unit processes. The mathematical relationships used in model development have been described in previous chapters of this manual. Using the relationships developed, modern computers and software packages are used to simultaneously solve multiple governing relationships, either directly or by iterations (Olsson & Newell, 1999).

Process models for WRRFs have developed over the years from basic activated sludge models, to the point where they are now used to develop whole facility simulations of all processes that are widely used for process design, facility evaluations, troubleshooting, and training. **Simulators** imitate the operation of a real-world process using models of the key characteristics of the process. There are general-purpose simulators supporting model development and application-specific simulators for total facility simulation, including models for other unit processes (e.g., biofilm processes). **Mathematical wastewater treatment models** are integrated into software algorithms or simulators to help provide a robust approach for design to achieve effluent quality goals (e.g., removing ammonia, nitrogen, and phosphorus), optimize process sizing with a reasonable safety factor, obtain mass balances (e.g., balance chemical oxygen demand [COD], solids, nitrogen [N], phosphorus [P], and sulfur), manage energy costs, control greenhouse gas emissions, and evaluate sizing needs for auxiliary systems (e.g., blowers, pumps). This chapter provides a general discussion and overview of model basics and applications. The reader is referred to *Wastewater Treatment Process Modeling* (Water Environment Federation [WEF], 2014) and Henze et al. (2000) for more in-depth model discussions.

Simulators are modeling tools designed using a series of models to simulate the performance of WRRFs using a key inputs (influent characteristics, temperature, etc.) processed through established relationships (kinetics and chemistry) and mass balances (using stoichiometric relationships) to generate or project outputs (effluent quality, aeration demand, offgas production).

Mathematical wastewater treatment models are mathematical relationships that predict key process reactions and stoichiometry in wastewater treatment reactions.

Steady-state refers to a model condition whereby the model inputs and outputs are balanced with rates reflecting no accumulation or loss in the system. Steady-state models are often used to represent long-term conditions with stable operation.

2.0 BASIS OF WASTEWATER MODEL DEVELOPMENT

The personal computer brought us the spreadsheet, a tool that completes calculations with each "enter" keystroke or on command at the completion of all entries. **Steady-state** spreadsheet models are often prepared by individual designers to determine tank volumes based on a set of input values and system requirements. Most models are set up to evaluate a steady-state condition, but the simulator may run the models for **dynamic** and peaking conditions in order to test worst-case design scenarios.

Dynamic refers to a model condition whereby model inputs and outputs do not need to be balanced with rates and accumulation or loss from the system can occur. Dynamic models are often used to evaluate short-term peaking or low flow/load conditions with possible instability.

The development of models of dynamic conditions was made possible by the advent of the digital computer and software to solve simulations involving simultaneous differential equations. Storer and Gaudy (1969) discussed computational analysis of transient responses to shock loading of heterogeneous populations of bacteria. Smith and Eilers (1970) at the Federal Water Pollution Control Administration prepared one of the first reports on the simulation of the activated sludge process. Andrews (1971) presented a discussion of kinetic models for biological waste treatment processes, and Jones (1971) reported on a mathematical model of the contact-stabilization process.

Graduate theses at Clemson University by Bryant (1972), Busby (1973), Stenstrom (1975), and Tracy (1973) led to publications by Busby and Andrews (1975) of a dynamic model using the Continuous System Modeling Program involving stored mass, active mass, and a settler model, whereby feed was introduced at the top of the sludge blanket. Tracy and Keinath (1974) presented a dynamic model of thickening activated sludge. Early models were also presented by Goodman and Englande (1974), Grady and Roper (1974), and Lawrence and McCarty (1970). This work was synthesized by Dold et al. (1980), who presented a general model for the aerobic activated sludge process. Modeling the nitrification process was not difficult: the model based on the Monod kinetics of Downing et al. (1964) could be directly incorporated to the general aerobic model. Following practical developments in biological denitrification using wastewater and endogenously generated organics as an electron donor by Ludzack and Ettinger (1962), Christensen-Henze and Harremoes (1977) presented a significant review of biological denitrification. This provided valuable information to Van Haandel et al. (1981) to extend the aerobic activated sludge model to include biological denitrification.

In 1982, on the recommendation of the International Water Association (IWA; then the International Association on Water Pollution Research and Control), a task group was formed to develop practical mathematical models for the design and operation of WRRFs. Members of the task group, Grady et al. (1986), presented details of the model, which became known as Activated Sludge Model No. 1; this work was later published by the same authors as the International Association on Water Quality *Scientific and Technical Report No. 1* (Henze et al., 1987). This was followed by Activated Sludge Model No. 2 (Henze et al., 1995) and Activated Sludge Model No. 2d (Henze et al., 1999). In separate efforts, Barker and Dold (1997) and Wentzel et al. (1989) developed biological phosphorus models that are used in some of the more popular simulator packages.

An additional task group was formed to prepare a comprehensive model of anaerobic digestion. The task group considered the breakdown of the various organics in sludge (proteins, fatty acids, and carbohydrates) and the bacteria attacking them leading to methanogenesis. The report is titled *Anaerobic Digestion Model No. 1 (ADM1)* (Batstone et al., 2002).

Mechanistic refers to a model based on the fundamental laws of natural sciences, including physical, biological, and chemical principles; sometimes called a physical model.

Empirical refers to a model relationship determined from statistical, or observed, relationships between variables.

3.0 MODEL TYPES

Most models used for design are termed **mechanistic models**, meaning that they are made up of a series of equations that describe the behavior of the process based on an understanding of the behavior of the system's components at its most fundamental levels. These models are also termed physics-based because they use equations that describe the physical behavior, or "physics," of the processes in the model. Another approach is the statistical model, which does not use a fixed set of equations, but instead develops a set of statistical rules based on historical data to determine how facility inputs affect facility outputs. These models are also known as black-box or empirical models because they typically are developed without the user considering the fundamental processes occurring in the system.

Empirical (experience-based) models express relationships between key measured variables such as solids retention time (SRT), sludge production, effluent quality, oxygen demand, and so on. Because of the complex nature of underlying processes and the different, often incompatible units of variables that can be readily measured by standard analytical methods (volatile suspended solids [VSS], 5-day biochemical oxygen demand [BOD_5], total suspended solids [TSS], etc.), empirical models typically are not mass-balance based. Rather, they are often expressed in correlations or in graphic format as nomograms or charts for ease of use. Empirical models and the knowledge they contain play an important role in engineering work, even with the detailed mechanistic models available today (WEF, 2010).

Mechanistic or structured models, on the other hand, are based on fundamental laws and principles that govern a process or the transformations occurring as part of the process. Mechanistic models use mass balances (Figure 16.1) and relate the mass balances to process kinetics and stoichiometry. (See Chapter 2, Section 2.1, for more information on the fundamental steps to developing a basic mass balance). With process simulations, the overall system is solved based on a consistent set of basic components (called **state variables** in mathematical modeling). Examples of state variables include pollutants (e.g., fractions of COD, ammonia), biomass (e.g., heterotrophic organisms, ammonia oxidizing organisms), and physicochemical factors (e.g., flow, calcium). State variables, in turn, are related to a number of **composite variables** more commonly measured in waste streams (e.g., BOD_5, TSS, VSS).

A state variable is an independent variable of model functions that is tracked through the model processes.

A composite variable is a model variable calculated from two or more state variables and a set of established relationships between the variables.

In general, mass balances in models can be expressed in the following two ways (WEF, 2010): (1) in steady state, and (2) dynamically. For a steady-state system, the mass input and output must balance with the reaction rate. For dynamic systems, masses can change, resulting in accumulation or loss from the system.

$$0 = \text{Input (kg/d)} - \text{Output (kg/d)} \pm \text{Reaction rate (kg/d)} \qquad (16.1)$$

$$\text{Mass change (kg/d)} = \text{Input (kg/d)} - \text{Output (kg/d)} \pm \text{Reaction rate (kg/d)} \qquad (16.2)$$

In Equation 16.1, the reaction rate is the change in the mass of the component as a result of chemical or biological reactions. Steady-state mechanistic models can only be used to calculate stable conditions, for example, annual or monthly averages. Models complete mass balance to be considered solved or converged. These mass balances provide value output information for the design engineer, and the mass balance acts as a basis of design for the WRRF. Mass balances may be exported and shared for COD or 5-day carbonaceous biochemical oxygen demand ($cBOD_5$), nitrogen, phosphorus, TSS, and so on. Mass balances apply to each individual unit process, and mass must be conserved for the system as a whole as well.

Keep in mind that the one of the goals of wastewater treatment is biodegradation of the pollutants in the waste stream. This means that some components (state variables) are incorporated into biomass and some mineralized from organic to inorganic forms or species. To close the mass balance, the transfer of components from the liquid stream to the gas stream may need to be taken into account. For example, a nitrogen balance on an anoxic zone results in a reduction in nitrogen from the inlet to the outlet water stream with the balance of nitrogen found in the offgas (Figure 16.1).

Dynamic models, on the other hand, express the mass change in time and thus can be used to calculate dynamic events such as diurnal variation, peak week dynamics, storms, peak oxygen demand, or similar events.

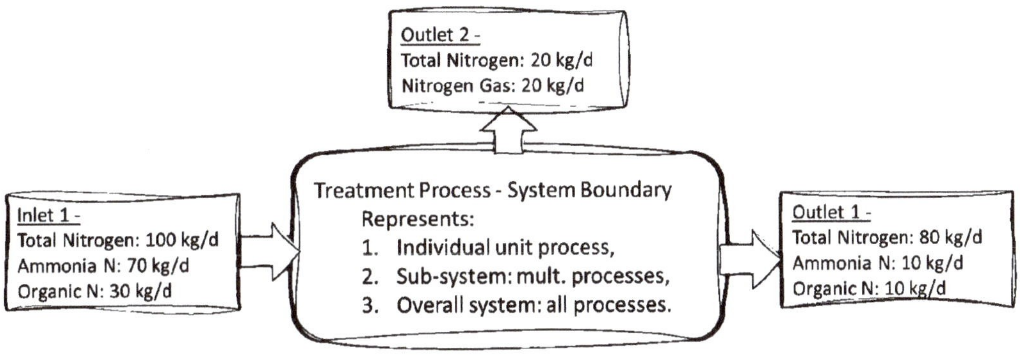

Figure 16.1 Example: Simple Nitrogen Mass Balance Over a System

The focus of this chapter is mechanistic, structured models that are used for calculation of facility-wide degradable organic carbon and nutrient transformation/removal performance and mass balances, typically implemented in simulation software packages. Models can be expanded and developed with a high level of complexity, accounting for a number of overlapping processes occurring in complex systems. However, model users are advised that the simplest model that gives satisfactory or useful results is preferred.

4.0 MODEL INPUTS AND OUTPUTS

Computer models, process or otherwise, attempt to represent the real world in the "virtual" world of the computer. Models and model systems represent an approximation of reality. Assumptions must be made in setting up the model, and the person or persons developing the model must decide which parameters and features to include. A model cannot be used to design or optimize a feature if it is not included in the model (i.e., a model of a fixed-volume, completely mixed reactor cannot be used to properly describe a system in which the reactor volume varies). Therefore, it is important to realize that models vary in complexity according to their intended use; it is also important that process engineers, as model users, realize the assumptions and limitations of the models they use to make good use of models for design or troubleshooting of wastewater treatment systems.

Figure 16.2 presents a diagram of the information flow between the process model and the real world. Different types of WRRF process data must be entered into the model or compared to the outputs from the model. Data from the real world include influent data, physical sizes of facilities, operating data, and effluent data. Within the model, these must be translated into numbers that can be used by the model through influent fractions and WRRF configuration interfaces. Model inputs and state variables are tracked through all the unit processes linked together within an overall WRRF model (Figure 16.3). Ultimately, the numbers evaluated by the model must be converted into outputs that can be compared to real-world data, such as facility effluent concentrations. An important use of model outputs is the comparison of effluent concentrations predicted by the model with effluent limits specified in facility permits to evaluate environmental permit and regulatory compliance.

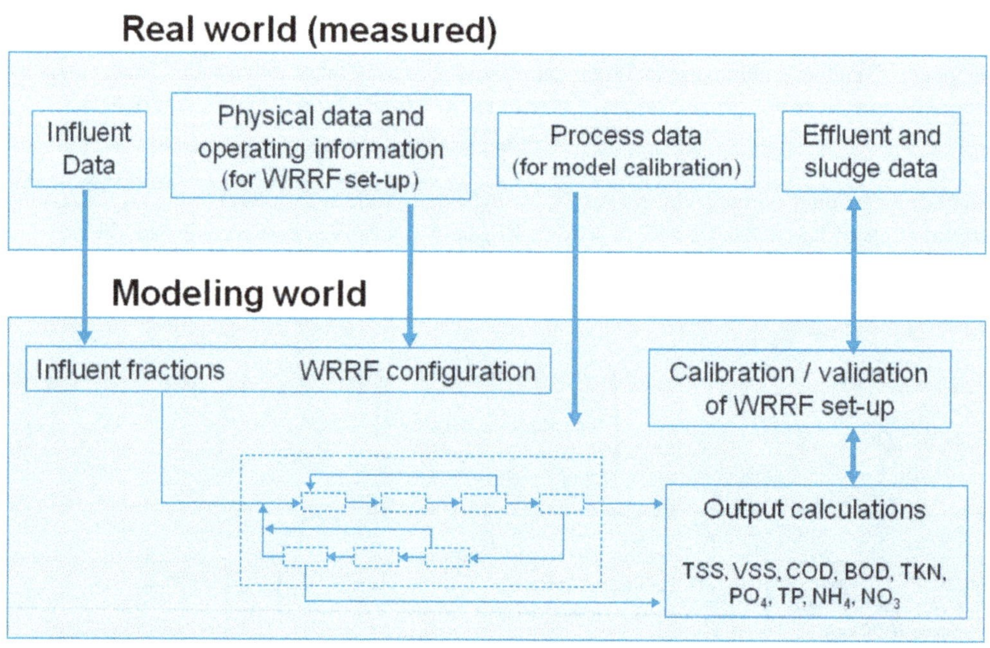

Figure 16.2 Diagram Showing Distinction Between the "Real World" and the "Modeling World" and the Interactions Between Them (Reprinted with permission from IWA GMP TG, 2008) (WRRF = water resource recovery facility; TKN = total Kjeldahl nitrogen; PO_4 = orthophosphate; NH_4 = ammonium; NO_3 = nitrate; and TP = total phosphorus)

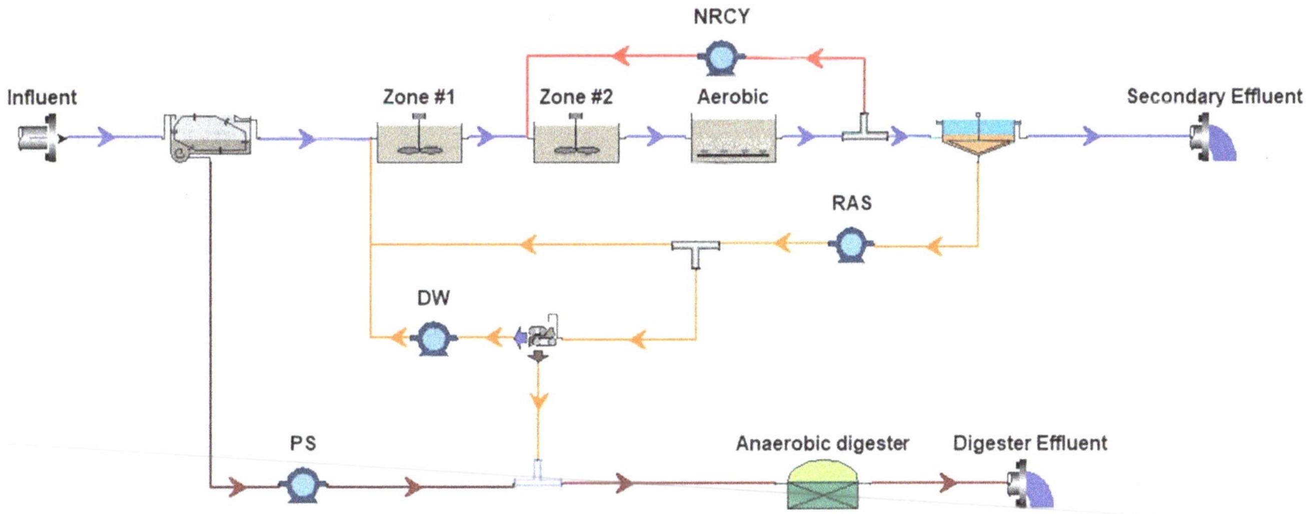

Figure 16.3 WRRF Configuration (Process Flow Schematic) With Linked Unit Processes

Considering the information flow in the model, the modeler must review the impacts of inputs on output quality. First, the less reliable inputs (influent data and physical conditions), make outputs from the model become less reliable. Therefore, it is important to develop the strongest input information possible within the constraints of the design effort. The modeler should clarify the level of reliability based on the calibration effort applied. Although reasonable default standards can be used, the reliability of the outputs is reduced due to the wide range of variability inherent to wastewater systems.

Calibration is the process of aligning a model relationships and predictions with actual WRRF observations.

A second consideration is that models can provide useful information even with less reliable inputs. While the range of outputs can vary, the trends or relationships associated with the model conditions applied give a consistent understanding of effects. For example, if the ammonia load increases and the temperature drops, a well-developed model shows that effluent ammonia concentration increases in response. Although this is a simple example, much more complex condition changes can be tested to yield value in understanding system response.

4.1 STATE AND COMPOSITE VARIABLES

Process engineers are accustomed to using influent parameters that are commonly and easily measured, such as BOD_5, COD, and TSS. These parameters are indicators of wastewater strength, which is important for facility design and operation, but they do not provide sufficient details about wastewater composition for many models. Although models do not require the exact chemical makeup of raw wastewater, they do require more information than just BOD_5 and TSS to predict the growth of microorganisms and the fate of pollutants within the model.

Model equations are based on basic building blocks called state variables. State variables are the simplest constituent parts within a model and are divided into two general categories: soluble and particulate (IWA nomenclature designation of "S" and "X", respectively [Corominas et al., 2010]). In addition, some models use state variables for colloidal material, which is difficult to define as truly soluble, or particulate material, which constitutes large molecules that may or may not be removed by filtration depending on the filter pore size. Examples of soluble state variables include soluble biodegradable substrate (expressed as COD), ammonia (expressed as N), and orthophosphate (expressed as P). When several state variables are combined, they constitute composite variables such as COD, $cBOD_5$, TSS, total Kjeldahl nitrogen (TKN), and total phosphorus. An example of the manner in which models combine state variables into composite variables is shown in Figure 16.4 (for COD), Figure 16.5 (for N), and Figure 16.6 (for P). The state variables are shown in white, while the black and gray boxes refer to composite variables. The examples shown for state and composite variables are typical for many wastewater models, although the exact state variables used are specific to the model selected. Chemical oxygen demand, rather than $cBOD_5$, is used as the basis for describing carbonaceous substrate in the IWA models.

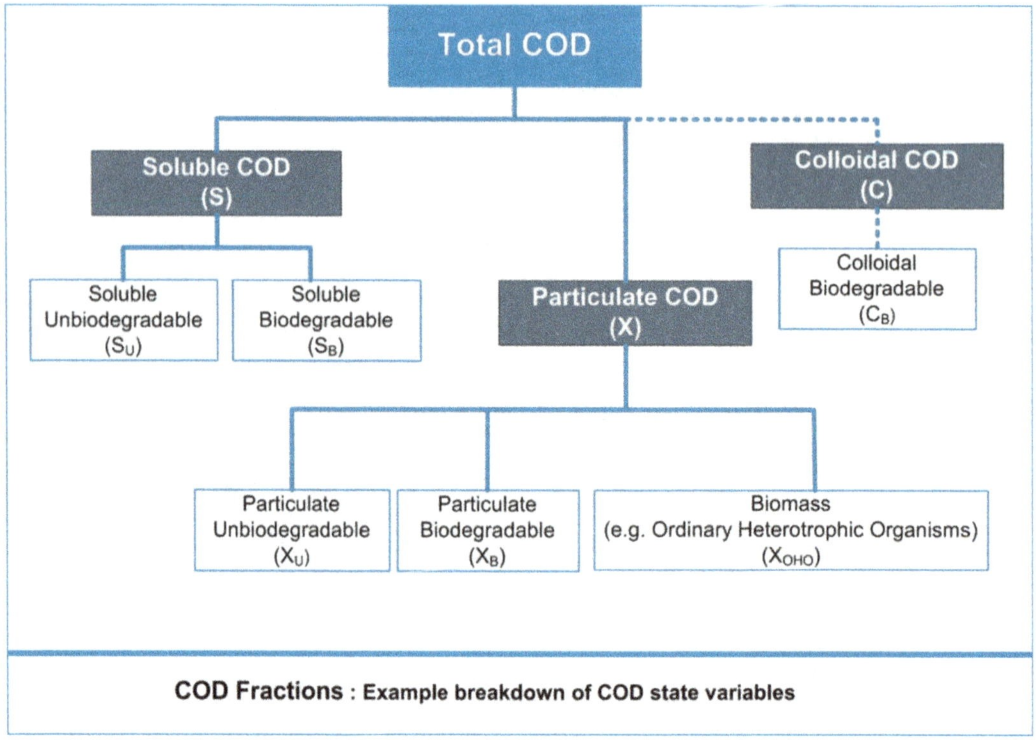

Figure 16.4 State Variables That Compose the Composite Variable, COD

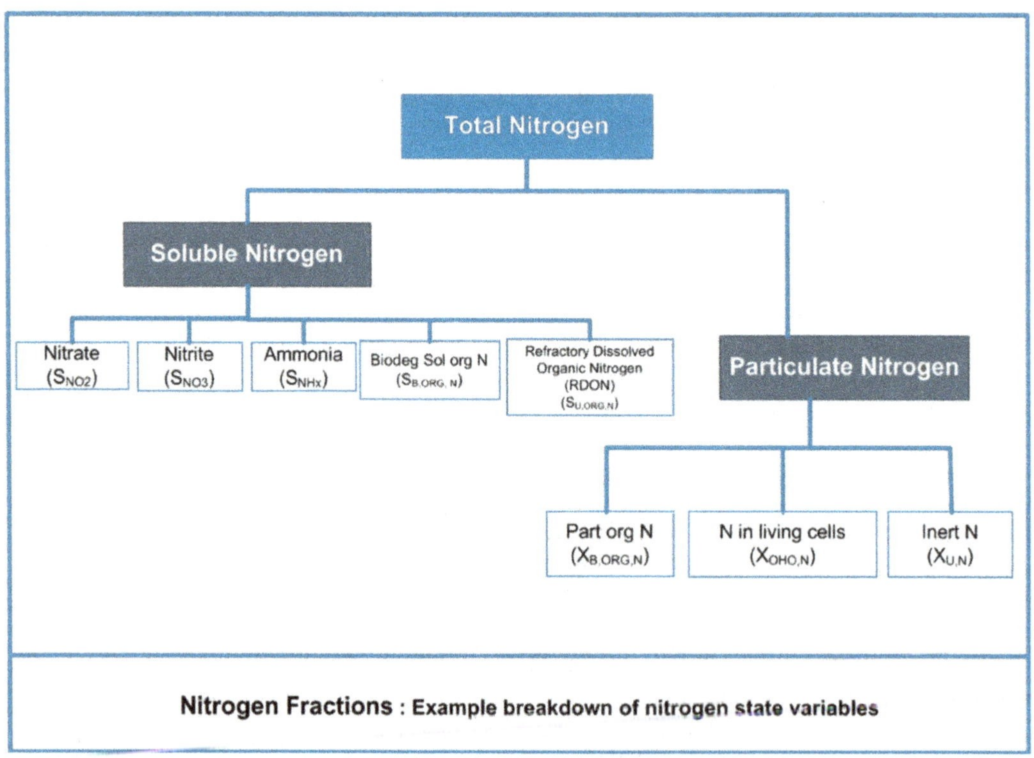

Figure 16.5 State Variables That Compose the Composite Variable, Total Nitrogen

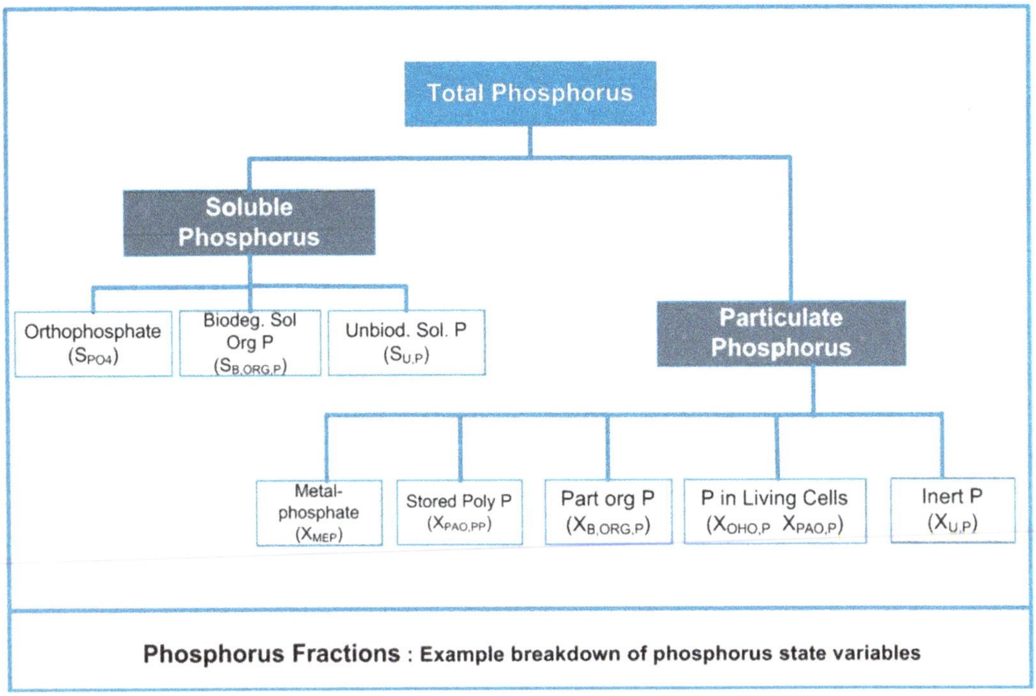

Figure 16.6 State Variables That Compose the Composite Variable, Total Phosphorus

The intent of this section is to provide a general overview of state variables and explain what they represent as an introduction to the more detailed discussion of models. Different mathematical models use similar state variables, although with some differences that correspond to their specific model structure. Table 16.1 illustrates the range of state variables considered in commonly used activated sludge models.

4.2 MODEL CALIBRATION AND VALIDATION

Setting up a model and working to better align inputs to real-world conditions is often termed calibrating the model. The calibration pathway, shown in Figure 16.7, reflects different levels of calibration (historical, intensive testing, kinetic measurements), and an iterative approach to achieving the right level of calibration depending on the model application:

- Level 1 calibration: Evaluation of influent quality based on assumptions and historical parameters
- Level 2 calibration: Intensive monitoring or supplemental sampling effort to align state variables
- Level 3 calibration: Most detailed level that involves testing of kinetic and stoichiometric parameters

Several factors should be considered when determining the appropriate calibration level. These factors include:

- WRRF size
- Treatment process planned
- Level of accuracy required (In other words, consider the margin for error. For example, if a discharge permit limits ammonia discharge to a concentration of 1.0 mg-N/L, then the model accuracy may need to be in the range from 0.1 to 0.5 mg-N/L rather than 1 to 5 mg-N/L.)
- Size of the project or upgrade
- Cost of the project or upgrade

As discussed previously, a higher-level calibration effort improves model accuracy. Increasing the calibration level requires additional effort from the modeler or model development team. The modelers need to organize sample collection, lab measurements, data analysis, and model development. As a result, this additional effort requires an investment to complete. The modeler or model team must balance the

Table 16.1 State Variables Commonly Considered in Activated Sludge Models

Description	Parameter	Units	ASM1 Symbol	ASM2d Symbol	ASM3 Symbol	ASM3 Bio-P Symbol	Barker and Dold Symbol	UCTPHO+ Symbol	ASM2d + TUD Symbol
COD soluble									
Soluble biodegradable organics	S_B	$g\ COD \cdot m^{-3}$	S_S		S_S	S_S			
Fermentable organic matter	S_F	$g\ COD \cdot m^{-3}$		S_F			S_{BSC}	S_F	S_F
Fermentation product (volatile fatty acids)	S_{VFA}	$g\ COD \cdot m^{-3}$		S_A			S_{BSA}	S_A	S_A
Soluble undegradable organics	S_U	$g\ COD \cdot m^{-3}$	S_I	S_I	S_I	S_I	S_{US}	S_1	S_1
Dissolved oxygen	S_{O2}	$-g\ COD \cdot m^{-3}$	S_O	S_{O2}	S_O	S_O	S_O	S_{O2}	S_O
COD particulate and colloidal									
Particulate biodegradable organics	X_C	$Bg\ COD \cdot m^{-3}$	X_S	X_S	X_S	X_S	S_{ENM}	X_{ENM}	X_S
Adsorbed slowly biodegradable substrate	X_{Ads}	$g\ COD \cdot m^{-3}$						X_{ADS}	
Particulate undegradable organics	X_U	$g\ COD \cdot m^{-3}$		X_I	X_I	X_I			X_I
Particulate undegradable organics from the influent	$X_{U,Inf}$	$g\ COD \cdot m^{-3}$	X_I				S_{UP}	X_I	
Particulate undegradable endogenous products	$X_{U,E}$	$g\ COD \cdot m^{-3}$	X_P				Z_E	X_E	
Nitrogen									
Ammonium and ammonia nitrogen ($NH_4 + NH_3$)	S_{NHx}	$g\ N \cdot m^{-3}$	S_{NH}	S_{NH4}	S_{NH}	S_{NH}	N_{N3}	S_{NH4}	S_{NH}
Nitrate and nitrite ($NO_3 + NO_2$) (considered to be NO_3 only for stoichiometry)	S_{NOx}	$g\ N \cdot m^{-3}$	S_{NO}	S_{NO3}	S_{NO}	S_{NO}	N_{O3}	S_{NO3}	S_{NO}
Dissolved nitrogen gas	S_{N2}	$g\ N \cdot m^{-3}$		S_{N2}	S_{N2}	S_{N2}			S_{N2}
Particulate biodegradable organic N	$XC_{B,N}$	$g\ N \cdot m^{-3}$	$X_{N,D}$			N_{BP}			
Soluble biodegradable organic N	$S_{B,N}$	$g\ N \cdot m^{-3}$	$S_{N,D}$			N_{BS}			
Soluble inert organic N	$S_{U,N}$	$g\ N \cdot m^{-3}$					N_{US}		
Phosphorus									
Soluble inorganic phosphorus	S_{PO4}	$g\ P \cdot m^{-3}$		S_{PO4}		S_{PO4}	P_{O4}	S_{PO4}	S_{PO}
Biomass									
Ordinary heterotrophic organisms (OHOs)	X_{OHO}	$g\ COD \cdot m^{-3}$	$X_{B,H}$	X_H	X_H	X_H	Z_H	X_H	X_H
Autotrophic nitrifying organisms (NH_4 to NO_3^-)	X_{ANO}	$g\ COD \cdot m^{-3}$	$X_{B,A}$	X_{AUT}	X_A	X_A	Z_A	X_{NIT}	Z_A
Phosphorus accumulating organisms (PAOs)	X_{PAO}	$g\ COD \cdot m^{-3}$		X_{PAO}		X_{PAO}	Z_E	X_{PAO}	X_{PAO}
Organisms (biomass)	X_{Bio}	$g\ COD \cdot m^{-3}$							

(continued)

Table 16.1 State Variables Commonly Considered in Activated Sludge Models (*Continued*)

Description	Parameter Units		ASM1 Symbol	ASM2d Symbol	ASM3 Symbol	ASM3 Bio-P Symbol	Barker and Dold Symbol	UCTPHO+ Symbol	ASM2d + TUD Symbol
Internal cell products									
Storage compound in OHOs	$X_{OHO,Stor}$	g COD · m^{-3}			X_{STO}	X_{STO}			
Storage compound in PAOs	$X_{PAO,Stor}$	g COD · m^{-3}		X_{PHA}		X_{PHA}	S_{PHB}	X_{PHA}	
Stored poly-β-hydroxyalkanoate in PAOs	$X_{PAO,PHA}$	g COD · m^{-3}							X_{PHA}
Stored glycogen in PAOs	$X_{PAO,Gly}$	g COD · m^{-3}							X_{GLY}
Stored polyphosphates in PAOs	$X_{PAO,PP}$	g P · m^{-3}		X_{PP}		X_{PP}		X_{PP}	X_{PP}
Releasable stored polyphosphates in PAOs (low molecular weight)	$X_{PAO,PP,Lo}$	g P · m^{-3}					$X_{PAO,PP,Lo}$		
Nonreleasable stored polyphosphates in PAOs (high molecular weight)	$X_{PAO,PP,Hi}$	g P · m^{-3}					$X_{PAO,PP,Hi}$		
Inorganics									
Metal hydroxide compounds	X_{MeOH}	g TSS · m^{-3}		X_{MeOH}					
Metal phosphate compounds	X_{MeP}	g TSS · m^{-3}		X_{MeP}					
Alkalinity (HCO_3^-)	S_{ALK}	mol HCO_3^- · m^{-3}	S_{ALK}	S_{ALK}	S_{ALK}	S_{HCO}			S_{HCO}
TSS									
Total suspended solids	X_{TSS}	g TSS · m^{-3}		X_{TSS}	X_{SS}	X_{SS}			X_{TSS}

level of calibration with the benefit of the model. When the model is used for a significant plant project, it is reasonable to expend additional effort and cost to improve the model calibration because it provides a return on the calibration investment in a more optimized or streamlined design.

Table 16.2 provides a list of information that may be collected when setting up a model. It is important to involve facility operators and laboratory managers in any discussions on data collection and analysis because they can provide insight regarding what is typical or unusual for the facility being modeled. The cost of sampling versus the benefit is also discussed, and an example of a sampling program is included.

When modeling a facility, it is impossible to have a complete set of data that describes all flows and concentrations throughout the entire facility. In reality, the modeler typically must deal with significant gaps in data. One powerful feature of process models is that they can bridge these gaps as they complete the mass balance. This is particularly useful in estimating flows and loads of sidestream returns, although the modeler must be careful in applying this information. It is always better to take real measurements, if possible, using good sampling and measurement techniques, than to blindly trust the output of a model. The IWA Task Group on Design and Operations Uncertainty was formed to develop a transparent method of addressing unknown factors in wastewater treatment modeling and to link efforts in academia with practice (Belia, Amerlinck et al., 2009; Belia, Benedetti et al., 2009).

Historical data collected at WRRFs are discussed in resources such as *Standards for Design of Water Resource Recovery Facilities* (WEF, 2023). Facilities typically monitor and record influent and effluent BOD_5 and TSS concentrations, and flow. They also may analyze for nutrients, pH, and dissolved oxygen. However, most models are based on, among other parameters, COD and inert suspended solids (ISS). Often, no

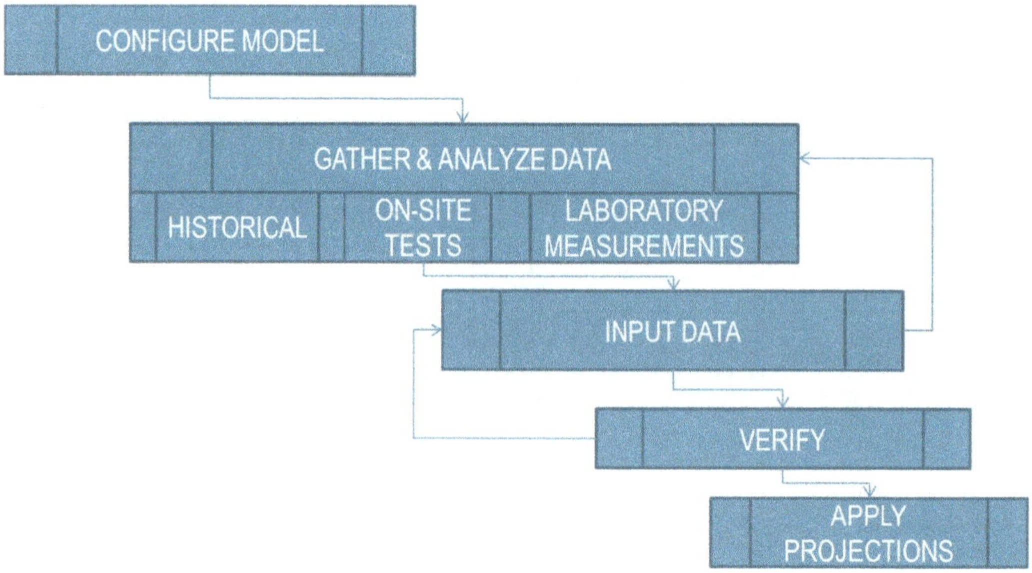

Figure 16.7 Calibration Approach and Philosophy

historical data exist for these parameters. Therefore, the modeler must clearly state his or her assumptions for the model calibration/validation efforts and his or her assumptions for the process evaluations—because they may differ—to account for future flow and load projections. This includes stating assumptions for solids processing sidestream contributions, including flow, quality, and return location.

Selecting the basis of the design is often a critical milestone in the project schedule, which means that the project manager may expedite the task to keep design efforts moving forward. There is significant risk in completing this task before special sampling results are in, however, because an incorrect assumption about wastewater characteristics can result in design failure. Although historical data are used to give a basic characterization of influent wastewater, there is often no record of COD, ISS, or even TKN or total phosphorus data. When designing nutrient-removal facilities in particular, the engineer must insist that

Table 16.2 Checklist of Information to Build a Facility Model

Information	Purpose
Most recent facility plan document	Confirm information is still current.
Recent facility influent data plots, and analyses of peaking factors	Discuss recent data trends vs observed long-term historical trends; discuss proposed modeling approach.
Additional sampling results	Determine if sampling occurred during a "typical" period or if additional sampling is required.
Recent facility operational data	Discuss trends and typical operating strategies.
Current process flow diagram	Determine locations of samplers, recycle flows, etc.
Current piping and instrumentation diagram	
Current as-built plan drawings	Determine surface area, volume, and configuration of all basins being modeled.
Current as-built hydraulic profiles	Determine sidewater depth of all basins being modeled.
Current inventory of equipment (blowers, diffusers, mixers, pumps etc.)	Determine firm capacity and discuss typical operation of equipment (constant speed or variable frequency drives for pumps, etc.).
List of calibration parameters	Discuss expectations and agree on level of accuracy.
Sample of model output and report	Discuss content and format; adjust based on feedback.

all necessary analyses be completed before any significant design efforts take place, or communicate the risk and associated cost of rework if data are not yet available. The engineer must also consider future changes in the service area when developing the design flow and loads, which serve as the basis for process evaluation.

When applicable, it is often possible to use historical BOD_5 and TSS concentrations to estimate historical COD and ISS concentrations, which are required model inputs. These parameters provide more detailed and useful characterization of influent wastewater needed in modeling and, as such, it is preferable to measure them directly if possible. Care must be taken when using BOD_5 data on nitrifying facilities with return liquors in the influent because the presence of nitrifying bacteria can cause significant errors in BOD_5 measurement. The $cBOD_5$ test can be used on influent to overcome this problem.

There are several methods for analyzing flow and load data, such as statistical methods, seasonal analyses, or applying published per capita factors in the case of a greenfield facility. It is common to size secondary treatment and solids processing facilities based on maximum month conditions or the highest 30-day average loads for a given set of data (Metcalf & Eddy, Inc./AECOM, 2014). Model calibration efforts might focus on a month of special sampling data, but a validation effort may be conducted using a different and wider set of data. Other modelers' efforts may calibrate a model using an entire year of data to account for dry- and wet-weather seasons. Regardless of the method used, it is important to analyze wastewater loads independent of concentration because wet-weather dilution can skew apparent observations in data. The modeler must also acknowledge that "worst-case" flows and loads in terms of capacity likely do not represent the worst case for nutrient-removal evaluations because BOD_5 is required for denitrification and biological phosphorus removal.

For a greenfield facility, or at WRRFs with smaller data sets, Tables 16.3, 16.4, and 16.5 provide abbreviated typical ranges for wastewater flowrates, per capita factors, and concentrations, respectively (see Chapter 1 for more information on flow and load development). For service areas that have significant commercial, institutional, recreational, or urban contributions, additional references may be consulted (Metcalf & Eddy, Inc./AECOM, 2014).

It is important to closely evaluate historical operating data before building a facility model because calibration efforts should focus on a period in which the facility was operating in a stable and consistent mode. Although this is not always possible, modeling efforts that attempt to calibrate around a time period for basins being taken in and out of service for maintenance, intermittent use of chemicals, or wide variations in recycle or wasting rates (SRT) may not yield useful results. In addition, it is important to have a good understanding of how operators control aeration, unless the process is run with effective aeration controls and dynamic data can be used directly. Dissolved oxygen readings may only be taken once or twice a day, and only at one location within the aeration basin. The modeler must have an understanding of how and when process control data are collected to accurately apply them to modeling efforts.

Another consideration is the effect of biosolids processing on liquid treatment performance because the two are directly related. It is common for facilities to dewater solids in shifts (8 hours per day, 5 days per week) and, if the filtrate (or centrate) is not equalized, this can introduce significant load variations on the facility that must be included in the modeling effort. Furthermore, if a facility stores digested biosolids

Table 16.3 Typical Wastewater Flowrates From Urban Residential Sources in the United States (Data from Metcalf & Eddy, Inc./AECOM, 2014)

Household Size	Flowrate Range (L/capita·d)
1	290–490
2	230–390
3	190–340
4	160–270

Table 16.4 Typical per Capita Factors of Untreated Domestic Wastewater (Data from Metcalf & Eddy, Inc./AECOM, 2014)

Parameter	Range of Values g/capita·d
BOD_5	50–120
COD	110–300
TSS	60–150
NH_3 as N	5–15
TKN	9–25
Total P	2.7–4.5

in the winter and uses drying beds only in the summer, the modeling effort must address this because nutrient loads from the beds in the summer is substantial. Even if the facility hauls sludge from the facility and sidestreams are not a concern, it is important to collect and evaluate waste activated sludge (WAS) data to ensure that the activated sludge model is predicting accurate sludge yields.

5.0 MODEL COMPONENTS

Mathematical models use state variables, kinetic expressions, and stoichiometric relationships to define processes. The wastewater treatment modeling community has "standardized" the use of **Gujer matrices** to summarize stoichiometry and kinetics associated with each model. Simulation tools are used to implement and solve models, either under steady-state or dynamic conditions. This section presents a general overview and background of modeling fundamentals. More detailed information on model development and application may be found in other references (e.g., Grady et al., 1999; WEF, 2014).

A Gujer matrix is a matrix relationship of models that relates processes with state variables, the corresponding stoichiometry of the variables, and the mathematical relationship for the process.

5.1 PROCESSES

There are a number of fundamental processes that occur in all biochemical systems. In mathematical models used for the simulation of wastewater treatment processes, simulator developers have elected to quantitatively describe processes that, based on current understanding, are the most important for the engineering application (nitrification, biological phosphorus removal, etc.). This section gives an overview of the processes included in the most commonly used simulators.

5.1.1 BIOLOGICAL REACTIONS

Wastewater treatment is centered on biological transformations of carbon, nitrogen, and phosphorus compounds, carried out by bacteria living in biological facilities. Mathematical representation of bacterial metabolism is a key element of modeling wastewater treatment processes, and it is an area of continuous development within the modeling community. Although different models may use slightly different

Table 16.5 Typical Composition of Untreated Domestic Wastewater (Data from Metcalf & Eddy, Inc./AECOM, 2014)

Parameter	Range of Values mg/L
TSS	120–400
VSS	95–315
BOD_5	110–350
COD	250–800
Total Nitrogen	20–70
NH_3 as N	12–45
Total Phosphorus	4–12

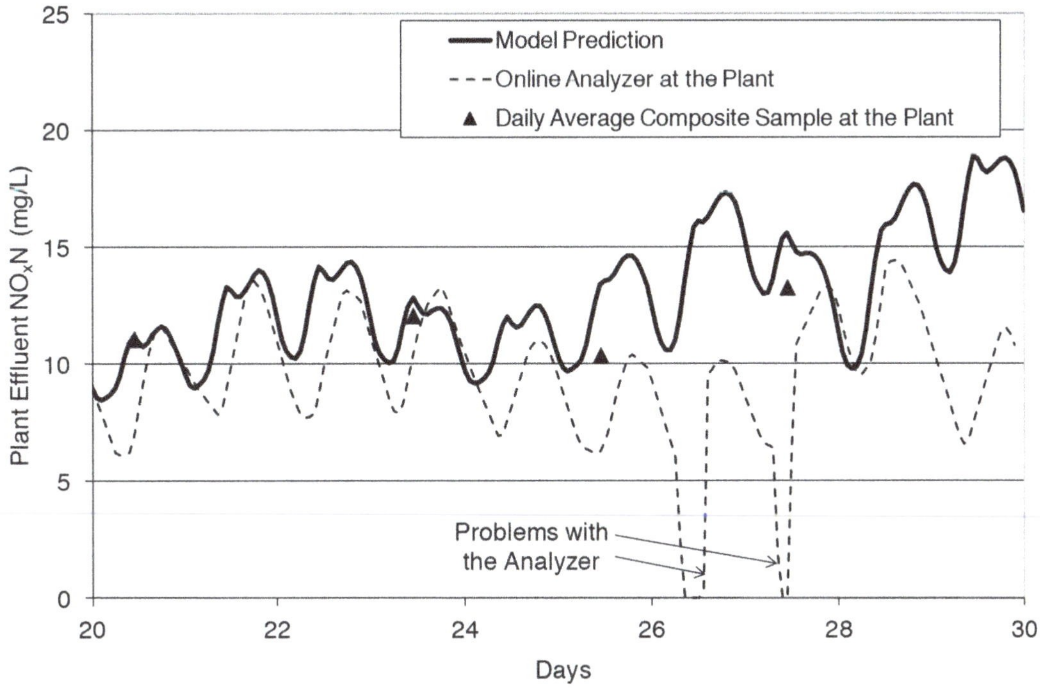

Figure 16.8 Dynamic Modeling Using Laboratory Data and Online Analyzer Data

approaches to mathematically describe biological reactions in wastewater treatment, the main biological processes of interest for modeling purposes are discussed in this subsection.

5.1.1.1 Biomass Growth, Substrate Utilization, and Yield

Biomass growth describes the growth of microorganisms using substances found in wastewater as energy and/or carbon sources. Biomass growth and substrate utilization are, in most cases, coupled. As a result, the removal of one unit of substrate results in production of Y units of biomass. The "Y" represents the true growth yield of the microbial community. True growth accounts for substrate utilization for cell maintenance purposes.

The description of growth in the models includes expressions of the four elements required for growth: carbon source, inorganic nutrients, energy source (electron donor), and a source of reducing power (electron acceptor). The mathematical expressions of the models describe the relationships of these four components for each process (Grady et al., 1999). For example, heterotrophic bacterial growths use biodegradable COD as carbon and energy sources, oxygen as an electron acceptor for reducing power, and nitrogen and phosphorus as limiting nutrients. Another example is autotrophic growth, wherein the energy sources are ammonia and nitrite, the carbon source is inorganic carbon from alkalinity, and nitrogen and phosphorus are the limiting nutrients.

5.1.1.2 Maintenance, Endogenous Metabolism, Decay, Lysis, and Death

Maintenance describes the energy required by microorganisms for functions other than growth. Such functions include motility, maintenance of ionic gradients, and molecular transport. The source of this energy can be external or internal. In the presence of sufficient external energy sources, the microorganisms use a portion of the available energy for maintenance and the rest for growth. With the availability of external energy diminishing, growth continuously decreases. If no external energy source is available for maintenance, internal energy sources are used, and maintenance needs are covered by endogenous metabolism.

Cell lysis results from the loss of a microorganism's ability to maintain the physical integrity of its cellular structure. As a result, the cell wall is ruptured, and the microorganism dies. Cell lysis is another process that describes reduced biomass growth as compared to the theoretical cell yield (Y).

Death describes the state of a microorganism that has lost the ability to metabolize substrates. There is debate among researchers regarding the number of microorganisms that are really dead in a facility. There is evidence to suggest that dead cells do not remain intact long, but instead break up into substrates and biomass debris. The inactive organic part of the mixed liquor in wastewater can be attributed to the accumulation of biomass debris rather than dead cells (Grady et al., 1999).

Mathematical models use different approaches to account for maintenance and decay mechanisms. Some models (e.g., ASM1) use a death regeneration approach, whereby bacterial decay produces biodegradable and nonbiodegradable material that is consumed by active bacteria or accumulated in the system. Other models (e.g., ASM3) use an endogenous respiration approach in which decay mechanisms for bacteria and their storage products account for decay and maintenance.

5.1.1.3 Storage

Bacteria often synthesize internal storage products that serve as energy or carbon sources. One of the more common storage inclusions involves poly-β-hydroxyalkanoate (PHA), which, in bacteria, functions as a carbon and energy storage product. Glycogen is another common carbon and energy storage product. Storage requires energy, which is obtained from respiration (Henze et al., 2000).

Given the right environmental conditions, some organisms can accumulate granules containing long chains of phosphate. Similar to PHA storage, polyphosphate storage requires energy. Poly-β-hydroxyalkanoate and polyphosphate storage and release are key processes in enhanced biological phosphorus removal.

Sulfur globules are also found in a variety of bacteria capable of oxidizing reduced sulfur compounds, such as hydrogen sulfide and thiosulfate. Oxidation of these compounds is linked to either energy metabolism or photosynthesis.

Models often consider storage products for phosphorus-accumulating organisms (PAOs) to represent biological phosphorus removal in a quantitative manner. Other storage products may be captured by some models.

5.1.1.4 Hydrolysis

Hydrolysis plays two key roles in biochemical systems. First, it releases cellular components following cell lysis. Second, through hydrolysis, complex organic materials are broken down by extracellular enzymes to produce low-molecular-weight compounds that can be transported inside the cell and metabolized. Current activated sludge and biofilm models include several hydrolysis reactions that cover the hydrolysis of complex organic molecules and entrapped organic nutrients, such as nitrogen and phosphorus. In anaerobic digestion models, particulate carbohydrates and proteins are hydrolyzed to produce simple sugars and amino acids; hydrolysis of lipids generates long chain fatty acids.

5.1.1.5 Fermentation, Acidogenesis, Acetogenesis, and Methanogenesis

Fermentation and acidogenesis describe processes by which, under strict anaerobic conditions, bacteria generate simple acids. During acidogenesis, amino acids and soluble sugars are degraded by fermentative reactions primarily to acetate, propionate, and butyrate products (Batstone et al., 2002). Acetogenesis describes the degradation of higher organic acids to acetate. During acetoclastic methanogenesis, acetate is used by bacteria to form methane and carbon dioxide.

5.1.2 CHEMICAL REACTIONS: PH AND PRECIPITATION

Chemical reactions are common in wastewater processes and occur between compounds produced from biological reactions and/or between compounds in the wastewater and chemicals added in the wastewater treatment process. These chemical reactions commonly include changes in pH through alkalinity production (denitrification) or alkalinity consumption (nitrification) or the effect of added chemicals and precipitation reactions such as phosphorus precipitation in anaerobic digesters. Because chemical precipitation is common in wastewater treatment, these reactions are commonly included in mathematical models, describing, for example, chemical precipitation of phosphorus with alum or ferric chloride.

5.1.3 PHYSICAL REACTIONS

Physical reactions are another important part of modeling wastewater treatment processes. Examples of physical reactions described here include mass transport of pollutants through processes such as dispersion and diffusion in biofilms, gas–liquid gas transfer (aeration and stripping), and mixing. These physical reactions are incorporated in process models, together with the biological reactions previously described in this chapter, to provide an overall representation of the treatment processes in facilities.

5.1.3.1 Gas–Liquid Mass Transfer

Mass transfer of compounds between gas and liquid phases is important in wastewater treatment. Several examples include aeration (i.e., oxygen transfer from air to mixed liquor), ammonia stripping (i.e., removal of ammonia from liquid solution), and carbon dioxide equilibrium.

Henry's law states that, at equilibrium, the ratio of the concentration of a volatile solute in a solvent to the solute's partial pressure in the vapor phase above the solvent–solute mixture is a constant. Henry's law is commonly used in mathematical models to describe the behavior of compounds that partition between liquid and gas phases (e.g., oxygen and methane).

Biological wastewater treatment systems require some type of aeration to drive aerobic biological processes. Aeration is the transfer of oxygen from air into the liquid phase so that bacteria can use oxygen dissolved in the liquid phase for aerobic growth. The rate of flux of oxygen from the gas to the liquid is given as

$$r_{O_2} = K_{La}(C_s - C) \tag{16.3}$$

Where r_{O_2} = the oxygen-transfer rate per unit volume, mg/(L · s);
K_{La} = the volumetric mass-transfer rate coefficient, L/s;
C_s = the saturation concentration or dissolved oxygen concentration (mg/L) in equilibrium with the bulk gas phase (governed by Henry's law constant); and
C = concentration in the bulk liquid, mg/L.

The mass transfer rate coefficient, K_{La}, is system-specific and depends on many factors, primarily the type of aeration system (surface aeration, fine bubble diffusers, coarse bubble diffusers, etc.) and size and shape of the aeration basins, among others. Other factors that need to be accounted for in wastewater aeration systems are the clean water versus wastewater factors, which affect oxygen transfer efficiency.

5.1.3.2 Diffusion

Diffusion is a physical process whereby mass is transported via gradients in concentration, temperature, or pressure gradients. In wastewater treatment processes, we are mainly concerned with diffusion because of concentration gradients in systems such as biofilms.

Biofilms are matrices of microorganisms and their extrapolymeric substances that grow attached to a fixed surface, such as rocks, plastic media, and so on. Concentration gradients are established within the biofilm because of the need to transport material throughout the biofilm. Mathematical models for biofilm systems must capture the existence of these concentration gradients to predict mass fluxes into and out of the biofilm.

5.1.3.3 Adsorption

Adsorption is a physical process in which substances adhere to a surface, resulting in the removal of such substances from the bulk liquid. Common examples of the use of adsorption processes in wastewater treatment include the use of granular or powdered activated carbon to remove residual chlorine, synthetic dyes, or metals from wastewater. Adsorption is also a factor in the removal of colloidal particulates by activated sludge and in chemical coagulation processes.

5.1.4 SOLIDS SEPARATION

Solids–liquid separation processes play an important role in modeling wastewater treatment systems and have a direct effect on the prediction of effluent quality. Solids–liquid separation processes include sedimentation (primary and secondary) and size exclusion filtration via media filtration and membranes.

In terms of modeling, primary and secondary sedimentations present different situations. In primary sedimentation, particle concentration is low enough that the particles typically do not have significant interactions. Moreover, the system can be modeled using Stoke's law, which governs the settling velocity of particles as a function of the densities of the particles and the fluid and the size of the particle. Primary sedimentation is typically modeled using an assumed solids capture efficiency, sometimes based on statistical correlations between solids capture efficiency and surface overflow rate.

In secondary sedimentation, mixed liquor with a relatively high concentration of particles needs to settle in the clarifiers to produce clarified effluent. Due to the high concentration of particles in the mixed liquor and interactions between particles, Stoke's law does not apply. Rather, a concept called zone settling is often used to model secondary sedimentation, whereby the velocity of the sludge relative to the liquid is a function of the local concentration. At high particle concentrations toward the bottom of the clarifier, flocs change size and shape and a phenomenon known as compaction occurs. In the compaction zone, water must be displaced so that the dense sludge can settle downward, producing a phenomenon that is complex to represent mathematically. Available clarifier models range from simple two-compartment models to complex computational fluid dynamic models suited for a range of modeling objectives.

Solids separation by size-exclusion filtration applies to filtration of secondary effluent or filtration of mixed liquor in membrane bioreactors. Filtration is based on size exclusion. Particles of a certain cutoff size will be retained in the filter, producing a clear effluent. Modeling filtration in wastewater treatment processes is typically limited to simple models involving an assumed solids capture efficiency or an assumed effluent quality. Backwash from filtration processes can represent a significant flow component and, therefore, it is desirable to include it in process models.

5.2 KINETICS

Kinetic expressions describe the rate of change of state variables in a model to describe the model processes. It is beyond the scope of this manual to give a detailed explanation of the different kinetic expressions used in process models. Model kinetics originally focused on development of biological reactions, but model simulators have advanced to account for kinetics of physical (diffusion and mass transfer) and chemical (precipitation) reactions as well. Key kinetic relationships used in models include the following:

- Zero order
- First order
- Monod (multiple Monod)
- Contois
- Inhibition
  - Substrate inhibition (Haldane-Andrews)
  - Competitive, non-competitive, uncompetitive

The most important kinetic relationship for the activated sludge process is based on the Monod equation, which was first proposed by Monod (1942) to describe the growth of bacteria. It should be noted that the Monod equation is identical to the Michaelis–Menten (Michaelis & Menten, 1913) formula used to describe enzyme reactions, although, unlike the Michaelis–Menten formula, the Monod equation does not have a fundamental biological or chemical basis but has proven to be a useful, simple expression to describe biological activity with a significant experience base.

In addition to using Monod expressions to describe how organisms use substrate to grow, similar equations can be used to adjust the rate based on environmental factors. These expressions are termed switching functions and have a value between 0 and 1, depending on the environmental condition and the concentration of the variable that defines that environmental condition. There are two common forms of switching functions: one that decreases the growth rate as the variable decreases, and one that decreases the growth rate as the variable increases. The equation form and curve shape for the first type is identical to the Monod equation previously described. The second form, which reduces the rate as the variable concentration increases, can be termed a reverse Monod, and its curve shape is shown in Figure 16.9. Figure 16.9 also shows that the rate is at half its maximum at the half-saturation concentration (K_s);

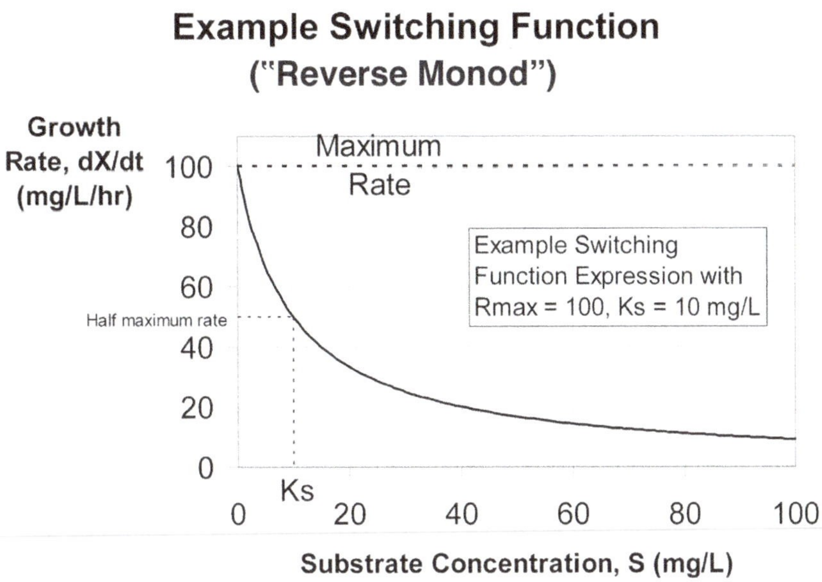

Figure 16.9 Reverse Monod Equation

however, when the variable concentration is zero, the rate is at its maximum and, at high concentrations, the rate approaches zero. Switching functions are commonly used to switch rates on and off, depending on the presence of compounds such as dissolved oxygen and nitrate (NO_3^-) to define whether conditions are aerobic, anoxic, or anaerobic. These formulas are more straightforward to handle in numerical solvers than functions that simply turn processes on or off at a particular threshold concentration. Care should be taken when making adjustments to switching function parameters because changing them slightly can significantly affect model outputs.

5.3 STOICHIOMETRY

All models contain **stoichiometric** parameters that are used to define mass relationships between state variables. Examples of stoichiometric parameters include biomass yields, nutrient fractions in biomass, and the ratio of COD to VSS for biomass or influent wastewater. Not to be confused with influent stoichiometric parameters, model yields and biomass fractions typically should not be adjusted from values reported in the literature as they describe fundamental relationships within the model. For example, a typical microorganism is a volatile solid that can be expressed chemically as $C_5H_7O_2N$ (molecular weight = 113) and yields a COD-to-VSS ratio of 1.42 based on the following equation:

Stoichiometry is the relationship between the relative quantities of substances taking part in a reaction.

$$C_5H_7O_2N + 5O_2 \rightarrow 5CO_2 + 2H_2O + NH_3 \quad (16.4)$$

Considering the stoichiometric coefficients of the aforementioned equation, the oxygen, O_2 (or COD), equivalent for the microorganism is as follows:

$$5 \times 32 \text{ g } O_2 \text{ } 1/113 \text{ l-}VSS = 1.42 \text{ g } O_2(g\text{-}VSS)^{-1} \quad (16.5)$$

5.4 GUJER MATRIX

Models are often presented by means of "Gujer matrices" (Henze et al., 2000) (previously referred to as "Petersen matrices" [Petersen, 1965]), which graphically describe the interactions of system components. The Gujer matrix for a given model can be used by the process designer to understand how state variables are affected by different process rates, and how state variables are related.

The Gujer matrix consists of three parts:

1. Stoichiometric matrix: This part of the matrix defines the changes in state variables for each of the processes included in the model.
2. Composition matrix: This matrix defines the COD, nitrogen, phosphorus, and charge in each state variable included in the model.

3. Kinetic rate expressions: Kinetics for each process included in the model is defined by a process rate equation.

Therefore, a Gujer matrix represents a comprehensive overview of processes, including reactions and stoichiometry. Table 16.6 shows an example Gujer matrix with processes versus state variables, the corresponding stoichiometry, and the rate relationship.

6.0 MODEL SIMULATORS

Models are often implemented in commercial simulators to produce an overall WRRF model. Many commercial simulators have been developed into software packages, or apps, during the past 20 years. The intent of this section is to describe how simulators use models and to discuss how the simulators link models together to develop a functioning representation of a WRRF.

Simulators vary in the number and complexity of the processes that are modeled and in the features included in each package. It should also be noted that the purpose of the simulator can vary as well. Some simulators are running models specifically focused on wastewater treatment design, while others use dynamic models to predict the effect of a design. This is a subtle difference but, in the former case, the simulator is used to calculate the size of a process and the required equipment given a user-defined required level of treatment. In the latter case, the model is used to predict the expected treatment behavior given user-defined process sizes and equipment.

Table 16.6 Activated Sludge Model 1: Gujer Matrix

		Soluble undegradable organics	Soluble biodegradable organics	Particulate undegradable organics from the influent	Particulate biodegradable organics	Ordinary heterotrophic organisms	Autotrophic nitrifying-organisms (NH_4 + to NO_3)	Particulate undegradable endogenous products
	Original parameter symbol	S_I	S_S	X_I	X_S	$X_{b,H}$	$X_{b,A}$	X_P
	Units	g cOD · m^{-3}	g cOD · m^{-3}	g cOD · m^{-3}	g cOD · m^{-3}	g cOD · m^{-3}	g cOD · m^{-3}	g cOD · m^{-3}
	Parameter	S_U	S_B	$X_{U,Inf}$	XC_B	X_{OHO}	X_{ANO}	$X_{U,E}$
1	Aerobic growth of heterotrophs		$-1/Y_{OHO}$			1		
2	Anoxic growth of heterotrophs		$-1/Y_{OHO}$			1		
3	Aerobic growth of autotrophs						1	
4	Decay of heterotrophs				$1 - f_{XU\_Bio,lys}$	-1		$1 - f_{XU\_Bio,lys}$
5	Decay of autotrophs				$1 - f_{XU\_Bio,lys}$		-1	$1 - f_{XU\_Bio,lys}$
6	Ammonification of soluble organic nitrogen							
7	Hydrolysis of entrapped organics		1		-1			
8	Hydrolysis of entrapped organic nitrogen							

[1] $\mu_{OHO,Max} *[S_B/K_{SB,OHO} + S_B] *[S_{O_2}/K_{O_2,OHO} + S_{O_2}] *[S_{NHx}/K_{NHx,OHO} + S_{NHx}] *X_{OHO}$
[2] $\mu_{OHO,Max} *[S_B/K_{SB,OHO} + S_B] *[K_{O_2,OHO}/(K_{O_2,OHO} + S_{O_2})] *(K_{NOx,OHO}/S_{NOx})] *[S_{NHx}]/(K_{NHx,OHO}/S_{NHx}) *n_{\mu OHO,Ax} *X_{OHO}$
[3] $\mu_{OHO,Max} *[S_{NHx}]/(K_{NHx,ANO}/S_{NHx})] *[S_{O_2}/K_{O_2,ANO} + S_{O_2}] *X_{ANO}$
[4] $b_{OHO} *X_{OHO}$

Table 16.6 Activated Sludge Model 1: Gujer Matrix (*Continued*)

Dissolved oxygen	Nitrate and nitrite ($NO_3 + NO_2$) (considered to be NO_3 only for stoichiometry)	Ammonia ($NH_4 + NH_3$)	Soluble biodegradable organic N	Particulate biodegradable organic N	Alkalinity (HcO_3)	Dissolved Nitrogen	Process rate
S_O	S_{NO}	S_{NH}	S_{ND}	X_{ND}	S_{ALK}		
$-g\ cOD \cdot m^{-3}$	$g\ N \cdot m^{-3}$	$g\ N \cdot m^{-3}$	$g\ N \cdot m^{-3}$	$g\ N \cdot m^{-3}$	$mol\ HcO_3^- \cdot m^{-3}$	$g\ N \cdot m^{-3}$	
S_{O_2}	S_{NOx}	S_{NHx}	$S_{b,N}$	$XC_{b,N}$	S_{ALK}	S_{N_2}	
$-(1 - Y_{OHO})/ Y_{OHO}$		$-i_{N,XBIO}$			$-i_{N,XBIO} * i_{Charge\_SNHx}$		(1)
	$-(1 - Y_{OHO})/ (i_{NO_3,N_2} * Y_{OHO})$	$-i_{N,XBIO}$			$-(1 - Y_{OHO})/ (i_{NO_3,N_2} * Y_{OHO}) i_{Charge\_SNOx} - i_{N,XBio} * i_{Charge\_SNHx}$	$(1 - Y_{OHO})/ (i_{NO_3,N_2} * Y_{OHO})$	(2)
$-(-i_{COD\_NO_3} - Y_{ANO})/ Y_{ANO}$	$1/Y_{ANO}$	$-i_{N,XBio} - 1/Y_{ANO}$			$-(i_{N,XBio} + (1/Y_{ANO}) * i_{Charge\_SNHx} + (1/Y_{ANO}) * i_{Charge\_SNOx}$		(3)
				$i_{N,XBio} - f_{XU\_Bio,lys} * i_{N,XUE}$	-1		(4)
				$i_{N,XBio} - f_{XU\_Bio,lys} * i_{N,XUE}$			(5)
		1	-1		i_{Charge_SNHx}		(6)
							(7)
			1	-1			(8)

[5] $b_{ANO} * X_{ANO}$

[6] $q_{am} * S_{B,N} * X_{OHO}$

[7] $q_{XCB\_SB,hyd} * [(XC_B/X_{OHO})/(K_{XCB,hyd} + XC_B/X_{OHO}) * ([S_{O_2}/(K_{O_2,OHO} + S_{O_2})] + n_{qhyd,Ax} * ([K_{O_2,OHO}/(K_{O_2,OHO} + S_{O_2})] * [S_{NOx}/(K_{NOx,OHO} + S_{NOx})] * X_{OHO}$

[8] $q_{XCB\_SB,hyd} * [(XC_{B,N}/XC_B)/(XC_B/X_{OHO})/(K_{XCB,hyd} + XC_B/X_{OHO}) * ([S_{O_2}/(K_{O_2,OHO} + S_{O_2})] + n_{qhyd,Ax} * ([K_{O_2,OHO}/(K_{O_2,OHO} + S_{O_2})] * [S_{NOx}/(K_{NOx,OHO} + S_{NOx})] * X_{OHO}$

Table 16.6 Activated Sludge Model 1: Gujer Matrix (*Continued*)

Stoichiometry	Parameter notation		Units	Value* T = 20°c
	Original	Standardized		
Yield for X_{OHO} growth	Y_H	Y_{OHO}	$g\ X_{OHO} \cdot g\ XC_B^{-1}$	0.67
Fraction of X_U generated in biomass decay	f_P	$f_{XU\_Bio,lys}$	$g\ X_U \cdot g\ X_{Bio}^{-1}$	0.08
Yield of X_{ANO} growth per S_{NO_3}	Y_A	Y_{ANO}	$g\ X_{AUT} \cdot g\ S_{NO_3}^{-1}$	0.24
N content of biomass (X_{OHO}, X_{PAO}, X_{ANO})	i_{XB}	i_{N_XBIO}	$g\ N \cdot g\ X_{Bio}^{-1}$	0.086
N content of products from biomass	i_{XP}	i_{N_XUE}	$g\ N \cdot g\ X_{UE}^{-1}$	0.06
Conversion factor for NO_3 reduction to N_2	i_{NO_3,N_2}	i_{NO_3,N_2}	$g\ COD \cdot g\ N^{-1}$	2.857
Conversion factor for NO_3 in COD	$i_{COD\_NO_3}$	$i_{COD\_NO_3}$	$g\ COD \cdot g\ N^{-1}$	−4.571
Conversion factor for N_2 in COD	$i_{COD\_N_2}$	$i_{COD\_N_2}$	$g\ COD \cdot g\ N^{-1}$	−1.714
Conversion factor for NHx in charge	i_{Charge_SNHx}	i_{Charge_SNHx}	$Charge \cdot g\ N^{-1}$	0.071
Conversion factor for NO_3 in charge	i_{Charge_SNOx}	i_{Charge_SNOx}	$Charge \cdot g\ N^{-1}$	−0.0714

*Henze, M.; Gujer, W.; Mino, T.; Matsuo, T.; Wentzel, M. C.; Marais, G. v. R.; van Loosdrecht, M. C. M. (2000) *Activated Sludge Model No. 2d*; Scientific and Technical Report No. 9; IWA Publishing: London, U.K.

Table 16.6 Activated Sludge Model 1: Gujer Matrix (*Continued*)

	Parameter notation			
Kinetics	**Original**	**Standardized**	**Units**	**Value* T = 20°c**
Maximum specific hydrolysis rate	k_h	$q_{XCB\_SB,hyd}$	$g\ XC_B \cdot g\ X_{OHO}^{-1} \cdot d^{-1}$	3
Saturation coefficient for X_B/X_{OHO}	K_x	$K_{XCB,hyd}$	$g\ XC_B \cdot g\ X_{OHO}^{-1}$	0.03
Correction factor for hydrolysis under anoxic conditions	η_h	$n_{qhyd,Ax}$		0.4
Maximum growth rate of X_{OHO}	μ_H	$\mu_{OHO,Max}$	d^{-1}	6
Reduction factor for anoxic growth of X_{OHO}	η_g	$n_{\mu OHO,Ax}$		0.8
Half-saturation coefficient for S_B	K_s	$K_{SB,OHO}$	$g\ S_{SB} \cdot m^{-3}$	20
Decay rate for X_{OHO}	b_h	b_{OHO}	d^{-1}	0.62
Half-saturation coefficient for S_{O_2}	$K_{O,H}$	$K_{O2,OHO}$	$g\ S_{O_2} \cdot m^{-3}$	0.2
Half-saturation coefficient for S_{NOx}	K_{NO}	$K_{NOx,OHO}$	$g\ S_{NOx} \cdot m^{-3}$	0.0.5
Half-saturation coefficient for NH_4	$K_{NH,H}$	$K_{NHx,OHO}$	$g\ S_{NHx} \cdot m^{-3}$	0.05
Maximum growth rate of X_{ANO}	μ_A	$\mu_{ANO,Max}$	d^{-1}	0.8
Decay rate for X_{ANO}	b_A	b_{ANO}	d^{-1}	0.15
Rate constant for ammonification	k_a	q_{am}	$m^3\ XC_{B,N}^{-1} \cdot d^{-1}$	0.08
Half-saturation coefficient for S_{O_2}	$K_{O,A}$	$K_{O2,ANO}$	$g\ S_{O_2} \cdot m^{-3}$	0.4
Half-saturation coefficient for S_{NHx}	K_{NH}	$K_{NHx,ANO}$	$g\ S_{NHx} \cdot m^{-3}$	1

*Same $K_{NH,H}$ value as ASM2d has been chosen.

Specific design simulators have their roots in the 1970s, and some use empirical design rules based on typically measured parameters such as BOD_5, food-to-microorganism ratio, and safety factors to come up with process sizing. For some of the original simulators, equipment was sized based on empirical curves and a defined database of available equipment options. In contrast, more recent, mechanistic simulators can be used for both design and to explore the behavior of a user-defined design to address questions such as: What is the expected effluent quality from a facility of a certain size and operated in a certain way? How does the effluent vary under dynamic loading conditions? How much oxygen is required for the process? How does oxygen demand vary throughout the day, week, and year? Unlike dedicated design software (where the output might be a predictor of tank size), dynamic simulators can be used to predict process dynamics and, with sufficient data, the dynamic risk associated with a particular design (i.e., how often the effluent ammonia will exceed a given limit, or whether dynamic control can change that frequency).

7.0 BUILDING A WATER RESOURCE RECOVERY FACILITY MODEL

During the definition phase of a project, the project scope and objectives are defined. The model is then developed and calibrated using operating data and any additional sampling data collected. The level of calibration effort depends on the objectives of the modeling project and available data. Once a calibrated model is developed, it can be used to assist with a wide range of activities throughout the entire life-cycle of a WRRF, such as:

- Treatment process selection
- Facility design and retrofitting
- Sensitivity and uncertainty analysis
- Control system design and tuning
- Operator training
- Facility startup planning
- Optimization

- Troubleshooting and "what if" analyses
- Facility capacity analysis and re-rating
- Facility auditing and benchmarking
- Planning and scheduling
- Biosolids management
- Design of experimental programs
- Data and sensor validation

The goal of the modeling task should be identified early in the project plan. The basic model is developed and calibrated with sufficient definition so that it can be used to answer key project questions. A WRRF is a complex system that undergoes changes in loading, operation, controls, and available treatment processes. The model can be used to determine performance in various process configurations, loading considerations, upset conditions, and operational changes.

Defining modeling scenarios requires consideration of key project objectives, operational conditions, and process configurations that are possible within the context of the project. This section provides background on whole-facility modeling, steady-state and dynamic modeling, scenario development, sensitivity analysis, handling uncertainty, and benchmarking of modeling results.

Whole-facility modeling refers to modeling of the entire WRRF, including biosolids treatment, thickening, dewatering, and the recycling of return liquors to the WRRF headworks. Before detailed anaerobic digestion and sidestream treatment process models became available in process simulators, sometimes return streams were ignored. Other times, return streams from biosolids treatment were modeled as separate influent streams with fixed flows and concentrations. With the realization that these streams can introduce a significant and variable nutrient loading to the WRRF and with the advancement of WRRF process models, the trend in recent years has been toward modeling the entire facility.

There are two general categories of wastewater treatment simulation: steady state and dynamic. A steady-state model does not include time variant inputs, while a dynamic model does. Most commercial simulators have the capability to do both. Both steady-state and dynamic simulations have their use in modeling projects. Selecting which method to use depends on specific requirements of the project.

For example, common design practice typically uses the maximum month flows and loads as the design basis for a facility, which may be modeled as a steady-state simulation. In contrast, a dynamic simulation would be warranted if the facility under design was required to achieve a daily limit on a nutrient component or if further information was needed on transient behavior in the facility, such as peak aeration demands or slug load effects of a dewatering liquor return. A modeler's decision about which type of simulation is best suited to meet project goals should be given careful consideration. In general, dynamic simulation requires significantly more time and effort to achieve a required level of accuracy than steady-state simulation.

Both steady-state and dynamic simulations require appropriate consideration of model uncertainty. Typically, this involves application of safety factors to model inputs, outputs, or both when using the models for design. Safety factors for steady-state simulations are larger than those required for dynamic simulations for a given level of calibration. Therefore, careful consideration must be given to what the appropriate safety factors are in a particular design. A key part of any simulation effort is capturing and documenting how the user applies safety factors in the design procedure and the effect on simulation results. The following sections discuss use of simulators and models in both steady-state and dynamic modes.

7.1 WHOLE-FACILITY MODELING

Simulators can be used for tasks ranging from unit-process evaluations to whole-facility studies. Models for biological unit processes and clarification/thickening unit operations are typically mechanistic models, while those for dewatering and wastewater solids filtration unit operations are often empirical models based on solids removal efficiency. Simulators typically may not be required for non-biological treatment unit processes, but are useful for evaluating the dynamic behavior of settling tanks. Flow equalization units, although not biological treatment units, can have a significant effect on process performance by storing wet-weather flow, leveling diurnal flow and load variations, and storing dewatering sidestream recycle flows for return during low-load periods.

Most wastewater treatment modeling tasks involving nitrogen or phosphorus should consider the need to model the entire facility, especially when the facility uses anaerobic digestion and has dewatering recycle sidestreams or reject streams. This approach captures the effects of recycling solids processing liquors within the system. The need to use a whole-facility model must be carefully balanced against project needs because whole-facility models are more complex and require additional time to solve.

7.2 STEADY-STATE MODELING

Steady-state simulation is typically used to analyze averaged process performance over an extended time period (e.g., two to three SRTs) over which the facility can be considered to be operating under steady conditions. For design, looking at averaged long-term performance is appropriate when developing the basic process configuration and tank sizes. Dynamic simulations can be used to further refine design, make the most of operating conditions, and ensure that the facility can handle anticipated peak loading conditions. One advantage of steady-state simulation analysis is the speed in both model solutions and in the time required by the user to produce needed results. The time required to achieve a model solution is related to the complexity of the model implemented in the simulator and the method the simulator uses to achieve the steady-state solution. There are two basic approaches to determining a steady-state solution. The first approach to solution, previously described, is to simultaneously solve the matrix of equations and mass balances for state variables. The second approach is to execute a dynamic simulation routine until a steady-state results are achieved; that is, a small change in parameters based on a defined tolerance on the derivative terms in the dynamic mass balances. The first method is often faster, more straightforward, and provides meaningful results. The second method is useful in accounting for dynamic factors driving steady-state conditions (such as airflow settings).

The data required to set up and run a steady-state simulation are identical to a dynamic model, except that the time-variant nature of the data is not required. This simplifies data needs in most instances, but adds an additional need to process the data set to be used for design that is consistent across the facility. Daily data from the facility must be averaged over the desired period. Desired periods typically are as follows:

- Average annual—the 365-day average value for various facility parameters.
- Average monthly—the 30-day average value for the various facility parameters.
- Shorter periods—careful consideration must be given by the modeler with the use of shorter periods of data.

As an example, maximum weekly data are sometimes used. However, it is likely that the facility is not at steady state for such a short period. In general, a facility approaches steady-state operation over periods of approximately two to three SRTs. In most facilities, the SRT of the bioreactor system is between 5 and 15 days, so a reliable steady-state period might range between 15 and 45 days. Shorter averaging periods can be used if desired, but the modeler needs to be aware that this is not a true steady-state application. Results from such runs can be either higher or lower than the simulator indicates. Dynamic simulation may be required if more definition is required.

7.3 DYNAMIC MODELING

Wastewater treatment processes are inherently dynamic and experience significant variations in flows and concentrations on an hourly, daily, weekly, and seasonal basis. To truly model the process performance and observe interactions between processes, it is necessary to run a dynamic simulation. Results from a dynamic simulation can be different from those obtained from a steady-state simulation, particularly for nutrient-removal systems. This is because biological nutrient removal (BNR) processes behave differently when they are subject to substrate and nutrient concentrations that repeatedly vary compared to behavior when they are subject to stable "average" conditions. Some processes, such as sequencing batch reactors (SBRs) and biological activated filters (BAFs) that operate in cycles, cannot be modeled using steady-state simulations and must be modeled dynamically if short-term control/performance information is needed.

Before running a dynamic simulation, it is useful to first run a steady-state simulation to obtain the approximate model variable values. Operating conditions used to run the initial steady-state simulation should be representative of the sludge inventory in the facility during an extended period preceding dynamic simulation (e.g., a period of two to three times the SRT that precedes the dynamic period being simulated). Alternatively, the dynamic model may be run for an extended period of time to reach a "pseudo" steady state for which model outputs have a repeatable pattern that matches measured or expected response in the real facility.

For time-varying processes such as SBRs, it is advisable to run the model dynamically until a pseudo-steady state is reached before running any investigative dynamic simulations. Simulation outputs typically are graphed, and averages, maximums, and minimums are obtained from the graphs. An infinite number of dynamic simulations can be run; some common simulations that are useful for design are described in the following subsections.

7.3.1 DIURNAL MODELING

One of the simplest dynamic simulations is to run the model over several days with a repeating diurnal flow and concentration pattern to see how the system responds to a typical diurnal variation. An example of this is shown in Figure 16.10 (input flows) and Figure 16.11 (output effluent ammonia and nitrate). Figure 16.12 shows dynamic output for a 1-week simulation showing the effect of dewatering during weekdays only.

7.3.2 WEEKLY, MONTHLY, AND SEASONAL VARIATIONS

An extension of diurnal modeling is to model a weekly pattern or to extend this to monthly or even seasonal variations. This can provide a better picture of the effect of different flows and loads or operating practices on weekends. An example of this is shown in Figure 16.13, which illustrates the 7-day output from a model that had different sludge liquor returns on the weekend than during the week; these had a direct effect on the effluent total nitrogen concentration. This model was subsequently used to demonstrate the benefit of running sludge dewatering operations 7 days a week.

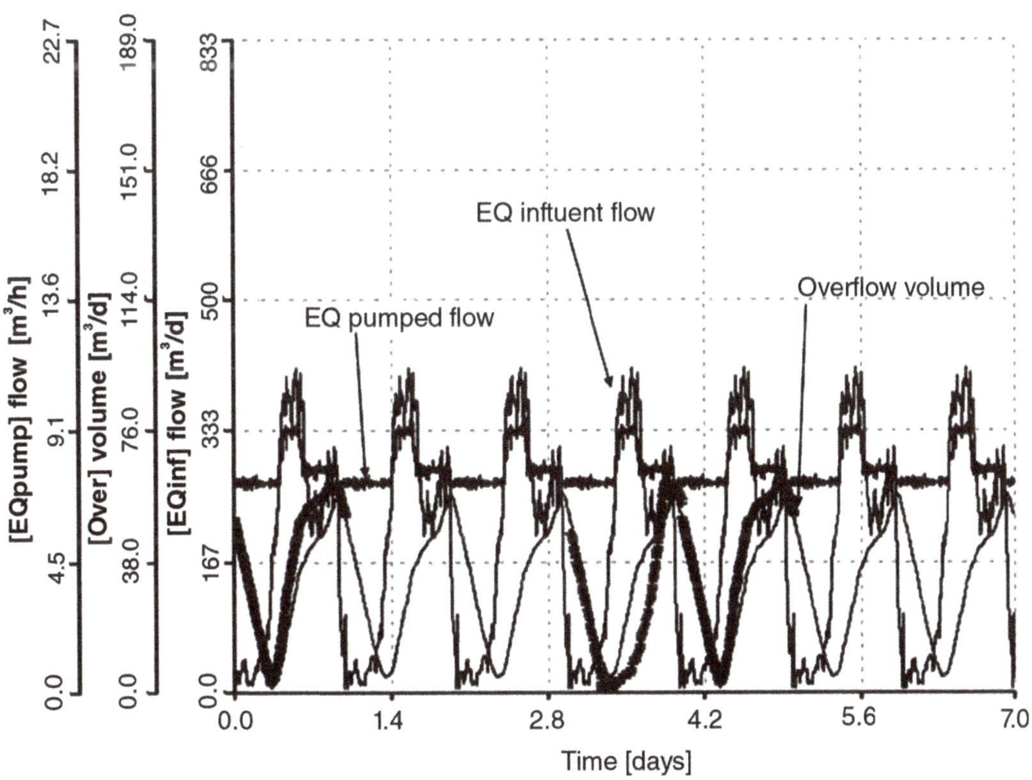

Figure 16.10 Influent Flows

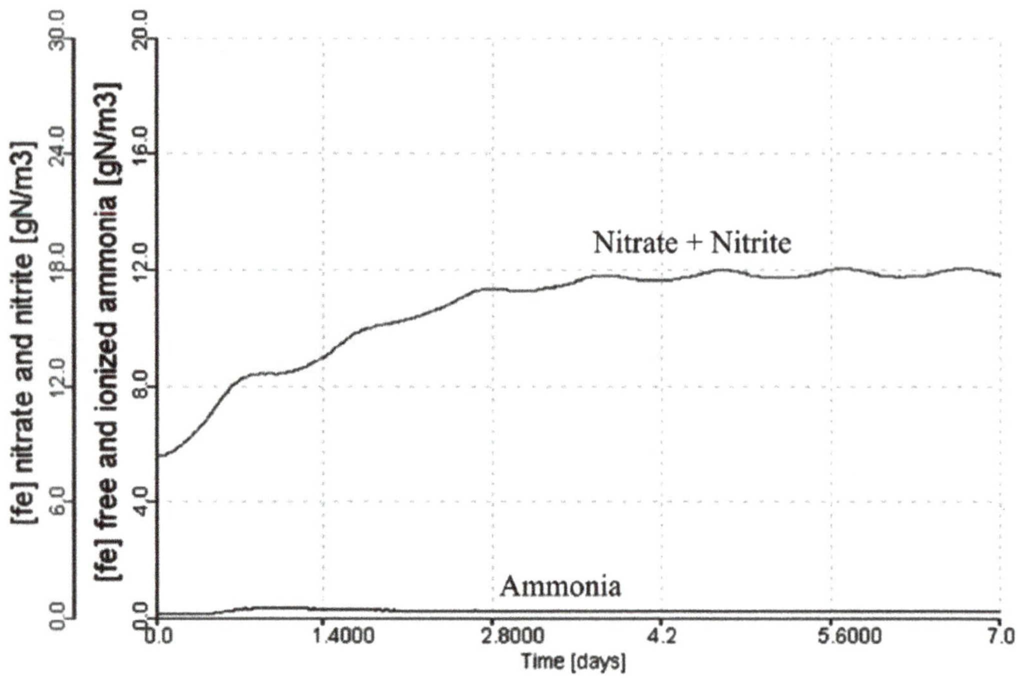

Figure 16.11 Output or Effluent Ammonia and Nitrate

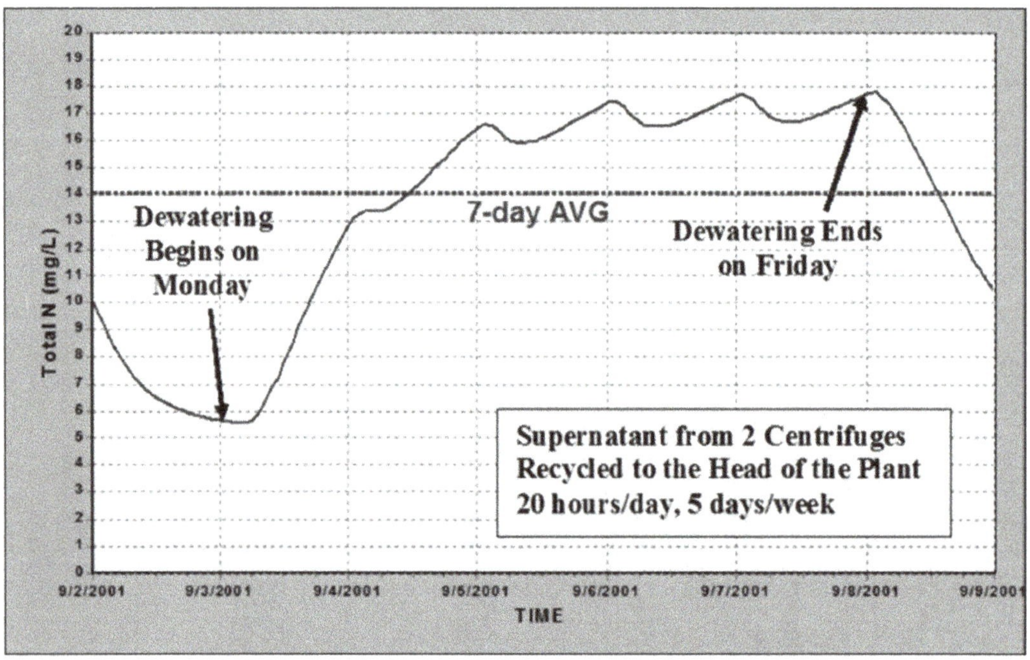

Figure 16.12 Dynamic Output

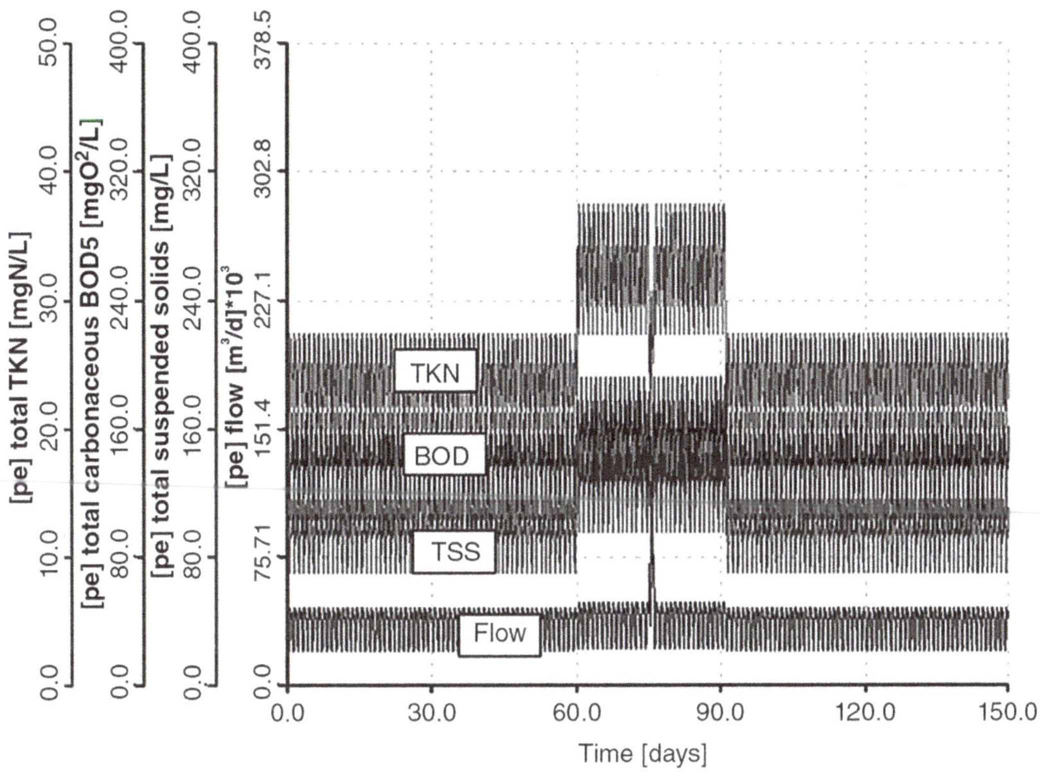

Figure 16.13 Dynamic Model Example: Birthday Cake Input (PE = primary effluent)

7.3.3 MAXIMUM MONTH ANALYSIS

A useful dynamic analysis for design is to run a simulation over several weeks and include the design loadings that are typically used to size equipment, including annual average, maximum monthly average, and peak day events. Ideally, long-term simulations would use actual facility data to describe the variation in influent characteristics. If the data are unavailable, an artificial loading pattern can be used that combines the annual average, maximum monthly average, and peak-day loading. The resulting flow input to the model sometimes resembles a "birthday cake" pattern. From this simulation, average and peak load conditions can be modeled, and the process response can be examined. Figures 16.13, 16.14, and 16.15 show typical dynamic monthly analysis from a model. From these types of figures, the expected average and peak performance can be evaluated.

7.3.4 DYNAMIC RESPONSES TO STORMS AND OTHER EVENTS

In addition to examining typical flow and load patterns throughout a facility, simulations can be run to observe how the model responds to storms or other unusual events. The simulation typically should be run for several days before the event and several days after the event to observe how long it takes for the outputs to return to "normal." Figures 16.16 and 16.17 show the dynamic input and response of a model that was subjected to 3 days of abnormal "special event" flows.

As allowances for frequency and volume of combined sewer overflows and sanitary sewer overflows are reduced or eliminated by regulatory agencies, these wet-weather flows are routed to WRRFs. A common solution to modeling wet-weather flows is to have a separate stormwater input, as shown in Figure 16.18.

7.4 SCENARIO DEVELOPMENT

Once a model has been developed and calibrated, there are many ways to use the model to answer "what if" questions. The modeler may choose to investigate several scenarios or model conditions to make the most of operations at an existing facility or to make the most of design of a new facility. Common scenarios used for design and operations are summarized in Table 16.7.

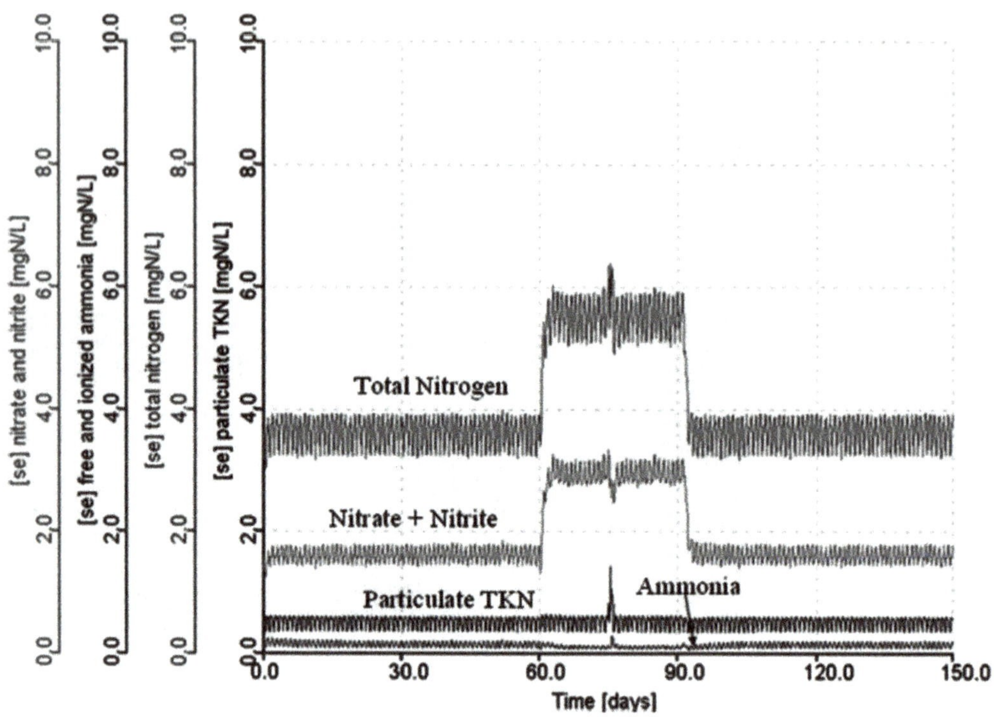

Figure 16.14 Dynamic Model: Modeled Effluent Nitrogen Species in Response to Figure 16.13 Input

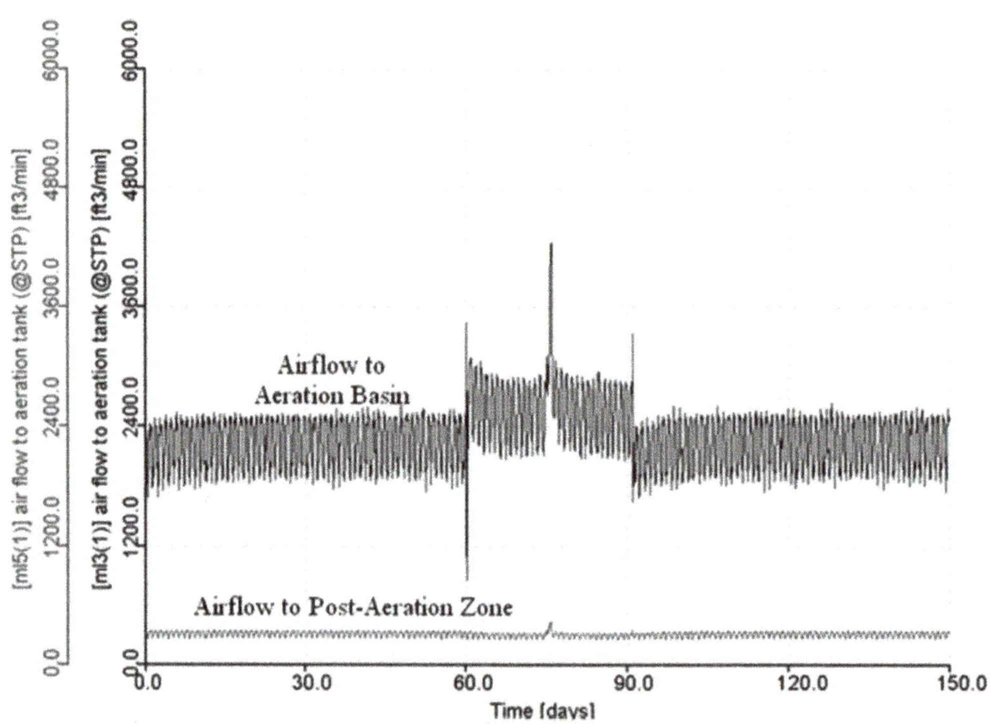

Figure 16.15 Dynamic Model: Modeled Airflow Required in Response to Figure 16.13 Input

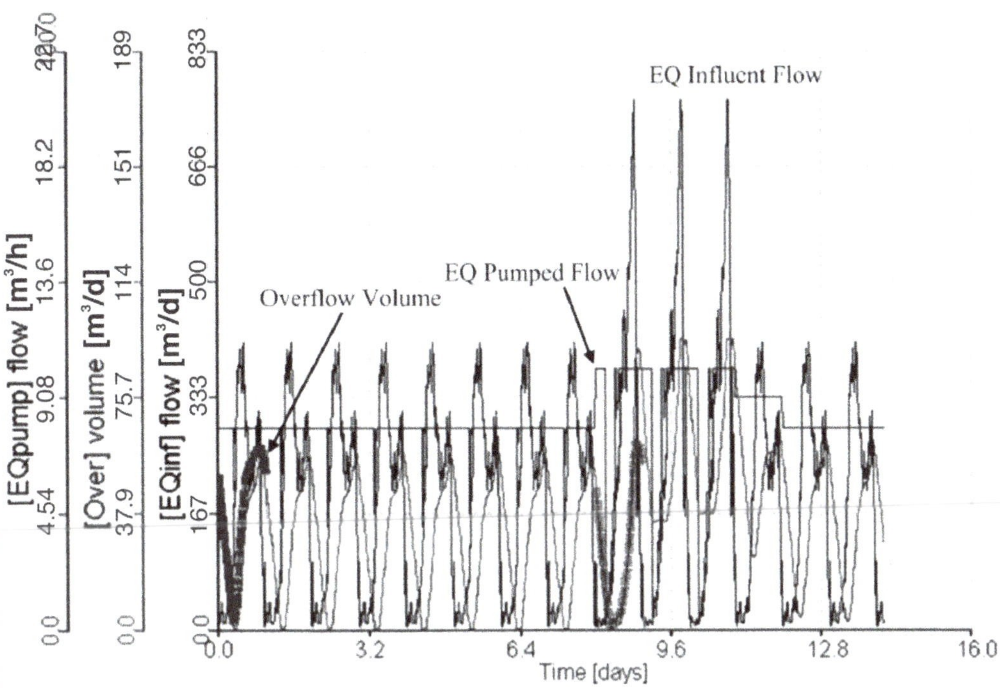

Figure 16.16 Dynamic Model: Simulation of a 3-Day Special Event (7-d EQ)

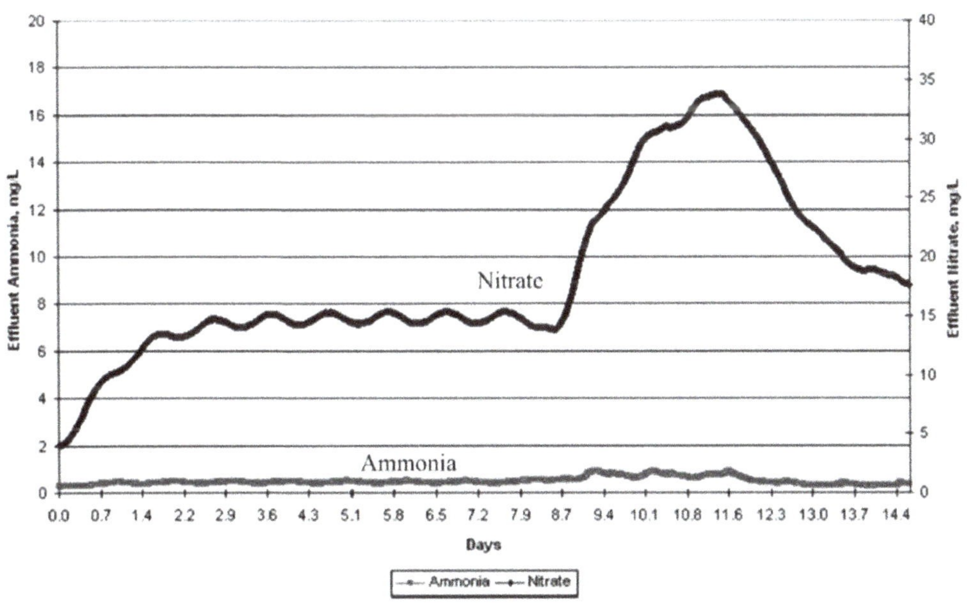

Figure 16.17 Dynamic Model: Modeled Effluent N Species in Response to Figure 16.16 Input

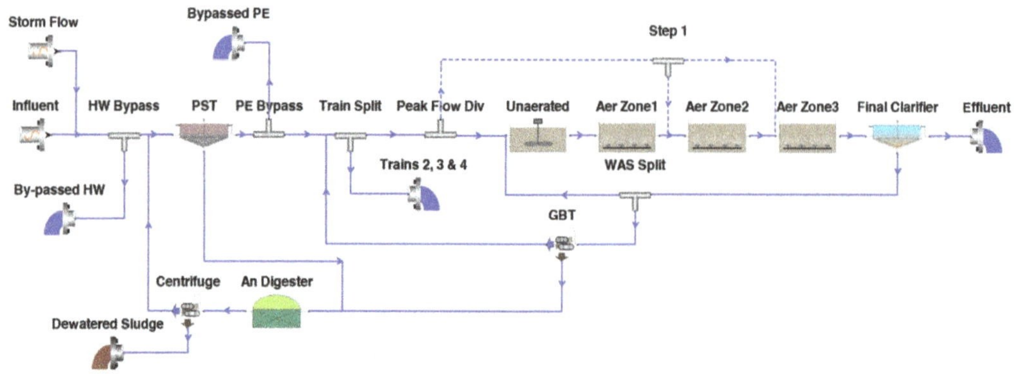

Figure 16.18 Model Build Example Using Stormwater Input and Wet-Weather Diversion

Table 16.7 Summary of Common Model Scenarios

Influent loading conditions	Effect on design
Design flow and loads, winter temperature	To size the aeration basins and clarifiers Determine effect on biological process at coldest temperature
Design flow and loads, summer temperature	To determine maximum oxygen demand at lowest oxygen transfer efficiency Determine effect on biological process at warmest temperature
Average flow and loads, average temperature	To determine average operations and maintenance costs
Minimum flow and loads, minimum	To determine minimum mixed liquor, aeration demand, pump turndown requirements, and options for taking basins out of service
Seasonal loading conditions	To determine the effect of seasonal loading from industrial sources
Operating conditions	
Basin out of service (maintenance or redundancy)	To determine capacity effect To determine effects on effluent quality To determine effect on aeration system (diffused air)
Higher (or lower) SRT	To determine effects on effluent nutrient concentration and capacity
Higher (or lower) RAS flowrates	To determine effects on secondary clarifier solids loading and effects to biological process performance
Higher (or lower) mixed liquor recycle rates	To determine effects on effluent quality
Digested sludge dewatering recycle return	To determine effects on effluent quality
Facility sizing scenarios	
Smaller (or larger) anoxic zones	To determine effects on effluent nitrates or the potential for swing zone operation
Smaller (or larger) anaerobic zones	To determine effects on effluent phosphorus or the potential for swing zone operation
More (or less) chemical addition	To make the most of performance, size pumps, and potential storage requirements

Evaluating different influent loading conditions is a relatively straightforward process; however, investigating some of the other scenarios listed in Table 16.7 can be time-consuming without an organized plan. Therefore, it is best to select design flows, loads, SRT, and approximate volumes before evaluating the effects of minor changes to recycle rates and other operating parameters.

7.5 SENSITIVITY ANALYSIS

Sensitivity analysis involves determining how sensitive the model outputs are to changes in the model inputs, which can include model parameters and initial conditions for a given scenario. Sensitivity analysis can be used to better understand system behavior, to determine the identifiability of model parameters, and to quantify uncertainty in modeling results. Sensitivity studies may include scenario analysis, local sensitivity analysis, and global sensitivity analysis.

Sensitivity analysis is an evaluation of the uncertainty in the outputs of a model in relation to specific inputs.

Scenario analysis (i.e., scenario-based sensitivity analysis) is used to help understand the behavior of the model under different operating conditions. This may involve specific "what if" studies, manually running a series of simulations over a range of parameters, or having the simulator automatically run a set of predefined simulations. These types of simulations are useful for assessing the relative importance of design criteria, loading conditions, operating conditions, facility capacity studies, development of operating strategies, and selection of sampling locations for model calibration because they will highlight the most sensitive locations.

To conduct a scenario-based sensitivity analysis, the modeler should change one model input (e.g., SRT) and observe changes in model output (e.g., effluent ammonia). The process is repeated for several different values of the input until a curve can be generated that shows the sensitivity of model output to changes in model input. It is important to vary only one parameter at a time to accurately show cause and effect. For example, by using a sensitivity analysis to compare effluent ammonia with SRT, the engineer can quantify the risk of nitrifier washout and select a design SRT with an appropriate safety factor, as shown in Figure 16.19. This figure was generated with a dynamic model, and the daily average, minimum, and maximum ammonia concentrations are shown as functions of SRT.

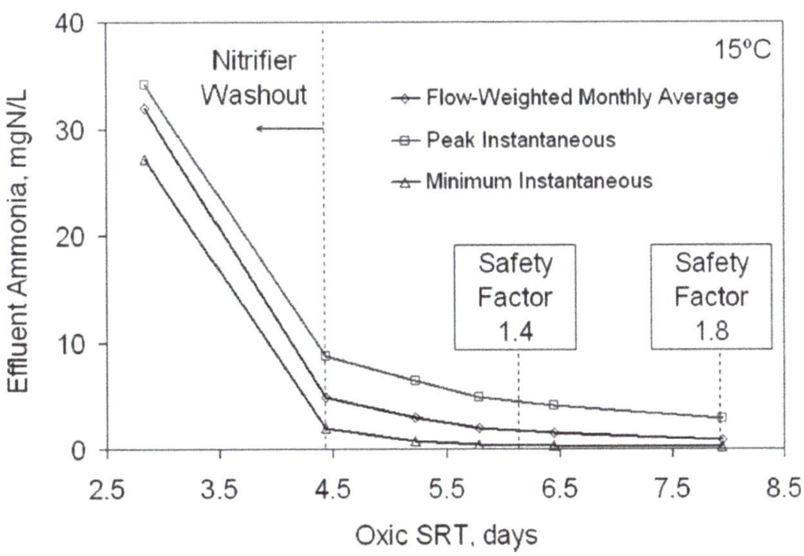

Figure 16.19 Sensitivity Analysis: Effluent Ammonia Versus Solids Retention Time

An important application of sensitivity analysis is to study the effect of important kinetic parameters on simulation results. For example, if the nitrifier growth rate was not experimentally measured at a particular WRRF and an upgrade to the facility was being designed to handle an increased influent ammonia loading, it would be important to study the effect of this parameter on the proposed design.

The aforementioned analysis of effluent ammonia versus SRT could be repeated over a range of nitrifier growth rates at the minimum expected system temperature to generate a series of effluent ammonia curves on the same plot. The range of the effluent ammonia curves at a given SRT shows the potential uncertainty in effluent ammonia at the selected loading conditions.

Another example analysis is to run a series of dynamic simulations with projected daily and diurnal variations over an entire year at a series of different nitrifier growth rates. This procedure could be repeated at a number of different target SRTs. Plots could then be generated at each SRT to show the dynamic effluent ammonia curves for different nitrifier growth rates. This analysis would help demonstrate how the upgraded facility would perform under dynamic loading conditions given potential uncertainty in the nitrifier growth rate, thereby helping to select appropriate SRT and tank sizes.

Sensitivity analyses can also be used for process optimization. Most activated sludge models have the ability to track residual dissolved oxygen throughout the process, which is essential for BNR facilities. Whether the engineer is selecting the size of mixed liquor recycle pumps or an operator is setting the pump speed, sensitivity analyses can be useful in predicting the points of diminishing return when recycled dissolved oxygen hinders denitrification, as shown in Figure 16.20.

7.6 DEALING WITH UNCERTAINTY

Table 16.8 summarizes the types of parameters available in WRRF models and provides examples of which parameters might be used in a sensitivity analysis to evaluate uncertainties that may persist. Physical parameters for study include reactor configuration (e.g., number of tanks in series, tanks out of service for maintenance, or use of swing zones); anaerobic/anoxic/aerobic mass fractions; and wastewater temperature. These parameters would typically be studied during design-based analyses.

Influent variables for study can include influent flow, diurnal patterns, concentrations, and composition (i.e., COD, nitrogen, and phosphorus fractionation among state variables). These parameters all relate to facility loading and are important when studying facility design, performance, and capacity.

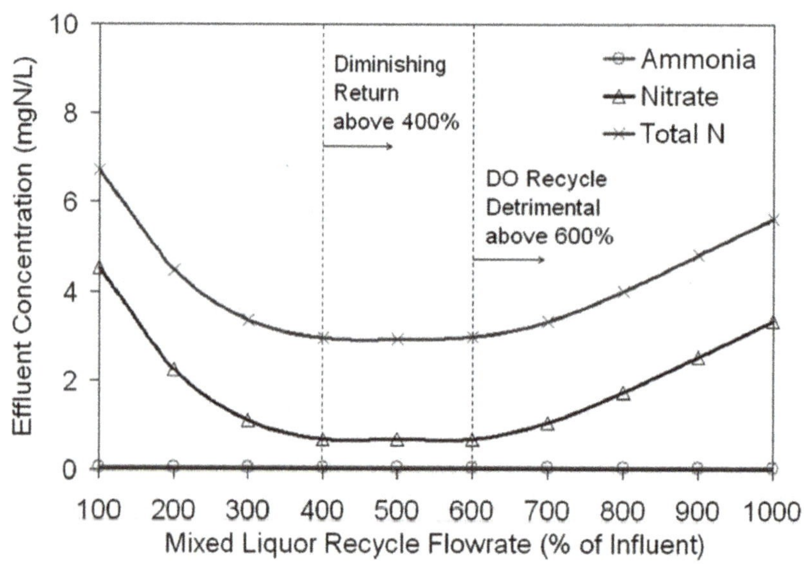

Figure 16.20 Sensitivity Analysis: Effluent N Species Versus Mixed Liquor Recycle Rate

Table 16.8 Parameters for Sensitivity Analysis (Adapted in part from Melcer et al. [2003])

Parameter type	Examples
Physical	■ Reactor configuration (e.g., number of tanks in series, effect of trains out of service, use of swing zones, and anaerobic, anoxic, aerobic volumes/mass fractions) ■ Wastewater temperature
Influent	■ Flow ■ Concentrations ■ Diurnal pattern ■ COD, nitrogen, and phosphorus fractionation among model state variables (e.g., readily biodegradable fraction of COD, fraction of un-biodegradable particulate COD)
Operational	■ Recycle flows ■ Wastage flow ■ Chemical dosage ■ Pumping schedules ■ Airflows ■ Aeration system constraints ■ Oxygen transfer efficiency ■ Aeration alpha factor ■ Controller setpoints (e.g., dissolved oxygen) ■ SRT, MLSS
Biological	■ Maximum specific growth rates (e.g., nitrifier growth rate, anoxic hydrolysis rate factor, polyphosphate accumulating organism maximum specific growth rate) ■ Half-saturation coefficients (e.g., half-saturation coefficient for poly-hydroxy-alkanoate) ■ Stoichiometric (e.g., nitrogen and phosphorus fractions in biomass)
Settling	■ Vesilind and double-exponential parameters
Initial conditions	■ Initial biomass concentrations

Specification of operational variables and strategies is an important aspect of troubleshooting and optimization studies. A sensitivity analysis can help identify the most important optimization variables. Potential parameters for analysis include recycle flows, wastage flows, airflows, chemical dosage, and controller setpoints (e.g., for dissolved oxygen, ammonia, SRT, and mixed liquor suspended solids [MLSS]).

Sensitivity of kinetic and stoichiometric parameters in activated sludge models has been well studied. In most instances, the majority of these parameters are left at published default values during modeling studies because of the difficulty in identifying them using typical facility data. Certain parameters may be site-specific and their effect can be studied using a sensitivity analysis. For example, Melcer et al. (2003) suggested that the following parameters may require adjustment during calibration: nitrifier growth rate, anoxic hydrolysis rate factor, polyphosphate-accumulating organism maximum specific growth rate, and the half-saturation coefficient for poly-β-hydroxyalkanoate. These parameters would be good candidates for further study using sensitivity analysis. Other parameters that are candidates for sensitivity analysis include the nitrogen and phosphorus fractions in biomass (which may need to be adjusted for industrial WRRFs) and settling model parameters (Vesilind and double-exponential parameters).

7.7 BENCHMARKING RESULTS

Models of existing facilities can be benchmarked against typical facility performance for similar facilities within the same region. The facility data (and model) can also be checked for consistency using basic data checks such as checking the influent flow expected based on population, influent concentration ratios (e.g., BOD_5/COD, volatile suspended solids/TSS, TKN/COD, and total phosphorus/COD), facility sludge production (based on flow, per capita, or BOD_5 removed), clarifier solids removal efficiency, clarifier solids mass balances, total phosphorus mass balances, oxygen consumption, digester volatile solids destruction, digester gas production (per capita, facility flow, or theoretical sludge production), and the hauled sludge concentrations. Models used for design can be benchmarked against performance for similar process configurations published within design manuals and for similar operating facilities in addition to the aforementioned basic data/model checks.

8.0 APPLICATION OF MODELS

Model setup and calibration is discussed in previous sections. Once set up, the model is ready to be applied to design and optimization, and each unit process should be evaluated to determine the validity of the model setup under a new set of conditions. The development of scenarios was discussed to evaluate various conditions, and unit processes should be validated with each scenario. During evaluation of alternative process scenarios, simulations should strive to conserve consistency in unit processes.

8.1 PRELIMINARY TREATMENT

In most modeling practices, preliminary treatment is not included in the overall facility model. Solids that are removed in the grit basin and screens typically are not included in any sampling used to characterize influent waste. The washing and compacting steps return most of the organic waste to the influent stream. The decision to include preliminary treatment steps should be based on the various scenarios considered in the process. Conditions for which preliminary treatment might be included are:

- Evaluation of scenarios where preliminary treatment currently does not exist at the treatment system. Initial calibration of the model may have included a greater mass of inert solids in the influent waste stream.
- Known effects from the preliminary treatment process such as dissolved oxygen effects from aerated grit removal facilities.

8.2 PRIMARY CLARIFICATION

In full-facility models, the primary clarification process is modeled on the basis of the percentage of TSS removal. Actual performance of the primary clarifiers may change based on changes in loading, wastewater characteristics, chemical addition, or co-thickening. The scenarios considered under this arrangement cannot be field-measured unless a pilot test is conducted. Particular care must be taken for configurations that include more sophisticated unit modeling than solids removal.

8.3 ACTIVATED SLUDGE

The majority of models available are centered on the operation and design of an activated sludge system. The various scenarios allow the designer to determine the comparative performance of process configurations, the effect of changes to zone sizing, and variations in operation options. In general, the process model will be used to

- Develop sizing of anaerobic, anoxic, and aerated zones;
- Predict solids generated for solids system sizing;
- Characterize aeration requirements;
- Evaluate alternate process configuration (e.g., step feed/Johannesburg/3-stage Phoredox) and swing zone use (i.e., to switch functions of zones between anaerobic, anoxic, and aerobic for seasonal variation);
- Determine effects from operating with a unit out of service; and
- Determine effects of changes to return activated sludge (RAS) and mixed liquor recycle flowrates.

The validity of results from scenario evaluation depends on the accuracy of the data that are available and the fit of the data during initial setup and calibration of the model. The models can only be as accurate as the conditions defined in each unit process. Several areas that should be considered during design or can be evaluated as part of troubleshooting are

- Actual flow splitting,
- Areas of possible air entrainment,
- Back-mixing potential,
- Requirements for air and mixing throughout the aerated zones, and
- Diffuser grid airflow control.

8.4 FIXED-FILM AND HYBRID PROCESSES

Fixed-film processes can be modeled using commercially available models as described in previous chapters. For optimization, similar parameters described under the calibration scenarios should be observed to ascertain that model predictions under various scenarios are equivalent between options.

In general, equipment suppliers should be consulted concerning operating characteristics to determine if the predicted results match installation experience. Scenarios to evaluate during design or optimization of a hybrid system are media-fill fraction and location of media in the treatment process. Scenarios to evaluate during design or optimization of a trickling filter system are media height, filters in series operation, and recycle ratios.

Although fixed-film treatment is not new to the industry, full-scale examples of integrated fixed-film activated sludge (IFAS) and moving bed bioreactor (MBBR) installations are limited. Parameters such as film thickness and detachment rates are generally empirically derived. The predictive ability of the models is directly tied to the quantity of full-scale operational data or pilot-testing results in which the biofilm thickness, detachment rates, and distribution of biomass population can be assessed.

Using a fixed-film process requires the modeler to consider metrics such as the mass of biomass per surface area to help assess the validity of model results. The approach to modeling typically has been to define maximum biofilm thickness and detachment rates to match the mass loading on the media to reported parameters. Modelers should note the purpose of the unit process because different mass loads are expected on media placed in different zones of the process.

8.5 SOLIDS HANDLING AND RECYCLE STREAMS

Recycle streams from solids processing can have a big effect on the overall treatment process. Simulating these streams provides loading characteristics to the rest of the process and allows the engineer to evaluate potential sidestream treatment options.

The solids processing simulation is dependent on liquid process stream characteristics. While developing scenarios, it is important to balance the conservatism in sizing needed for the solids system with actual return streams. The following actions should be taken for each simulation:

- Review solids concentrations after each unit process,
- Watch solids processing flow and mass load splits under the various scenarios,
- Consider batch operation of certain components, and
- Consider feed tanks and equalization tanks.

8.6 SLUDGE PRETREATMENT

Solids processes often include holding tanks that allow hydraulic buffering to downstream processes. These tanks should be included in the model, both to determine dynamic interactions when solids processing is not constant and for biological effects within the holding tanks. Combining streams such as RAS and primary sludge in an unaerated environment could cause release of phosphorus from a biological phosphorus removal facility. This additional phosphorus load is sent back to the activated sludge process.

8.7 ANAEROBIC DIGESTION

Anaerobic digesters seek stabilization of sludge through anaerobic processes. When modeling anaerobic digesters, a separate fundamental model is used (often, the ADM1). The integration point between the ADM and the ASM models should be evaluated as part of the overall model. Like the ASM model, completely mixed tanks are typically assumed. Modelers should evaluate the actual mixing and effective digester volume. Geometry or incomplete mixing can result in inert debris accumulation (e.g., grit and struvite) in the corners of the tank, which, in turn, reduces the effective digester volume.

8.8 AEROBIC DIGESTION

Aerobic digestion seeks stabilization of the sludge through aerobic processes. When modeling an aerobic digestion process, it is important to assess potential unaerated operation of the digester when decanting from the digester back to the head of the facility. During the unaerated time, the contents will shift to either anoxic or anaerobic processes depending on the presence of nitrate and nitrite.

8.9 SIDESTREAM TREATMENT AND BIOAUGMENTATION

Return streams from the solids process can present a large mass loading on the liquid processing stream. Although models are typically set up for continuous operation, sidestream treatment should consider use of different holding tank steps that are used for equalization, combining and mixing streams, or for hold time. Dynamic modeling is required for these facilities to determine the effect on noncontinuous dewatering practices.

Sidestreams also present an opportunity for sidestream treatment to either reduce the loads returning to the liquid stream process or to grow biomass that will augment the main liquid stream biomass. Multiple sidestream processes have been developed. The modeler should consider the environment under which the sidestream treatment occurs. In bioaugmentation, additional biomass could be generated under higher temperatures than the main liquid stream process. The kinetics that result from adding this "seed" biomass should be considered when modeling. The seed biomass population that grows at higher temperatures will not be fully acclimated to the temperature of the main liquid treatment train. A large shift in temperature can reduce the biological activity of the seed biomass by a greater amount than suggested by conventional Arrhenius coefficients. This is a structural uncertainty in most commercially available models.

8.10 TERTIARY TREATMENT

Tertiary treatment processes consist of advanced treatment processes that can include coagulation, flocculation, clarification, and filtration units. As part of a whole-facility model, these unit processes are often approximated strictly as a solids separation device.

Backwashing and countercurrent continuous backwash systems may not be accurately represented in standard solids separation modules. The designer must determine which module or combination of modules best describe the treatment step being modeled.

9.0 USING MODEL RESULTS

It is important to keep in mind that models are not reality. Even the best calibrated model has a level of uncertainty associated with many aspects. In general, there are three categories of uncertainty:

1. Quantifiable uncertainty: Parameters that are known to be uncertain and for which the degree of uncertainty can be reduced. An example of this is uncertainty related to model structure. The model structure may have inherent assumptions that are only approximations of reality, such as the assumption that the autotrophic half-saturation constant for oxygen is a constant. These could potentially be made more accurate through further study and model development, provided the model allows for variable values.
2. Irreducible uncertainty: Uncertainty that cannot be reduced by any degree of study. An example of this is uncertainty related to future flows and loads. No amount of study would be able to quantify future flows and loads beyond a certain point.
3. Total ignorance: Parameters that the modeler does not know are uncertain.

These items could be either reducible or not reducible. Any parameter might fall in this category, and the only method of reducing this is through learning and experience. All of these sources of uncertainty are present within models and contribute to each model's variation from reality. Project team members (stakeholders) need to be aware of this potential during projects so that appropriate measures can be taken to address uncertainty in the modeling results.

However, this does not mean that simulation results cannot be used. As discussed earlier, simulations give valuable insight to how WRRFs function, although the accuracy of the output must be viewed in light of the uncertainties involved.

9.1 INTERPRETING MODEL RESULTS

Reliability of modeling results is often directly dependent on the degree of effort expended by the modeler in reducing uncertainty. Once the model/simulator is selected, calibration and validation are the most common methods of reducing uncertainty of modeling results. The amount of calibration and validation performed on a model should be selected according to use of the model and available budget.

In general, time-variant data add additional uncertainty to simulation results. As raw wastewater is inherently variable, this uncertainty is present in all modeling results that examine short-term effects. Predictions of future performance also are more uncertain because they are extended further into the future. It is up to the modeler to determine how best to interpret modeling results. Irreducible uncertainty ensures that the modeling results never exactly match reality.

9.2 COMMUNICATING MODEL RESULTS AND UNCERTAINTY

Management of a project team member's expectations regarding the reliability and accuracy of modeling results is a key aspect of any modeling project. It is not uncommon for non-modelers to have unrealistic expectations concerning level of effort and modeling accuracy; these expectations must be addressed early in the project and continue throughout the effort.

Presentation of modeling results can take many different routes. In general, the degree of accuracy that is presented in modeling results should reflect the level of calibration. For example, a facility option comparison with little calibration might only report the percentage difference between the options. If a highly calibrated model was developed to look only a short time into the future, it may be possible to report actual results with more significant digits in some of the results. Model results and limitations are as follows:

- Mass balance data presentation: Detailed mass balance data can be summarized and presented by a number of methods. In general, composite variables such as BOD_5, COD, TSS, TKN, and total phosphorus should be presented in data summaries. Detailed mass balance data typically are only suitable for presentation in appendices.
- Model/simulator setup data: Similar to mass balance data, only significant parameters should be presented in the main body of a project report or presentation. Detailed model parameters should be included in an appendix so that future users can reproduce the modeling results if necessary. Commercial simulators typically have reporting features that can produce such detailed information automatically.
- Model/simulator assumptions: A critical component of documenting a model includes assumptions that were used in the model/simulator. It is important to model results for the user to document the degree of conservatism used in determining model parameters.
- Identification of physical modifications necessary: The modeler should incorporate the need to discuss other modifications that would go along with the modeling effort.
- Graphic display of dynamic model results is an effective presentation method.

Care should be taken to not present too much information in a single graph. Comparing different time-series data sets together is an effective method of displaying cause and effect. Figure 16.21 shows the effect of limited aeration air quantity under maximum day demand on aeration basin dissolved oxygen, airflow rate per diffuser, and the effect on final effluent quality.

9.3 QUALITY ASSURANCE

Modeling of wastewater treatment processes is a complex task with many variables. As a result, **quality assurance/quality control** (QA/QC) is a critical part of any modeling project. Although QA/QC can take many different forms, all of these include participation by a person other than the modeler. Quality assurance/quality control should include:

- A "reality check" of modeling results by an external reviewer. This reality check can be through heuristic rules (general rules), by senior reviewers, or through simple spreadsheet models.
- A review of the model files and simulation results.

Regardless of the type of review selected, the following items should be reviewed as part of the QA/QC process:

- Modeling project structure
- Flows and loads analysis
- Interpretation of additional data collected through extra sampling and analysis
- Model default parameters
- Documentation of results
- Conclusions drawn from modeling results

Quality Assurance/Quality Control (QA/QC) is a process that maintains model quality by reviewing model development against well established standards, guidelines, and procedures.

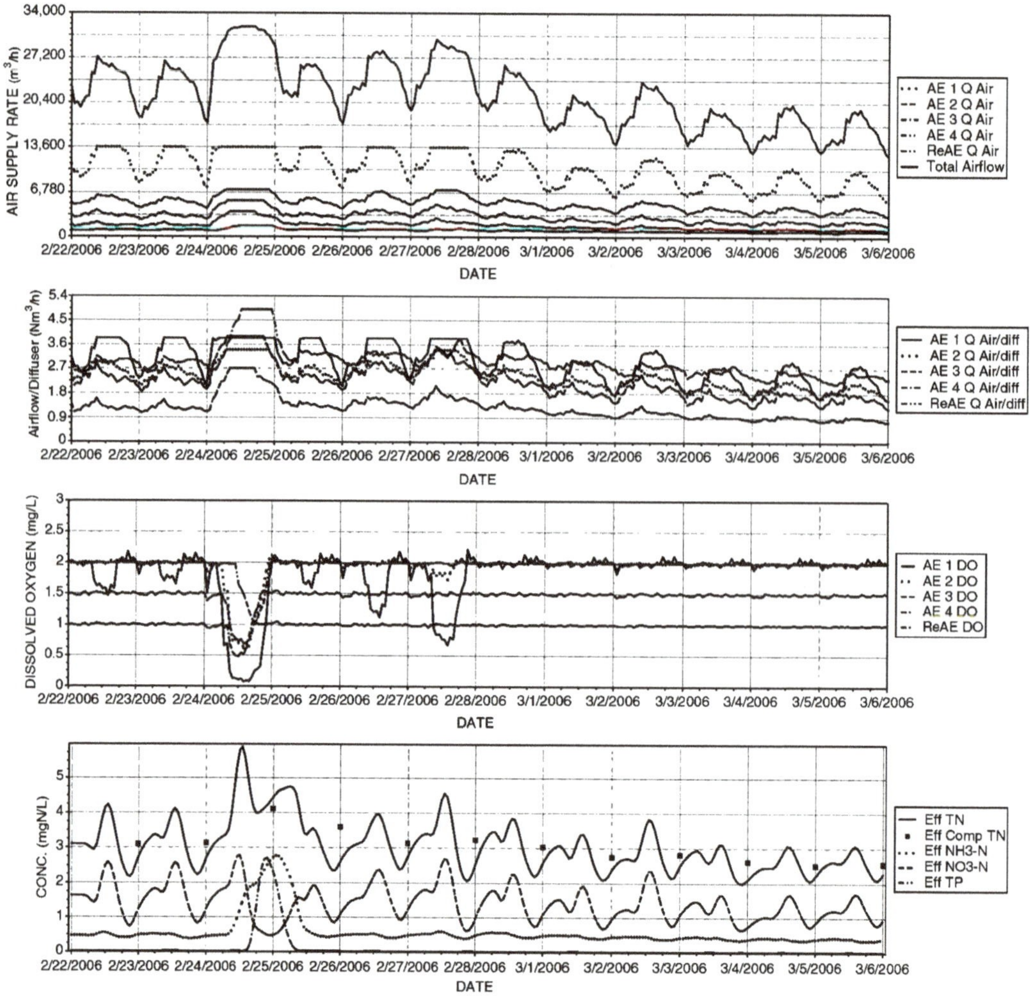

Figure 16.21 Graphical Dynamic Results Showing Cause and Effect

9.3.1 *DIRECT QUALITY CHECKS*

As the modeler documents decisions made throughout the modeling process, a reviewer can backtrack the results to review assumptions and standards along the way. This method requires that the reviewer be an equally qualified modeler. The benefit of this method is direct identification of any changes required in the modeling effort. However, limitations of the model can be missed during review.

9.3.2 *INDEPENDENT VERIFICATION*

There are many different design and modeling tools available to a modeler. Using a second method may identify issues in assumptions used in the modeling effort. These checks could include different simulators or comparisons with typical design values. Discrepancies that are found during this effort should be evaluated against the original goal of the project. The chosen modeling method may better represent the conditions being evaluated.

10.0 REPORTING AND DOCUMENTATION

At the end of a modeling task, a report must be generated that captures the purpose and results of the simulations. The report can provide both historical documentation of assumptions and results from modeling to help decision makers understand the results of the various scenarios that were simulated, along with background information for future users of the model.

10.1 REPORT PREPARATION

The report should contain information from each of the steps described in this manual. This may include information as presented in previously issued memorandums or reports. The report as a whole should follow and document the steps of the project and include the following information:

- Objectives of the modeling effort
- Description of the existing facility
- Data collection
- Data analysis
- Setup and description of model
- Calibration and validation results
- Description of simulation scenarios
- Presentation of results
- Uncertainty analysis

During each of these steps, input and output files used in the model should be provided. Assumptions for each unit process should be tabulated and any deviation from default parameters should be documented. The goal of this documentation is to accurately describe the decision-making process that went into the project.

10.2 DOCUMENTATION FOR FUTURE USE

The report described in the preceding section focuses mainly on results of the modeling effort. Appendices also contain information describing modeling parameters (kinetic, stoichiometric, etc.) that are used for each simulation. After the model has been used for design or optimization, it may be used again in the future for forecasting capabilities, to change operating or influent characteristics to match changes in the facility, or to evaluate options developed later.

It is important to document all the operating parameters of the system, including default parameters that were not changed. As modeling tools progress, the capabilities of the model improves and some parameters may be changed based on further research.

11.0 MODEL LIFE CYCLES

As more wastewater facilities are using models for operation, it is important that consultants coordinate their design efforts with the operators using the model. In addition, because different consultants are likely be involved in planning and design efforts over the years, it is important that modeling information is transferred to all parties involved. This prevents information from being lost and consultants having to "reinvent the wheel" with every new project. In addition, it provides an opportunity to refine model accuracy as new data are collected. Special sampling campaigns should be repeated for different seasons and loading conditions to validate model accuracy over a model's life cycle. At the end of modeling efforts, consultants should provide model reports that document all inputs, including kinetic parameters and influent characteristics. Whether or not facility staff use the model, they retain ownership of all reports and files and should pass this information on to the next project team.

Another factor to consider in a model's life cycle is the issue of new versions of software. Before a new model is used for design or operation, a scenario from the old model should be run in the new model to compare predictions. Changes in model fundamentals or default parameters may cause results to be different, and the model may need to be recalibrated. Latimer et al. (2008) describe a specific challenge they encountered when upgrading a model with single-step nitrification to two-step nitrification. Along with software developers, they identified that the model was "looping" nitrite back to nitrate in a simultaneous nitrification–denitrification process, which resulted in higher effluent nitrate predictions.

11.1 MAINTAINING AND REFINING THE MODEL OVER TIME

Once a modeling effort is complete, conclusions are used to make process or capital investment decisions. In some instances, the model may be reused to evaluate implemented changes to the process, troubleshoot performance issues, or re-evaluate the process based on changes to loading conditions.

Assumptions made in the initial modeling effort should be re-evaluated if project budget allows. Modelers should be aware of safety factors that were used in the initial evaluation. In practice, the performance may be better than anticipated based on conservative parameters initially used. If project budget allows, the model should be updated to reflect current performance, which allows it to be used as a tool for future activities. This may include some additional sampling and analysis of the operating WRRF to update the model.

11.2 MODEL OWNERSHIP

Modeling projects may be conducted for a utility by an independent consulting firm. At the end of the evaluation, ownership of modeling files can come into question if not otherwise stated in the contract. Unless specifically documented in the contract, an owner owns the modeling files if he or she has paid for the work. Modeling files developed as part of the effort become the property of the client once the work is complete. In some circumstances, the company may not want to release a tool that they have developed in-house. Even if it states in the contract that these files are not to be transferred to the utility, it is the responsibility of the modeler to provide a comprehensive set of all calculations that support the conclusions reached as part of the modeling effort.

Models are often reused by other entities after a modeling project is completed. Models may be reused by another modeler within the same consulting company, by modelers from another consulting company, or by a member of a utility using a model developed by a consultant or developed in-house at the utility. Regardless of the circumstance, it is the responsibility of the modeler using a previously developed model to take charge of the model in much the same way a professional engineer stamps drawings based on work from another engineer. When using modeling work performed by others, it is important to understand the goals of the project under which it was developed and to identify assumptions that may not apply to the current modeling effort. As a modeler who may need to transition his or her work to another modeling effort, the reporting and documentation sections of this chapter are important to facilitate this exchange of information.

11.3 DOCUMENTATION FOR FUTURE USE

The documentation process provides information to re-create models for future projects. When preparing documentation, parameters and assumptions should be clearly presented to allow future users (or the current modeler) to understand what decisions were made. Use of a previously developed model does not serve as a guide to future users, and it is the responsibility of future users to verify their own assumptions in implementing the model. The model was developed to be used for conditions defined in the original project. Deviations from these conditions, changes in industry understanding, or goals of the project are not necessarily transferable to any future modeling project.

The accuracy of a model varies greatly with the quality of data used to calibrate and create the model. As models are revisited during construction and operation of a new facility, model accuracy increases as physical effects are further understood and influent waste characteristics can be better documented.

12.0 COMMON PITFALLS

There are many pitfalls that should be avoided when producing and using a process model. Shaw et al. (2007) described some of the most common pitfalls and suggested ways to avoid them. These pitfalls are summarized in the following sections.

12.1 SAFETY FACTORS

Most engineering equations and typical practices used to design and operate wastewater facilities include safety factors. Process models, on the other hand, have no safety factors. Users must apply their own safety factors and good judgment in determining how close to the wire to push the facility based on modeling predictions. Actual facility operation can result in discrepancies in sludge age, temperature, and influent loading when compared to assumptions made during modeling.

12.2 GARBAGE IN EQUALS GARBAGE OUT

The accuracy of any model is highly dependent on the quality of the data used to develop the model. The accuracy of the model then directly affects confidence in the results it produces. This, in turn, affects the safety factor that should be applied to its predictions. Special sampling for influent characteristics and direct kinetic measurements are important steps to ensure that input data are sound. Data checking also is an important step in this process.

12.3 ADJUSTING THE APPROPRIATE PARAMETERS

Expert judgment must be used to adjust the proper parameters in the model. Often, it is possible to fit a model to facility data by adjusting any one of several parameters.

In modeling terms, the process models used in wastewater are considerably "over-parameterized," meaning that there are too many parameters than can be adjusted to obtain the same result, making it almost impossible to obtain a unique solution using purely numerical methods. It is possible for a modeler to adjust one parameter to get a model to fit the data. However, this may not properly represent mechanisms occurring in the facility, which renders the model useless for exploring conditions other than those used during calibration. In principle, the model can be forced to show anything the user wants. In practice, knowledge and modeling expertise must be used to discern correct adjustments to make to the model. In many instances, influent characteristics must be reevaluated and corrected, or the mixing characteristics of the reactor need to be investigated if a model does not fit the facility data for a domestic WRRF.

Calibration of the model to reasonably match the quality of the input source and the facility solids balance, and to identify the source of any discrepancies, are critical initial steps that must be completed before other parameters are adjusted. Kinetics or sludge yields should only be adjusted from established default parameters as a last resort. Additional calibration information for critical parameters can be obtained by conducting facility-specific tests for parameters such as nitrifier growth rates.

12.4 MODEL LIMITATIONS

Models have traditionally been based on activated sludge models whose focus is on solids generation, effluent quality, and biomass growth. During operation, physical geometry and biology add additional complicating factors such as foaming, bulking, poor flow splitting, and short-circuiting. In the clarifiers, a one-dimensional model is often used to describe the settling characteristics of the activated sludge but cannot predict physical clarifier limitations. In these instances, computational fluid dynamic modeling can be implemented to help describe the clarifiers; however, this can be a costly task that does not directly integrate the rest of the facility model. Use of uncertainty analysis can help identify possible effects to population dynamics and overall process performance with changes in return streams.

Although simulators do not predict conditions such as bulking, the modeler should use data that indicate that these conditions may exist. Excessively long or short sludge ages favor some filamentous organisms, as would inadequate aeration at the front of an aerobic zone.

12.5 CONSIDERING MINIMUM CONDITIONS

In addition to considering the overall treatment capacity of the system, the modeler should consider how the facility performs when first started up and during seasonal and diurnal minimum conditions. Under these conditions, the aeration system and pumping system are often oversized for loading. For example, the effect of too much basin volume could result in thin mixed liquor that does not settle well or long sludge ages that result in nitrification and aeration costs not necessary to meet permit requirements. The designer should focus on the range of scenarios that can be seen at the WRRF to determine treatment capabilities throughout the year.

12.6 FALSE PRECISION IN CONTROL CAPABILITIES

Whether or not the model is used to assess effects from controls, the engineer must consider actual operation of the facility including how precisely dissolved oxygen must be controlled in each zone, flowrate control for influent splitting and recycle flow, and control of mixed liquor age based on limited solids analysis.

12.7 DEFINING WORST-CASE CONDITIONS

The worst-case scenario is considered to be the highest loading to the system under the most adverse environmental conditions. The maximum month or maximum week used for design to meet effluent quality may not be the same maximum month or maximum week used for design of the aeration system. These maximum loading conditions may also be the most advantageous for influent carbon with respect to nutrient removal. Worst-case conditions should examine potential combinations of loading characteristics, such as high nitrogen and low influent BOD_5 and extremes in temperature. Table 16.7 suggests a variety of operating conditions to be considered. Optimizing the design for a worst case could result in an inefficient facility under average and startup loading conditions.

12.8 CONFIGURATIONS THAT CONFORM TO MODELING CAPABILITIES

Certain process configurations are inherently easier to model. Focusing on these treatment options because your predictive capabilities are highest could eliminate a more suitable process configuration. Examples of this include

- Focusing on the suspended growth-activated sludge process instead of fixed-film processes such as MBBR, IFAS, BAF, or trickling filters; and
- Using chemical phosphorus removal because of uncertainties in implementation of biological phosphorus removal.

Take care not to limit the alternatives that may work well for a facility in order to make modeling work simpler.

13.0 CONCLUSION

The use of models and models implemented in simulators offers a tremendous tool for supporting design and operation of WRRFs. Models should be developed with appropriate care to link the real facility with the model facility to the extent necessary in order to maximize the benefit from the model. Modern mechanistic models are becoming increasingly complex, and it is important to understand the model mechanics in order apply the model properly.

As discussed earlier, models are only as useful as the input used to build the model. Data used to develop models should be reviewed carefully, and data that are based on higher confidence should be the priority for model alignment. The quality of the output from the model are limited to the quality of the input and, therefore, the quality of the data. All inputs and outputs from the model, and the model files themselves, need to be reviewed for quality and must be properly documented.

Looking to the future of WRRF modeling and simulation, utilities may seek to create "digital twins" of processes and whole facilities. These digital twins may be represented by current simulators, or they may be advanced to mimic the facility based on real-time data collection fed directly into the digital twin or model (e.g., flow, dissolved oxygen concentration, MLSS concentration, effluent ammonia concentration). Online, or linked, digital twins may be used by operations and engineering staff to better understand WRRF treatment variables and responses. One way to use the output from digital twins is to maximize treatment capacity. Another way to use the output is to optimize the facility for more efficient and effective treatment (more treatment at reduced cost; e.g., energy and chemical usage).

Digital twins may advance even further in the future, with the linkages between the WRRF and the model providing opportunities for real-time operational control. In this case, the digital twin may be used to directly optimize performance by making real-time changes to operational setpoints. For example, the model may predict a need for increasing the dissolved oxygen setpoint due to increased loading at the WRRF. With real-time control system in place, the change in setpoint may be made automatically. Model development and integration in this case requires an even higher standard of care.

PRACTICE PROBLEMS

Concepts and Fundamentals:

1. Why are models (and simulators) used for evaluating and designing WRRFs?
2. What is an empirical model, and what are the limitations of an empirical model?
3. What is a mechanistic model, and what are the limitations of a mechanistic model?
4. What is the difference between a state variable and a composite variable?
5. Identify three state variables.
6. Identify three composite variables.
7. What is the difference between a model calibration and a model validation?
8. What are three steps to calibrating a model?
9. Identify three biological processes or mechanisms that make up typical model components.
10. Describe the components of a Gujer matrix and explain why a Gujer matrix is useful.
11. Identify three physical processes or mechanisms that make up typical model components.
12. What is the difference between a model and a simulator?
13. What kind of evaluation or design scenario would be modeled with a steady-state model?
14. Identify four evaluation or design scenarios that would be modeled with a dynamic model.

Modeling Problems (Note: access to and use of a simulator is needed for problems 15 and 16):

15. Build a WRRF into a wastewater simulator (e.g., BioWin, GPS-X, Sumo, Simba, Stoat) based on the following:
    - Influent:
        - Flow = 100 000 m^3/d
        - COD = 500 mg/L
        - TKN = 50 mg-N/L
        - Total phosphorus = 8 mg-P/L
        - Temperature = 16 °C
        - Alkalinity = 250 mg-$CaCO_3$/L
        - pH = 7.1
    - Activated sludge: Modified Ludzack-Ettinger Process
        - Anoxic volume = 25 000 m^3
        - Aeration volume = 52 500 m^3
        - Final clarifier: Area = 4000 m^2; Depth = 4 m
        - Mixed liquor recycle flow = 300% (influent)
        - SRT = 10 d

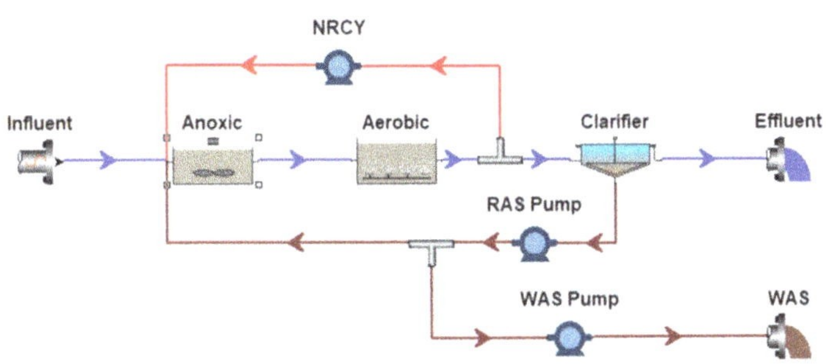

Configuration for Problem 15

After configuring, run the model and answer the following questions:

a. What is the MLSS concentration?

b. What is the clarifier solids loading rate?

c. Are there any model alarms or limitations on the process (e.g., high oxygen uptake rate, high airflow per diffuser, high clarifier surface overflow rate)?

d. Does the effluent show high nitrification efficiency?

e. Does the effluent show total nitrogen and/or total phosphorus removal?

Run the model at several different SRTs and mass loading rates. Describe the effects on the following:

a. MLSS concentration

b. Airflow rate

c. Effluent ammonia concentration

d. Effluent total nitrogen concentration

16. Build a WRRF into a wastewater simulator (e.g., BioWin, GPS-X, Sumo, Simba) based on the following:
    - Influent:
        - Flow = 150 000 m^3/d
        - COD = 550 mg/L
        - TKN = 45 mg-N/L
        - Total phosphorus = 8.5 mg-P/L
        - Temperature = 22 °C
        - Alkalinity = 200 mg-$CaCO_3$/L
        - pH = 7.2
    - Activated sludge: Modified Ludzack-Ettinger Process
        - Anaerobic volume = 12 500 m^3
        - Anoxic volume = 25 000 m^3
        - Aeration volume = 50 000 m^3
        - Final clarifier: Area = 5000 m^2; Depth = 4 m
        - Mixed liquor recycle flow = 250% (influent)
        - SRT = 8 d

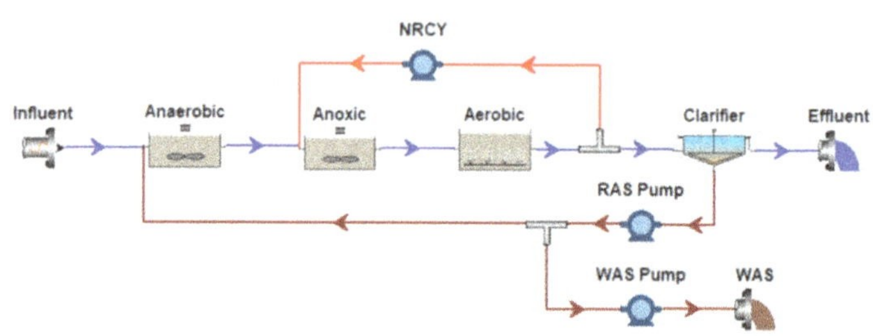

Configuration for Problem 16

After configuring, run the model and answer the following questions:

a. What is the MLSS concentration?

b. What is the clarifier solids loading rate?

c. Are there any model alarms or limitations on the process (e.g., high oxygen uptake rate, high airflow per diffuser, high clarifier surface overflow rate)?

d. Does the effluent show high nitrification efficiency?

e. Does the effluent show total nitrogen and/or total phosphorus removal?

Run the model at several different SRTs, mass loading rates, and RAS rates. Describe the effects on the following:

a. MLSS concentration

b. Airflow rate

c. Effluent ammonia concentration

d. Effluent total nitrogen concentration

e. Effluent total phosphorus concentration

14.0 REFERENCES

Andrews, J. F. (1971). Kinetic models of biological waste treatment processes. *Biotechnology Bioengineering Symposium 2*. Wiley & Sons.

Barker, P. S., & Dold, P. L. (1997). General model for biological nutrient removal activated sludge systems: Model presentation. *Water Environment Research, 69*(5), 969–984.

Batstone, D. J., Keller, J., Angelidaki, L., Kalyuzhnvi, S., Pavlostathis, S. G., Rozzi, A., Sanders, W., Siegrist, H., & Vavilin, V. (2002). *Anaerobic Digestion Model No. 1 (ADM1)*. IWA Publishing.

Belia, E., Amerlinck, Y., Benedetti, L., Johnson, B., Sin, G., Vanrolleghem, P. A., Gernaey, K. V., Gillot, S., Neumann, M. B., Rieger, L., Shaw, A., & Villez, K. (2009). Wastewater treatment modeling: Dealing with uncertainties. *Water Science and Technology, 60*, 1929–1941.

Belia, E., Benedetti, L., Johnson, B., Murthy, S., Vanrolleghem, P. A., Gernaey, K. V., Gillot, S., Neumann, M. B., Weijers, S., & Copp, J. B. (2009). Wastewater treatment modeling: Discussing uncertainty. *Influents, 4*, 81–83.

Bryant, J. O. (1972). *Continuous time simulation of the conventional activated sludge wastewater renovation system* [Doctoral dissertation, Clemson University].

Busby, J. B. (1973). *Dynamic modeling and control strategies for the activated sludge process.* [Doctoral dissertation, Clemson University].

Busby, J. B., & Andrews, J. F. (1975). Dynamic modeling and control strategies for the activated sludge process. *Journal of the Water Pollution Control Federation, 47*, 1055–1080.

Christensen-Henze, M., & Harremoes, P. (1977). Biological denitrification of sewage: A literature review. *Progress in Water Technology, 8*, 509–555.

Corominas, L., Rieger, L., Takacs, I., Ekama, G., Hauduc, H., Vanrolleghem, P. A., Oehmen, A., Gernaey, K. V., van Loosdrecht, M. C. M., & Comeau, Y. (2010). New framework for standardized notation in wastewater treatment modeling. *Water Science and Technology, 61*(4), 841–857.

Dold, P. L., Ekama, G. A., & Marais, G. V. R. (1980). A general model for the activated sludge process. *Progress in Water Technology, 12*, 44–77.

Downing, A. L., Painter, H. A., & Knowles, G. (1964). Nitrification in the activated sludge process. *Journal and Proceedings of the Institute of Sewage Purification, 63*, 130–153.

Goodman, B. L., & Englande, A. J., Jr. (1974). A unified model of the activated sludge process. *Journal of the Water Pollution Control Federation, 46*, 312–332.

Grady, C. P. L., Jr., & Roper, R. E. (1974). A model for the bio-oxidation process which incorporates the viability concept. *Water Research, 8*(7), 471–483.

Grady, L., Jr., Daigger, G. T., & Lim, H. C. (1999). *Biological wastewater treatment* (2nd ed.). Marcel Dekker.

Grady, C. P. L., Jr., Gujer, W., Henze, M., Marais, G. V. R., & Matsuo, T. (1986). A model for single sludge wastewater treatment systems. *Water Science and Technology, 18*, 47–61.

Henze, M., Grady, C. P. L., Gujer, W., Marais, G. V. R., & Matsuo, T. (1987). *Activated sludge model no. 1* (IAWPRC Scientific and Technical Report No. 1). International Association on Water Pollution Research and Control.

Henze, M., Gujer, W., Mino, T., Matsuo, T., Wentzel, M. C., & Marais, G. V. R. (1995). *Activated sludge model no. 2* (IAWQ Scientific and Technical Report No. 3). International Association on Water Quality.

Henze, M., Gujer, W., Mino, T., Matsuo, T., Wentzel, M. C., Marais, G. V. R., & van Loosdrecht, M. C. M. (1999). Activated sludge nodel no. 2d, ASM2d. *Water Science and Technology, 39*(1), 165–182.

Henze, M., Gujer, W., Mino, T., & van Loosdrecht, M. (2000). *Activated sludge models ASM1, ASM2, ASM2D and ASM3*. IWA Publishing.

International Water Association Task Group on Good Modeling Practice. (2008, September 5–6). Plant model set-up and calibration. *Proceedings of the International Water Association Task Group on Good Modeling Practice Course on Modeling Activated Sludge Plants*, Vienna, Austria.

Jones, P. H. (1971). A mathematical model for contact stabilization-modification of the activated sludge process. In *Advances in Water Pollution Research: Proceedings of the 5th International Conference on Water Pollution Research*. Pergamon Press.

Latimer, R. J., Pitt, P., Dold, P., Takács, I., & Lynch, T. J. (2008, June 1–3). Kinetic parameters for modeling two-step nitrification and denitrification—Case study. *Proceedings of the 1st International Water Association/Water Environment Federation Wastewater Treatment Modeling Seminar*, Mont-Sainte-Anne, Québec, Canada. International Water Association.

Lawrence, A. W., & McCarty, P. L. (1970). Unified basis for biological treatment design and operation. *American Society of Civil Engineers, Journal of the Sanitary Engineering Division, 96*, 757–778.

Ludzack, F. J., & Ettinger, M. B. (1962). Controlling operation to minimize activated sludge effluent nitrogen. *Journal of the Water Pollution Control Federation, 34*, 920–931.

Melcer, H., Dold, P. L., Jones, R. M., Bye, C. M., Táakacs, I., Stensel, H. D., Wilson, A. W., Sun, P., & Bury, S. (2003). *Methods for wastewater characterization in activated sludge modeling* (Report 99-WWF-3). Water Environment Research Foundation.

Metcalf & Eddy, Inc./AECOM (2014). *Wastewater engineering: Treatment and resource recovery* (5th ed.). McGraw-Hill.

Michaelis, L., & Menten, M. (1913). Die Kinetik der Invertinwirkung. *Biochem. Z., 49*, 333–369.

Monod, J. (1942). Recherches sur la Croissance des Cultures Bactériennes; Librairie Scientifique.

Olsson, G., & Newell, B. (1999). *Wastewater treatment systems, modeling, diagnosis, and control*. IWA Publishing.

Petersen, E. E. (1965). *Chemical research analysis*. Prentice-Hall.

Shaw, A., Phillips, H. M., Sabherwal, B., & deBarbadillo, C. (2007). Succeeding at simulation. *Water Environment Technology, 19*(4), 54–58.

Smith, R., & Eilers, R. G. (1970). *Simulation of the time-dependent performance of the activated sludge process using digital computer*. U.S. Department of the Interior, Federal Water Pollution Control Administration.

Stenstrom, M. K. (1975). *A dynamic model and computer compatible control strategies for wastewater treatment plants* [Doctoral dissertation, Clemson University].

Storer, F. F., & Gaudy, A. F., Jr. (1969). Computational analysis of transient response to quantitative shock loadings of heterogeneous populations in continuous culture. *Environmental Science and Technology, 3*, 143–149.

Tracy, K. D. (1973). *Mathematical modeling of unsteady state thickening of compressible slurries* [Doctoral dissertation, Clemson University].

Tracy, K. D., & Keinath, T. M. (1974). Dynamic model for thickening of activated sludge. *AIChE Symposium Series, 136*(70), 291–308.

Van Haandel, A. C., Ekama, G. A., & Marais, G. V. R. (1981). A general model for the activated sludge process, Part 3: Single sludge denitrification. *Water Research, 15*, 1135–1152.

Water Environment Federation. (2014). *Wastewater treatment process modeling* (Manual of Practice No. 31). McGraw-Hill.

Water Environment Federation. (2010). *Nutrient removal* (Manual of Practice No. 34). McGraw-Hill.

Water Environment Federation. (2023). *Standards for design of water resource recovery facilities* (WEF 8).

Wentzel, M. C., Dold, P. L., Ekama, G. A., & Marais, G. V. R. (1989). Enhanced polyphosphate organism cultures in activated sludge systems, Part III: Kinetic model. *Water SA, 15*(2), 89–102.

Index